The Living Earth and Earth at Night

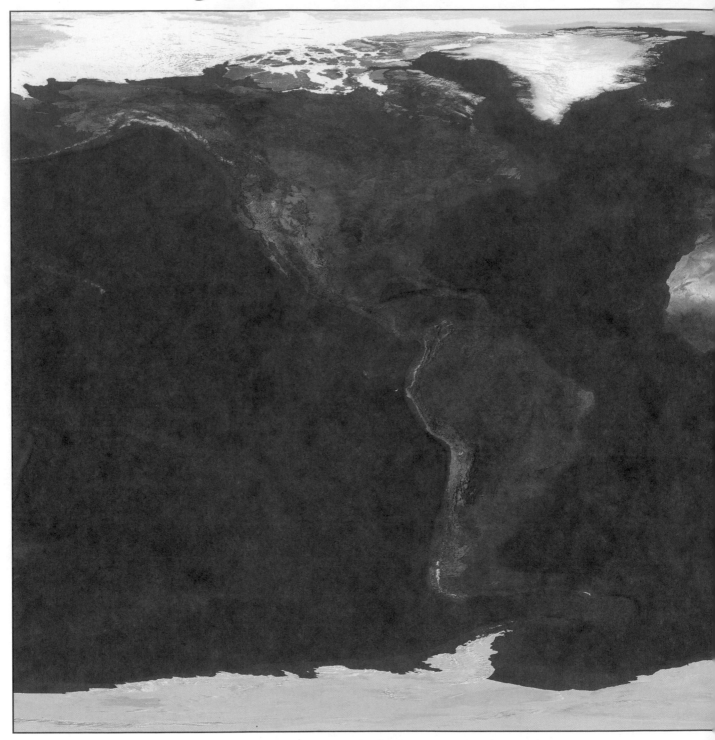

Hundreds of NOAA satellite images went into the making of "The Living Earth." Computers translate the image data into natural true color of local summer with cloudless conditions. Enhancement of elevation data was done with Gtopo30data—a composite of bathymetric (ocean floor) and topographic information. "Earth at Night" inset is derived from Defense Meteorological Satellite Program (DMSP), using Operational Linescan System in low-altitude polar orbit, 2001. Please see an animation of these two images in the World Map References section on the Student Animations CD-ROM that accompanies this text.

YOUR ACCESS TO SUCCESS

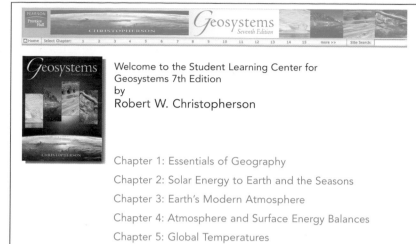

Welcome to the Student Learning Center for
Geosystems 7th Edition
by
Robert W. Christopherson

Chapter 1: Essentials of Geography
Chapter 2: Solar Energy to Earth and the Seasons
Chapter 3: Earth's Modern Atmosphere
Chapter 4: Atmosphere and Surface Energy Balances
Chapter 5: Global Temperatures

Students with new copies of Christopherson, *Geosystems*, Seventh Edition have full access to the book's Student Learning Center—a 24/7 study tool with review exercises, web destinations, and other features designed to help you make the most of your study time.

Just follow the easy website registration steps listed below...

Registration Instructions for
www.prenhall.com/christopherson

1. Go to *www.prenhall.com/christopherson*
2. Click the cover for Christopherson, *Geosystems*, Seventh Edition.
3. Click "Student Learning Center" link, then the "Register" button.
4. Using a coin (not a knife) scratch off the metallic coating below to reveal your Access Code.
5. Complete the online registration form, choosing your own personal Login Name and Password.
6. Enter your pre-assigned Access Code exactly as it appears below.
7. Complete the online registration form by entering your School Location information.
8. After your personal Login Name and Password are confirmed by e-mail, go back to *www.prenhall.com/christopherson*, click your book's cover, enter your new Login Name and Password, and click "Log In".

Your Access Code is:

PSWCHR-SLUNK-SIDLE-SEWAN-MIMIR-PORES

If there is no metallic coating covering the access code above, the code may no longer be valid and you will need to purchase online access using a major credit card to use the website. To do so, go to *www.prenhall.com/christopherson*, click the cover for Christopherson, *Geosystems*, Seventh Edition, then click the "Student Learning Center" link and the "Get Access" button, and follow the instructions under "Students".

Minimum system requirements

PC Operating Systems:

Windows 2000, XP

Pentium II 233 MHz processor. 64 MB RAM In addition to the minimum memory required by your OS.

Internet Explorer 6.0, Netscape Navigator 7.2, or Firefox 1.0.x

Macintosh Operating Systems:

Macintosh Power PC with OS X (10.2, 10.3)

In addition to the RAM required by your OS, this product requires 64 MB RAM, with 40MB Free RAM, with Virtual Memory enabled

Netscape Navigator 7.2, Safari 1.0, 1.3, or Firefox 1.0.x

Macromedia Shockwave(TM)

8.50 release 326 plugin

Macromedia Flash Player 6.0.79 & 7.0

Acrobat Reader 6.0.1

800 x 600 pixel screen resolution

Technical Support
Visit our support site at
http://247.pearsoned.custhelp.com
E-mail support is available 24/7.

GEOSYSTEMS

Combination of satellite *Terra* MODIS sensor image and *GOES* satellite image produces a true-color view of North and South America from 35,000 km (22,000 mi) in space. [Courtesy of MODIS Science Team, Goddard Space Flight Center, NASA and NOAA.]

Along Hudson Bay, Manitoba, Canada, in November, these two males are play fighting, a form of exercise, with plenty of mock attacks. Dramatically, both bears stand on their hind legs and throw jabs at their sparring partner, then into a clinch and back on the ground, each taking his turn as the lead aggressor. You can see movie clips made by the author on the *Student Animation CD* featuring polar bears. Climate change is reducing sea-ice extent, this reduces food availability for the bears. Forecasts predict an ice-free Arctic Ocean in as little as a decade; devastating for the polar bears since they depend on the ice. [Photo by Bobbé Christopherson.]

SEVENTH EDITION

GEOSYSTEMS

An Introduction to Physical Geography

Robert W. Christopherson

PEARSON

Prentice
Hall

Pearson Education International

CIP data available

C.3

Publisher, Geosciences and Environment: *Dan Kaveney*
Editor-in-Chief, Science: *Nicole Folchetti*
Project Manager: *Tim Flem*
Assistant Managing Editor, Science: *Gina M. Cheselka*
In-House Production Liasions: *Beth Sweeten, Wendy Perez*
Media Editor: *Andrew Sobel*
Associate Editor: *Amanda Brown*
Production Editors: *Cindy Miller, Suganya Karuppasamy*
Editorial Assistant: *Jessica Neumann*
Marketing Manager: *Amy Porubsky*
Marketing Assistant: *Ilona Samouha*
Production Assistant to the Author: *Bobbé Christopherson*
Director of Operations: *Barbara Kittle*
Senior Operations Supervisor: *Alan Fischer*
Art Director: *Heather Scott*
Interior Designer: *Tamara Newnam*
Cover Designer: *Bobbé Christopherson, Tamara Newnam*
Media Project Manager: *Rich Barnes*
AV Production Liaison: *Connie Long*
Art Studio: *Precision Graphics*
Project Manager, Precision Graphics: *JC Morgan*
Lead Artist: *Ron Kempke*
Copy Editor: *Pamela Rockwell*
Proofreader: *Jeff Georgeson*
Director, Image Resource Center: *Melinda Patelli*
Image Permission Coordinator: *Zina Arabia*
Composition: GGS Book Services
Cover Credits: *Iceberg near Isispynten Island, Arctic Ocean; mountain reflection in Hornsund, southwest Spitsbergen; weathered sandstone in Valley of Fire State Park, Nevada; forest ecosystem in central Scotland,* by Bobbé Christopherson; Astronaut photograph from the International Space Station courtesy of NASA, JSC.
Dedication Page Quote: *Barbara Kingsolver, Small Wonder (New York: Harper Collins Publishers, 2002), p. 39*

ISBN-13: 978-0-13-715496-8
ISBN-10: 0-13-715496-8

Pearson Education LTD. , London
Pearson Education Australia PTY, Limited
Pearson Education Singapore, Pte. Ltd
Pearson Education North Asia Ltd
Pearson Education Canada, Ltd.
Pearson Educación de Mexico, S.A. de C.V.
Pearson Education -- Japan
Pearson Education Malaysia, Pte. Ltd
Pearson Education, Upper Saddle River, New Jersey

— *Dedication* —

To all the students and teachers of Earth,
our home planet, and a sustainable future.

For the polar bears and all the grandchildren who will
enjoy them, may there be sea ice.

The land still provides our genesis,
however we might like to forget
that our food comes from dank,
muddy Earth, that the oxygen in our lungs was
recently inside a leaf, and that every newspaper or
book we may pick up is made from the hearts of
trees that died for the sake of our imagined lives.
What you hold in your hands right now,
beneath these words,
is consecrated air and time and sunlight.
—Barbara Kingsolver

Environmental Statement

This book is carefully crafted to minimize environmental impact. Pearson Prentice Hall is proud to report that the materials used to manufacture this book originated from sources committed to sustainable forestry practices, tree harvesting, and associated land management. The binding, cover, and paper come from facilities that minimize waste, energy usage, and the use of harmful chemicals.

Equally important, Pearson Prentice Hall closes the loop by recycling every out-of-date text returned to our warehouse. We pulp the books, and the pulp is used to produce other items such as paper coffee cups or shopping bags.

The future holds great promise for reducing our impact on Earth's environment, and Pearson Prentice Hall is proud to be leading the way in this initiative. From production of the book to putting a copy in your hands, we strive to publish the best books with the most up-to-date and accurate content, and to do so in ways that minimize our impact on Earth.

PEARSON
Prentice
Hall

Brief Contents

Contents

PART II
The Water, Weather, and Climate Systems 174

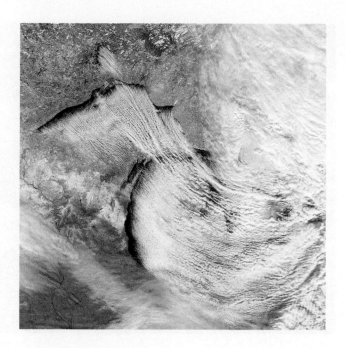

PART III
The Earth–Atmosphere Interface 320

11 The Dynamic Planet 323

12 Tectonics, Earthquakes, and Volcanism 358

15 Eolian Processes and Arid Landscapes 468

13 Weathering, Karst Landscapes, and Mass Movement 400

14 River Systems and Landforms 430

PART IV
Soils, Ecosystems, and Biomes   572

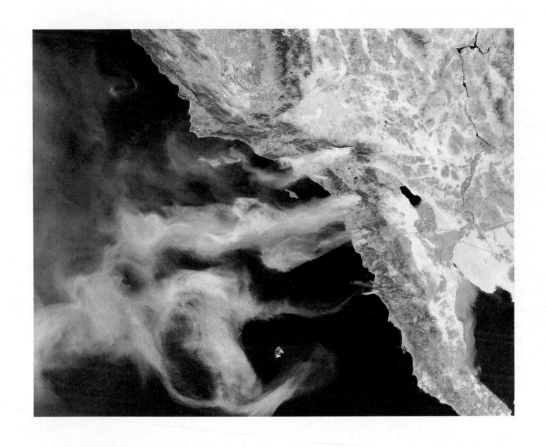

Preface

Welcome to the Seventh Edition of *Geosystems, An Introduction to Physical Geography*! Much has happened since the sixth edition of *Geosystems* (© 2006). The International Polar Year (IPY; http://www.ipy.org/) began in 2007 and concludes March 2009. The Intergovernmental Panel on Climate Change (IPCC; http:www.ipcc.ch/) completed its Fourth Assessment Report and published four supporting volumes of its consensus science regarding human-forced climate change.

In every chapter, *Geosystems* presents updated science, information, and Internet URLs to enhance your understanding of Earth systems. The currency of *Geosystems* makes it important for students and teachers alike to take advantage of the updated content by using this latest edition.

The goal of physical geography is to explain the spatial dimension of Earth's dynamic systems—its energy, air, water, weather, climate, tectonics, landforms, rocks, soils, plants, ecosystems, and biomes. Understanding human–Earth relations is part of our challenge—to create a holistic (or complete) view of the planet and its inhabitants. Armed with the spatial analysis tools of geographic science and our systems approach, physical geographers are well equipped to participate in a planetary understanding of environmental conditions.

Geosystems presents introductory physical geography as a contemporary spatial science, integrating the latest research across the broad sweep of this discipline. This edition builds on the widespread success of the first six editions in the United States, Canada, and elsewhere, as well as its companion texts: *Elemental Geosystems*, now in its fifth edition, and *Geosystems, Canadian Edition*. Students and teachers continue to communicate their appreciation of the up-to-date content, systems organization, scientific accuracy, clarity of the summary and review sections, and overall relevancy. Welcome to physical geography!

New to the Seventh Edition

There is not room in this Preface to list all the changes and additions to *Geosystems*, 7/e, which you will see as you work through the book. Begin by examining all 21 chapter-opening and 4 part-opening photos and images and read their captions—all but two are new. You will see the devastation of an EF-5 tornado that swept away Greensburg, Kansas; photos of the midnight Sun in the Arctic, the Mauna Loa Observatory, a blizzard in Canada, Katrina damage photos, a 2003 and 2007 comparative photo update of declining Lake Mead at Hoover Dam, a glacier as it calves into the sea, and harvesting peat, among others. These are content-specific and begin the learning process from the start of the chapter.

The 2005 Atlantic tropical storm season set many records, including 27 named storms (average is 10); 15 hurricanes (average is 5); and 7 intense hurricanes, category 3 or higher (average is 2). *Geosystems* takes the approach that the 2005 storm was named Katrina but the disaster that followed was of human origins due to poorly conceived planning and flawed engineering. A new aerial photo from August 30, 2005, shows the Inner Harbor Navigation Canal levee break. A new section in Chapter 14, "Engineering Failures in New Orleans, 2005," with satellite before-and-after *Landsat* images of the city with a detailed locator map illustrates the catastrophe.

The latest climate-change science is woven throughout the book and into Chapter 10 with 10 new graphs and charts from the IPCC and results from the ice cores. All temperature data have been updated through the latest available measurements. Focus Study 17.1 has expanded coverage of Dome C, Antarctica, and the ice-core analysis that now stretches the climate record back to 800,000 years ago, confirming present levels of greenhouse gases in the atmosphere as highest in the record.

Throughout the text hundreds of new photos, satellite images, illustrations, and maps support content and learning. *Geosystems* presents many current examples for discussion as new applied topics, such as: the extinction of 67% of all harlequin frog species as an example of applied systems analysis; an updated air pressure section including the 2005 records; the unexpected hurricanes that have hit Brazil, Spain, the Persian Gulf, and mid-Pacific since 2004; the losses of sand inventory from the Chandeleur Islands in the Gulf of Mexico in a new before-and-after Katrina photo comparison; new U.S. drought monitor maps for 2007 and 2006–2007 comparing moisture changes; new High Plains aquifer maps showing its status through 2002; a climate comparison of Death Valley, California, and Baghdad, Iraq; a photo and description of the new Ocean Drilling Program research ship; a new Pu'u O'o crater aerial photo at Kīlauea volcano, Hawai'i, from summer 2007 placed next to an aerial photo made at roughly the same location in 1999 showing huge collapses on the west side; a new sediment-accumulation-model illustration added to the deflation-model art for desert pavement formation, representing both hypotheses; the latest information on the troubled Colorado River system and western drought, a subject present in all my books since the first edition in 1992; a new image of the record low sea-ice extent in the Arctic Ocean, reached September 2007; the latest soil science in the 2006 edition of *Keys to Soil Taxonomy*, 10/e; USDA 1990 map of Hardiness Zones compared to a 2006 revised map, along with a map showing how climate change is forcing the zones to migrate northward; and the endangered status of polar bears and the USGS forecast of a two-thirds loss of their population by 2050. Again, this is but a brief sample of what you will discover in the pages ahead.

Systems Organization Makes *Geosystems* Flow

Each section of this book is organized around the flow of energy, materials, and information. *Geosystems* presents subjects in the same sequence in which they occur in nature. In this way you and your teacher logically progress through topics which unfold according to the flow of individual systems, or in accord with time and the flow of events. See Figure 1.7 in the text for an illustration of this systems organization.

A good example of the systems organization is the eruption of Mount Pinatubo in the Philippines. The global implications of this major event (one of the largest eruptions in the twentieth century) are woven through seven chapters of the book (see Figure 1.6, p. 12, and page 393 for a summary) and not just in the chapter dealing with volcanoes. Another example is the updated coverage on global climate change and its related potential effects included as part of the fabric of many chapters, not just the temperature or climate chapters, since numerous systems are impacted.

For flexibility, *Geosystems* is divided into four parts, each containing chapters that link content in logical groupings. The diagram on the facing page, illustrates our part structure. A quick check of the table of contents and this illustration shows you the order of chapters within these four parts.

The text culminates with Chapter 21, "Earth and the Human Denominator," a unique capstone chapter that summarizes physical geography as an important discipline in understanding Earth's present status and possible future. Think of the world's population and the totality of our impact as the *human denominator*. Just as the denominator in a fraction tells how many parts a whole is divided into, so the growing human population and the increasing demand for resources and their rising planetary impact suggest how much the whole Earth system must adjust. This chapter is sure to stimulate further thought and discussion, dealing as it does with the most profound issue of our time, Earth's stewardship.

Geosystems Is a Text That Teaches

Teaching and learning begin with the front cover photographs. An astronaut provides the background photo from the International Space Station over the tropical Pacific Ocean, the shadows of cumulonimbus thunderheads are cast on the ocean. The four photos dramatically express Earth's diversity: energy, atmsophere, weather, landscapes and weathering, and the complexity of life in biogeography. Thus is the scope of our discipline of physical geography.

To assist you in becoming acquainted with physical geography, this edition of *Geosystems* features more than 550 photographs from across the globe, and more than 125 remote-sensing images from a wide variety of orbital platforms. For spatial analysis and location support, 265 maps are utilized, and more than 400 illustrations locate and explain concepts.

Three heading levels are used throughout the text, and precise topic sentences begin each paragraph to help you outline and review material. **Boldface** words are defined where they first appear in the text. The Glossary collects these terms and concepts alphabetically, with chapter-number references. Our complete Glossary appears on the *Student Animation CD* that accompanies this text. *Italics* are used in the text to emphasize other words and phrases of importance. Every figure has a title that summarizes the caption. Also, in the introduction to each chapter, a feature called "In this chapter" gives you an overview.

An important continuing feature is a list of *Key Learning Concepts* that opens each chapter, stating what you should be able to do upon completing it. These objectives are keyed to the main headings in the chapter. At the end of each chapter is a unique *Summary and Review* section that corresponds to the *Key Learning Concepts*. Grouped under each learning concept is a narrative review that defines the boldfaced terms, a key terms list with page numbers, and specific review questions for that concept. You can conveniently review each concept, test your understanding with review questions, and check key terms in the glossary, then return to the chapter and the next learning concept. In this way, the chapter content is woven together using specific concepts.

A *Critical Thinking* section ends each chapter, challenging you to take the next step with information from the chapter. The key learning concepts help you determine what you want to learn, the text guides you in developing information and more questions, the summary and review assesses what you have learned and what more you might want to know about the subject, and the critical thinking section provokes action and application.

"Focus Study" essays, some completely revised and several new to this edition, provide additional explanation of key topics. "News Reports" relate topics of special interest.

We now live on a planet served by the Internet and its World Wide Web, a resource that weaves threads of information from around the globe into a vast fabric. The fact that we have Internet access to almost all the compartments aboard Spaceship Earth is clearly evident in *Geosystems*. Many entry points link directly from the words in a chapter to an Internet source, allowing you to be up-to-the minute. You will find more than 200 URLs (Internet addresses) in the body of the text (printed in blue color and boldface). Given the fluid nature of the Internet, URLs were rechecked at press time for accuracy. If some URLs have changed since publication, you can most likely find the new location using elements of the old

Geosystems: Our "Sphere of Contents"

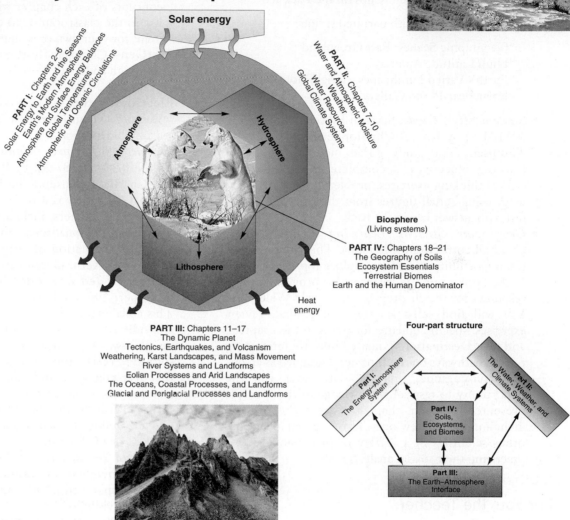

Solar energy

PART I: Chapters 2–6
Solar Energy to Earth and the Seasons
Earth's Modern Atmosphere
Atmosphere and Surface Energy Balances
Global Temperatures
Atmospheric and Oceanic Circulations

PART II: Chapters 7–10
Water and Atmospheric Moisture
Weather
Water Resources
Global Climate Systems

Atmosphere

Hydrosphere

Lithosphere

Biosphere
(Living systems)

PART IV: Chapters 18–21
The Geography of Soils
Ecosystem Essentials
Terrestrial Biomes
Earth and the Human Denominator

Heat energy

PART III: Chapters 11–17
The Dynamic Planet
Tectonics, Earthquakes, and Volcanism
Weathering, Karst Landscapes, and Mass Movement
River Systems and Landforms
Eolian Processes and Arid Landscapes
The Oceans, Coastal Processes, and Landforms
Glacial and Periglacial Processes and Landforms

Four-part structure

Part I:
The Energy–Atmosphere System

Part II:
The Water, Weather, and Climate Systems

Part IV:
Soils, Ecosystems, and Biomes

Part III:
The Earth–Atmosphere Interface

FIGURE 1.8 Earth's four spheres.
Each sphere is a model of vast Earth systems. This four-part structure is the organizational framework for *Geosystems*. [Photos by Bobbé Christopherson.]
  Part I—Atmosphere: The Energy–Atmosphere System
  Part II—Hydrosphere: The Water, Weather, and Climate Systems
  Part III—Lithosphere: The Earth–Atmosphere Interface
  Part IV—Biosphere: Soils, Ecosystems, and Biomes

address. This Internet link begins with a new Table 1.1, presenting the URLs for major geography organizations. The *Geosystems* texts remain the only ones that have embedded URLs for your use.

The Geosystems Learning/ Teaching Package

The seventh edition provides a complete physical geography program for you and your teacher.

For You, the Student:

- ***Student Animation CD***, by Robert Christopherson, is included in each copy of this book.
- 67 animations of key text figures and concepts, with self-tests
- 18 satellite loops, including 4 Notebook features
- 9 items in a World Map Reference library, including:

  "Living Earth/Earth at Night" morphing maps
  "The Blue Marble" rotating Earth
  7 maps with grabber and zoom control

- 5 High Latitude Connection video clips
- 4 new movies:

  "Niagara Falls Close Up"
  "1891–2006 Temperature Anomalies Movie"
  "27 Storms Arlene to Zeta—The 2005 Season"
  "Polar Bear Mom and Cubs on the Pack Ice"

- 4 photo galleries with narrated scripts:

  "Geographic Scenes: East Greenland
  "High Latitude Animals"
  "Earth's Varied Landscapes, Aerial Photos"
  "Polar Bear Photo Gallery"

- *Student Study Guide*, Seventh Edition (ISBN 0-13-601184-5) by Robert Christopherson and Charlie Thomsen. The study guide includes additional learning objectives, a complete chapter outline, critical thinking exercises, problems and short essay work using actual figures from the text, and a self-test with answer key in the back.
- *Geosystems Student Learning Center* (**www.prenhall.com/christopherson**). This site gives you the opportunity to further explore topics presented in the book using the Internet and provides on-line resources for the chapters on the World Wide Web. You will find self-tests that are graded, review exercises, specific updates for items in the chapters, and in "Destinations" many links to interesting related pathways on the Internet. Designed as an on-line study guide, this web site gives you the opportunity to test your knowledge of the concepts presented in each chapter. *Career Link* essays, including several new ones, feature geographers and other scientists in a variety of professional fields practicing their spatial analysis craft.

For You, the Teacher:

Geosystems is designed to give you flexibility in presenting your course. The text is comprehensive in that it is true to each scientific discipline from which it draws subject matter. This diversity is a strength of physical geography, yet makes it difficult to cover the entire book in a school term. *Geosystems* is organized to help you customize your presentation. You should feel free to use the text based on your specialty or emphasis, rearranging parts and chapters as desired. The four-part structure of chapters, systems organization within each chapter, and focus study and news report features, all will assist you in sampling some chapters while covering others in greater depth. The following materials are available to assist you—have a great class!

- *Instructor Resource Center DVD (IRC)*, (ISBN: 0-13-601182-9) contains everything you need for efficient course preparation. Find all your digital resources in one, well-organized, easy-to-access place. Included on this DVD are:

Figures—JPEGs of all illustrations and, new to this edition, all the photos from the book.
Animations—All the animations ready to use in your classroom presentations as PPT slides or Flash files.
PowerPoint™—Preauthored slides outline the concepts of each chapter with embedded art for use in the classroom or to customize with other items for your own presentation needs.
TestGen—Complete TestGen software featuring a thorough and revised test bank, with questions and answers.
Electronic files—The complete Instructor's Resource Manual and Test Item File in one easily accessible location.

- *Instructor's Resource Manual*, Seventh Edition (ISBN: 0-13-601187-X), by Robert Christopherson and Charlie Thomsen. The Instructor's Resource Manual, intended as a resource for both new and experienced teachers, includes lecture outlines and key terms, additional source materials, teaching tips, complete annotation of chapter review questions, and a list of overhead transparencies.
- *Geosystems Test Item File* (ISBN: 0-13-601186-1), by Robert Christopherson and Charlie Thomsen. This collaboration has produced the most extensive and fully revised test item file available in physical geography. This test bank employs **TestGen** software. **TestGen** is a computerized test generator that lets you view and edit test bank questions, transfer questions to tests, and print customized formats. Visit the Prentice Hall Instructor Resource Center catalog at **www.prenhall.com** to download in MAC or PC format.
- *Overhead Transparencies* (ISBN: 0-13-601185-3) including hundreds of illustrations from the text on 175 transparencies, all enlarged for excellent classroom visibility.
- *Applied Physical Geography—Geosystems in the Laboratory*, Seventh Edition (ISBN: 0-13-601181-0), by Robert Christopherson and Charlie Thomsen of American River College. The new seventh edition is the result of a careful revision. Twenty lab exercises, divided into logical sections, allow flexibility in presentation. Each exercise comes with a list of learning concepts. We integrate *Google Earth*™ mapping service links to KMZ files into the exercises so that students can actually fly through and experience 3-D landscapes as they work problems. Our manual comes with its own complete glossary and stereolenses and stereomaps for viewing photo stereopairs in the manual. A complete *Solutions and Answers Manual*, 7/e, is available to teachers (ISBN: 0-13-601180-2).
- *Geosystems 18-month Calendar* highlights important dates relevant to physical geography and our text. Seventy content-specific photos by Bobbé Christopherson grace the pages of our first calendar, which covers January 2008–June 2009.

Acknowledgments

As in all past editions, I recognize my family, for they never waver in supporting *Geosystems'* goals—mom, my sister Lynne, brothers Randy and Marty, and our children Keri, Matt, Reneé, and Steve. And now the next generation: Chavon, Bryce, Payton, Brock, Trevor, Blake, Chase, Téyenna, and our newest, Cade. When I look into our grandchildren's faces and think of their lives this century, that tells me why we need to work toward a sustainable future, one for the children. We honor our grandchildren in the Dedication to this edition. I feel gratitude to all the students over my 30 years at American River College for defining the importance of Earth's future, for their questions, and their enthusiasm.

My thanks go to the many authors and scientists who published research, articles, and books that have enriched my work; to all the colleagues who served as reviewers on one or more editions of each book, or who offered helpful suggestions in conversations at our national and regional geography meetings; and to all the correspondence received from students and teachers from across the globe who, although unnamed here, shared with me through e-mail, FAX, and phone—a continuing appreciated dialogue. I am grateful to all for the generosity of ideas and sacrifice of time. Here is a master list of all our reviewers of the *Geosystems* books, including those that reviewed the *Student Animation CD*. I appreciate several of the reviewers—such as Bob Rohli, Dorothy Sack, Paul Larson, and Brent Yarnal, among others—who have assisted on many editions over the years.

Ted J. Alsop, *Utah State University*
Ward Barrett, *University of Minnesota*
Steve Bass, *Mesa Community College*
Stefan Becker, *University of Wisconsin–Oshkosh*
Daniel Bedford, *Weber State University*
David Berner, *Normandale Community College*
Franco Biondi, *University of Nevada, Reno*
Peter D. Blanken, *University of Colorado, Boulder*
David R. Butler, *Southwest Texas State University*
Mary-Louise Byrne, *Wilfred Laurier University*
Ian A. Campbell, *University of Alberta–Edmonton*
Randall S. Cerveny, *Arizona State University*
Fred Chambers, *University of Colorado, Boulder*
Muncel Chang, *Butte College, Emeritus*
Andrew Comrie, *University of Arizona*
C. Mark Cowell, *Indiana State University*
Richard A. Crooker, *Kutztown University*
Armando M. da Silva, *Towson State University*
Dirk H. de Boer, *University of Saskatchewan*

Dennis Dahms, *University of Northern Iowa*
Mario P. Delisio, *Boise State University*
Joseph R. Desloges, *University of Toronto*
Lee R. Dexter, *Northern Arizona University*
Don W. Duckson, Jr., *Frostburg State University*
Christopher H. Exline, *University of Nevada–Reno*
Michael M. Folsom, *Eastern Washington University*
Mark Francek, *Central Michigan University*
Glen Fredlund, *University of Wisconsin–Milwaukee*
William Garcia, *University of North Carolina–Charlotte*
Doug Goodin, *Kansas State University*
David E. Greenland, *University of North Carolina–Chapel Hill*
Duane Griffin, *Bucknell University*
John W. Hall, *Louisiana State University–Shreveport*
Barry N. Haack, *George Mason University*
Roy Haggerty, *Oregon State University*
Vern Harnapp, *University of Akron*
John Harrington, *Kansas State University*
Jason "Jake" Haugland, *University of Colorado, Boulder*
Gail Hobbs, *Pierce College*
Thomas W. Holder, *University of Georgia*
David A. Howarth, *University of Louisville*
Patricia G. Humbertson, *Youngstown State University*
David W. Icenogle, *Auburn University*
Philip L. Jackson, *Oregon State University*
J. Peter Johnson, Jr., *Carleton University*
Guy King, *California State University–Chico*
Ronald G. Knapp, *SUNY–The College at New Paltz*
Peter W. Knightes, *Central Texas College*
Thomas Krabacher, *California State University–Sacramento*
Richard Kurzhals, *Grand Rapids Junior College*
Hsiang-te Kung, *University of Memphis*
Steve Ladochy, *California State University–Los Angeles*
Charles W. Lafon, *Texas A & M University*
Paul R. Larson, *Southern Utah University*
Robert D. Larson, *Southwest Texas State University*
Elena Lioubimtseva, *Grand Valley State University*
Joyce Lundberg, *Carleton University*

W. Andrew Marcus, *Montana State University*

Nadine Martin, *University of Arizona*

Elliot G. McIntire, *California State University–Northridge*

Norman Meek, *California State University–San Bernardino*

Leigh W. Mintz, *California State University–Hayward, Emeritus*

Sherry Morea-Oaks, *Boulder, CO*

Patrick Moss, *University of Wisconsin, Madison*

Lawrence C. Nkemdirim, *University of Calgary*

Andrew Oliphant, *San Francisco State University*

John E. Oliver, *Indiana State University*

Bradley M. Opdyke, *Michigan State University*

Richard L. Orndorff, *University of Nevada, Las Vegas*

Patrick Pease, *East Carolina University*

James Penn, *Southeastern Louisiana University*

Greg Pope, *Montclair State University*

Robin J. Rapai, *University of North Dakota*

Philip D. Renner, *American River College, Emeritus*

William C. Rense, *Shippensburg University*

Leslie Rigg, *Northern Illinois University*

Dar Roberts, *University of California–Santa Barbara*

Wolf Roder, *University of Cincinnati*

Robert Rohli, *Louisiana State University*

Bill Russell, *L.A. Pierce College*

Dorothy Sack, *Ohio University*

Randall Schaetzl, *Michigan State University*

Glenn R. Sebastian, *University of South Alabama*

Daniel A. Selwa, *U.S.C. Coastal Carolina College*

Peter Siska, *Austin Peay State University*

Thomas W. Small, *Frostburg State University*

Daniel J. Smith, *University of Victoria*

Stephen J. Stadler, *Oklahoma State University*

Michael Talbot, *Pima Community College*

Paul E. Todhunter, *University of North Dakota*

Susanna T.Y. Tong, *University of Cincinnati*

Liem Tran, *Florida Atlantic University*

Suzanne Traub-Metlay, *Front Range Community College*

David Weide, *University of Nevada–Las Vegas*

Thomas B. Williams, *Western Illinois University*

Brenton M. Yarnal, *Pennsylvania State University*

Stephen R. Yool, *University of Arizona*

Susie Zeigler-Svatek, *University of Minnesota*

I extend my continuing gratitude to the editorial, production, and sales staff of Pearson Prentice Hall. Thanks to Dan Kaveney, Geosciences Publisher, for the willingness to risk new ideas and technologies, for his role in both *Geosystems* and *Elemental Geosystems*, and in geographic education in general; my continuing compliments to the expertise of Associate Editor Amanda Brown, who oversees my ancillaries and *Applied Physical Geography* lab manual, for all the years of our friendship; and warm recognition to Charlie Thomsen for his collaboration on the lab manual and ancillaries; to Media Editor Andrew Sobel for skill and creativity on the CD and web site projects; and thanks to Project Manager Tim Flem, who operates the organizational switching yard, for all his coordination effort, help, and conversation. To Marketing Manager Amy Porubsky and the many sales representatives who spend months in the field communicating the *Geosystems* approach, as always, thanks and safe travels to them—may the wind be at their backs!—and to all for allowing us to participate in the entire publishing process.

My thanks to our colleague Production Editor Cindy Miller of GGS Book Services for such friendship and managerial ability through several editions, and to Suganya Karuppasamy of GGS for her skills and oversight of pages, converting my materials into this textbook. Thanks to the art and design team at Prentice Hall for carrying out Bobbé's cover design, for their beautiful text design, and for letting us in on many decisions; and to our son with pride, a Pearson Regional Sales Director who has risen to the top of his field!

At the heart of *Geosystems* and of this life's work is a partner, photographer, and special editorial assistant, my wife Bobbé Christopherson. She has worked tirelessly on all the *Geosystems*' projects. You see more than 370 of her photographs in the seventh edition and on the cover, all carefully made to illustrate physical geography concepts. Please visit the four photo galleries on the *Student Animation CD* to see her artistry with a camera. She prepares our extensive figure and photo tracking logs and processes satellite imagery and all photos, cropping and sizing them for pages. She produced 70 photos for use in our new *Geosystems 18-month Calendar*. Bobbé's love for the smallest of living things and her strength guide me forward. And, she continues as my best friend, colleague, and expedition companion, by my side on our nine polar-region expeditions.

Physical geography teaches us a holistic view of the intricate supporting web that is Earth's environment and our place in it. Dramatic changes that demand our understanding are occurring in many human–Earth relations, as we alter physical, chemical, and biological systems. Our attention to the polar regions is in recognition to the linkages between all planetary systems and the high latitudes—the polar bears become our early warning system, our miner's Canary. All things considered, this is a critical time for you to be enrolled in a physical geography course! The best to you in your studies—and *carpe diem!*

Robert W. Christopherson
P. O. Box 128
Lincoln, California 95648–0128
E-mail: bobobbe@aol.com
Web site: **http://www.prenhall.com/christopherson/**

Greensburg, Kansas, essentially leveled by an EF-5 tornado, as rated on the Enhanced Fujita Scale (see Chapter 8 for more on this). Winds exceeded 330 kmph (205 mph) in the 2.7-km (1.7-mi) wide funnel that locked onto the ground along a 35-km (22-mi) track. The 8:45 P.M. CDT, May 4, 2007, twister destroyed 95% of all public, commercial, and private structures in this town of 1500 people. With 30 minutes notice from sirens, the loss of life totaled 10 killed. [Image by satellite *IKONOS* May 14, courtesy of MJ Harden/GeoEye; lower-left photo made May 12 by Orlin Wagner, Associated Press; lower-right photo made May 12, courtesy of MJ Harden/GeoEye. All rights reserved.]

Essentials of Geography

■ **Key Learning Concepts**

After reading the chapter, you should be able to:

- *Define* geography and physical geography in particular.

- *Describe* systems analysis, open and closed systems, and feedback information, and *relate* these concepts to Earth systems.

- *Explain* Earth's reference grid: latitude and longitude and latitudinal geographic zones and time.

- *Define* cartography and mapping basics: map scale and map projections.

- *Describe* remote sensing, and *explain* the geographic information system (GIS) as tools used in geographic analysis.

Welcome to the Seventh Edition of *Geosystems: An Introduction to Physical Geography*! Much has happened since the sixth edition of *Geosystems* (© 2006). The International Polar Year (IPY, http://www.ipy.org/), begun in 2007 and concluding March 2009, has provided new research, data, and discoveries from the changing polar regions. The Intergovernmental Panel on Climate Change (IPCC, http:www.ipcc.ch/) completed its Fourth Assessment Report and published four volumes of supporting science on human-forced climate change. In every chapter, *Geosystems* presents updated science and information to enhance your understanding of Earth systems.

Imagine the challenge society as a whole and we in physical geography face to understand how Earth's systems operate and then to forecast changes in these systems to determine what future environments we may expect. The scope of this challenge was outlined by scientist Jack Wilson:

> By the end of the 21st century, large portions of the Earth's surface may experience climates not found at present, and some 20th-century climates may disappear....Novel climates are projected to develop primarily in the tropics and subtropics.*

Physical geography deals with powerful Earth systems that influence our lives and the many ways humans alter Earth's systems. Therefore, as we near the end of this first decade of the twenty-first century, a century that will see many changes to the natural environment, now is a scientifically exciting time to be enrolled in a physical geography course, studying the building blocks that form the landscapes and seascapes upon which we depend. Perhaps you will consider physical geography for a major as we move through the chapters together—we'll see.

Scientists and governments conduct research to understand the worldwide impact of global changes. Former United Nations Secretary General and recipient of the 2001 Nobel Peace Prize, Kofi Annan, speaking to the Association of American Geographers (AAG) annual meeting in March 2001, offered this assessment:

> As you know only too well the signs of severe environmental distress are all around us. Unsustainable practices are woven deeply into the fabric of modern life. Land degradation threatens food security. Forest destruction threatens biodiversity. Water pollution threatens public health, and fierce competition for freshwater may well become a source of conflict and wars in the future ... the overwhelming majority of scientific experts have concluded that climate change is occurring, that humans are contributing, and that we cannot wait any longer to take action ... environmental problems build up over time, and take an equally long time to remedy.

Examine the chapter-opening image. This devastation resulted from the strongest of 66 tornadoes that hit the central United States in a 3-day period in 2007. By June 1, 2007, tornado occurrences exceeded the historical annual average of 787—in less than half the year. The remainder of the year contuned this record trend. In Chapter 8 we study weather and the nature of the air masses that interacted to produce such a remarkable storm.

A U.N. Environment Programme (UNEP, http://www.unep.org/) report in 2005 that included data from 15 of the largest financial institutions estimated that weather-related damage losses (drought, floods, hail, tornadoes, derechos, tropical systems, monsoon intensity, storm

surges, blizzard and ice storms, and wildfires) will exceed $1 trillion by A.D. 2040 (adjusted to current dollars)—up from $210 billion in 2005 (Hurricane Katrina alone produced $125 billion in losses). Global climate change is driving this increasing total, at a pace that is accelerating. Physical geography is the one course you will take that explains the operations of Earth systems from a spatial perspective and gives you the tools to understand what is happening.

Why do all of these conditions occur? How are these events different from past experience? Why does the environment vary from equator to midlatitudes, between deserts and polar regions? How does solar energy influence the distribution of trees, soils, climates, and lifestyles? What about the record level of wildfires we are experiencing? Why are earthquakes and volcanoes active in certain regions and what is the nature of the risk? What produces the patterns of wind, weather, and ocean currents? Why are global sea levels on the rise? Why did the countries of the world join in the IPY study of the polar regions? How do natural systems affect human populations, and, in turn, what impact are humans having on natural systems? In this book, we explore those questions, and more, through geography's unique perspective. Welcome to an exploration of physical geography!

We live in an extraordinary era of **Earth systems science**. This science contributes to our emerging view of Earth as a complete entity—an interacting set of physical, chemical, and biological systems that produce a whole Earth. Physical geography is at the heart of Earth systems science as we answer the *spatial* questions concerning Earth's physical systems and their interaction with living things.

In this chapter: Our study of geosystems—Earth systems—begins with a look at the science of physical geography and the geographic tools we use. Physical geography is key to studying entire Earth systems because of its integrative spatial approach.

Physical geographers analyze systems to study the environment. Therefore, we discuss systems and the feedback mechanisms that influence system operations. We then consider location, a key theme of geographic inquiry—the latitude, longitude, and time coordinates that inscribe Earth's surface and the new technologies in use to measure them. The study of longitude and a universal time system provides us with interesting insights into geography. Next, we examine maps as critical tools that geographers use to portray physical and cultural information. This chapter concludes with an overview of the technology that is adding exciting new dimensions to geography: remote sensing from space and computer-based geographic information systems (GIS).

The Science of Geography

Geography (from *geo*, "Earth," and *graphein*, "to write") is the science that studies the relationships among natural systems, geographic areas, society, cultural activities, and the interdependence of all of these *over space*. The term **spatial** refers to the nature and character of physical space, its measurement, and the distribution of things within it.

*Jack Williams et al., "Projected distributions of novel and disappearing climates by A.D. 2100," *Proceedings of the National Academy of Sciences*, April 3, 2007.

For example, think of your own route to the classroom or library today or the way you get to work and how you used your knowledge of street patterns, traffic trouble spots, one-way streets, parking spaces, or bike rack locations to minimize walking distance. All these are spatial considerations. Humans are spatial actors. We profoundly influence vast areas because of our mobility and access to energy and technology. In turn, Earth's systems influence our activities in a most obvious way—these systems give us life.

To help in this definition, we separate geographic science into five spatial themes: **location**, **region**, **human–Earth relationships**, **movement**, and **place**, each illustrated and defined in Figure 1.1. *Geosystems* draws on each theme. For a listing of some important geography organizations and their URLs, see Table 1.1.

Geographic Analysis

Within these five geographic themes, a *method* rather than a specific body of knowledge governs geography. The method is **spatial analysis**. Using this method, geography synthesizes (brings together) topics from many fields, integrating information to form a whole-Earth concept. Geographers view phenomena as occurring in spaces, areas, and locations. The language of geography reflects this spatial view: *space, territory, zone, pattern, distribution, place, location, region, sphere, province,* and *distance*. Geographers analyze the differences and similarities among places.

Process, a set of actions or mechanisms that operate in some special order, is central to geographic analysis. As examples in *Geosystems*, numerous processes are involved in Earth's vast water–atmosphere–weather system, or in continental crust movements and earthquake occurrences, or in ecosystem functions, or the dynamics in a river's drainage basin. Geographers use spatial analysis to examine how Earth's processes interact over space or area.

Therefore, **physical geography** is the *spatial analysis of all the physical elements and processes that make up the environment: energy, air, water, weather, climate, landforms, soils,* *animals, plants, microorganisms, and Earth itself*. As a science, physical geographers employ the **scientific method**. Focus Study 1.1, beginning on page 8, explains this essential process of science.

The Geographic Continuum

Geography is eclectic, integrating a wide range of subject matter from diverse fields; virtually any subject can be examined geographically. Figure 1.2 shows a continuous distribution—a continuum—along which the content of geography is arranged. Disciplines in the physical and life sciences are at one end, and those in the human and cultural sciences are at the other. As the figure shows, various specialties within geography draw from these subject areas.

The continuum in Figure 1.2 reflects a basic duality, or split, within geography—*physical geography* versus *human/cultural geography*. Society parallels this duality. Humans sometimes think of themselves as exempt from physical Earth processes—like actors not paying attention to their stage, props, and lighting. We all depend on Earth's systems to provide oxygen, water, nutrients, energy, and materials to support life. The growing complexity of the human–Earth relationship requires that we shift our study of geographic processes toward the center of the continuum in Figure 1.2 to attain a more balanced perspective—such is the thrust of *Geosystems*.

Earth Systems Concepts

The word *system* pervades our lives daily: "Check the car's cooling system"; "Is a broadband system available for my Internet browser?"; "How does the grading system work?"; "There is a weather system approaching." Systems of many kinds surround us. *Systems analysis* techniques began with studies of energy and temperature (thermodynamics) in the nineteenth century and were further developed in engineering during World War II. Today, geographers use systems methodology as an analytical tool. In this book's 4

Table 1.1	
A Few Geography Organizations	**URL Addresses**
American Geographical Society	http://www.amergeog.org/
Association of American Geographers*	http://www.aag.org/
National Council for Geographic Education	http://www.ncge.org/
National Geographic Society	http://www.nationalgeographic.com/
Canadian Association of Geographers	http://www.cag-acg.ca/en/
Royal Canadian Geographical Society	http://www.rcgs.org/
Institute of Australian Geographers	http://www.iag.org.au/
Australian Geography Teachers Association	http://www.agta.asn.au/
European Geography Association	http://www.egea.eu/
Royal Geographical Society, Institute of British Geographers	http://www.rgs.org/
Complete global listing of geography organizations	http://www.geoggeol.fau.edu/Links/GeographicalSites.htm

*Includes nine regional divisions: East Lakes, Great Plains/Rocky Mountain, Middle Atlantic, Middle States, New England–St. Lawrence Valley, Pacific Coast Regional, Southeastern, Southwestern, West Lakes.

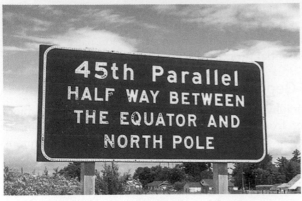

Location

Absolute and relative location on Earth. Location answers the question *Where?*—the specific planetary address of a location. This road sign is posted on the Interstate 5 freeway in Oregon telling drivers their position on Earth.

Region

Areas having uniform characteristics; how they form and change; their relation to other regions. The edge of the northern boreal forest west of Hudson Bay, Canada, is a region undergoing climatic change and increasing temperatures. Here the pioneer spruce trees take root as the forest moves northward into the tundra biome.

Human–Earth Relationships

Humans and the environment: resource exploitation, hazard perception, and environmental pollution and modification. The western U.S. drought, into its eighth year is going to alter agricultural activities in the Imperial Valley, California. Here, flowers are grown, harvested, and placed on a waiting cargo jet for distant markets. The lack of irrigation water and concern over fossil fuel use might eventually phase out this unique desert agriculture.

Geographic Science

Place

Tangible and intangible living and nonliving characteristics that make each place unique. No two places on Earth are exactly alike. The coast of Cornwall in southwest England is a place of myth and legend. Here the ruins of Tintagel Castle sit starkly against the Celtic Sea. This was the site of Uther Pendragon's castle, and in myth where King Arthur was born. These coastal headlands have a rich and varied history, including the mining of slate from the bedrock beneath.

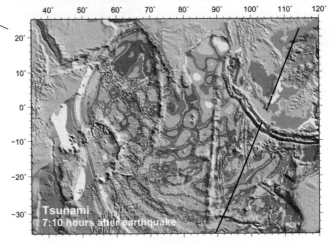

Tsunami
7:10 hours after earthquake

Movement

Communication, movement, circulation, migration, and diffusion across Earth's surface. Global interdependence links all regions and places. The radiating waves of energy in an Indian Ocean tsunami generated by a magnitude 9.3 earthquake off the coast of Sumatra, Indonesia, on December 26, 2004. Damage and loss of life exceeding 170,000 people, resulted when waves hit distant shores.

FIGURE 1.1 Five themes of geographic science.
Definitions of five fundamental themes in geographic science with examples of each—location, region, human–Earth relationships, movement, and place. Drawing from your own experience, can you think of several examples of each theme? [Regions, Human–Earth relationships, Place photos by Bobbé Christopherson; Location photo by author; and Movement *GFO-Satellite* image courtesy of NOAA, Laboratory for Satellite Altimetry.]

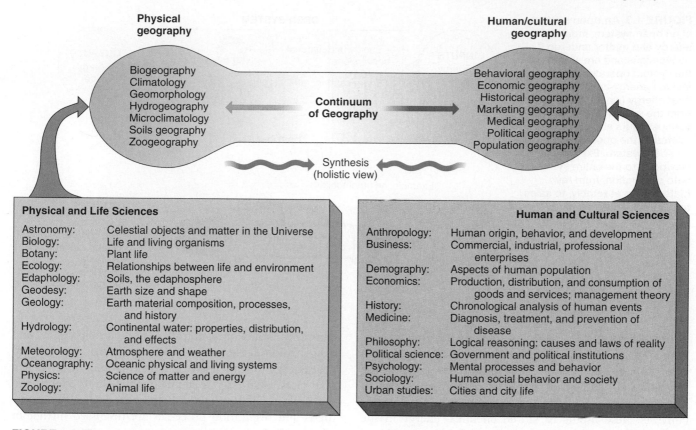

FIGURE 1.2 The content of geography.
Geography derives subject matter from many different sciences. The focus of this book is physical geography, but we also integrate some human and cultural components. Movement toward the middle of the continuum suggests a synthesis of Earth topics and human topics. Check the two boxes for any courses you have completed.

parts and 21 chapters, content is organized along logical flow paths consistent with systems thinking.

Systems Theory

Simply stated, a **system** is any ordered, interrelated set of things and their attributes, linked by flows of energy and matter, as distinct from the surrounding environment outside the system. The elements within a system may be arranged in a series or interwoven with one another. A system comprises any number of subsystems. Within Earth's systems, both matter and energy are stored and retrieved and energy is transformed from one type to another. (Remember: *Matter* is mass that assumes a physical shape and occupies space; *energy* is a capacity to change the motion of, or to do work on, matter.)

Open Systems Systems in nature are generally not self-contained: Inputs of energy and matter flow into the system, and outputs of energy and matter flow from the system. Such a system is an **open system**. Within a system, the parts function in an interrelated manner, acting together in a way that gives each system its character. Earth is an open system *in terms of energy*, because solar energy enters freely and heat energy leaves, going back into space. Most natural systems are open in terms of energy. Figure 1.3 schematically illustrates an open

system and presents the inputs and outputs of an automobile as an example.

Most Earth systems are dynamic (energetic, in motion) because of the tremendous infusion of radiant energy from thermonuclear reactions deep within the Sun. This energy comes through the outermost edge of Earth's atmosphere to power terrestrial systems, being transformed along the way into various forms of energy, such as *kinetic energy* (of motion), *potential energy* (of position), or chemical or mechanical energy—setting the fluid atmosphere and ocean in motion. Eventually, Earth radiates this energy back to the cold vacuum of space as heat energy. Earth's interior is likewise a dynamic system, with a fluid core and physically convecting mantle that influence crustal plates set in motion as energy flows toward the surface.

Closed Systems A system that is shut off from the surrounding environment so that it is self-contained is a **closed system**. Although such closed systems are rarely found in nature, Earth is essentially a closed system *in terms of physical matter and resources*—air, water, and material resources. The only exceptions are the slow escape of lightweight gases (such as hydrogen) from the atmosphere into space and the input of frequent but tiny meteors and cosmic and meteoric dust. The fact that Earth is a closed material system makes recycling efforts inevitable if we want a sustainable global economy.

FIGURE 1.3 An open system. In an open system, inputs of energy and matter undergo conversions and are stored as the system operates. Outputs include energy and matter and heat energy (waste) that flow from the system. See how the various inputs and outputs are related to the operation of a car—an open system. Expand your viewpoint to the entire system of auto production, from raw materials, to assembly, to sales, to car accidents, to junkyards. Can you identify other open systems that you encounter in your daily life?

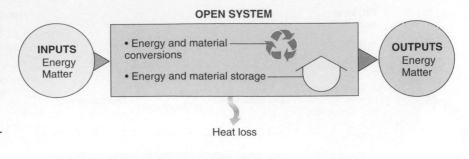

System Example   Figure 1.4 illustrates a simple open-flow system, using plant photosynthesis and respiration as an example. In *photosynthesis* (Figure 1.4a), plants use sunlight as an energy input and material inputs of water, nutrients, and carbon dioxide. The photosynthetic process converts these inputs to stored chemical energy in the form of plant sugars (carbohydrates). The process also releases an output from the plant system: the oxygen we breathe.

Reversing the process, plants derive energy for their operations from respiration. In *respiration*, the plant consumes inputs of chemical energy (carbohydrates) and oxygen and releases outputs of carbon dioxide, water, and heat energy into the environment (Figure 1.4b). Thus, a plant acts as an open system, in which both energy and materials freely flow into and out of the plant. (Photosynthesis and respiration processes are discussed further in Chapter 19.)

System Feedback   As a system operates, it generates outputs that influence its own operations. These outputs function as "information" that is returned to various

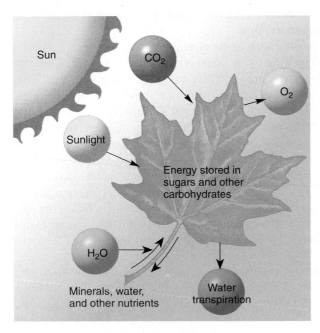

(a) Plant photosynthesis

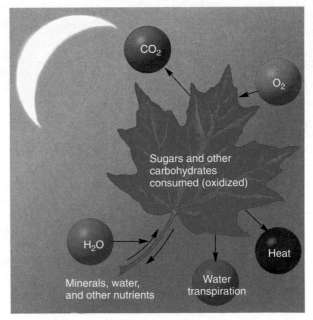

(b) Plant respiration

FIGURE 1.4 A leaf is a natural open system.
A plant leaf provides an example of a natural open system. (a) In the process of photosynthesis, plants consume light, carbon dioxide (CO_2), nutrients, and water (H_2O), and produce outputs of oxygen (O_2) and carbohydrates (sugars) as stored chemical energy. (b) Plant respiration, illustrated here at night, approximately reverses this process and produces outputs of carbon dioxide (CO_2) and water (H_2O) using oxygen (O_2) and consuming (oxidizing) carbohydrates to produce energy for cell operations.

points in the system via pathways called **feedback loops**. Feedback information can guide, and sometimes control, further system operations. In the plant's photosynthetic system (see Figure 1.4), any increase or decrease in daylength (sunlight availability), carbon dioxide, or water produces feedback that causes specific responses in the plant. For example, decreasing the water input slows the growth process; increasing daylength increases the growth process, within limits.

If the feedback *information* discourages response in the system, it is **negative feedback**—like bad reviews affecting ticket sales for a film or concert. *Further production in the system decreases the growth of the system.* Such negative feedback causes self-regulation in a natural system, stabilizing the system. Think of a weight-loss diet— the human body is an open system. When you stand on the scales, the good or bad news reported about your weight acts as negative feedback to you. This, in turn, provides you with guidance (negative feedback) as to how inputs such as food Calories need to be adjusted, or how exercise and increased metabolism are needed to burn excess Calories.

If feedback *information* encourages increased response in the system, it is **positive feedback**—like good reviews affecting ticket sales for a movie. *Further production in the system stimulates the growth of the system.* In finance, a compound-interest-bearing account provides an example:

The larger the account becomes, the more interest it earns, thus the larger the account becomes, and so on. Unchecked positive feedback in a system can create a runaway ("snowballing") condition. In natural systems, such unchecked growth can reach a critical limit, leading to instability, disruption, or death of organisms. Global climate change creates an example of positive feedback in our first High Latitude Connection feature.

Think of the numerous devastating wildfires that occurred globally in the last half decade, including area-record wildfires in the western United States, as examples of positive feedback. As the fires burned, they dried wet shrubs and green wood around the fire, thus providing more fuel for combustion. The greater the fire, the greater becomes the availability of fuel, and thus more fire is possible—a positive feedback for the fire. Control of the input of flammable fuel and oxygen is the key to extinguishing such fires. Knowing this process allows us to design strategies for controlling fuel availability, regulating excessive landscaping in vulnerable urban areas and practicing "control burns."

Can you describe possible negative and positive feedback in the automobile illustration in Figure 1.3? For instance, how would the inputs and outputs be affected if tailpipe exhaust emission standards are weakened or strengthened? What if the automobile is operated at high elevation? Or if the car is a sports utility vehicle

High Latitude Connection 1.1

Meltponds as Positive Feedback

Another example of positive feedback is the significant increase in meltponds in the polar regions—on icebergs, ice shelves, and the Greenland ice sheet. As explained in News Report 17.2, Chapter 17, meltponds are darker and reflect less sunlight (have a lower albedo) and therefore absorb more solar energy, which in turn melts more ice, which forms more meltponds, and so forth. Thus, a positive feedback system is in operation, further enhancing the effects of higher temperatures and warming trends.

Based on satellite images and aerial surveys, scientists determined a spatial increase in meltpond distribution of 400% between 2001 and 2003 across the Arctic, then slightly retreating through 2005 (Figure 1.1.1). Overall ice-mass losses from Greenland more than doubled between 1996 and 2005. The meltponds are symptomatic of climatic warming.

Note that throughout *Geosystems* you find these brief High Latitude Connection features. I continue this feature in the book because the polar regions are experiencing dynamic changes in system opera-

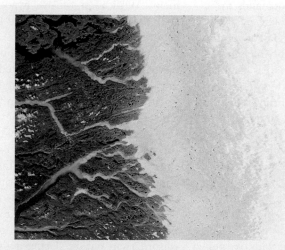

FIGURE 1.1.1 High-latitude meltpond increase results from positive feedback.
Meltponds on the western region of the Greenland ice sheet in 2003. The darker water-saturated surface layer from melt and blue-colored meltponds indicate significant increase in temperatures. [Image by MODIS sensor, *Terra* satellite, courtesy of MODIS Rapid Response Team, NASA/GSFC.]

tions that appear to be accelerating. Many people think these polar environs are acting as our "miner's canary," that is, our

early warning system. Watch for these points of applied geographic science in the chapters ahead.

The Scientific Method

The term *scientific method* may have an aura of complexity, but it shouldn't. The scientific method is simply the application of common sense in an organized and objective manner. A scientist observes, makes a general statement to summarize the observations, formulates a hypothesis, conducts experiments to test the hypothesis, and develops a theory and governing scientific laws. Sir Isaac Newton (1642–1727) developed this method of discovering the patterns of nature, although the term *scientific method* was applied later.

Scientists are curious about nature and appreciate the challenge of problem solving. Society depends on the discoveries of science to decipher mysteries, find solutions, and foster progress. *Complexity* dominates nature, making several outcomes possible as a system operates. Science serves an important function in reducing such uncertainty. Yet, the more knowledge we have, the more the uncertainty and awareness of other possible scenarios (outcomes and events) increase. This in turn demands more precise and aggressive science.

A problem for society occurs when scientific uncertainty fuels an antiscience viewpoint, whereas defining uncertainty is a by-product of discovery. As we realize scientific principles of complexity and chaos in natural and human-made systems, the need for critical thinking and the scientific method deepens in all aspects of life.

Steps in the Scientific Method

Follow the scientific method illustration in Figure 1.1.1 as you read. The scientific method begins with our perception of the real world and a determination of what we know, what we want to know, and the many unanswered questions that exist. Scientists who study the physical environment turn to nature for clues that they can observe and measure. They figure out what data are needed and begin to collect those data. Then, these observations and data are analyzed to identify coherent patterns that may be present. This search for patterns requires *inductive reasoning*, or the process of drawing generalizations from specific facts. This step is important in modern Earth systems science, in which the goal is to understand *a whole functioning Earth*, rather than isolated, small compartments of information. Such understanding allows the scientist to construct models that simulate general operations of Earth systems.

If patterns are discovered, the researcher may formulate a *hypothesis*—a formal generalization of a principle. Examples include the planetesimal hypothesis, nuclear-winter hypothesis, and moisture-benefits-from-hurricanes hypothesis. Further observations are related to the general principles established by the hypothesis. Further data gathering may support or refute the hypothesis, or predictions made according to it may prove accurate or inaccurate. All these findings provide feedback to adjust data collection and model building and to refine the hypothesis statement. Verification of the hypothesis after exhaustive testing may lead to its elevation to the status of a *theory*.

A theory is constructed on the basis of several extensively tested hypotheses. Theories represent truly broad general principles—unifying concepts that tie together the laws that govern nature (for example, the theory of relativity, theory of evolution, atomic theory, Big Bang theory, stratospheric ozone depletion theory, or plate tectonics theory). A theory is a powerful device with which to understand both the order and chaos (disorder) in nature. Using a theory allows predictions to be made about things not yet known, the effects of which can be tested and verified or disproved through tangible evidence. The value of a theory is the continued observation, testing, understanding, and pursuit of knowledge that the theory stimulates. A general theory reinforces our perception of the real world, acting as positive feedback.

Pure science does not make value judgments. Instead, pure science provides people and their institutions with objective information on which to base their own value judgments. Social and political judgments about the applications of science are increasingly critical as Earth's natural systems respond to the impact of modern civilization. From Jane Lubchenco's 1997 AAAS Presidential Address:

> Science alone does not hold the power to achieve the goal of greater sustainability, but scientific knowledge and wisdom are needed to help inform decisions that will enable society to move toward that end.

The Method and Applied Science

The growing awareness that human activity is producing global change places increasing pressure on scientists to participate in decision making. Numerous editorials in scientific journals have called for such *applied science* involvement. F. Sherwood Rowland and Mario Molina provide an example, discussed in Chapter 3, for they first proposed a hypothesis that certain human-made chemicals caused damaging reactions in the stratosphere. They hypothesized in 1974—along with Paul Crutzen, who also contributed to these discoveries—that chlorine-containing products such as chlorofluorocarbons (CFCs), commonly used as aerosol propellants, refrigerants, foaming agents, and cleaning solvents, were depleting our protective stratospheric ozone (O_3) layer.

Subsequently, surface, atmosphere, and satellite measurements confirmed the photochemical reactions and provided data to map the real losses that were occurring and led to international treaties and agreements to ban the disruptive chemicals. In 1995, the Royal Swedish Academy of Sciences awarded these three scientists the Nobel Prize for Chemistry for their pioneering research and for their direct involvement in later political change.

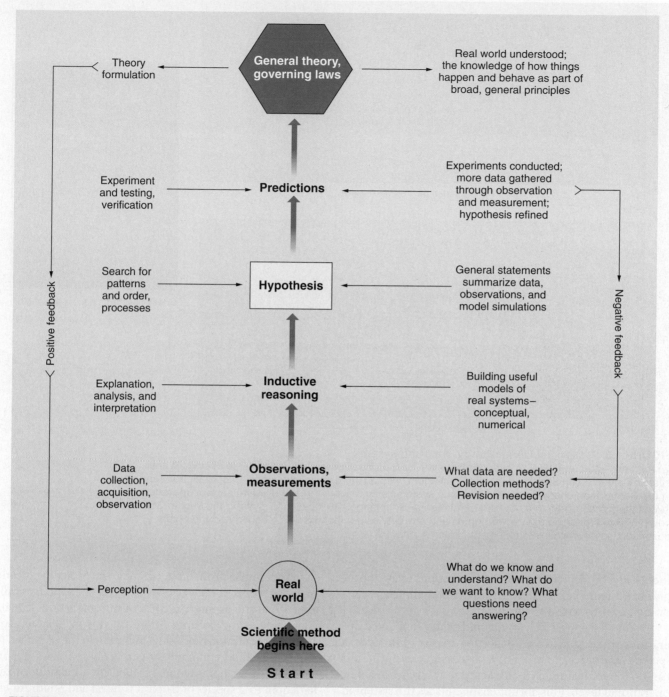

FIGURE 1.1.1 Scientific method flow chart.
The scientific method process: from perceptions, to observations, reasoning, hypothesis, predictions, and possibly to general theory and natural laws.

(SUV) that weighs 2300 kg (5070 lb)? Or if the price of gasoline increases or decreases? Or if the car is a fuel-efficient hybrid gas–electric model? Take a moment to assess the effect of such changes on automobile system operations.

System Equilibrium Most systems maintain structure and character over time. An energy and material system that remains balanced over time, in which conditions are constant or recur, is in a *steady-state condition*. When the rates of inputs and outputs in the system are equal and the amounts of energy and matter in storage within the system are constant (or more realistically, as they fluctuate around a stable average, such as a person's body weight), the system is in **steady-state equilibrium**.

However, a steady-state system may demonstrate a changing trend over time, a condition described as **dynamic equilibrium**. These changing trends of either

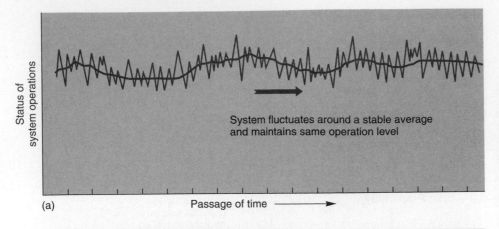

Coastal systems pass a threshold.

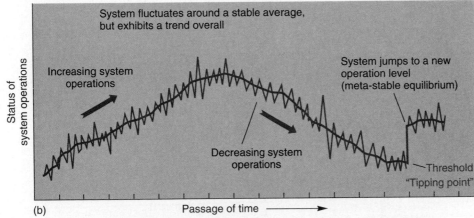

FIGURE 1.5 System equilibria: steady-state and dynamic.
(a) A steady-state equilibrium over time; system operations fluctuate around a stable average. Other systems are in a condition of dynamic equilibrium, with an increasing or decreasing operational trend as in (b). Rather than changing gradually, some systems may reach a *threshold* where system operations lurch (change abruptly) to a new set of relations—a "tipping point." Landslides along the Pacific Coast south of San Francisco destroyed homes and property as a threshold was reached and the bluffs collapsed. [Inset photo by Sam Sargent/Liaison Agency Inc.]

increasing or decreasing system operations may appear gradual. Figure 1.5 illustrates these two equilibrium conditions, steady-state and dynamic.

Note that systems try to maintain their functional operations; they tend to resist abrupt change. However, a system may reach a **threshold**, or *tipping point*, where it can no longer maintain its character, so it lurches to a new operational level. This abrupt change places the system in a *metastable equilibrium*. An example of such a condition is a landscape, such as a hillside or coastal bluff, which adjusts after a sudden landslide. A new equilibrium is eventually achieved among slope, materials, and energy over time and may not be as desirable or supportive to us as previous conditions.

This threshold concept raises concern in the scientific community, especially if some natural systems reach their tipping-point limits. The relatively sudden collapse of ice shelves surrounding a portion of Antarctica, or the sudden crack-up of the Ward Hunt, in 2003, and Ayles, in 2005, ice shelves on the north coast of Ellesmere Island, Canada, serve as examples of systems at a threshold, changing to a new status, that of disintegration (more on this in Chapter 17).

As examples from biogeography, the bleaching (death) of living coral reefs worldwide accelerated dramatically after 1997, and as much as 50% of the corals on Earth are ailing or nearing collapse (more on this in Chapter 16). Warming conditions and some pollution in the ocean led to such a threshold and coral system collapse. News Report 1.1 looks at a tipping point reached by a majority of the harlequin frog species of tropical Central and South America. In this report, note the power of systems analysis to explain what is causing these extinctions.

Mount Pinatubo—Global System Impact   A dramatic example of interactions between volcanic eruptions and Earth systems illustrates the strength of spatial analysis in physical geography and the organization of this textbook. Mount Pinatubo in the Philippines erupted violently in 1991, injecting 15–20 million tons of ash and sulfuric acid mist into the upper atmosphere (Figure 1.6). This was the second greatest eruption during the twentieth century; Mount Katmai in Alaska (1912) was the only one greater. The eruption materials from Mount Pinatubo affected Earth systems in several ways, which are noted on the map.

News Report 1.1

Frogs at a Tipping Point—Check the System

Global climate change is pushing systems to their threshold. Many amphibian species are threatened by these changes. The "Declining Amphibian Population Task Force," founded in 1991 and merged with the IUCN (World Conservation Union) Amphibian Group in 2006, is a response to this growing crisis. Scientific understanding of frog population declines from habitat loss and pollution, and now rising temperatures, is emerging.

The key to unraveling what was happening to harlequin frog (*Atelopus varius*, Figure 1) and golden toad (*Bufo periglenes*) species in the Monteverde Cloud Forest Reserve, Costa Rica, and elsewhere in Central and South America was a consideration of climate change and how this alters optimum conditions for the transmission of diseases and pathogens. Research had to uncover complex synergistic interactions (combined actions that magnify results) in physical and biological systems. One infectious agent, the *chytrid fungus* (*Bactrachochytrium dendrobactidis*), was spreading as thermal-optimum conditions were reached in its mid-elevation range (1000 m to 2400 m; 621 mi−1491 mi) in the mountain-cloud forest. Here is how system operations and feedbacks produced this species failure.

Human-forced climate change is increasing the temperature of the ocean and atmosphere. Higher temperatures cause higher evaporation rates. However, even though humidity (water-vapor content of the air) is higher, relative humidity is lower because the air can absorb more, which affects the condensation level (see Chapter 7). As the warm, moist air moves onshore and reaches the mountains, the air mass lifts

FIGURE 1.1.1 Harlequin frog extinction alarm.
About two-thirds of harlequin frog species went extinct over the last two decades as a result of climate change, specifically increasing temperatures. The frog is nearing extinction due to a pathogen that thrives in the climate-altered ecosystem.

and cools and condensation occurs at higher elevations than before (see Chapter 8). Increasing levels of air pollution provide more condensation nuclei for cloud droplets to form (see Chapters 3 and 7).

Increased clouds affect the daily temperature range: Clouds at night, acting like insulation, raise nighttime minimum temperatures, whereas clouds during the day act as reflectors, sending sunlight back into space and lowering daytime maximum temperatures (see Chapter 4). Nighttime minimum temperatures are increasing more rapidly than daytime maximums. The chytrid fungus does best when temperatures are between 17°C and 25°C, stops growing at 28°C, and dies at 30°C. The cloud cover keeps the daytime maximum below 25°C (77°F) at the leaf-litter and moss-ground cover microscale; without cloud cover, the surface reaches 30°C. In these more favorable conditions the disease pathogen flourishes.

Harlequin frogs have moist, porous skin that the fungus penetrates, killing the frog. Between 1986 and 2006, approximately 67% of the 110 known species of harlequin frogs went extinct—extinction is forever. The IUCN Red List classifies the remaining population as "Critically Endangered."

From this example can you see the power of systems analysis? Global warming intensifies the hydrologic cycle, raises air and water temperatures, and increases evaporation, all of which changes the nature of cloud formations and alters daily temperature patterns. These changes impact ecological systems by increasing the range and number of infectious disease pathogens, which attack species and ultimately end their existence. A summary of the research by J. Alan Pounds, and his team of scientists, concluded with a sobering thought:

The powerful synergy between pathogen transmission and climate change should give us cause for concern about human health in a warmer world.... The frogs are sending an alarm call to all concerned about the future of biodiversity and the need to protect the greatest of all open-access resources—the atmosphere.[*]

*A. Blaustein and A. Dobson, "A message from the frogs," summary of J. A. Ponds et al., "Widespread amphibian extinctions from epidemic disease driven by global warming," *Nature* 439 (January 12, 2006): p. 144.

As you progress through this book, you see the story of Mount Pinatubo and its implications woven through eight chapters: Chapter 1 (systems theory), Chapter 4 (effects on energy budgets in the atmosphere), Chapter 6 (satellite images of the spread of debris by atmospheric winds), Chapter 10 (temporary effect on global atmospheric temperatures), Chapters 11 and 12 (volcanic processes), Chapter 17 (past climatic effects of volcanoes), and Chapter 19 (effects on net

Eruption cloud spread to 500-kilometer diameter, 35-km high.

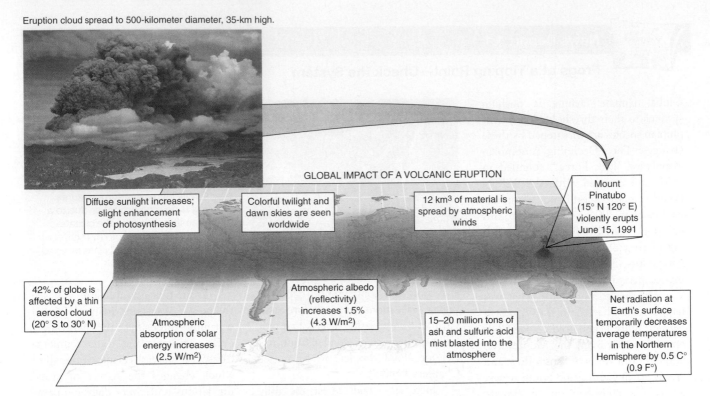

FIGURE 1.6 The eruption of Mount Pinatubo.
The 1991 Mount Pinatubo eruption affected the Earth–atmosphere system on a global scale.
Geographers and other scientists use the latest technology to study how such eruptions affect the
atmosphere's dynamic equilibrium. For a summary of the impacts from this eruption, refer to Chapter 12.
[Inset photo by Van Cappellen/REA/SABA.]

photosynthesis). This illustrates the value of a systems approach; instead of simply describing the eruption, we see the linkages and global impacts of such a volcanic explosion.

Systems in *Geosystems* To further illustrate the systems organization in this textbook, notice its organization around the flow of energy, materials, and information throughout the chapters and portions of chapters (Figure 1.7). This book presents subjects in the same sequence in which they occur in nature. In this way, topics follow a logical progression that unfolds according to the energy and material flows within individual systems or with time and the flow of events.

The sequence (Figure 1.7a)—*input* (components and driving force), *actions* (movements, processes, and storage changes), *outputs* (results and consequences), and *human impacts/impact on humans* (measure of relevance)—is seen in Part 1 as an example. The Sun begins Chapter 2: The energy flows across space to the top of the atmosphere and through the atmosphere to the surface and surface energy budgets (Chapters 3 and 4). Then we look at the outputs of temperature (Chapter 5) and winds and ocean currents (Chapter 6). Note the same logical systems flow in the other three parts and in each chapter of this text—Part 2 and Chapter 16 also are shown as examples.

Models of Systems A **model** is a simplified, idealized representation of part of the real world. Models are designed with varying degrees of generalization. In your

life, you may have built or drawn a model of something that was a simplification of the real thing—such as a model airplane or model house. Physical geographers use simple system models to demonstrate complex things in the environment.

The simplicity of a model makes a system easier to understand and to simulate in experiments (for example, the leaf model in Figure 1.4). A good example is a model of the *hydrologic system*, which models Earth's entire water system, its related energy flows, and the atmosphere, surface, and subsurface environments through which water moves (see Figure 9.1 in Chapter 9).

Adjusting the variables in a model produces differing conditions and allows predictions of possible system operations. A general circulation model (GCM) of the atmosphere (see Figure 10.33), operated by the Goddard Institute, accurately predicted the effect of Mount Pinatubo's ash on the atmosphere, the lowering and subsequent recovery of global air temperatures, and continues to produce accurate temperature forecasts for the future. However, predictions are only as good as the assumptions and accuracy built into the model. It is best to view a model for what it is—a simplification to help us understand complex processes.

We discuss many *system models* in this text, including the hydrologic system, water balance, surface energy budgets, earthquakes and faulting as outputs of Earth systems, river drainage basins, glacier mass budgets, soil profiles, and various ecosystems. Computer-based models are in use to study most natural systems—from the upper atmosphere, to

Systems in *Geosystems*

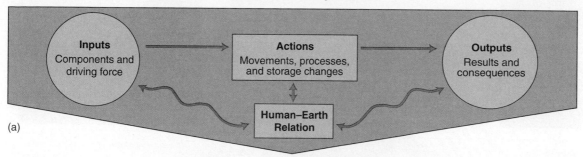

(a)

Examples of systems organization in text:

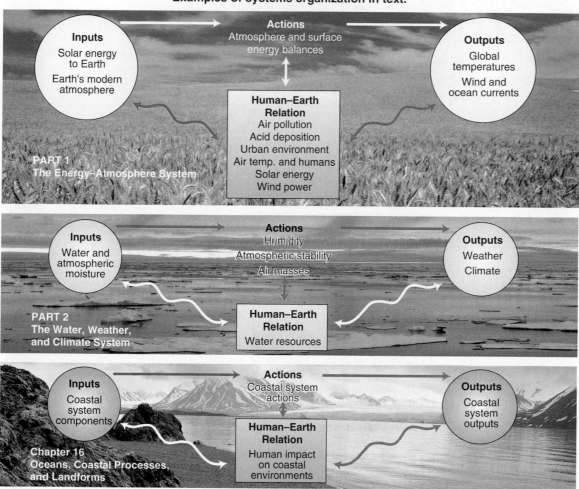

(b)

FIGURE 1.7 The systems in *Geosystems*.
(a) Chapters, sections, and topics are organized around simple flow systems, or around time and the flow of events, at various scales. (b) This systems structure as it is applied to Parts 1 and 2. As a chapter example, turn to Chapter 16 on coastal processes and see how the chapter is organized using the systems flow.

climate change, to Earth's interior. Each of Earth's four "spheres" represents such a model of Earth's systems.

Earth's Four "Spheres"

Earth's surface is a vast area of 500 million square kilometers (193 million square miles) where four immense open systems interact. Figure 1.8 shows a simple model of three **abiotic** (nonliving) systems overlapping to form the realm of the **biotic** (living) system. The abiotic spheres are the

atmosphere, *hydrosphere*, and *lithosphere*. The biotic sphere is the *biosphere*. Because these four systems are not independent units in nature, their boundaries are really transition zones rather than sharp barriers.

As noted in the figure, these four spheres form the part structure in which chapters are grouped in this book. The various arrows in the figure show that content in each part and among chapters interrelates as we build our discussion of Earth's abiotic and biotic systems.

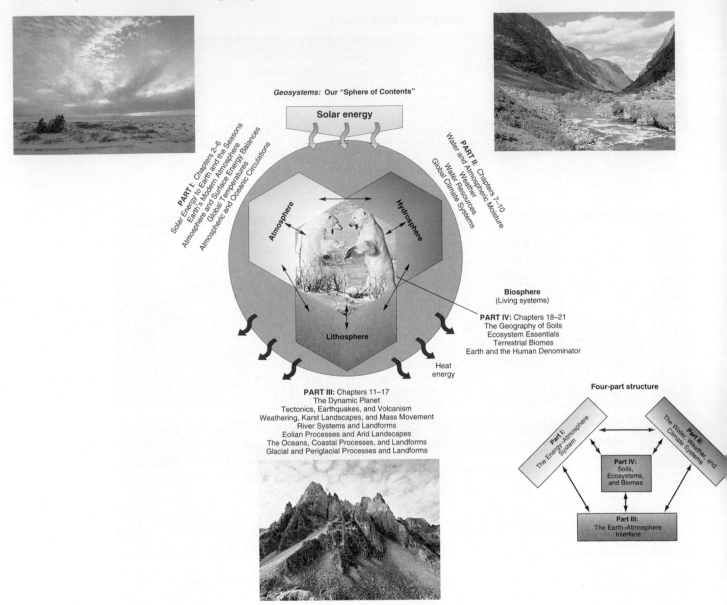

Geosystems: Our "Sphere of Contents"

Solar energy

PART I: Chapters 2–6
Solar Energy to Earth and the Seasons
Earth's Modern Atmosphere
Atmosphere and Surface Energy Balances
Global Temperatures
Atmospheric and Oceanic Circulations

PART II: Chapters 7–10
Water and Atmospheric Moisture
Water, Weather,
Water Resources
Global Climate Systems

Atmosphere

Hydrosphere

Lithosphere

Biosphere
(Living systems)

PART IV: Chapters 18–21
The Geography of Soils
Ecosystem Essentials
Terrestrial Biomes
Earth and the Human Denominator

Heat
energy

PART III: Chapters 11–17
The Dynamic Planet
Tectonics, Earthquakes, and Volcanism
Weathering, Karst Landscapes, and Mass Movement
River Systems and Landforms
Eolian Processes and Arid Landscapes
The Oceans, Coastal Processes, and Landforms
Glacial and Periglacial Processes and Landforms

Four-part structure

Part I:
The Energy–Atmosphere
System

Part II:
The Water, Weather, and
Climate Systems

Part IV:
Soils,
Ecosystems,
and Biomes

Part III:
The Earth–Atmosphere
Interface

FIGURE 1.8 Earth's four spheres.
Each sphere is a model of vast Earth systems. This four-part structure is the organizational framework
for *Geosystems*. [Photos by Bobbé Christopherson.]
Part I—Atmosphere: The Energy–Atmosphere System
Part II—Hydrosphere: The Water, Weather, and Climate Systems
Part III—Lithosphere: The Earth–Atmosphere Interface
Part IV—Biosphere: Soils, Ecosystems, and Biomes

Atmosphere (Part I, Chapters 2–6)   The **atmosphere**
is a thin, gaseous veil surrounding Earth, held to the planet
by the force of gravity. Formed by gases arising from
within Earth's crust and interior and the exhalations of all
life over time, the lower atmosphere is unique in the Solar
System. It is a combination of nitrogen, oxygen, argon,
carbon dioxide, water vapor, and trace gases.

Hydrosphere (Part II, Chapters 7–10)   Earth's waters exist in the atmosphere, on the surface, and in the
crust near the surface. Collectively, these waters form
the **hydrosphere**. That portion of the hydrosphere that
is frozen is the **cryosphere**—ice sheets, ice caps and
fields, glaciers, ice shelves, sea ice, and subsurface

ground ice—and topics related to it are covered in chapters in all four parts of the book. Water of the hydrosphere exists in all three states: liquid, solid (the frozen
cryosphere), and gaseous (water vapor). Water occurs in
two general chemical conditions, fresh and saline (salty).
It exhibits important heat-storage properties. Water is
an extraordinary solvent. Water is the medium of life.
Among the planets in the Solar System, only Earth possesses surface water in such quantity, adding to Earth's
uniqueness.

Lithosphere (Part III, Chapters 11–17)   Earth's crust
and a portion of the upper mantle directly below the crust
form the **lithosphere**. The crust is quite brittle compared

with the layers deep beneath the surface, which move slowly in response to an uneven distribution of heat energy and pressure. In a broad sense, the term *lithosphere* sometimes refers to the entire solid planet. The soil layer is the *edaphosphere* and generally covers Earth's land surfaces. In this text, soils represent a bridge between the lithosphere (Part III) and biosphere (Part IV) and are featured in the first chapter in Part IV, "Soils, Ecosystems, and Biomes."

Biosphere (Part IV, Chapters 18–21)   The intricate, interconnected web that links all organisms with their physical environment is the **biosphere**, or **ecosphere**. The biosphere is the area in which physical and chemical factors form the context of life. The biosphere exists in the overlap among the abiotic, or nonliving, spheres, extending from the seafloor, the upper layers of the crustal rock, to about 8 km (5 mi) into the atmosphere. Life is sustainable within these natural limits. In turn, life processes have powerfully shaped the other three spheres through interactive processes. The biosphere evolves, reorganizes itself at times, faces extinctions, and manages to flourish. Earth's biosphere is the only one known in the Solar System; thus, life as we know it is unique to Earth.

A Spherical Planet

We all have heard that some people believed Earth to be flat. Yet Earth's *sphericity*, or roundness, is not as modern a concept as many think. For instance, more than two millennia ago, the Greek mathematician and philosopher Pythagoras (ca. 580–500 B.C.) determined through observation that Earth is spherical. We do not know what observations led Pythagoras to this conclusion. Can you guess at what he saw to deduce Earth's roundness?

He might have noticed ships sailing beyond the horizon and apparently sinking below the water's surface, only to arrive back at port with dry decks. Perhaps he noticed Earth's curved shadow cast on the lunar surface during an eclipse of the Moon. He might have deduced that the Sun and Moon are not just the flat disks they appear to be in the sky, but are spherical, and that Earth must be a sphere as well.

Earth's sphericity was generally accepted by the educated populace as early as the first century A.D. Christopher Columbus, for example, knew he was sailing around a sphere in 1492; that is one reason why he thought he had arrived in the East Indies.

Earth as a Geoid   Until 1687, the spherical-perfection model was a basic assumption of **geodesy**, the science that determines Earth's shape and size by surveys and mathematical calculations. But in that year, Sir Isaac Newton postulated that Earth, along with the other planets, could not be perfectly spherical. Newton reasoned that the more rapid rotational speed at the equator—the equator being farthest from the central axis of the planet and therefore moving faster—would produce an equatorial bulge as centrifugal force pulls Earth's surface outward. He was convinced that Earth is slightly misshapen into what he termed an *oblate spheroid*, or more correctly, an *oblate ellipsoid* (*oblate* means "flattened"), with the oblateness occurring at the poles.

Earth's equatorial bulge and its polar oblateness are universally accepted and confirmed with tremendous precision by satellite observations. The "geoidal epoch" is our modern era of Earth measurement because Earth is a **geoid**, meaning literally that "the shape of Earth is Earth-shaped." Imagine Earth's geoid as a sea-level surface that is extended uniformly worldwide, beneath the continents. Both heights on land and depths in the oceans are measured from this hypothetical surface. Think of the geoid surface as a balance among the gravitational attraction of Earth's mass, the distribution of water and ice along its surface, and the centrifugal pull caused by Earth's rotation.

Figure 1.9 gives Earth's average polar and equatorial circumferences and diameters. The Greek geographer, astronomer, and librarian Eratosthenes (ca. 276–195 B.C.) first measured Earth's polar circumference more than

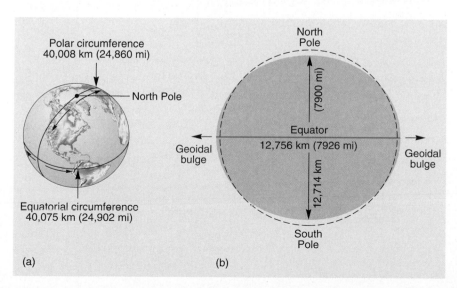

Polar circumference
40,008 km (24,860 mi)

North Pole

Equatorial circumference
40,075 km (24,902 mi)

North Pole

(7900 mi)

Equator
12,756 km (7926 mi)

Geoidal bulge

Geoidal bulge

12,714 km

South Pole

(a)   (b)

FIGURE 1.9 Earth's dimensions. Earth's equatorial and polar circumference (a) and diameter (b). The dashed line is a perfect circle for reference to Earth's geoid.

2200 years ago. The ingenious reasoning by which he arrived at his calculation follows.

Measuring Earth in 247 B.C.

Foremost among early geographers, Eratosthenes served as the librarian of Alexandria in Egypt during the third century B.C. He was in a position of scientific leadership, for Alexandria's library was the finest in the ancient world. Among his achievements was calculating Earth's polar circumference to a high level of accuracy, quite a feat for 247 B.C. Here's how he did it. As you read, follow along on Figure 1.10.

Travelers told Eratosthenes that on June 21 they saw the Sun's rays shine directly to the bottom of a well at Syene, the location of present-day Aswan, Egypt. This meant that the Sun was directly overhead on that day. North of Syene in Alexandria, Eratosthenes knew from his own observations that the Sun's rays never were directly overhead, even at noon on June 21. This day is the longest day of the year and the day on which the noon Sun is at its northernmost position in the sky. He knew objects in Alexandria, unlike objects in Syene, always cast a noontime shadow. Using his knowledge of geometry, Eratosthenes conducted an experiment to explain the different observations.

In Alexandria at noon on June 21, he measured the angle of a shadow cast by an obelisk (a perpendicular column). Knowing the height of the obelisk and measuring the length of the Sun's shadow from its base, he solved the triangle for the angle of the Sun's noon rays, which he determined to be 7.2° off from directly overhead. However, at Syene on the same day, the angle of the Sun's rays was 0° from a perpendicular, arriving from directly overhead.

Geometric principles told Eratosthenes that the distance on the ground between Alexandria and Syene formed an arc of Earth's circumference equal to the angle of the Sun's rays at Alexandria. Since 7.2° is roughly 1/50 of the 360° (360° ÷ 7.2° = 50) in Earth's total circumference, he knew the distance between Alexandria and Syene must represent approximately 1/50 of Earth's total polar circumference.

Next, Eratosthenes determined the surface distance between the two cities as 5000 stadia. A *stadium*, a Greek unit of measure, equals approximately 185 m (607 ft). He then multiplied 5000 stadia by 50 to determine that Earth's polar circumference is about 250,000 stadia. Eratosthenes' calculations convert to roughly 46,250 km (28,738 mi), which is remarkably close to the correct value of 40,008 km (24,860 mi) for Earth's polar circumference. Not bad for 247 B.C.! (Several values can be used for the distance of a Greek stadium—the 185 m used here represents an average value.)

Location and Time on Earth

An essential for geographic science is an internationally accepted, coordinated grid system to determine location on Earth. You can see the importance in Figure 1.16 showing the Cape Hatteras lighthouse location change. The terms *latitude* and *longitude* were in use on maps as early as the first century A.D., with the concepts themselves dating back to Eratosthenes and others.

The geographer, astronomer, and mathematician Ptolemy (ca. A.D. 90–168) contributed greatly to the development of modern maps, and many of his terms and configurations are still used today. Ptolemy divided the circle into 360 degrees (360°), with each degree having 60 minutes (60′) and each minute having 60 seconds (60″) in a manner adapted from the ancient Babylonians. He located places using these degrees, minutes, and seconds. However, the precise length of a degree of latitude and a degree of longitude remained unresolved for the next 17 centuries.

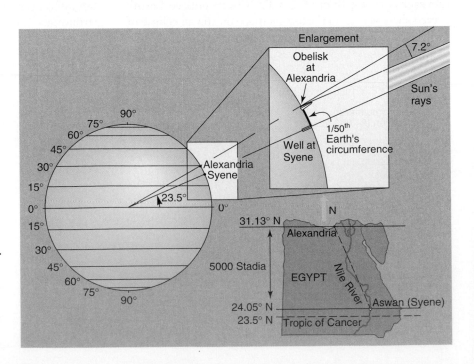

FIGURE 1.10 Eratosthenes' calculation. Eratosthenes' work teaches the value of observing carefully and integrating all observations with previous learning. Calculating Earth's circumference required application of his knowledge of Earth–Sun relationships, geometry, and geography.

Latitude

Latitude *is an angular distance north or south of the equator*, measured from the center of Earth (Figure 1.11a). On a map or globe, the lines designating these angles of latitude run east and west, parallel to the equator (Figure 1.11b). Because Earth's equator divides the distance between the North Pole and the South Pole exactly in half, it is assigned the value of 0° latitude. Thus, latitude increases from the equator northward to the North Pole, at 90° north latitude, and southward to the South Pole, at 90° south latitude.

A line connecting all points along the same latitudinal angle is a **parallel**. In the figure, an angle of 49° north latitude is measured, and, by connecting all points at this latitude, we have the 49th parallel. Thus, *latitude* is the name of the angle (49° north latitude), *parallel* names the line (49th parallel), and both indicate distance north of the equator.

Latitude is readily determined by observing *fixed celestial objects* such as the Sun or the stars, a method dating to ancient times. During daylight hours, the angle of the Sun above the horizon indicates the observer's latitude, after adjustment is made for the season and for the time of day. Because Polaris (the North Star) is almost directly overhead at the North Pole, persons anywhere in the Northern Hemisphere can determine their latitude at night simply by sighting Polaris and measuring its angle above the local horizon. The angle of elevation of Polaris above the horizon equals the latitude of the observation point. Go to this chapter on our web site to see an illustration of how sighting on Polaris determines latitude.

In the Southern Hemisphere, Polaris cannot be seen, because it is below the horizon. Instead, latitude measurement south of the equator is accomplished by sighting on a constellation that points to a celestial location above the South Pole. This indicator constellation is the Southern Cross (*Crux Australis*).

Latitudinal Geographic Zones Natural environments differ dramatically from the equator to the poles. These differences result from the amount of solar energy received, which varies by latitude and season of the year. As a convenience, geographers identify *latitudinal geographic zones* as regions with fairly consistent qualities. "Lower latitudes" are those nearer the equator, whereas "higher latitudes" are those nearer the poles. Make a note of the geographic zone in which each of the five cities whose temperature graphs appear in Figure 5.4 is located.

Figure 1.12 portrays these zones, their locations, and their names: *equatorial* and *tropical*, *subtropical*, *midlatitude*, *subarctic* or *subantarctic*, and *arctic* or *antarctic*. These generalized latitudinal zones are useful for reference and comparison, but they do not have rigid boundaries; rather, think of them as transitioning one to another.

The *Tropic of Cancer* (about 23.5° north parallel) and the *Tropic of Capricorn* (about 23.5° south parallel) are the most extreme northern and southern parallels that experience perpendicular (directly overhead) rays of the Sun at local noon. When the Sun arrives overhead at these tropics, it marks the first day of summer in each hemisphere. (Further discussion of the tropics is in Chapter 2.) The *Arctic Circle* (about 66.5° north parallel) and the *Antarctic Circle* (about 66.5° south parallel) are the parallels farthest from the poles that still experience 24 uninterrupted hours of night, including dawn and twilight, during local winter or of day during local summer.

Longitude

Longitude *is an angular distance east or west of a point on Earth's surface*, measured from the center of Earth (Figure 1.13a). On a map or globe, the lines designating these angles of longitude run north and south (Figure 1.13b). A line connecting all points along the same longitude is a **meridian**. In the figure, a longitudinal angle of 60° E is measured. These meridians run at right angles (90°) to all parallels, including the equator.

Thus, *longitude* is the name of the angle, *meridian* names the line, and both indicate distance east or west of an arbitrary **prime meridian**—a meridian designated as 0° (Figure 1.13b). Earth's prime meridian passes through the

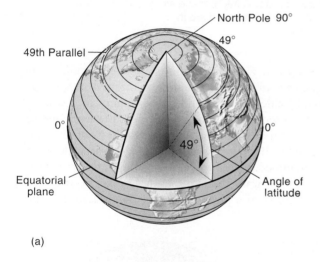

(a)

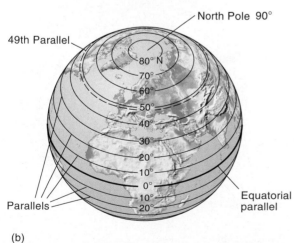

(b)

FIGURE 1.11 Parallels of latitude.
(a) Latitude is measured in degrees north or south of the equator, which is 0°. Earth's poles are at 90°. Note the measurement of 49° latitude. (b) These angles of latitude determine parallels along Earth's surface. Do you know your present latitude?

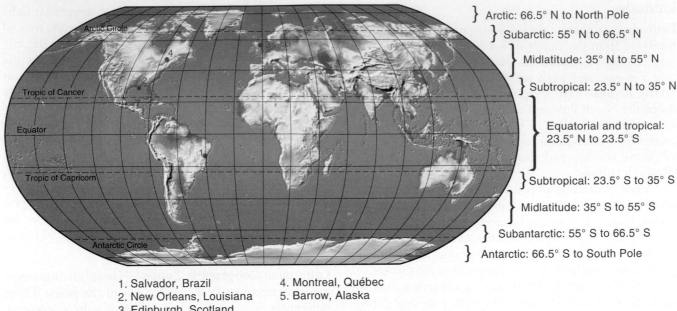

Arctic: 66.5° N to North Pole

Subarctic: 55° N to 66.5° N

Midlatitude: 35° N to 55° N

Subtropical: 23.5° N to 35° N

Equatorial and tropical: 23.5° N to 23.5° S

Subtropical: 23.5° S to 35° S

Midlatitude: 35° S to 55° S

Subantarctic: 55° S to 66.5° S

Antarctic: 66.5° S to South Pole

1. Salvador, Brazil
2. New Orleans, Louisiana
3. Edinburgh, Scotland
4. Montreal, Québec
5. Barrow, Alaska

FIGURE 1.12 Latitudinal geographic zones.
Geographic zones are generalizations that characterize various regions by latitude. Think of these as transitioning into one another over broad areas. Temperature graphs for the five cities noted are in Figure 5.4. Reference that figure as you locate each city in its geographic zone.

old Royal Observatory at Greenwich, England, as set by treaty—the *Greenwich prime meridian.*

Determination of Latitude and Longitude Table 1.2 compares the length of latitude and longitude degrees. Because meridians of longitude converge toward the poles, the actual distance on the ground spanned by a degree of longitude is greatest at the equator (where meridians separate to their widest distance apart) and

diminishes to zero at the poles (where meridians converge). In comparison, note the consistent distance represented by a degree of latitude from equator to poles.

We noted that latitude is determined easily by sighting the Sun or the North Star or by using the Southern Cross as a pointer. In contrast, a method of accurately determining longitude, especially at sea, remained a major difficulty in navigation until the late 1700s. The key to measuring the longitude of a place lies in accurately knowing time.

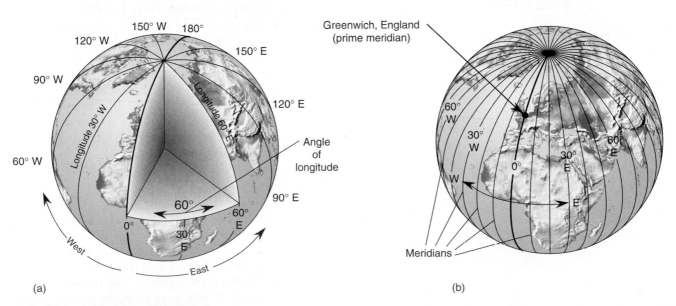

(a)

(b)

FIGURE 1.13 Meridians of longitude.
(a) Longitude is measured in degrees east or west of a 0° starting line, the prime meridian. Note the measurement of 60° E longitude. (b) Angles of longitude measured from the prime meridian determine other meridians. The prime meridian is drawn from the North Pole through the Royal Observatory in Greenwich, England, to the South Pole. North America is west of Greenwich; therefore, it is in the Western Hemisphere. Do you know your present longitude?

Table 1.2	Physical Distances Represented by Degrees of Latitude and Longitude				
	Latitude Degree		**Longitude Degree**		
	Length		**Length**		
Latitudinal Location	**km**	**(mi)**	**km**	**(mi)**	
90° (poles)	111.70	(69.41)	0	(0)	
60°	111.42	(69.23)	55.80	(34.67)	
50°	111.23	(69.12)	71.70	(44.55)	
40°	111.04	(69.00)	85.40	(53.07)	
30°	110.86	(68.89)	96.49	(59.96)	
0° (equator)	110.58	(68.71)	111.32	(69.17)	

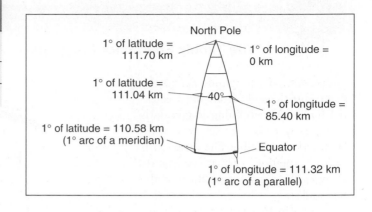

The relation between time and longitude and an exciting chapter in human discovery is the topic of Focus Study 1.2.

Today, using a handheld instrument that reads radio signals from satellites, you can accurately calibrate latitude, longitude, and elevation—and even know where your car is as you drive along. This is the **Global Positioning System (GPS)** technology. News Report 1.2 discusses how GPS works and the dramatic applications of GPS.

Great Circles and Small Circles

Great circles and small circles are important concepts that help summarize latitude and longitude (Figure 1.14). A **great circle** is any circle of Earth's circumference whose center coincides with the center of Earth. An infinite number of great circles can be drawn on Earth. Every meridian is one-half of a great circle that passes through the poles. On flat maps, airline and shipping routes appear to arch their way across oceans and landmasses. These are *great circle routes*, the shortest distance between two points on Earth (discussion with Figure 1.22 is just ahead).

In contrast to meridians, only one parallel is a great circle—the *equatorial parallel*. All other parallels diminish in length toward the poles and, along with any other non-great circles that one might draw, constitute **small circles**. These circles have centers that do not coincide with Earth's center.

(text continued on page 22)

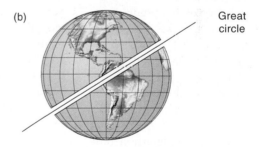

(b)

Great circle

A plane intersecting the globe along a great circle divides the globe into equal halves and passes through its center

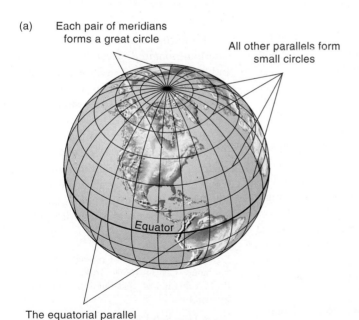

(a) Each pair of meridians forms a great circle

All other parallels form small circles

Equator

The equatorial parallel is a great circle

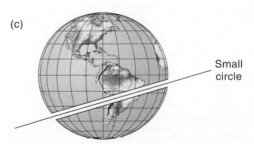

(c)

Small circle

A plane that intersects the globe along a small circle splits the globe into unequal sections—this plane does not pass through the center of the globe

FIGURE 1.14 Great circles and small circles.
(a) Examples of great circles and small circles on Earth. (b) Any plane that divides Earth into equal halves will intersect the globe along a great circle; this great circle is a full circumference of the globe and is the shortest distance between any two surface points. (c) Any plane that splits the globe into unequal portions will intersect the globe along a small circle.

Focus Study 1.2

The Timely Search for Longitude

Unlike latitude, longitude cannot be determined readily from fixed celestial bodies. Determining longitude is particularly critical at sea, where no landmarks are visible. The problem is Earth's rotation, which constantly changes the apparent position of the Sun and stars.

In his historical novel *Shogun*, author James Clavell expressed the frustration of the longitude problem through his pilot, Blackthorn: "Find how to fix longitude and you're the richest man in the world.... The Queen, God bless her, 'll give you ten thousand pound and a dukedom for the answer to the riddle.... Out of sight of land you're always lost, lad."*

In the early 1600s, Galileo explained that longitude could be measured by using two clocks. Any point on Earth takes 24 hours to travel around the full 360° of one rotation (1 day). If you divide 360° by 24 hours, you find that any point on Earth travels through 15° of longitude every hour. Thus, if there were a way to measure time accurately at sea, a comparison of two clocks could give a value for longitude.

One clock would indicate the time back at home port (Figure 1.2.1). The other clock would be reset at local noon each day, as determined by the highest Sun position in the sky (solar zenith). The time difference between the ship and home port would indicate the longitudinal difference traveled: 1 hour for each 15° of longitude. The principle was sound; all that was needed were accurate clocks. Unfortunately, the pendulum clock invented by Christian Huygens in 1656, and accurate on land, did not work on the rolling deck of a ship at sea!

In 1707, the British lost four ships and 2000 men in a sea tragedy that was blamed specifically on the longitude problem. In response, Parliament passed an act in 1714—"Publik Reward...to Discover the Longitude at

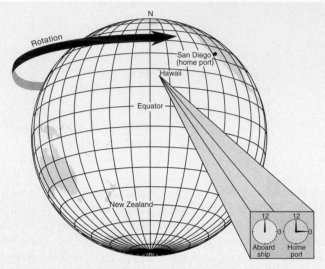

FIGURE 1.2.1 Clock time determines longitude.
Using two clocks on a ship to determine longitude. For example, if the shipboard clock reads local noon and the clock set for home port reads 3:00 P.M., ship time is 3 hours earlier than home time. Therefore, calculating 3 hours at 15° per hour puts the ship at 45° west longitude from home port.

Sea"—and authorized a prize worth more than $2 million in today's dollars to the first successful inventor of an accurate seafaring clock. The Board of Longitude was established to judge any devices submitted.

John Harrison, a self-taught country clockmaker, began work on the problem in 1728 and finally produced his brilliant marine chronometer, known as *Number 4*, in 1760. The clock was tested on a voyage to Jamaica in 1761. When taken ashore and compared to land-based longitude, Harrison's ingenious *Number 4* was only 5 seconds slow, an error that translates to only 1.25′ or 2.3 km (1.4 mi), well within Parliament's standard. After many delays, Harrison finally received most of the prize money in his last years of life.

With his marine clocks, John Harrison tested the waters of space–time. He succeeded, against all odds, in using the fourth—temporal—dimension to link points on the three-dimensional globe. He wrested

the world's whereabouts from the stars, and locked the secret in a pocket watch.†

From that time it was possible to determine longitude accurately on land and sea, as long as everyone agreed upon a meridian to use as a reference for time comparisons—after more than a century that became the Royal Observatory in Greenwich, England. In this modern era of atomic clocks and satellites in mathematically precise orbits, we have far greater accuracy available for the determination of longitude on Earth's surface and a basis for precise navigation.

*From J. Clavell, *Shogun*, Copyright © 1975 by James Clavell (New York: Delacorte Press, a division of Dell Publishing Group, Inc.), p. 10.

†From Dava Sobel, *Longitude, The Story of a Lone Genius Who Solved the Greatest Scientific Problem of His Time* (New York: Walker and Co., 1995), p. 175.

News Report 1.2

GPS: A Personal Locator

The Global Positioning System (GPS) comprises 24 orbiting satellites, in six orbital planes, that transmit navigational signals for Earth-bound use (backup GPS satellites are in orbital storage as replacements). Originally devised in the 1970s by the Department of Defense for military purposes, the present system is commercially available worldwide. Think of the satellites as a constellation of navigational beacons with which you interact to determine your unique location. Every possible square meter of Earth's surface has its own address relative to the latitude–longitude grid.

A receiver, some about the size of a pocket radio or smaller, senses signals from three or more satellites at the same time and triangulates the receiver's location. The key is the time lag the signals take, travelling at the speed of light (Figure 1.2.1). The receiver reports a calculatation of latitude and longitude within 10-m accuracy (33 ft) and elevation within 15 m (49 ft). *Differential GPS (DGPS)* increases accuracy by comparing readings with another base station (reference receiver) for a differential correction (Figure 1.2.2).

The GPS satellite measurements help refine our knowledge of Earth's exact shape and how this shape is changing. GPS is useful for diverse applications, such as ocean navigation, land surveying, tracking small changes in Earth's crust, managing the movement of fleets of trucks, mining and resource mapping, tracking wildlife migration and behavior, police and security work, and environmental planning. Measuring and mapping an ominous bulge on an Oregon volcano is another present-day application. GPS also is useful to the backpacker and sportsperson.

Relative to earthquakes in southern California, the USGS, Jet Propulsion Laboratory (JPL), NASA, and NSF, among other partners, operate the GPS Observation Office that mon-

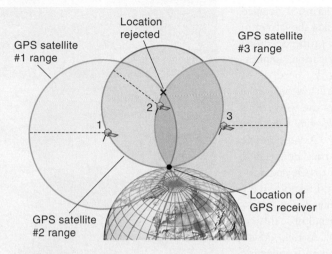

FIGURE 1.2.1 Using satellites to triangulate location through GPS.
Imagine a ranging sphere around each of three GPS satellites. These spheres intersect at two points, one easily rejected because it is some distance above Earth and the other at the true location of the GPS receiver. The three spheres can only intersect at two points. The key is gauging the distance between the satellites and the GPS receiver—signals travel at the speed of light.

FIGURE 1.2.2 GPS in action.
A GPS unit in operation for surveying and location analysis in an environmental study. [Photo courtesy of Trimble Navigation Ltd., Sunnyvale, California.]

itors a network of 250 continuously operating GPS seismic stations—the Southern California Integrated GPS Network (SCIGN, http://www.scign.org/). This system can record fault movement as small as 1 mm (0.04 in.).

Commercial airlines use GPS to improve accuracy of routes flown and thus increase fuel efficiency. Scientists used GPS to accurately determine the height of Mount Everest in the Himalayan Mountains—now 8850 m compared to the former 8848 m (29,035 ft; 29,028 ft) (Figure 1.2.3). In contrast, GPS measurements of Mount Kilimanjaro lowered its summit from 5895 m to 5892 m (19,340 ft; 19,330 ft).

(continued)

News Report 1.2 (continued)

In *precision agriculture*, farmers use GPS to determine crop yields on specific parts of their farms. A detailed plot map is made to guide the farmer in adding more fertilizer, proper seed distribution, precise irrigation applications, or other work needed. A computer and GPS unit on board the farm equipment guides the work. This is the science of *variable-rate technology*, made possible by GPS. As an example among others, California Polytechnic State University, San Luis Obispo, California State University, Fresno, and the University of California, Davis, established Precision Agriculture (http://www.precisionag.org/). A full, downloadable curriculum with supporting faculty is posted on their web site.

The importance of GPS to geography is obvious because this precise technology reduces the need to maintain ground control points for location, mapping, and spatial analysis. Instead, geographers working in the field can determine their position accurately as they work. Boundaries and data points in a study area are easily determined and entered into a database, reducing the need for traditional surveys.

For this and myriad other applications, GPS sales are exceeding $10 billion a year. Additional frequencies

FIGURE 1.2.3 GPS used to measure Everest's summit height. Installation of a Trimble 4800 GPS unit at Earth's highest benchmark, only 18 m (60 ft) below the 8850-m (29,035-ft) summit of Mount Everest. Scientists using GPS data accurately analyzed the height of the world's tallest mountain and the rate that the mountain range is moving due to tectonic forces. [Photo and GPS installation by adventurer/climber Wally Berg, May 20, 1998.]

were added in 2003 and 2006, which increased accuracy significantly. For a GPS overview, see http://www.colorado.edu/geography/gcraft/notes/gp/gps_f.html.

Figure 1.15 combines latitude and parallels with longitude and meridians to illustrate Earth's complete coordinate grid system. Note the dot that marks our measurement of 49° N and 60° E, a location in western Kazakstan. Next time you look at a world globe, follow the parallel and meridian that converge on your location.

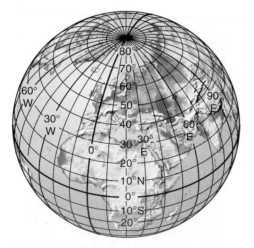

FIGURE 1.15 Earth's coordinate grid system. Latitude and parallels, longitude and meridians allow all places on Earth to be precisely located. The red dot is at 49° N latitude and 60° E longitude.

Where Is the Lighthouse?

A century of storm surges and a rising sea level increased losses of sand from coastal barrier islands. This deteriorating condition necessitated the dramatic 1999 move of the Cape Hatteras lighthouse more than 884 m (2900 ft) inland. Ironically, this is about the same distance the lighthouse was from the ocean when first built in 1870 (the lower-left photo insert in Figure 1.16 shows the path along which it was moved).

This move placed the lighthouse in a new location relative to Earth's surface coordinates. The new location is 12" (seconds) south and 48" (seconds) west from the former site. See coverage of this vulnerable barrier island chain in Chapter 16, Figure 16.17.

A quick survey of coastlines begs the question about how many more "relocations" of thousands of coastal structures we might see in the decades ahead as global sea level continues to rise from the thermal expansion of warmer water and the glacial and ice-sheet meltdown underway. Location is important in geographic science.

Prime Meridian and Standard Time

Coordination of international trade, airline schedules, business and agricultural activities, and daily living depends on a worldwide time system. Today, we take

FIGURE 1.16 Lighthouse at a new planetary address. Hatteras Lighthouse is moved a safer distance from rising seas and storm surges. The inset photos show the path taken and the massive lighthouse with its indicator spiral paint design. Can you tell the direction Hatteras Lighthouse was moved by comparing the new coordinates (on the white card) and old coordinates (etched in marble) in the photo? [All photos by Bobbé Christopherson.]

for granted standard time zones and an agreed-upon prime meridian, but such a standard is a relatively recent development.

Setting time was not a great problem in small European countries, most of which are less than 15° wide. But in North America, which spans more than 90° of longitude (the equivalent of six 15° time zones), the problem was serious. In 1870, railroad travelers going from Maine to San Francisco made 22 adjustments to their watches to stay consistent with local time.

In Canada, Sir Sanford Fleming led the fight for standard time and for an international agreement on a prime meridian. His struggle led the United States and Canada to adopt a standard time in 1883. Today, only three adjustments are needed in the continental United States—from Eastern Standard Time to Central, Mountain, and Pacific—and four changes across Canada.

Twenty-seven countries attended the 1884 International Meridian Conference in Washington, DC. Before that year, most nations used their own national capital as a prime meridian for their land maps, whereas more than 70% of the world's merchant ships were using Greenwich as a prime meridian on marine charts. After lengthy debate at the conference, most participating nations chose the highly respected Royal Observatory at Greenwich, London, England, as the place for the prime meridian of 0° longitude for all maps. Thus, a world standard was set—**Greenwich Mean Time (GMT)**—and a consistent Universal Time was established (see http://www.mt2000.co.uk/ meridian/place/plco0a1.htm).

The basis of time is that Earth revolves 360° every 24 hours, or 15° per hour (360° ÷ 24 = 15°). Thus, a time zone of 1 hour is established, spanning 7.5° on either side of a *central meridian.* Assuming it is 9:00 P.M. in Greenwich, then it is 4:00 P.M. in Baltimore (+5 hr), 3:00 P.M. in Oklahoma City (+6 hr), 2:00 P.M. in Salt Lake City (+7 hr), 1:00 P.M. in Seattle and Los Angeles (+8 hr), noon in Anchorage (+9 hr), and 11:00 A.M. in Honolulu (+10 hr). To the east, it is midnight in Ar Riyāḍ, Saudi Arabia (−3 hr). The designation A.M. is for *ante meridiem,* "before noon," whereas P.M. is for *post meridiem,* "after noon." A 24-hour clock avoids the use of these designations: 3 P.M. is stated as 15:00 hours; 3 A.M. is 3:00 hours.

As you can see from the modern international time zones in Figure 1.17, national boundaries and political considerations distort time boundaries. For example, China spans four time zones, but its government decided to keep the entire country operating at the same time. Thus, in some parts of China clocks are several hours off from what the Sun is doing. In the United States, parts of Florida and west Texas are in the same time zone.

International Date Line   An important corollary of the prime meridian is the 180° meridian on the opposite side of the planet. This meridian is the **International Date Line** and marks the place where each day officially begins (at 12:01 A.M.). From this "line" the new day sweeps westward. This *westward* movement of time is created by Earth's turning *eastward* on its axis.

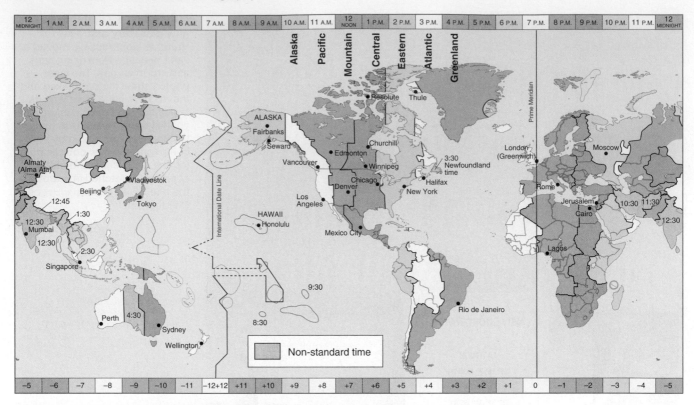

FIGURE 1.17 Modern international standard time zones.
Numbers along the bottom of the map indicate how many hours each zone is earlier (plus sign) or later (minus sign) than Coordinated Universal Time (UTC) at the prime meridian. If it is 7 P.M. in Greenwich, determine the present time in Moscow, London, Halifax, Chicago, Winnipeg, Denver, Los Angeles, Fairbanks, Honolulu, Tokyo, and Singapore. [Adapted from Standard Time Zone Chart of the World, Defense Mapping Agency, Bethesda, Maryland. See http://aa.usno.navy.mil/faq/docs/world_tzones.html.]

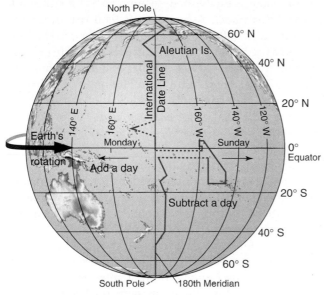

FIGURE 1.18 International Date Line.
International Date Line location, approximately along the 180th meridian (see the IDL location on Figure 1.17). Officially, you gain 1 day crossing the IDL from east to west.

At the International Date Line, the west side of the line is always one day ahead of the east side. No matter what time of day it is when the line is crossed, the calendar changes a day (Figure 1.18). Note in the illustration the departures from the IDL and the 180° meridian; this deviation is due to local administrative and political preferences.

In 1995 when Kiribati moved the IDL designation to east of its islands, placing them west of the IDL, it could claim "first dawn" on Earth. This curious political move did not apply to the ocean between its territorial waters and the 180th meridian—note the dotted lines on the map—as political control of island nations extends only 3.5 nautical miles (4 mi) offshore.

Locating the date line in the sparsely populated Pacific Ocean minimizes most local confusion. However, early explorers before the date-line concept were "lost." For example, Magellan's crew returned from the first circumnavigation of Earth in 1522, confident from their ship's log that the day of their arrival home was Wednesday, September 7. They were shocked when informed by insistent local residents that it was actually Thursday, September 8. Of course, without an International Date Line they had no idea that they must advance a day somewhere when sailing around the world in a westward direction. Imagine the confusion as the arriving crew knelt down on the dock accounting for each day in their log!

Coordinated Universal Time For decades, Greenwich Mean Time from the Royal Observatory's astronomical clocks was the world's Universal Time (UT) standard for accuracy. GMT was broadcast using radio time signals as early as 1910. Taking the initiative in 1912, the French government called a gathering of nations to better coordinate the various radio time signals from many countries. At this conference, GMT was made standard, and a new organization was set up to be the custodian of "exact" time—the International Bureau of Weights and Measures (BIPM) (see **http://www.bipm.fr/**). Progress was made in accurately measuring time with the invention of a quartz clock in 1939 and atomic clocks in the early 1950s.

Accurate time is not a matter of simply keeping track of Earth's rotation, since it is slowing down from the tug of tidal forces and rearrangement of water. Note that 150 million years ago a "day" was 22 hours long, and 150 million years in the future a "day" will be approaching 27 hours in length.

The **Coordinated Universal Time (UTC*)** time-signal system replaced GMT universal time in 1972 and became the legal reference for official time in all countries. Although the prime meridian still runs through Greenwich, UTC is based on average time calculations collected worldwide from 55 locations by the BIPM near Paris and broadcast worldwide. There are 100 clocks reporting from the U.S. Naval Observatory alone. You might still see official UTC referred to as GMT or Zulu time.

Regular vibrations (natural frequency) of cesium atoms in primary standard clocks measure the length of a second and UTC—accuracies now range down to 15 decimal places, or nanoseconds of time. The newest clocks are so accurate they sense where they are in space and time relative to Earth's geoid and mass! Time and Frequency Services of the National Institute for Standards and Technology (NIST), U.S. Department of Commerce, operates several of the most advanced clocks. (For the time, call 303-499-7111 or 808-335-4363, or see **http://nist.time.gov/** for UTC.) In Canada, the Institute for Measurement Standards, National Research Council Canada, participates in determining UTC (English, 613-745-1578; French, 613-745-9426; for more see **http://www.nrc.ca/**).

Daylight Saving Time   In 70 countries, mainly in the temperate latitudes, time is set ahead 1 hour in the spring and set back 1 hour in the fall—a practice known as **daylight saving time**. The idea to extend daylight for early evening activities (at the expense of daylight in the morning) was first proposed by Benjamin Franklin. It was not adopted until World War I and again in World War II, when Great Britain, Australia, Germany, Canada, and the United States used the practice to save energy (1 less hour of artificial lighting needed).

In 1986 and again in 2007, the United States and Canada increased daylight saving time. Time "springs forward" 1 hour on the second Sunday in March and "falls back" an hour on the first Sunday in November, except in

a few places that do not use daylight saving time (Hawai'i, Arizona, and Saskatchewan). In Europe, the last Sundays in March and October are used to begin and end what is called "summer-time period." (See **http://webexhibits.org/daylightsaving/**.)

Maps, Scales, and Projections

The earliest known graphic maps date to 2300 B.C., when the Babylonians used clay tablets to record information about the region of the Tigris and Euphrates Rivers (the area of modern-day Iraq). Today, the making of maps and charts is a specialized science as well as an art, blending aspects of geography, engineering, mathematics, graphics, computer science, and artistic specialties. It is similar in ways to architecture, in which aesthetics and utility combine to produce a useful product.

A **map** is a generalized view of an area, usually some portion of Earth's surface, as seen from above and greatly reduced in size. **Cartography** is that part of geography that involves mapmaking. Maps are critical tools with which geographers depict spatial information and analyze spatial relationships.

We all use maps at some time to visualize our location in relation to other places, or maybe to plan a trip, or to coordinate commercial and economic activities. Have you found yourself looking at a map, planning real and imagined adventures to far-distant places? Maps are wonderful tools! Understanding a few basics about maps is essential to our study of physical geography.

The Scale of Maps

Architects, toy designers, and mapmakers have something in common: They all create scale models. They reduce real things and places to the more convenient scale of a drawing; a model car, train, or plane; a diagram; or a map. An architect renders a blueprint of a structure to guide the building contractors, selecting a scale so that a centimeter (or inch) on the drawing represents so many meters (or feet) on the proposed building. Often, the drawing is 1/50 to 1/100 of real size.

The cartographer does the same thing in preparing a map. The ratio of the image on a map to the real world is **scale**; it relates a unit on the map to a similar unit on the ground. A 1:1 scale means that a centimeter on the map represents a centimeter on the ground (although this is an impractical map scale, for the map is as large as the area mapped!). A more appropriate scale for a local map is 1:24,000, in which 1 unit on the map represents 24,000 identical units on the ground.

Cartographers express map scale as a representative fraction, a graphic scale, or a written scale (Figure 1.19). A *representative fraction* (*RF*, or *fractional scale*) is expressed with either a colon or a slash, as in 1:125,000 or 1/125,000. No actual units of measurement are mentioned because any unit is applicable as long as both parts of the fraction are in the same unit: 1 cm to 125,000 cm,

---

*UTC is in use because agreement was not reached on whether to use the English word order, CUT, or the French order, TUC. UTC was the compromise and is recommended for all timekeeping applications; use of the term GMT is discouraged.

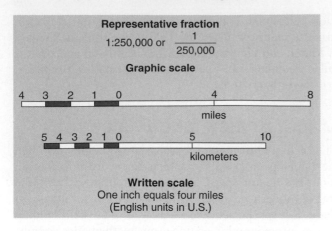

FIGURE 1.19 Map scale.
Three common expressions of map scale—representative fraction, graphic scale, and written scale.

1 in. to 125,000 in., or even 1 arm length to 125,000 arm lengths, and so on.

A *graphic scale*, or *bar scale*, is a bar graph with units to allow measurement of distances on the map. An important advantage of a graphic scale is that if the map is enlarged or reduced, the graphic scale enlarges or reduces along with the map. In contrast, written and fractional scales become incorrect with enlargement or reduction. As an example, you can shrink a map from 1:24,000 to 1:63,360, but the scale will still say "1 in. to 2000 ft," instead of the new correct scale of 1 in. to 5280 ft (1 mi).

Scales are *small*, *medium*, and *large*, depending on the ratio described. In relative terms, a scale of 1:24,000 is a large scale, whereas a scale of 1:50,000,000 is a small scale. The greater the denominator in a fractional scale (or the number on the right in a ratio expression), the smaller the scale and the more abstract the map is in relation to what is being mapped. Examples of selected representative

Table 1.3	Sample Representative Fractions and Written Scales for Small-, Medium-, and Large-Scale Maps		
System	**Scale Size**	**Representative Fraction**	**Written System Scale**
English	Small	1:3,168,000	1 in. = 50 mi
		1:1,000,000	1 in. = 16 mi
		1:250,000	1 in. = 4 mi
	Medium	1:125,000	1 in. = 2 mi
		1:63,360	1 in. = 1 mi
		(or 1:62,500)	
	Large	1:24,000	1 in. = 2000 ft
System		**Representative Fraction**	**Written System Scale**
Metric		1:1,000,000	1 cm = 10.0 km
		1:25,000	1 cm = 0.25 km
		1:10,000	1 cm = 0.10 km

fractions and written scales are listed in Table 1.3 for small-, medium-, and large-scale maps.

If there is a globe or map available in your library or classroom, check to see the scale at which it was drawn. See if you can find examples of representative, graphic, and written scales on wall maps, highway maps, and in atlases. In general, do you think a world globe is a small- or a large-scale map of Earth's surface?

Map Projections

A globe is not always a helpful map representation of Earth. When you go on a trip, you need more detailed information than a globe can provide. Consequently, to provide local detail, cartographers prepare large-scale *flat maps*, which are two-dimensional representations (scale models) of our three-dimensional Earth. Unfortunately, such conversion from three dimensions to two causes distortion.

A globe is the only true representation of *distance*, *direction*, *area*, *shape*, and *proximity*. A flat map distorts these properties. Therefore, in preparing a flat map, the cartographer must decide which characteristic to preserve, which to distort, and how much distortion is acceptable. To understand this problem, consider these important properties of a *globe*:

- Parallels always are parallel to each other, always are evenly spaced along meridians, and always decrease in length toward the poles.
- Meridians converge at both poles and are evenly spaced along any individual parallel.
- The distance between meridians decreases toward poles, with the spacing between meridians at the 60th parallel equal to one-half the equatorial spacing.
- Parallels and meridians always cross each other at right angles.

The problem is that all these qualities cannot be reproduced on a flat surface. Simply taking a globe apart and laying it flat on a table illustrates the challenge faced by cartographers (Figure 1.20). You can see the empty spaces that open up between the sections, or gores, of the globe. This reduction of the spherical Earth to a flat surface is a **map projection**. Thus, no flat map projection of Earth can ever have all the features of a globe. Flat maps always possess some degree of distortion—much less for large-scale maps representing a few kilometers; much more for small-scale maps covering individual countries, continents, or the entire world.

Properties of Projections There are many map projections, four of which are shown in Figure 1.21. *The best projection is always determined by its intended use.* The major decisions in selecting a map projection involve the properties of **equal area** (equivalence) and **true shape** (conformality). A decision for one property sacrifices the other, for they cannot be shown together on the same flat map.

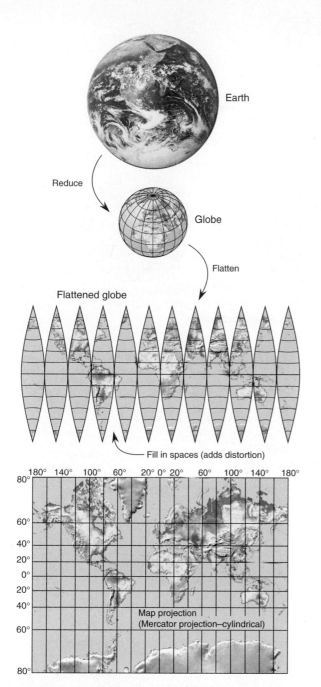

FIGURE 1.20 From globe to flat map.
Conversion of the globe to a flat map projection requires decisions about which properties to preserve and the amount of distortion that is acceptable. [NASA astronaut photo from *Apollo 17*, 1972.]

If a cartographer selects equal area as the desired trait—for example, for a map showing the distribution of world climates—then true shape must be sacrificed by *stretching* and *shearing*, which allow parallels and meridians to cross at other than right angles. On an equal-area map, a coin covers the same amount of surface area no matter where you place it on the map. In contrast, if a cartographer selects the property of true shape, such as for a map used for navigational purposes, then equal area must be sacrificed and the scale will actually change from one region of the map to another.

The Nature and Classes of Projections Despite the fact that modern cartographic technology uses mathematical constructions and computer-assisted graphics, the word *projection* is still used. The term comes from times past, when geographers actually projected the shadow of a wire-skeleton globe onto a geometric surface. The wires represented parallels, meridians, and outlines of the continents. A light source then cast a shadow pattern of latitude and longitude lines from the globe onto various geometric surfaces, such as a *cylinder*, *plane*, or *cone*.

Figure 1.21 illustrates the general classes of map projections and the perspectives from which they are generated. The classes shown include the cylindrical, planar (or azimuthal), and conic. Another class of projections, which cannot be derived from this physical-perspective approach, is the nonperspective oval shape. Still other projections derive from purely mathematical calculations.

With projections, the contact line or contact point between the wire globe and the projection surface—a *standard line* or *standard point*—is *the only place where all globe properties are preserved.* Thus, a *standard parallel* or *standard meridian* is a standard line true to scale along its entire length without any distortion. Areas away from this critical tangent line or point become increasingly distorted. Consequently, this line or point of accurate spatial properties should be centered by the cartographer on the area of interest.

The commonly used **Mercator projection** (from Gerardus Mercator, A.D. 1569) is a cylindrical projection (Figure 1.21a). The Mercator is a true-shape projection, with meridians appearing as equally spaced straight lines and parallels appearing as straight lines that are spaced closer together near the equator. The poles are infinitely stretched, with the 84th north parallel and 84th south parallel fixed at the same length as that of the equator. Note in Figures 1.20 and 1.21a that the Mercator projection is cut around the 80th parallel in each hemisphere because of the severe distortion at higher latitudes.

Unfortunately, Mercator classroom maps present false notions of the size (area) of midlatitude and poleward landmasses. A dramatic example on the Mercator projection is Greenland, which looks bigger than all of South America. In reality, Greenland is an island only one-eighth the size of South America and is actually 20% smaller than Argentina alone.

The advantage of the Mercator projection is that a line of constant direction, known as a **rhumb line**, or *loxodrome*, is straight and thus facilitates plotting directions between two points (see Figure 1.22). Thus, the Mercator projection is useful in navigation and is the standard for nautical charts prepared by the National Ocean Service since 1910 (formerly, U.S. Coast and Geodetic Survey).

The *gnomonic*, or *planar*, *projection* in Figure 1.21b is generated by projecting a light source at the center of a globe onto a plane that is tangent to (touching) the

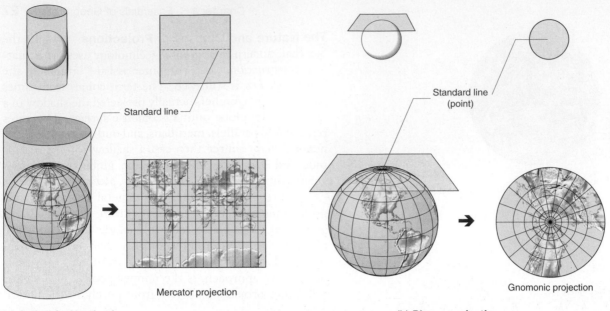

(a) **Cylindrical projection**

Mercator projection

Standard line

(b) **Planar projection**

Standard line (point)

Gnomonic projection

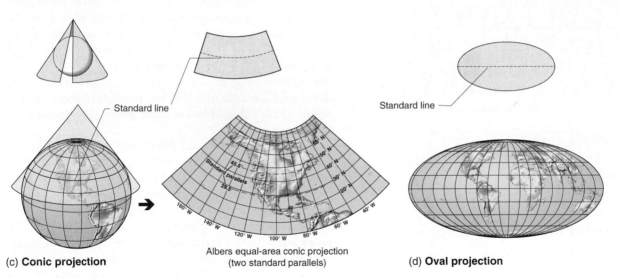

(c) **Conic projection**

Standard line

Albers equal-area conic projection
(two standard parallels)

(d) **Oval projection**

Standard line

FIGURE 1.21 Classes of map projections.
Four general classes and perspectives of map projections—(a) cylindrical, (b) planar, (c) conic, and (d) oval projections.

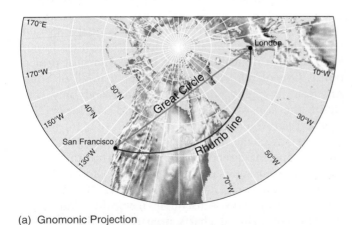

(a) Gnomonic Projection

(b) Mercator Projection (conformal, true shape)

FIGURE 1.22 Determining great circle routes.
A gnomonic projection (a) is used to determine the shortest distance—great circle route—between San Francisco and London, because on this projection the arc of a great circle is a straight line. This great circle route is then plotted on a Mercator projection (h), which has true compass direction. Note that straight lines of constant direction (or bearing) on a Mercator projection—rhumb lines—are not the shortest route in terms of distance.

globe's surface. The resulting severe distortion prevents showing a full hemisphere on one projection. However, a valuable feature is derived: *All great circle routes, which are the shortest distance between two points on Earth's surface, are projected as straight lines* (Figure 1.22a). The great circle routes plotted on a gnomonic projection then can be transferred to a true-direction projection, such as the Mercator, for determination of precise compass headings (Figure 1.22b).

For more information on maps used in this text and standard map symbols, turn to Appendix A, Maps in This Text and Topographic Maps. Topographic maps are essential tools of landscape analysis. Geographers, other scientists, travelers, and anyone visiting the outdoors may use topographic maps. U.S. Geological Survey (USGS) topographic maps appear in several chapters of this text. Perhaps you have used a "topo" map in planning a hike. *The National Map* that the USGS says is "The Nation's Topographic Map for the 21st Century," discussed in the appendix, also is available on the Internet. Go to **http://nationalmap.gov/**, click through "The National Map Viewer," and find your campus and your home town; the topographic shading shows you the texture of the landscape.

Remote Sensing and GIS

Geographers probe, analyze, and map our home planet through remote sensing and geographic information systems (GIS). These technologies enhance our understanding of Earth. Among many subjects, geographers use remote-sensing data to study humid and arid lands, natural and economic vegetation, snow and ice, Earth energy budgets, seasonal variation of atmospheric and oceanic circulation, sea-level measurements, atmospheric chemistry, geologic features and events, changes in the timing of seasons, and the human activities that produce global change.

Remote Sensing

In this era of observations from orbit outside the atmosphere, from aircraft within it, and from remote submersibles in the oceans, scientists obtain a wide array of remotely sensed data (Figure 1.23). Remote sensing is nothing new to humans; we do it with our eyes as we scan the environment, sensing the shape, size, and color of objects from a distance, registering energy from the visible-wavelength portion of the electromagnetic spectrum. Similarly, when a camera views the wavelengths for which its film or sensor is designed (visible light or infrared), it remotely senses energy that is reflected or emitted from a scene.

Our eyes and cameras are familiar means of obtaining **remote-sensing** information about a distant subject without having physical contact. Aerial photographs have been used for years to improve the accuracy of surface maps faster and more cheaply than can be done by on-site

surveys. Deriving accurate measurements from photographs is the realm of **photogrammetry**, an important application of remote sensing.

Remote sensors on satellites, the International Space Station, and other craft sense a broader range of wavelengths than can our eyes. They can be designed to "see" wavelengths shorter than visible light (ultraviolet) and wavelengths longer than visible light (infrared and microwave radar).

Following launch, satellites can be set in several types of orbits: geostationary, polar, and Sun-synchronous, illustrated in Figure 1.24. Figure 1.24a shows a geostationary (geosynchronous) orbit, typically 35,790 km (22,239 mi) altitude, where satellites pace Earth's rotation speed and therefore remain "parked" above a specific location, usually the equator, which provides a full hemisphere and is good for recording weather; an example is *GOES*.

Figure 1.24b shows a polar-orbiting satellite typically residing between 200 km and 1000 km (124 mi−621 mi) altitude and passing over every portion of the globe as Earth rotates beneath it, completing each orbit in 90 minutes. Such satellites are good for difficult regions such as the poles, for environmental studies, and for spy surveillance; examples include *Terra* and *Aqua*.

Figure 1.24c demonstrates a Sun-synchrous orbit, shifting its 600 km to 800 km (373 mi to 497 mi) altitude, near-polar track, about 1° per day (365 days in a year, 360° in a circle), providing a surface below in constant sunlight for visible-wavelength images or in constant darkness for longwave radiation images.

Satellites do not take conventional-film photographs. Rather, they record *images* that are transmitted to Earth-based receivers in a way similar to television satellite transmissions or a digital camera. A scene is scanned and broken down into pixels (*pic*ture *el*ements), each identified by coordinates named *lines* (horizontal rows) and *samples* (vertical columns). For example, a grid of 6000 lines and 7000 samples forms 42,000,000 pixels, providing great detail. The large amount of data needed to produce a single image requires computer processing and data storage at ground stations.

Digital data are processed in many ways to enhance their utility: simulated natural color, "false" color to highlight a particular feature, enhanced contrast, signal filtering, and different levels of sampling and resolution. Active and passive are two types of remote-sensing systems.

Active Remote Sensing   Active systems direct a beam of energy at a surface and analyze the energy reflected back. An example is *radar* (*ra*dio *d*etection *a*nd *r*anging). A radar transmitter emits short bursts of energy that have relatively long wavelengths (0.3 to 10 m) toward the subject terrain, penetrating clouds and darkness. Energy reflected back to a radar receiver for analysis is known as *backscatter*. An example is the computer image of wind and sea-surface patterns over the Pacific in Figure 6.6a,

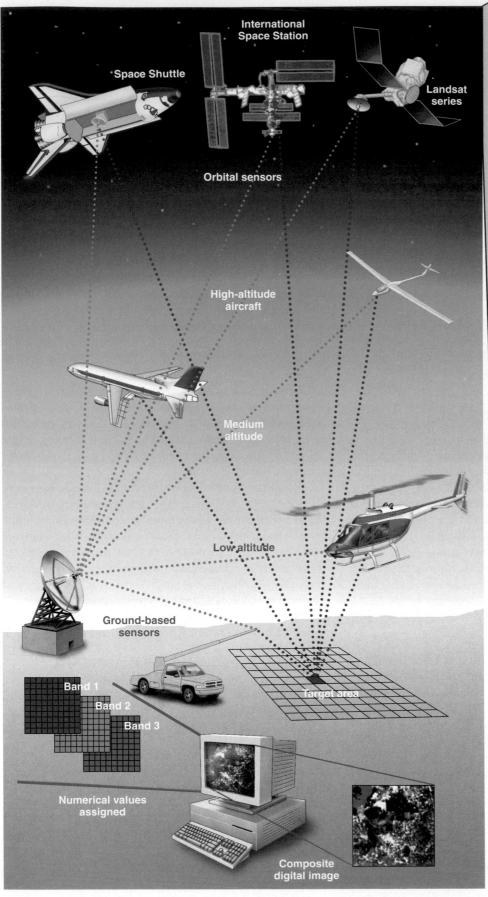

A sample of orbital platforms

Envisat: ESA environment-monitoring satellite; 10 sensors, including a next-generation radar.

ERBS: Earth Radiation Budget Satellite.

GOES: Weather monitoring and forecasting.

JASON: Measures sea level heights.

Landsat: First in 1972 to *Landsat-7* in 1999, millions of images for Earth systems science and global change.

NOAA: First in 1978 through *NOAA-15, -16*, -17, and -18 now operating, global data, short- and long-term weather forecasts.

RADARSAT-1, -2: Synthetic Aperature Radar in near polar orbit operated by Canada Space Agency.

SciSat-1: Analyzes trace gases, thin clouds, atmospheric aerosols with Arctic focus.

SeaStar: Carries the SeaWiFS (Sea-viewing Wide Field-of-View instrument) to observe Earth's oceans and microscopic marine plants.

Terra and **Aqua:** Environmental change, error-free surface images, cloud properties, through five instrument packages.

TOMS-EP: Total Ozone Mapping Spectrometer, monitoring stratospheric ozone, similar instruments on *NIMBUS-7* and *Meteor-3*.

TOPEX/POSEIDON: Measures sea level heights.

TRMM: Tropical Rainfall Measuring Mission, includes lightning detection, and global energy budget measurements.

UARS: Since 1991 measuring atmospheric chemistry and ozone layer changes.

For more info see:
http://www.nasa.gov/centers/goddard/mission/index.html

FIGURE 1.23 Remote-sensing technologies.
Remote-sensing technology measures and monitors Earth's systems from orbiting spacecraft, aircraft in the atmosphere, and ground-based sensors. Various wavelengths (bands) are collected from sensors. Computers process the data to produce digital images for analysis. A sample of remote-sensing platforms is in the margin. The Space Shuttle is shown in its inverted orbital flight mode. (Illustration is not to scale.)

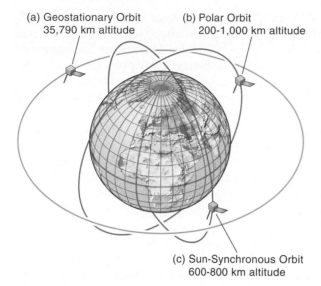

FIGURE 1.24 Three satellite orbital paths.
(a) Geostationary (geosynchronous) orbit, typically 35,790 km altitude. (b) Polar orbiting satellites typically are between 200 and 1000 km altitude. (c) Sun-synchronous orbits shift their near-polar track at 600 to 800 km altitude.

(a) Geostationary Orbit 35,790 km altitude
(b) Polar Orbit 200-1,000 km altitude
(c) Sun-Synchronous Orbit 600-800 km altitude

developed from 150,000 radar-derived measurements made on a single day by the *Seasat* satellite.

In addition, NASA sent imaging radar systems into orbit on several Space Shuttles. Shuttle missions in 1994 by *Endeavour* and *Atlantis* made dramatic contributions to

Earth observations using radar and other sensors to study stratospheric ozone, weather, oceanography, volcanic activity, earthquakes, water resources, and biogeography, among many subjects.

One Space Shuttle mission in September 1994 was appropriately loaded with radar sensors to study volcanoes. Only 8 hours after launch, the Kliuchevskoi Volcano on the Kamchatka Peninsula of Russia erupted unexpectedly. Previously, this volcano had erupted in 1737 and 1945. The Shuttle radar was able to see through ash and smoke and expose lava flows and the volcanic eruption in dramatic images (Figure 1.25). Kliuchevskoi Volcano again erupted in May 2007. Astronaut Mission Specialist Dr. Thomas Jones operated the radar and camera to make the image and photo in the figure. He is profiled in a Career Link on the *Geosystems* web site. Chapter 12 discusses volcanic processes.

Several radar-imaging satellites are operating. *QuikSCAT* (for quick scatterometer) was launched in 1999 into a Sun-synchronous polar orbit covering 90% of Earth's surface a day. In addition to measuring wind speed and direction, *QuikSCAT* detected massive snow melt in Antarctica in January 2005, when surface temperatures remained at 5°C (41°F) for about a week. The satellite monitors snowfall and melt in Greenland as well.

Canada operates *RADARSAT–1*, and –2 and *SCISAT–1*, and the Japanese have *JERS–1*, that image in radar wavelengths (Figure 1.26). The European Space Agency (ESA) now operates two Earth resource satellites

Visible light Radar

FIGURE 1.25 A volcanic eruption seen from orbit.
Photograph (passive, visible light) and image (active, radar) of the eruption of the Kliuchevskoi Volcano on the Kamchatka Peninsula of Siberia, Russia, as captured by the Space Shuttle *Endeavour*, September 1994. [Image and photo by JPL/NASA; made by Astronaut Thomas Jones.]

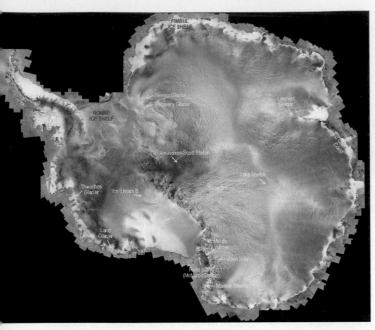

FIGURE 1.26 *RADARSAT–1* composite of Antarctica. A mosaic of *RADARSAT–1* images made over 18-day period. Resolution can be as fine as 8 m (26 ft) with this microwave Synthetic Aperture Radar (SAR) system. The platform is placed in a near-polar orbit, approximately 800 km (497 mi) above Earth. [Image courtesy of Canadian Space Agency.]

(*ERS–1* and *–2*). They work in tandem, producing a spectacular 10-cm (3.9-in.) resolution. Imaging the same area at different times, pairs of images produce a digital three-dimensional data set.

ESA's environment-monitoring satellite, *Envisat*, went into service in 2002. In addition to a more sophisticated radar than *ERS*, *Envisat* carries nine passive sensor packages, making it a multitalented craft. It monitors ocean temperatures, sea level, and wave patterns; polar ice; forests; biological activity in the oceans and on land; cloud heights; atmospheric ozone and pollution; and carbon dioxide concentrations (see **http://envisat.esa.int/**).

Passive Remote Sensing Passive remote-sensing systems record energy radiated from a surface, particularly visible light and infrared. Our own eyes are passive remote sensors, as was the *Apollo 17* astronaut camera that made the picture of Earth on the back cover of this book from a distance of 37,000 km (23,000 mi).

Passive remote sensors on five *Landsat* satellites, launched by the United States, provide a variety of data, as shown in images of the Appalachian Mountains and Mount St. Helens in Chapter 12, river deltas and New Orleans before and after Katrina in Chapter 14, and the Malaspina and Kuskulana Glaciers in Alaska in Chapter 17. Two *Landsats* remain operational, *Landsat–5* and *–7*. See **http://landsat.gsfc.nasa.gov/** and **http://geo.arc.nasa.gov/sge/landsat/landsat.html** for more information and additional links.

A NASA, National Oceanic and Atmospheric Administration (NOAA, see **http://www.noaa.gov/**), and U.K. joint venture, the *Polar Operational Environmental Satellites* (*POES*), completes nearly polar orbits 14 times a day, carrying the *advanced very high resolution radiometer* (*AVHRR*). These are in Sun-synchronous orbits aboard *NOAA-15* and *NOAA-*17 for morning hours, and *NOAA-16* and *NOAA-18* for afternoon hours. These sensors are sensitive to visible and infrared wavelengths. They are used for weather analysis and forecasting, sea-surface temperature measurement, ozone concentrations, and temperatures in the stratosphere, among many other functions. See **http://www.oso.noaa.gov/poesstatus/**.

Key to NASA's Earth Observing System (EOS) is satellite *Terra*, which began beaming back data and images in 2000 (see **http://terra.nasa.gov/**), followed by another satellite in the series called *Aqua*. Five instrument packages observe Earth systems in detail, exploring the atmosphere, landscapes, oceans, environmental change, and climate, among other things. Of more than 100 remote-sensing images presented throughout this text, watch for those from the *Terra* satellite, such as the Chapter 21-opening image of southern California wildfires in 2007. The image of Earth on the half-title page of this book includes a cloud snapshot from *GOES* added to *Terra* MODIS images in 2000.

The *Geostationary Operational Environmental Satellites*, known as *GOES*, became operational in 1994, providing frequent infrared and visible images—the ones you see on television weather reports. Geostationary satellites stay in semipermanent positions because they keep pace with Earth's rotational speed at their altitude of 35,400 km (22,000 mi).

GOES-9, launched in 1995, is now above 160° E in a backup orbital mode. *GOES-10*, active in 1998, was moved in December 2006 to monitor South America from above 60° W. *GOES-12* sits above 75° W longitude to monitor central and eastern North America and the western Atlantic (Figure 1.27). Think of these satellites as hovering over these meridians for continuous coverage, using visual wavelengths for daylight hours and infrared for nigthtime views. *GOES-11* operates above 135° W longitude for western North America and eastern Pacific Ocean weather. *GOES-13*, launched in May 2006, is parked in orbit to become next in line when needed. (See the Geostationary Satellite Server at **http://www.goes.noaa.gov/**.) The *GOES* Project Science appears at **http://rsd.gsfc.nasa.gov/goes/**, or see **http://www.ghcc.msfc.nasa.gov/GOES/**. For the future, *GOES-O*, launched in July 2007, will become *GOES-14*, and *GOES-P*, after a successful 2008 launch, will be operational as *GOES-15*.

Other satellites used for weather include Japan's *MTSAT-1R* (Multi-Functional Transport Satellite, 2005) and *MTSAT-2* (2006), which now serve weather and communication needs over the western Pacific (*GOES-9* serves as backup). China's *Feng Yun-2B* and *Feng Yun-1D* above 105° E cover the Far East, and *METEOSAT-7*,

FIGURE 1.27 *GOES–12* first image.
The first image from *GOES–12* demonstrates excellent quality from its 37,000 km (23, 300 mi) orbital post. This satellite "hovers" above 75° W longitude for the Atlantic, central, and eastern coverage. [Image courtesy of NOAA.]

FIGURE 1.28 EROS Data Center (EDS).
USGS operates the EROS (Earth Resources Observation Systems) Data Center, outside Sioux Falls, South Dakota. This center operates one of the largest computer complexes dedicated to science (see http://edc.usgs.gov/). [Photo by Bobbé Christopherson.]

above 58° E, covers Europe and Africa and is operated by the European Space Agency. (See the Remote Sensing Virtual Library at **http://virtual.vtt.fi/space/rsvlib/** for remote-sensing links; click on "Satellite Data" for specific coverage.)

One of two commercial systems includes the three French satellites (1, 2, and 4) called *SPOT* (Systeme Probatoire d'Observation de la Terre) that resolve objects on Earth down to 10 to 20 m (33 to 66 ft), depending on which sensors are used. Other commercial systems are operated by GeoEye, formerly Space Imaging, which was acquired by ORBIMAGE. Its satellites offer 0.41-m to 4-m resolution from *Ikonos–2*, *OrbView-2*, *-3*, and *–5*, and its new *GeoEye-1* satellite, all in Sun-synchronous orbits at 680 km (420 mi) altitude (**http://www.geoeye.com/**). See the Chapter 1 opening image for an example of an image from *Ikonos–2*.

The USGS *EROS* (Earth Resources Observation and Science) Data Center, near Sioux Falls, South Dakota, is the national repository for geophysical and geospatial data—remote-sensing imagery, photos, maps, the Global Land Inventory System, NASA's EOS Gateway, and more (Figure 1.28). *EROS* is where we gather all the data and imagery in one place that will guide future generations about Earth systems operations and what conditions were like in this era.

For a sampling of global remote-sensing coverage from a variety of satellite platforms go to the USGS Global Visualization Viewer, select the satellite and click on the map for a location, and then watch (**http://glovis. usgs.gov/**).

Geographic Information Systems (GIS)

Techniques such as remote sensing acquire large volumes of spatial data that must be stored, processed, and retrieved in useful ways. A **geographic information system (GIS)** is a computer-based, data-processing tool for gathering, manipulating, and analyzing geographic information. Today's sophisticated computer systems allow the integration of geographic information from direct surveys (on-the-ground mapping) and remote sensing in complex ways never before possible. Through a GIS, Earth and human phenomena are analyzed over time. This can include the study and forecasting of diseases such as West Nile virus, or the population displacement caused by Hurricane Katrina, or interactive land-cover maps, to name a few examples (Figure 1.29).

GIS is a rapidly expanding career field in many sectors of the economy. Regardless of your academic major, the ability to analyze data spatially is important. The annual ESRI International Users Conference, whose attendance is approaching 15,000 people, is where GIS professionals share the latest applications of the GIS technology (see **http://www.esri.com/**). Be sure to check some of the URLs listed in News Report 1.3 for information on professional career directions in this exciting field or how to acquire this tool for use in your chosen career.

The beginning component for any GIS is a coordinate system such as latitude–longitude, which establishes reference points against which to position data. The coordinate system is digitized, along with all areas, points, and lines. Remotely sensed imagery and data are then added on the coordinate system.

A GIS is capable of analyzing patterns and relationships within a single data plane, such as the floodplain or

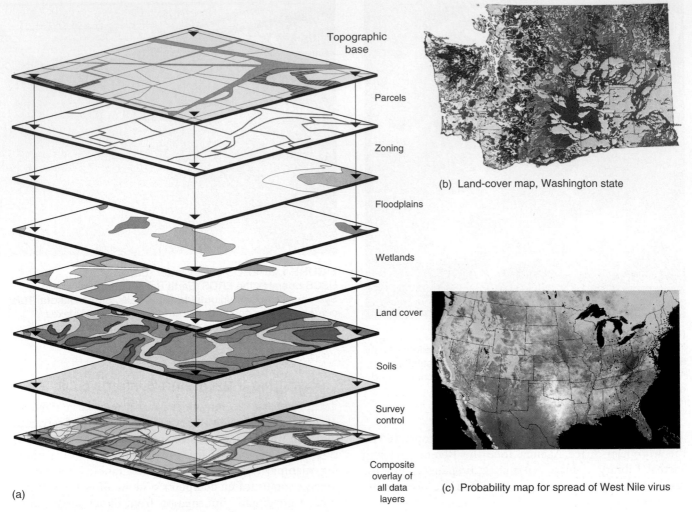

Topographic
base

Parcels

Zoning

Floodplains

Wetlands

Land cover

Soils

Survey
control

Composite
overlay of
all data
layers

(a)

(b) Land-cover map, Washington state

(c) Probability map for spread of West Nile virus

FIGURE 1.29 A geographic information system (GIS) model.
(a) Layered spatial data in a geographic information system (GIS) format. (b) This comprehensive
land-cover map is an important component of a statewide GIS analysis for Washington state.
The goal is to evaluate the protection of species and biodiversity. (c) A NASA GIS study uses
satellite-derived temperature, humidity, and vegetation variables in a GIS overlay of environmental
conditions for a probability map of the spread of West Nile virus. Various colors indicate
temperatures and temperature ranges, moisture or dryness, and vegetation as an indicator
of humidity. [(a) After USGS. (b) GIS map courtesy of Dr. Kelly M. Cassidy, *Gap Analysis of Washington State*,
v. 5, Map 3, University of Washington, 1997. (c) Composite image from *NOAA-15* AVHRR data, NASA/GSFC.]

soil layer in Figure 1.28a. A GIS also can generate an overlay analysis where two or more data planes interact. When the layers are combined, the resulting synthesis—a *composite overlay*—is a valuable product, ready for use in analyzing complex problems. A research study may follow specific points or areas through the complex of overlay planes. The utility of a GIS compared with that of a fixed map is the ability to manipulate the variables for analysis and to constantly change the map—the map is alive!

Before the advent of computers, an environmental-impact analysis required someone to gather data and painstakingly hand-produce overlays of information to determine positive and negative impacts of a project or event. Today, this layered information is handled by a computer-driven GIS, which assesses the complex interconnections among different components. In this way,

subtle changes in one element of a landscape may be identified as having a powerful impact elsewhere.

The applications of GIS are wide-ranging and too numerous to detail here. For GIS resources and information, see **http://www.geo.ed.ac.uk/home/giswww.html** and **http://erg.usgs.gov/isb/pubs/gis_poster/**.

USGS Geography (see **http://geography.usgs.gov/**; formerly *Mapping*) monitors and analyzes natural resource and environmental issues confronting us. The program is in charge of building and maintaining an innovative national map. The value of the on-line, interactive system is that geospatial data are portrayed in a manner that can be readily updated—a dynamic mapping program—involving the principles of GIS. For more on this work in progress, see **http://nationalmap. usgs.gov/**.

News Report 1.3

Careers in GIS

Geographic information system (GIS) methodology offers great career opportunities in industry, government, business, marketing, teaching, sales, military, and other fields. Right now, geographers trained in geospatial science analyze ozone depletion, natural hazards, deforestation, soil erosion, and acid deposition, among many possible applications. They map ecosystems and monitor the declining diversity of plant and animal species. Geographers plan, design, and survey urban developments, follow the trends of global warming, study the impact of human population, and analyze air and water pollution. Police cruisers in many cities are equipped with laptops running GIS maps developed by geographers.

We are in the midst of a GIS revolution. Potent new careers are emerging in business, marketing, and academia. For a list of current trends in GIS, simply enter this topic in your search engine; for an alphabetized GIS resources list, check out **http://www.geo.ed.ac.uk/home/giswww.html.**

GIS degree programs are available at many colleges and universities. GIS curriculum and certificate programs are now available at many community colleges. For example, at American River College, Sacramento, California, a curriculum offering 15 GIS courses drives programs leading to a GIS certificate or an A.S. degree. To look at this program, go to **http://www.arc.losrios.edu/~earthsci/** and click "GIS."

A consortium of three universities forms the National Center for Geographic Information and Analysis (NCGIA) for GIS education, research, outreach, and model generation. These GIS centers are: Department of Geography, University of California Santa Barbara, Santa Barbara, CA 93106; Department of Surveying and Engineering, University of Maine, Orono, ME 04669; and State University of New York–Buffalo, Buffalo, NY 14260. (See **http://www.ncgia.ucsb.edu/** or **http://www.ncgia.maine.edu/** or **http://www.geog.buffalo.edu/ncgia/.**)

Summary and Review—Essentials of Geography

To assist you, here is a review in handy summary form of the Key Learning Concepts listed on this chapter's title page. Each concept review concludes with a list of the key terms from the chapter, their page numbers, and review questions. Such summary and review sections follow each chapter in the book.

■ *Define* geography and physical geography in particular.

Geography brings together disciplines from the physical and life sciences with the cultural and human sciences to attain a holistic view of Earth—physical geography is an essential aspect of the emerging **Earth systems sciences**. **Geography** is a science of method, a special way of analyzing phenomena over space; **spatial** refers to the nature and character of physical space. Geography integrates a wide range of subject matter, and geographic education recognizes five major themes: **location**, **region**, **human–Earth relationships**, **movement**, and **place**. Geography's method is **spatial analysis**, used to study the interdependence among geographic areas, natural systems, society, and cultural activities over space or area. **Process**—that is, analyzing a set of actions or mechanisms that operate in some special order—is central to geographic synthesis.

Physical geography applies spatial analysis to all the physical elements and processes that make up the environment: energy, air, water, weather, climate, landforms, soils, animals, plants, microorganisms, and Earth itself. Understanding the complex relations among these elements is important to human survival because Earth's physical systems and human society are so intertwined. The development of hypotheses and theories about the Universe, Earth, and life involves the **scientific method**.

Earth systems science (p. 2)
geography (p. 2)
spatial (p. 2)
location (p. 3)
region (p. 3)
human–Earth relationships (p. 3)
movement (p. 3)
place (p. 3)
spatial analysis (p. 3)
process (p. 3)
physical geography (p. 3)
scientific method (p. 3)

1. What is unique about the science of geography? On the basis of information in this chapter, define physical geography and review the geographic approach.
2. In general terms, how might a physical geographer analyze water pollution in the Great Lakes?
3. Assess your geographic literacy by examining atlases and maps. What types of maps have you used—political? Physical? Topographic? Do you know what projections they employed? Do you know the names and locations of the four oceans, seven continents, and most individual countries? Can you identify the new countries that have emerged since 1990?
4. Suggest a representative example for each of the five geographic themes and use that theme in a sentence.
5. Have you made decisions today that involve geographic concepts discussed within the five themes presented? Explain briefly.

■ *Describe* systems analysis, open and closed systems, and feedback information, and *relate* these concepts to Earth systems.

A **system** is any ordered, related set of things and their attributes, as distinct from their surrounding environment. Systems analysis is an important organizational and analytical tool used by geographers. Earth is an **open system** in terms of energy, receiving energy from the Sun, but it is essentially a **closed system** in terms of matter and physical resources.

As a system operates, "information" is returned to various points in the system via pathways called **feedback loops**. If the feedback information discourages response in the system, it is called **negative feedback**. (Further production in the system decreases the growth of the system.) If feedback information encourages response in the system, it is called **positive feedback**. (Further production in the system stimulates the growth of the system.) When the rates of inputs and outputs in the system are equal and the amounts of energy and matter in storage within the system are constant (or as they fluctuate around a stable average), the system is in **steady-state equilibrium**. A system that demonstrates a steady increase or decrease in system operations—a trend over time—is in **dynamic equilibrium**. A **threshold** is the moment at which a system can no longer maintain its character, so it lurches to a new operational level, one which may not be compatible with previous conditions. Geographers often construct a simplified **model** of natural systems to better understand them.

Four immense open systems powerfully interact at Earth's surface: three nonliving **abiotic** systems [**atmosphere**, **hydrosphere** (including the **cryosphere**), and **lithosphere**] and a living **biotic** system (**biosphere**, or **ecosphere**).

system (p. 5)
open system (p. 5)
closed system (p. 5)
feedback loops (p. 7)
negative feedback (p. 7)
positive feedback (p. 7)
steady-state equilibrium (p. 9)
dynamic equilibrium (p. 9)
threshold (p. 10)
model (p. 12)
abiotic (p. 13)
biotic (p. 13)
atmosphere (p. 14)
hydrosphere (p. 14)
cryosphere (p. 14)
lithosphere (p. 14)
biosphere (p. 15)
ecosphere (p. 15)

6. Define systems theory as an organizational strategy. What are open systems, closed systems, and negative feedback? When is a system in a steady-state equilibrium condition? What type of system (open or closed) is the human body? A lake? A wheat plant?
7. Describe Earth as a system in terms of both energy and matter; use simple diagrams to illustrate your description.
8. What are the three abiotic spheres (nonliving) that make up Earth's environment? Relate these to the biotic (living) sphere: the biosphere.

■ *Explain* Earth's reference grid: latitude and longitude and latitudinal geographic zones and time.

The science that studies Earth's shape and size is **geodesy**. Earth bulges slightly through the equator and is oblate (flattened) at the poles, producing a misshapen spheroid called a **geoid**. Absolute location on Earth is described with a specific reference grid of **parallels** of **latitude** (measuring distances north and south of the equator) and **meridians** of **longitude** (measuring distances east and west of a prime meridian). A historic breakthrough in navigation and timekeeping occurred with the establishment of an international **prime meridian** (0° through Greenwich, England) and the invention of precise chronometers that enabled accurate measurement of longitude. Latitude, longitude, and elevation are accurately calibrated using a handheld **global positioning system (GPS)** instrument that reads radio signals from satellites. A **great circle** is any circle of Earth's circumference whose center coincides with the center of Earth. Great circle routes are the shortest distance between two points on Earth. **Small circles** are those whose centers do not coincide with Earth's center.

The prime meridian provided the basis for **Greenwich Mean Time (GMT)**, the world's first universal time system. A corollary of the prime meridian is the 180° meridian, the **International Date Line**, which marks the place where each day officially begins. Today, **Coordinated Universal Time (UTC)** is the worldwide standard and the basis for international time zones. **Daylight saving time** is a seasonal change of clocks by 1 hour in summer months.

geodesy (p. 15)
geoid (p. 15)
latitude (p. 17)
parallel (p. 17)
longitude (p. 17)
meridian (p. 17)
prime meridian (p. 17)
Global Positioning System (GPS) (p. 19)
great circle (p. 19)
small circles (p. 19)
Greenwich Mean Time (GMT) (p. 23)
International Date Line (p. 23)
Coordinated Universal Time (UTC) (p. 25)
daylight saving time (p. 25)

9. Draw a simple sketch describing Earth's shape and size.
10. How did Eratosthenes use Sun angles to figure out that the 5000-stadia distance between Alexandria and Syene was 1/50 of Earth's circumference? Once he knew this fraction of Earth's circumference, how did he calculate the entirety of Earth's circumference?
11. What are the latitude and longitude coordinates (in degrees, minutes, and seconds) of your present location? Where can you find this information?
12. Define *latitude* and *parallel* and define *longitude* and *meridian* using a simple sketch with labels.
13. Define a great circle, great circle routes, and a small circle. In terms of these concepts, describe the equator, other parallels, and meridians.
14. Identify the various latitudinal geographic zones that roughly subdivide Earth's surface. In which zone do you live?
15. What does timekeeping have to do with longitude? Explain this relationship. How is Coordinated Universal Time (UTC) determined on Earth?
16. What and where is the prime meridian? How was the location originally selected? Describe the meridian that is opposite the prime meridian on Earth's surface.
17. What is GPS and how does it assist you in finding location and elevation on Earth? Give a couple of examples where it was utilized to correct heights for some famous mountains.

■ *Define* cartography and mapping basics: map scale and map projections.

A **map** is a generalized view of an area, usually some portion of Earth's surface, as seen from above and greatly reduced in size. The science and art of mapmaking is called **cartography**. For the spatial portrayal of Earth's physical systems geographers use maps. **Scale** is the ratio of the image on a map to the real world; it relates a unit on the map to an identical unit on the ground. Cartographers create **map projections** for specific purposes, selecting the best compromise of projection for each application. Compromise is always necessary because Earth's round, three-dimensional surface cannot be exactly duplicated on a flat, two-dimensional map. **Equal area** (equivalence), **true shape** (conformality), true direction, and true distance are all considerations in selecting a projection. The **Mercator projection** is in the cylindrical class; it has true-shape qualities and straight lines that show constant direction. A **Rhumb line** denotes constant direction and appears as a straight line on the Mercator.

> map (p. 25)
> cartography (p. 25)
> scale (p. 25)
> map projections (p. 26)
> equal area (p. 26)
> true shape (p. 26)
> Mercator projection (p. 27)
> rhumb line (p. 27)

18. Define cartography. Explain why it is an integrative discipline.

19. What is map scale? In what three ways may it be expressed on a map?

20. State whether the following ratios are large scale, medium scale, or small scale: 1:3,168,000; 1:24,000; 1:250,000.

21. Describe the differences between the characteristics of a globe and those that result when a flat map is prepared.

22. What type of map projection is used in Figure 1.12? In Figure 1.17? (See Appendix A.)

■ *Describe* remote sensing, and *explain* geographic information system (GIS) as tools used in geographic analysis.

Orbital and aerial **remote sensing** analyzes the operation of Earth's systems. Satellites do not take photographs, but record images that are transmitted to Earth-based receivers. Satellite images are recorded in digital form for later processing, enhancement, and generation. Aerial photographs have been used for years to improve the accuracy of surface maps. This is the realm of **photogrammetry**, an important application of remote sensing.

The mountain of data collected is put to work in **geographic information system (GIS)** technology. Computers process geographic information from direct surveys and remote sensing in complex ways never before possible. GIS methodology is an important step in better understanding Earth's systems and is a vital career opportunity for geographers.

The science of physical geography is in a unique position to synthesize the spatial, environmental, and human aspects of our increasingly complex relationship with our home planet—Earth.

> remote sensing (p. 29)
> photogrammetry (p. 29)
> geographic information system (GIS) (p. 33)

23. What is remote sensing? What are you viewing when you observe a weather satellite image on TV or in the newspaper? Explain.

24. Describe the *Terra*, *Landsat*, *GOES*, and *NOAA* satellites and explain them using several examples.

25. If you were in charge of planning the development of a large tract of land, how would GIS methodologies assist you? How might planning and zoning be affected if a portion of the tract in the GIS is a floodplain or prime agricultural land?

 NetWork

The *Geosystems Student Learning Center* provides on-line resources for this chapter on the World Wide Web. To begin: Once at the Center, click on the cover of this textbook, scroll the Table of Contents menu, and select this chapter. You will find self-tests that are graded, review exercises, specific updates for items in the chapter, and in "Destinations" many links to interesting related pathways on the Internet. *Geosystems Student Learning Center* is found at **http://www.prenhall.com/** christopherson/.

Critical Thinking

A. Select a location (for example, your campus, home, workplace, a public place, or a city) and determine the following: latitude, longitude, and elevation. Describe the resources you used to gather this geographic information. Have you ever used a GPS unit to determine these aspects of your location?

B. Let's say there is a world globe in the library or geography department that is 61 cm (24 in.) in diameter. We know that Earth has an equatorial diameter of 12,756 km (7926 mi), so the scale of the globe is the ratio of 61 cm to 12,756 km. We divide Earth's actual diameter by the globe's diameter (12,756 km ÷ 61 cm) and determine that 1 cm of the globe's diameter equals about _____ cm of Earth's actual diameter. Thus, the representative fraction for the globe is expressed in centimeters as 1/_____. (*Hint*: 1 km = 1000 m, 1 m = 100 cm; therefore, Earth's diameter of 12,756 km represents 1,275,600,000 cm.)

C. The Chapter 1 opening feature shows the May 4, 2007, EF-05 monster tornado that leveled Greenburg, KS. As you take time examing the aerial photos and orbital image, think through each of the five themes of geographic science detailed in Figure 1.1 and describe how each theme might be implemented to analyze this storm and its destruction—how the theme applies. Use you search browser to explore the tornado, turn ahead to the related section in Chapter 8, and go to some of the sources listed. Be sure to include any URLs for the sites you visit.

The Energy–Atmosphere System

PART I

Sunlight powers this field of wheat in eastern Washington State. Of the visible light spectrum, chlorophyll absorbs orange-red and violet blue wavelengths for photochemical operations within leaves. Wheat, along with maize (corn) and rice, provides more than 50% of the world grain supply. [Photo by Bobbé Christopherson.]

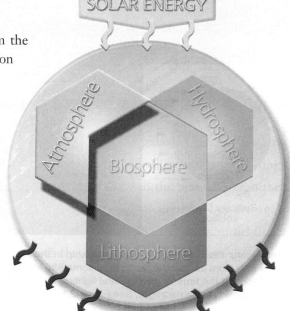

Our planet and our lives are powered by radiant energy from the star that is closest to Earth—the Sun. For more than 4.6 billion years, solar energy has traveled across interplanetary space to Earth, where a small portion of the solar output is intercepted. Because of Earth's curvature, the energy at the top of the atmosphere is unevenly distributed, creating imbalances from the equator to each pole—the equatorial region experiences energy surpluses; the polar regions experience energy deficits. Also, the annual pulse of seasonal change varies the distribution of energy during the year.

Earth's atmosphere acts as an efficient filter, absorbing most harmful radiation, charged particles, and space debris so that they do not reach Earth's surface. In the lower atmosphere the unevenness of daily energy receipt empowers atmospheric and surface energy budgets, giving rise to global patterns of temperature and circulation of wind and ocean currents. We see this in the part-opening photo, wheat plants almost aglow with the infusion of this radiant energy. Each of us depends on many systems that are set into motion by energy from the Sun. These systems are the subject of Part I.

In this chapter: Incoming solar energy that arrives at the top of Earth's atmosphere establishes the pattern of energy input that drives Earth's physical systems and which daily influences our lives. This solar energy input to the atmosphere, plus Earth's tilt and rotation, produce daily, seasonal, and annual patterns of changing daylength and Sun angle. The Sun is the ultimate energy source for most life processes in our biosphere.

The Solar System, Sun, and Earth

Our Solar System is located on a remote, trailing edge of the **Milky Way Galaxy**, a flattened, disk-shaped mass estimated to contain nearly 400 billion stars (Figure 2.1a, b). Our Solar System is embedded more than halfway out from the galactic center, in one of the Milky Way's spiral arms, called the Orion Arm. A supermassive black hole some 2 million solar masses in size, named *Sagittarius A** (pronounced "*Sagittarius A Star*"), sits in the center. From our Earth-bound perspective in the Milky Way, the Galaxy appears to stretch across the night sky like a narrow band of hazy light. On a clear night, the unaided eye can see only a few thousand of these billions of stars gathered about us in our "neighborhood."

Solar System Formation and Structure

According to prevailing theory, our Solar System condensed from a large, slowly rotating and collapsing cloud of dust and gas called a *nebula*. **Gravity**, the mutual attracting force exerted by the mass of an object upon all other objects, was the key force in this condensing solar nebula. As the nebular cloud organized and flattened into a disk shape, the early *protosun* grew in mass at the center, drawing more matter to it. Small accretion (accumulation) eddies swirled at varying distances from the center of the solar nebula; these were the *protoplanets*.

The beginnings of the Sun and its Solar System are estimated to have occurred more than 4.6 billion years ago. The early protoplanets, called *planetesimals*, orbited at approximately the same distances from the Sun as the planets are today. The **planetesimal hypothesis**, or *dust-cloud hypothesis*, explains how suns condense from nebular clouds with planetesimals forming in orbits about the central mass. Astronomers study this formation process in other parts of the Galaxy, where planets are observed orbiting distant stars. By the end of 2007, astronomers had discovered more than 250 planets orbiting other stars. Closer to home in our Solar System, some 165 moons (planetary satellites) are in orbit about the eight planets. As of 2007, the new satellite count for the four outer planets was: Jupiter, 63 moons; Saturn, 60 moons; Uranus, 27 moons; and Neptune, 13 moons.

Dimensions and Distances The **speed of light** is 300,000 kmps (kilometers per second), or 186,000 mps (miles per second)—in other words, about 9.5 trillion kilometers per year, or nearly 6 trillion miles per year. (In more precise numbers, light speed is 299,792 kmps, or 186,282 mps.) This tremendous distance that light travels in a year is known as a *light-year*, and it is used as a unit of measurement for the vast Universe.

For spatial comparison, our Moon is an average distance of 384,400 km (238,866 mi) from Earth, or about 1.28 seconds in terms of light speed—for the *Apollo* astronauts this was a 3-day space voyage. Our entire Solar System is approximately 11 hours in diameter, measured by light speed (Figure 2.1c). In contrast, the Milky Way is about 100,000 light-years from side to side, and the known Universe that is observable from Earth stretches approximately 12 billion light-years in all directions. (See a Solar System simulator at **http://space.jpl.nasa.gov/**.)

Earth's Orbit Earth's orbit around the Sun is presently *elliptical*—a closed, oval path (Figure 2.1d). Earth's average distance from the Sun is approximately 150 million km (93 million mi), which means that light reaches Earth from the Sun in an average of 8 minutes and 20 seconds. Earth is at **perihelion** (its closest position to the Sun) during the Northern Hemisphere winter (January 3 at 147,255,000 km, or 91,500,000 mi). It is at **aphelion** (its farthest position from the Sun) during the Northern Hemisphere summer (July 4 at 152,083,000 km, or 94,500,000 mi). This seasonal difference in distance from the Sun causes a slight variation in the solar energy incoming to Earth but is not an immediate reason for seasonal change.

The structure of Earth's orbit is not a constant, but instead exhibits change over long periods. As shown in Chapter 17, Figure 17.31, Earth's distance from the Sun varies more than 17.7 million km (11 million mi) during a 100,000-year cycle, placing it closer or farther at different periods in the cycle. This variation is thought to be one of several factors that create Earth's long-term cyclical pattern of glaciations (colder) and interglacial (warmer) periods.

Solar Energy: From Sun to Earth

Our Sun is unique to us but is a commonplace star in the Galaxy. It is only average in temperature, size, and color when compared with other stars, yet it is the ultimate energy source for most life processes in our biosphere.

The Sun captured about 99.9% of the matter from the original solar nebula. The remaining 0.1% of the matter formed all the planets, their satellites, asteroids, comets, and debris. Consequently, the dominant object in our region of space is the Sun. In the entire Solar System, it is the only object having the enormous mass needed to sustain a nuclear reaction and produce radiant energy.

The solar mass produces tremendous pressure and high temperatures deep in its dense interior. Under these conditions, the Sun's abundant hydrogen atoms are forced together and pairs of hydrogen nuclei are joined in the process of **fusion**. In the fusion reaction, hydrogen nuclei form helium, the second-lightest element in nature, and enormous quantities of energy are liberated—literally, disappearing solar mass becomes energy.

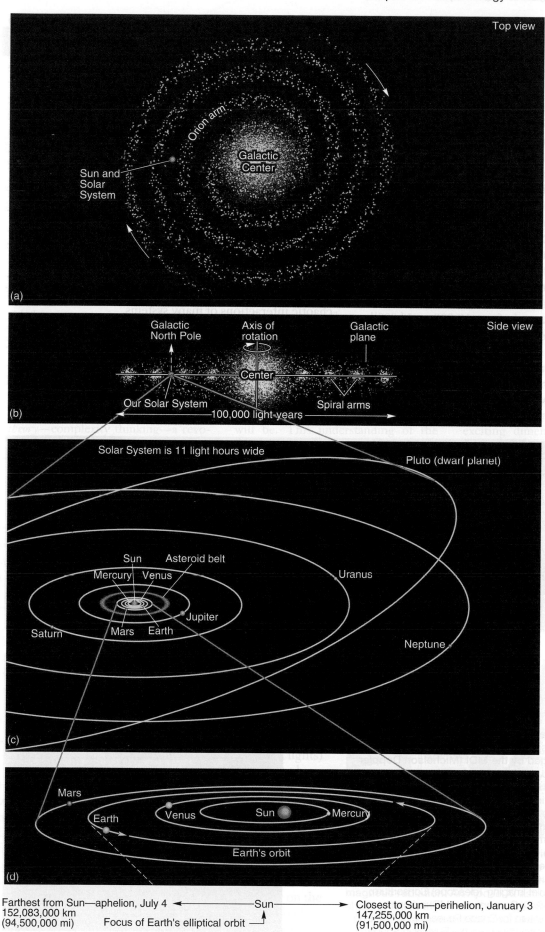

ANIMATION

Nebular Hypothesis

FIGURE 2.1 Milky Way Galaxy, Solar System, and Earth's orbit.
The Milky Way Galaxy viewed from above (a) and cross-section side view (b). Our Solar System of eight planets, four dwarf planets, and asteroids is some 30,000 light-years from the center of the Galaxy. In 2006, Pluto was reclassified a "dwarf planet" and part of the Kuiper Asteroid Belt. All of the planets except Pluto have orbits closely aligned to the plane of the ecliptic (c). The four inner terrestrial planets and the structure of Earth's elliptical orbit, illustrating perihelion (closest) and aphelion (farthest) positions during the year, are shown in (d). Have you ever observed the Milky Way Galaxy in the night sky?

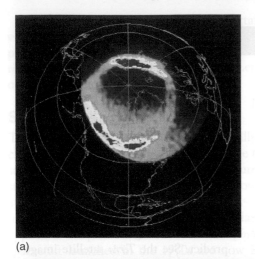

(a)

(b)

FIGURE 2.4 Auroras from an orbital perspective and from the ground in Alaska.
(a) *Polar* satellite false-color image of auroral halo over Earth's North Pole region from April 1996—the UVI sensor is able to capture the aurora on the day and night sides of Earth.
(b) Surface view of an aurora borealis in the night sky over Alaska. [(a) Image from the Ultraviolet Imager (UVI) aboard *Polar* satellite courtesy of NASA; (b) Aurora photo by Johnny Johnson/Tony Stone Images, Inc.]

links between the sunspot cycle and patterns of drought and periods of wetness—the sunspot–weather connection.

Our understanding of the solar wind is increasing dramatically as data are collected by a variety of satellites: the *SOHO* (*Solar and Heliospheric Observatory*), *FAST* (*Fast Auroral Snapshot*), *WIND*, *Ulysses*, the *Dynamics Explorer*, and *Genesis* with solar wind collectors, among others. All satellite data are available on the Internet. (For auroral activity, see **http://www.sel.noaa.gov/pmap/**; and for forecasts, see **http://www.gi.alaska.edu**; or for auroral photographs, see **http://northernlightsnome.homestead.com/**.)

Electromagnetic Spectrum of Radiant Energy

The key essential solar input to life is electromagnetic energy of various wavelengths. Solar radiation occupies a portion of the **electromagnetic spectrum** of radiant energy. This radiant energy travels at the speed of light to Earth. The total spectrum of this radiant energy is made

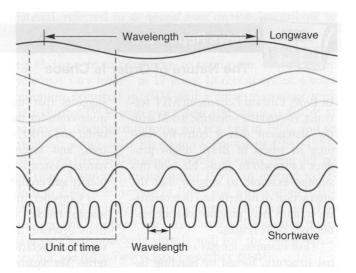

FIGURE 2.5 Wavelength and frequency.
Wavelength and frequency are two ways of describing the same phenomenon—electromagnetic wave motion. More short wavelengths pass a given point during a unit of time, so they are higher in frequency; whereas fewer long wavelengths pass a point in a unit of time, so they are lower in frequency.

up of different wavelengths. Figure 2.5 shows that a **wavelength** is the distance between corresponding points on any two successive waves. The number of waves passing a fixed point in 1 second is the *frequency*.

The Sun emits radiant energy composed of 8% ultraviolet, X-ray, and gamma-ray wavelengths; 47% visible light wavelengths; and 45% infrared wavelengths. A portion of the electromagnetic spectrum is illustrated in Figure 2.6, with wavelengths increasing from the top of the illustration to the bottom. Note the wavelengths at which various phenomena and human applications of energy occur.

An important physical law states that all objects radiate energy in wavelengths related to their individual surface temperatures: the hotter the object, the shorter the wavelengths emitted. This law holds true for the Sun and Earth. Figure 2.7 shows that the hot Sun radiates shorter wavelength energy, concentrated around 0.5 μm (micrometer).

The Sun's surface temperature is about 6000 K (6273°C, 11,459°F), and its emission curve shown in the figure is similar to that predicted for an idealized 6000 K surface, or *blackbody radiator* (shown in Figure 2.7). An ideal blackbody absorbs all the radiant energy that it receives and subsequently emits all that it receives. A hotter object like the Sun emits a much greater amount of energy per unit area of its surface than does a similar area of a cooler object like Earth. Shorter wavelength emissions are dominant at these higher temperatures.

Although cooler than the Sun, Earth radiates nearly all that it absorbs and acts as a blackbody (also shown in Figure 2.7). Because Earth is a cooler radiating body, longer wavelengths are emitted as radiation mostly in the infrared portion of the spectrum, centered around 10.0 μm. Some of the atmospheric gases vary in their response to radiation received, being transparent to some while absorbing others.

Figure 2.8 illustrates the flows of energy into and out of Earth systems. To summarize, the Sun's radiated energy

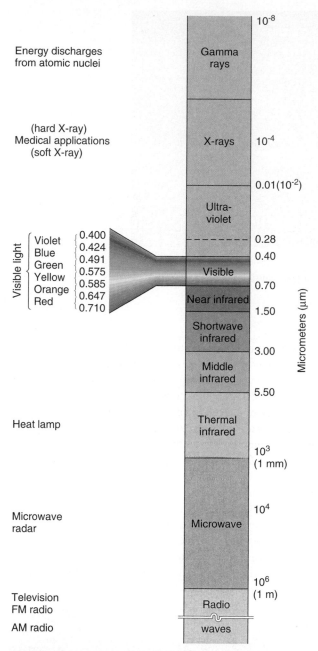

FIGURE 2.6 A portion of the electromagnetic spectrum of radiant energy.
The spectrum is oriented with shorter wavelengths toward the top and longer wavelengths toward the bottom.

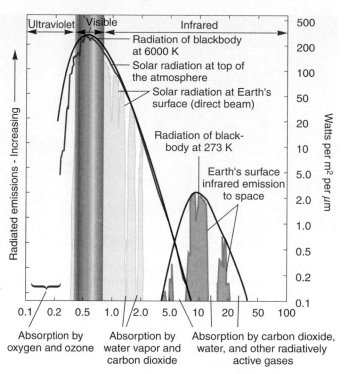

FIGURE 2.7 Solar and terrestrial energy distribution by wavelength.
The solar output peaks in shorter wavelengths of visible light in relation to its higher surface temperature, whereas Earth's emissions are concentrated in the infrared portion of the spectrum in relation to its lower surface temperature. Ideal blackbody curves for a radiating body like the Sun (hotter) and Earth (cooler) illustrate the concept. The dropouts in the plot lines of solar and terrestrial radiation represent absorption bands of water vapor, water, carbon dioxide, oxygen, ozone (O_3), and other gases. [Adapted from W. D. Sellers, *Physical Climatology* (Chicago: University of Chicago Press, 1965), p. 20. Used by permission.]

 ANIMATION   Electromagnetic Spectrum and Plants

is *shortwave radiation* that peaks in the short visible wavelengths, whereas Earth's radiated energy is *longwave radiation* concentrated in infrared wavelengths. In Chapter 4, we see that Earth, clouds, sky, ground, and things that are terrestrial radiate longer wavelengths in contrast to the Sun.

Incoming Energy at the Top of the Atmosphere

The region at the top of the atmosphere, approximately 480 km (300 mi) above Earth's surface, is the **thermopause**. It is the outer boundary of Earth's energy system and provides a useful point at which to assess the arriving solar radiation before it is diminished by scattering and absorption in passage through the atmosphere.

Earth's distance from the Sun results in its interception of only one two-billionth of the Sun's total energy output. Nevertheless, this tiny fraction of energy from the Sun is an enormous amount of energy flowing into Earth's systems. Solar radiation that reaches a horizontal plane at Earth is **insolation**, derived from the words *in*coming *sol*ar *radi*ation. Insolation specifically applies to radiation arriving at Earth's atmosphere and surface. Insolation at the top of the atmosphere is expressed as the *solar constant*.

Solar Constant   Knowing the amount of insolation incoming to Earth is important to climatologists and other scientists. The **solar constant** is the average insolation received at the thermopause when Earth is at its average distance from the Sun, a value of 1372 W/m² (watts per square meter).*

*A *watt* is equal to 1 joule (a unit of energy) per second and is the standard unit of power in the International System of Units (SI). (See the conversion tables in Appendix C of this text for more information on measurement conversions.) In nonmetric *calorie* heat units, the solar constant is expressed as approximately 2 calories per square centimeter per minute, or 2 *langleys* per minute (a langley being 1 cal/cm²). A calorie is the amount of energy required to raise the temperature of 1 gram of water (at 15°C) 1 degree Celsius and is equal to 4.184 joules.

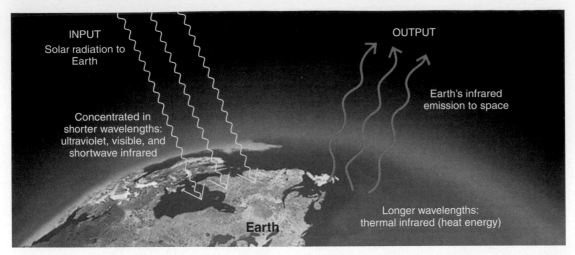

INPUT
Solar radiation to
Earth

OUTPUT

Earth's infrared
emission to space

Concentrated in
shorter wavelengths:
ultraviolet, visible, and
shortwave infrared

Longer wavelengths:
thermal infrared (heat energy)

Earth

FIGURE 2.8 Earth's energy budget simplified.
Inputs of shorter wavelengths arrive at Earth from the Sun. Outputs of longer wavelengths of infrared
radiate to space from Earth.

As we follow insolation through the atmosphere to Earth's surface (Chapters 3 and 4), we see that the value of the solar constant is reduced by half or more through reflection, scattering, and absorption of shortwave radiation.

Uneven Distribution of Insolation Earth's curved surface presents a continually varying angle to the incoming parallel rays of insolation (Figure 2.9). Differences in the angle of solar rays at each latitude result in an uneven distribution of insolation and heating. The only point receiving insolation perpendicular to the surface (from directly overhead) is the **subsolar point**. During the year this point occurs at lower latitudes between the tropics (23.5° north and 23.5° south latitudes), where the energy

received is more concentrated. All other places away from the subsolar point receive insolation at an angle less than 90° and thus experience more diffuse energy; this effect is pronounced at higher latitudes.

The thermopause above the equatorial region receives 2.5 times more insolation annually than the thermopause above the poles. Of lesser importance is the fact that the lower-angle solar rays toward the poles must pass through a greater thickness of atmosphere, resulting in greater losses of energy due to scattering, absorption, and reflection.

Figure 2.10 illustrates the daily variations throughout the year of energy at the top of the atmosphere for various latitudes in watts per square meter (W/m²). The chart shows a decrease in insolation from the equatorial regions northward and southward toward the poles. However, in

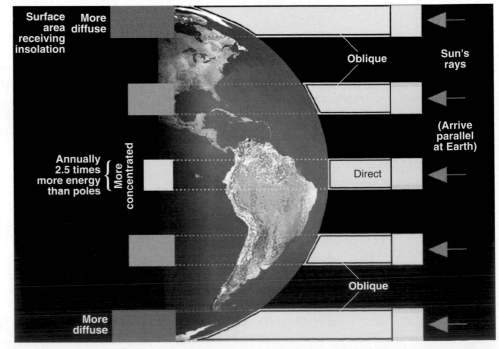

Surface area receiving insolation

More diffuse

Oblique

Sun's rays

(Arrive parallel at Earth)

Annually 2.5 times more energy than poles

More concentrated

Direct

Oblique

More diffuse

FIGURE 2.9 Insolation receipts and Earth's curved surface. Solar insolation angles determine the concentration of energy receipts by latitude. Note the area covered by identical columns of solar energy arriving at Earth's surface at higher latitudes (more diffuse, larger area covered) and at lower latitudes (more concentrated and direct, smaller area covered).

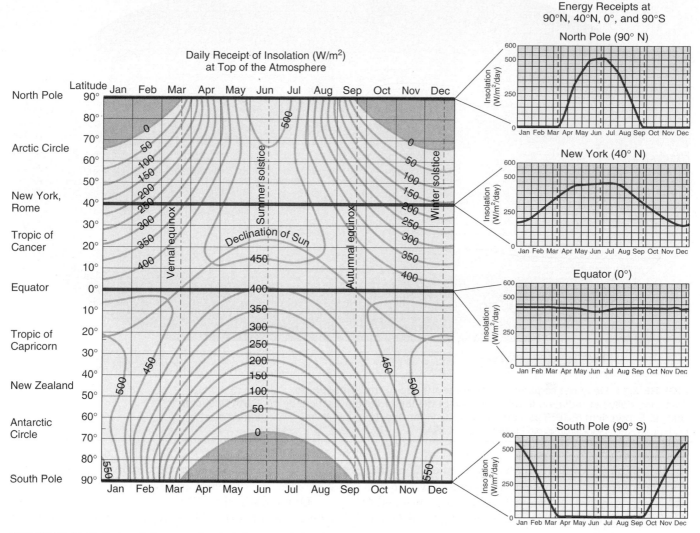

FIGURE 2.10 Daily insolation received at the top of the atmosphere.
The total daily insolation received at the top of the atmosphere is charted in watts per square meter
per day by latitude and month (1 W/m²/day = 2.064 cal/cm²/day). [Reproduced by permission of the
Smithsonian Institution Press from *Smithsonian Miscellaneous Collections: Smithsonian Meteorological Tables*,
vol. 114, 6th ed., Robert List, ed. (Washington, DC: Smithsonian Institution, 1984), p. 419, Table 134.]

June, the North Pole receives more than 500 W/m² per
day, which is more than is ever received at 40° N latitude or
at the equator. Such high values result from long daylengths
at the poles in summer: 24 hours a day, compared with only
15 hours of daylight at 40° N latitude and 12 hours at the
equator. However, at the poles the summertime Sun at
noon is low in the sky, so a daylength twice that of the
equator yields only about a 100 W/m² difference.

In December, the pattern reverses. Note that the top
of the atmosphere at the South Pole receives even more
insolation than the North Pole does in June (more than
500 W/m²). This is a function of Earth's closer location to
the Sun at perihelion (January 3 in Figure 2.1d).

Along the equator, two maximum periods of approximately 430 W/m² occur at the spring and fall equinoxes,
when the subsolar point is at the equator. Find your latitude on the graph and follow across the months to determine the seasonal variation of insolation where you live.
The four graphs to the right show plots of the energy

received at the North Pole, along 40° N, along the equator, and at the South Pole during the year to give you
energy profiles to compare.

Global Net Radiation Earth Radiation Budget Experiment instruments aboard several satellites measured
shortwave and longwave flows of energy at the top of the
atmosphere. ERBE sensors collected the data used to develop the map in Figure 2.11. This map shows *net radiation*,
or the balance between incoming shortwave and outgoing
longwave radiation—energy inputs minus energy outputs.
See News Report 2.2 for more on these experiments.

Note the latitudinal energy imbalance in net radiation on the map—positive values in lower latitudes and
negative values toward the poles. In middle and high latitudes, approximately poleward of 36° north and south latitudes, net radiation is negative. This occurs in these
higher latitudes because Earth's climate system loses more
energy to space than it gains from the Sun, as measured at

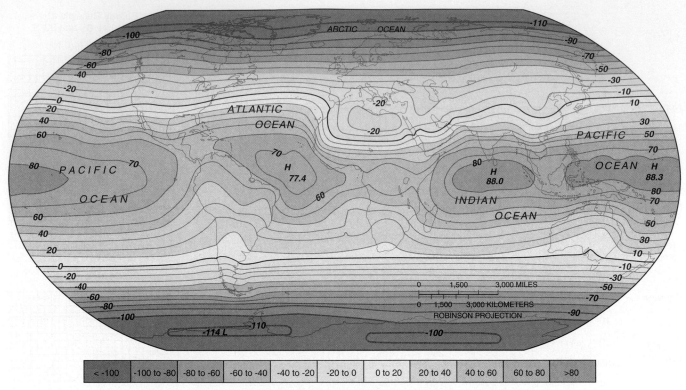

FIGURE 2.11 Daily net radiation patterns at the top of the atmosphere.
Averaged daily net radiation flows measured at the top of the atmosphere by the Earth Radiation
Budget Experiment (ERBE) aboard the *Nimbus-7* satellite. Units are W/m². [Data for map courtesy of
Dr. H. Lee Kyle, Goddard Space Flight Center, NASA.]

the top of the atmosphere. In the lower atmosphere, these polar energy deficits are offset by flows of energy from tropical energy surpluses (as we will see in Chapters 4 and 6). The largest net radiation values, averaging 80 W/m², are

above the tropical oceans along a narrow equatorial zone. Net radiation minimums are lowest over Antarctica.

Of interest is the –20 W/m² area over the Sahara region of North Africa. Here, typically clear skies—which

News Report 2.2

Monitoring Earth's Radiation Budget

Earth's weather and climate are a direct result of the balance between sunlight received from space and energy reflected and radiated to space. Patterns of insolation absorption and energy reradiation by Earth systems produce an energy budget. Since 1978, the Earth–atmosphere energy budget has been monitored by the Earth Radiation Budget (ERB) package on board the *Nimbus*-7 satellite, and originally aboard NOAA-9 and NOAA-10, with later energy budget monitoring by the *NOAA* series of satellites, now with NOAA-15 through NOAA-18 active. The original program satellites collected more than a decade of data. ERB sensors mapped the complex exchanges of

energy among atmosphere, ocean, and land.

Measurements include monthly energy budget and variations, seasonal shifts in energy between equator and poles, and daily energy budgets at the regional scale. The map in Figure 2.11 is a direct result of ERBE, showing regions that absorb more energy than radiated (a heat-energy source, rose color on map in Figure 2.11) and that radiate more energy than received (a heat-energy sink, purple color).

A new tool in understanding Earth's radiation budget is the CERES (Clouds and Earth Radiant Energy System) sensor aboard satellite *Terra*. *CERES* monitors shortwave (light) and

longwave (heat energy) radiation and, thus, Earth's energy balance of incoming and outgoing radiation—a pair of CERES images are in Chapter 4.

In addition, ERB measurements determined that solar irradiance, or the luminous brightness of solar radiation, is directly correlated to the sunspot cycle discussed earlier. During solar maximums, the solar constant rises slightly above the 1372 W/m² average, whereas during solar minimums, it shifts slightly lower, below the average. The ERBE findings help scientists better understand the systems described in Part I of this textbook (see **http://asd-www.larc.nasa. gov/erbe/ASDerbe.html**).

permit great longwave radiation losses from Earth's surface—and light-colored reflective surfaces work together to reduce net radiation values at the thermopause. In other regions, clouds and atmospheric pollution in the lower atmosphere also affect net radiation patterns at the top of the atmosphere by reflecting more shortwave energy to space.

The atmosphere and ocean form a giant heat engine, driven by differences in energy from place to place and causing major circulations within the lower atmosphere and in the ocean. These circulations include global winds, ocean currents, and weather systems—subjects to follow in Chapters 6 and 8. As you go about your daily activities, let these dynamic natural systems remind you of the constant flow of solar energy through the environment.

Having examined the flow of solar energy to Earth and the top of the atmosphere, let us now look at how seasonal changes affect the distribution of insolation as Earth orbits the Sun during the year.

The Seasons

Earth's periodic rhythms of warmth and cold, dawn and daylight, twilight and night, have fascinated humans for centuries. In fact, many ancient societies demonstrated an intense awareness of seasonal change and formally commemorated these natural energy rhythms with festivals, monuments, ground markings, and calendars (Figure 2.12). Such seasonal monuments and calendar markings are found worldwide, including thousands of sites in North America, demonstrating an ancient awareness of seasons and astronomical relations. Unfortunately, the 12-month Gregorian calendar in common use has no relation to natural rhythms of Sun, Earth, the Moon, and the march of the seasons.

Seasonality

Seasonality refers to both the seasonal variation of the Sun's position above the horizon and changing daylengths during the year. Seasonal variations are a response to changes in the Sun's **altitude**, or the angle between the horizon and the Sun. At sunrise or sunset, the Sun is at the horizon, so its altitude is 0°. If during the day, the Sun reaches halfway between the horizon and directly overhead, it is at 45° altitude. If the Sun reaches the point directly overhead, it is at 90° altitude. The Sun is directly overhead (90° altitude, or *zenith*) only at the *subsolar point*, where insolation is at a maximum. At all other surface points, the Sun is at a lower altitude angle, producing more diffuse insolation.

The Sun's **declination** is the latitude of the subsolar point. Declination annually migrates through 47° of latitude, moving between the Tropic of Cancer at 23.5° N and the Tropic of Capricorn at 23.5° S latitude. Other than Hawai'i, which is between 19° N and 22° N, the subsolar point does not reach the continental United States or Canada; all other states and provinces are too far north.

Also, seasonality means changing **daylength**, or duration of exposure to insolation. Daylength varies during the year, depending on latitude. The equator always receives equal hours of day and night: If you live in Ecuador, Kenya, or Singapore, every day and night is 12 hours long, year-round. People living along 40° N latitude (Philadelphia, Denver, Madrid, Beijing) or 40° S latitude (Buenos Aires, Capetown, Melbourne) experience about 6 hours' difference in daylight between winter (9 hours) and summer (15 hours). At 50° N or S latitude (Winnipeg, Paris, Falkland, or Malvinas Islands), people experience almost 8 hours of annual daylength variation.

At the North and South Poles, the range of daylength is extreme with a 6-month period of no insolation, beginning with weeks of twilight to darkness then on to weeks of pre-dawn. Following sunrise, daylight covers a 6-month period of continuous 24-hour insolation—literally the poles experience one long day and one long night each year! The chapter-opening photo illustrates the shock to a midlatitude person at seeing the Midnight Sun.

FIGURE 2.12 Stonehenge on the Salisbury Plain, England.
Neolithic peoples beginning about 5000 years ago constructed this complex of standing rock—sarsen uprights, topped by lintel capstones. Inherent in the layout is an indication of important seasonal dates and precise lunar-cycle predictions. [Photo by Bobbé Christopherson.]

Reasons for Seasons

Seasons result from variations in the Sun's *altitude* above the horizon, the Sun's *declination* (latitude location of the subsolar point), and *daylength* during the year. These in turn are created by several physical factors that operate in concert: Earth's *revolution* in orbit around the Sun, its daily *rotation* on its axis, its *tilted axis*, the unchanging *orientation of its axis*, and its *sphericity* (Table 2.1). Of course, the essential ingredient is having a single source of radiant energy—the Sun. We now look at each of these factors individually. As we do, please note the distinction between revolution—Earth's travel around the Sun—and rotation—Earth's spinning on its axis (Figure 2.13).

Revolution Earth's orbital **revolution** about the Sun is shown in Figure 2.1d and 2.13. Earth's speed in orbit averages 107,280 kmph (66,660 mph). This speed, together with Earth's distance from the Sun, determines the time required for one revolution around the Sun and, therefore, the length of the year and duration of the seasons. Earth completes its annual revolution in 365.2422 days. This number is based on a *tropical year*, measured from equinox to equinox, or the elapsed time between two crossings of the equator by the Sun.

The Earth-to-Sun distance (aphelion to perihelion) might seem a seasonal factor, but it is not significant. It varies about 3% (4.8 million km, or 3 million mi) during the year, amounting to a 50 W/m² difference between local polar summers. Remember that the Earth–Sun distance averages 150 million km (93 million mi).

Rotation Earth's **rotation**, or turning on its axis, is a complex motion that averages slightly less than 24 hours in duration. Rotation determines daylength, creates the apparent deflection of winds and ocean currents, and produces the twice-daily rise and fall of the ocean tides in relation to the gravitational pull of the Sun and the Moon.

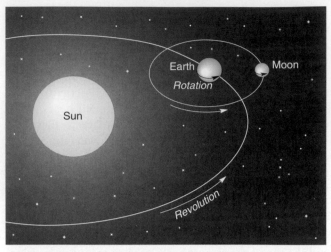

FIGURE 2.13 Earth's revolution and rotation. Earth's *revolution* about the Sun and *rotation* on its axis, viewed from above Earth's orbit. Note the Moon's rotation on its axis and revolution are counterclockwise about Earth, as well.

Earth rotates about its **axis**, an imaginary line extending through the planet from the geographic North Pole to the South Pole. When viewed from above the North Pole, Earth rotates counterclockwise around this axis. Viewed from above the equator, Earth rotates west to east, or eastward. This eastward rotation creates the Sun's *apparent* westward daily journey from sunrise in the east to sunset in the west. Of course, the Sun actually remains in a fixed position in the center of the Solar System. (Note in Figure 2.13 that the Moon both revolves around Earth and rotates on its axis in a counterclockwise direction.)

Although every point on Earth takes the same 24 hours to complete one rotation, the linear velocity of rotation at any point on Earth's surface varies dramatically with latitude. The equator is 40,075 km (24,902 mi) long; therefore, rotational velocity at the equator must be approximately 1675 kmph (1041 mph) to cover that distance in one day. At 60° latitude, a parallel is only half the length of the equator, or 20,038 km (12,451 mi) long, so the rotational velocity there is 838 kmph (521 mph). At the poles, the velocity is 0. (This variation in rotational velocity establishes the effect of the Coriolis force, discussed in Chapter 6.) Table 2.2 lists the speed of rotation for several selected latitudes.

Earth's rotation produces the diurnal (daily) pattern of day and night. The dividing line between day and night is called the **circle of illumination** (as illustrated in Figure 2.15). Because this day-night dividing circle of illumination intersects the equator, *daylength at the equator is always evenly divided*—12 hours of day and 12 hours of night. All other latitudes experience uneven daylength through the seasons, except for 2 days a year, on the equinoxes. (Any two great circles on a sphere bisect one another.)

A true day varies slightly from 24 hours, but by international agreement a day is defined as exactly 24 hours, or

Table 2.1 Five Reasons for Seasons

Factor	Description
Revolution	Orbit around the Sun; requires 365.24 days to complete at 107,280 kmph (66,660 mph)
Rotation	Earth turning on its axis; takes approximately 24 hours to complete at 1675 kmph (1041 mph) at the equator
Tilt	Axis is aligned at a 23.5° angle from a perpendicular to the plane of the ecliptic (the plane of Earth's orbit)
Axial parallelism	Remains in a fixed alignment, with Polaris directly overhead at the North Pole throughout the year
Sphericity	Appears as an oblate spheroid to the Sun's parallel rays; the geoid

Table 2.2	Speed of Rotation at Selected Latitudes		
Latitude	Speed kmph	(mph)	Cities at Approximate Latitudes
90°	0	(0)	North Pole
60°	838	(521)	Seward, Alaska; Oslo, Norway; Saint Petersburg, Russia
50°	1078	(670)	Chibougamau, Québec; Kyyîv (Kiev), Ukraine
40°	1284	(798)	Valdivia, Chile; Columbus, Ohio; Beijing, China
30°	1452	(902)	Pôrto Alegre, Brazil; New Orleans, Louisiana
0°	1675	(1041)	Quito, Ecuador; Pontianak, Indonesia

86,400 seconds. This average, called *mean solar time*, eliminates predictable variations in rotation and revolution that cause the solar day to change slightly in length throughout the year.

The complexity of Earth's rotation is now exactly measured by satellites that are in precise mathematical orbits: GPS (global positioning system, see News Report 1.2, Chapter 1), SLR (satellite-laser ranging), and VLBI (very-long baseline interferometry). All contribute to our knowledge of Earth's rotation. The International Earth Rotation Service at **http://www.iers.org/** issues monthly and annual reports. Earth's rotation is gradually slowing, partially owing to the drag of lunar tidal forces. A "day" on Earth today is many hours longer than it was 4 billion years ago.

Tilt of Earth's Axis To understand Earth's **axial tilt**, imagine a plane (a flat surface) that intersects Earth's elliptical orbit about the Sun, with half of the Sun and Earth above the plane and half below. Such a plane touching all points of Earth's orbit is the **plane of the ecliptic**. Earth's tilted axis remains fixed relative to this plane as Earth revolves around the Sun. The plane of the ecliptic is important to our discussion of Earth's seasons. Now, imagine a perpendicular (at a 90° angle) line passing through the plane. From this perpendicular, Earth's axis is tilted 23.5°. It forms a 66.5° angle from the plane itself (Figure 2.14). The axis through Earth's two poles points just slightly off Polaris, which is named, appropriately, the *North Star*.

This text uses "about" with the tilt angle just described because Earth's axial tilt changes over a complex 41,000-year cycle (see Figure 17.31). The axial tilt ranges roughly between 22° and 24.5° from a perpendicular to the plane of the ecliptic. The present tilt is 23° 27′ (or 66° 33′ from the plane). In decimal numbers, 23° 27′ is approximately 23.45°. For convenience, this is rounded off to a 23.5° tilt (or 66.5° from the plane) in our discussion. Scientific evidence shows that the angle of tilt is lessening in its 41,000-year cycle.

Hypothetically, if Earth were tilted on its side, with its axis parallel to the plane of the ecliptic, we would experience a maximum variation in seasons worldwide. In contrast, if Earth's axis were perpendicular to the plane of its orbit—that is, with no tilt—we would experience no seasonal changes, with something like a perpetual spring or fall season, and all latitudes would experience 12-hour days and nights.

Axial Parallelism Throughout our annual journey around the Sun, Earth's axis *maintains the same alignment* relative to the plane of the ecliptic and to Polaris and other stars. You can see this consistent alignment in Figure 2.15. If we compared the axis in different months, it would always appear parallel to itself, a condition known as **axial parallelism**.

Sphericity Although Earth is not a perfect sphere, as discussed in Chapter 1, we can refer to Earth's *sphericity* as a part of seasonality, for it produces the uneven receipt of insolation from pole to pole shown in Figures 2.9 and 2.10. All five reasons for seasons are summarized in Table 2.1: revolution, rotation, tilt, axial parallelism, and sphericity. Now, considering all these factors operating together, let us examine the annual march of the seasons.

Annual March of the Seasons

During the annual march of the seasons on Earth, daylength is the most obvious way of sensing changes in season at latitudes away from the equator. Daylength is the interval between **sunrise**, the moment when the disk of the Sun first appears above the horizon in the east, and **sunset**, that moment when it totally disappears below the horizon in the west. Table 2.3 lists the average times of sunrise and sunset and the daylength for various latitudes and seasons in the Northern Hemisphere. (For the Southern Hemisphere, merely switch the solstice column headings and switch the equinox column headings.)

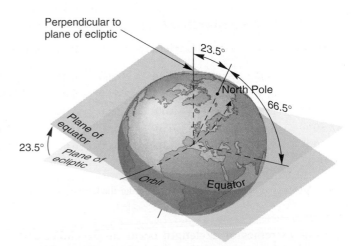

FIGURE 2.14 The plane of Earth's orbit—the ecliptic—and Earth's axial tilt.
Note on the illustration that the plane of the equator is inclined to the plane of the ecliptic at about 23.5°.

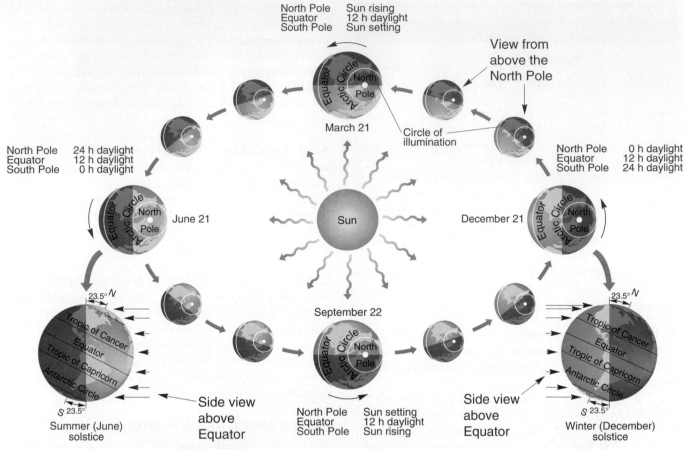

FIGURE 2.15 Annual march of the seasons.
Annual march of the seasons as Earth revolves about the Sun. Shading indicates the changing position of the circle of illumination. Note the hours of daylight for the equator and the poles. To follow the text, begin on the right side at December 21 and move counterclockwise. Given this illustration, turn to the back cover of this textbook and see whether you can determine the month during which the *Apollo* astronaut made the Earth photo. (The answer is on this book's copyright page.)

ANIMATION

Earth-Sun Relations, Seasons

Table 2.3	Daylength Times (Sunrise and Sunset) at Selected Latitudes (Northern Hemisphere)											
	Winter Solstice (December Solstice) December 21–22			**Vernal Equinox (March Equinox) March 20–21**			**Summer Solstice (June Solstice) June 20–21**			**Autumnal Equinox (September Equinox) September 22–23**		
Latitude	A.M.	P.M.	Daylength	A.M.	P.M.	Daylength	A.M.	P.M.	Daylength	A.M.	P.M.	Daylength
0°	6:00	6:00	12:00	6:00	6:00	12:00	6:00	6:00	12:00	6:00	6:00	12:00
30°	6:58	5:02	10:04	6:00	6:00	12:00	5:02	6:58	13:56	6:00	6:00	12:00
40°	7:30	4:30	9:00	6:00	6:00	12:00	4:30	7:30	15:00	6:00	6:00	12:00
50°	8:05	3:55	7:50	6:00	6:00	12:00	3:55	8:05	16:10	6:00	6:00	12:00
60°	9:15	2:45	5:30	6:00	6:00	12:00	2:45	9:15	18:30	6:00	6:00	12:00
90°	No sunlight			Rising Sun			Continuous sunlight			Setting Sun		

Note: All times are standard and do not consider the local option of daylight saving time.

The extremes of daylength occur in December and June. The times around December 21 and June 21 are termed *solstices*. Strictly speaking, the solstices are specific points in time at which the Sun's declination is at its position farthest north at the **Tropic of Cancer**, 23.5° N, or south at the **Tropic of Capricorn**, 23.5° S. "Tropic" is from *tropicus*, meaning a turn or change, so a tropic latitude is where the Sun's declination appears to stand still briefly (Sun stance, or *sol stice*); then it "turns" and heads toward the other tropic.

Table 2.4   Annual March of the Seasons

Approximate Date	Northern Hemisphere Name	Location of the Subsolar Point
December 21–22	Winter solstice (December solstice)	23.5° S latitude (Tropic of Capricorn)
March 20–21	Vernal equinox (March equinox)	0° (Equator)
June 20–21	Summer solstice (June solstice)	23.5° N latitude (Tropic of Cancer)
September 22–23	Autumnal equinox (September equinox)	0° (Equator)

Table 2.4 presents the key seasonal anniversary dates during which the specific time of the equinoxes or solstices occur, their names, and the subsolar point location (declination). During the year, places on Earth outside of the equatorial region experience a continuous but gradual shift in daylength, a few minutes each day, and the Sun's altitude increases or decreases a small amount. You may have noticed that these daily variations become more pronounced in spring and autumn, when the Sun's declination changes at a faster rate.

Figure 2.15 demonstrates the annual march of the seasons and illustrates Earth's relationship to the Sun during the year. Let us begin with December, on the right side of the illustration. On December 21 or 22, at the moment of the **winter solstice** ("winter Sun stance"), or **December solstice**, the circle of illumination excludes the North Pole region from sunlight but includes the South Pole region. The subsolar point is at 23.5° S latitude, the Tropic of Capricorn parallel. The Northern Hemisphere is tilted away from these more direct rays of sunlight—our northern winter—thereby creating a lower angle for the incoming solar rays and thus a more diffuse pattern of insolation.

From 66.5° N latitude to 90° N (the North Pole), the Sun remains below the horizon the entire day. This latitude (66.5° N) marks the **Arctic Circle**, the southernmost parallel (in the Northern Hemisphere) that experiences a 24-hour period of darkness. During this period, twilight and dawn provide some lighting for more than a month at the beginning and end of the Arctic night. During the following 3 months, daylength and solar angles gradually increase in the Northern Hemisphere as Earth completes one-fourth of its orbit.

The sunset photo in Figure 2.16 was made approaching the Antarctic coast, December 14, 2003, at 11:30 P.M. local time, just a week before the December solstice. We were crossing the Branfield Strait at 63° S at the time. Sunrise was only a couple of hours later, at 2 A.M. local time—making the sunless twilight/dawn period only a little over 2 hours!

The moment of the **vernal equinox**, or **March equinox**, occurs on March 20 or 21. At that time, the circle of illumination passes through both poles so that all locations on Earth experience a 12-hour day and a 12-hour night. People living around 40° N latitude (New York, Denver) have gained 3 hours of daylight since the December solstice. At the North Pole, the Sun peeks above the

FIGURE 2.16 Antarctic sunset, 11:30 P.M.!
Approaching the continent of Antarctica at 63° S latitude in December. The low Sun angle produces incredible sunset colors that persist for almost an hour. Note a large iceberg and ice-covered mountains ahead. [Photo by Bobbé Christopherson.]

horizon for the first time since the previous September; at the South Pole, the Sun is setting—a dramatic 3-day "moment" for the people working the Scott–Amundson Base.

From March, the seasons move on to June 20 or 21 and the moment of the **summer solstice**, or **June solstice**. The subsolar point migrates from the equator to 23.5° N latitude, the Tropic of Cancer. Because the circle of illumination now includes the North Polar region, everything north of the Arctic Circle receives 24 hours of daylight—the *Midnight Sun*. Figure 2.17 is a multiple-image photo of the Midnight Sun as seen north of the Arctic Circle. In contrast, the region from the **Antarctic Circle** to the South Pole (66.5°–90° S latitude) is in darkness the entire 24 hours. Those working in Antarctica call this *Midwinter's Day*.

September 22 or 23 is the moment in time of the **autumnal equinox**, or **September equinox**, when Earth's orientation is such that the circle of illumination again passes through both poles so that all parts of the globe experience a 12-hour day and a 12-hour night. The subsolar point returns to the equator, with days growing shorter to the north and longer to the south. Researchers stationed at the South Pole see the disk of the Sun just

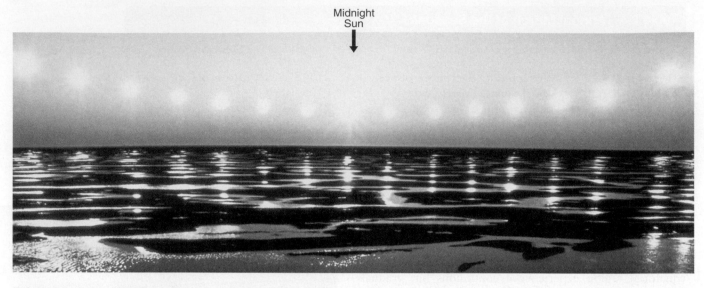

FIGURE 2.17 The Midnight Sun.
The Midnight Sun north of the Arctic Circle captured in a series of 18 exposures on the same piece of film. The camera is facing due north. Midnight is the exposure showing the Sun closest to the horizon. The photographer removed the lens cap at regular intervals to make the multiple exposures. The middle exposure is marked *Midnight Sun*. [Photo by Gary Braasch/Tony Stone Images.]

rising, ending their 6 months of darkness. In the Northern Hemisphere, autumn arrives, a time of many colorful changes in the landscape, whereas in the Southern Hemisphere it is spring.

Dawn and Twilight   *Dawn* is the period of diffused light that occurs before sunrise. The corresponding evening time after sunset is *twilight*. During both periods, light is scattered by molecules of atmospheric gases and reflected by dust and moisture in the atmosphere. The duration of both is a function of latitude, because the angle of the Sun's path above the horizon determines the thickness of the atmosphere through which the Sun's rays must pass. The illumination may be enhanced by the presence of pollution aerosols and suspended particles from volcanic eruptions or forest and grassland fires.

At the equator, where the Sun's rays are almost directly above the horizon throughout the year, dawn and twilight are limited to 30–45 minutes each. These times increase to 1–2 hours each at 40° latitude, and at 60° latitude they each range upward from 2.5 hours, with little true night in summer. The poles experience about 7 weeks of dawn and 7 weeks of twilight, leaving only 2.5 months of "night" during the 6 months when the Sun is completely below the horizon.

Seasonal Observations   In the midlatitudes of the Northern Hemisphere, the position of sunrise on the horizon migrates from day to day, from the southeast in December to the northeast in June. Over the same period, the point of sunset migrates from the southwest to the northwest. The Sun's altitude at local noon at 40° N latitude increases from a 26° angle above the horizon at the winter

(December) solstice to a 73° angle above the horizon at the summer (June) solstice—a range of 47° (Figure 2.18).

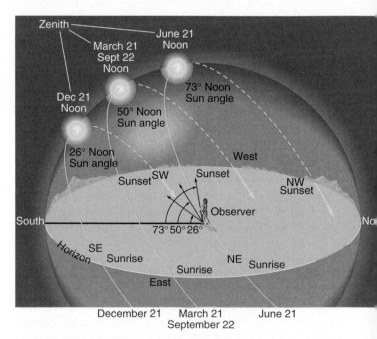

FIGURE 2.18 Seasonal observations—sunrise, noon, and sunset through the year.
Seasonal observations are at 40° N latitude for the December solstice, March equinox, June solstice, and September equinox. The Sun's altitude increases from 26° in December to 73° above the horizon in June—a difference of 47°. Note the changing position of sunrise and sunset along the horizon during the year. For a useful sunrise and sunset calculator for any location, go to http://www.srrb.noaa.gov/highlights/sunrise/sunrise.html.

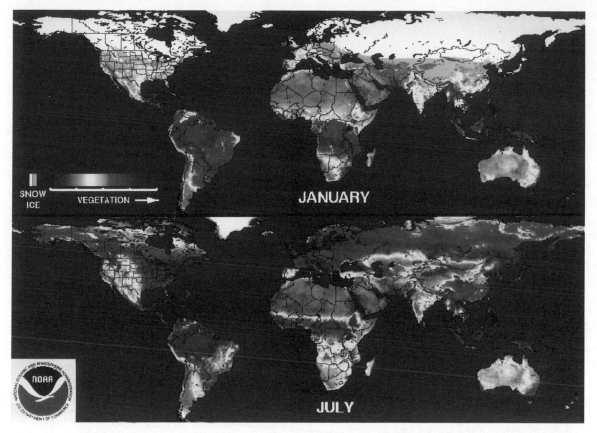

FIGURE 2.19 Seasonal change from orbit.
Seasonal change is monitored and measured by sensors aboard polar-orbiting satellites. Compare and contrast similar regions for January and July: the American Midwest, China, Argentina, and Europe. Orbital remote sensing is tracking an earlier spring and later fall seasons worldwide. [Images courtesy of Garik Gutman, NOAA, National Environmental Satellite, Data, and Information Service (NESDIS).]

Seasonal change is quite noticeable across the landscape away from the equator. Think back over the past year. What seasonal changes have you observed in vegetation, temperatures, and weather? Figure 2.19 presents two composite images of vegetation cover in winter and late summer as recorded by the AVHRR sensors aboard polar-orbiting satellites. Recently, the timing of the seasons is changing as global climates shift in middle and high latitudes. Spring and leafing out is occurring as much as 3 weeks earlier than expected from average conditions. Likewise, the timing of fall is happening later. Ecosystems are changing in response.

Summary and Review—Solar Energy to Earth and the Seasons

■ Distinguish among galaxies, stars, and planets, and **locate** Earth.

Our Solar System—the Sun and nine planets—is located on a remote, trailing edge of the **Milky Way Galaxy**, a flattened, disk-shaped mass estimated to contain up to 400 billion stars. **Gravity**, the mutual attracting force exerted by the mass of an object upon all other objects, is an organizing force in the Universe. The process of stars (like our Sun) condensing from nebular clouds with planetesimals (protoplanets) forming in orbits around their central masses is the **planetesimal hypothesis**.

Milky Way Galaxy (p. 42)
gravity (p. 42)
planetesimal hypothesis (p. 42)

1. Describe the Sun's status among stars in the Milky Way Galaxy. Describe the Sun's location, size, and relationship to its planets.
2. If you have seen the Milky Way at night, briefly describe it. Use specifics from the text in your description.
3. Briefly describe Earth's origin as part of the Solar System.
4. Compare the locations of the nine planets of the Solar System.

■ Overview the origin, formation, and development of Earth, and *construct* Earth's annual orbit about the Sun.

The Solar System, planets, and Earth began to condense from a nebular cloud of dust, gas, debris, and icy comets approximately 4.6 billion years ago. Distances in space are so vast that the speed of light (300,000 kmps, or 186,000 mps, which is about 9.5 trillion km, or nearly 6 trillion mi, per year) is used to express distance.

In its orbit, Earth is at **perihelion** (its closest position to the Sun) during our Northern Hemisphere winter (January 3 at 147,255,000 km, or 91,500,000 mi). It is at **aphelion** (its farthest position from the Sun) during our Northern Hemisphere summer (July 4 at 152,083,000 km, or 94,500,000 mi). Earth's average distance from the Sun is approximately 8 minutes and 20 seconds in terms of light speed.

  speed of light (p. 42)
  perihelion (p. 42)
  aphelion (p. 42)

5. How far is Earth from the Sun in terms of light speed? In terms of kilometers and miles?
6. Briefly describe the relationship among these concepts: Universe, Milky Way Galaxy, Solar System, Sun, and Earth.
7. Diagram in a simple sketch Earth's orbit about the Sun. How much does it vary during the course of a year?

■ *Describe* the Sun's operation, and *explain* the characteristics of the solar wind and the electromagnetic spectrum of radiant energy.

The **fusion** process—hydrogen nuclei forced together under tremendous temperature and pressure in the Sun's interior—generates incredible quantities of energy. Solar energy in the form of charged particles of **solar wind** travels out in all directions from magnetic disturbances on the Sun, such as from large **sunspots**. Solar wind is deflected by Earth's **magnetosphere**, producing various effects in the upper atmosphere, including spectacular **auroras**, the northern and southern lights, which surge across the skies at higher latitudes. Another effect of the solar wind in the atmosphere is its possible influence on weather.

The **electromagnetic spectrum** of radiant energy travels outward in all directions from the Sun. The total spectrum of this radiant energy is made up of different **wavelengths**—the distance between corresponding points on any two successive waves of radiant energy. Eventually, some of this radiant energy reaches Earth's surface.

  fusion (p. 42)
  solar wind (p. 44)
  sunspots (p. 44)
  magnetosphere (p. 44)
  auroras (p. 45)
  electromagnetic spectrum (p. 46)
  wavelength (p. 46)

8. How does the Sun produce such tremendous quantities of energy?
9. What is the sunspot cycle? At what stage is the cycle in the year 2008?
10. Describe Earth's magnetosphere and its effects on the solar wind and the electromagnetic spectrum.
11. Summarize the presently known effects of the solar wind relative to Earth's environment.
12. Describe the various segments of the electromagnetic spectrum, from shortest to longest wavelength. What are the main wavelengths produced by the Sun? Which are principally radiated by Earth to space?

■ *Portray* the intercepted solar energy and its uneven distribution at the top of the atmosphere.

Electromagnetic radiation from the Sun passes through Earth's magnetic field to the top of the atmosphere—the **thermopause**, at approximately 500 km (300 mi) altitude. Incoming solar radiation that reaches a horizontal plane at Earth is **insolation**, a term specifically applied to radiation arriving at Earth's surface and atmosphere. The term **solar constant** describes insolation at the top of the atmosphere: the average insolation received at the thermopause when Earth is at its average distance from the Sun. The solar constant is measured as an average of 1372 W/m² (2.0 cal/cm²/min; 2 langleys/min). The place receiving maximum insolation is the **subsolar point**, where solar rays are perpendicular to the surface (radiating from directly overhead). All other locations away from the subsolar point receive slanting rays and more diffuse energy.

  thermopause (p. 47)
  insolation (p. 47)
  solar constant (p. 47)
  subsolar point (p. 48)

13. What is the solar constant? Why is it important to know?
14. Select 40° or 50° N latitude on Figure 2.10 and plot the amount of energy in watts per square meter (W/m²) per day on a graph for each month throughout the year. Compare this with the amount at the North Pole and at the equator.
15. If Earth were flat and oriented perpendicularly to incoming solar radiation (insolation), what would be the latitudinal distribution of solar energy at the top of the atmosphere?

■ *Define* solar altitude, solar declination, and daylength, and *describe* the annual variability of each—Earth's seasonality.

The angle between the Sun and the horizon is the Sun's **altitude**. The Sun's **declination** is the latitude of the subsolar point. Declination annually migrates through 47° of latitude, moving between the *Tropic of Cancer* at 23.5° N (June) and the *Tropic of Capricorn* at 23.5° S latitude (December). Seasonality means an annual change in the Sun's altitude and changing **daylength**, or duration of exposure.

Earth's distinct seasons are produced by interactions of **revolution** (annual orbit about the Sun), **rotation** (Earth's turning on its **axis**), **axial tilt** (at about 23.5° from a perpendicular to the **plane of the ecliptic**) and **axial parallelism** (the parallel alignment of the axis throughout the year), and *sphericity*. Earth rotates about its axis, an imaginary line extending through the planet from the geographic North Pole to the South Pole. In the Solar System, an imaginary plane touching all points of Earth's orbit is termed the **plane of the ecliptic**. As Earth rotates, the boundary that divides daylight and darkness, called the **circle of illumination**, travels. Daylength is the interval between **sunrise**, the moment when the disk of the Sun first appears above the horizon in the east, and **sunset**, that moment when it totally disappears below the horizon in the west. The **Tropic of Cancer** parallel marks the farthest north the subsolar point migrates during the year, 23.5° N latitude. The **Tropic of Capricorn** parallel marks the farthest south the subsolar point migrates during the year, 23.5° S latitude.

On December 21 or 22, at the moment of the **winter solstice** ("winter Sun stance"), or **December solstice**, the circle of illumination excludes the North Pole but includes the South Pole. The subsolar point is at 23.5° S latitude, the parallel called the Tropic of Capricorn. The Sun remains below the horizon the entire day. This latitude (66.5° N) marks the **Arctic Circle**, the southernmost parallel (in the Northern Hemisphere) that experiences a 24-hour period of darkness, or on June 21, a 24-hour period of daylight. The moment of the **vernal equinox**, or **March equinox**, occurs on March 20 or 21. At that time, the circle of illumination passes through both poles so that all locations on Earth experience a 12-hour day and a 12-hour night.

June 20 or 21 is the moment of the **summer solstice**, or **June solstice**. The subsolar point migrates from the equator to 23.5° N latitude, the Tropic of Cancer. Because the circle of illumination now includes the North Polar region, everything north of the Arctic Circle receives 24 hours of daylight—the "Midnight Sun." In contrast, the area from the **Antarctic Circle** to the South Pole (66.5°–90° S latitude) is in darkness the entire 24 hours, or on December 21, in constant daylight. September 22 or 23 is the time of the **autumnal equinox**, or **September equinox**, when Earth's orientation is such that the circle of illumination again passes through both poles so that all parts of the globe experience a 12-hour day and a 12-hour night.

altitude (p. 51)
declination (p. 51)
daylength (p. 51)
revolution (p. 52)
rotation (p. 52)
axis (p. 52)
circle of illumination (p. 52)
axial tilt (p. 53)
plane of the ecliptic (p. 53)
axial parallelism (p. 53)
sunrise (p. 53)

sunset (p. 53)
Tropic of Cancer (p. 54)
Tropic of Capricorn (p. 54)
winter solstice, December solstice (p. 55)
Arctic Circle (p. 55)
vernal equinox, March equinox (p. 55)
summer solstice, June solstice (p. 55)
Antarctic Circle (p. 55)
autumnal equinox, September equinox (p. 55)

16. Assess the 12-month Gregorian calendar, with its months of different lengths, and leap years, and its relation to the annual seasonal rhythms—the march of the seasons. What do you find?
17. The concept of seasonality refers to what specific phenomena? How do these two aspects of seasonality change during a year at 0° latitude? At 40°? At 90°?
18. Differentiate between the Sun's altitude and its declination at Earth's surface.
19. For the latitude at which you live, how does daylength vary during the year? How does the Sun's altitude vary? Does your local newspaper publish a weather calendar containing such information?
20. List the five physical factors that operate together to produce seasons.
21. Describe Earth's revolution and rotation and differentiate between them.
22. Define Earth's present tilt relative to its orbit about the Sun.
23. Describe seasonal conditions at each of the four key seasonal anniversary dates during the year. What are the solstices and equinoxes, and what is the Sun's declination at these times?

 NetWork

The *Geosystems Student Learning Center* provides on-line resources for this chapter on the World Wide Web. To begin: Once at the Center, click on the cover of this textbook, scroll the Table of Contents menu, and select this chapter. You will find self-tests that are graded, review exercises, specific

updates for items in the chapter, and in "Destinations" many links to interesting related pathways on the Internet. *Geosystems Student Learning Center* is found at **http://www.prenhall.com/ christopherson/**.

Critical Thinking

A. Using the concepts in Figure 2.18, use a protractor and stick or ruler to measure the angle of the Sun's altitude at noon (or 1 P.M., if in daylight saving time). Do not look at the Sun; rather, with your back to the Sun, align the stick so that it casts no shadow as you measure its rays against the protractor. Place this measurement in your notebook and affix a Post-It® in the textbook to remind you to repeat the measurement near the end of the semester. Compare and analyze seasonal change using your different measurements of the Sun's altitude.

B. Also in reference to the concepts in Figure 2.18, post a reminder in your notebook to check the position of sunrise and sunset at least twice during the semester. If you have a magnetic compass, note the degrees from north for sunrise

and sunset (*azimuth* is read from north in a clockwise direction—0° and 360° being the same point, north). Find some place where you can see the horizon. If new to such observations, the degree of location change over the span of months in a school term will surprise you. Be sure and ask your teacher about the magnetic declination of your location to adjust your compass readings.

C. The variability of Earth's axial tilt, orbit about the Sun, and a wobble to the axis is described in Figure 17.31. Please refer to this figure and compare these changing conditions to the information in Table 2.1 and the related figures in this chapter. Speculate on the effects of each of these changes on the annual march of the seasons.

In this chapter: We examine the modern atmosphere using criteria of composition, temperature, and function. Our look at the atmosphere also includes the spatial aspects of both natural and human-produced air pollution. We all participate in the atmosphere with each breath we take, the energy we consume, the traveling we do, and the products we buy. Human activities caused the stratospheric ozone predicament and the blight of acid deposition on ecosystems. These topics are essential to physical geography, for we are influencing the atmospheric composition for the future.

Atmospheric Composition, Temperature, and Function

The *modern* atmosphere probably is the fourth general atmosphere in Earth's history, making the modern atmosphere a gaseous mixture of ancient origin, the sum of all the exhalations and inhalations of life on Earth throughout time. The principal substance of this atmosphere is air, the medium of life as well as a major industrial and chemical raw material. **Air** is a simple mixture of gases that is naturally odorless, colorless, tasteless, and formless, blended so thoroughly that it behaves as if it were a single gas.

In his book *The Lives of a Cell*, the late physician and self-styled "biology watcher" Lewis Thomas compared the atmosphere of Earth to an enormous cell membrane.

The membrane around a cell regulates the interactions between the cell's delicate inner workings and the potentially disruptive outer environment. Each cell membrane is selective as to what it will allow to pass through. The modern atmosphere acts as Earth's protective membrane, as Thomas described so vividly (Figure 3.1).

As a practical matter, we consider the top of our atmosphere to be around 480 km (300 mi) above Earth's surface, the same altitude we used in Chapter 2 for measuring the solar constant and insolation receipt. Beyond that altitude is the **exosphere**, which means "outer sphere," where the rarefied atmosphere is nearly a vacuum. It contains scarce lightweight hydrogen and helium atoms, weakly bound by gravity as far as 32,000 km (20,000 mi) from Earth.

Atmospheric Profile

As critical as the atmosphere is to us, it represents only the thinnest envelope, amounting to less than one-millionth of Earth's total mass. Think of Earth's modern atmosphere as a series of imperfectly shaped concentric "shells" or "spheres" that grade into one another, all bound to the planet by gravity. To study the atmosphere let's view it in layers, each with distinctive properties and purposes. Figure 3.2 charts the atmosphere in a vertical cross-section profile, or side view, and is a useful reference for the following discussion. To simplify we use three atmospheric

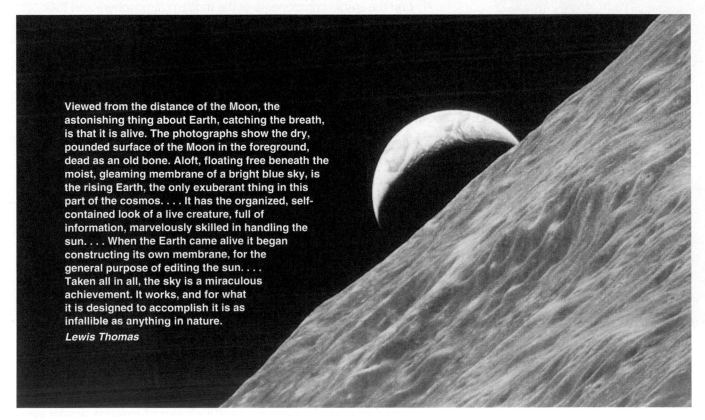

Viewed from the distance of the Moon, the astonishing thing about Earth, catching the breath, is that it is alive. The photographs show the dry, pounded surface of the Moon in the foreground, dead as an old bone. Aloft, floating free beneath the moist, gleaming membrane of a bright blue sky, is the rising Earth, the only exuberant thing in this part of the cosmos. . . . It has the organized, self-contained look of a live creature, full of information, marvelously skilled in handling the sun. . . . When the Earth came alive it began constructing its own membrane, for the general purpose of editing the sun. . . . Taken all in all, the sky is a miraculous achievement. It works, and for what it is designed to accomplish it is as infallible as anything in nature.

Lewis Thomas

FIGURE 3.1 Earthrise.
Earthrise over the stark and lifeless lunar surface. [Photo by NASA; quotation from "The World's Biggest Membrane" in *The Lives of a Cell* by Lewis Thomas. Copyright © 1973, Massachusetts Medical Society. Reprinted by permission of Viking Penguin, Penguin Books USA, Inc.]

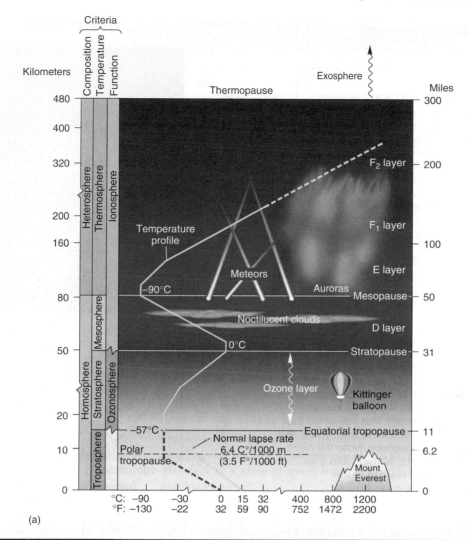

(a)

(b)

FIGURE 3.2 Profile of the modern atmosphere.
(a) The integrated chart of our modern atmosphere is from Earth's surface to the top of the atmosphere at 480 km (300 mi). The columns along the left side show the division of the atmosphere by composition, temperature, and function. The plot of temperature by altitude relates to the scale along the bottom axis. (The small balloon shows the height achieved by Kittinger, discussed in News Report 3.1.) (b) Space Shuttle astronauts captured a dramatic sunset through various atmospheric layers across the "edge" of our planet—*Earth's limb*. A silhouetted cumulonimbus thunderhead cloud is topping out at the tropopause. [Space Shuttle photo from NASA.]

criteria detailed in Table 3.1: *composition*, *temperature*, and *function*, noted along the left side of Figure 3.2a.

Earth's atmosphere exerts its weight, pressing downward under the pull of gravity. Air molecules create air pressure through their motion, size, and number. Pressure is exerted on all surfaces in contact with the air. The weight (force over a unit area) of the atmosphere, or **air pressure**, pushes in on all of us. Fortunately, that same

Table 3.1	Atmospheric Criteria	
Composition	**Temperature**	**Function**
Heterosphere	Thermosphere	Ionosphere
	Mesosphere	
Homosphere	Stratosphere	Ozonosphere
	Troposphere	

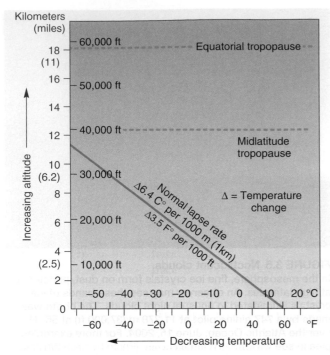

FIGURE 3.6 The temperature profile of the troposphere. During daytime, temperature decreases with increased altitude at a rate known as the *normal lapse rate*. Scientists use a concept called the *standard atmosphere* as an accepted description of air temperature and pressure changes with altitude. Note the approximate locations of the equatorial and midlatitude tropopauses.

Figure 3.6 illustrates the normal temperature profile within the troposphere during daytime. As the graph shows, temperatures decrease rapidly with increasing altitude at an average of 6.4 C° per kilometer (3.5 F° per 1000 ft), a rate known as the **normal lapse rate**. This temperature plot also appears in the troposphere portion of Figure 3.2.

The normal lapse rate is an average. The actual lapse rate at any particular time and place, which may deviate considerably because of local weather conditions, is the **environmental lapse rate**. This variation in temperature gradient in the lower troposphere is central to our discussion of weather processes in Chapters 7 and 8.

In the stratosphere, the marked warming with increasing altitude causes the tropopause to act like a lid, essentially preventing whatever is in the cooler (denser) air below from mixing into the warmer (less dense) stratosphere. However, the tropopause may be disrupted wherever jet streams produce vertical turbulence and an interchange between the troposphere and the stratosphere, as we see in Chapter 6. Also, hurricanes occasionally inject moisture above this temperature-inversion layer at the tropopause, and powerful volcanic eruptions may loft ash and sulfuric acid mists into the stratosphere, as did the Mount Pinatubo eruptions in 1991.

Atmospheric Function Criterion

Looking at our final atmospheric criterion of *function*, the atmosphere has two specific zones, the *ionosphere* and the *ozonosphere* (*ozone layer*), which remove most of the harmful wavelengths of incoming solar radiation and charged particles

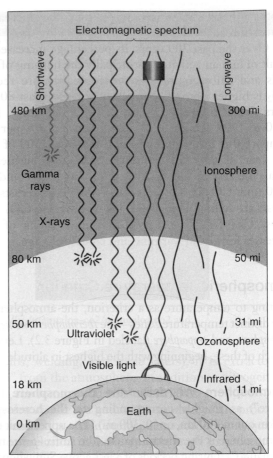

FIGURE 3.7 The atmosphere protects Earth's surface. As solar energy passes through the atmosphere, the shortest wavelengths are absorbed. Only a fraction of the ultraviolet radiation, and most of the visible light and infrared, reaches Earth's surface.

ANIMATION Ozone Breakdown / Ozone Hole

(Focus Study 3.1). Figure 3.7 depicts in a general way the absorption of radiation by the various functional layers of the atmosphere.

Ionosphere The outer functional layer, the **ionosphere**, extends throughout the thermosphere and into the mesosphere below (Figure 3.2). The ionosphere absorbs cosmic rays, gamma rays, X-rays, and shorter wavelengths of ultraviolet radiation, changing atoms to positively charged ions and giving the ionosphere its name. The glowing auroral lights discussed in Chapter 2 occur principally within the ionosphere.

Figure 3.2 shows the average daytime altitudes of four regions within the ionosphere, known as the D, E, F_1 and F_2 layers. These are important to broadcast communications, for they reflect certain radio wavelengths, including AM radio and other shortwave broadcasts, especially at night. However, during the day the ionosphere actively absorbs arriving radio signals, preventing distant reception. Normally unaffected are FM or television broadcast wavelengths, which pass through to space, requiring the use of communication satellites.

Ozonosphere That portion of the stratosphere that contains an increased level of ozone is the **ozonosphere**,

(text continued on page 72)

Stratospheric Ozone Losses: A Worldwide Health Hazard

Consider these news reports and conditions:

- The stratospheric ozone above Antarctica during the months of Antarctic spring (September to November) suffered its most severe loss in 2006, covering an area three times larger than the United States. Each year, the "ozone hole" (actually a thinning) widens and deepens, even though the rate of accumulation of offending chemicals has leveled off.
- As stratospheric ozone flows into the depleted region from lower latitudes, the protective layer thins over southern South America, southern Africa, Australia, and New Zealand. In Ushuaia, Argentina, surface UVB intensity increased 236% compared to normal levels.
- An international scientific consensus confirmed previous assessments of the anthropogenic, or human-caused, disruption of the ozone layer—chlorine atoms and chlorine monoxide molecules in the stratosphere are of human origin. (See *Scientific Assessment of Ozone Depletion*, by NASA, NOAA, United Nations Environment Programme, and World Meteorological Organization.)
- At Earth's opposite pole, a similar ozone depletion over the Arctic annually exceeds 30%, to a record 45% loss during spring 1997. The Canadian and U.S. governments report an "ultraviolet index" to help the public protect themselves (http://www.ec.gc.ca/ozone/).
- Increased ultraviolet radiation is affecting atmospheric chemistry, biological systems, oceanic phytoplankton (small photosynthetic organisms that form the basis of the ocean's primary food production), fisheries, crop yields, and human skin, eye tissues, and immunity.

More ultraviolet radiation than ever before is breaking through Earth's protective ozone layer. What

is going on in the stratosphere everywhere and in the polar regions specifically? Why is this happening, and how are people and their governments responding? What effects does this have on you personally?

Monitoring Earth's Fragile Safety Screen

A sample from the ozone layer's densest part (at 29 km, or 18 mi altitude) contains only 1 part ozone per 4 million parts of air—compressed to surface pressure, the ozone layer would be only 3 mm thick. Yet, this rarefied layer was in steady-state equilibrium for several hundred million years, absorbing intense ultraviolet radiation and permitting life to proceed safely on Earth.

The ozone layer has been monitored since the 1920s. Ground stations with instrumented balloons, aircraft, orbiting satellites, and a 30-station ozone-monitoring network (mostly in North America) observe stratospheric ozone. The total ozone mapping spectrometer (TOMS) began operations in 1978 aboard various satellites. A September 2006 image by the OMI (Ozone Monitoring Instrument) on NASA's *Aura* satellite is in Figure 3.1.1, and shows the all-time record ozone depletion. These measures mean that the ozone column over Antarctica was

virtually gone. *Aura* and its OMI package became operational in 2004. (See http://www.nasa.gov/mission_pages/aura/main/index.html.)

Ozone Losses Explained

What is causing the decline in stratospheric ozone? In 1974, two atmospheric chemists, F. Sherwood Rowland and Mario Molina, hypothesized that some synthetic chemicals were releasing chlorine atoms that decompose ozone. These **chlorofluorocarbons**, or **CFCs**, are synthetic molecules of chlorine, fluorine, and carbon. (See Rowland and Molina's report: "Stratospheric sink for chlorofluoromethanes: Chlorine atom catalyzed destruction of ozone," *Nature* 249 (1974): 810.)

CFCs are stable, or inert, under conditions at Earth's surface and they possess remarkable heat properties. Both qualities made them valuable as propellants in aerosol sprays and as refrigerants. Also, some 45% of CFCs were solvents in the electronics industry and used as foaming agents. Being inert, CFC molecules do not dissolve in water and do not break down in biological processes. (In contrast, chlorine compounds derived from volcanic eruptions and the ocean are water soluble and rarely reach the stratosphere.)

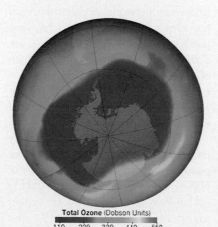

FIGURE 3.1.1 The Antarctic ozone hole.
OMI image for September 24, 2006, shows the largest ozone hole observed, nearly 29.5 million km^2 (11.4 million mi^2). Measurements dropped below 100 Dobson units in the region of blues and purples; greens, yellows, and reds denote more ozone. The ozone "hole" has grown larger since 1979, although the accumulation of damaging chemicals has leveled. [OMI instrument aboard *Aqua* courtesy of GSFC/NASA.]

(continued)

Focus Study 3.1 (continued)

Researchers Rowland and Molina hypothesized that stable CFC molecules slowly migrate into the stratosphere, where intense ultraviolet radiation splits them, freeing chlorine (Cl) atoms. This process produces a complex set of reactions that breaks up ozone molecules (O_3) and leaves oxygen gas molecules (O_2) in their place. The effect is severe, for a single chlorine atom decomposes more than 100,000 ozone molecules. The long residence time of chlorine atoms in the ozone layer (40 to 100 years) is likely to produce long-term consequences through this century from the chlorine already in place. More than 22 million metric tons (24 million tons) of CFCs were sold worldwide and subsequently released into the atmosphere.

Of many identified reactions, here are simplified equations that show chlorine monoxide-produced (ClO) and chlorine-produced ozone losses. Note that chlorine monoxide (ClO) reacts with ozone (O_3) to produce a free chlorine atom (Cl) and two oxygen gas molecules (O_2). The free chlorine atom then reacts with an ozone molecule (O_3) to produce more chlorine monoxide (ClO) and oxygen gas (O_2)

Ozone depletion by ClO and by Cl:

$$ClO + O_3 \rightarrow Cl + 2O_2$$

$$Cl + O_3 \rightarrow ClO + O_2$$

The chlorine monoxide is then available to go through the cycle all over again. The *net reaction*, with the chlorine catalyst removed, is a loss of ozone (O_3) and its conversion to oxygen gas (O_2) molecules. The problem then is that O_2, unlike O_3, is transparent to ultraviolet radiation. (See http://www.epa.gov/ozone/index.html.)

Political Realities—An International Response and the Future

As the science became established, sales and production of CFCs declined until a March 1981 presidential order permitting the export and sales of banned products. Sales increased and hit a new peak in 1987 at 1.2 million metric tons (1.32 million tons), when an international agreement went into effect halting further sales growth.

Chemical manufacturers once claimed that no hard evidence existed to prove the ozone-depletion model, and they successfully delayed remedial action for 15 years. Today, with extensive scientific evidence and verification of losses, even the CFC manufacturers admit that the problem is serious.

The *Montreal Protocol on Substances That Deplete the Ozone Layer* (1987), as amended five times in 1990, 1992, 1995, 1997, and 1999, aims to reduce and eliminate CFC damage to the ozone layer. With 189 signatory countries the protocol is regarded as probably the most successful international agreement in history.

CFC sales continue their decline and all production of harmful CFCs will cease by 2010, although there is concern over some of the substitute compounds and a robust black market for banned CFCs, rivaling drug trafficking at some ports. If the protocol is fully enforced, it is estimated that the stratosphere will return to more normal conditions in a century. (See the United Nations Ozone Secretariat at http://ozone.unep.org/ and Montreal Protocol at http://www.ec.gc.ca/international/multilat/ozone1_e.htm.)

The protocol also bans other chemicals, such as bromine, that are damaging to the atmosphere, using different target dates; progress is slow. Bromine sources include halon-charged fire extinguishers and methyl bromide pesticides and soil fumigants. They are about 50 times more effective than chlorine as catalysts causing reactions that destroy ozone, although they have a residence time in the atmosphere of only a couple of years.

The world without this treaty might prove challenging. Imagine possible stratospheric ozone losses of 50% in the midlatitudes, producing 1.5 million more cases of malignant melanoma (a skin cancer). We owe much to doctors Rowland and Molina.

Mario Molina stated, "It was frustrating for many years, but it really paid off with the Protocol, which was a marvelous example of what the international community can do working together. We can see from atmospheric measurements that it is already working" (*Nature* 389 (September 18, 1997): 219). For their work, Rowland, Molina, and another colleague, Paul Crutzen, received the 1995 Nobel Prize for Chemistry. In making the award, the Royal Swedish Academy of Sciences said, "By explaining the chemical mechanism that affects the thickness of the ozone layer, the three researchers have contributed to our salvation from a global environmental problem that could have catastrophic consequences."

Ozone Losses over the Poles

How do Northern Hemisphere CFCs become concentrated over the South Pole? Evidently, chlorine freed in the Northern Hemisphere midlatitudes concentrates over Antarctica through the work of atmospheric winds. Persistent cold temperatures over the South Pole and the presence of thin, icy clouds in the stratosphere promote development of the ozone hole. Over the North Pole, conditions are more changeable, so the hole is smaller, although growing each year.

Polar stratospheric clouds (PSCs) are thin clouds that are important catalysts in the release of chlorine for ozone-depleting reactions. During the long, cold winter months, a tight circulation pattern forms over the Antarctic continent—the polar vortex. Chlorine that is freed from droplets in PSCs triggers the breakdown of the otherwise inert molecules and frees the chlorine for catalytic reactions. Figure 3.1.2a graphs the negative correlation between ClO presence and O_3 losses over the Antarctic continent within the polar vortex. Figure 3.1.2b graphs the progressive loss of ozone during the year to its usual September low, and Figure 3.1.2c presents satellite data on the areal size of the ozone depleted region during the year.

Atmospheric science is active in Antarctica. The U.S. scientific base, Palmer Station (64° 46′S) on Anvers Island, Antarctic Peninsula, experiences a 285% increase in surface UVB radiation above normal levels in the fall. Nearby at Ukraine's Akademik Vernadsky Station, the former British Faraday Base in the Argentine Islands (65° 9′S), one of the original ozone spectrophotometer instruments is actively measuring ultraviolet radiation (Figure 3.1.3). Ukraine scientists are studying UVA (320–400 nm) and UVB (290–320 nm) wavelengths using a variety of instruments, including *biodosimeters* that use biological material to measure radiation impact.

UV Index Helps Save Your Skin

Newspaper and television weather reports regularly include the *UV Index* (UVI) in forecasts, as reported by the National Weather Service (NWS) and the Environmental Protection Agency (EPA) since 1994. A revision in 2004 aligned the UVI with guidelines adopted by the World Health Organization and World Meteorological Organization.

Note the 2004 UVI scale does not rely on "burn times." People were misinterpreting time-to-burn estimates thinking there was some safe level of exposure, or safe tanning time—which was the wrong message. Table 3.1.1 presents the revised UVI. Fair-skinned people (melano-compromised) must exercise more caution than moderate to darker skin types (melano-competent and melano-protected), but all skin types should feel warned, especially for babies and children.

As stratospheric ozone levels continue to thin, surface exposure to cancer-causing radiation climbs. The public now is alerted to take extra precautions in the form of sunscreens, hats, protective clothing, and sunglasses. Remember, damage is cumulative and it may be decades before you experience the ill effects triggered by this summer's sunburn. (For more information, contact the American Cancer Society, 800-227-2345, http://www.cancer.org/; or write the American Academy of Dermatology, P.O. Box 681069, Schaumburg, IL 60168-1069.)

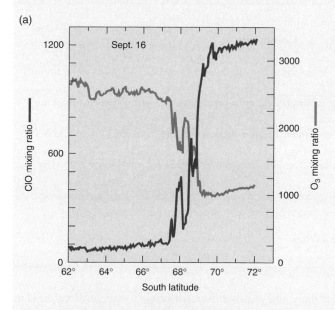

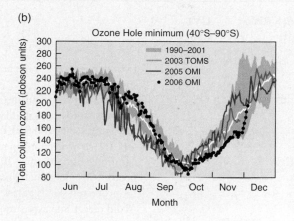

Southern Hemisphere
Ozone 2002–2003

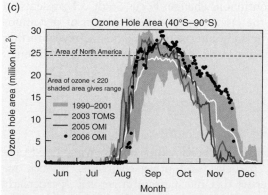

FIGURE 3.1.2 Chemical evidence and the timing and extent of ozone damage.
(a) The negative correlation between ClO and O_3 over the Antarctic continent poleward of 68° S latitude. The data were collected from flights during September 1987 at stratospheric altitudes. (b) Total ozone column data from 1990 to 2006 illustrate the seasonal loss. (c) The area of ozone hole depletion generally peaks in late September, reaching a record low 85 Dobson units in 2006. [(a) Data from NASA; (b and c) TOMS aboard various satellites and OMI on *Aqua*, collected data courtesy of GSFC/NASA.]

(continued)

Focus Study 3.1 (continued)

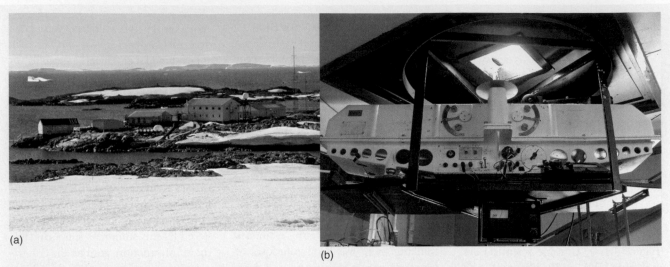

(a)

(b)

FIGURE 3.1.3 Atmospheric science in Antarctica.
(a) Ukraine's Vernadsky Station operates year-round performing valuable scientific research. (b) This ozone spectrophotometer instrument measures the ultraviolet component of insolation arriving through the opening in the science station's roof. [Photos by Bobbé Christopherson.]

Table 3.1.1   UV Index (EPA, NWS, WHO, WMO)

Exposure Risk Category	UVI Range	Comments
Low	less than 2	Low danger for average person. Wear sunglasses on bright days. Watch out for reflection off snow.
Moderate	3–5	Take covering precautions, such as sunglasses, sunscreen, hats, and protective clothing, and stay in shade during midday hours.
High	6–7	Use sunscreens with SPF of 15 or higher. Reduce time in the Sun between 11 A.M. and 4 P.M. Use protections mentioned above.
Very High	8–10	Minimize Sun exposure 10 A.M. to 4 P.M. Use sunscreens with SPF ratings of over 15. Use protections mentioned above.
Extreme	11+	Unprotected skin is at risk of burn. Sunscreen application every two hours if out-of-doors. Avoid direct Sun exposure during midday hours. Use protections mentioned above.

See http://www.epa.gov/sunwise/uvindex.html for a UVI forecast for your location.

or **ozone layer**. Ozone is a highly reactive oxygen molecule made up of three oxygen atoms (O_3) instead of the usual two atoms (O_2) that make up oxygen gas. Ozone absorbs certain wavelengths of ultraviolet (principally, all the UVC, 100–290 nm, and about 90% of UVB, 290–320 nm).* This absorbed energy subsequently reradiates at longer wavelengths as infrared radiation. This process converts most harmful ultraviolet radiation, effectively "filtering" it and safeguarding life at Earth's surface. UVA, at 320–400 nm, makes up about 98% of all UV radiation that reaches Earth's surface.

The ozone layer is presumed to have been relatively stable over the past several hundred million years (allowing

*Nanometer (nm) = one-millionth of a millimeter; 1 nm = 10^{-9}. For comparison, a micrometer, or micron (μm), one-millionth of a meter; 1 μm = 10^{-6} m. A millimeter (mm) = one-thousandth of a meter; 1 mm = 10^{-3}.

for daily and seasonal fluctuations). Today, however, it is in a state of continuous change. Focus Study 3.1 presents an analysis of the crisis in this critical portion of our atmosphere. Fortunately, international treaties to prevent further ozone losses are working, although the damage continues.

Variable Atmospheric Components

The troposphere contains natural and human-caused variable gases, particles, and other chemicals. In fact, researchers report that the increase in aerosols, particularly soot, has produced a "dimming" of sunlight reaching the surface—globally, a measured 4% to 8% reduction in surface receipts. The World Health Organization estimated that air pollution kills approximately 4.3 million people worldwide. Spatial aspects of these variables are important applied topics in physical geography.

Air pollution is not a new problem. Romans complained more than 2000 years ago about the foul air of their cities. Filling Roman air was the stench of open sewers, smoke from fires, and fumes from ceramic-making kilns and smelters (furnaces) that converted ores into metals. In human experience, cities are always the place where the environment's natural ability to process and recycle waste is stressed. Air pollution is closely linked to our production and consumption of energy and resources.

Solutions require regional, national, and international strategies, because the pollution sources often are distant from the observed impact—moving across political boundaries and even crossing oceans. Regulations to curb human-caused air pollution have achieved great success, although much remains to be done. Before we discuss these topics, let's examine some natural pollution sources.

Natural Sources

Natural air pollution sources produce a greater quantity of pollutants—nitrogen oxides, carbon monoxide, hydrocarbons from plants and trees, and carbon dioxide—than do human-made sources. Table 3.3 lists some of these natural sources and the substances they contribute to the air. However, any attempt to dismiss the impact of human-made air pollution through a comparison with natural sources is irrelevant, for we evolved with and adapted to the natural ingredients in the air. We did not evolve in relation to the comparatively recent concentrations of *anthropogenic* (human-caused) contaminants in our metropolitan regions.

A dramatic natural source of pollution was the 1991 eruption of Mount Pinatubo in the Philippines (15° N 120° E), perhaps the twentieth century's second-largest eruption. This event injected nearly 20 million tons of sulfur dioxide (SO_2) into the stratosphere. The spread of these emissions is shown in a sequence of satellite images in Figure 6.1, Chapter 6.

Devastating wildfires on several continents, including widespread fires across the western United States, produce natural air pollution as each year's fire season breaks the previous year's record. Soot, ash, and gases darken skies and damage health in affected regions. Wind

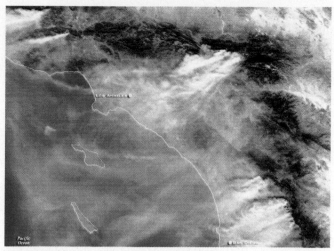

Southern California, 2003 (compare to Chapter 21 opener)

FIGURE 3.8 Southern California wildfires fill the atmosphere with smoke.
Terra image of wildfires in Southern California consuming thousands of acres of drought-damaged forests; image made October 28, 2003. [Image is *Terra* MODIS sensor, courtesy of the MODIS Land Rapid Response Team, NASA/GSFC.]

patterns spread the pollution from the fires to nearby cities, closing airports and forcing evacuations to avoid the health-related dangers. Wildfire smoke contains particulate matter (dust, smoke, soot, ash), nitrogen oxides, carbon monoxide, and volatile organic compounds.

Scientists established the connection between climate change and wildfire occurrence, meaning this source of smoke is on the increase as is the release of enormous quantities of carbon dioxide to the atmosphere. A sample of findings related to climate-change impacts and the western United States:

> Higher spring and summer temperatures and earlier snowmelt are extending the wildfire season and increasing the intensity of wildfires in the western US... large wildfire activity increased suddenly and markedly in the mid-1980s, with higher large-wildfire frequency... longer wildfire durations, and longer wildfire seasons.*

For example, in drought-plagued southern California, dozens of wildfires across the region burned more than 600,000 acres (243,000 ha) in October 2003 and again in October 2007 (Figure 3.8). Each year since 2002 across the country, wildfire acreage broke the record of the previous year—2006 was triple the acreage lost in 2000 (see **http://www.usgs.gov/hazards/wildfires/**).

Natural Factors That Affect Air Pollution

The problems resulting from both natural and human-made atmospheric contaminants are made worse by several important natural factors. Among these, wind, local, and regional landscape characteristics and temperature inversions in the troposphere dominate.

Table 3.3	Sources of Natural Variable Gases and Materials
Source	**Contribution**
Volcanoes	Sulfur oxides, particulates
Forest fires	Carbon monoxide and dioxide, nitrogen oxides, particulates
Plants	Hydrocarbons, pollens
Decaying plants	Methane, hydrogen sulfides
Soil	Dust and viruses
Ocean	Salt spray and particulates

*A. L. Westerling et. al., "Warming and earlier spring increase western U.S. forest wildfire activity," *Science* 313, 5789 (Aug. 18, 2006): 940.

Winds Winds gather and move pollutants, sometimes reducing the concentration of pollution in one location while increasing it in another. Wind can produce dramatic episodes of dust movement. (Dust is defined as particles less than 62 μm, microns, or 0.0025 in.). Traveling on prevailing winds, dust from Africa contributes to the soils of South America and Europe, and Texas dust ends up across the Atlantic. Such movement is confirmed by chemical analysis, used by scientists to track dust to its source area (Figure 3.9). Imagine at any one moment a billion tons of dust aloft in the atmospheric circulation, carried by the winds!

Such air movements make the atmosphere's condition an international issue. For example, prevailing winds transport air pollution from the United States to Canada, causing much complaint and negotiation between the two governments. Pollution from North America adds to European air problems. In Europe, the cross-boundary drift of pollution is a major issue because of the proximity of countries. This issue was a factor in the formation of the 27-country European Union (EU). See "Arctic Haze," High Latitude Connection 3.1, for another aspect of pollution drift, as winds carry pollution into the remote northern polar region.

Local and Regional Landscapes   Local and regional landscapes are another important factor in air pollution. Surrounding mountains and hills can form barriers to air movement or can direct pollutants from one area to another. Some of the worst incidents result when local landscapes trap and concentrate air pollution.

Places with volcanic landscapes such as Iceland and Hawai'i have their own natural pollution with which to deal. During periods of sustained volcanic activity at Kīlauea, some 2000 metric tons (2200 tons) of sulfur dioxide are produced a day. Concentrations are sometimes high enough to merit broadcast warnings about health concerns. The resulting acid rain and volcanic smog, called *vog* by Hawaiians (for *vo*lcanic smog), causes losses to agriculture as well as other economic impacts.

(a)

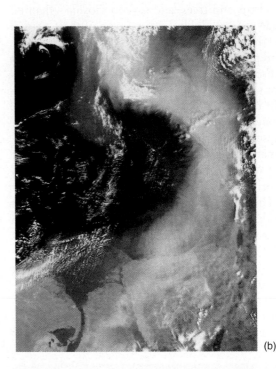
(b)

(c)

FIGURE 3.9 Natural variable dust in the atmosphere.
(a) A dust storm in central Nevada. (b) A *Terra* orbital view of a wind-blown dust plume flowing northward from Africa, across the Mediterranean Sea, toward Turkey. Can you find the Nile Delta? (c) Alkali dust, a serious air pollutant, rises from the exposed shorelines of Mono Lake, California. [(a) Photo by author; (b) *Terra* image courtesy of the MODIS Land Rapid Response Team NASA/GSFC; (c) photo by Bobbé Christopherson.]

High Latitude Connection 3.1

Arctic Haze

Arctic haze is a term from the 1950s, when pilots noticed decreased visibility, either horizontally or at a slant angle from their aircraft. Since there is no heavy industry at these latitudes and sparse population, this seasonal haze (concentration of microscopic particles and air pollution that diminishes air clarity) was a mystery at first. When an observer is looking toward the Sun, the haze has a reddish-brown tint, especially when the polar air mass is dominated by stable high-pressure systems and calm weather conditions.

A remarkable attribute of industrialization in the Northern Hemisphere is the haze across the Arctic region, especially from November to April. This winter haze is worse toward Alaska's North Slope. Although the haze can extend up to 8 km (5 mi) in altitude, it sometimes appears layered since sources are at varying distances from the Arctic and the warmer transporting winds are variously lifted over the cold air mass that hugs the surface.

Simply, winds of atmospheric circulation transport pollution to sites far distant from points of origin. There is no comparable haze over the Antarctic continent. Based on this analysis, can you think of why Antarctica lacks such a condition?

Temperature Inversion   Vertical temperature and atmospheric density in the troposphere also can worsen pollution conditions. A **temperature inversion** occurs when the normal temperature decrease with altitude (normal lapse rate) begins to *increase* at some point. This can happen at any elevation from ground level to several thousand meters. Figure 3.10 compares a normal temperature profile with that of a temperature inversion. The normal profile (Figure 3.10a) permits warmer (less dense) air at the surface to rise, ventilating the valley and moderating surface pollution. But the warm-air inversion (Figure 3.10b) prevents the rise of cooler (denser) air beneath, halting the vertical mixing of pollutants with other atmospheric gases. Thus, instead of being carried away, pollutants are trapped under the *inversion layer*.

Inversions most often result from certain weather conditions, such as when the air near the ground is radiatively cooled on clear nights, or from topographic situations that produce cold-air drainage into valleys. In addition, the air above snow-covered surfaces or beneath subsiding air in a high-pressure system may cause a temperature inversion. (The concepts of high- and low-pressure systems are in Chapter 6.)

Anthropogenic Pollution

Anthropogenic, or human-caused, air pollution remains most prevalent in urbanized regions. Approximately 2% of annual deaths in the United States are attributable to air pollution. Comparable risks are identified in Canada, Europe, Mexico, Asia—especially China and India—and elsewhere.

The human population is moving to cities and is therefore coming in increasing contact with air pollution. By the year 2010, approximately 3.3 billion people

(a)

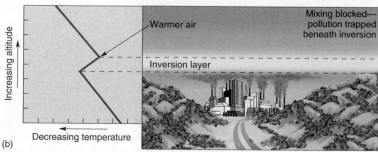

(b)

(c)

FIGURE 3.10 Normal and inverted temperature profiles.
A comparison of a normal temperature profile in the atmosphere (a) with a temperature inversion in the lower atmosphere (b). Note how the warmer air layer prevents mixing of the denser (cooler) air below the inversion, thereby trapping pollution. (c) An inversion layer is visible in the morning hours over a valley.
[Photo by Bobbé Christopherson.]

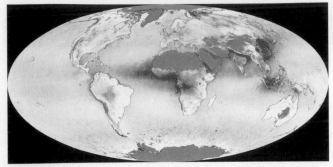

FIGURE 3.11 Portrait of global aerosols, 2006.
The MODIS (Moderate Resolution Imaging
Spectroradiometer) aboard NASA's *Terra* and *Aqua* satellites
captured the average annual AOD for 2006, a composite of
daily measurements. Darker tones represent higher AOD—
dust, soot, smoke, and pollution suspended in the
atmosphere. Areas where data could not be measured are in
gray. [Image MODIS aboard *Terra* and *Aqua* satellites compiled
by Reto Stockli, Earth Observatory, courtesy of NASA/GSFC.]

(48% of world population) will live in metropolitan regions, some one-third with unhealthful levels of air pollution. This represents a potentially massive public health issue in this century.

Small particles of dust, soot, and pollution suspended in the air comprise **aerosols**. Remote sensing now provides a rough global portrait of aerosols, as described by the atmosphere's *optical depth*, or AOD. Figure 3.11 shows a composite MODIS image from NASA's *Terra* and *Aqua* satellites for 2006. Note the high aerosol concentrations in Indonesia (wildfires and burning), northern India and China (urban and industrial pollution), western and central Africa (Saharan dust and agricultural fire smoke), and Russia (drought-plagued Siberian boreal forest wildfires). Bright surfaces reflecting sunlight such as snow and ice, desert, clouds, or sun-glint reflections limit accurate measurements. The limits of MODIS are obvious over the Amazon Basin where cloudiness causes an underestimation of AOD.

Table 3.4 lists the names, chemical symbols, principal sources, and impacts of variable anthropogenic components in the air. The first seven pollutants in the table result from combustion of fossil fuels in transportation (specifically automobiles) and at stationary sources such as power plants and factories. Overall, automobiles contribute more than 60% of United States and 50% of Canadian human-caused air pollution. The apparent manageability of this transportation-pollution problem is obvious to an informed public. Reducing air pollution from the transportation sector involves known technologies and strategies, monetary savings for the consumer, and significant health benefits.

Stationary sources, such as electric power plants and industrial plants that use fossil fuels, contribute the most sulfur oxides and particulates. For this reason, concentrations are focused in the Northern Hemisphere and the industrial, developed countries. The last three gases shown in Table 3.4 are discussed elsewhere in this text: Water vapor is examined with water and weather (Chapters 7

and 8); carbon dioxide and methane are discussed with greenhouse gases and climate (Chapters 4, 5, and 10).

Carbon Monoxide Pollution Carbon monoxide **(CO)** is a combination of one atom of carbon and one of oxygen. Incomplete combustion (burning with limited oxygen) of fuels or other carbon-containing substances produces carbon monoxide. A log decaying in the woods produces carbon monoxide, as does a forest fire or other organic decomposition. Natural sources produce up to 90% of existing carbon monoxide, whereas anthropogenic sources, principally transportation, produce the other 10%. A dangerous point source of carbon monoxide for individuals is from primary and secondary tobacco smoke. Generally, the carbon monoxide from human sources is from the transportation sector and is concentrated in urban areas, where it directly affects human health (see Table 3.4).

In the tropics of south-central Africa, the Amazon region of South America, and Indonesia, a large source of carbon monoxide is the widespread burning of biomass (trees, grasses, brush) from August to October. Figure 3.12a presents carbon monoxide emissions from both human sources and wildfire biomass burning, measured by satellite during November 2006. Note the highest levels are from Indonesia, where a prolonged drought caused exceptionally dry conditions and thus wildfire conditions (Figure 3.12b). Carbon monoxide, soot, and dust spread across the Southern Hemisphere. Such conditions coupled with widespread deforestation are introducing more carbon to the atmosphere. The former forests once provided a carbon sink, or removal mechanism; Chapter 20 presents data on forest losses.

Photochemical Smog Pollution Photochemical smog was not generally experienced in the past, but developed with the advent of the automobile. Today, it is the major component of anthropogenic air pollution (Figure 3.13). **Photochemical smog** results from the interaction of sunlight and the combustion products in automobile exhaust (nitrogen oxides and VOCs). Although the term *smog*—a combination of the words *smoke* and *fog*—is misnamed, it is generally used to describe this pollution. Smog is responsible for the hazy sky and reduced sunlight in many of our cities. Figure 3.14 shows such conditions in China and the U.S. Mid-Atlantic states—both regions under an air pollution attack.

The connection between automobile exhaust and smog was not determined until 1953 in Los Angeles, long after society had established its dependence upon individualized transportation. Despite this discovery, widespread mass transit declined, the railroads dwindled, and the polluting, inefficient individual automobile remains America's preferred transportation. Continuing poor gas economy efficiency is principally due to gas-guzzling sport-utility vehicles and pickups that account for more than half of new sales. Efficiency standards for this class of vehicles have not changed since 1975; they are set lower than for cars.

Table 3.4   Anthropogenic Gases and Materials in the Lower Atmosphere

Name	Symbol	Sources	Description and Effects of Criteria Pollutants (Main sources listed for U.S.)
Carbon monoxide	CO	Incomplete combustion of fuels	Odorless, colorless, tasteless gas Toxicity; affinity for hemoglobin Displaces O_2 in bloodstream; 50 to 100 ppm causes headaches and vision and judgment losses *Source:* 78% transportation
Nitrogen oxides	NO_x (NO, NO_2)	High temperature/pressure combustion	Reddish-brown choking break gas Inflames respiratory system, destroys lung tissue Damages plants; 3 to 5 ppm is dangerous *Source:* 55% transportation
Volatile organic compounds	VOC	Incomplete combustion of fossil fuels such as gasoline; cleaning and paint solvents	Prime agents of ozone formation Sources: 45% industrial 47% transportation
Ozone	O_3	Photochemical reactions	Highly reactive, unstable gas Oxidizes surfaces, dries rubber and elastic Damages plants at 0.01 to 0.09 ppm Agricultural losses at 0.1 ppm; 0.3 to 1.0 ppm irritates eyes, nose, throat
Peroxyacetyl nitrates	PAN	Photochemical reactions	Produced by NO + VOC photochemistry No human health effects Major damage to plants, forests, crops
Sulfur oxides	SO_x (SO_2, SO_3)	Combustion of sulfur-containing fuels	Colorless; irritating smell 0.1 to 1 ppm; impairs breathing, taste threshold Human asthma, bronchitis, emphysema Leads to acid deposition *Source:* 85% fuel combustion
Particulate matter	PM	Dust, dirt, soot, salt, metals, organics; agriculture, construction, roads	Complex mixture of solid and aerosol particles Dust, smoke, and haze affect visibility Various health effects: bronchitis, pulmonary function PM_{10} negative health effects established by researchers *Sources:* 42% industrial 33% fuel combustion 25% transportation
Carbon dioxide	CO_2	Complete combustion, mainly from fossil-fuel consumption	Principal greenhouse gas Atmospheric concentration increasing; 64% of greenhouse warming effect
Methane	CH_4	Organic processes	Secondary greenhouse gas Atmospheric concentration increasing; 19% of greenhouse warming effect
Water vapor	H_2O vapor	Combustion processes, steam	See Chapter 7 for more on the role of water vapor in the atmosphere.

Figure 3.15 summarizes how car exhaust is converted into major air pollutants—PAN, ozone, and nitric acid. The high temperatures in modern automobile engines produce nitrogen dioxide (NO_2). Also, to a lesser extent nitrogen dioxide is emitted from power plants. Worldwide, the problem with **nitrogen dioxide** production is its concentration in metropolitan regions. North American urban areas may have from 10 to 100 times higher nitrogen dioxide concentrations than nonurban areas.

Nitrogen dioxide interacts with water vapor to form nitric acid (HNO_3), a contributor to acid deposition by precipitation, the subject of Focus Study 3.2.

In the photochemistry reaction in the illustration, ultraviolet radiation liberates atomic oxygen (O) and a nitric oxide (NO) molecule from the NO_2. The free oxygen atom combines with an oxygen molecule, O_2, to form the oxidant ozone, O_3. In addition, the nitric oxide (NO) molecule reacts with VOCs to produce *peroxyacetyl nitrates (PAN)*.

FIGURE 3.12 Forests on fire—a source of global carbon monoxide (CO).
(a) Biomass burning and drought-assisted wildfires are imaged by satellite. CO is primarily from the burning of rain forest in Indonesia, Africa, and the Amazon Basin. The soot and ash is lifted high in the atmosphere by strong cloud convection. Higher levels of CO in pink, lower levels in blue. (b) A November 5, 2006, image captures these Indonesian human and natural wildfires. [Images from (a) MOPITT (Measurement of Pollution in the Troposphere) sensor aboard the *Terra* satellite courtesy of MOPITT Science Team NCAR/University of Toronto; (b) MODIS sensor aboard *Aqua* satellite courtesy of NASA/GSFC.]

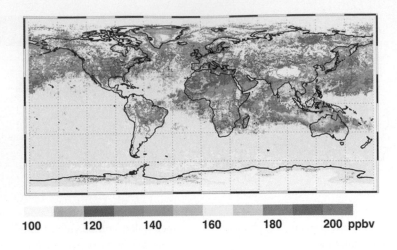

100   120   140   160   180   200 ppbv

FIGURE 3.13 Photochemical smog near Denver.
Photochemical reactions in the skies over Denver, Colorado, produce the blanket of smog that moves over mountains and forest to the west. What do you notice about air pollution conditions where you live or go to college? [Photo by Bobbé Christopherson.]

Peroxyacetyl nitrates (PAN) produce no known health effect in humans, but they are particularly damaging to plants, including both agricultural crops and forests. Damage in California is estimated to exceed $1 billion a year, and it is estimated to be several billion dollars nationwide in the farming and forestry sectors.

Ozone is the primary ingredient in photochemical smog. (This is the same gas that is beneficial to us in the stratosphere in absorbing ultraviolet radiation.) The reactivity of ozone is a health threat, for it damages biological tissues. For several reasons, children are at greatest risk from ozone pollution—one in four children in U.S. cities is at risk of developing health problems from ozone pollution. This ratio is significant; it means that more than 12 million children are vulnerable in those metropolitan regions with the worst polluted air (Los Angeles, Sacramento to Fresno to Bakersfield, Houston, Knoxville, Dallas–Fort Worth, Washington, DC, and Philadelphia; for more information and city listings of the worst and the best,

FIGURE 3.14 Serious air pollution in China and the United States.
(a) Smog and haze blanket most of eastern China. Beijing is about 150 km (93 mi) west of the coast where the heaviest pollution exists. Poor air pollution controls and the burning of coal contribute to the problem.
(b) A serious air pollution alert masks most of the Mid-Atlantic region of the United States—from northern Georgia to New York. Prevailing winds were blowing the mass offshore over the Atlantic Ocean.
[(a) *Terra* MODIS sensor images courtesy of MODIS Land Rapid Response Team NASA/GSFC; (b) *Terra* MODIS image courtesy of Liam Gumley, Space Science and Engineering Center, University of Wisconsin–Madison and NASA.]

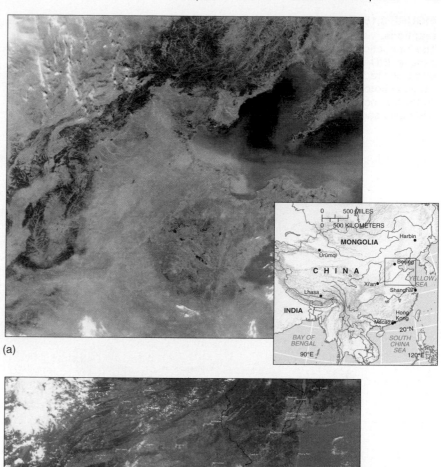

(a)

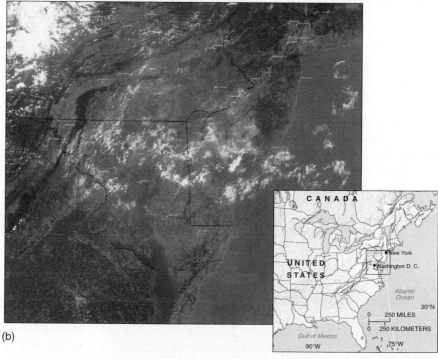

(b)

see http://www.lungusa.org). The **volatile organic compounds (VOCs)**, including hydrocarbons from gasoline, surface coatings, and combustion at electric utilities, are important factors in ozone formation. States such as California base their standards for control of ozone pollution on VOC emission controls—a scientifically accurate emphasis.

Industrial Smog and Sulfur Oxides   Over the past 300 years, except in some developing countries, coal slowly replaced wood as the basic fuel used by society. The Industrial Revolution required high-grade energy to run machines. The changes involved conversion from *animate* energy (energy from animal sources, such as animal-powered farm equipment) to *inanimate* energy (energy from nonliving sources, such as coal, steam, and water). Air pollution associated with coal-burning industries is known as **industrial smog** (Figure 3.16). The term *smog* was coined by a London physician in A.D. 1900 to describe the combination of fog and smoke containing sulfur gases (sulfur is an impurity in fossil fuels).

Once in the atmosphere, **sulfur dioxide (SO₂)** reacts with oxygen (O) to form sulfur trioxide (SO₃), which is highly reactive and, in the presence of water or water vapor, forms **sulfate aerosols**, tiny particles about 0.1 to 1 μm in

FIGURE 3.15 Photochemical reactions.
The interaction of automobile exhaust (NO$_2$, VOCs, CO) and ultraviolet radiation in sunlight causes photochemical reactions. To the left, note the formation of nitric acid and acid deposition.

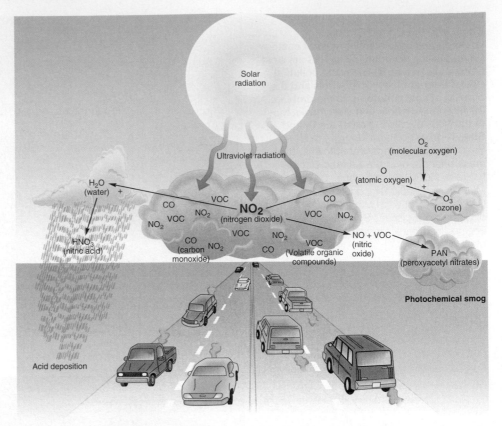

diameter. Sulfuric acid (H$_2$SO$_4$) can form even in moderately polluted air at normal temperatures. Sulfur dioxide–laden air is dangerous to health, corrodes metals, and deteriorates stone building materials at accelerated rates. Focus Study 3.2 discusses this vital atmospheric issue.

In the United States, coal-burning electric utilities and steel manufacturing are the main sources of sulfur dioxide, principally in the East and Midwest. And, because

FIGURE 3.16 Typical industrial smog.
Pollution generated by industry differs from that produced by transportation. Industrial pollution has high concentrations of sulfur oxides, particulates, and carbon dioxide. [Photo by author.]

of the prevailing movement of air masses, they are the main sources of sulfur dioxide in adjacent Canadian regions. A majority of Canadian sulfur dioxide is initiated within the United States.

Particulates Particulate matter (PM) is a diverse mixture of fine particles, both solid and aerosol, that impact human health—refer to Figure 3.11. Haze, smoke, and dust are visible reminders of particulate material in the air we breathe. PM$_{10}$, particulates smaller than 10 microns (10 μm or less) in diameter, was designated a matter for concern in 1987. PM$_{2.5}$ is currently being debated as an appropriate standard for human health. New studies are implicating even smaller particles, known as *ultrafines*, which are PM$_{0.1}$. These are many times more potent than the larger PM$_{2.5}$ and PM$_{10}$ particles because they can get into smaller channels in lung tissue and cause scarring, abnormal thickening, and damage called *fibrosis*.

Studies in Provo and Orem, Utah (1989), Philadelphia (1992), and other cities and by the American Cancer Society (1995), and more recently the WHO, established links between PM pollution and health. Nationally, a study of major cities disclosed a 26% greater risk of premature death due to respirable particulate pollution as compared with nonpolluted air—further driving up medical costs and related expenses. As science gains more information, the public should take needed action.

We are now contributing significantly to the creation of the **anthropogenic atmosphere**, a tentative label for

(text continued on page 83)

Focus Study 3.2

Acid Deposition: Damaging to Ecosystems

Acid deposition is a major environmental problem in some areas of the United States, Canada, Europe, and Asia. Such deposition is most familiar as "acid rain," but it also occurs as "acid snow" and in dry form as dust or aerosols—tiny liquid droplets or solid particles. In addition, winds can carry the acid-producing chemicals many kilometers from their sources before they settle on the landscape, where they enter streams and lakes as runoff and groundwater flows.

Acid deposition is causally linked to serious problems: declining fish populations and fish kills in the northeastern United States, southeastern Canada, Sweden, and Norway; widespread forest damage in these same places and Germany; widespread changes in soil chemistry; and damage to buildings, sculptures, and historic artifacts. In New Hampshire's Hubbard Brook Experimental Forest (http://www.hubbardbrook.org/), a study covering 1960 to the present found half the nutrient calcium and magnesium base cations (see Chapter 18) were leached from the soil. Excess acids are the cause of the decline.

Despite scientific agreement about the problem, which the U.S. General Accounting Office calls a "combination of meteorological, chemical, and biological phenomena," corrective action was delayed by its complexity and politics.

The acidity of precipitation is measured on the pH scale, which expresses the relative abundance of free hydrogen ions (H^+) in a solution. Free hydrogen ions in a solution are what make an acid corrosive, for they easily combine with other ions. The pH scale is logarithmic: Each whole number represents a tenfold change. A pH of 7.0 is neutral (neither acidic nor basic). Values less than 7.0 are increasingly *acidic*, and values greater than 7.0 are increasingly *basic*, or *alkaline*. (A pH scale for soil acidity and alka-

linity is portrayed graphically in Chapter 18.)

Natural precipitation dissolves carbon dioxide from the atmosphere to form carbonic acid. This process releases hydrogen ions and produces an average pH reading for precipitation of 5.65. The normal range for precipitation is 5.3–6.0. Thus, normal precipitation is always slightly acidic.

Some anthropogenic gases are converted to acids in the atmosphere and then are removed by wet and dry deposition processes. Specifically, nitrogen and sulfur oxides released in the combustion of fossil fuels can produce nitric acid (HNO_3) and sulfuric acid (H_2SO_4) in the atmosphere.

Acid Precipitation Damage

Precipitation as acidic as pH 2.0 has fallen in the eastern United States, Scandinavia, and Europe. By comparison, vinegar and lemon juice register slightly less than 3.0. Aquatic plant and animal life perishes when lakes drop below pH 4.8.

More than 50,000 lakes and some 100,000 km (62,000 mi) of streams in the United States and Canada are at a pH level below normal (that is, below 5.3 pH), with several hundred lakes incapable of supporting any aquatic life. Acid deposition causes the release of aluminum and magnesium from clay minerals in the soil, and both of these are harmful to fish and plant communities.

Also, relatively harmless mercury deposits in lake-bottom sediments convert in acidified lake waters into highly toxic *methylmercury*, which is deadly to aquatic life. Local health advisories in two provinces and 22 U.S. states are regularly issued to warn those who fish of the methylmercury problem. Mercury atoms rapidly bond with carbon and move through biological systems as an *organometallic compound*.

Damage to forests results from the rearrangement of soil nutrients, the death of soil microorganisms, and an aluminum-induced calcium defi-

ciency that is currently under investigation. The most advanced impact is seen in forests in Europe, especially in eastern Europe, principally because of its long history of burning coal and the density of industrial activity. In Germany and Poland, up to 50% of the forests are dead or damaged; in Switzerland, 30% are afflicted (Figure 3.2.1a).

In the United States, regional-scale decline in forest cover is significant, especially red spruce and sugar maples. In some maples, aluminum is collecting around rootlets; in spruce, acid fogs and rains leach calcium from needles directly. Affected trees are susceptible to winter cold, insects, and droughts. In New England, some stands of spruce are as much as 75% affected, as evidenced through analysis of tree-growth rings, which become narrower in adverse growing years. An indicator of forest damage is the reduction by almost half of the annual production of U.S. and Canadian maple sugar. Acid-laden cloud cover damages trees at higher elevations in the Appalachians (Figure 3.2.1b).

Government estimates of damage in the United States, Canada, and Europe exceed $50 billion annually. Because wind and weather patterns are international, efforts at reducing acidic deposition also must be international in scope. Figure 3.2.2 maps the reduction in sulfate deposition between the 1990–1994 and 1996–2000 periods. However, this progress is only a beginning. According to a study in *BioScience*, power plants and other sources must cut sulfur dioxide emissions 80% beyond the Clean Air Act mandate to truly reverse damage trends.

Ten leading acid-deposition researchers reported in an extensive study in *BioScience*:

Model calculations suggest that the greater the reduction in atmospheric sulfur deposition, the

(continued)

Focus Study 3.2 (continued)

FIGURE 3.2.1 The blight of acid deposition.
The harm done to forests and crops by acid deposition is well established, especially in Europe, such as in the Czech Republic (a), and in the forests of the Appalachian Mountains in the United States, such as in the forests on Mount Mitchell (b). [(a) Photo by Simon Fraser/Science Photo Library/Photo Researchers, Inc.; (b) Will and Deni McIntyre/Photo Researchers, Inc.]

(a)

(b)

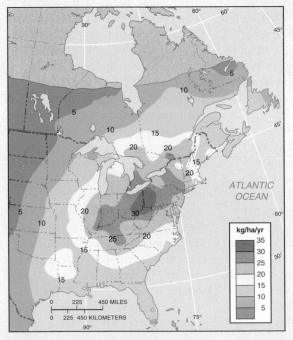

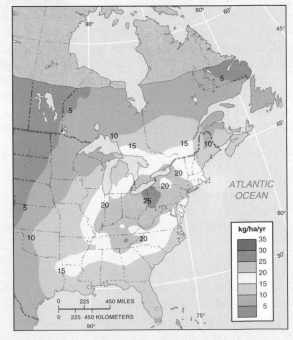

FIGURE 3.2.2 Improvement in sulfate wet deposition rate.
Spatial portrayal of annual sulfate (principally SO_4) wet deposition on the landscape, (a) 1990 to 1994 and (b) 1996 to 2000, in kilograms per hectare per year. The Canadian Environmental Protection Act, the Canada–United States Border Air Quality Strategy, and the U.S. Clean Air Act regulations lowered levels of emissions that add acid to the environment. [Maps from *Cleaner Air through Cooperation, Canada—United States, Progress under the Air Quality Agreement—2003,* Environment Canada, 2003. Reproduced with the permission of the Minister of Public Works and Government Services Canada.]

greater the magnitude and rate of chemical recovery. Less aggressive proposals for controls of sulfur emissions will result in slower chemical and biological recovery and in delays in regaining the services of a fully functional ecosystem. . . . North America and Europe are in the midst of a large-scale experiment. Sulfuric and nitric acids have acidified soils, lakes, and streams, thereby stressing or killing terrestrial and aquatic biota.*

Acid deposition is an issue of global spatial significance for which science is providing strong incentives for action. Reductions in troublesome emissions are closely tied to energy conservation and therefore directly related to production of greenhouse gases and global warming concerns—thus, linking these environmental issues.

*C. T. Driscoll et al., "Acidic deposition in the northeastern United States: Sources and inputs, ecosystems effects, and management strategies," *BioScience* 51 (March 2001): 195.

Earth's next atmosphere. The urban air we breathe today may be just a preview. What is the air quality like where you live, work, and go to college? Where could you go to find out its status?

Benefits of the Clean Air Act

The concentration of many air pollutants declined over the past several decades because of Clean Air Act (CAA) legislation (1970, 1977, 1990), resulting in the avoidance of trillions of dollars in health, economic, and environmental losses. Despite this reality, air pollution controls are subject to a continuing political debate, a weakening of some standards, and proposals that will increase pollution emissions.

Since 1970 and the CAA, there were significant reductions in atmospheric concentrations of carbon monoxide (–45%), nitrogen oxides (–22%), volatile organic compounds (–48%), PM_{10} particulates (–75%), sulfur oxides (–52%; see Focus Study 3.2, Figure 3.2.2), and lead (–98%). Prior to the CAA, lead was added to gasoline, emitted in the exhaust, dispersed over great distances, and settled in living tissues, especially in children. These all represent remarkable reductions, a successful linking of science and public policy.

To be justified, abatement (mitigation and prevention) costs must not exceed the financial benefits derived from reducing pollution damage. Compliance with the CAA affected patterns of industrial production, employment, and capital investment. Although these expenditures must be viewed as investments that generated benefits, the dislocation and job loss in some regions were difficult—reductions in high-sulfur coal mining and cutbacks in polluting industries such as steel, for example.

In 1990, Congress requested the Environmental Protection Agency (EPA) to answer the question: How do the overall health, welfare, ecological, and economic benefits of CAA programs compare with the costs of these programs? In response, the EPA performed an exhaustive cost-benefit analysis and published a report in 1997. *The Benefits of the Clean Air Act, 1970 to 1990* (Office of Policy, Planning, and Evaluation, U.S. EPA) calculated the following:

- The *total direct cost* to implement all the federal, state, and local Clean Air Act rules from 1970 to 1990 was *$523 billion* (in 1990 dollars). This cost was borne by businesses, consumers, and government entities.
- The estimate of *direct monetized benefits* from the Clean Air Act from 1970 to 1990 falls in a range from $5.6 to $49.4 trillion, with a central mean of *$22.2 trillion*. (The uncertainty of the assessment is indicated by the range of benefit estimates.)
- Therefore, the estimated *net financial benefit* of the Clean Air Act is *$21.7 trillion*! "The finding is overwhelming. The benefits far exceed the costs of the CAA in the first 20 years," said Richard Morgenstern, associate administrator for policy planning and evaluation at the EPA.

The benefits to society, directly and indirectly, were widespread across the entire population, including improved health and environment, less lead to harm children, lowered cancer rates, less acid deposition, and an estimated 206,000 fewer deaths related to air pollution in 1990 alone. These benefits took place during a period in which the U.S. population grew by 22% and the economy expanded by 70%. The benefits continued between 1990 and 2000 as air quality continued to improve. Recent efforts to weaken or rescind CAA regulations seem counterproductive to the progress made. Given a choice, which benefits from the CAA might the public choose to lose?

As you reflect on this chapter and our modern atmosphere, the treaties to protect stratospheric ozone, and the EPA study of benefits from the CAA, you should feel encouraged. Scientists did the research and society knew what to do, took action, and reaped enormous economic and health benefits. An important role for physical geographers is to explain these global impacts through spatial analysis and to guide an informed citizenry toward a better understanding.

Summary and Review—Earth's Modern Atmosphere

■ *Construct* a general model of the atmosphere based on three criteria, composition, temperature, and function, and *diagram* this model in a simple sketch.

Our modern atmosphere is a gaseous mixture so evenly mixed it behaves as if it were a single gas. The principal substance of this atmosphere is air—the medium of life. **Air** is naturally odorless, colorless, tasteless, and formless.

Above 480 km (300 mi) altitude, the atmosphere is rarefied (nearly a vacuum) and is the **exosphere**, which means "outer sphere." The weight (force over a unit area) of the atmosphere, exerted on all surfaces, is **air pressure**. It decreases rapidly with altitude.

By *composition*, we divide the atmosphere into the **heterosphere**, extending from 480 km (300 mi) to 80 km (50 mi), and the **homosphere**, extending from 80 km to Earth's surface. Within the heterosphere and using *temperature* as a criterion, we identify the **thermosphere**. Its upper limit, called the **thermopause**, is at approximately 480 km altitude. **Kinetic energy**, the energy of motion, is the vibrational energy that we measure and call temperature. The actual heat produced in the thermosphere is very small. The density of the molecules is so low that little actual **heat**, the flow of kinetic energy from one body to another because of a temperature difference between them, is produced. Nearer Earth's surface, the greater number of molecules in the denser atmosphere transmits their kinetic energy as **sensible heat**, meaning that we can feel it.

The homosphere includes the **mesosphere, stratosphere**, and **troposphere**, as defined by temperature criteria. Within the mesosphere, cosmic or meteoric dust act as nuclei around which fine ice crystals form to produce rare and unusual night clouds called **noctilucent clouds**.

The normal temperature profile within the troposphere during the daytime decreases rapidly with increasing altitude at an average of 6.4 C° per kilometer (3.5 F° per 1000 ft), a rate known as the **normal lapse rate**. The top of the troposphere is wherever a temperature of –57°C (–70°F) is recorded, a transition known as the **tropopause**. The actual lapse rate at any particular time and place may deviate considerably because of local weather conditions and is called the **environmental lapse rate**.

We distinguish a region in the heterosphere by its *function*. The **ionosphere** absorbs cosmic rays, gamma rays, X-rays, and shorter wavelengths of ultraviolet radiation and converts them into kinetic energy. A functional region within the stratosphere is the **ozonosphere**, or **ozone layer**, which absorbs life-threatening ultraviolet radiation, subsequently raising the temperature of the stratosphere.

air (p. 62)
exosphere (p. 62)
air pressure (p. 63)
heterosphere (p. 64)
homosphere (p. 65)
thermosphere (p. 67)
thermopause (p. 67)
kinetic energy (p. 67)
heat (p. 67)
sensible heat (p. 67)
mesosphere (p. 67)
noctilucent clouds (p. 67)
stratosphere (p. 67)
troposphere (p. 67)
tropopause (p. 67)
normal lapse rate (p. 68)
environmental lapse rate (p. 68)
ionosphere (p. 68)
ozonosphere, ozone layer (p. 68)

1. What is air? Where did the components in Earth's present atmosphere originate?
2. In view of the cell analogy by Lewis Thomas, characterize the various functions the atmosphere performs that protect the surface environment.
3. What three distinct criteria are employed in dividing the atmosphere for study?
4. Describe the overall temperature profile of the atmosphere and list the four layers defined by temperature.
5. Describe the two divisions of the atmosphere on the basis of composition.
6. What are the two overall primary functional layers of the atmosphere and what does each do?

■ *List* the stable components of the modern atmosphere and their relative percentage contributions by volume, and *describe* each.

Even though the atmosphere's density decreases with increasing altitude in the homosphere, the blend (proportion) of gases is nearly uniform. This stable mixture of gases has evolved slowly.

The homosphere is a vast reservoir of relatively inert *nitrogen*, originating principally from volcanic sources and from bacterial action in the soil; *oxygen*, a by-product of photosynthesis; *argon*, constituting about 1% of the homosphere and completely inert; and *carbon dioxide*, a natural by-product of life processes and fuel combustion.

7. Name the four most prevalent stable gases in the homosphere. Where did each originate? Is the prevalence of any of these changing at this time?

■ *Describe* conditions within the stratosphere; specifically, *review* the function and status of the ozonosphere (ozone layer).

The overall reduction of the stratospheric ozonosphere, or ozone layer, during the past several decades represents a hazard for society and many natural systems and is caused by chemicals introduced into the atmosphere by humans. Since World War II, quantities of human-made **chlorofluorocarbons (CFCs)** and bromine-containing compounds have made their way into the stratosphere. The increased ultraviolet light at those altitudes breaks down these stable chemical compounds, thus freeing chlorine and bromine atoms. These atoms act as catalysts in reactions that destroy ozone molecules.

chlorofluorocarbons (CFCs) (p. 69)

8. Why is stratospheric ozone so important? Describe the effects created by increases in ultraviolet light reaching the surface.
9. Summarize the ozone predicament, and describe trends and any treaties that intend to protect the ozone layer.

10. Evaluate Crutzen, Rowland, and Molina's use of the scientific method in investigating stratospheric ozone depletion, and public reaction to their findings.

■ *Distinguish* between natural and anthropogenic variable gases and materials in the lower atmosphere.

Within the troposphere, both natural and human-caused variable gases, particles, and other chemicals are part of the atmosphere. We coevolved with natural "pollution" and thus are adapted to it. But we are not adapted to cope with our own anthropogenic pollution. It constitutes a major health threat, particularly where people are concentrated in cities.

Vertical temperature and atmospheric density distribution in the troposphere can worsen pollution conditions. A **temperature inversion** occurs when the normal temperature decrease with altitude (normal lapse rate) reverses. In other words, temperature begins to increase at some altitude.

temperature inversion (p. 75)

11. Why are anthropogenic gases more significant to human health than are those produced from natural sources?
12. In what ways does a temperature inversion worsen an air pollution episode? Why?

■ *Describe* the sources and effects of carbon monoxide, nitrogen dioxide, and sulfur dioxide, and *construct* a simple equation that illustrates photochemical reactions that produce ozone, peroxyacetyl nitrates, nitric acid, and sulfuric acid.

Small particles of dust, soot, and pollution suspended in the air are **aerosols**. Odorless, colorless, and tasteless, **carbon monoxide (CO)** is produced by incomplete combustion (burning with limited oxygen) of fuels or other carbon-containing substances. Transportation is the major human-caused source for carbon monoxide. The toxicity of carbon monoxide is due to its affinity for blood hemoglobin, which is the oxygen-carrying pigment in red blood cells. In the presence of carbon monoxide, the oxygen is displaced and the blood becomes deoxygenated (Table 3.4).

Photochemical smog results from the interaction of sunlight and the products of automobile exhaust, the single largest contributor of pollution that produces smog. Car exhaust, containing *nitrogen dioxide* and *volatile organic compounds* (*VOCs*), in the presence of ultraviolet light in sunlight converts into the principal photochemical by-products—*ozone, peroxyacetyl nitrates* (*PAN*), and *nitric acid.*

Ozone (O_3) causes negative health effects, oxidizes surfaces, and kills or damages plants. **Peroxyacetyl nitrates (PAN)** produce no known health effects in humans but are particularly damaging to plants, including both agricultural crops and forests. **Nitrogen dioxide (NO_2)** damages and inflames human respiratory systems, destroys lung tissue, and damages plants. Nitric oxides participate in reactions that form nitric acid (HNO_3) in the atmosphere, forming both wet and dry acidic deposition. The **volatile organic compounds (VOCs)**, including hydrocarbons from gasoline, surface coatings, and electric utility combustion, are important factors in ozone formation.

The distribution of human-produced **industrial smog** over North America, Europe, and Asia is related to transportation and electrical production. Such characteristic pollution contains **sulfur dioxide**. Sulfur dioxide in the atmosphere reacts to produce **sulfate aerosols**, which produce sulfuric acid (H_2SO_4) deposition, which can be dangerous to health and affect Earth's energy budget by scattering and reflecting solar energy. **Particulate matter (PM)** consists of dirt, dust, soot, and ash from industrial and natural sources.

Energy conservation and efficiency and reducing emissions are key to reducing air pollution. Earth's next atmosphere most accurately may be described as the **anthropogenic atmosphere** (human-influenced atmosphere).

aerosols (p. 76)
carbon monoxide (CO) (p. 76)
photochemical smog (p. 76)
nitrogen dioxide (p. 77)
peroxyacetyl nitrates (PAN) (p. 78)
volatile organic compounds (VOCs) (p. 79)
industrial smog (p. 79)
sulfur dioxide (p. 79)
sulfate aerosols (p. 79)
particulate matter (PM) (p. 80)
anthropogenic atmosphere (p. 80)

13. What is the difference between industrial smog and photochemical smog?
14. Describe the relationship between automobiles and the production of ozone and PAN in city air. What are the principal negative impacts of these gases?
15. How are sulfur impurities in fossil fuels related to the formation of acid in the atmosphere and acid deposition on the land?
16. In summary, what are the results from the first 20 years under Clean Air Act regulations?

NetWork

The *Geosystems* Student Learning Center provides on-line resources for this chapter on the World Wide Web. To begin: Once at the Center, click on the cover of this textbook, scroll the Table of Contents menu, and select this chapter. You will find self-tests that are graded, review exercises, specific updates for items in the chapter, and in "Destinations" many links to interesting related pathways on the Internet. *Geosystems* Student Learning Center is found at **http://www.prenhall.com/ christopherson/.**

Critical Thinking

A. A 'scientific study, *The Benefits of the Clean Air Act, 1970 to 1990* (Office of Policy, Planning, and Evaluation, U.S. EPA, October 1997), determined that the Clean Air Act provided the American people with benefits. Health, welfare, ecological, and economic benefits rated 42 to 1 over the costs of CAA implementation (estimated at $21.7 trillion in net benefits compared to $523 billion in costs). Do you think this is significant? Should this be part of the debate about future weakening or strengthening of the Clean Air Act? Do you think such benefits might result from other regulations such as the Clean Water Act or groundwater protection measures, or action on global climate change?

B. Relative to item A: In your opinion, why is the public generally unaware of these details? What are the difficulties in informing the public? Why is there such media attention given to antiscience and nonscientific opinions? Take a moment and brainstorm recommendations for action, education, and public awareness on these issues.

C. To determine the total ozone column at your present location, go to **http://toms.gsfc.nasa.gov/teacher/ozone_overhead.html**, "What was the total ozone column at your house?" Select a point on the map or enter your latitude and longitude and the date you want to check. The ozone column is measured by the Ozone Monitoring Instrument (OMI) sensor aboard the *Aqua* satellite and is mainly sensitive to stratospheric ozone. (Note also the limitations listed on the extent of data availability.) For several different dates, when do the lowest values occur? The highest values? Briefly explain what your results mean. How do you interpret the values found?

Jet condensation trails stimulate high-cirrus cloud formation. Inset photos and image show thin, newer contrails in contrast to older contrails widening over time as they develop. Note the density of contrails over the Midwest and Great Lakes region in the satellite image. Studies found that such contrails affect daily temperature ranges between daytime maximum and nighttime minimum—another connection between transportation systems and Earth's atmospheric energy budget. [Photos by Bobbé Christopherson; inset image *Terra* satellite courtesy of MODIS Land Rapid Response Team, NASA/GSFC.]

Atmosphere and Surface Energy Balances

■ Key Learning Concepts

After reading the chapter, you should be able to:

■ *Identify* the pathways of solar energy through the troposphere to Earth's surface: transmission, scattering, diffuse radiation, refraction, albedo (reflectivity), conduction, convection, and advection.

■ *Describe* what happens to insolation when clouds are in the atmosphere, and *analyze* the effect of clouds and air pollution on solar radiation received at ground level.

■ *Review* the energy pathways in the Earth–atmosphere system, the greenhouse effect, and the patterns of global net radiation.

■ *Plot* the daily radiation curves for Earth's surface, and *label* the key aspects of incoming radiation, air temperature, and the daily temperature lag.

■ *Portray* typical urban heat island conditions, and *contrast* the microclimatology of urban areas with that of surrounding rural environments.

E arth's biosphere pulses with flows of solar energy that sustain our lives and empower natural systems. Changing seasons, daily weather conditions, and Earth's variety of climates remind us of this constant flow of energy cascading through the atmosphere. Earth's shifting seasonal rhythms are described in Chapter 2, and a profile of the atmosphere in Chapter 3. Energy and moisture exchanges between Earth's surface and the atmosphere are essential elements of weather and climate, discussed in later chapters.

In this chapter: We follow solar energy through the troposphere to Earth's surface. The *outputs* of reflected light and emitted longwave energy from the atmosphere and surface environment counter the *input* of insolation. Together, this input and output determine the net energy available to perform work. We examine surface energy budgets and analyze how net radiation is spent. The chapter concludes with a look at the unique energy and moisture environment in our cities. The view across a hot parking lot of cars and pavement in summer is all too familiar—the air shimmers as heat energy radiates skyward. The climate of our urban areas differs measurably from that of surrounding rural areas. Focus Study 4.1 discusses solar energy, an applied aspect of surface energy and a renewable energy resource of great potential.

Energy Essentials

When you look at a photograph of Earth taken from space, you clearly see the surface receipts of incoming insolation (see Earth photo on the back cover of this book). Land and water surfaces, clouds, and atmospheric gases and dust intercept solar energy. The flows of energy are manifest in swirling weather patterns, powerful oceanic currents, and the varied vegetation of land cover. Specific energy patterns differ for deserts, oceans, mountaintops, plains, rain forests, and ice-covered landscapes. In addition, the presence or absence of clouds may make a 75% difference in the amount of energy that reaches the surface, because clouds reflect incoming energy.

Energy Pathways and Principles

Solar energy that heats Earth's atmosphere and surface is unevenly distributed by latitude and fluctuates seasonally. Figure 4.1 is a simplified flow diagram of shortwave and longwave radiation in the Earth–atmosphere system, discussed in the pages that follow. You will find it helpful to refer to this figure, and the more detailed energy balance illustration in Figure 4.12, as you read through the following section. We first look at some important pathways and principles for insolation as it passes through the atmosphere to Earth's surface.

Transmission refers to the passage of shortwave and longwave energy through either the atmosphere or water. Our budget of atmospheric energy comprises shortwave radiation *inputs* (ultraviolet light, visible light, and near-infrared wavelengths) and longwave radiation *outputs* (thermal infrared) that pass through the atmosphere by transmission.

Insolation Input   *Insolation* is the single energy input driving the Earth–atmosphere system. The world map in Figure 4.2 shows the distribution of average annual solar energy received at Earth's surface. It includes all the radiation that arrives at Earth's surface, both direct and diffuse (scattered by the atmosphere).

Several patterns are notable on the map. Insolation decreases poleward from about 25° latitude in both the Northern and Southern Hemispheres. Consistent daylength and high Sun altitude produce average annual values of 180–220 W/m² throughout the equatorial and tropical latitudes. In general, greater insolation of 240–280 W/m² occurs in low-latitude deserts worldwide because of frequently cloudless skies. Note this energy pattern in the cloudless subtropical deserts in both hemispheres (for example, the Sonoran, Saharan, Arabian, Gobi, Atacama, Namib, Kalahari, and Australian deserts).

Scattering (Diffuse Radiation)   Insolation encounters an increasing density of atmospheric gases as it travels toward the surface. Atmospheric gases and dust physically

ANIMATION   Global Warming, Climate Change

ANIMATION   Earth-Atmosphere Energy Balance

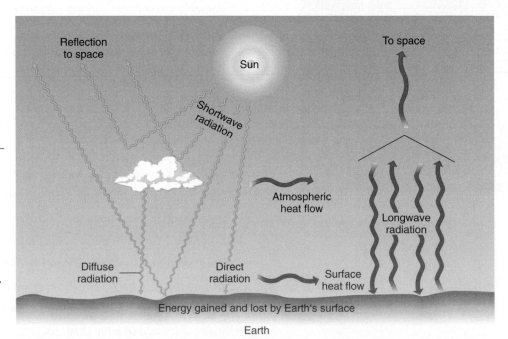

FIGURE 4.1 Energy gained and lost by Earth's surface and atmosphere.
Simplified view of the Earth–atmosphere energy system. Circuits include incoming shortwave insolation, reflected shortwave radiation, and outgoing longwave radiation.

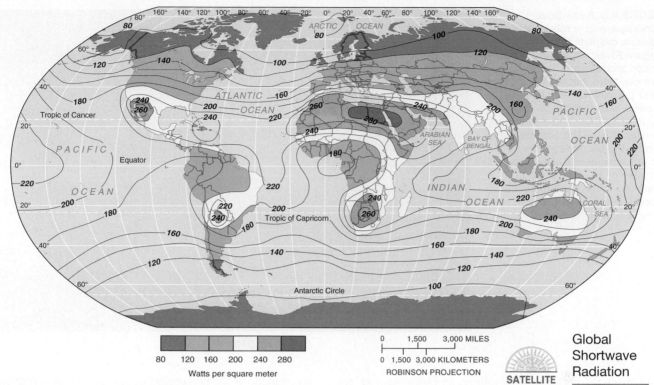

FIGURE 4.2 Insolation at Earth's surface.
Average annual solar radiation received on a horizontal surface at ground level in watts per square meter
(100 W/m² = 75 kcal/cm²/year). [After M. I. Budyko, *The Heat Balance of the Earth's Surface* (Washington, DC:
U.S. Department of Commerce, 1958), p. 99.]

interact with insolation. The gas molecules redirect radiation, changing the direction of the light's movement *without altering its wavelengths*. **Scattering** describes this phenomenon and represents 7% of Earth's reflectivity, or albedo (see Figure 4.12). Dust particles, pollutants, ice, cloud droplets, and water vapor produce further scattering.

Have you wondered why Earth's sky is blue? And why sunsets and sunrises are often red? These simple questions have an interesting explanation, based upon a principle known as Rayleigh scattering (named for English physicist Lord Rayleigh, 1881). This principle relates wavelength to the size of molecules or particles that cause the scattering. The general rule is: *The shorter the wavelength, the greater the scattering; and the longer the wavelength, the lesser the scattering.* Small gas molecules in the air scatter shorter wavelengths of light. Thus, the shorter wavelengths of visible light—the blues and violets—scatter the most and dominate the lower atmosphere. And, because there are more blue than violet wavelengths in sunlight, a blue sky prevails. A sky filled with smog and haze appears almost white because the larger particles associated with air pollution act to scatter all wavelengths of visible light.

The angle of the Sun's rays determines the thickness of atmosphere they must pass through to reach the surface. Therefore, direct rays (from overhead) experience less scattering and absorption than do low, oblique-angle rays that must travel farther through the atmosphere. Insolation from the low-altitude Sun undergoes more scattering of shorter wavelengths, leaving only the residual oranges and reds to reach the observer at sunset or sunrise.

Some incoming insolation is diffused by clouds and atmosphere and transmits to Earth as **diffuse radiation**, the downward component of scattered light (labeled in Figure 4.12). Such diffuse light is multidirectional and thus casts shadowless light on the ground.

Refraction   As insolation enters the atmosphere, it passes from one medium to another, from virtually empty space into atmospheric gases, or from air into water. This transition subjects the insolation to a change of speed, which also shifts its direction, a bending action called **refraction**. In the same way, a crystal or prism refracts light passing through it, bending different wavelengths to different angles, separating the light into its component colors to display the spectrum. A rainbow is created when visible light passes through myriad raindrops and is refracted and reflected toward the observer at a precise angle (see Figure 4.3).

Another example of refraction is a **mirage**, an image that appears near the horizon where light waves are refracted by layers of air at different temperatures (and consequently of different densities) on a hot day. The atmospheric distortion of the setting Sun in Figure 4.4 is also a product of refraction—light from the Sun low in the sky must penetrate more air than when the Sun is high; it is refracted through air layers of different densities on its way to the observer.

An interesting function of refraction is that it adds approximately 8 minutes of daylight that we would lack if Earth had no atmosphere. Sunlight refracts in its passage

FIGURE 4.3 A rainbow. Raindrops, and in this photo moisture droplets from the Niagara River, refract and reflect light to produce a primary rainbow. Note that the color order in the primary rainbow has the shortest wavelengths on the inside of the bow and the longest wavelengths on the outside of the bow. In the secondary bow note the color sequence is reversed because of an extra angle of reflection within each moisture droplet. [Photo by Bobbé Christopherson.]

from space through the atmosphere, and so, at sunrise, we see the Sun's image about 4 minutes before the Sun actually peeks over the horizon. Similarly, as the Sun actually sets at sunset its refracted image is visible from over the horizon for about 4 minutes afterward. To this day, science cannot predict the exact time of visible sunrise or sunset given the degree of refraction, which varies with atmospheric temperature, moisture, and pollutants.

Albedo and Reflection   A portion of arriving energy bounces directly back into space without being absorbed or performing any work—this is the **reflection** process. **Albedo** is the reflective quality, or intrinsic brightness, of a surface. It is an important control over the amount of insolation that is available for absorption by a surface. We state albedo as the percentage of insolation that is reflected (0% is total absorption; 100% is total reflectance). Examine the different surfaces and their albedo values in Figure 4.5.

In the visible wavelengths, darker colors have lower albedos, and lighter colors have higher albedos. On water surfaces, the angle of the solar rays also affects albedo values; lower angles produce a greater reflection than do higher angles. In addition, smooth surfaces increase albedo, whereas rougher surfaces reduce it.

Specific locations experience highly variable albedo values during the year in response to changes in cloud and ground cover. Earth Radiation Budget (ERB) sensors aboard the *Nimbus*-7 satellite measured average albedos of 19%–38% for all surfaces between the tropics (23.5° N to 23.5° S), whereas albedos for the polar regions measured as high as 80%.

Earth and its atmosphere reflect 31% of all insolation when averaged over a year. Looking ahead to Figure 4.12, you can see that Earth's average albedo is a combination of 21% reflected by clouds, 3% reflected by the ground (combined land and ocean surfaces), and 7% reflected and scattered by the atmosphere. By comparison, a full Moon, which is bright enough to read by under clear skies, has only a 6%–8% albedo value. Thus, with *earthshine* being 4 times brighter than moonlight (4 times the albedo), and with Earth 4 times greater in diameter than the Moon, it is no surprise that astronauts report how startling our planet looks from space. News Report 4.1 presents information on earthshine as a climate status indicator.

Figure 4.6 portrays total albedos for July and January as measured by the Earth Radiation Budget experiments aboard several satellites. These patterns are typical of most years. As compared with July albedos, January albedos are

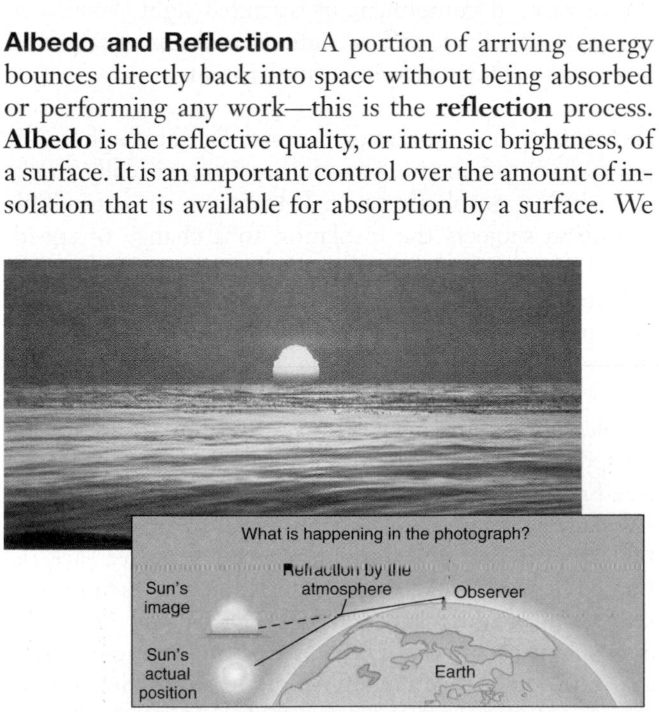

What is happening in the photograph?

Sun's image — Refraction by the atmosphere — Observer

Sun's actual position — Earth

FIGURE 4.4 Sun refraction. The distorted appearance of the Sun nearing sunset over the ocean is produced by refraction of the Sun's image in the atmosphere. Have you ever noticed this effect? [Photo by author.]

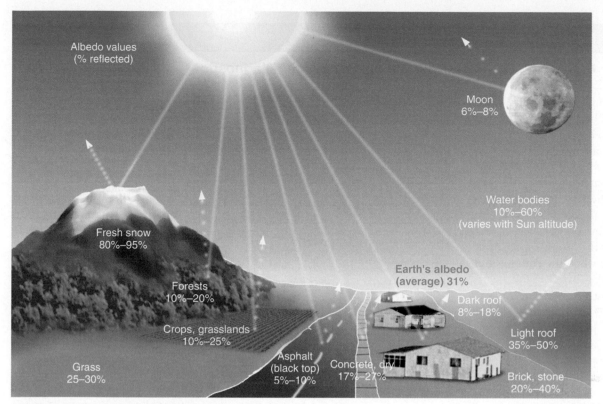

FIGURE 4.5 Various albedo values.
Different surfaces have different albedo values. In general, light surfaces are more reflective than dark surfaces and thus have higher albedo values. [Data from M. I. Budyko, *The Heat Balance of the Earth's Surface* (Washington, DC: U.S. Department of Commerce, 1958), p. 36.]

SATELLITE

Global Albedo Values,
Global Shortwave
Radiation

higher poleward of 40° N because of the snow and ice that cover the ground. Tropical forests are characteristically low in albedo (15%), whereas generally cloudless deserts have high albedos (35%). The southward-shifting cloud cover over equatorial Africa is quite apparent on the January map. What other seasonal changes do you see on these ERB albedo maps?

Clouds, Aerosols, and the Atmosphere's Albedo   An unpredictable factor in the tropospheric energy budget, and therefore in refining climatic models, is the role of clouds. Clouds reflect insolation and thus cool Earth's surface. The term **cloud-albedo forcing** is an *increase in albedo* caused by clouds. Yet clouds act as insulation, trapping longwave radiation from Earth and raising minimum

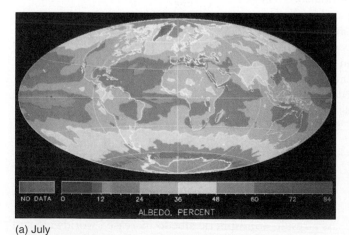

(a) July

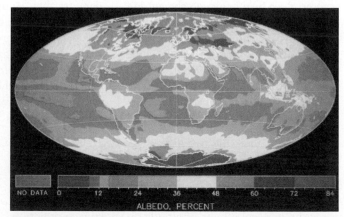

(b) January

FIGURE 4.6 Total albedos for July and January.
Total albedos for a July (a) and a January (b) as measured by the Earth Radiation Budget experiments (ERB) aboard satellites *Nimbus-7, NOAA-9,* and *ERBS (ERB Satellite)*. Each color represents a 12% interval in albedo values. [Courtesy of Radiation Services Branch, Langley Research Center, NASA.]

News Report 4.1

Earthshine Studies—Possible Energy Budget Diagnostic

Earth's reflection of light to space might prove to be a useful tool in analyzing energy budgets and climate. Earth's average albedo is 31%—that is, 31% of all incoming insolation returns to space without being absorbed. We have all seen the effect of *earthshine*—Earth's total albedo reflectance to space—as a faint illumination of the dark portion of the Moon, especially in crescent phase, when the dark portion of the lunar disk is barely visible.

In contrast, the Moon reflects only 6% to 8% of the sunlight that hits its surface, producing the "moonlight" familiar to us all. Despite the bright, white appearance of the Moon, the lunar surface albedo is about the same as that of asphalt pavement. Because Earth reflects about 400% as much light as the Moon and is about 400% greater in diameter, Earth is a startling presence in space.

Earthshine, if properly analyzed, is a proxy (substitute, representative) measure of Earth's shortwave reflectance and an important diagnostic tool in climate change studies. The significance of measuring earthshine is this: An approximate 1% change in albedo will produce about a 1 C° (2 F°) change in average atmospheric temperature. Thus, an albedo change can produce a temperature change, which can produce further albedo change and further alter climate. The Sun's average 11-year activity cycle produces about a 0.1% variation in earthshine and is not considered a cause for present global warming.

Cloud cover is the significant factor that drives changes in Earth's short-

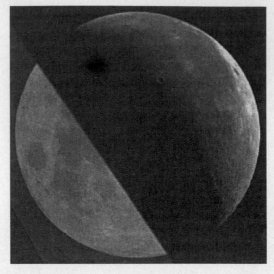

FIGURE 4.1.1 Earthshine illuminates the Moon. Our Moon intercepts some of Earth's shortwave reflectance to space. Earthshine is most visible when the Moon is in a crescent phase. Check this out next time you see the Moon in its crescent to half phase. The combined photos show the moonshine segment (lower-left, dimmed by a filter because it is 10,000 times brighter than earthshine) and the earthshine segment (upper-right). [Composite photos courtesy of Earthshine Project, Big Bear Solar Observatory.]

wave reflectance, although surface changes and atmospheric pollution also are important. Increases in cloud cover are an indicator of increasing atmospheric temperatures. Relative to surface changes, imagine the impact on global albedo values if there were an increase in snow-free landscapes.

Scientists from the New Jersey Institute of Technology and California Institute of Technology now measure earthshine at the Big Bear Solar Observatory during the Moon's crescent phase when Earth's sunlight hemisphere is toward the Moon (see: **http://www.bbso.njit.edu/**). These data are then compared to the International Satellite Cloud Climatology Project (ISCCP) record. André Danjon of France began earthshine studies in 1925. Current research shows Earth's average albedo was on a steady decline from 1984 to 2000 (less cloudy), whereas 2001–2004 (more

clouds) demonstrated a complete reversal with increasing Earth reflectance.

These large variations in reflectance imply climatologically significant cloud-driven changes in Earth's radiation budget, which is consistent with the large tropospheric warming that has occurred over the most recent decades. Moreover, if the observed reversal in Earth's reflectance trend is sustained during the next few years, it might also play a very important role in future climate change.*

A lot remains to be learned about this yardstick for Earth's energy balance. Think about earthshine the next time you take a moonlit walk.

*E. Pallé et al., "Changes in Earth reflectance over the past two decades," *Science* 304 (May 28, 2004): 1299–1301.

temperatures. **Cloud-greenhouse forcing** is an *increase in greenhouse warming* caused by clouds. Figure 4.7 illustrates the general effects of clouds on shortwave radiation and longwave radiation. Chapter 7 details more on clouds.

The eruption of Mount Pinatubo in the Philippines, which began explosively during June 1991, illustrates how Earth's internal processes affect the atmosphere. Approximately 15–20 megatons of sulfur dioxide were injected

into the stratosphere; winds rapidly spread these aerosols (tiny droplets) worldwide (see the images in Figure 6.1). As a result, atmospheric albedo increased worldwide and produced a temporary average cooling of 0.5 C° (0.9 F°).

Other mechanisms affect atmospheric albedo and therefore atmospheric and surface energy budgets. Industrialization is producing a haze of pollution that is increasing the reflectivity of the atmosphere, including sulfate

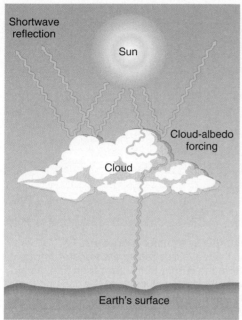
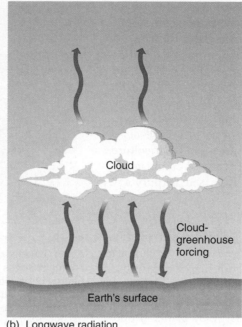

(a) Shortwave radiation

(b) Longwave radiation

FIGURE 4.7 The effects of clouds on shortwave and longwave radiation. (a) Clouds reflect and scatter shortwave radiation; a high percentage is returned to space. (b) Clouds absorb and reradiate longwave radiation emitted by Earth; some longwave energy returns to space and some is radiated toward the surface.

aerosols, soot and fly ash, and black carbon. Emissions of sulfur dioxide and subsequent chemical reactions in the atmosphere form *sulfate aerosols*. These aerosols act as an insolation-reflecting haze in clear-sky conditions.

Scientists are actively analyzing such pollution-dimmed sunlight reductions. The pollution causes both an atmospheric warming through absorption by the pollutants and a surface cooling through reduction in insolation reaching ground and water surfaces. An overall term has emerged that describes this decline in insolation making it

to Earth's surface—**global dimming**. In determining present temperature increases in global warming, this solar dimming is perhaps forcing an underestimation of the actual amount of warming.

Figure 4.8 is from the CERES (Clouds and the Earth Radiant Energy System) sensors aboard satellite *Terra* and shows pollution-haze and soot effects on the atmospheric energy budget in the Indian Ocean region. The increased aerosols (1), containing black carbon, increase reflection off the atmosphere, thus raising atmospheric albedos (2).

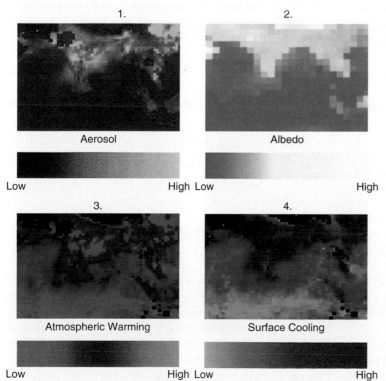

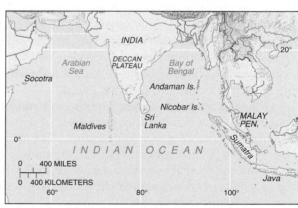

FIGURE 4.8 Aerosols impact Earth–atmosphere energy budgets.
Aerosols including black carbon both reflect and absorb incoming insolation as identified in these four images by the CERES sensors aboard satellite *Terra*, made between January and March 2001 over southern Asia and the Indian Ocean. Note the color scale for each image rating from low to high values. [CERES images aboard *Terra* courtesy of Goddard Space Flight Center, NASA.]

This causes an increase in atmospheric warming (as aerosols absorb energy) (3) and an increase in surface cooling as less insolation reaches Earth's surface (4). These aerosols reduce surface insolation by 10% and increase energy absorption in the atmosphere by 50%.

Such pollution effects in southern Asia influence the dynamics of the Asian monsoon through alteration of the region's Earth–atmosphere energy budget (see monsoons in Chapter 6). A weakening monsoonal flow will negatively impact regional water resources and agriculture. This research is part of the international, multiagency Indian Ocean Experiment (INDOEX, see **http://www.indoex.ucsd.edu/**).

Absorption   **Absorption** is the *assimilation* of radiation by molecules of matter and its conversion from one form of energy to another. Insolation, both direct and diffuse, that is not part of the 31% reflected from Earth's surface and atmosphere is absorbed. It is converted into either longwave radiation or chemical energy by plants in photosynthesis. The temperature of the absorbing surface is raised in the process, and warmer surfaces radiate more total energy at shorter wavelengths—thus, *the hotter the surface, the shorter the wavelengths that are emitted.* Examples are an electric burner beginning to glow as it heats, or molten metal glowing at high temperatures. In addition to absorption by land and water surfaces (about 45% of incoming insolation), absorption also occurs in atmospheric gases, dust, clouds, and stratospheric ozone (about 24% of incoming insolation). Figure 4.12 summarizes the pathways of insolation and the flow of heat in the atmosphere and at the surface.

Conduction, Convection, and Advection   There are several means of transferring heat energy in a system. **Conduction** is the molecule-to-molecule transfer of heat energy as it diffuses through a substance. As molecules warm, their vibration increases, causing collisions that produce motion in neighboring molecules, thus transferring heat from warmer to cooler materials.

Different materials (gases, liquids, and solids) conduct sensible heat directionally from areas of higher temperature to those of lower temperature. This heat flow transfers energy through matter at varying rates, depending on the conductivity of the material—Earth's land surface is a better conductor than air; moist air is a slightly better conductor than dry air.

Gases and liquids also transfer energy by movements called **convection** when the physical mixing involves a strong vertical motion. When horizontal motion dominates, the term **advection** applies. In the atmosphere or in bodies of water, warmer (less dense) masses tend to rise and cooler (denser) masses tend to sink, establishing patterns of convection. Sensible heat transports physically through the medium in this way.

You commonly experience such energy flows in the kitchen: Energy is conducted through the handle of a pan, or boiling water bubbles in the saucepan in convective

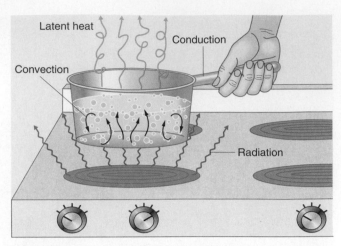

FIGURE 4.9 Heat energy transfer processes. A pan of water on the stove illustrates heat transfer. Infrared energy *radiates* from the burner to the saucepan and the air. Energy *conducts* through the molecules of the pan and the handle. The water physically mixes, carrying heat energy by *convection.* The energy in the water and handle is measurable as *sensible heat.* The vapor leaving the surface of the water contains the *latent heat* absorbed in the change of water to water vapor.

motions (Figure 4.9). Also in the kitchen you may use a convection oven that has a fan to circulate heated air to uniformly cook food.

In physical geography, we find many examples of each physical transfer mechanism:

- *Conduction* examples include surface energy budgets, temperature differences between land and water bodies and between darker and lighter surfaces, the heating of surfaces and overlying air, and soil temperatures;
- *Convection* includes atmospheric and oceanic circulation, air mass movements and weather systems, internal motions deep within Earth that produce a magnetic field, and movements in the crust; and
- *Advection* involves horizontal movement of winds from land to sea and back, fog that forms and moves to another area, and air mass movements from source regions.

Now with these pathways and principles in mind, we put it all together to describe the energy budgets of the lower atmosphere.

Energy Balance in the Troposphere

The Earth–atmosphere energy system budget naturally balances itself in a steady-state equilibrium. The atmosphere and surface eventually radiate longwave energy back to space, and this energy, together with reflected shortwave energy, equals the initial solar input—think of cash flows into and out of a checking account and the balance when deposits and cash withdrawals are equal.

Greenhouse gases in the atmosphere effectively delay losses to space and act to warm the lower atmosphere. We examine this effect and then develop an overall budget for the troposphere.

The Greenhouse Effect and Atmospheric Warming

Previously, we characterized Earth as a cool-body radiator, emitting energy in longer wavelengths from its surface and atmosphere toward space. (In contrast, the Sun is a hot-body radiator, emitting shorter wavelengths from its surface.) However, some of this longwave radiation is absorbed by carbon dioxide, water vapor, methane, nitrous oxide, chlorofluorocarbons (CFCs), and other gases in the lower atmosphere and then emitted back toward Earth (detailed in Figure 2.7). This absorption and emission delays energy loss to space and is an important factor in warming the troposphere. The rough similarity between this process and the way a greenhouse operates gives the process its name—the **greenhouse effect**.

In a greenhouse, the glass is transparent to shortwave insolation, allowing light to pass through to the soil, plants, and materials inside, where absorption and conduction take place. The absorbed energy is then radiated as longwave radiation back toward the glass, but the glass physically traps both the longer wavelengths and the warmed air inside the greenhouse. Thus, the glass acts as a one-way filter, allowing shortwave energy in but not allowing the longwave energy out, except through conduction. You experience the same process in a car parked in direct sunlight.

Opening the greenhouse roof vent, or the car windows, allows the air inside to mix with the outside environment, thereby removing heat by moving air physically from one place to another—the process of convection. It is surprising how hot the interior of a car gets, even on a day with mild temperatures outside. Many people place a sunscreen across the windshield to prevent shortwave energy from entering the car to begin the greenhouse process. Have you taken steps to reduce the insolation input into your car's interior or through windows at home?

Overall, the atmosphere behaves a bit differently. In the atmosphere, the greenhouse analogy does not fully apply because longwave radiation is not trapped as in a greenhouse. Rather, its passage to space is *delayed* as the longwave radiation is absorbed by certain gases, clouds, and dust in the atmosphere and is emitted to Earth's surface. According to scientific consensus, today's increasing carbon dioxide concentration is forcing more longwave radiation absorption in the lower atmosphere, thus producing a warming trend and changes in the Earth–atmosphere energy system.

Clouds and Earth's "Greenhouse"

Clouds affect the heating of the lower atmosphere in several ways, depending on cloud type. Not only is the percentage of cloud cover important, but the cloud type, height, and thickness (water content and density) also have an effect. High-altitude, ice-crystal clouds reflect insolation with albedos of about 50%, whereas thick, lower cloud cover reflects about 90% of incoming insolation.

To understand the actual effects on the atmosphere's energy budget, we must consider both transmission of shortwave and longwave radiation and cloud type. Figure 4.10a portrays the *cloud-greenhouse forcing* caused by high clouds (warming, because their greenhouse effects exceed their albedo effects); and Figure 4.10b portrays the *cloud-albedo forcing* produced by lower, thicker clouds (cooling, because albedo effects exceed greenhouse effects).

Jet contrails (*condensation trails*) produce high cirrus clouds (Figure 4.10c and chapter-opening photo inset)—sometimes called *false cirrus clouds*. The chapter-opening photos and satellite image illustrate how cloud development is stimulated by aircraft exhaust. Scientists were suspicious that these clouds affected atmospheric and surface temperatures. As shown in Figure 4.10a, these high-altitude ice-crystal clouds are efficient at reducing outgoing longwave radiation and therefore contribute to global heating. However, these contrail-seeded cirrus clouds have a higher density and higher albedo than normal cirrus clouds, thus reducing some surface insolation.

The tragedy that struck the World Trade Center, and humanity, on September 11, 2001, inadvertently provided researchers with a chance to study these effects. Following 9/11, there was a 3-day grounding of all commercial airline traffic and therefore no contrails across the United States. Weather data from 4000 stations over 30 years were compared with data during the 3-day flight shutdown. *Diurnal temperature range* (DTR)—the difference between daytime maximum and nighttime minimum temperatures—increased during the 3 days. Specifically, maximum temperatures were more responsive to the missing clouds than minimums.

Later study of synoptic conditions (air pressure, humidity, air mass data—all gathered at a specific time), analysis of Canadian air space, a check on conditions during adjacent 3-day periods, and a look at the few military flights that took place during the halt of air travel served to verify the initial findings and strengthen the conclusion.[*] The elimination of contrail-stimulated clouds produced an increase in DTR—slightly higher afternoon and slightly lower nighttime temperature readings across the United States. The largest effects were in those regions where the greatest number of flights occurred. This ongoing study is helpful in the spatial analysis of possible impacts of future aircraft design and timing of flight schedules on the Earth–atmosphere energy budget.

To better understand the role of natural clouds as well as jet contrails, NASA operates the CERES program, an

---

[*]See David J. Travis et al., "Contrails reduce daily temperature range," *Nature* 418 (August 8, 2002): 601, from the Department of Geography and Geology, University of Wisconsin–Whitewater; and, D. J. Travis, A. M. Carleton, and R. G. Lauritsen, "Regional variations in U.S. diurnal temperature range for the 9/11–14/2001 aircraft groundings: Evidence of jet contrail influence on climate," *Journal of Climate* 17 (March 1, 2004): 1123–34.

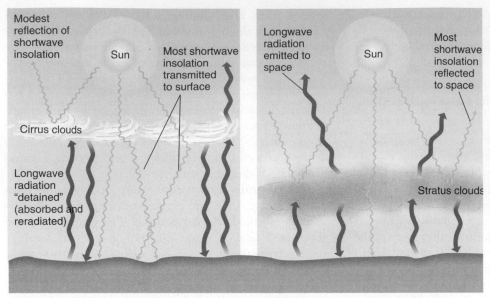

Modest reflection of shortwave insolation

Sun

Most shortwave insolation transmitted to surface

Cirrus clouds

Longwave radiation "detained" (absorbed and reradiated)

(a) High clouds: net greenhouse forcing and atmospheric warming

Longwave radiation emitted to space

Sun

Most shortwave insolation reflected to space

Stratus clouds

(b) Low clouds: net albedo forcing and atmospheric cooling

FIGURE 4.10 Energy effects of two cloud types.
Cloud effects vary depending on cloud type. (a) High, ice-crystal clouds (called *cirrus*) transmit most of the insolation. However, they absorb and delay losses of outgoing longwave radiation, producing a greater greenhouse forcing and a net warming of Earth. (b) Low, thick clouds (*stratus*) reflect most of the incoming insolation and radiate longwave radiation to space, producing a greater albedo forcing and a net cooling of Earth. (c) Jet aircraft exhaust triggers denser cirrus-cloud development. Newer, thinner contrails spread to form cirrus clouds. [(c) Astronaut photo May 15, 2002 courtesy of NASA/JSC.]

(c)

acronym for Clouds and Earth's Radiant Energy System. CERES sensors aboard satellites began operating in 1997 (*TRMM*), in addition to *Terra* and *Aqua* satellites. CERES data are an important part of the study of aerosols and the monsoons in Figure 4.8. (See **http://asd-www.larc.nasa. gov/ceres/**.)

Data collected by CERES sensors cover both reflected and emitted radiation. Figure 4.11 shows the first global monthly images from satellite *Terra* for March 2000 in W/m² flux (meaning energy "flow"). In Figure 4.11a, lighter regions indicate where more sunlight is reflected into space than is absorbed—for example, by lighter-colored land surfaces such as deserts or by cloud cover such as over tropical lands. Green and blue areas illustrate where less light was reflected.

In Figure 4.11b, orange and red pixels indicate regions where more longwave radiation was absorbed and emitted

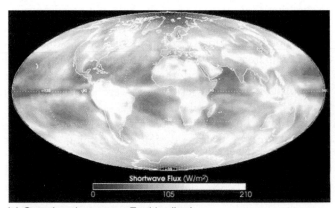

Shortwave Flux (W/m²)

0   105   210

(a) Outgoing shortwave – Earth's albedo.

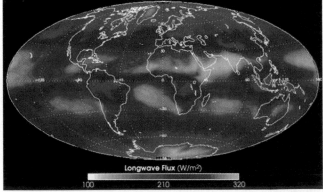

Longwave Flux (W/m²)

100   210   320

(b) Longwave energy flux to space.

FIGURE 4.11 Shortwave and longwave images show Earth's radiation budget components.
The CERES sensors aboard *Terra* made these portraits, capturing (a) outgoing shortwave energy flux reflected from clouds, land, and water—Earth's albedo; and (b) longwave energy flux emitted by these surfaces back to space. The scale beneath each image displays value gradations in watts per square meter. [Images courtesy of CERES Instrument Team, Langley Research Center, NASA]

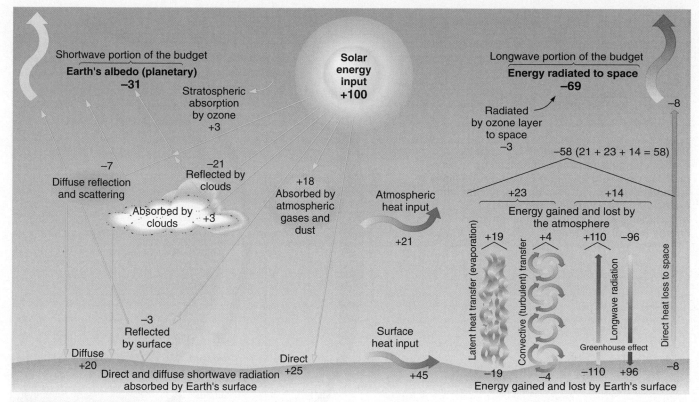

FIGURE 4.12 Detail of the Earth–atmosphere energy balance.
Solar energy cascades through the lower atmosphere (left-hand portion of the illustration), where it is absorbed, reflected, and scattered. Clouds, atmosphere, and the surface reflect 31% of this insolation back to space. Atmospheric gases and dust and Earth's surface absorb energy and radiate longwave radiation (right-hand portion of the illustration). Over time, Earth emits, on average, 69% of incoming energy to space. When added to Earth's average albedo (31%, reflected energy), this equals the total energy input from the Sun (100%).

Global Warming,
Climate Change
ANIMATION   Earth-Atmosphere
Energy Balance

to space; less longwave energy is escaping in the blue and purple regions. Those blue regions of lower longwave emissions over tropical lands are due to tall, thick clouds along the equatorial region (the Amazon, equatorial Africa, and Indonesia); these clouds also caused higher shortwave reflection. Subtropical desert regions exhibit greater longwave radiation emissions owing to the presence of little cloud cover and greater radiative energy losses from surfaces that absorbed a lot of energy, as we saw in Figure 4.2.

Earth–Atmosphere Energy Balance

If Earth's surface and its atmosphere are considered separately, neither exhibits a balanced radiation budget. The average annual energy distribution is positive (an energy surplus) for Earth's surface and negative (an energy deficit) for the atmosphere as it radiates energy to space. Considered together, these two equal each other, making it possible for us to construct an overall energy balance. However, regionally and seasonally Earth absorbs more energy in the tropics and less in the polar regions, establishing the imbalance that drives global circulation patterns.

Figure 4.12 summarizes the Earth–atmosphere radiation balance. It brings together all the elements discussed to this point in the chapter by following 100% of arriving insolation through the troposphere. The shortwave portion of the budget is on the left in the illustration; the longwave part of the budget is on the right.

Out of 100% of the solar energy arriving, Earth's average albedo is 31%. Figure 4.11a shows this shortwave radiation reflected from the Earth–atmosphere system. Absorption by atmospheric clouds, dust, and gases involves another 21% and accounts for the atmospheric heat input. Stratospheric ozone absorption and radiation account for another 3% of the atmospheric budget. About 45% of the incoming insolation transmits through to Earth's surface as direct and diffuse shortwave radiation.

The natural energy balance occurs through longwave energy transfers from the surface, which are both *nonradiative* (physical motion) and *radiative*. Nonradiative transfers include convection, conduction, and the latent heat of evaporation (energy that is absorbed and dissipated by water as it evaporates and condenses). Radiative transfer is by longwave radiation between the surface, the atmosphere, and space, as illustrated to the right in Figure 4.12's depiction of the greenhouse effect.

To summarize, Earth eventually emits the longwave radiation part of the budget into space: 21% (atmosphere heating) + 45% (surface heating) + 3% (ozone emission) = 69%. This longwave radiation from the Earth–atmosphere system (net outgoing longwave) is shown in Figure 4.11b.

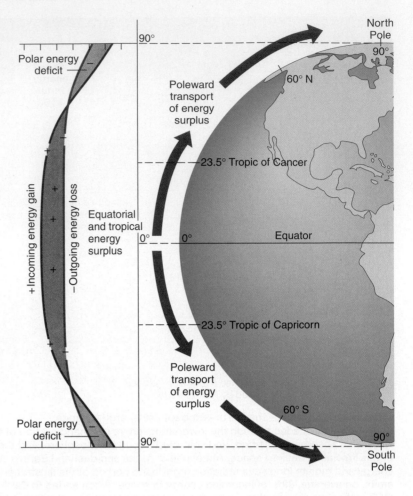

FIGURE 4.13 Energy budget by latitude. Earth's energy surpluses and deficits by latitude produce poleward transport of energy and mass in each hemisphere—atmospheric circulation and ocean currents. Outside of the tropics, atmospheric winds dominate the transport of energy toward each pole.

Review, from Chapter 2, Figure 2.11 showing "Daily Net Radiation" at the top of the atmosphere. The map clearly shows the latitudinal energy imbalance in net radiation—positive values in lower latitudes and negative values toward the poles. Figure 4.13 summarizes the Earth–atmosphere energy balance for all shortwave and longwave energy by latitude:

- Between the tropics, the angle of incoming insolation is high and daylength is consistent, with little seasonal variation, so more energy is gained than lost—*energy surpluses dominate*.
- In the polar regions, the Sun is low in the sky, surfaces are light (ice and snow) and reflective, and for up to 6 months during the year no insolation is received, so more energy is lost than gained—*energy deficits prevail*.
- At around 36° latitude, a balance exists between energy gains and losses for the Earth–atmosphere system.

The imbalance of net radiation from *tropical surpluses* to the *polar deficits* drives a vast global circulation of both energy and mass. The meridional (north–south) transfer agents are winds, ocean currents, dynamic weather systems, and related phenomena. Dramatic examples of such energy and mass transfers are tropical cyclones—hurricanes and typhoons. Forming in the tropics, these powerful storms mature and migrate to higher latitudes, carrying with them energy, water, and water vapor.

Having established the Earth–atmosphere energy balance, we next focus on energy characteristics at Earth's surface and examine energy budgets along the ground level, as portrayed in Figure 4.12. The surface environment is the final stage in the Sun to Earth energy system.

Energy Balance at Earth's Surface

Solar energy is the principal heat source at Earth's surface. Figure 4.12 illustrates the direct and diffuse shortwave radiation and longwave radiation arriving at the ground surface. These radiation patterns at Earth's surface are of great interest to geographers and should be of interest to everyone because these are the surface environments where we live.

Daily Radiation Patterns

Figure 4.14 shows the fluctuating daily pattern of incoming shortwave energy absorbed and resultant air temperature. This graph represents idealized conditions for bare soil on a cloudless day in the middle latitudes. Incoming energy arrives during daylight, beginning at sunrise, peaking at noon, and ending at sunset.

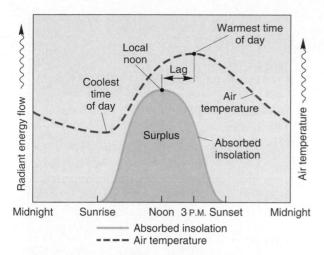

FIGURE 4.14 Daily radiation curves.
Sample radiation plot for a typical day shows the changes in insolation (orange line) and air temperature (dashed line). Comparing the curves demonstrates a lag between local noon (the insolation peak for the day) and the warmest time of day.

The shape and height of this insolation curve varies with season and latitude. The highest trend for such a curve occurs at the time of the summer solstice (around June 21 in the Northern Hemisphere and December 21 in the Southern Hemisphere). The air temperature plot also responds to seasons and variations in insolation input. Within a 24-hour day, air temperature generally peaks between 3:00 and 4:00 P.M. and dips to its lowest point right at or slightly after sunrise.

The relationship between the insolation curve and the air temperature curve on the graph is interesting—they do not align; there is a *lag*. The warmest time of day occurs not at the moment of maximum insolation, but at the moment when a *maximum of insolation is absorbed* and emitted to the atmosphere from the ground. As long as the incoming energy exceeds the outgoing energy, air temperature continues to increase, not peaking until the incoming energy begins to diminish in the afternoon as the Sun's altitude decreases. In contrast, if you have ever gone camping in the mountains, you no doubt experienced the coldest time of day with a wake-up chill at sunrise!

The annual pattern of insolation and air temperature exhibits a similar lag. For the Northern Hemisphere, January is usually the coldest month, occurring after the December solstice and the shortest days. Similarly, the warmest months of July and August occur after the June solstice and the longest days.

Simplified Surface Energy Balance

Energy and moisture are continually exchanged at the surface, creating worldwide "boundary-layer climates" of great variety. **Microclimatology** is the science of physical conditions at or near Earth's surface. The following discussion is more meaningful if you visualize an actual surface—perhaps a park, a front yard, or a place on campus.

The surface receives visible light and longwave radiation and it reflects light and radiates longwave according to the following simple scheme:

$$\underset{\text{(Insolation)}}{+\text{SW}\downarrow} \quad \underset{\text{(Reflection)}}{-\text{SW}\uparrow} \quad \underset{\text{(Infrared)}}{+\text{LW}\downarrow} \quad \underset{\text{(Infrared)}}{-\text{LW}\uparrow} = \underset{\text{(Net Radiation)}}{\text{NET R}}$$

We use SW for shortwave and LW for longwave for simplicity. You may come across other symbols in the microclimatology literature, such as "Q^*" for NET R (net radiation), "K" for shortwave, and "L" for longwave.

Figure 4.15 shows these components of a surface energy balance. The soil column shown continues to a depth at which energy exchange with surrounding materials or with the surface becomes negligible, usually less than a meter. Sensible heat transfer in the soil is through conduction, predominantly downward during the day or in summer and toward the surface at night or in winter. Energy from the atmosphere that is moving toward the surface is a positive (a gain), and energy that is moving away from the surface, through sensible and latent heat transfers, is a negative (a loss) to the surface account.

Adding and subtracting the energy flow at the surface completes the calculation of **net radiation (NET R)**, or the balance of all radiation at Earth's surface. As the components of this simple equation vary with daylength through the seasons, cloudiness, and latitude, NET R varies. Figure 4.16 illustrates the components of a surface energy balance for a typical summer day at a midlatitude location, showing the daily change in the components of net radiation. The items plotted are direct and diffuse insolation (+SW ↓) reflected energy (surface albedo value, −SW ↑), and longwave radiation arriving (+LW ↓) and leaving (−LW ↑) from the surface.

Surface albedo values (−SW ↑) dictate the amount of insolation reflected, and therefore not absorbed, at the

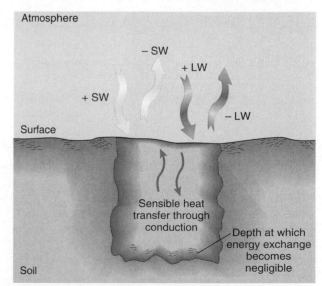

FIGURE 4.15 Surface energy budget.
Idealized input and output energy budget components for a surface and a soil column. Sensible heat transfer in the soil is through conduction, predominantly downward during the day or in summer and toward the surface at night or in winter (SW = shortwave, LW=longwave).

surface. As an example, imagine a snow-covered landscape as a surface energy system. Snow has high reflectivity, sending sunlight back to space and reducing the amount of insolation absorbed. Consequently, surfaces and air temperatures are lower, and thus the snow cover does not melt. This is a system with positive feedback.

Remember from Chapter 1 that *positive feedback* increases response in the system: more cooling–more snow–increasing albedo–more cooling–then colder, drier air–then less snow–and so on. Likewise, in High Latitude Connection 1.1 we see the increasing presence of melt-ponds on ice as a positive feedback in ice loss—the darker colors of meltponds with lower albedo, more absorption of insolation, thus more melting, forming more melt-ponds, and so on.

At night, the net radiation value in Figure 4.16 becomes negative because the SW component ceases at sunset and the surface continues to lose longwave radiation to the atmosphere. The surface rarely reaches a zero net radiation value—a perfect balance—at any one moment. However, over time, Earth's surface naturally balances incoming and outgoing energies.

Net Radiation   The net radiation (NET R of all wavelengths) available at Earth's surface is the final outcome of the entire energy-balance process discussed in this chapter. Figure 4.17 displays the global mean annual net radiation at ground level. The abrupt change in radiation balance from ocean to land surfaces is evident on the map. Note that all values are positive; negative values probably

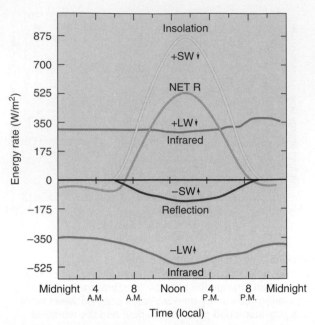

FIGURE 4.16 A day's radiation budget.
Radiation budget components and resulting net radiation (NET R) on a typical July summer day at a midlatitude location (Matador in southern Saskatchewan, about 51° N). [Adapted by permission from T. R. Oke, *Boundary Layer Climates* (New York: Methuen & Co., 1978), p. 21]

occur only over ice-covered surfaces poleward of 70° latitude in both hemispheres. The highest net radiation occurs north of the equator in the Arabian Sea at 185 W/m² per year. Aside from the obvious interruption caused by

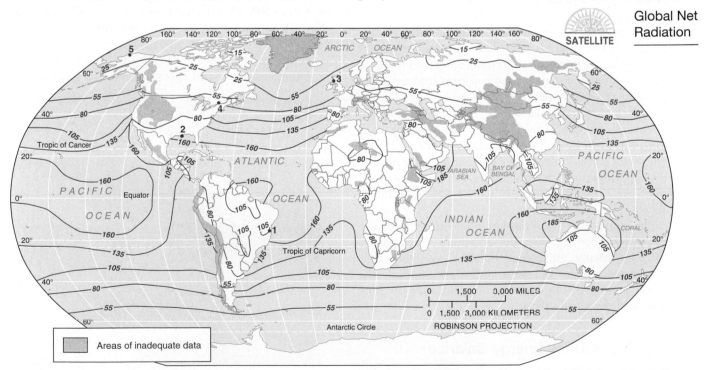

FIGURE 4.17 Global net radiation (NET R).
Distribution of global mean annual net radiation (NET R) at surface level in watts per square meter (100 W/m² = 75 kcal/cm²/year). Temperature graphs for the five cities noted on the map and in Figures 4.18 and 4.19, are graphed in Figure 5.4. The latitude geographic zone in which each city is located is in Figure 1.12. [After M. I. Budyko, *The Heat Balance of the Earth's Surface* (Washington, DC: U.S. Department of Commerce, 1958), p. 106.]

1. Salvador, Brazil
2. New Orleans, Louisiana
3. Edinburgh, Scotland
4. Montreal, Canada
5. Barrow, Alaska
   (see figure 5.4)

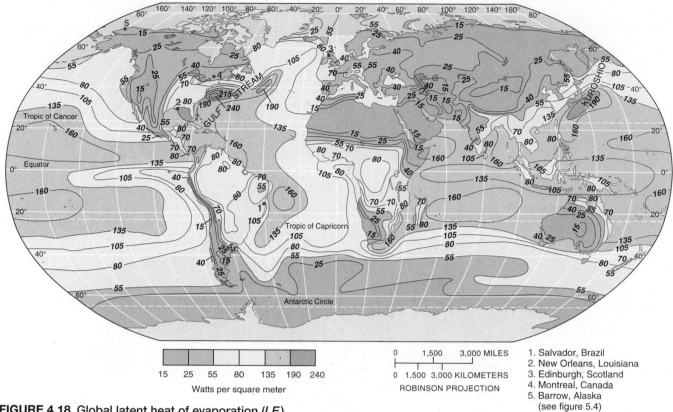

FIGURE 4.18 Global latent heat of evaporation (*LE*).

Distribution of annual energy expenditure as the latent heat of evaporation (*LE*) at surface level in watts per square meter (100 W/m² = 75 kcal/cm²/year). Note the high values associated with high sea-surface temperatures in the area of the Gulf Stream and Kuroshio currents. [Adapted by permission from M. I. Budyko, *The Earth's Climate Past and Future* (New York: Academic Press, 1982), p. 56.]

Global Latent Heat Flux
SATELLITE

1. Salvador, Brazil
2. New Orleans, Louisiana
3. Edinburgh, Scotland
4. Montreal, Canada
5. Barrow, Alaska
   (see figure 5.4)

landmasses, the pattern of values appears generally zonal, or parallel, decreasing away from the equator.

Net radiation is expended from a nonvegetated surface through three pathways:

- *LE*, or *latent heat of evaporation*, is the energy that is stored in water vapor as water evaporates. Water absorbs large quantities of this *latent heat* as it changes state to water vapor, thereby removing this heat energy from the surface. Conversely, this heat energy releases to the environment when water vapor changes state back to a liquid (discussed in Chapter 7). Latent heat is the dominant expenditure of Earth's entire NET R, especially over water surfaces.

- *H*, or *sensible heat*, is the back-and-forth transfer between air and surface in turbulent eddies through convection and conduction within materials. This activity depends on surface and boundary-layer temperatures and on the intensity of convective motion in the atmosphere. About one-fifth of Earth's entire NET R is mechanically radiated as sensible heat from the surface, especially over land.

- *G*, or *ground heating and cooling*, is the energy that flows into and out of the ground surface (land or water) by conduction. During a year, the overall *G* value is zero because the stored energy from spring and summer is equaled by losses in fall and winter. Another factor in ground heating is energy absorbed

at the surface to melt snow or ice. In snow- or ice-covered landscapes, most available energy is in sensible and latent heat used in the melting and warming process.

On land, the highest annual values for latent heat of evaporation (*LE*) occur in the tropics and decrease toward the poles (Figure 4.18). Over the oceans, the highest *LE* values are over subtropical latitudes, where hot, dry air comes into contact with warm ocean water.

The values for sensible heat (*H*) are distributed differently, being highest in the subtropics (Figure 4.19). Here, vast regions of subtropical deserts feature nearly waterless surfaces, cloudless skies, and almost vegetation-free landscapes. The bulk of NET R is expended as sensible heat in these dry regions. Moist and vegetated surfaces expend less in *H* and more in *LE*, as you can see by comparing the maps in Figures 4.18 and 4.19.

Understanding net radiation is essential to solar energy technologies that concentrate shortwave energy for use. Solar energy offers great potential worldwide and is presently the fastest-growing form of energy conversion by humans, although it is still decades away from its potential. Focus Study 4.1 briefly reviews such direct application of surface energy budgets.

Sample Stations   A couple of real locations help complete our discussion. Variation in the expenditure of NET R among sensible heat (*H*, energy we can feel), latent heat (*LE*,

(text continued on page 107)

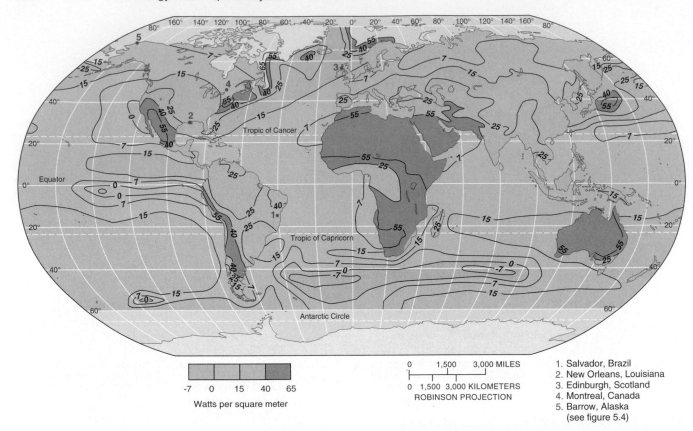

FIGURE 4.19 Global sensible heat (*H*).
Distribution of annual energy expenditure as sensible heat (*H*) at surface level in watts per square meter (100 W/m² = 75 kcal/cm²/year). [Adapted by permission from M. I. Budyko, *The Earth's Climate Past and Future* (New York: Academic Press, 1982), p. 59.]

Global Sensible Heat

SATELLITE

Focus Study 4.1

Solar Energy Collection and Concentration

Consider the following:

- The insolation receipt in just 35 minutes at the surface of the United States exceeds the amount of energy derived from the burning of fossil fuels (coal, oil, natural gas) in a year.
- An average building in the United States receives 6 to 10 times more energy from the Sun hitting its exterior than is required to heat the inside.
- Old photographs of homes in southern California and Florida show solar (flat-plate) water heaters on many rooftops. Early twentieth-century newspaper ads and merchandise catalogues featured the Climax solar water heater (1905) and the Day and Night water heater (1909). Low-priced natural gas and oil displaced many of these

applied solar energy technologies. In 2005, installed solar water heating rose to over 125 million m² of installed surface area, or 40 million households worldwide.

- Installed photovoltaic capacity increased 45% in 2005, to nearly 6 times the installed level in 2000. Solar-electric power is the fastest growing energy source in the world, rising to over 6000 MW.
- The U.S. share of total wind-power production, which was at 44% in 1996, fell to only 9% in 2005—Japan and Germany lead the world. The U.S. administration continued with budget cuts in its photovoltaic program in 2003, 2004, and 2005. (All statistics from Worldwatch Institute, **http://www.worldwatch. org**.)

Not only does insolation warm Earth's surface, it also provides an inexhaustible supply of energy far into the future for humanity. Sunlight is direct and widely available, and solar power installations are decentralized, labor-intensive, and renewable. Sunlight has been collected for centuries through various technologies. Yet it is underutilized. The utility energy industry appears to prefer indirect, centralized, power-plant-specific, capital-intensive, and nonrenewable sources.

Rural villages in developing countries could benefit greatly from the simplest, most cost-effective solar application—the *solar-panel cooker*. For example, people in Kenya walk many kilometers collecting fuel wood for cooking fires (Figure 4.1.1a). Each

(a)

(b)

FIGURE 4.1.1 The solar-cooking solution.
(a) Five women haul firewood many miles to the Dadaab refugee camp in northeastern Kenya. (b) Kenyan women in training to use their solar-panel cookers, which do not require scavenging the countryside for scarce fuel wood. These simple cookers collect direct and diffuse insolation through transparent glass or plastic and trap longwave radiation in an enclosed box or cooking bag (c). This is a small-scale, efficient application of the greenhouse effect. Construction is easy, using cardboard components. Temperatures easily exceed 107°C (225°F) for baking, boiling, purifying water, and sterilizing instruments. [Photos by: (a) and (b) Solar Cookers International, Sacramento, California; (c) Bobbé Christopherson.]

(c)

village and refugee camp is surrounded by impoverished land, stripped of wood. Using solar cookers, villagers are able to cook meals and sanitize their drinking water without scavenging for wood (Figure 4.1.1b). These solar devices in Figure 4.1.1c are simple yet efficient, reaching temperatures between 107°C–127°C (225°F–260°F). See Solar Cookers International at **http://solarcooking.org/** for more information.

In less-developed countries, the money for electrification (a *centralized* technology) is not available despite the push from more developed countries and energy corporations for large capital-intensive power projects in those countries. In reality, the pressing need is for *decentralized* energy sources, appropriate in scale to everyday needs, such as cooking, heating water, and pasteurization. Net per capita (per person) cost for solar cookers is far less than for centralized electrical production, regardless of fuel source.

Collecting and Concentrating Solar Energy

Any surface that receives light from the Sun is a *solar collector*. But the diffuse nature of solar energy received at the surface requires that it be collected, concentrated, transformed, and stored to be useful. Space heating is the simplest application. Windows that are carefully designed and placed allow sunlight to shine into a building, where it is absorbed and converted into sensible heat—an everyday application of the greenhouse effect.

A *passive solar system* captures heat energy and stores it in a "thermal mass," such as water-filled tanks, adobe, tile, or concrete. An *active solar system* involves heating water or air in a collector and then pumping it through a plumbing system to a tank, where it can provide hot water for direct use or for space heating.

Solar energy systems can generate heat energy of an appropriate scale for approximately half the present domestic applications in the United States, which includes space heating

and water heating. In marginal climates, solar-assisted water and space heating is feasible as a backup; even in New England and the Northern Plains states, solar collection systems prove effective. (Among many sources of information, see National Solar Radiation Data Base User's Manual at **http://rredc.nrel.gov/solar/**.)

Focusing (concentrating) mirrors, such as Fresnel lenses, or parabolic (curved surface) troughs and dishes can be used to attain very high temperatures to heat water or other heat-storing fluids. Kramer Junction, California, about 225 km (140 mi) northeast of Los Angeles, in the Mojave Desert near Barstow, has the world's largest operating solar electric-generating facility, with a capacity of 150 MW (megawatts; 150 million watts). Long troughs of computer-guided curved mirrors concentrate sunlight to create temperatures of 390°C (735°F) in vacuum-sealed tubes filled with synthetic oil. The heated oil heats water; the heated water produces steam that rotates

(continued)

Foucs Stude 4.1 (continued)

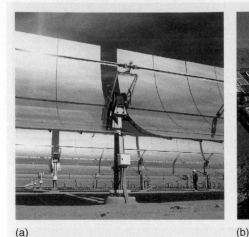

(a)

(b)

(c)

FIGURE 4.1.2 Solar thermal and photovoltaic energy production.
(a) Kramer Junction solar-thermal energy installation in southern California. (b) NREL Outdoor Test Facility in Golden, Colorado, where a variety of photovoltaic cell arrays successfully convert sunlight directly into electricity. (c) A residence with a 7380-W rooftop solar photovoltaic array generates electricity. Excess electricity is fed to the grid for credits that offset 100% of the home's electric bill. [Photos by (a) Kramer Junction Operating Company, Los Angeles; (b) and (c) Bobbé Christopherson.]

turbines to generate cost-effective electricity. The facility converts 23% of the sunlight it receives into electricity during peak hours (Figure 4.1.2a), and operation and maintenance costs continue to decrease.

The National Renewable Energy Laboratory (NREL, http://www.nrel.gov) and the National Center for Photovoltaics, http://www.nrel.gov/ncpv/, were established in 1974 to coordinate solar energy research, development, and testing in partnership with private industry. NREL's headquarters building in Golden, Colorado, features the latest in passive and active solar design. At NREL's Outdoor Test Facility, successful tests continue on arrays of prototype solar cells. Some of these panels have been in operation for more than 25 years (Figure 4.1.2b).

Electricity Directly from Sunlight
Producing electricity by *photovoltaic cells*

(PVs) is a technology used in spacecraft since 1958. Familiar to us all are the solar cells in pocket calculators (hundreds of millions now in use). When light shines upon a semiconductor material in these cells, it stimulates a flow of electrons (an electrical current) in the cell. PV cells are arranged in modules that can be assembled in large arrays.

The efficiency of these cells has improved to the level that they are generally cost-competitive, especially if government policies and subsidies were to be balanced evenly among all energy sources. The residential installation in Figure 4.1.2c features 36 panels, generating 205 W each, 7380 W total, at a 21.5% conversion efficiency. People are able to run their electric meters in reverse and supply electricity to the power grid. In experiments, NREL developed high-efficiency concentrator solar cells that broke the 40% conversion efficiency barrier!

Rooftop photovoltaic electrical generation is now cheaper than power line construction to rural sites. For example, hundreds of thousands of homes in Mexico, Indonesia, Philippines, South Africa, India, Norway, and elsewhere have PV roof systems. (See the Department of Energy's "Photovoltaic Home Page," at http://www.eere.energy.gov/solar/.)

Obvious drawbacks of both solar-heating and solar-electric systems are periods of cloudiness and night, which inhibit operations. Research is under way to enhance energy-storage technologies, such as hydrogen fuel production (using energy to extract hydrogen from water for later use in producing more energy), and to improve battery technology.

The Promise of Solar Energy
Solar energy is a wise choice for the future. It is directly available to the consumer; it is based on a renewable energy source of an appropriate scale for end-use needs; and most solar strategies are labor-intensive, rather than capital-intensive as in centralized, nonrenewable power production. Solar is economically preferable to further development of our decreasing fossil-fuel reserves, further increases in oil imports and tanker spills, investment in foreign military incursions, or adding more troubled nuclear power, especially in a world with security issues.

Whether or not we follow the alternative path of solar energy is a matter of political control and not technological innovation. Much of the technology is ready for installation and is cost-effective when all direct and indirect costs are considered for the alternatives.

NREL asserts in its mission statement that the laboratory wants to "lead the nation toward a sustainable energy future by developing renewable energy technologies, improving energy efficiency, advancing related science, and engineering commercialization." Unfortunately, political decision-making slows progress, cuts budgets, and creates uneven investment incentives weighted toward fossil fuels.

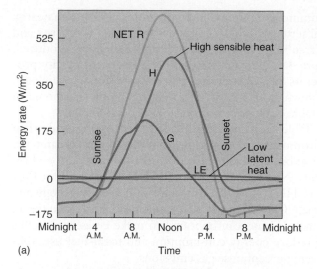

(a)

(b)

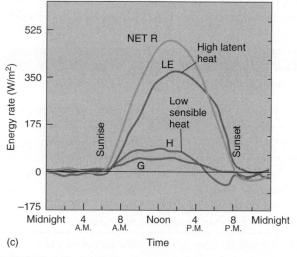

(c)

(d)

FIGURE 4.20 Radiation budget comparison for two stations.
(a) The graph shows the daily net radiation expenditure for El Mirage, California, east of Los Angeles, at about 35° N, and (b) a typical desert landscape near the hot, dry site. (c) The daily net radiation expenditure for Pitt Meadows in southern British Columbia, at about 49° N, and (d) the irrigated blueberry orchards characteristic of agricultural activity in this moist environment of moderate temperatures. (*H* = turbulent sensible heat transfer; *LE* = latent heat of evaporation; *G* = ground heating and cooling.) [(a) Adapted by permission from William D. Sellers, *Physical Climatology,* Fig. 33, copyright © 1965 by The University of Chicago. All rights reserved. (c) Adapted by permission from T. R. Oke, *Boundary Layer Climates* (New York: Methuen & Co., 1978), p. 23. Photos by author.]

energy for evaporation), and ground heating and cooling (*G*) produces the variety of environments we experience in nature. Let us examine the daily energy balance at two locations, El Mirage in California and Pitt Meadows in British Columbia.

El Mirage, at 35° N, is a hot desert location characterized by bare, dry soil with sparse vegetation (Figure 4.20a, b). Our sample is on a clear summer day, with a light wind in the late afternoon. The NET R value is lower than might be expected, considering the Sun's position close to zenith (June solstice) and the absence of clouds. But the income of energy at this site is countered by surfaces of higher albedo than forest or cropland and by hot soil surfaces that radiate longwave energy back to the atmosphere throughout the afternoon.

El Mirage has little or no energy expenditure for evaporation (*LE*). With little water and sparse vegetation,

most of the available radiant energy dissipates through turbulent transfer of sensible heat (*H*), warming air and soil to high temperatures. Over a 24-hour period, *H* is 90% of NET R; the remaining 10% is for ground heating (*G*). The *G* component is greatest in the morning, when winds are light and turbulent transfers are lowest. In the afternoon, heated air rises off the hot ground and convective heat expenditures are accelerated as winds increase.

Compare graphs for El Mirage (Figure 4.20a) and Pitt Meadows (Figure 4.20c). Pitt Meadows is midlatitude (49° N), vegetated, and moist, and its energy expenditures differ greatly from those at El Mirage. The Pitt Meadows landscape is able to retain much more of its energy because of lower albedo values (less reflection), the presence of more water and plants, and lower surface temperatures than those of El Mirage.

The energy balance data for Pitt Meadows is on a cloudless summer day. Higher *LE* values result from the moist environment of rye grass and irrigated mixed-orchard ground cover for the sample area (Figure 4.20d), contributing to the more moderate sensible heat (*H*) levels during the day.

The Urban Environment

An urban landscape produces the temperatures most of you reading this book feel each day. Urban microclimates generally differ from those of nearby nonurban areas. In fact, the surface energy characteristics of urban areas are similar to desert locations. Because almost 50% of the world's population lives in cities, urban microclimatology and other specific environmental effects related to cities are important topics for physical geographers.

The physical characteristics of urbanized regions produce an **urban heat island** that has on average both maximum and minimum temperatures higher than nearby rural settings. Table 4.1 lists five urban characteristics and the resulting temperature and moisture effects produced. Figure 4.21 illustrates these traits. Every major city produces its own **dust dome** of airborne pollution, which can be blown from the city in elongated plumes; as noted in the table, such domes affect urban energy budgets.

Table 4.2 compares climatic factors of rural and urban environments. The worldwide trend toward greater urbanization is placing more and more people on urban heat islands. NASA launched its Urban Heat Island Pilot Project (UHIPP) in 1997, through its Global Hydrology and Climate Center, to better understand the role of cities in climate. If ways can be found to make cities cooler, this will reduce energy consumption and fossil-fuel use, thus reducing greenhouse gas emissions.

In 1998, as part of UHIPP, NASA's research jet made longwave (thermal infrared) radiation measurements over Atlanta, Georgia; Sacramento, California; Baton Rouge,

Table 4.1 Urban Physical Characteristics and Conditions

Urban Characteristics	Results and Conditions
Urban surfaces typically are metal, glass, asphalt, concrete, or stone, and their energy characteristics respond differently from natural surfaces.	Albedos of urban surfaces are lower, leading to higher net radiation values. Urban surfaces expend more energy as sensible heat than do nonurban areas (70% of the net radiation to *H*). Surfaces conduct up to 3 times more energy than wet, sandy soil and thus are warmer. During the day and evening, temperatures above urban surfaces are higher than those above natural areas.
Irregular geometric shapes in a city affect radiation patterns and winds.	Incoming insolation is caught in mazelike reflection and radiation "canyons." Delayed energy is conducted into surface materials, thus increasing temperatures. Buildings interrupt wind flows, diminishing heat loss through advective (horizontal) movement. Maximum heat island effects occur on calm, clear days and nights.
Human activity alters the heat characteristics of cities.	In summer, urban electricity production and use of fossil fuels release energy equivalent to 25%–50% of insolation. In winter, urban-generated sensible heat averages 250% greater than arriving insolation, reducing winter heating requirements.
Many urban surfaces are sealed (built on and paved), so water cannot reach the soil.	Central business district surfaces average 50% sealed, producing more water runoff than the suburbs that average 20% sealed surfaces. Urban areas respond as a desert landscape: A storm may cause a flash flood over the hard, sparsely vegetated surfaces, to be followed by dry conditions a few hours later.
Air pollution, including gases and aerosols, greater in urban areas than in comparable natural settings; increases possible convection and precipitation.	Pollution increases the atmosphere's reflectivity above a city, reducing insolation and absorbing infrared radiation, reradiating infrared downward. Increased particulates in pollution are condensation nuclei for water vapor, increasing cloud formation and precipitation. Urban-stimulated increases in precipitation may occur downwind from cities.

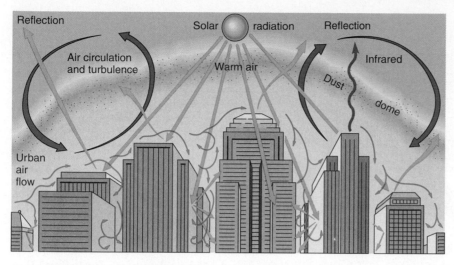

FIGURE 4.21 The urban environment. Insolation, wind movements, and dust dome in city environments.

Table 4.2 Average Differences in Climatic Elements Between Urban and Rural Environments

Element	Urban Compared with Rural Environs	Element	Urban Compared with Rural Environs
Contaminants		**Precipitation (cont.)**	
Condensation nuclei	10 times more	Snowfall, downwind (lee) of city	10% more
Particulates	10 times more		
Gaseous admixtures	5–25 times more	Thunderstorms	10%–15% more
Radiation		**Temperature**	
Total on horizontal surface	0%–20% less	Annual mean	0.5–3 C° (0.9–5.4 F°) more
Ultraviolet, winter	30% less	Winter minima (average)	1.0–2 C° (1.8–3.6 F°) more
Ultraviolet, summer	5% less	Summer maxima	1.0–3 C° (1.8–3.0 F°) more
Sunshine duration	5%–15% less	Heating degree days	10% less
Cloudiness		**Relative Humidity**	
Clouds	5%–10% more	Annual mean	6% less
Fog, winter	100% more	Winter	2% less
Fog, summer	30% more	Summer	8% less
Precipitation		**Wind Speed**	
Amounts	5%–15% more	Annual mean	20%–30% less
Days with 5mm (0.2 in.)	10% more	Extreme gusts	10%–20% less
Snowfall, inner city	5%–10% less	Calm	5%–20% more

Source: H. E. Landsberg, *The Urban Climate*, International Geophysics Series, vol. 28 (1981), p. 258. Reprinted by permission from Academic Press.

Louisiana; and Salt Lake City, Utah. Teachers and students on the ground assisted by making temperature measurements at the same time as the flights. Instruments carried by balloons tracked vertical temperature, relative humidity, and air pressure profiles. (See **www.ghcc.msfc. nasa.gov/urban/** for an update.)

Figure 4.22 illustrates a generalized cross section of a typical urban heat island, showing increasing temperatures toward the downtown central business district. Note that temperatures drop over areas of trees and parks.

Sensible heat is lessened because of the latent heat of evaporation and plant effects such as transpiration and shade. Urban forests are important factors in cooling cities. NASA measurements in an outdoor mall parking lot found temperatures of 48°C (118°F), but a small planter with trees in the same lot was significantly cooler at 32°C (90°F)—16 C° (29 F°) lower in temperature! In Central Park in New York City, daytime temperatures average 5–10 C° (9–18 F°) cooler than in urban areas outside the park.

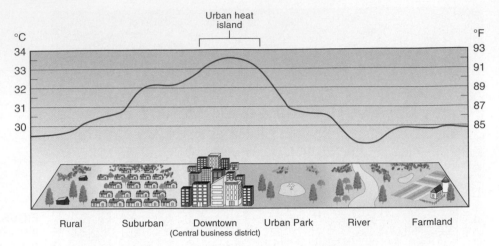

FIGURE 4.22 Typical urban heat island profile.
Cross section of a typical urban heat island. The trend of the temperature gradient from rural to downtown is measurable. Temperatures steeply rise in urban settings, plateau over the suburban built-up area, and peak where temperature is highest at the urban core. Note the cooling over the park area and river.

Figure 4.23 is a thermal infrared image of a portion of Sacramento, California, taken at midday. The false colors of red and white are relatively hot areas (60°C, 140°F), blues and greens are relatively cool areas (29°–36°C, 85°–96°F); clearly buildings are the hottest objects in the scene. As you look across this landscape, what strategies do you think would reduce the urban heat island effect? Possibilities include lighter-colored surfaces for buildings, streets, and parking lots; more reflective roofs; and more trees, parks, and open space. (The images from all the test cities are available at **http://science.nasa.gov/newhome/headlines/essd01jul98%5F1.htm**.)

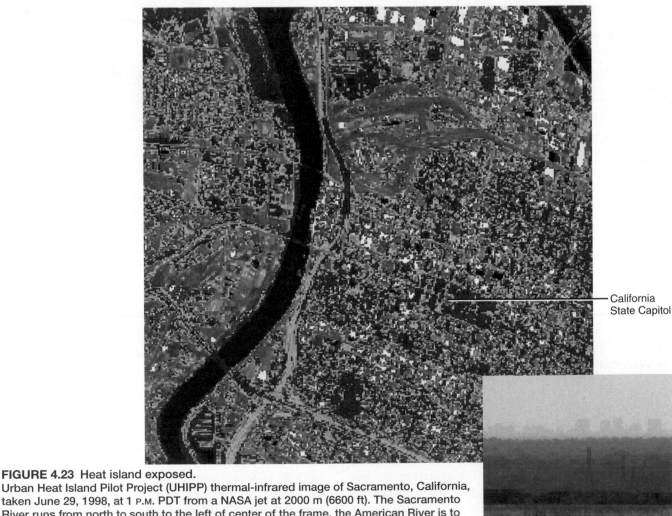

FIGURE 4.23 Heat island exposed.
Urban Heat Island Pilot Project (UHIPP) thermal-infrared image of Sacramento, California, taken June 29, 1998, at 1 P.M. PDT from a NASA jet at 2000 m (6600 ft). The Sacramento River runs from north to south to the left of center of the frame, the American River is to the upper right, and the California state capitol building is the red area within the green rectangle right of center from the river. False color tags white and red as relatively hot sites; greens and blues are relatively cool. The white rectangles are buildings and the hottest spots. [Image courtesy of UHIPP, Marshall Space Flight Center, Global Hydrology and Climate Center, Huntsville, Alabama.]

Sacramento urban heat island in the distance

Summary and Review—Atmosphere and Surface Energy Balances

■ *Identify* the pathways of solar energy through the troposphere to Earth's surface: transmission, scattering, diffuse radiation, refraction, albedo (reflectivity), conduction, convection, and advection.

Radiant energy from the Sun that cascades through complex circuits to the surface powers Earth's biosphere. Our budget of atmospheric energy comprises shortwave radiation *inputs* (ultraviolet light, visible light, and near-infrared wavelengths) and longwave radiation *outputs* (thermal infrared).

Transmission refers to the passage of shortwave and longwave energy through either the atmosphere or water. The gas molecules redirect radiation, changing the direction of the light's movement *without altering its wavelengths*. This phenomenon is known as **scattering** and represents 7% of Earth's reflectivity, or albedo. Dust particles, pollutants, ice, cloud droplets, and water vapor produce further scattering. Some incoming insolation is diffused by clouds and atmosphere and is transmitted to Earth as **diffuse radiation**, the downward component of scattered light. The speed of insolation entering the atmosphere changes as it passes from one medium to another; the change of speed causes a bending action called **refraction**. **Mirage** is a refraction effect when an image appears near the horizon where light waves are refracted by layers of air at different temperatures (and consequently of different densities).

A portion of arriving energy bounces directly back into space without being converted into heat or performing any work. **Albedo** is the reflective quality (intrinsic brightness) of a surface; this returned energy is **reflection**. Albedo is an important control over the amount of insolation that is available for absorption by a surface. We state albedo as the percentage of insolation that is reflected. Earth and its atmosphere reflect 31% of all insolation when averaged over a year.

An increase in albedo and reflection of shortwave radiation caused by clouds is **cloud-albedo forcing**. Also, clouds can act as insulation, thus trapping longwave radiation and raising minimum temperatures. An increase in greenhouse warming caused by clouds is **cloud-greenhouse forcing**. **Global dimming** describes the decline in sunlight reaching Earth's surface owing to pollution, aerosols, and clouds and is perhaps masking the actual degree of global warming occurring.

Absorption is the assimilation of radiation by molecules of a substance and its conversion from one form to another—for example, visible light to infrared radiation. **Conduction** is the molecule-to-molecule transfer of energy as it diffuses through a substance.

Energy also is transferred in gases and liquids by **convection** (when the physical mixing involves a strong vertical motion) or **advection** (when the dominant motion is horizontal). In the atmosphere or bodies of water, warmer portions tend to rise (they are less dense) and cooler portions tend to sink (they are more dense), establishing patterns of convection.

transmission (p. 90)
scattering (p. 91)
diffuse radiation (p. 91)
refraction (p. 91)
mirage (p. 91)
reflection (p. 92)
albedo (p. 92)
cloud-albedo forcing (p. 93)

cloud-greenhouse forcing (p. 94)
global dimming (p.95)
absorption (p. 96)
conduction (p. 96)
convection (p. 96)
advection (p. 96)

1. Diagram a simple energy balance for the troposphere. Label each shortwave and longwave component and the directional aspects of related flows.
2. What would you expect the sky color to be at 50 km (30 mi) altitude? Why? What factors explain the lower atmosphere's blue color?
3. Define *refraction*. How is it related to daylength? To a rainbow? To the beautiful colors of a sunset?
4. List several types of surfaces and their albedo values. Explain the differences among these surfaces. What determines the reflectivity of a surface?
5. Using Figure 4.6, explain the seasonal differences in albedo values for each hemisphere. Be specific, using the albedo values given in Figure 4.5 where appropriate.
6. Define the concepts transmission, absorption, diffuse radiation, conduction, and convection.

■ *Describe* what happens to insolation when clouds are in the atmosphere, and *analyze* the effect of clouds and air pollution on solar radiation received at ground level.

Clouds reflect *insolation*, thus cooling Earth's surface—*cloud-albedo forcing*. Yet clouds also act as insulation, thus trapping longwave radiation and raising minimum temperatures—*cloud-greenhouse forcing*. Clouds affect the heating of the lower atmosphere, depending on cloud type, height, and thickness (water content and density). High-altitude, ice-crystal clouds reflect insolation with albedos of about 50% and redirect infrared to the surface, producing a net cloud-greenhouse forcing (warming); thick, lower cloud cover reflects about 90%, producing a net cloud-albedo forcing (cooling). **Jet contrails**, or *condensation trails*, are produced by aircraft exhaust, particulates, and water vapor and can form high cirrus clouds, sometimes called *false cirrus clouds*.

Emissions of sulfur dioxide and the subsequent chemical reactions in the atmosphere form *sulfate aerosols*, which act either as insolation-reflecting haze in clear-sky conditions or as a stimulus to condensation in clouds that increases reflectivity.

jet contrails (p. 97)

7. What role do clouds play in the Earth–atmosphere radiation balance? Is cloud type important? Compare high, thin cirrus clouds and lower, thick stratus clouds.
8. Jet contrails affect the Earth–atmosphere balance in what ways? Describe the recent scientific findings.
9. In what way does the presence of sulfate aerosols affect solar radiation received at ground level? How does it affect cloud formation?

■ *Review* the energy pathways in the Earth–atmosphere system, the greenhouse effect, and the patterns of global net radiation.

The Earth–atmosphere energy system naturally balances itself in a steady-state equilibrium. It does so through energy transfers that are *nonradiative* (convection, conduction, and the latent heat of evaporation) and *radiative* (by longwave radiation between the surface, the atmosphere, and space).

Carbon dioxide, water vapor, methane, CFCs (chloroflurocarbons), and other gases in the lower atmosphere absorb infrared radiation that is then emitted to Earth, thus delaying energy loss to space. This process is the **greenhouse effect**. In the atmosphere, longwave radiation is not actually trapped, as it would be in a greenhouse, but its passage to space is delayed (heat energy is detained in the atmosphere) through absorption and counterradiation.

Between the tropics, high insolation angle and consistent daylength cause more energy to be gained than lost (there are energy surpluses). In the polar regions, an extremely low insolation angle, highly reflective surfaces, and up to 6 months of no insolation annually cause more energy to be lost (there are energy deficits). This imbalance of net radiation from tropical surpluses to the polar deficits drives a vast global circulation of both energy and mass.

Surface energy balances are used to summarize the energy expenditure for any location. Surface energy measurements are used as an analytical tool of **microclimatology**. Adding and subtracting the energy flow at the surface lets us calculate **net radiation (NET R)**, or the balance of all radiation at Earth's surface available to do work—shortwave (SW) and longwave (LW).

greenhouse effect (p. 97)
microclimatology (p. 101)
net radiation (NET R) (p. 101)

10. What are the similarities and differences between an actual greenhouse and the gaseous atmospheric greenhouse? Why is Earth's greenhouse effect changing?
11. In terms of energy expenditures for latent heat of evaporation, describe the annual pattern as mapped in Figure 4.18.
12. Generalize the pattern of global net radiation. How might this pattern drive the atmospheric weather machine? (See Figures 4.13 and 4.17.)
13. In terms of surface energy balance, explain the term *net radiation* (NET R).
14. What are the expenditure pathways for surface net radiation? What kind of work is accomplished?
15. What is the role played by latent heat in surface energy budgets?
16. Compare the daily surface energy balances of El Mirage, California, and Pitt Meadows, British Columbia. Explain the differences.

■ *Plot* the daily radiation curves for Earth's surface, and *label* the key aspects of incoming radiation, air temperature, and the daily temperature lag.

The greatest insolation input occurs at the time of the summer solstice in each hemisphere. Air temperature responds to seasons and variations in insolation input. Within a 24-hour day, air temperature peaks between 3:00 and 4:00 P.M. and dips to its lowest point right at or slightly after sunrise.

Air temperature lags behind each day's peak insolation. The warmest time of day occurs not at the moment of maximum insolation, but at that moment when a maximum of insolation is absorbed.

17. Why is there a temperature lag between the highest Sun altitude and the warmest time of day? Relate your answer to the insolation and temperature patterns during the day.

■ *Portray* typical urban heat island conditions, and *contrast* the microclimatology of urban areas with that of surrounding rural environments.

A growing percentage of Earth's people live in cities and experiences a unique set of altered microclimatic effects: increased conduction, lower albedos, higher NET R values, increased water runoff, complex radiation and reflection patterns, anthropogenic heating, and the gases, dusts, and aerosols of urban pollution. Urban surfaces of metal, glass, asphalt, concrete, and stone conduct up to 3 times more energy than wet, sandy soil and thus are warmed, producing an **urban heat island**. Air pollution, including gases and aerosols, is greater in urban areas than in rural ones. Every major city produces its own **dust dome** of airborne pollution.

urban heat island (p. 108)
dust dome (p. 108)

18. What is the basis for the urban heat island concept? Describe the climatic effects attributable to urban as compared with nonurban environments. What did NASA determine from the UHIPP overflight of Sacramento (Figure 4.23)?
19. Which of the items in Table 4.2 have you yourself experienced? Explain.
20. Assess the potential for solar energy applications in our society. What are some negatives? What are some positives?

NetWork

The *Geosystems* Student Learning Center provides online resources for this chapter on the World Wide Web. To begin: Once at the Center, click on the cover of this textbook, scroll the Table of Contents menu, and select this chapter. You will find self-tests that are graded, review exercises, specific updates for items in the chapter, and in "Destinations" many links to interesting related pathways on the Internet. *Geosystems* Student Learning Center is found at **http://www.prenhall.com/christopherson/**.

Critical Thinking

A. Given what you now know about reflection, albedo, absorption, and net radiation expenditures, assess your wardrobe (fabrics and colors), house or apartment (colors of exterior walls, especially south and west facing; or, in the Southern Hemisphere, north and west facing), and roof (orientation relative to the Sun), automobile (color, use of sun shades), bicycle seat (color), and other aspects of your environment to determine a personal "Energy IQ." What grade do you give yourself? Be cool!

B. On the *Geosystems* Student Learning Center website, in Chapter 4, under the "Destinations," section, as well as in this chapter of the text, there are several listings of URLs relating to solar energy applications. Take some time and explore the Internet for a personal assessment of these necessary technologies (solar-thermal, solar-electric, photovoltaic cells, solar-box cookers, and the like). As we near the climatic limitations of the fossil-fuel era, and the depletion of the resource itself, these available technologies are destined to become part of the fabric of our lives. Briefly describe your search results. Given these findings, determine if solar technology is available in your area.

C. Examine the photograph to the right. In Antarctica, on Petermann Island off the Graham Land Coast, we noticed this piece of kelp (a seaweed) dropped by a passing bird. The kelp was resting about 10 cm (4 in.) deep in the snow, in a hole about the shape of the kelp. In your opinion, what energy factors operated to make this scene? Now, expand your conclusion to the issue of mining in Antarctica. There are coal deposits and other minerals in Antarctica. The international Antarctic Treaty (1959, additional measures in 1972, 1982, 1991) blocks mining exploitation, although reconsideration is possible after 2048. Given your energy budget analysis of the kelp in the snow, information on surface energy budgets in this chapter, the dust and particulate output of mining, and the importance of a stable sea level, construct a case opposed to such mining activities based on these factors.

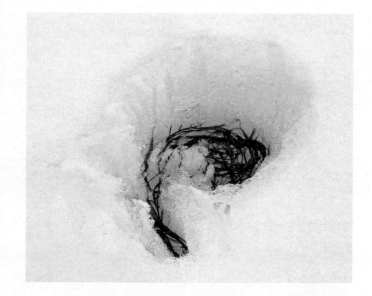

A November blizzard whistles across the tundra and a frozen pond in northern Canada. Note the downwind accumulation of snow at each rock. At the moment of this photo it didn't seem possible, but increasing temperatures are affecting the Arctic region at more than double the rate of the rest of the globe. The polar regions are acting like a "miners' Canary" early warning system about the state of global temperatures and climate change. [Photo by Bobbé Christopherson.]

Global Temperatures

■ Key Learning Concepts

After reading the chapter, you should be able to:

■ *Define* the concepts of temperature, kinetic energy, and sensible heat, and *distinguish* among Kelvin, Celsius, and Fahrenheit temperature scales and how they are measured.

■ *List* and *review* the principal controls and influences that produce global temperature patterns.

■ *Review* the factors that produce different marine effects and continental effects as they influence temperatures, and *utilize* several pairs of cities to illustrate these differences.

■ *Interpret* the pattern of Earth's temperatures from their portrayal on January and July temperature maps and on a map of annual temperature ranges.

■ *Contrast* wind chill and heat index, and *determine* human response to these apparent temperature effects.

What is the temperature now—both indoors and outdoors—as you read these words? How is it measured, and what does the value mean? How is air temperature influencing your plans for the day? Our bodies subjectively sense temperature and judge comfort and react to changing temperature. Air temperature plays a remarkable role in our lives, both at the micro level and at the macro and global levels.

We read of heat waves and cold spells affecting people, crops, events, and energy consumption. For example, in 1995 devastating summer heat and high humidity killed more than 1000 people in the Midwest and East; Europe lost more than 20,000 people in 2003 record heat-index conditions; more than 150 died in a California heat wave in July 2006; and hundreds died in heat waves across southern Europe and the southeastern United States in 2007 (see Focus Study 5.1 on "Air Temperature and the Human Body").

Record warmth is reaching into the Arctic. Ice-core data and other proxy measures indicate that present temperatures across the region are higher than at any time during the past 125,000 years. In response, since satellite observations began 3 decades ago, the Arctic Ocean sea-ice extent melted to its lowest area during September 2007. This warming trend in air temperature is the subject of much scientific, geographic, and political action. Temperature of land, air, and water is a key *output* of the energy–atmosphere system.

In this chapter: A variety of temperature regimes affect cultures, decision-making, and resources consumed across the globe. Understanding some temperature concepts helps us begin our study of these Earth systems. We look at the principal temperature controls of latitude, altitude and elevation, cloud cover, and land–water heating differences as they interact to produce Earth's temperature patterns. We end by examining the effect of temperature on the human body and the resultant spatial aspects of hot and cold conditions on human experience.

Temperature Concepts and Measurement

Heat and temperature are not the same. *Heat* is a form of energy that flows from one system or object to another because the two are at different temperatures. **Temperature** is a measure of the average kinetic energy (motion) of individual molecules in matter. We feel the effect of temperature as the *sensible heat* transfer from warmer objects to cooler objects. For instance, when you jump into a cool lake, you can sense the heat transfer from your skin to the water as kinetic energy leaves your body and flows to the water; a chill develops.

Temperature and heat are related because changes in temperature are caused by the absorption or emission (gain or loss) of heat energy. The term *heat energy* is frequently used to describe energy that is added to or removed from a system or substance.

Temperature Scales

The temperature at which atomic and molecular motion in matter completely stops is *absolute zero, "0° absolute temperature."* Its value on the different temperature-measuring scales is –273° Celsius (C), –459.4° Fahrenheit (F), and 0 Kelvin (K). Figure 5.1 compares these three scales. Formulas for converting between Celsius, SI (Système International), and English units are in Appendix C.

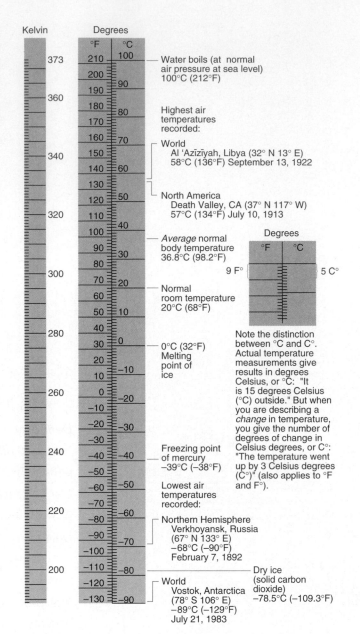

FIGURE 5.1 Temperature scales. Scales for expressing temperature in Kelvin (K) and degrees Celsius (°C) and Fahrenheit (°F), with details of what we experience. Note the distinction between temperature and units of temperature expression, and the placement of the degree symbol.

The Fahrenheit scale places the melting point of ice at 32°F, with 180 subdivisions to the boiling point of water at 212°F. The scale is named for its developer Daniel G. Fahrenheit, a German physicist (1686–1736). Note that there is only one melting point for ice, but there are many freezing points for water, ranging from 32°F down to –40°F, depending on its purity, volume, and certain conditions in the atmosphere.

About a year after the adoption of the Fahrenheit scale, Swedish astronomer Anders Celsius (1701–1744) developed the Celsius scale (formerly called centigrade). He placed the melting point of ice at 0°C and boiling temperature of water at sea level at 100°C, dividing his scale into 100 degrees using a decimal system.

British physicist Lord Kelvin (born William Thomson, 1824–1907) proposed the Kelvin scale in 1848. Science uses this scale because it starts at absolute zero temperature and thus readings are proportional to the actual kinetic energy in a material. The Kelvin scale's melting point for ice is 273 K, and its boiling point of water is 373 K.

Most countries use the Celsius scale to express temperature. The United States remains the only major country still using the Fahrenheit scale. The continuing pressure from the scientific community and corporate interests, not to mention the rest of the world, makes adoption of Celsius and SI units inevitable in the U.S. This textbook presents Celsius (with Fahrenheit equivalents in parentheses) throughout to help bridge this transitional era in the United States.

Measuring Temperature

A *mercury thermometer* or *alcohol thermometer* is a sealed glass tube that measures outdoor temperatures. Fahrenheit invented both the alcohol and mercury thermometers. Cold climates demand alcohol thermometers because alcohol freezes at –112°C (–170°F), whereas mercury freezes at –39°C (–38.2°F). The principle of these thermometers is simple: When fluids are heated, they expand; upon cooling, they contract. A thermometer stores fluid in a small reservoir at one end and is marked with calibrations to measure the expansion or contraction of the fluid, which reflects the temperature of the thermometer's environment.

Figure 5.2 shows a mercury minimum–maximum thermometer. It preserves readings of the day's highest

FIGURE 5.3 Instrument shelter.
This standard thermometer shelter is white (for high albedo) and louvered (for ventilation), replacing the traditional Cotton-region, or Stephenson screen, shelters for temperature measurement. [Photo by Bobbé Christopherson.]

and lowest temperatures until reset (by moving the markers with a magnet). Another type, the recording thermometer, creates an inked record on a turning drum, usually set for one full rotation every 24 hours or every 7 days.

Thermometers for standardized official readings are placed outdoors in small shelters that are white (for high albedo) and louvered (for ventilation) to avoid overheating of the instruments. They are placed at least 1.2–1.8 m (4–6 ft) above a surface, usually on turf; in the United States 1.2 m (4 ft) is standard. Official temperature measurements are made in the shade to prevent the effect of direct insolation. If you install a thermometer for outdoor temperature reading, be sure to avoid direct sunlight on the instrument and place it in an area of good ventilation and at least 1.2 m off the ground.

The instrument shelter in Figure 5.3 contains a *thermistor* that measures temperature by sensing the electrical resistance of a semiconducting material. Temperature is sensed because resistance changes at 4% per C°, which is reported electronically to the weather station.

Temperature readings are taken daily, sometimes hourly, at more than 16,000 weather stations worldwide. Some stations with recording equipment also report the duration of temperatures, rates of rise or fall, and temperature variation over time throughout the day and night. One of the goals of the Global Climate Observing System (GCOS, **http://www.wmo.ch/pages/prog/gcos/**) is

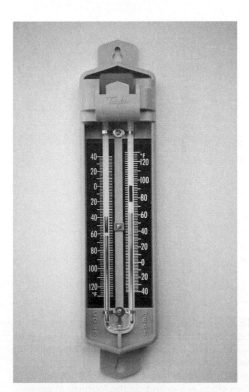

FIGURE 5.2 A minimum–maximum thermometer.
[Photo by Bobbé Christopherson.]

to establish a reference network of one station per 250,000 km² (95,800 mi²). (See the World Meteorological Organization at **http://www.wmo.ch/**.)

The *daily mean temperature* is an average of daily minimum–maximum readings. The *monthly mean temperature* is the total of daily mean temperatures for the month divided by the number of days in the month. An *annual temperature range* expresses the difference between the lowest and highest monthly mean temperatures for a given year.

Principal Temperature Controls

The interaction of complex physical controls produces Earth's temperature patterns. These principal influences upon temperatures include latitude, altitude and elevation, cloud cover, and land–water heating differences.

Latitude

Insolation is the single most important influence on temperature variations. Figure 2.9 shows how insolation intensity decreases as one moves away from the subsolar point—a point that migrates annually between the Tropic of Cancer and Tropic of Capricorn (between 23.5° N and 23.5° S). In addition, daylength and Sun angle change throughout the year, increasing seasonal effect with increasing latitude. The five cities graphed in Figure 5.4 demonstrate the effects of latitudinal position on temperature. From equator to poles, Earth ranges from continually warm, to seasonally variable, to continually cold.

Altitude/Elevation

Within the troposphere, temperatures decrease with increasing altitude above Earth's surface. (Recall that the *normal lapse rate* of temperature change with altitude is 6.4 C°/1000 m, or 3.5 F°/1000 ft; see Figure 3.6). The density of the atmosphere also diminishes with increasing altitude. In fact, the density of the atmosphere at an elevation of 5500 m (18,000 ft) is about half that at sea level. As the atmosphere thins, there is a loss in its ability to absorb and radiate sensible heat. Thus, worldwide, mountainous areas experience lower temperatures than do regions nearer sea level, even at similar latitudes. Note that "altitude" refers to airborne objects above Earth's surface, whereas "elevation" usually refers to the height of a point on Earth's surface above some plane of reference.

The consequences are that, at high elevations, average air temperatures are lower, nighttime cooling is greater, and the temperature range between day and night is greater than at low elevations. The temperature difference between areas of sunlight and shadow is greater than at sea level. You may have felt temperatures decrease noticeably in the shadows and shortly after sunset when you were in the mountains. Surfaces both gain energy rapidly and lose energy rapidly to the thinner atmosphere.

Also, at higher elevations, the insolation received is more intense because of the reduced mass of atmospheric gases. As a result of this intensity, the ultraviolet energy component makes sunburn a distinct hazard.

The snowline seen in mountain areas indicates where winter snowfall exceeds the amount of snow lost through

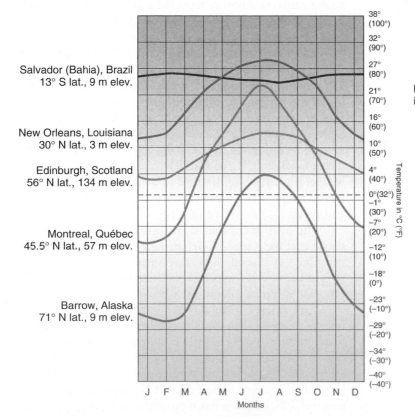

Located on maps in:
Figure 1.12 – latitudinal zone
Figure 4.17 – net radiation
Figure 4.18 – latent heat
Figure 4.19 – sensible heat

FIGURE 5.4 Latitude affects temperatures.
A comparison of five cities from near the equator to north of the Arctic Circle demonstrates changing seasonality and increasing differences between average minimum and maximum temperatures. These five cities are noted on the latitudinal zone map in Figure 1.12 and on three energy maps in Figures 4.17, 4.18, and 4.19.

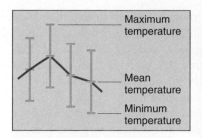

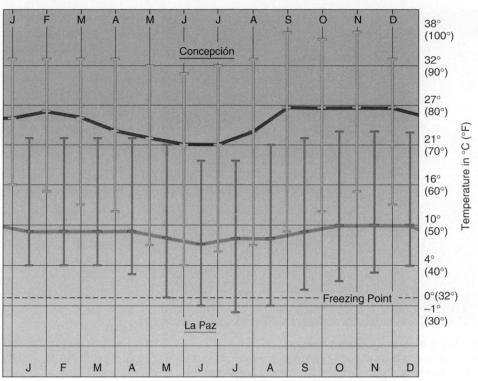

	Station	
	Concepción, Bolivia	**La Paz, Bolivia**
Latitude/longitude	16° 15′ S 62° 03′ W	16° 30′ S 68° 10′ W
Elevation	**490 m (1608 ft)**	**4103 m (13,461 ft)**
Avg. ann. temperature	24°C (75.2°F)	9°C (48.2°F)
Ann. temperature range	5 C° (9 F°)	3 C° (5.4 F°)
Ann. precipitation	121.2 cm (47.7 in.)	55.5 cm (21.9 in.)
Population	10,000	810,300 (Administrative division 1.6 million)

FIGURE 5.5 Effects of latitude and elevation.
Comparison of temperature patterns in two Bolivian cities: cooler, drier La Paz (4103 m; 13,461 ft) and warmer, wetter Concepción (490 m; 1608 ft).

summer melting and evaporation. The snowline's location is a function both of elevation and latitude, so glaciers can exist even at equatorial latitudes if the elevation is great enough. Permanent ice fields and glaciers exist on equatorial mountain summits in the Andes and East Africa. In equatorial mountains, the snowline occurs at approximately 5000 m (16,400 ft) because of the latitude (more insolation in the tropics produces temperatures that place the snowline at higher elevation). With increasing latitude, snowlines gradually lower in elevation from 2700 m (8850 ft) in the midlatitudes to lower than 900 m (2950 ft) in southern Greenland.

Two cities in Bolivia illustrate the interaction of the two temperature controls, latitude and elevation. Figure 5.5 displays temperature data for the cities of Concepción and La Paz, which are near the same latitude (about 16°S). Note the elevation, average annual temperature, and precipitation for each location noted on the figure.

The hot, humid climate of Concepción at its much lower elevation stands in marked contrast to the cool, dry climate of highland La Paz. People living around La Paz actually grow wheat, barley, and potatoes—crops characteristic of the cooler midlatitudes—despite the fact that La Paz is 4103 m (13,461 ft) above sea level (Figure 5.6). (For comparison, the summit of Pikes Peak in Colorado is at 4301 m, or 14,111 ft, and Mount Rainier, Washington, is at 4392 m, or 14,410 ft.)

The combination of elevation and low-latitude location guarantees La Paz nearly constant daylength and moderate temperatures, averaging about 9°C (48°F) every month. Such moderate temperature and moisture conditions lead to the formation of more fertile soils than those found in the warmer, wetter climate of Concepción, with an annual average temperature of 24°C (75°F).

Cloud Cover

Orbiting satellites reveal that approximately 50% of Earth is cloud covered at any given moment. Clouds moderate temperature, and their effect varies with cloud type, height, and density. Their moisture reflects, absorbs, and liberates large amounts of energy released upon condensation and cloud formation.

FIGURE 5.6 High-elevation farming.
People in these high-elevation villages grow potatoes and wheat in view of the permanent ice-covered peaks of the Andes Mountains. The combination of low latitude and high elevation creates consistent, moderate temperatures throughout the year, averaging about 9°C (48°F). [Photo by Mireille Vautier/Woodfin Camp & Associates.]

At night, clouds act as insulation and radiate long-wave energy, preventing rapid energy loss. During the day, clouds reflect insolation as a result of their high albedo values. In general, they lower daily maximum temperatures and raise nighttime minimum temperatures. In the last chapter, Figures 4.7 and 4.10 portrayed cloud-albedo forcing and cloud-greenhouse forcing as they relate to the presence of clouds and cloud types. Clouds also reduce latitudinal and seasonal temperature differences.

Clouds are the most variable factor influencing Earth's radiation budget, making them the subject of much simulation in computer models of atmospheric behavior. The International Satellite Cloud Climatology Project (**http://isccp.giss.nasa.gov/**), part of the World Climate Research Programme (**http://www.wmo.ch/pages/prog/wcrp/**), is presently in the midst of such research. Clouds and the Earth Radiant Energy System (CERES, **http://asd-www.larc.nasa.gov/ceres/**) sensors aboard the *TRMM*, *Terra*, and *Aqua* satellites are assessing cloud effects on longwave, shortwave, and net radiation patterns as never before possible.

Land–Water Heating Differences

Another major control over temperature is the different ways in which land and water respond to insolation. Earth presents these two surfaces in an irregular arrangement of continents and oceans. Each absorbs and stores energy differently than the other and therefore contributes to the global pattern of temperature. Water bodies tend to produce moderate temperature patterns, whereas inland in continental interiors more extreme temperatures occur.

The physical nature of land (rock and soil) and water (oceans, seas, and lakes) is the reason for **land–water heating differences**—land heats and cools faster than water. Figure 5.7 visually summarizes the following discussion of the five land–water temperature controls: evaporation, transparency, specific heat, movement and ocean currents, and sea-surface temperatures.

Evaporation Evaporation consumes more of the energy arriving at the ocean's surface than is expended over a comparable area of land, simply because so much water is available. An estimated 84% of all evaporation on Earth is from the oceans. When water evaporates and thus changes to water vapor, *heat energy is absorbed in the process and is stored in the water vapor as latent heat.* We saw this process in the map of energy expended for the latent heat of evaporation in Figure 4.18.

You experience this evaporative heat loss (cooling) by wetting the back of your hand and then blowing on the

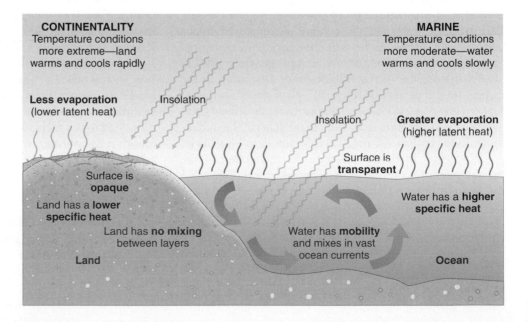

FIGURE 5.7 Land–water heating differences.
The differential heating of land and water produces contrasting marine (more moderate) and continental (more extreme) temperature regimes.

moist skin. Sensible heat energy is drawn from your skin to supply some of the energy for evaporation, and you feel the cooling. You probably appreciate the water "misters" at an outdoor market or café that lower temperatures. Similarly, as surface water evaporates, it absorbs substantial energy from the immediate environment, resulting in a lowering of temperatures. Temperatures over land, with far less water, are not as moderated by evaporative cooling as are marine locations.

Transparency   The transmission of light obviously differs between soil and water: Solid ground is opaque, water is transparent. Consequently, light striking a soil surface does not penetrate but is absorbed, heating the ground surface. That energy is accumulated during times of exposure and is rapidly lost at night or in shadows.

Figure 5.8 shows the profile of diurnal (daily) temperatures for a column of soil and the atmosphere above it at a midlatitude location. You can see that maximum and minimum temperatures generally are experienced right at ground level. Below the surface, even at shallow depths, temperatures remain about the same throughout the day. You encounter this at a beach, where surface sand may be

FIGURE 5.9 Ocean is transparent.
The transparency of the ocean at Taveuni Reef near Fiji permits insolation to penetrate to average depths of 60 m (200 ft), greatly increasing the volume of water that absorbs energy. [Photo by Copr. F. Stuart Westmorland/Photo Researchers, Inc.]

painfully hot to your feet, but as you dig in your toes and feel the sand a few centimeters below the surface, it is cooler, offering relief.

In contrast, when light reaches a body of water, it penetrates the surface because of water's **transparency**— water is clear and light transmits through it to an average depth of 60 m (200 ft) in the ocean (Figure 5.9). This illuminated zone is the *photic layer* and has been recorded in some ocean waters to depths of 300 m (1000 ft). This characteristic of water results in the distribution of available heat energy over a much greater depth and volume, forming a larger energy reservoir than that of the surface layers of the land.

Specific Heat   When comparing equal volumes of water and land, water requires far more energy to increase its temperature than does land. In other words, *water can hold more heat than can soil or rock*, and therefore water is said to have a higher **specific heat**, the heat capacity of a substance. On average, the specific heat of water is about 4 times that of soil. A given volume of water represents a more substantial energy reservoir than an equal volume of soil or rock, so changing the temperature of the oceanic

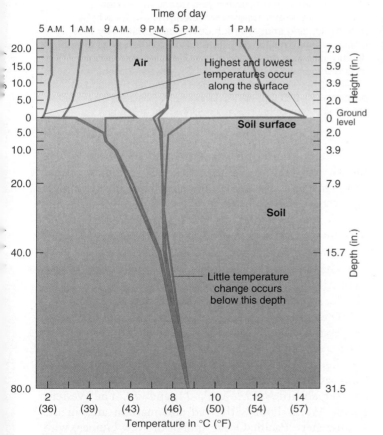

FIGURE 5.8 Land is opaque.
Profile measurements of air and soil temperatures in Seabrook, New Jersey. Note that little temperature change occurs throughout the year at depths beyond a meter, whereas the daily extremes of temperature register along the surface, where insolation is absorbed. [Adapted by permission from John R. Mather, *Climatology: Fundamentals and Applications* (New York: McGraw-Hill, 1974), p. 36.]

energy reservoir is a slower process than changing the temperature of land.

Likewise, for that oceanic heat reservoir to lose heat energy requires more time than would a similar volume of land. The temperature response of water bodies is "sluggish" in comparison with land surfaces. For this reason, day-to-day temperatures near a substantial body of water tend to be moderated.

Movement Land is a rigid, solid material, whereas water is a fluid and is capable of movement. Differing temperatures and currents result in a mixing of cooler and warmer waters, and that mixing spreads the available energy over an even greater volume than if the water were still. Surface water and deeper waters mix, redistributing energy. Both ocean and land surfaces radiate longwave radiation at night, but land loses its energy more rapidly than does the greater mass of the moving oceanic energy reservoir.

Ocean Currents and Sea-Surface Temperatures

Warm water adds energy to overlying air through high evaporation rates and transfers of latent heat. Thus, in an air mass, the amount of water vapor that ocean temperature affects forms an interesting *negative feedback* mechanism.

Across the globe, ocean water is rarely found warmer than 31°C (88°F), although hurricanes Katrina, Rita, and Wilma intensified as they moved over 33.3°C (92°F) sea-surface temperatures in the Gulf of Mexico in 2005. Higher ocean temperatures produce higher evaporation rates and more energy is lost from the ocean as latent heat. As the water vapor content of the overlying air mass increases, the ability of the air to absorb longwave radiation also increases. Therefore, the air mass becomes warmer. The warmer the air and the ocean become, the more evaporation will occur, increasing the amount of water vapor entering the air mass. More water vapor leads to cloud formation, which reflects insolation and produces lower temperatures. Lower temperatures of air and ocean reduce evaporation rates and the ability of the air mass to absorb water vapor.

As a specific example, the **Gulf Stream** moves northward off the east coast of North America, carrying warm water far into the North Atlantic (Figure 5.10). As a result, the southern third of Iceland experiences much milder temperatures than would be expected for a latitude of 65° N, just south of the Arctic Circle (66.5°). In Reykjavík, on the southwestern coast of Iceland, monthly temperatures average above freezing during all months of the year. Similarly, the Gulf Stream moderates temperatures in coastal Scandinavia and northwestern Europe. In the western Pacific Ocean, the Kuroshio, or Japan Current, similar to the Gulf Stream, functions much the same in its warming effect on Japan, the Aleutians, and the northwestern margin of North America.

In contrast, along midlatitude and subtropical west coasts, cool ocean currents flowing toward the equator influence air temperatures. When conditions in these regions are warm and moist, fog frequently forms in the chilled air over the cooler currents. Chapter 6 discusses ocean currents.

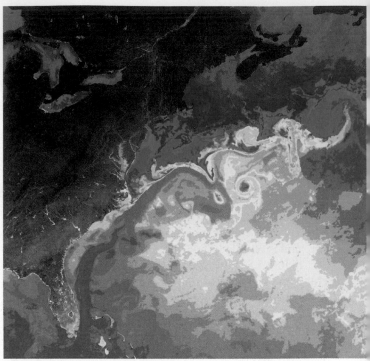

FIGURE 5.10 The Gulf Stream.
Satellite image of the warm Gulf Stream as it flows northward along the North American eastern coast. Instruments sensitive to thermal infrared wavelengths produced this remote-sensing image, which covers approximately 11.4 million km² (4.4 million mi²). Temperature differences are distinguished by computer-enhanced false coloration: reds/oranges = 25°–29°C (76°–84°F), yellows/greens = 17°–24°C (63°–75°F); blues = 10°–16°C (50°–61°F); and purples = 2°–9°C (36°–48°F). [Imagery by Rosenstiel School of Marine and Atmospheric Science, University of Miami.]

Remote sensing from satellites is providing sea-surface temperature (SST) data that are well correlated with actual sea-surface measurements. This correlation permits a thorough global assessment of SSTs in programs such as Tropical Ocean Global Atmosphere (TOGA) and Coupled Ocean-Atmosphere Response Experiment (COARE, **http://lwf.ncdc.noaa.gov/oa/coare/index.html**).

The Physical Oceanography Distributed Active Archive Center (PODAAC, **http://podaac.jpl.nasa.gov/**) is responsible for storage and distribution of data relevant to the physical state of the ocean. Sea-level height, currents, and ocean temperatures are measured as part of the effort to understand oceans and climate interactions.

Figure 5.11 displays SST data for February and July from satellites in the NOAA/NASA Pathfinder AVHRR data set, aboard *NOAA*-7, *-9*, *-11*, and *-14*. The Western Pacific Warm Pool is the red and maroon area in the southwestern Pacific Ocean (north of New Guinea) with temperatures above 30°C (86°F). This region has the highest average ocean temperatures in the world. Note the seasonal change in ocean temperatures; for example, compare the waters around Australia on the two images.

Mean annual SSTs increased steadily from 1982 through 2006 to record-high levels. Increasing warmth is measured at depths to 1000 m (3200 ft), and even slight

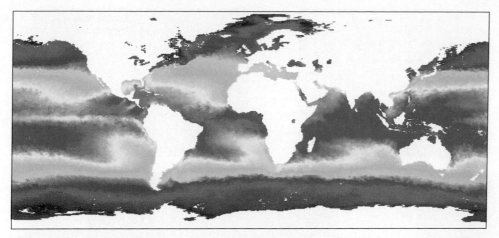

(a) February

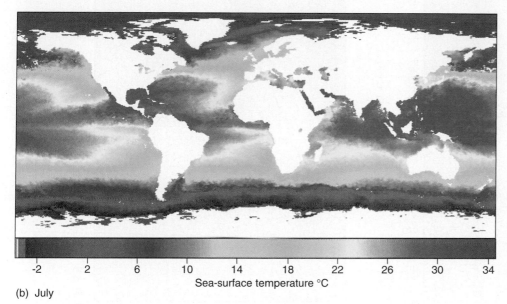

Global
Sea-surface
Temperatures

SATELLITE

FIGURE 5.11 Sea-surface
temperatures.
**Average annual sea-surface
temperatures for February (a) and
July (b) 1999 from satellites in the
Pathfinder AVHRR data set.**
[Satellite image data from the NASA
PODAAC at the Jet Propulsion
Laboratory, California Institute of
Technology.]

-2   2   6   10   14   18   22   26   30   34
Sea-surface temperature °C

(b) July

increases were found in the temperature of deep bottom water as reported in 2004. Scientists suggest that the ocean's ability to absorb excess heat energy from the atmosphere may be nearing its capacity.

Summary of Marine Effects vs. Continental Effects

As noted, Figure 5.7 summarizes the operation of the land–water temperature controls presented: evaporation, transparency, specific heat, movement, ocean currents, and sea-surface temperatures. The term **marine effect**, or *maritime*, describes locations that exhibit the moderating influences of the ocean, usually along coastlines or on islands. **Continental effect**, a condition of *continentality*, refers to areas less affected by the sea and therefore having a greater range between maximum and minimum temperatures on both a daily and yearly basis.

The Canadian cities of Vancouver, British Columbia, and Winnipeg, Manitoba, exemplify marine and continental conditions (Figure 5.12). Both cities are at approximately 49° N latitude. However, Vancouver has a more moderate pattern of average maximum and minimum temperatures. Vancouver's annual range of 16.0 C° (28.8 F°) is far less than Winnipeg's 38.0 C° (68.4 F°) range. In fact, Winnipeg's

continental temperature pattern is more extreme in every aspect than that of maritime Vancouver.

A similar comparison of San Francisco, California, and Wichita, Kansas, provides us another example of marine and continental conditions (Figure 5.13). Both cities are at approximately 37° 40′ N latitude. The cooling waters of the Pacific Ocean and San Francisco Bay surround San Francisco on three sides. Summer fog helps delay the warmest summer month in San Francisco until September. In 100 years of weather records, marine effects moderated temperatures so that only a few days a year have summer maximums that exceed 32.2°C (90°F). Winter minimums rarely drop below freezing.

In contrast, Wichita is susceptible to freezes from late October to mid-April, with daily variations slightly increased by its elevation, experiencing –30°C (–22°F) as a record low. January 2006 was 13 C° (23.4 F°) above normal when it hit an average of 6.2°C (43.2°F) for the month. West of Wichita, winters increase in severity with increasing distance from the moderating influences of invading air masses from the Gulf of Mexico. Wichita's temperature reaches 32.2°C (90°F) or higher more than 65 days each year, with 46°C (114°F) as a record high.

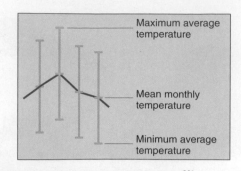

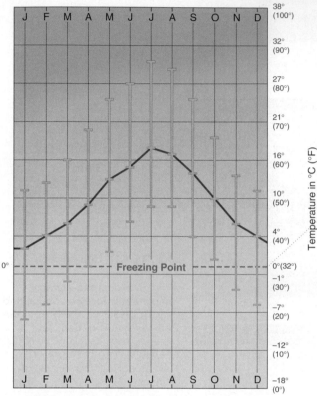

Station: Vancouver, British Columbia
Lat/long: 49° 11′ N 123° 10′ W
Avg. ann. temp.: 10°C (50°F)
Total ann. precip.:
   104.8 cm (41.3 in.)

Elevation: sea level
Population: 520,000
Ann. temp. range:
   16 C° (28.8 F°)

Station: Winnipeg, Manitoba
Lat/long: 49° 54′ N 97° 14′ W
Avg. ann. temp.: 2°C (35.6°F)
Total ann. precip.:
   51.7 cm (20.3 in.)

Elevation: 248 m (813.6 ft)
Population: 620,000
Ann. temp. range:
   38 C° (68.4 F°)

FIGURE 5.12 Marine and continental cities—Canada.
Comparison of temperatures in coastal Vancouver, British Columbia, and continental Winnipeg, **Manitoba.** [Vancouver waterfront photo by author; Winnipeg photo by John Eastcott/Yva Momatiak/The Image Works.]

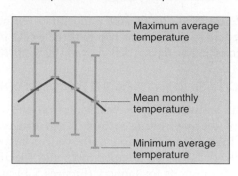

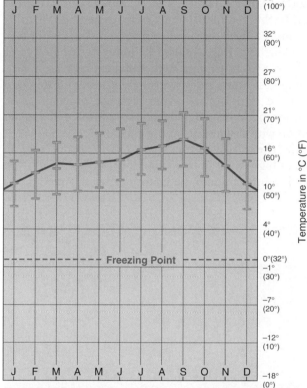

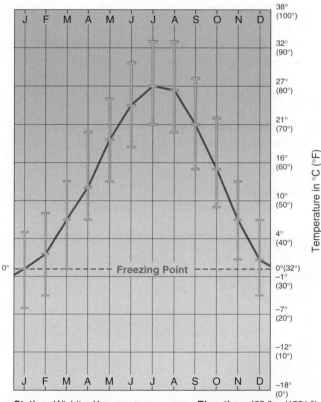

Station: San Francisco, California
Lat/long: 37° 37′ N 122° 23′ W
Avg. ann. temp.: 14.6°C (58.3°F)
Total ann. precip.:
 56.6 cm (22.3 in.)

Elevation: 5 m (16.4 ft)
Population: 777,000
Ann. temp. range:
 11.4 C° (20.5 F°)

Station: Wichita, Kansas
Lat/long: 37° 39′ N 97° 25′ W
Avg. ann. temp.: 13.7°C (56.6°F)
Total ann. precip.:
 72.2 cm (28.4 in.)

Elevation: 402.6 m (1321 ft)
Population: 327,000
Ann. temp. range:
 27 C° (48.6 F°)

FIGURE 5.13 Marine and continental cities—United States.
Comparison of temperatures in coastal San Francisco, California, and continental Wichita, Kansas. [San Francisco photo by Bobbé Christopherson; Wichita region photo by Philip Gould/Corbis.]

In 1980, 1990, 2000, 2005, and 2006 temperatures exceeded 38°C (100°F) for two to three weeks in a row—23 nonconsecutive days reached over 100°C in 2003.

Earth's Temperature Patterns

Earth's temperature patterns result from the combined effect of the controlling factors in our discussion. Let us now look at temperatures portrayed on maps that show worldwide mean air temperatures and specific maps for the polar regions for January (Figures 5.14 and 5.15) and July (Figures 5.17 and 5.18). To complete the analysis, Figure 5.19 presents the temperature range differences between the January and July maps, or the difference between averages of the coolest and warmest months.

Maps are for January and July instead of the solstice months of December and June because a lag occurs between insolation received and maximum or minimum temperatures experienced, as explained earlier in Chapter 4. The U.S. National Climate Data Center provided the data for these maps. Some ship reports go back to 1850 and land reports to 1890, although the bulk of the record is representative of conditions since 1950. In using these maps, remember that their small scale permits only generalizations about actual temperatures at specific locations.

The lines on temperature maps are known as *isotherms*. An **isotherm** is an isoline—a line along which there is a constant value—that connects points of equal temperature and portrays the temperature pattern, just as a contour line on a topographic map illustrates points of equal elevation. Isotherms help with the spatial analysis of temperatures.

January Temperature Maps

Figure 5.14 maps January's mean temperatures for the world. In the Southern Hemisphere, the higher Sun altitude causes longer days and summer weather conditions; in the Northern Hemisphere, the lower Sun angle causes the short days of winter. Isotherms generally are zonal, trending east–west, parallel to the equator, and they appear to be interrupted by the presence of landmasses. Isotherms mark the general decrease in insolation and net radiation with distance from the equator.

The **thermal equator** is an isotherm connecting all points of highest mean temperature, roughly 27°C (80°F); it trends southward into the interior of South America and

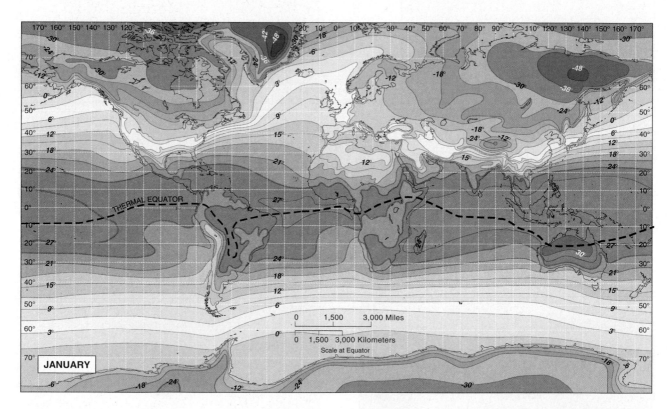

SATELLITE

Global Surface Temperatures, Land and Ocean

°F -50 -40 -30 -20 -10 0 10 20 30 40 50 60 70 80 90 100 110 120 °F

°C -40 -30 -20 -10 0 10 20 30 40 50 °C

FIGURE 5.14 Global mean temperatures for January. Temperatures are in Celsius (convertible to Fahrenheit by means of the scale) as taken from separate air-temperature databases for ocean and land. Note the inset map of North America and the equatorward-trending isotherms in the interior. Compare with **Figure 5.17.** [Adapted and redrawn from National Climatic Data Center, *Monthly Climatic Data for the World*, 47 (January 1994), in cooperation with the WMO and NOAA.]

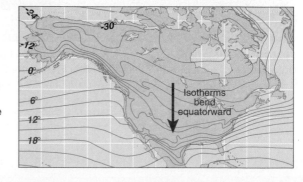

January

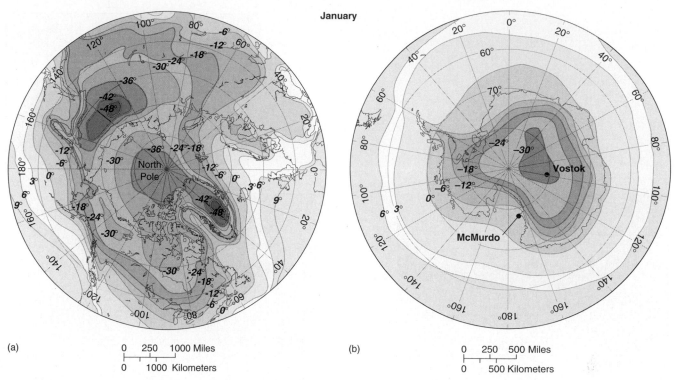

(a)

```
0   250   1000 Miles
0      1000 Kilometers
```

(b)

```
0   250   500 Miles
0       500 Kilometers
```

FIGURE 5.15 January mean temperatures for Polar Regions.
January temperatures in (a) North Polar region and (b) South Polar region. Use temperature conversions in Figure 5.14. Note that each map is at a different scale. [Author-prepared maps using same sources as Figures 5.14 and 5.17.]

Africa, indicating higher temperatures over landmasses. In the Northern Hemisphere, isotherms shift equatorward as cold air chills the continental interiors. The oceans, in contrast, are more moderate, with warmer conditions extending farther north than over land at comparable latitudes.

As an example, in Figure 5.14, follow along 50° N latitude (the 50th parallel) and compare isotherms: 3° to 6°C in the North Pacific and 3° to 9°C in the North Atlantic, as contrasted to –18°C in the interior of North America and –24°C to –30°C in central Asia. Also, note the orientation of isotherms over areas where there are mountain ranges and how they illustrate the cooling effects of elevation. Check the South American Andes as an example of these elevation effects.

The North Polar and South Polar regions (Figure 5.15a and b) are featured on two maps. Remember: the North Polar region is an ocean surrounded by land, in contrast to the South Polar region, which is the Antarctic continent surrounded by ocean. On the northern map (a), the island of Greenland has a summit elevation of 3240 m (10,630 ft) on Earth's second-largest ice sheet, the highest elevation north of the Arctic Circle. Two-thirds of the island is north of the Arctic Circle. Its north shore is only 800 km (500 mi) from the North Pole. This combination of high latitude and interior high elevation on the ice sheet produces cold midwinter temperatures.

In Antarctica (b), "summer" is underway in December. This is Earth's coldest and highest average elevation landmass. January average temperatures for the three scientific bases noted on the map are: McMurdo Station, at the coast on Ross Island, –3°C; the Admundsen–Scott

Station at the South Pole (elevation 2835 m, 9301 ft) –28°C; and the Russian Vostok Station at its more continental location (elevation 3420 m, 11,220 ft) –32°C; respectively, 26.6°F, –18.4°F, and –25.6°F.

For a continental region other than Antarctica, Russia is the coldest area on the map, specifically northeastern Siberia. The intense cold results from winter conditions of consistent clear, dry, calm air; small insolation input; and an inland location far from moderating maritime effects. Verkhoyansk, Russia (located within the –48°C isotherm on the map), actually recorded a minimum temperature of –68°C (–90°F) and experiences a daily average of –50.5°C (–58.9°F) for January. Verkhoyansk (Figure 5.16) has 7 months of temperatures below freezing, including at least 4 months below –34°C (–30°F).

In contrast, Verkhoyansk was hit with a maximum temperature of +37°C (+98°F) in July—an incredible 105°C (189 F°) min–max range! People do live and work in Verkhoyansk, which has a population of 1400; the town has been occupied continuously since 1638 and is today a secondary mining district.

Trondheim, Norway, is near the latitude of Verkhoyansk and at a similar elevation. But Trondheim's coastal location moderates its annual temperature regime (Figure 5.15). January minimum and maximum temperatures range between –17°C and +8°C (+1.4°F and +46°F), and the minimum–maximum range for July is from +5°C to +27°C (+41°F to +81°F). The lowest minimum and highest maximum temperatures ever recorded in Trondheim are –30°C and +35°C (–22°F and +95°F)—quite a difference from the extremes at Verkhoyansk.

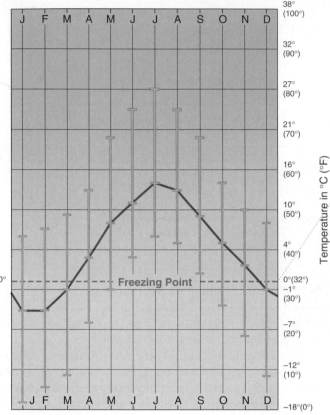

Station: Trondheim, Norway **Elevation:** 115 m (377.3 ft)
Lat/long: 63° 25′ N 10° 27′ E **Population:** 139,000
Avg. ann. temp.: 5°C (41°F) **Ann. temp. range:**
Total ann. precip.: 17 C° (30.6 F°)
  85.7 cm (33.7 in.)

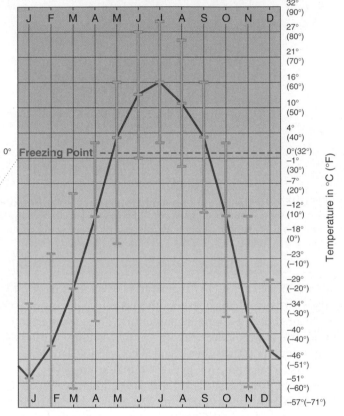

Station: Verkhoyansk, Russia **Elevation:** 137 m (449.5 ft)
Lat/long: 67° 35′ N 135° 23′ E **Population:** 1400
Avg. ann. temp.: −15°C (5°F) **Ann. temp. range:**
Total ann. precip.: 63 C° (113.4 F°)
  15.5 cm (6.1 in.)

FIGURE 5.16 Marine and continental cities—Eurasia.
Comparison of temperatures in coastal Trondheim, Norway, and continental Siberian Russia. Note that the freezing levels on the two graphs are positioned differently to accommodate the contrasting data.
[Trondheim photo by Norman Benton/Peter Arnold, Inc.; Verkhoyansk photo by TASS/Sovfoto/Eastfoto.]

July Temperature Map

Average July temperatures are presented in Figure 5.17 for the world. The longer days of summer and higher Sun altitude are in the Northern Hemisphere. Winter dominates the Southern Hemisphere, although it is milder than winters north of the equator because continental landmasses are smaller and the dominant oceans and seas store and release more energy. The *thermal equator* shifts northward with the high summer Sun and reaches the Persian Gulf–Pakistan–Iran area. The Persian Gulf is the site of the highest recorded sea-surface temperature of 36°C (96°F), difficult to imagine for a large water body.

During July in the Northern Hemisphere, isotherms shift poleward over land, as higher temperatures dominate continental interiors. July temperatures in Verkhoyansk average more than 13°C (56°F), which represents a 63 C° (113 F°) seasonal range between winter and summer averages. The Verkhoyansk region of Siberia is probably Earth's most dramatic example of continental effects on temperature.

The hottest places on Earth occur in Northern Hemisphere deserts during July. The reasons are simple: clear skies, strong surface heating, virtually no surface water, and few plants. Prime examples are portions of the Sonoran Desert of North America and the Sahara of Africa. Africa recorded a shade temperature higher than 58°C (136°F), a record set on September 13, 1922, at Al 'Azīzīyah, Libya (32° 32′ N, 112 m, or 367 ft elevation).

The highest maximum and annual average temperatures in North America occurred in Death Valley, California, where the Greenland Ranch Station reached 57°C (134°F) in 1913. The station is at 37° N and is –54.3 m, or

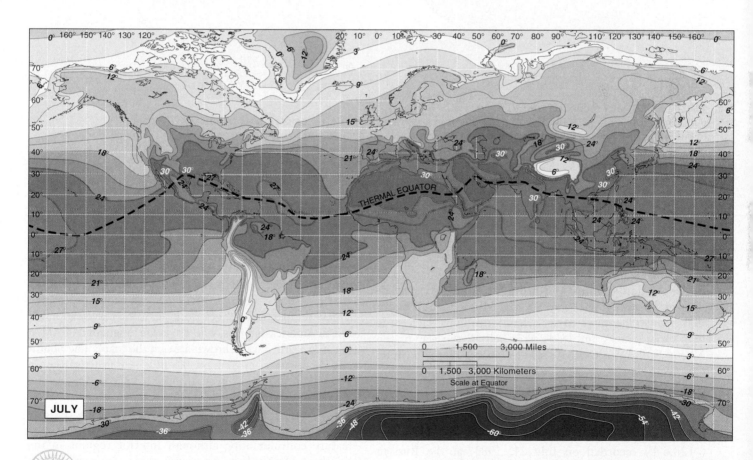

SATELLITE

Global Surface
Temperatures,
Land and Ocean

FIGURE 5.17 Global mean temperatures for July. Temperatures are in Celsius (convertible to Fahrenheit by means of the scale) as taken from separate air-temperature databases for ocean and land. Note the inset map of North America and the poleward-trending isotherms in the interior. Compare with Figure 5.14. [Adapted and redrawn from National Climatic Data Center, *Monthly Climatic Data for the World*, 47 (July 1994), in cooperation with the WMO and NOAA.]

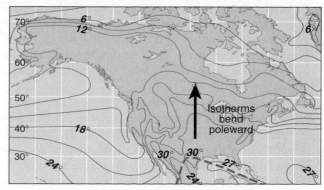

July

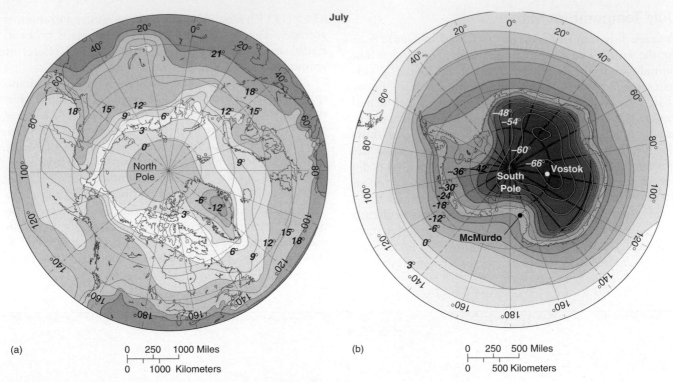

(a)

```
0    250  1000 Miles
0         1000  Kilometers
```

(b)

```
0    250   500 Miles
0          500 Kilometers
```

FIGURE 5.18 July mean temperatures for Polar Regions.
July temperatures in (a) North Polar region and (b) South Polar region. Use temperature conversions in Figure 5.17. Note that each map is at a different scale. [Author-prepared maps using same sources as Figures 5.14 and 5.17.]

−178 ft, below sea level. Chapter 15 discusses such hot, arid lands.

The North Polar and South Polar regions (Figure 5.18a and b) are featured on two maps. July is "summer" in the Arctic Ocean (a) and brings a thinning of permanent and seasonal pack ice. See the Midnight Sun photo that opens Chapter 2 to see what continuous daylight is like. Open cracks and leads stretched all the way to the North Pole over the past several years. High Latitude Connection 5.1 presents an overview of temperature trends in the polar regions.

Figure 5.18 shows when nights in Antarctica are 24 hours long. This lack of insolation results in the lowest natural temperature reported on Earth, a frigid −89.2°C (−128.6°F) recorded on July 21, 1983, at the Russian Vostok Station, Antarctica (noted on the map). Such a temperature is 11 C° (19.8 F°) colder than the freezing point of dry ice (solid carbon dioxide)! If the concentration of carbon dioxide were large enough, such a cold temperature would theoretically freeze tiny carbon dioxide dry-ice particles out of the sky.

The average July temperature around Vostok is −68°C (−90.4°F). For average temperature comparisons, the Admundsen–Scott Station at the South Pole experiences −60°C (−76°F), and McMurdo Station has a −26°C (−14.8°F) reading. Note that the coldest temperatures in Antarctica are usually in August, not July, just before the equinox sunrise in September at the end of the long polar night.

Annual Temperature Range Map

The temperature range map helps identify areas that experience the greatest annual extremes (continental locations) and the most moderate temperature regimes (marine locations), as demonstrated in Figure 5.19. As you might expect, the largest temperature ranges occur in subpolar locations in North America and Asia, where average ranges of 64 C° (115 F°) are recorded (dark brown area on the map). The Southern Hemisphere, in contrast, has little seasonal variation in mean temperatures, owing to the lack of large landmasses and the vast expanses of water to moderate temperature extremes.

Southern Hemisphere temperature patterns are generally maritime and Northern Hemisphere patterns feature continentality; although interior regions in the Southern Hemisphere experience some continentality effects. For example, in January (Figure 5.14), Australia is dominated by isotherms of 20°–30°C (68°–86°F), whereas in July (Figure 5.17), Australia is crossed by the 12°C (54°F) isotherm. The Northern Hemisphere, with greater land area overall, registers a slightly higher average surface temperature than does the Southern Hemisphere.

Imagine living in some of these regions and the degree to which you would need to adjust your personal wardrobe and make other comfort adaptations. See Focus Study 5.1 for more on air temperature and the human body. The recent summer heat-wave deaths remain a powerful reminder of the role of temperatures in our lives.

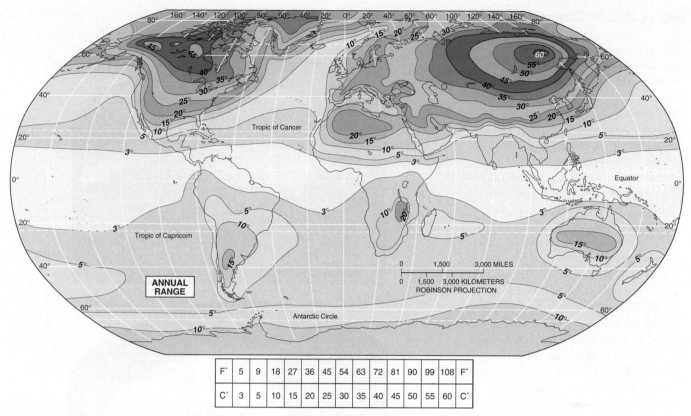

FIGURE 5.19 Global annual temperature ranges.
A generalized portrait of the annual range of global temperatures in Celsius degrees, C°, with conversions to Fahrenheit degrees, F°, shown on scale. The mapped data show the difference between January and July temperature means.

F°	5	9	18	27	36	45	54	63	72	81	90	99	108	F°
C°	3	5	10	15	20	25	30	35	40	45	50	55	60	C°

High Latitude Connection 5.1

Overview of Trends in the Polar Regions

Climate models used by scientists to predict and direct measurements show that the polar regions are experiencing some dramatic global climate-change impacts. These changes lead some to say that the high latitudes are providing the global community an early warning signal. The question is, how do we read changing temperature signals, how do they relate to natural circulation cycles, and what are the spatial implications to the rest of the planet?

Arctic

Many proxy and direct measurements confirm that the twentieth century was a time of increasing temperatures at both middle and low latitudes and in the polar regions, especially during the last half century. Daily maximum and minimum temperatures show there is warming in the Arctic at a rate of about 5 C° (9 F°) per century. Since 1978, warming increased to a rate of 1.2 C° (2.2 F°) per decade, which means the last 20 years warmed at nearly 7 times the rate of the last 100 years.

In response to rising air and water temperatures, Arctic sea ice is on a significant decline. The sea ice is melting at a rate of about 9% per decade since 1978, especially in the eastern and central Arctic Ocean. Atmospheric pressure patterns cause sea ice to shift as well. There have been record losses since 2000, as sea ice reached its smallest spatial distribution in the satellite record. Figure 5.1.1 is a composite satellite image for 1979 and 2007 comparing sea-ice coverage in the Arctic. September 14, 2007 marked the lowest sea-ice extent on record during the era of satellite surveillance. The sea ice thins as warmer air melts the surface and warmer waters attack the base of the ice. The summer melt period is increasing as spring arrives earlier and fall is delayed. Since 1970, about 43% of the Arctic sea ice has disappeared. More open water means darker surfaces that absorb more insolation that leads to more warming.

Chapters 1 and 17 discuss the 400% increase in meltponds on the surface of the Greenland ice sheet, ice shelves, and glaciers. This increase represents a significant positive feedback mechanism. In response to the increased discharge of freshwater in the Arctic,

(*continued*)

High Latitude Connection 5.1 (continued)

the northern oceans are freshening—actually, lowering in salinity. There is valid concern that these meltwaters riding atop saline water can disrupt important circulation systems in the northern oceans.

In Alaska, temperatures rose over the past 30 years at a 3 C° (5 F°) per decade rate. All 11 mountain ranges experienced glacial ice losses. Glacial thinning reached 1.8 m per year by the late 1990s across Alaska. In response to this unloading of ice mass, Earth's crust is rebounding upward. A News Report in Chapter 17 explains the negative net mass balance of 67 Alaskan glaciers.

All of these changes are affecting living systems as specific high-latitude ecosystems shift across the landscape poleward at 6.1 km (3.8 mi) per decade and spring arrives earlier at a pace of 2.3 days per decade. Changes in tundra soils and the increased depth of active-layer melting in permafrost indicate regional warming, as discussed in Chapter 17. These thawing lands provide less nutrition to caribou and other grazing animals, worsen forest fire conditions, and are an additional source of carbon dioxide to the atmosphere.

By 2080, the Intergovernmental Panel on Climate Change (IPCC) forecasts increasing warmth in the Arctic ranging from 4.0° to 7.5 C° in summer and 2.5° to 14.0 C° in winter over land, and 0.5° to 4.5 C° in summer and 3.0° to 16.0 C° in winter over the Arctic Ocean. With these temperature changes precipitation will increase since warmer air can absorb more moisture.

Antarctica

In Antarctica and the Southern Ocean, the temperature records of the permanently occupied research stations over the last 45 years demonstrate an overall warming. This record includes periods of slight cooling within this time frame. The present warming rate is less than that of the Arctic, at about a 0.9° to 1.2 C° increase per century.

However, along the Antarctic Peninsula, the location of numerous

1979

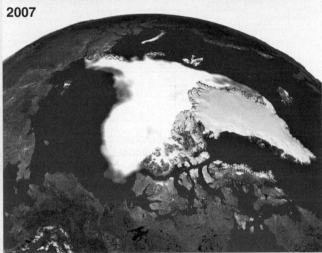

2007

FIGURE 5.1.1 Loss of Arctic sea ice.
A composite of satellite data and surface data from 1979 and 2007 shows remarkable loss of Arctic sea ice. Each composite is a portrait of minimum sea-ice concentrations during summer months. For scale, in just 2 years between the 2005 record low and the new 2007 record, the amount of sea-ice loss exceeded the area of California in size (423,950 km²; 163,700 mi²), or 24% below 2005. Not only did the atmosphere record increased warmth but the Arctic Ocean warmed as well. In addition to sea-ice losses, changing pressure patterns cause sea ice to shift. [Images courtesy of Scientific Visualization Studio, GSFC/NASA.]

ice-shelf collapses and disintegrations, the rate of warming is more than twice this rate, as the January 0° isotherm continues to shift southward. The exposed land is open to exotic plants and animals that move into these altered ecosystems. In terms of sea ice, the Antarctic trend is not as pronounced as that of the Arctic, with sea-ice extent reduced by 0.4°–1.8° of latitude over the last century. Continued changes will affect marine life and birds.

By 2080, the IPCC forecasts increasing warmth in the Antarctic

ranging from 1.0° to 4.8 C° in summer and 1.0° to 5.0 C° in winter over the continent; and 0.0° to 2.8 C° in summer and 0.5° to 5.0 C° in winter over the Southern Ocean.

For the latest science on both polar regions, please see: "Polar Regions (Arctic and Antarctic)" in *Climate Change 2007: Climate Change Impacts, Adaptation, and Vulnerability,*" Working Group II Contribution to the Fourth Assessment Report of the Intergovernmental Panel on Climate—the web site is http://www.ipcc.ch/.

Focus Study 5.1

Air Temperature and the Human Body

We describe our perception of temperature with the term *apparent temperature*, or *sensible temperature*. This perception varies among individuals and cultures. Through complex mechanisms, our bodies maintain an average internal temperature at about 36.8°C (98.2°F),[*] slightly lower in the morning or in cold weather and slightly higher at emotional times or during exercise and work.

Taken together, the water vapor content of air, wind speed, and air temperature affect each individual's sense of comfort. The most discomfort comes with high temperatures, high humidity, and low winds. Modern heating and cooling systems, where available, reduce the impact of extreme temperatures indoors. Deaths still happen, especially among groups of people who can't afford to alter temperatures in their living spaces.

When there are changes in the surrounding air, the human body reacts to maintain its core temperature and to protect the brain at all costs. Table 5.1.1 summarizes the human body's response to stress induced by exposure to *hypothermia* (low temperature) and *hyperthermia* (high temperature). Carefully review these two lists of responses for future reference.

Wind Chill

The wind-chill index is important to people who experience winters with freezing temperatures and was first proposed by Antarctic explorer Paul Siple in 1939. The *wind-chill factor* indicates the enhanced rate at which body heat

[*]The traditional value for "normal" body temperature, 37°C (98.6°F), was set in 1868 using old methods of measurement. According to Dr. Philip Mackowiak of the University of Maryland School of Medicine, a more accurate modern assessment places normal at 36.8°C (98.2°F), with a range of 2.7 C°(4.8 F°) for the human population (*Journal of the American Medical Association*, September 23–30, 1992).

Table 5.1.1   The Human Body's Response to Temperature Stress

At Low Temperatures (Heat-gaining mechanism)	At High Temperatures (Heat-dissipation mechanism)
Temperature-Regulation Methods	
Constriction of surface blood vessels	Dilation of surface blood vessels
Concentration of blood	Dilution of blood
Flexing to reduce surface exposure	Extending to increase exposure
Increased muscle tone	Decreased muscle tone
Decrease in sweating	Sweating
Inclination to increase activity	Inclination to decrease activity
Shivering	
Increased cell metabolism	
Consequent Disturbances	
Increased urine volume	Decreased urine volume
Danger of inadequate blood supply to exposed parts; frostbite	Reduced blood supply to brain; dizziness, nausea, fainting
Discomfort leading to neuroses	Discomfort leading to neuroses
Increased appetite	Decreased appetite
	Mobilization of tissue fluid
	Thirst and dehydration
	Reduced chloride balance; heat cramps
Failure of Regulation	
Falling body temperature	Rising body temperature
Drowsiness	Impaired heat-regulating center
Cessation of heartbeat and respiration	Failure of nervous regulation; cessation of breathing

Source: Climatology—Arid Zone Research X. © UNESCO. Reproduced by permission of UNESCO.

is lost to the air and at best represents an estimate of heat-energy loss. As wind speeds increase, heat loss from the skin increases. A formula for calculating these relationships was developed in 1945. This older index tended to overestimate heat loss from skin.

The National Weather Service (NWS, **http://www.nws.noaa.gov/om/windchill/**) and the Meteorological Services of Canada (MSC, **http://www.msc.ec.gc.ca/education/windchill/index_e.cfm**) revised the wind-chill formula for the 2001–2002 winter season. The new *Wind Chill*

Temperature Index (WCT) is an effort to improve the accuracy of heat-loss calculations. Figure 5.1.1 presents the new version that began service in November 2001.

For example, using the new chart, if the air temperature is –7°C (20°F) and the wind is blowing at 32 kmph (20 mph), skin temperatures will be at –16°C (4°F). The lower wind-chill values present a serious freezing hazard to exposed flesh. Imagine the wind chill experienced by a downhill ski racer going 130 kmph (80 mph)—frostbite is a definite

(continued)

Focus Study 5.1 (continued)

Actual Air Temperature in °C (°F)

Wind speed, kmph (mph) Calm	4° (40°)	−1° (30°)	−7° (20°)	−12° (10°)	−18° (0°)	−23° (−10°)	−29° (−20°)	−34° (−30°)	−40° (−40°)
8 (5)	2° (36°)	−4° (25°)	−11° (13°)	−17° (1°)	−24° (−11°)	−30° (−22°)	−37° (−34°)	−43° (−46°)	−49° (−57°)
16 (10)	1° (34°)	−6° (21°)	−13° (9°)	−20° (−4°)	−27° (−16°)	−33° (−28°)	−41° (−41°)	−47° (−53°)	−54° (−66°)
24 (15)	0° (32°)	−7° (19°)	−14° (6°)	−22° (−7°)	−28° (−19°)	−36° (−32°)	−43° (−45°)	−50° (−58°)	−57° (−71°)
32 (20)	−1° (30°)	−8° (17°)	−16° (4°)	−23° (−9°)	−30° (−22°)	−37° (−35°)	−44° (−48°)	−52° (−61°)	−59° (−74°)
40 (25)	−2° (29°)	−9° (16°)	−16° (3°)	−24° (−11°)	−31° (−24°)	−38° (−37°)	−46° (−51°)	−53° (−64°)	−61° (−78°)
48 (30)	−2° (28°)	−9° (15°)	−17° (−1°)	−24° (−12°)	−32° (−26°)	−39° (−39°)	−47° (−53°)	−55° (−67°)	−62° (−80°)
56 (35)	−2° (28°)	−10° (14°)	−18° (0°)	−26° (−14°)	−33° (−27°)	−41° (−41°)	−48° (−55°)	−56° (−69°)	−63° (−82°)
64 (40)	−3° (27°)	−11° (13°)	−18° (−1°)	−26° (−15°)	−34° (−29°)	−42° (−43°)	−49° (−57°)	−57° (−71°)	−64° (−84°)
72 (45)	−3° (26°)	−11° (12°)	−19° (−2°)	−27° (−16°)	−34° (−30°)	−42° (−44°)	−50° (−58°)	−58° (−72°)	−66° (−86°)
80 (50)	−3° (26°)	−11° (12°)	−19° (−3°)	−27° (−17°)	−35° (−31°)	−43° (−45°)	−51° (−60°)	−59° (−74°)	−67° (−88°)

Frostbite times: 30 min.   10 min.   5 min.

FIGURE 5.1.1 Wind Chill Temperature Index.
WCT Index factor for various temperatures and wind speeds. For quick calculations, see
http://www.srh.noaa.gov/elp/wxcalc/windchill.html (in English units the new formula is: Wind chill
$(F°) = 35.74 + 0.6125T − 35.75(V^{0.16}) + 0.4275T(V^{0.16})$, where T = air temperature in °F, V = wind speed in
mph. In Canada, heat loss is expressed in watts per square meter.) [Adapted from the National
Weather Service and Meteorological Services of Canada, version 11/01/01.]

possibility during a 2-minute run. The wind chill does omit consideration of sunlight intensity, a person's physical activity, and the use of protective clothing, such as a windbreaker, that prevents the wind from gaining access to your skin.

Heat Index

The *heat index* (*HI*) indicates the human body's reaction to the combination of air temperature and water vapor. The heat index, combining the effects of temperature and humidity, indicates how hot the air feels to an average person. Water vapor in air is expressed as relative humidity, a concept presented in Chapter 7. For now, it is sufficient to say that the amount of water vapor in the air affects the evaporation rate of perspiration on the skin; a lower evaporation rate means

less evaporative cooling of the skin and increasing discomfort. We simply feel stuffy in warm, humid air.

Figure 5.1.2 is an abbreviated version of the heat index that the NWS now includes in its daily weather summaries during summer months. The table beneath the graph describes the effects of heat-index categories on higher-risk groups. A combination of high temperature and high humidity can severely reduce the body's natural ability to regulate internal temperature (see http://www.nws.noaa.gov/om/heat/index.shtml); a heat-index calculator is at http://www.crh.noaa.gov/jkl/?n=heat_index_calculator.

Summer Heat-Index Deaths In an average year fewer than 200 people die in the United States from heat-related

causes, totaling nearly 20,000 people since the mid-1930s. With air temperatures on the increase, more years are surpassing the average. For nearly a week during July 1995, Chicago's heat-index values went to category I in dwellings that lacked air-conditioning. Chicago had never experienced a 48-hour period during which temperatures did not go below 31.5°C (89°F).

With high pressure (hot, stable air) dominating the Midwest to the Atlantic and moist air from the Gulf of Mexico, all the ingredients for this disaster were present. Afternoon temperatures were above 32°C (90°F) for a week in Chicago, 38°C (100°F) in South Dakota, and 40°C (104°F) in Toledo, Ohio, and New York City; the temperature hit 43°C (109°F) in Omaha, Nebraska. Cities in New England that had reached 37.7°C

(100°F) only twice in their history exceeded that for 4 or 5 days during July 1995.

This combination of high temperatures and water vapor in the air produced stifling conditions for everyone, the sick and elderly in particular. Some apartments without air-conditioning exceeded indoor heat-index temperatures of 54°C (130°F) when the official temperature at Midway Airport reached a record 41°C (106°F). The heat-index death toll totaled 700 people in Chicago. Nearly 1000 people died overall in the Midwest and East from these conditions.

Europe was paralyzed in the summer of 2003 with high heat-index conditions when temperatures topped 40°C (104°F) in June, July, and August. Official estimates across the continent placed the death count at more than 20,000, but this triggered a great undercount controversy, as governments were ill-prepared to adequately deal with such a catastrophe.

France was hardest hit, with more than 14,000 deaths, some 11,000 in the first two weeks of August alone. Although the heat wave began August 1, a full emergency response was not begun until August 13—the day temperatures began to fall.

The rush for medical help in the Midwest and in Europe quickly swamped hospitals, which had to turn away hundreds of patients. An added problem in France was that many doctors, nurses, and fellow citizens were out of town as this was the beginning of the typical holiday season.

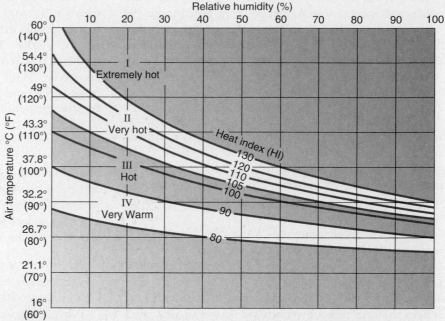

Level of concern	Category	Heat Index Apparent Temperature	General Effect of Heat Index on People in High-Risk Groups
Extreme danger	I	54°C (130°F) or higher	Heat/sunstroke highly likely with continued exposure
Danger	II	41° – 54°C (105° – 130°F)	Sunstroke, heat cramps, or heat exhaustion likely and heatstroke possible with prolonged exposure and/or physical activity
Extreme caution	III	32° – 41°C (90° – 105°F)	Sunstroke, heat cramps, and heat exhaustion possible with prolonged exposure and/or physical activity
Caution	IV	27° – 32°C (80° – 90°F)	Fatigue possible with prolonged exposure and/or physical activity

FIGURE 5.1.2 Heat index.
Heat-index graph for various temperatures and relative humidity levels. A heat-index calculator is at http://www.crh.noaa.gov/jkl/?n=heat_index_calculator. [Courtesy of the National Weather Service.]

In 2005, for four days in the aftermath of Hurricane Katrina, some 100,000 stranded and homeless New Orleans and Gulf Coast residents suffered from heat-index values exceeding 38°C (100°F), made worse by a lack of drinking water and medical aid—no one will ever know how many perished simply from the heat.

Record Temperatures and Greenhouse Warming

Future humans will experience greater temperature-related challenges owing to complex changes now under way in the lower atmosphere. Scientists agree that human activities, principally the burning of fossil fuels, are increasing greenhouse gases that absorb longwave radiation, delaying losses of heat energy to space. We are experiencing human-forced climate change, as Earth's natural greenhouse is enhanced.

The Dome C ice core taken from the East Antarctic Plateau shows that present levels of carbon dioxide, methane, and nitrous oxides, and resulting air temperatures, are higher in the present decade than in the entire 800,000-year record!

These are remarkable times in which we live. Twelve of the last 13 years, 1995–2007, rank among the warmest years in instrumental records stretching back to 1850.

The Intergovernmental Panel on Climate Change (IPCC), through major reports in 1990, 1992, 1995, 2001, and its *Fourth Assessment Report* (*AR4*) in 2007, confirms that global warming is occurring. The IPCC *AR4* reached a consensus, in part stating:

Warming of the climate system is unequivocal, as is now evident from observations of increases in global average air and ocean temperatures, widespread melting of snow and ice, and rising global average sea level.... Most of the observed increase in globally averaged temperatures since the mid-20th century is *very likely* [>90%] due to the observed increases in anthropogenic greenhouse gas concentrations...human influences now extend to other aspects of climate, including ocean warming, continental-average temperatures, temperature extremes and wind patterns...and, *very likely* (<90%) that this is not due to known natural causes alone.*

The Arctic Climate Impact Assessment (ACIA) reported confirmation of conditions in the Arctic in their science report released in late 2004. ACIA stated, "The Arctic is extremely vulnerable to observed and projected climate change and its impacts. The Arctic is now experiencing some of the most rapid and severe climate change on Earth."

Our own Association of American Geographers (AAG) at the 2006 Chicago meeting passed the "AAG Resolution Requesting Action on Climate Change." This resolution places the AAG in line with other professional organizations and national academies of sciences that are now on record for action to slow rates of climate change.

*Intergovernmental Panel on Climate Change (IPCC) *Fourth Assessment Report* (2007), Working Group I, *The Physical Science Basis*, "Summary for Policy Makers," (February 5, 2007), IPCC Secretariat, Geneva, Switzerland, pp. 5, 10. See: http://www.ipcc.ch/ and download the three Working Group summaries free.

ANIMATION   Global Warming, Climate Change

An array of sophisticated satellites, remote-sensing capabilities, and powerful computers equip scientists to run global circulation models and decipher the climate trends. Actual field measurements and observations confirm these remotely sensed conclusions and a scientific consensus. With the cause being the exhaust from our fossil-fuel consumption and our modern technological society, the political fallout and debate is intense.

Chapter 10 discusses global warming and climate change and Chapter 17 the vulnerable polar regions and the ice-core record of climate. We address what is known and unknown, how global models work, the radiatively active gases that are forcing the warming, the consequences to natural systems, the uncertainties, and what international action is underway to slow the effects of such change. We should be aware of what scientists are discovering:

> By the end of the 21st century, large portions of Earth's surface may experience climates not found at present, and some 20th century climates may disappear.... Novel climates are projected to develop primarily in the tropics and subtropics.[†]

[†]Jack Williams, et al., "Projected distributions of novel and disappearing climates by A.D. 2100," *Proceedings of the National Academy of Sciences*, April 3, 2007: 5738.

NOTEBOOK   GISS Surface Temperature Analysis, 1891 to 2006 (Chapter 10)

Summary and Review—Global Temperatures

■ *Define* the concepts of temperature, kinetic energy, and sensible heat, and *distinguish* among Kelvin, Celsius, and Fahrenheit temperature scales and how they are measured.

Temperature is a measure of the average kinetic energy (motion) of individual molecules in matter. We feel the effect of temperature as the sensible heat transfer from warmer objects to cooler objects when these objects are touching. Temperature scales include

- Kelvin scale: 100 units between the melting point of ice (273 K) and the boiling point of water (373 K).
- Celsius scale: 100 degrees between the melting point of ice (0°C) and the boiling point of water (100°C).
- Fahrenheit scale: 180 degrees between the melting point of ice (32°F) and the boiling point of water (212°F).

The Kelvin scale is used in scientific research because temperature readings start at absolute zero and thus are proportional to the actual kinetic energy in a material.

temperature (p. 116)

1. Distinguish between sensible heat and sensible temperature.

2. What does air temperature indicate about energy in the atmosphere?
3. Compare the three scales that express temperature. Find "normal" human body temperature on each scale and record the three values in your notes.
4. What is your source of daily temperature information? Describe the highest temperature you have experienced and the lowest temperature. From what we have discussed in this chapter, can you identify the factors that may have contributed to these temperatures?

■ *List* and *review* the principal controls and influences that produce global temperature patterns.

Principal controls and influences upon temperature patterns include *latitude* (the distance north or south of the equator), *altitude* and *elevation*, *cloud cover* (reflection, absorbtion, and re-radiation of energy), and *land–water heating differences* (the nature of evaporation, transparency, specific heat, movement, and ocean currents and sea-surface temperatures).

5. Explain the effect of altitude/elevation on air temperature. Why is air at higher altitude/elevation lower in temperature? Why does it feel cooler standing in shadows at higher elevation than at lower elevation?

6. What noticeable effect does air density have on the absorption and radiation of energy? What role does elevation play in that process?

7. How is it possible to grow moderate-climate-type crops such as wheat, barley, and potatoes at an elevation of 4103 m (13,460 ft) near La Paz, Bolivia, so near the equator?

8. Describe the effect of cloud cover with regard to Earth's temperature patterns. From the last chapter, review the cloud-albedo forcing and cloud-greenhouse forcing of different cloud types and relate the concepts with a simple sketch.

■ *Review* the factors that produce different marine effects and continental effects as they influence temperatures, and *utilize* several pairs of cities to illustrate these differences.

The physical nature of land (rock and soil) and water (oceans, seas, and lakes) is the reason for **land–water heating differences**, the fact that land heats and cools faster than water. Moderate temperature patterns are associated with water bodies, and extreme temperatures occur inland. The controls that cause differences between land and water surfaces are *evaporation*, *transparency*, *specific heat*, *movement*, and *ocean currents and sea-surface temperatures*.

Light penetrates water because of its **transparency**. Water is clear and light transmits to an average depth of 60 m (200 ft) in the ocean. This penetration distributes available heat energy over a much greater volume than could occur on opaque land; thus, a larger energy reservoir is formed. When equal volumes of water and land are compared, water requires far more energy to increase its temperature than does land. In other words, water can hold more energy than can soil or rock, so water has a higher **specific heat**, the heat capacity of a substance, averaging about 4 times that of soil.

Ocean currents affect temperature. An example of the effect of ocean currents is the **Gulf Stream**, which moves northward off the east coast of North America, carrying warm water far into the North Atlantic. As a result, the southern third of Iceland experiences much milder temperatures than would be expected for a latitude of 65° N, just below the Arctic Circle (66.5°).

Marine effect, or maritime, describes locations that exhibit the moderating influences of the ocean, usually along coastlines or on islands. **Continental effect** refers to the condition of areas that are less affected by the sea and therefore have a greater range between maximum and minimum temperatures diurnally and yearly.

land–water heating differences (p. 120)
transparency (p. 121)
specific heat (p. 121)
Gulf Stream (p. 122)
marine effect (p. 123)
continental effect (p. 123)

9. List the physical aspects of land and water that produce their different responses to heating from absorption of insolation. What is the specific effect of transparency in a medium?

10. What is specific heat? Compare the specific heat of water and soil.

11. Describe the pattern of sea-surface temperatures (SSTs) as determined by satellite remote sensing. Where is the warmest ocean region on Earth?

12. What effect does sea-surface temperature have on air temperature? Describe the negative feedback mechanism created by higher sea-surface temperatures and evaporation rates.

13. Differentiate between marine and continental temperatures. Give geographic examples of each from the text: Canada, the United States, Norway, and Russia.

■ *Interpret* the pattern of Earth's temperatures from their portrayal on January and July temperature maps and on a map of annual temperature ranges.

Maps for January and July instead of the solstice months of December and June are used for temperature comparison because of the natural lag that occurs between insolation received and maximum or minimum temperatures experienced. Each line on these temperature maps is an **isotherm**, an isoline that connects points of equal temperature. Isotherms portray temperature patterns.

Isotherms generally are zonal, trending east–west, parallel to the equator. They mark the general decrease in insolation and net radiation with distance from the equator. The **thermal equator** (isoline connecting all points of highest mean temperature) trends southward in January and shifts northward with the high summer Sun in July. In January it extends farther south into the interior of South America and Africa, indicating higher temperatures over landmasses.

In the Northern Hemisphere in January, isotherms shift equatorward as cold air chills the continental interiors. The coldest area on the map is in Russia, specifically northeastern Siberia. The intense cold experienced there results from winter conditions of consistent clear, dry, calm air; small insolation input; and an inland location far from any moderating maritime effects.

isotherm (p. 126)
thermal equator (p. 126)

14. What is the thermal equator? Describe its location in January and in July. Explain why it shifts position annually.

15. Observe trends in the pattern of isolines over North America and compare the January average temperature map with the July map. Why do the patterns shift locations?

16. Describe and explain the extreme temperature range experienced in north-central Siberia between January and July.

17. Where are the hottest places on Earth? Are they near the equator or elsewhere? Explain. Where is the coldest place on Earth?

18. From the maps in Figures 5.14, 5.17, and 5.19, determine the average temperature values and annual range of temperatures for your present location.

19. Compare the maps in Figures 5.15 and 5.18: (a) Describe what you find in central Greenland between January and July; (b) look at the South polar region and describe seasonal changes between the two maps. Characterize conditions along the Antarctic Peninsula between January and July (around 60° W longitude).

■ *Contrast* **wind chill and the heat index, and** *determine* **human response to these apparent temperature effects.**

The *wind-chill factor* indicates the enhanced rate at which body heat is lost to the air. As wind speeds increase, heat loss from the skin increases. The *heat index* (HI) indicates the human body's reaction to air temperature and water vapor. The level of humidity in the air affects our natural ability to cool through evaporation from skin.

20. What is the wind-chill temperature on a day with an air temperature of –12°C (10°F) and a wind speed of 32 kmph (20 mph)?
21. On a day when temperature reaches 37.8°C (100°F), how does a relative humidity reading of 50% affect apparent temperature?

NetWork

The *Geosystems* Student Learning Center provides online resources for this chapter on the World Wide Web. To begin: Once at the Center, click on the cover of this textbook, scroll the Table of Contents menu, and select this chapter. You will find self-tests that are graded, review exercises, specific updates for items in the chapter, and in "Destinations" many links to interesting related pathways on the Internet. *Geosystems* Student Learning Center is found at **http://www.prenhall.com/ christopherson/**.

Critical Thinking

A. With each temperature map (Figures 5.14, 5.17, and 5.19), begin by finding your own city or town and noting the temperatures indicated by the isotherms for January and July and the annual temperature range. Record the information from these maps in your notebook. The scale of all these maps allows you to make only a rough approximation for your location. As you work through the different maps throughout this text, note atmospheric pressure and winds, annual precipitation, climate type, landforms, soil orders, vegetation, and terrestrial biomes. By the end of the course you will have recorded a complete physical geography profile for your regional environment.

B. Have you ever experienced any of the different responses of the human body to the low-temperature or high-temperature stress noted in Focus Study 5.1, Table 5.1.1? Where were you at the time of each experience? Make copies of the *wind-chill* and *heat-index* charts given in the Focus Study for later reference. On an appropriate day, check air temperature and wind speed and determine their combined effect on skin chilling. In contrast, on an appropriate day, check air temperature and relative humidity values and determine their combined effect on your personal comfort. The *Geosystems* Home Page presents several interesting temperature links under "Destinations."

The largest wind turbines in the UK operate at Burger Hill Wind Farm, Birsay Moors, in the Orkney Islands, northeast Scotland. The inset photo shows a maintenance ladder for scale on the *nacelle* that houses the machinery. The towers are 68 m (223 ft) tall with a rotor diameter of 72 m (236 ft); each turbine is rated at 2 MW.
[Photo by Bobbé Christopherson.]

Atmospheric and Oceanic Circulations

■ Key Learning Concepts

After reading the chapter, you should be able to:

- ■ **Define** the concept of air pressure, and **describe** instruments used to measure air pressure.

- ■ **Define** wind, and **describe** how wind is measured, how wind direction is determined, and how winds are named.

- ■ **Explain** the four driving forces within the atmosphere—gravity, pressure gradient force, Coriolis force, and friction force—and **describe** the primary high- and low-pressure areas and principal winds.

- ■ **Describe** upper-air circulation and its support role for surface systems, and define the jet streams.

- ■ **Overview** several multiyear oscillations of air temperature, air pressure, and circulation in the Arctic, Atlantic, and Pacific oceans.

- ■ **Explain** several types of local winds: land–sea breezes, mountain–valley breezes, katabatic winds, and the regional monsoons.

- ■ **Discern** the basic pattern of Earth's major surface ocean currents and deep thermohaline circulation.

Early in April 1815, on the island Sumbawa in present-day Indonesia, Tambora erupted violently. The volcano spewed an estimated 150 km² (36 mi²) of material, 25 times the volume produced by the 1980 Mount St. Helens eruption in Washington State. Global atmospheric circulation carried Tambora debris worldwide, creating a stratospheric veil of dust and acid mist. Remarkable optical and meteorological effects lasted in Earth's atmosphere years after the eruption—beautiful sunrises and sunsets and a temporary lowering of temperatures worldwide. Scientists in 1815 lacked remote-sensing satellite capabilities and had no way of knowing the global impact of Tambora's eruption.

(a)

Luzon Island, Philippines

6/15 – 6/19/91

(b)

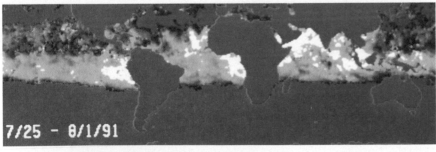

7/4 – 7/10/91

(c)

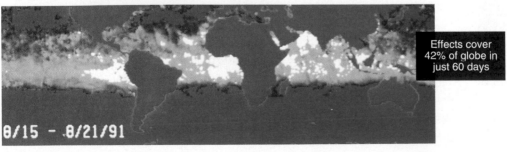

7/25 – 8/1/91

(d)

7/25 – 8/1/91

8/15 – 8/21/91

(e)

Effects cover 42% of globe in just 60 days

FIGURE 6.1 Volcanic eruption effects spread worldwide by winds. Satellite image captures Mount Pinatubo eruption (a). False-color images (b–e) of aerosols from Mount Pinatubo, smoke from fires, and dust storms, all swept about the globe by the general atmospheric circulation. The false color shows aerosol concentration as its aerosol optical thickness (AOT): White is densest, dull yellow indicates medium values, and brown areas have the lowest aerosol concentration. In (b) note the dust moving westward from Africa, smoke from Kuwaiti oil well fires set during the first Persian Gulf War, smoke from forest fires in Siberia, and haze off the east coast of the United States. [(a) *AVHRR* satellite image courtesy of U.S. Geological Survey, EROS Data Center. (b-d) *AVHRR* satellite AOT images courtesy of Dr. Larry L. Stowe, NESDIS/NOAA. Used by permission.]

After 635 years of dormancy, Mount Pinatubo in the Philippines erupted in 1991 (Figure 6.1a). This event had tremendous atmospheric impact, lofting 15–20 million tons of ash, dust, and sulfur dioxide (SO_2) into the atmosphere. As the sulfur dioxide rose into the stratosphere, it quickly formed sulfuric acid (H_2SO_4) aerosols, which concentrated at 16–25 km (10–15.5 mi) altitude. This debris increased atmospheric albedo about 1.5%, giving scientists an estimate of the aerosol volume generated by the eruption (see Figure 6.1).

The AVHRR instrument aboard *NOAA-11* monitored the reflected solar radiation from Mount Pinatubo's aerosols as global winds swept them around Earth. Figure 6.1b–d shows images made at three-week intervals that clearly track the spread of the debris worldwide. Some 60 days after the eruption (the last satellite image in the sequence), the aerosol cloud spanned about 42% of the globe, from 20° S to 30° N. Again, for almost two years colorful sunrises and sunsets and a small lowering of average temperatures followed. The eruption provided a unique insight into the dynamics of atmospheric circulation.

Today, technology permits a depth of analysis unknown in the past—satellites now track the atmospheric effects from dust storms, forest fires, industrial haze, warfare, and the dispersal of volcanic explosions, among many others. As an example, millions of tons of dust from African soils cross the Atlantic each year, borne by the winds of atmospheric circulation. This dust sometimes increases the iron content of the waters off Florida, promoting the toxic algal blooms (*Karenia brevis*) known as "Red Tides."

Global winds are certainly an important reason why the United States, the former Soviet Union, and Great Britain signed the 1963 Limited Test Ban Treaty. That treaty banned above-ground testing of nuclear weapons because atmospheric circulation spread radioactive contamination worldwide. Such agreements illustrate how the fluid movement of the atmosphere socializes humanity more than perhaps any other natural or cultural factor. Our atmosphere makes the world a spatially linked society—one person's or country's exhalation is another's inhalation.

In this chapter: We begin with a discussion of wind essentials, consisting of air pressure and its measurement and a description of wind. The driving forces that produce surface winds are pressure gradient, Coriolis, and friction. We examine the circulation of Earth's atmosphere and the patterns of global winds, including principal pressure systems and winds. We consider Earth's wind-driven oceanic currents and explain multiyear oscillations in atmospheric flows, local winds, and the powerful monsoons. The energy driving all this movement comes from one source: the Sun.

Wind Essentials

Earth's atmospheric circulation transfers both energy and mass on a grand scale. In the process, the imbalance between equatorial energy surpluses and polar energy deficits is resolved, Earth's weather patterns formed, and ocean currents produced. Air pollutants, whether natural or human-caused, are spread worldwide by atmospheric circulation, far from their point of origin.

Atmospheric circulation is generally categorized at three levels: *primary circulation* consisting of general worldwide circulation, *secondary circulation* of migratory high-pressure and low-pressure systems, and *tertiary circulation* that includes local winds and temporal weather patterns. Winds that move principally north or south along meridians are known as *meridional flows*. Winds moving east or west along parallels of latitude are called *zonal flows*.

Air Pressure and Its Measurement

Air pressure, its measurement and expression, is key to understanding wind. The molecules that constitute air create **air pressure** through their motion, size, and number—the factors that determine the temperature and density of the air. Air pressure, then, is a product of the temperature and density of a mass of air. Pressure is exerted on all surfaces in contact with the air.

In 1643, Galileo's pupil Evangelista Torricelli was working on a mine-drainage problem. His work led him to discover a method for measuring air pressure (Figure 6.2a). He knew that pumps in the mine were able to "pull" water upward about 10 m (33 ft), but no higher. He did not know why. Careful observation led him to discover that this was caused not by weak pumps but by the atmosphere itself. Torricelli noted that the water level in the vertical pipe fluctuated from day to day. He figured out that air pressure, the weight of the air, varies with weather conditions.

To simulate the problem at the mine, Torricelli devised an instrument, at Galileo's suggestion, using a much denser fluid than water—mercury (Hg)—in a glass tube 1 m high. Torricelli sealed the glass tube at one end, filled it with mercury, and inverted it into a dish of mercury (Figure 6.2b). He determined that the average height of the column of mercury in the tube was 760 mm (29.92 in.) and that it did vary day-to-day as the weather changed. He concluded that the mass of surrounding air was exerting pressure on the mercury in the dish that counterbalanced the column of mercury.

Using similar instruments, scientists set a standard of *normal sea-level pressure* at 1013.2 mb (*millibar*, or *mb*, which expresses force per square meter of surface area), or 29.92 in. mercury (Hg). In Canada and other countries, normal air pressure is expressed as 101.32 kPa (*kilopascal*; 1 kPa = 10 mb).

Any instrument that measures air pressure is a *barometer* (from the Greek *baros*, meaning "weight"). Torricelli developed a **mercury barometer**. A more compact barometer design, which works without a meter-long tube of mercury, is the aneroid barometer, shown in Figure 6.2c. *Aneroid* means "using no liquid." The **aneroid barometer** principle is simple: Imagine a small chamber, partially emptied of air, which is sealed and connected to a

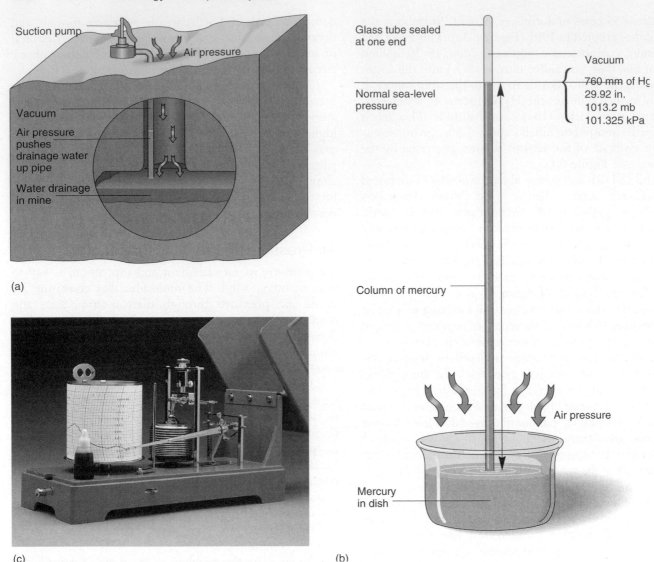

FIGURE 6.2 Developing the barometer.
(a) Torricelli developed the barometer to measure air pressure while trying to solve a mine-drainage problem. Two types of instruments are used to measure atmospheric pressure: (b) a mercury barometer (idealized sketch) and (c) an aneroid barometer. Have you used a barometer? If so, what type is it? Have you tried to reset it using a local weather information source?

mechanism attached to a needle on a dial. As air pressure increases, it presses on the chamber; as air pressure decreases, it relieves pressure on the chamber. The chamber responds to these changes in air pressure and moves the needle. An aircraft altimeter is a type of aneroid barometer. It accurately measures altitude because air pressure diminishes with altitude above sea level. For accuracy the altimeter must be adjusted for temperature changes.

Figure 6.3 illustrates comparative scales in millibars and inches used to express air pressure and its relative force. The normal range of Earth's atmospheric pressure from strong high pressure to deep low pressure is about 1050 to 980 mb (31.00 to 29.00 in.). The figure also indicates the extreme highest and lowest pressures ever recorded in the United States, Canada, and on Earth.

Hurricane Wilma is now the record holder from 2005. For comparison, note hurricanes Gilbert (1988), Rita, and Katrina (2005), and their lowest central pressures.

Wind: Description and Measurement

Simply stated, **wind** is generally the horizontal motion of air across Earth's surface. Turbulence adds wind updrafts and downdrafts and a vertical component to this definition. *Differences in air pressure (density) between one location and another produce wind.* Wind's two principal properties are speed and direction, and instruments measure each. An **anemometer** measures wind speed in kilometers per hour (kmph), miles per hour (mph), meters per second (mps), or knots. (A *knot* is a nautical mile per hour, covering 1 minute of Earth's arc in an hour, equivalent to

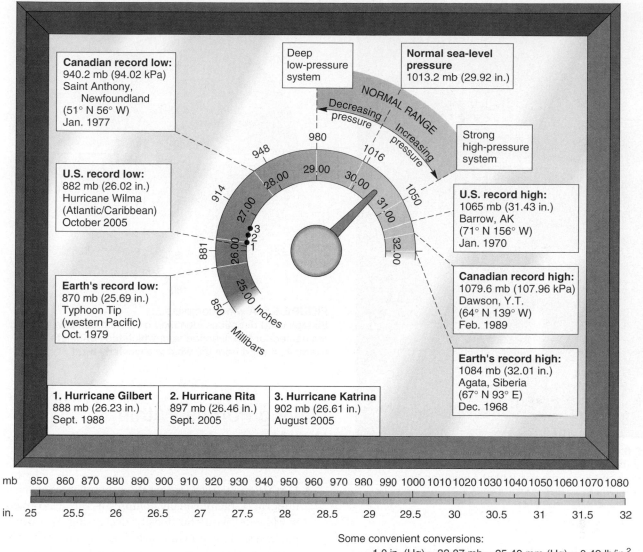

FIGURE 6.3 Air pressure readings and conversions.
Scales express barometric air pressure in millibars and inches, with average air pressure values and recorded pressure extremes. Canadian values include kilopascal equivalents, 10 mb = 1 kPa. The intensity of the 2005 hurricane season is evident; note the positions of Katrina, Rita, and Wilma on the pressure dial.

1.85 kmph, or 1.15 mph.) A **wind vane** determines wind direction; the standard measurement is taken 10 m (33 ft) above the ground to reduce the effects of local topography on wind direction (Figure 6.4).

Winds are named for the direction *from which they originate*. For example, a wind from the west is a *westerly* wind (it blows eastward); a wind out of the south is a *southerly* wind (it blows northward). Figure 6.5 illustrates a simple wind compass, naming 16 principal wind directions used by meteorologists.

The traditional *Beaufort wind scale* is a descriptive scale useful in visually estimating wind speed (Table 6.1). Admiral Beaufort of the British Navy introduced his wind scale in 1806. In 1926, G. C. Simpson expanded Beaufort's scale to include wind speeds on land. The National Weather Service (old Weather Bureau) standardized

the scale in 1955 (see **http://www.hpc.ncep.noaa.gov/html/beaufort.shtml**). The scale is still referenced on ocean charts, enabling estimation of wind speed without instruments. Although most ships use sophisticated equipment to perform such measurements, you find the Beaufort scale posted on the bridge of many ships.

Using no instruments, try the Beaufort scale in Table 6.1 to estimate wind speed as you walk across campus today. Then, moisten your finger, hold it in the air, and sense evaporative cooling on one side of your finger to indicate from which direction the wind is blowing.

Global Winds

The primary circulation of winds across Earth has fascinated travelers, sailors, and scientists for centuries, although only in the modern era is a true picture emerging

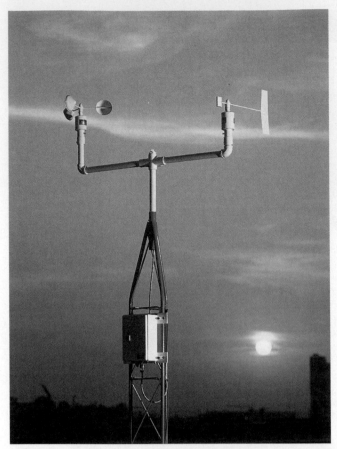

FIGURE 6.4 Wind vane and anemometer.
Instruments used to measure wind direction (wind vane, right) and wind speed (anemometer, left) at a weather station installation. [Photo by Belfort Instruments.]

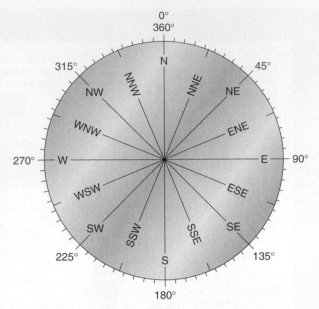

FIGURE 6.5 A wind compass.
Sixteen wind directions identified on a wind compass. Winds are named for the direction from which they originate. For example, a wind from the west is a westerly wind.

of global winds. Breakthroughs in space-based observations and Earth-bound computer technology are refining models that simulate total atmospheric and oceanic circulation.

Scientists at the National Center for Atmospheric Research determined that atmospheric flows of global winds redistribute much of the energy between the tropics and the polar regions. The atmosphere is the dominant medium from about 35° latitude to the poles in each hemisphere, whereas ocean currents redistribute more heat in a zone straddling the equator between the 17th parallels in each hemisphere.

Scientists at the Jet Propulsion Laboratory and the University of California, Los Angeles, painstakingly assembled a remarkable portrait of surface winds across the Pacific Ocean (Figure 6.6). The *Seasat* satellite produced the image using radar to measure the motion and direction of ocean waves. Because wind drives waves on the ocean surface, wave patterns indicate winds.

The patterns in the figure are the result of specific forces at work in the atmosphere: *pressure gradient force, Coriolis force, friction force,* and *gravity.* These forces are our next topic. As we progress through this chapter, you may want to refer to this *Seasat* image to identify the winds, eddies, and vortexes it portrays.

Driving Forces Within the Atmosphere

Four forces determine both speed and direction of winds:

- Earth's *gravitational force* on the atmosphere is virtually uniform. Gravity equally compresses the atmosphere worldwide, with the density decreasing as altitude increases. The gravitational force counteracts the outward *centrifugal force* acting on Earth's spinning surface and atmosphere. Without gravity, there would be no atmospheric pressure—or atmosphere, for that matter.

- **Pressure gradient force** drives air from areas of higher barometric pressure (more dense air) to areas of lower barometric pressure (less dense air), thereby causing winds. Without a pressure gradient force, there would be no wind.

- The **Coriolis force**, a deflective force, makes wind that travels in a straight path appear to be deflected in relation to Earth's rotating surface. The Coriolis force deflects wind to the right in the Northern Hemisphere and to the left in the Southern Hemisphere. Without Coriolis force, winds would move along straight paths between high- and low-pressure areas.

- The **friction force** drags on the wind as it moves across surfaces; it decreases with height above the surface. Without friction, winds would simply move in paths parallel to isobars and at high rates of speed.

All four of these forces operate on moving air and ocean currents at Earth's surface and influence global wind circulation patterns. The following sections describe the

Table 6.1   Beaufort Wind Scale

Wind Speed			Beaufort Wind Scale			
kmph	Mph	Knots	Beaufort Number	Wind Description	Observed Effects at Sea	Observed Effects on Land
<1	<1	<1	0	Calm	Glassy calm, like a mirror	Calm, no movement of leaves
1–5	1–3	1–3	1	Light air	Small ripples; wavelet scales; no foam on crests	Slight leaf movement; smoke drifts; wind vanes still
6–11	4–7	4–6	2	Light breeze	Small wavelets; glassy look to crests, which do not break	Leaves rustling; wind felt; wind vanes moving
12–19	8–12	7–10	3	Gentle breeze	Large wavelets; dispersed whitecaps as crests break	Leaves and twigs in motion; small flags and banners extended
20–29	13–18	11–16	4	Moderate breeze	Small, longer waves; numerous whitecaps	Small branches moving; raising dust, paper, litter, and dry leaves
30–38	19–24	17–21	5	Fresh breeze	Moderate, pronounced waves; many whitecaps; some spray	Small trees and branches swaying; wavelets forming on inland waterways
39–49	25–31	22–27	6	Strong breeze	Large waves, white foam crests everywhere; some spray	Large branches swaying; overhead wires whistling; difficult to control an umbrella
50–61	32–38	28–33	7	Moderate (near) gale	Sea mounding up; foam and sea spray blown in streaks in the direction of the wind	Entire trees moving; difficult to walk into wind
62–74	39–46	34–40	8	Fresh gale (or gale)	Moderately high waves of greater length; breaking crests forming sea spray; well-marked foam streaks	Small branches breaking; difficult to walk; moving automobiles drifting and veering
75–87	47–54	41–47	9	Strong gale	High waves; wave crests tumbling and the sea beginning to roll; visibility reduced by blowing spray	Roof shingles blown away; slight damage to structures; broken branches littering the ground
88–101	55–63	48–55	10	Whole gale (or storm)	Very high waves and heavy, rolling seas; white appearance to foam-covered sea; overhanging waves; visibility reduced	Uprooted and broken trees; structural damage; considerable destruction; seldom occurring
102–116	64–73	56–63	11	Storm (or violent storm)	White foam covering a breaking sea of exceptionally high waves; small- and medium-sized ships lost from view in wave troughs; wave crests frothy	Widespread damage to structures and trees, a rare occurrence
>117	>74	>64	12–17	Hurricane	Driving foam and spray filling the air; white sea; visibility poor to nonexistent	Severe to catastrophic damage; devastation to affected society

actions of the pressure gradient, Coriolis, and friction forces. (The gravitational force operates uniformly worldwide.)

Pressure Gradient Force

High- and low-pressure areas exist in the atmosphere principally because Earth's surface is unequally heated. For example, cold, dense air at the poles exerts greater pressure than warm, less dense air along the equator. These pressure differences establish a *pressure gradient force*.

An **isobar** is an *isoline* (a line along which there is a constant value) plotted on a weather map to connect points of equal pressure. A pattern of isobars on a weather map provides a portrait of the pressure gradient between an area of higher pressure and one of lower pressure. The spacing between isobars indicates the intensity of the pressure difference, or *pressure gradient*.

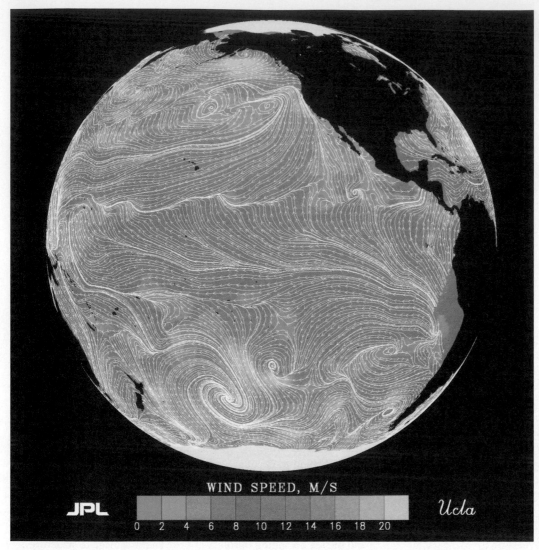

(a)　　　　　　　　　　　　　　　　　　　　　　　　　　　　(b)

FIGURE 6.6 Wind portrait of the Pacific Ocean.
(a) Surface wind measured by radar scatterometer aboard the *Seasat* satellite on a September day. Scientists analyzed 150,000 measurements for this image. Colors correlate with wind speeds, and the white arrows denote wind direction. Compare the wind patterns with a visible-light image of the same region (b). Can you identify the pattern of trade winds, westerlies, high-pressure cells, and low-pressure cells from the cloud patterns on the image? [(a) Courtesy of Dr. Peter Woiceshyn, Jet Propulsion Laboratory, Pasadena, California. (b) Satellite image from Laboratory of Planetary Studies, Cornell University. Used by permission.]

Just as closer contour lines on a topographic map indicate a steeper slope on land, so closer isobars denote steepness in the pressure gradient. In Figure 6.7a, note the spacing of the isobars (green lines). A steep gradient causes faster air movement from a high-pressure area to a low-pressure area. Isobars spaced wider apart from one another mark a more gradual pressure gradient, one that creates a slower airflow. Along a horizontal surface, the pressure gradient force alone acts at right angles to the isobars, so wind blows across isobars at right angles. Note the location of steep ("strong winds") and gradual ("light winds") pressure gradients and their relationship to wind intensity on the map in Figure 6.7b.

Figure 6.8 illustrates the forces that direct winds. Figure 6.8a shows the pressure gradient force acting alone. A field of subsiding, or sinking, air develops in a high-pressure area. Air descends in a high-pressure area and diverges outward at the surface in all directions. On the other hand, in a low-pressure area, as air rises, it pulls air from all directions, converging into the area of lower pressure at the surface.

Coriolis Force

You might expect surface winds to move in a straight line from areas of higher pressure to areas of lower pressure. On a nonrotating Earth, they would. But on our rotating planet, the Coriolis force deflects anything that flies or flows across Earth's surface—wind, an airplane, or ocean currents—from a straight path. In simplest terms, this

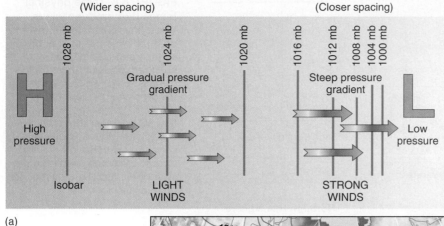

(Wider spacing)   (Closer spacing)

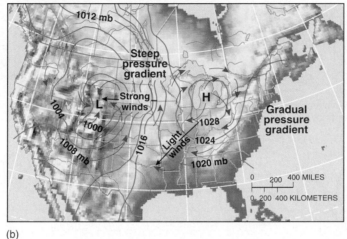

(b)

FIGURE 6.7 Pressure gradient determines wind speed. Pressure gradient (a) portrayed on a weather map (b). The closer spacing of isobars represents a steeper pressure gradient producing stronger winds; wider spacing of isobars denotes a gradual pressure gradient for lighter winds. Here we see surface winds spiraling clockwise out of a high-pressure system and spiraling counterclockwise into a low-pressure system.

force is an effect of Earth's rotation. Earth's rotational speed varies with latitude, increasing from 0 kmph at the poles (surface is at Earth's axis) to 1675 kmph (1041 mph) at the equator (surface is farthest from Earth's axis). Table 2.2 lists rotational speeds by latitude.

The deflection occurs regardless of the direction in which the object is moving. Because Earth rotates eastward, objects that move in an absolute straight line over a distance (such as winds and ocean currents) appear to curve to the right in the Northern Hemisphere and to the left in the Southern Hemisphere (Figure 6.9a). The Coriolis force is zero along the equator, increases to half the maximum deflection at 30° N and 30° S latitudes, and reaches maximum deflection flowing away from the poles. The effect of the Coriolis force increases as the speed of the moving object increases; thus, the faster the wind speed, the greater its apparent deflection. The Coriolis force does not normally affect small-scale motions that cover insignificant distance and time (see News Report 6.1).

We work through a simple explanation here focused on the observed phenomena. If you choose, you can use physics and identify the variables that interplay to produce these observed results. Mathematical formulas explain how the conservation of angular and linear momentum, as the distance to Earth's axis changes with latitudinal location, operates to produce the deflection, as well as the interplay of the opposing gravitational and centrifugal forces.

With this in mind, we observe the Coriolis force from different viewpoints. From the viewpoint of an airplane that is passing over Earth's surface, the surface is seen to rotate slowly below. But, looking from the surface at the airplane, the surface seems stationary, and the airplane appears to curve off course. The airplane does not actually deviate from a straight path, but it *appears* to do so because we are standing on Earth's rotating surface beneath the airplane. Because of this apparent deflection, the airplane must make constant corrections in flight path to maintain its "straight" heading relative to a rotating Earth.

As an example of the effect of this force, see Figure 6.9b. A pilot leaves the North Pole and flies due south toward Quito, Ecuador. If Earth were not rotating, the aircraft would simply travel along a meridian of longitude and arrive at Quito. But Earth is rotating eastward beneath the aircraft's flight path. If the pilot does not allow for this increase in rotational speed, the plane will reach the equator over the ocean along an apparently curved path, far to the west of the intended destination. Likewise, flying northward on a return flight, the plane maintains its eastward speed as it makes its way northward over slower rotational speeds of Earth. Without correction, the plane ends up to the east of the pole, in a right-hand deflection. Pilots must correct for this Coriolis deflection in their navigational calculations.

This effect occurs regardless of the direction of the moving object. A flight from California to New York

Top view and side view of air movement in an idealized high-pressure area and low-pressure area on a nonrotating Earth.

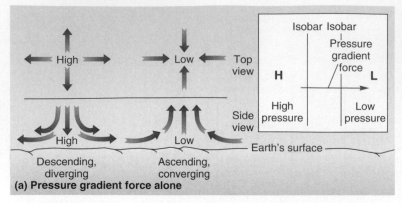

(a) Pressure gradient force alone

FIGURE 6.8 Three physical forces that produce winds. Three physical forces interact to produce wind patterns at the surface and aloft: (a) the pressure gradient force; (b) the Coriolis force, which counters the pressure gradient force, producing a geostrophic wind flow in the upper atmosphere; and (c) the friction force, which, combined with the other two forces, produces characteristic surface winds. The gravitational force is assumed.

Earth's rotation adds the Coriolis force and a "twist" to air movements. High-pressure and low-pressure areas develop a rotary motion, and wind flowing between highs and lows flows parallel to isobars.

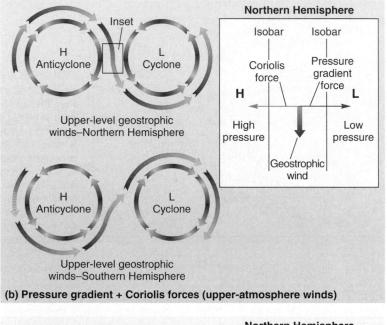

(b) Pressure gradient + Coriolis forces (upper-atmosphere winds)

Wind Pattern Development

ANIMATION

(b)   N.H.

→ Pressure gradient
⟹ Wind
→ Coriolis force
← Friction force

Surface friction adds a countering force to Coriolis, producing winds that spiral out of a high-pressure area and into a low-pressure area. Surface winds cross isobars at an angle. Air flows into low pressure cyclones and turns to the left, because of deflection to the right.

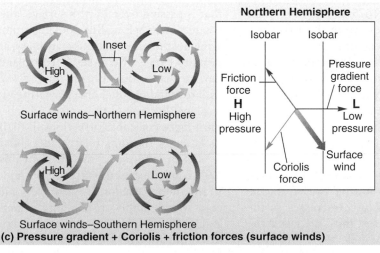

(c) Pressure gradient + Coriolis + friction forces (surface winds)

(c)   N.H.

provides an example in Figure 6.9c. The Coriolis deflection occurs as the airplane flew to New York because Earth continued to rotate eastward; this increased the centrifugal force acting on the plane (Earth rotation speed + plane speed), so the plane moves toward the equator in a right-hand deflection (moving farther from Earth's axis). Unless the pilot corrected for this

deflective force, the flight would end up somewhere in North Carolina. Likewise, flying westward on a return flight opposite Earth's rotation direction, which decreases the centrifugal force (Earth rotation speed − plane speed), the plane moves away from the equator, in a right-hand deflection (moving closer to Earth's axis).

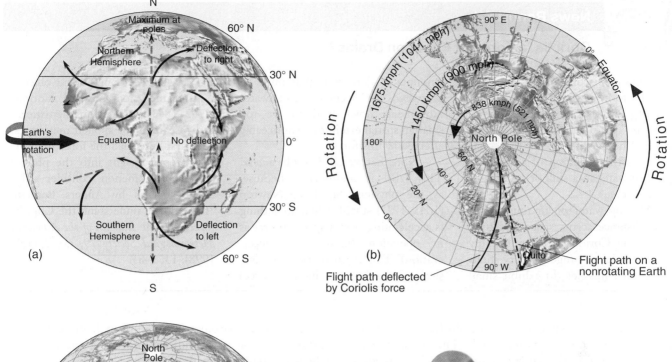

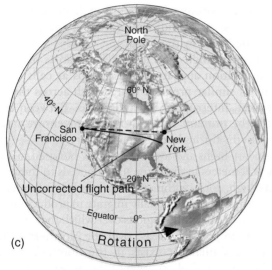

During an eastward flight, in the same direction as Earth's rotation, the centrifugal force acting on the plane in flight increases to counter the gravitational force pulling toward Earth's axis, so the plane's path is deflected toward the equator. Wind and ocean currents experience the same deflection.

FIGURE 6.9 The Coriolis force—an apparent deflection.
Distribution of the Coriolis force on Earth: (a) apparent deflection to the right of a straight line in the Northern Hemisphere and apparent deflection to the left in the Southern Hemisphere; (b) Coriolis deflection of a flight path between the North Pole and Quito, Ecuador, which is on the equator; (c) deflection of a flight path between San Francisco and New York. Deflection from a straight path occurs regardless of the direction of movement.

Coriolis Force and Winds   How does the Coriolis force affect wind? As air rises from the surface through the lowest levels of the atmosphere, it leaves the drag of surface friction behind and increases speed. This increases the Coriolis force, spiraling the winds to the right in the Northern Hemisphere or to the left in the Southern Hemisphere, generally producing upper-air westerly winds from the subtropics to the poles. In the upper troposphere, the Coriolis force just balances the pressure gradient force. Consequently, the winds between higher-pressure and lower-pressure areas in the upper troposphere flow parallel to the isobars.

Figure 6.8b illustrates the combined effect of the pressure gradient force and the Coriolis force on air currents aloft. Together, they produce winds that do not flow directly from high to low, but *around* the pressure areas, remaining parallel to the isobars. Such winds are called **geostrophic winds** and are characteristic of upper tropospheric circulation. (The suffix *-strophic* means "to

turn.") Geostrophic winds produce the characteristic pattern shown on the upper-air weather map just ahead in Figure 6.15. Note the inset illustration showing the effects of the pressure gradient and Coriolis forces that produce a geostrophic flow of air.

Friction Force

Figure 6.8c adds the effect of friction to the Coriolis and pressure gradient forces on wind movements; combining all three forces produces the wind patterns we see along Earth's surface. The effect of surface friction extends to a height of about 500 m (around 1600 ft) and varies with surface texture, wind speed, time of day and year, and atmospheric conditions. In general, rougher surfaces produce more friction.

Near the surface, friction disrupts the equilibrium established in geostrophic wind flows between the pressure gradient and Coriolis forces—note the inset illustration in Figure 6.8c. Because surface friction decreases wind

downward and heats by compression on its descent to the surface, as Figure 6.12 shows. Warmer air has a greater capacity to absorb water vapor than does cooler air, making this descending warm air relatively dry (large water-vapor capacity, low water-vapor content). The air is also dry because heavy precipitation along the equatorial portion of the circulation removes moisture.

Surface air diverging from the subtropical high-pressure cells generates Earth's principal surface winds: the westerlies and the trade winds. The **westerlies** are the dominant surface winds from the subtropics to high latitudes. They diminish somewhat in summer and are stronger in winter in both hemispheres.

As you examine the global pressure maps in Figure 6.10, you find several high-pressure areas. In the Northern Hemisphere, the Atlantic subtropical high-pressure cell is called the **Bermuda high** (in the western Atlantic) or the **Azores high** (when it migrates to the eastern Atlantic in winter). The area in the Atlantic under this subtropical high features clear, warm waters and large quantities of *Sargassum* (a seaweed) that gives the area its name—the Sargasso Sea. The **Pacific high**, or *Hawaiian high*, dominates the Pacific in July, retreating southward in January. In the Southern Hemisphere, three large high-pressure centers dominate the Pacific, Atlantic, and Indian oceans, especially in January, and tend to move along parallels of latitude in shifting zonal positions.

The entire high-pressure system migrates with the summer high Sun, fluctuating about 5°–10° in latitude. The *eastern sides* (right-hand side) of these anticyclonic systems are drier and more stable (less convective activity) and feature cooler ocean currents than do the *western sides* (left-hand side). The drier eastern side of these systems and dry-summer conditions influence climate along subtropical and midlatitude west coasts (Figure 6.13). In fact, Earth's major deserts generally occur within the subtropical belt and extend to the west coast of each continent except Antarctica. In Figures 6.13 and Figure 6.21, note the desert regions of Africa come right to the shore in both hemispheres, with the cool, southward-flowing *Canaries Current* offshore in the north and the cool, northward-flowing *Benguela* current offshore in the south.

Because the subtropical belts are near 25° N and 25° S latitudes, these areas sometimes are known as the *calms of Cancer* and the *calms of Capricorn*. These zones of windless, hot, dry desert air, so deadly in the era of sailing ships, earned the name *horse latitudes*. The origin of this term is popularly attributed to becalmed and stranded sailing crews of past centuries, who destroyed the horses on board, not wanting to share food or water with the livestock. The term's true origin may never be known; the *Oxford English Dictionary* calls it "uncertain."

Subpolar Low-Pressure Cells: Cool and Moist Air

In January, two low-pressure cyclonic cells exist over the oceans around 60° N latitude, near their namesakes: the North Pacific **Aleutian low** and the North Atlantic **Icelandic low** (see Figure 6.10a). Both cells are dominant in winter and weaken or disappear in summer with the

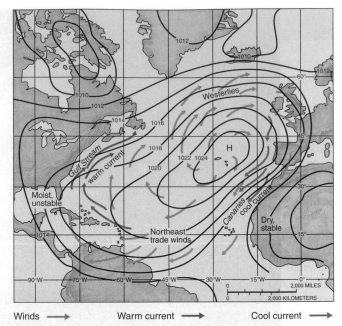

Winds ⟶   Warm current ⟶   Cool current ⟶

FIGURE 6.13 Subtropical high-pressure system in the Atlantic.
Characteristic circulation and climate conditions related to the Atlantic subtropical high-pressure anticyclone in the Northern Hemisphere. Note deserts extending to the shores of Africa with offshore cool currents, whereas the southeastern United States is moist and humid, with offshore warm currents.

strengthening of high-pressure systems in the subtropics. The area of contrast between cold (from higher latitudes) and warm (from lower latitudes) air forms the **polar front**, where masses of air with different characteristics battle. This front encircles Earth and is focused in these low-pressure areas.

Figure 6.12 illustrates this confrontation between warm, moist air from the westerlies and cold, dry air from the polar and Arctic regions. Cooling and condensation in the lifted air occurs because of upward displacement of the warm air. Low-pressure cyclonic storms migrate out of the Aleutian and Icelandic frontal areas and may produce precipitation in North America and Europe, respectively. Northwestern sections of North America and Europe generally are cool and moist as a result of the passage of these cyclonic systems onshore—consider the weather in British Columbia, Washington, Oregon, Ireland, and the United Kingdom.

In the Southern Hemisphere, a discontinuous belt of subpolar low-pressure systems surrounds Antarctica. The spiraling cloud patterns produced by these cyclonic systems are visible on the spacecraft image in Figure 6.14. Severe cyclonic storms can cross Antarctica, producing strong winds and new snowfall. How many cyclonic systems can you identify around Antarctica on the image?

Polar High-Pressure Cells: Frigid, Dry Deserts

Polar high-pressure cells are weak. The polar atmospheric mass is small, receiving little energy from the Sun to put it into motion. Variable winds, cold and dry, move

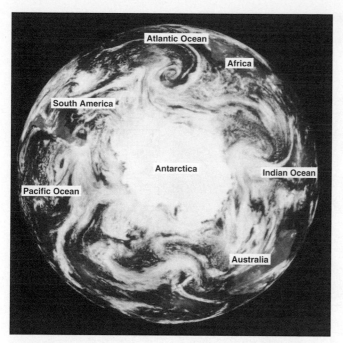

FIGURE 6.14 Clouds portray subpolar and polar circulation patterns.
Centered on Antarctica, this image shows a series of subpolar low-pressure cyclones in the Southern Hemisphere. Antarctica is fully illuminated by a midsummer Sun as the continent approaches the December solstice. The *Galileo* spacecraft made this image during its December 1990 flyby of Earth. [The Solid State Imaging instrument (violet, green, and red filters) image courtesy of Dr. W. Reid Thompson, Laboratory of Planetary Studies, Cornell University. Used by permission.]

away from the polar region in an anticyclonic direction. They descend and diverge clockwise in the Northern Hemisphere (counterclockwise in the Southern Hemisphere) and form weak, variable winds named **polar easterlies**.

Of the two polar regions, the **Antarctic high** is stronger and more persistent, forming over the Antarctic landmass. Over the Arctic Ocean, a polar high-pressure cell is less pronounced. When it does form, it tends to locate over the colder northern continental areas in winter (Canadian and Siberian highs) rather than directly over the relatively warmer Arctic Ocean.

Upper Atmospheric Circulation

Circulation in the middle and upper troposphere is an important component of the atmosphere's general circulation. Just as sea level is a reference datum for evaluating air pressure at the surface, we use a pressure level such as 500 mb as a **constant isobaric surface** for a pressure-reference datum in the upper atmosphere.

On upper-air pressure maps, we plot the height above sea level at which an air pressure of 500 mb occurs. The isobaric chart and illustration for an April day in Figures 6.15a and b illustrate such an undulating isobaric surface, upon which all points have the same pressure. In contrast, on surface weather maps we plot different pressures at the fixed elevation of sea level—a *constant height surface*.

We use this 500-mb level to analyze upper-air winds and the possible support they provide for surface weather conditions. Similar to surface maps, closer spacing of the isobars indicates faster winds; wider spacing indicates slower winds. On this constant isobaric pressure surface, altitude variations from the reference datum are called *ridges* for high pressure (with isobars on the map bending poleward) and *troughs* for low pressure (with isobars on the map bending equatorward). Looking at the figure, can you identify such ridges and troughs in the isobaric surface?

The pattern of ridges and troughs in the upper-air wind flow is important in sustaining surface cyclonic (low-pressure) and anticylonic (high-pressure) circulation. Near ridges in the isobaric surface, winds slow and converge (pile up), whereas winds near the area of maximum wind speeds along the troughs in the isobaric surface accelerate and diverge (spread out). Note the wind-speed indicators and labels in Figure 6.15a near the ridge (over Alberta, Saskatchewan, Montana, and Wyoming); now compare these with the wind-speed indicators around the trough (over Kentucky, West Virginia, the New England states, and the Maritimes). Also, note the wind relationships off the Pacific Coast.

Now look at Figure 6.15c. Divergence in the upper-air flow is important to cyclonic circulation at the surface because it creates an outflow of air aloft that stimulates an inflow of air into the low-pressure cyclone (such as what happens when you open a chimney damper to create an upward draft). Convergence aloft, on the other hand, drives descending airflows and divergent winds at the surface, moving out from high-pressure anticyclones.

Rossby Waves   Within the westerly flow of geostrophic winds are great waving undulations called **Rossby waves**, named for meteorologist Carl G. Rossby, who first described them mathematically in 1938. The polar front is the line of conflict between colder air to the north and warmer air to the south (Figure 6.16a). Rossby waves bring tongues of cold air southward, with warmer tropical air moving northward. The development of Rossby waves in the upper-air circulation is shown in the three-part figure. As these disturbances mature, distinct cyclonic circulation forms, with warmer air and colder air mixing along distinct fronts. These wave-and-eddy formations and upper-air divergences support cyclonic storm systems at the surface. Rossby waves develop along the flow axis of a jet stream.

Jet Streams   The most prominent movement in these upper-level westerly wind flows is the **jet stream**, an irregular, concentrated band of wind occurring at several different locations that influences surface weather systems. (Figure 6.12a shows the location of four jet streams.) Rather flattened in vertical cross section, the jet streams normally are 160–480 km (100–300 mi) wide by 900–2150 m (3000–7000 ft) thick, with core speeds that can exceed 300 kmph (190 mph). Jet streams in each hemisphere tend to weaken during the hemisphere's summer and strengthen

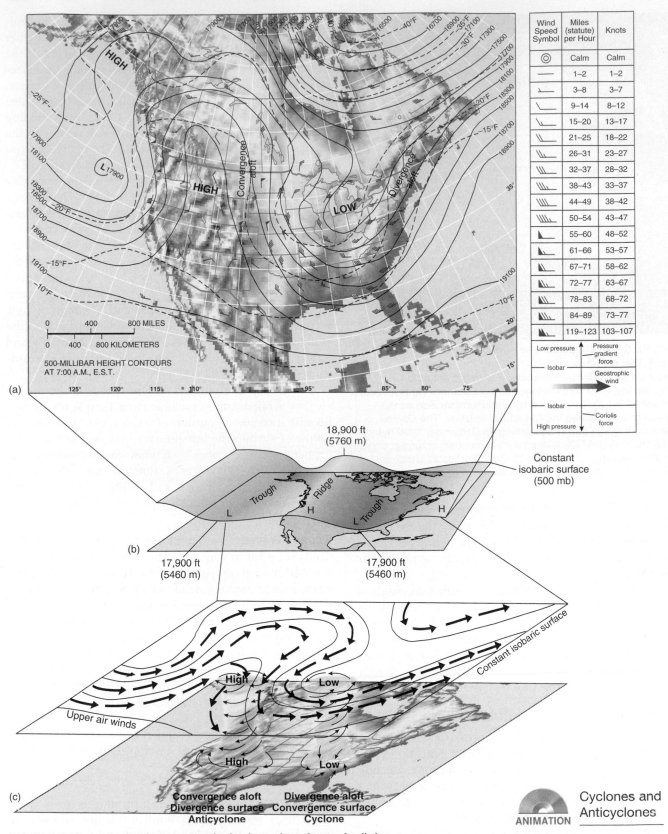

Wind Speed Symbol	Miles (statute) per Hour	Knots
◎	Calm	Calm
—	1–2	1–2
⊾	3–8	3–7
⊿	9–14	8–12
	15–20	13–17
	21–25	18–22
	26–31	23–27
	32–37	28–32
	38–43	33–37
	44–49	38–42
	50–54	43–47
	55–60	48–52
	61–66	53–57
	67–71	58–62
	72–77	63–67
	78–83	68–72
	84–89	73–77
	119–123	103–107

FIGURE 6.15 Analysis of a constant isobaric surface for an April day.
(a) Contours show elevation (in feet) at which 500-mb pressure occurs—a constant isobaric surface. The pattern of contours reveals a 500-mb isobaric surface and geostrophic wind patterns in the troposphere ranging from 16,500 to 19,100 ft elevation. (b) Note the "ridge" of high pressure over the Intermountain West, at 5760 m altitude, and the "trough" of low pressure over the Great Lakes region and off the Pacific Coast, also at 5460 m altitude, on the map and in the sketch beneath the chart. (c) Note areas of convergence aloft (corresponding to surface divergence) and divergence aloft (corresponding to surface convergence)—upper-atmosphere conditions at the 500-mb level support surface cyclones and anticyclones. [Data for map in (a) from the National Weather Service, NOAA.]

Cyclones and Anticyclones

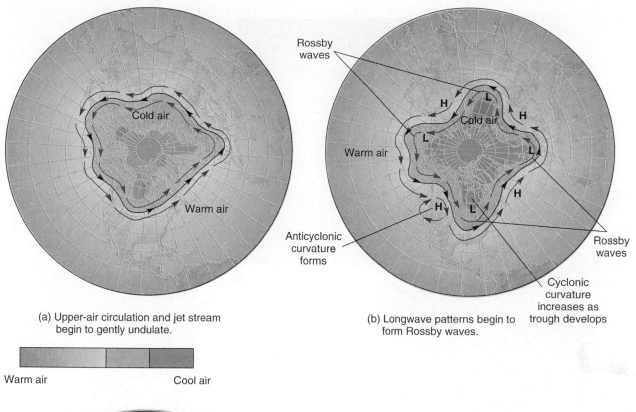

(a) Upper-air circulation and jet stream begin to gently undulate.

(b) Longwave patterns begin to form Rossby waves.

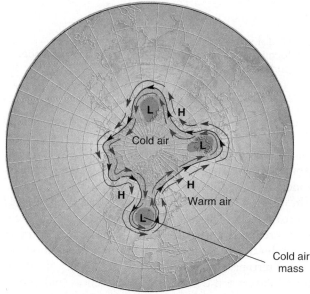

(c) Strong development of waves produces cells of cold and warm air—high-pressure ridges and low-pressure troughs.

Jet Stream, Rossby Waves

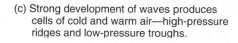

FIGURE 6.16 Rossby upper-atmosphere waves. Development of waves in the upper-air circulation, first described by C. G. Rossby in 1938 and expanded by J. Namias in 1952. Note in (b) undulations in Rossby waves are labeled "L" for troughs and "H" for ridges.

during winter as the streams shift closer to the equator. The pattern of ridges and troughs causes variation in jet stream speeds (convergence and divergence).

The *polar jet stream* meanders between 30° and 70° N latitude, at the tropopause along the polar front, at altitudes between 7600 and 10,700 m (24,900 and 35,100 ft). The polar jet stream can migrate as far south as Texas, steering colder air masses into North America and influencing surface storm paths traveling eastward. In the

summer, the polar jet stream exerts less influence on storms by staying over higher latitudes. Figure 6.17 shows a stylized view of a polar jet stream and two jet streams over North America. An experiment in 1998 inadvertently provided confirming information on the flow of the jet stream; see News Report 6.2 for details.

In subtropical latitudes, near the boundary between tropical and midlatitude air, the *subtropical jet stream* flows near the tropopause (see Figure 6.12). The subtropical jet

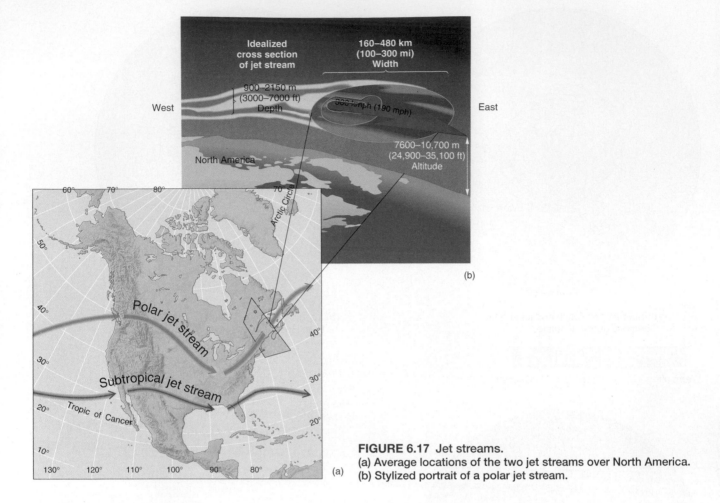

FIGURE 6.17 Jet streams.
(a) Average locations of the two jet streams over North America.
(b) Stylized portrait of a polar jet stream.

stream meanders from 20° to 50° latitude and may occur over North America simultaneously with the polar jet stream—sometimes the two will actually merge for brief episodes.

Multiyear Oscillations in Global Circulation

Several system fluctuations that occur in multiyear or shorter periods are important in the global circulation picture, yet an understanding of them is just emerging. The most famous of these is the El Niño–Southern Oscillation (ENSO) phenomenon discussed later in Chapter 10. Multiyear oscillations affect temperatures, air masses, and air pressure patterns and thus affect global winds and climates. Here we describe three of these hemisphere-scale oscillations in the briefest terms. Please consult the three URLs for illustrations and further descriptions.

North Atlantic Oscillation   A north–south fluctuation of atmospheric variability marks the North Atlantic Oscillation (NAO) as pressure differences between the Icelandic low and the Azores high in the Atlantic alternate in strength. The *NAO Index* is in its *positive phase* when the Icelandic low-pressure system is lower than normal and the Azores (west of Portugal) high-pressure cell is higher than normal. Strong westerly winds and jet stream cross the eastern Atlantic. Northerly winds blow in the

Labrador Sea (west of Greenland) and southerly winds move through the Norwegian Sea (east of Greenland). In the eastern United States, winters tend to be less severe in contrast to the strong, warm, wet storms hitting northern Europe; however, the Mediterranean region is dry.

In its *negative phase*, the NAO features a weaker pressure gradient than normal between the Azores and Iceland and reduced westerlies and jet stream. Storm tracks shift southward in Europe, bringing moist conditions to the Mediterranean and cold, dry winters to northern Europe. The eastern United States experiences cold, snowy winters as Arctic air masses plunge to lower latitudes.

With unpredictable flips between positive and negative phases, sometimes changing from week-to-week, a trend to a more positive phase than negative has emerged since 1960. Since 1980, the NAO has been predominantly positive (see **http://www.ldeo.columbia.edu/NAO/**).

Arctic Oscillation   Variable fluctuations between middle- and high-latitude air mass conditions over the Northern Hemisphere produce the Arctic Oscillation (AO). The AO is associated with the NAO, especially in winter, and the two phases of the *AO Index* correlate with the *NAO Index*. In the *warm phase* of the AO Index (positive NAO), the pressure gradient is affected by lower pressure than normal over the North Pole region and relatively higher pressures at lower latitudes. This sets up stronger westerly

News Report 6.2

Mantra's Big Balloon—Blowing in the Wind

A 25-story balloon named MANTRA (Middle Atmosphere Nitrogen Trend Assessment) launched from Vanscoy, Saskatchewan, August 24, 1998. The unpiloted mission was intended to measure concentrations of human-made chemicals and ozone in the stratosphere using a sophisticated instrument array.

After its 3:25 A.M. lift off, the balloon rose to a planned 38-km altitude and collected valuable data throughout the day. Upon mission completion, explosive charges were to release the instrument payload to parachute for a safe landing in nearby Manitoba. However, a malfunction left the payload attached to the balloon, adrift in the stratosphere in the tow of upper wind streams. An unexpected voyage was underway.

The balloon's path traced upper-atmospheric wind circulation and the flowing polar jet stream across Manitoba, northern Ontario and Québec, Labrador, the tip of Newfoundland, and out over the Atlantic (Figure 6.2.1). Realizing the potential threat to commercial jets, Canadian military jets dispatched to shoot MANTRA down. Thousands of rounds of ammunition were fired into the balloon to no avail because MANTRA was a zero-pressure balloon.

The helium was not under pressure in the plastic, so the holes created by the ammunition merely produced

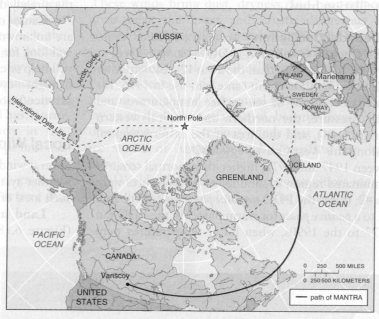

FIGURE 6.2.1 Flight path of Mantra 1998's big balloon. Westerly winds carried the balloon east from Vanscoy, Saskatchewan, across the globe, eventually landing in Finland. [Used by permission of the Minister of Public Works and Government Services Canada; Natural Resources Canada, Geological Survey of Canada.]

slow leaks. American and British military radar tracked MANTRA's continuing voyage. Imagine this balloon adrift in the Rossby waves depicted in Figure 6.16.

The upper winds drove MANTRA northeastward over the Denmark Straits between Greenland and Iceland, across the Norwegian Svalbard Islands, poleward of 80° N, and over the Russian Franz Josef Land and Novaya Zemlya island groups. Then winds took the balloon

south over the Russian coast and westward across Finland. The depleted balloon began its descent September 2, after a 9-day odyssey, eventually settling to Earth in Mariehamn, a city on Aland Island, in the Gulf of Bothnia. Finnish meteorologists gathered the payload and shipped it back to Canada. MANTRA's 1998 unscheduled voyage to Europe dramatically demonstrated the course of the polar jet stream within Earth's upper-atmospheric circulation.

---

winds and a flow of warmer Atlantic water currents into the Arctic Ocean. Cold air masses do not migrate as far south in winter, whereas winters are colder than normal in Greenland.

In the *cold phase* of the AO Index (negative NAO), the pattern reverses, with higher-than-normal pressure over the North Pole region and relatively lower pressure in the central Atlantic. Here the air masses of winter bring cold conditions to the eastern United States, northern Europe, and Asia, and sea ice in the Arctic Ocean becomes a bit thicker. Greenland is warmer than normal.

Since the 1980s, the NAO and AO indices have been in a strong positive, warm phase. In response, the ice drift

in the eastern Arctic Ocean is counterclockwise and air pollution from Europe and Russia moves across the region to Canada. Since the expected variability in these oscillations is essentially random, the emergence of a strong positive, warm phase is challenging to understand. Scientists are examining correlations with changes in sea-surface temperatures to understand what is happening and perhaps develop a forecast capability. For instance, researchers have discovered linkages between the behavior of the NAO and the increasingly warm Indian Ocean.

Given increasing ocean temperatures and the present trends of freshening in the northern ocean and increasing salinity in the tropical ocean, there is concern about a

Monsoonal Winds

Some regional wind systems seasonally change direction. Intense, seasonally shifting wind systems occur in the tropics over Southeast Asia, Indonesia, India, northern Australia, and equatorial Africa. A milder version of such monsoonal-type flow affects the extreme southwestern United States. These winds involve an annual cycle of returning precipitation with the summer Sun. The Arabic word for season, *mausim*, or **monsoon**, names them. (Specific monsoonal weather, associated climate types, and vegetation regions are discussed in Chapters 8, 10, and 20.) The location and size of the Asian landmass and its proximity to the Indian Ocean drive the monsoons of southern and eastern Asia (Figure 6.20). Also important to the generation of monsoonal flows are wind and pressure patterns in the upper-air circulation.

The extreme temperature range from summer to winter over the Asian landmass is due to its continentality (isolation from the modifying effects of the ocean). An intense high-pressure anticyclone dominates this continental landmass in winter (see Figure 6.10a and Figure 6.20a), whereas the equatorial low-pressure trough (ITCZ) dominates the central area of the Indian Ocean. This pressure gradient produces cold, dry winds from the Asian interior over the Himalayas and across India. Average temperatures range between 15° and 20°C (60°–68°F) at lower elevations. These winds desiccate, or dry out, the

landscape and then give way to hot weather from March through May. The second photo of the GPS unit in News Report 1.2 (Chapter 1) was made 18 m (60 ft) below the summit of Mount Everest on May 20, 1998—the monsoons dictate climbing schedules.

During the June–September wet period, the subsolar point (direct overhead rays of sunlight) shifts northward to the Tropic of Cancer, near the mouths of the Indus and Ganges rivers. The ITCZ shifts northward over southern Asia, and the Asian continental interior develops a thermal low pressure, associated with high average temperatures (remember the summer warmth in Verkhoyansk, Siberia, from Chapter 5). Meanwhile, subtropical high pressure dominates the Indian Ocean, with a surface temperature of 30°C (86°F). As a result of this reversed pressure gradient, hot subtropical air sweeps over the warm ocean, producing extremely high evaporation rates (Figure 6.20b).

By the time this air mass and the convergence zone reach India, it is laden with moisture in thunderous, dark clouds. When the monsoonal rains arrive from June to September, they are welcome relief from the dust, heat, and parched land of Asia's springtime. Likewise, the annual monsoon is an integral part of Indian music, poetry, and life. World-record rainfalls of the wet monsoon drench India. Cherrapunji, India, received both the second-highest average annual rainfall (1143 cm, or 450 in.) and the highest single-year rainfall (2647 cm, or 1042 in.) on Earth.

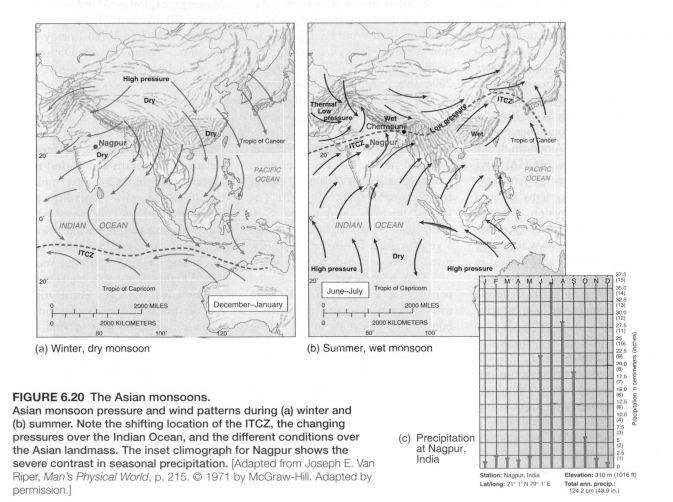

(a) Winter, dry monsoon

(b) Summer, wet monsoon

(c) Precipitation at Nagpur, India

Station: Nagpur, India   Elevation: 310 m (1016 ft)
Lat/long: 21° 1′ N 79° 1′ E   Total ann. precip.: 124.2 cm (48.9 in.)

FIGURE 6.20 The Asian monsoons.
Asian monsoon pressure and wind patterns during (a) winter and (b) summer. Note the shifting location of the ITCZ, the changing pressures over the Indian Ocean, and the different conditions over the Asian landmass. The inset climograph for Nagpur shows the severe contrast in seasonal precipitation. [Adapted from Joseph E. Van Riper, *Man's Physical World*, p. 215. © 1971 by McGraw-Hill. Adapted by permission.]

The fact that the monsoons of southern Asia involve vast global pressure systems leads us to ask whether future climate change might affect present monsoonal patterns. Researchers are conducting a field experiment in the Indian Ocean, the Indian Ocean Experiment (INDOEX), to examine this question (**http://www-indoex.ucsd.edu/ProjDescription.html**). Concern increases following events such as July 27, 2005, when a monsoonal deluge drowned Mumbai (Bombay)—in only a few hours, 94.2 cm (37.1 in.) of rain fell, causing widespread flooding. Again in August 2007, extensive flooding from an intense monsoon struck across India.

Their model projections forecast higher rainfall if consideration is given to increases in carbon dioxide and its greenhouse warming alone. However, when consideration is given to increases in aerosols—principally sulfur compounds and black carbon—then a drop in precipitation of 7%–14% appears in the model. Air pollution reduces surface heating and therefore decreases the pressure differences at the heart of monsoonal flows. Considering that 70% of the annual precipitation for the entire region comes during the wet monsoon, the actual occurrence of such changes would force difficult societal adjustments to changing water resources. (See the satellite images in Figure 4.8 and review the text discussion.)

Oceanic Currents

The driving force for ocean currents is the frictional drag of the winds, thus linking the atmospheric and oceanic systems. Also important in shaping these currents is the interplay of the Coriolis force, density differences caused by temperature and salinity, the configuration of the continents and ocean floor, and astronomical forces (the tides).

Surface Currents

Figure 6.21 portrays the general patterns of major ocean currents. Because ocean currents flow over distance and through time, the Coriolis force deflects them. However, their pattern of deflection is not as tightly circular as that of the atmosphere. Compare this ocean-current map with the map showing Earth's pressure systems (see Figure 6.10) and you can see that ocean currents are driven by the circulation around subtropical high-pressure cells in both hemispheres. These circulation systems are known as *gyres* and generally appear to be offset toward the western side of each ocean basin. (Remember, in the Northern Hemisphere, winds and ocean currents move *clockwise* about high-pressure cells—note the currents in the North Pacific and North Atlantic on the map—see Figure 6.13. In the Southern Hemisphere, circulation is *counterclockwise* around high-pressure cells, evident on the map.)

In sailing days, the Spanish galleons would leave San Blas and Acapulco, Mexico (16.5° N), and sail southwest, catching the northeast trade winds across the Pacific to the Philippines and Manila (14° N). Goods would be traded and new commodities obtained, and the ships would move northward to catch the westerlies in the midlatitudes to be blown across the ocean to the shores of present-day Alaska or British Columbia and sail along the coast to Northern California. The galleons would sail south, fighting the rain, frequent fog, and the right-hand Coriolis deflection pushing them away from the coast. The journey would end back in Mexico with cargo from the Orient. Thus the history of the *Manilla Galleons* is a lesson in oceanic currents around the Pacific gyre. News Report 6.3 dramatically portrays this clockwise circulation around the Pacific Ocean.

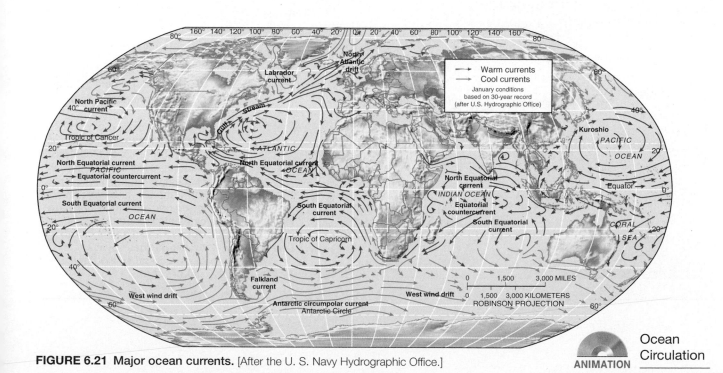

FIGURE 6.21 Major ocean currents. [After the U. S. Navy Hydrographic Office.]

Ocean Circulation

News Report 6.3

A Message in a Bottle and Rubber Duckies

To give you an idea of the dynamic circulation of the ocean, consider the following examples. A 9-year-old child at Dana Point, California (33.5° N), a small seaside community south of Los Angeles, placed a letter in a glass juice bottle in July 1992 and tossed it into the waves. Thoughts of distant lands and fabled characters filled the child's imagination as the bottle disappeared. The vast circulation around the Pacific high, clockwise-circulating gyre now took command (Figure 6.3.1).

Three years passed before ocean currents carried the message in a bottle to the coral reefs and white sands of Mogmog, a small island in Micronesia (7° N). A 7-year-old child there had found a pen pal from afar and immediately sent a photo and card to the sender of the message. Imagine the journey of that note from California—traveling through storms and calms, clear moonlit nights and typhoons, as it floated on ancient currents as the galleons once had.

In January 1994, a large container ship from Hong Kong loaded with toys and other goods was ravaged by a powerful storm. One of the containers on board split apart in the wind off the coast of Japan, dumping nearly 30,000 rubber ducks, turtles, and frogs into the North Pacific. Westerly winds and the North Pacific current swept this floating cargo at up to 29 km (18 mi) a day across the ocean to the coast of Alaska, Canada, Oregon, and California. Other toys, still adrift, went through the Bering Sea and into the Arctic Ocean (this route is indicated by the dashed line in Figure 6.3.1). Their drifting migration took a decade, but some objects made it around the Arctic Ocean to the Labrador Sea and will end up in the Atlantic Ocean.

The high-floating message bottle and rubber ducks offered a much better opportunity to study winds than did an earlier spill of 60,000 athletic shoes near Japan. The low-floating shoes tracked across the Pacific Ocean until they landed in the Pacific Northwest. By the way, the shoes that did not make landfall headed (or footed) back around the Pacific gyre into the tropics and westward, back toward Japan! Scientists are using incidents such as the message in a bottle and rubber duckies to learn more about wind systems and ocean currents.

On August 19, 2006, Tropical Storm Ioke formed about 1285 km (800 mi) south of Hawai'i. The storm's track moved across the Pacific Ocean over Wake Island and Johnston Atoll, becoming the strongest Super Typhoon in recorded history—a category 5, as we discuss in Chapter 8. Turning northward east of Sakhalin Island of Japan and the Kamchatka Peninsula of Russia, the storm moved into higher latitudes. Typhoon Ioke eventually swung northeastward and crossed the Aleutian Islands, reaching 55° N as an extratropical depression, dying out September 7. Much coastal erosion resulted in Alaska. The storm roughly followed the path around the Pacific gyre and along the track of the rubber duckies (Figure 6.3.1).

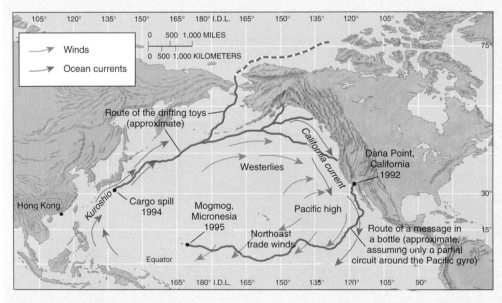

FIGURE 6.3.1 Pacific Ocean currents transport human artifacts.
The approximate route of a message in a bottle from Dana Point, California, to Mogmog, Micronesia, assumes a partial circuit around the Pacific gyre. Given the 3-year travel time, we do not know if the message circumnavigated the Pacific Ocean more than once. The rubber duckies voyaged eastward across the Pacific and beyond.

Equatorial Currents   In Figure 6.21, you can see that trade winds drive the ocean surface waters westward in a concentrated channel along the equator. These currents, called *equatorial currents*, are kept near the equator by the Coriolis force, which diminishes to zero at the equator. As these surface currents approach the western margins of the oceans, the water actually piles up against the eastern shores of the continents. The average height of this pileup is 15 cm (6 in.). This phenomenon is the **western intensification**.

The piled-up ocean water then goes where it can, spilling northward and southward in strong currents, flowing in tight channels along the eastern shorelines. In the Northern Hemisphere, the *Gulf Stream* and the *Kuroshio* (current east of Japan) move forcefully northward as a result of western intensification. Their speed and depth are increased by the constriction of the area they occupy. The warm, deep, clear water of the ribbon-like Gulf Stream (Figure 5.10) usually is 50–80 km (30–50 mi) wide and 1.5–2.0 km (0.9–1.2 mi) deep, moving at 3–10 kmph (1.8–6.2 mph). In 24 hours, ocean water can move 70–240 km (40–150 mi) in the Gulf Stream.

Upwelling and Downwelling Flows   Where surface water is swept away from a coast, either by surface divergence (induced by the Coriolis force) or by offshore winds, an **upwelling current** occurs. This cool water generally is nutrient-rich and rises from great depths to replace the vacating water. Such cold upwelling currents exist off the Pacific coasts of North and South America and the subtropical and midlatitude west coast of Africa. These areas are some of Earth's prime fishing regions.

In other regions where there is an accumulation of water—such as the western end of an equatorial current, or the Labrador Sea, or along the margins of Antarctica—

the excess water gravitates downward in a **downwelling current**. These are the deep currents that flow vertically and along the ocean floor and travel the full extent of the ocean basins, carrying heat energy and salinity.

Thermohaline Circulation— The Deep Currents

Differences in temperatures and salinity produce density differences important to the flow of deep currents. This is Earth's **thermohaline circulation**, and it is different from wind-driven surface currents. Traveling at slower speeds than surface currents, the thermohaline circulation hauls larger volumes of water. Figure 16.4 illustrates the ocean's physical structure and profiles of temperature, salinity, and dissolved gases; note the temperature and salinity differences with depth in this illustration.

To picture such a deep, salty current, imagine a continuous channel of water beginning with cold-water downwelling in the North Atlantic and along Antarctica and upwelling in the Indian Ocean and North Pacific (Figure 6.22). Here it warms and then is carried in surface currents back to the North Atlantic. A complete circuit of the current may require 1000 years from downwelling, to deep cold current covering thousands of kilometers, to its reemergence at the surface elsewhere. Even deeper Antarctic bottom water flows northward in the Atlantic Basin beneath these currents.

These current systems appear to play a profound role in global climate; in turn, global warming has the potential to disrupt the downwelling in the North Atlantic and the thermohaline circulation. If this sinking of surface waters slowed or stopped, the ocean's ability to redistribute absorbed heat energy would be disrupted and surface accumulation of heat energy would add to the warming

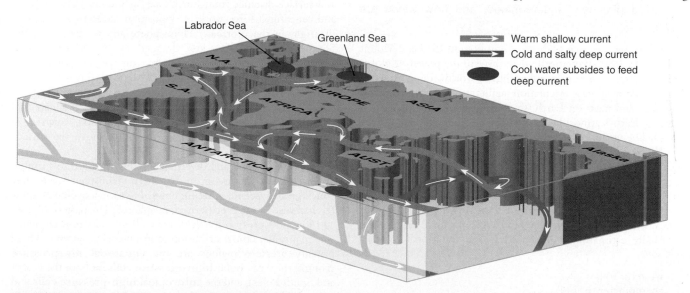

FIGURE 6.22 Deep-ocean thermohaline circulation.
Scientists are deciphering centuries-long deep circulation in the oceans, sluggishly driven by temperature- and salinity-caused density differences. This global circulation is like a vast conveyor belt of water, drawing heat energy from surface warm, shallow currents and transporting it for release in the depths of the ocean basins in deep, cold, salty currents. Four blue areas at high latitudes are where surface water cools, sinks, and feeds the deep circulation.

ANIMATION   Ocean Circulation

already underway. There is a freshening of ocean surface waters in both polar regions, contrasted with large increases in salinity in the surface waters at lower latitudes. Increased rates of glacier, sea-ice, and ice-sheet melting produce these fresh, lower-density surface waters that ride on top of the denser saline water. The concern is that such changes in ocean temperature and salinity could dampen the rate of the North Atlantic Deep Water downwelling in the North Atlantic.

This is a vital research frontier because Earth's hydrologic cycle is still little understood. Research groups such as Scripps Institution of Oceanography (**http://sio. ucsd.edu/**), Woods Hole Oceanographic Institution (**http://www.whoi.edu/**), Institute for the Study of Earth, Oceans, and Space (**http://www.eos.sr.unh.edu/**), and Bedford Institute of Oceanography **http://www.bio.gc.ca/ welcome-e.html**), to name only a few of many, are places to periodically check for updated information.

Summary and Review—Atmospheric and Oceanic Circulations

■ *Define* the concept of air pressure, and *describe* instruments used to measure air pressure.

The weight (created by motion, size, and number of molecules) of the atmosphere is **air pressure**, which exerts an average force of approximately 1 kg/cm^2 (14.7 lb/in.^{2}). A **mercury barometer** measures air pressure at the surface (mercury in a tube—closed at one end and open at the other, with the open end placed in a vessel of mercury—that changes level in response to pressure changes) or an **aneroid barometer** (a closed cell, partially evacuated of air, that detects changes in pressure).

air pressure (p. 143)
mercury barometer (p. 143)
aneroid barometer (p. 143)

1. How does air exert pressure? Describe the basic instrument used to measure air pressure. Compare the operation of two different types of instruments discussed.
2. What is normal sea-level pressure in millimeters? Millibars? Inches? Kilopascals?

■ *Define* wind, and *describe* how wind is measured, how wind direction is determined, and how winds are named.

Volcanic eruptions such as those of Tambora in 1815 and Mount Pinatubo in 1991 dramatically demonstrate the power of global winds to disperse aerosols and pollution worldwide in a matter of weeks. Atmospheric circulation facilitates important transfers of energy and mass on Earth, thus maintaining Earth's natural balances. Earth's atmospheric and oceanic circulations represent a vast heat engine powered by the Sun.

Wind is the horizontal movement of air across Earth's surface; turbulence adds wind updraft and downdrafts and a vertical component to the definition. Its speed is measured with an **anemometer** (a device with cups that are pushed by the wind) and its direction with a **wind vane** (a flat blade or surface that is directed by the wind). A descriptive scale useful in visually estimating wind speed is the traditional *Beaufort wind scale*.

wind (p. 144)
anemometer (p. 144)
wind vane (p. 145)

3. What is a possible explanation for the beautiful sunrises and sunsets during the summer of 1992 in North America? Relate your answer to global circulation.

4. Explain this statement: "The atmosphere socializes humanity, making all the world a spatially linked society." Illustrate your answer with some examples.
5. Define wind. How is it measured? How is its direction determined?
6. Distinguish among primary, secondary, and tertiary general classifications of global atmospheric circulation.
7. What is the purpose of the Beaufort wind scale? Characterize winds given Beaufort numbers of 4, 8, and 12, and describe specific effects over both water and land.

■ *Explain* the four driving forces within the atmosphere—gravity, pressure gradient force, Coriolis force, and friction force—and *describe* the primary high- and low-pressure areas and principal winds.

Earth's gravitational force on the atmosphere operates uniformly worldwide. Winds are driven by the **pressure gradient force** (air moves from areas of high pressure to areas of low pressure), deflected by the **Coriolis force** (an apparent deflection in the path of winds or ocean currents caused by the rotation of Earth and surface distance from Earth's axis as the gravitational force and centrifugal force work in opposition, deflecting objects to the right in the Northern Hemisphere and to the left in the Southern Hemisphere), and dragged by the **friction force** (Earth's varied surfaces exert a drag on wind movements in opposition to the pressure gradient). Maps portray air pressure patterns using the **isobar**—an isoline that connects points of equal pressure. A combination of the pressure gradient and Coriolis forces alone produces **geostrophic winds**, which move parallel to isobars, characteristic of winds above the surface frictional layer.

Winds descend and diverge, spiraling outward to form an **anticyclone** (clockwise in the Northern Hemisphere), and they converge and ascend, spiraling upward to form a **cyclone** (counterclockwise in the Northern Hemisphere). The pattern of high and low pressures on Earth in generalized belts in each hemisphere produces the distribution of specific wind systems. These primary pressure regions are the **equatorial low-pressure trough**, the weak **polar high-pressure cells** (at both the North and South Poles), and the **subtropical high-pressure cells** and **subpolar low-pressure cells**.

All along the equator winds converge into the equatorial low, creating the **intertropical convergence zone (ITCZ)**. Air rises along the equator and descends in the subtropics in each hemisphere. The winds returning to the ITCZ from the northeast in the Northern Hemisphere and from the southeast in the Southern Hemisphere produce the **trade winds**.

Winds flowing out of the subtropics to higher latitudes produce the **westerlies** in either hemisphere. The subtropical high-pressure cells on Earth, generally between 20° and 35° in either hemisphere, are variously named the **Bermuda high**, **Azores high**, and **Pacific high**.

Along the polar front and the series of low-pressure cells, the **Aleutian low** and **Icelandic low** dominate the North Pacific and Atlantic, respectively. This region of contrast between colder air toward the poles and warmer air toward the equator is the **polar front**. The weak and variable **polar easterlies** diverge from the polar high-pressure cells, particularly the **Antarctic high**.

pressure gradient force (p. 146)
Coriolis force (p. 146)
friction force (p. 146)
isobar (p. 147)
geostrophic winds (p. 152)
anticyclone (p. 152)
cyclone (p. 152)
equatorial low-pressure trough (p. 152)
polar high-pressure cells (p. 152)
subtropical high-pressure cells (p. 152)
subpolar low-pressure cells (p. 152)
intertropical convergence zone (ITCZ) (p. 154)
trade winds (p. 154)
westerlies (p. 156)
Bermuda high (p. 156)
Azores high (p. 156)
Pacific high (p. 156)
Aleutian low (p. 156)
Icelandic low (p. 156)
polar front (p. 156)
polar easterlies (p. 157)
Antarctic high (p. 157)

8. What does an isobaric map of surface air pressure portray? Contrast pressures over North America for January and July.
9. Describe the effect of the Coriolis force. Explain how it apparently deflects atmospheric and oceanic circulations.
10. What are geostrophic winds, and where are they encountered in the atmosphere?
11. Describe the horizontal and vertical air motions in a high-pressure anticyclone and in a low-pressure cyclone.
12. Construct a simple diagram of Earth's general circulation; begin by labeling the four principal pressure belts or zones, then add arrows between these pressure systems to denote the three principal wind systems.
13. How is the intertropical convergence zone (ITCZ) related to the equatorial low-pressure trough? How does it appear on a satellite image of accumulated precipitation in Figure 6.11?
14. Characterize the belt of subtropical high pressure on Earth: Name the specific cells. Describe the generation of westerlies and trade winds. Discuss sailing conditions.
15. What is the relation among the Aleutian low, the Icelandic low, and migratory low-pressure cyclonic storms in North America? In Europe?

■ *Describe* upper-air circulation and its support role for surface systems, and *define* the jet streams.

A **constant isobaric surface**, or surface along which the same pressure, such as 500 mb, occurs regardless of altitude, describes air pressure in the middle and upper troposphere. The height of this surface above the ground forms ridges and troughs that support the development of and help sustain surface pressure systems; surface lows are sustained by divergence aloft, and surface highs are sustained by convergence aloft.

Vast, flowing, longwave undulations in these upper-air westerlies form wave motions, the **Rossby waves**. Prominent streams of high-speed westerly winds in the upper-level troposphere are the **jet streams**. Depending on their latitudinal position in either hemisphere, they are termed the *polar jet stream* or the *subtropical jet stream*.

constant isobaric surface (p. 157)
Rossby waves (p. 157)
jet streams (p. 157)

16. What is the relation between wind speed and the spacing of isobars?
17. How is the constant isobaric surface (ridges and troughs) related to surface pressure systems? To divergence aloft and surface lows? To convergence aloft and surface highs?
18. Relate the jet-stream phenomenon to general upper-air circulation. How is the presence of this circulation related to airline schedules from New York to San Francisco and the return trip to New York?

■ *Overview* several multiyear oscillations of air temperature, air pressure, and circulation in the Arctic, Atlantic, and Pacific oceans.

Several system fluctuations that occur in multiyear or shorter periods are important in the global circulation picture. The most famous of these is the El Niño–Southern Oscillation (ENSO) phenomenon. A north–south fluctuation of atmospheric variability marks the *North Atlantic Oscillation (NAO)* as pressure differences between the Icelandic low and the Azores high in the Atlantic alternate in strength. The *Atlantic Oscillation (AO)* is the variable fluctuation between middle- and high-latitude air mass conditions over the Northern Hemisphere. The AO is associated with the NAO especially in winter, and the two phases of the *AO Index* correlate to the two-phased *NAO Index*.

Across the Pacific Ocean the *Pacific Decadal Oscillation (PDO)* involves variability between two regions of sea-surface temperatures and related air pressure: the northern and tropical western Pacific (region #1) and the area of the eastern tropical Pacific, along the West Coast (region #2). The PDO switches between positive and negative phases in 20- to 30-year cycles.

19. What phases are identified for the AO and NAO indices? What winter weather conditions generally affect the eastern United States during each phase?
20. What is the apparent relation between the PDO and the strength of El Niño events? Between PDO phases and the intensity of drought in the southwestern United States?

■ *Explain* several types of local winds: land–sea breezes, mountain–valley breezes, katabatic winds, and the regional monsoons.

Different heating characteristics of land and water surfaces create **land–sea breezes**. Temperature differences during the day and evening between valleys and mountain summits cause **mountain–valley breezes**. **Katabatic winds**, or gravity drainage winds, are of larger regional scale and are usually stronger than mountain–valley breezes, under certain conditions. An elevated plateau or highland is essential, where layers of air at the surface cool, become denser, and flow downslope.

Intense, seasonally shifting wind systems occur in the tropics over Southeast Asia, Indonesia, India, northern Australia, equatorial Africa, and southern Arizona. These winds involve an annual cycle of returning precipitation with the summer Sun and carry the Arabic word for season, *mausim*, or **monsoon**. The location and size of the Asian landmass and its proximity to the Indian Ocean and the seasonally shifting ITCZ drive the monsoons of southern and eastern Asia.

land–sea breezes (p. 162)
mountain–valley breezes (p. 163)
katabatic winds (p. 163)
monsoon (p. 166)

21. People living along coastlines generally experience variations in winds from day to night. Explain the factors that produce these changing wind patterns.
22. The arrangement of mountains and nearby valleys produces local wind patterns. Explain the day and night winds that might develop.
23. Describe the seasonal pressure patterns that produce the Asian monsoonal wind and precipitation patterns. Contrast January and July conditions.

■ *Discern* the basic pattern of Earth's major surface ocean currents and deep thermohaline circulation.

Ocean currents are primarily caused by the frictional drag of wind and occur worldwide at varying intensities, temperatures, and speeds, both along the surface and at great depths in the oceanic basins. The circulation around subtropical high-pressure cells in both hemispheres is notable on the ocean circulation map—these *gyres* are usually offset toward the western side of each ocean basin.

The trade winds converge along the ITCZ and push enormous quantities of water that pile up along the eastern shore of continents in a process known as the **western intensification**. Where surface water is swept away from a coast, either by surface divergence (induced by the Coriolis force) or by offshore winds, an **upwelling current** occurs. This cool water generally is nutrient-rich and rises from great depths to replace the vacating water. In other portions of the sea where there is an accumulation of water, the excess water gravitates downward in a **downwelling current**. These currents generate important mixing currents that flow along the ocean floor and travel the full extent of the ocean basins, carrying heat energy and salinity.

Differences in temperatures and salinity produce density differences important to the flow of deep, sometimes vertical, currents; this is Earth's **thermohaline circulation**. Traveling at slower speeds than wind-driven surface currents, the thermohaline circulation hauls larger volumes of water. There is scientific concern that increased surface temperatures in the ocean and atmosphere, coupled with climate-related changes in salinity, can alter the rate of thermohaline circulation in the oceans.

western intensification (p. 169)
upwelling current (p. 169)
downwelling current (p. 169)
thermohaline circulation (p. 169)

24. What is the relationship between global atmospheric circulation and ocean currents? Relate oceanic gyres to patterns of subtropical high pressure.
25. Define the western intensification. How is it related to the Gulf Stream and Kuroshio currents?
26. Where on Earth are upwelling currents experienced? What is the nature of these currents? Where are the four areas of downwelling that feed these dense bottom currents?
27. What is meant by deep-ocean thermohaline circulation? At what rates do these currents flow? How might this circulation be related to the Gulf Stream in the western Atlantic Ocean?
28. Relative to Question 27, what relation do these deep currents have to global warming and possible climate change?

NetWork

The *Geosystems Student Learning Center* provides online resources for this chapter on the World Wide Web. To begin: Once at the Center, click on the cover of this textbook, scroll the Table of Contents menu, and select this chapter. You will find self-tests that are graded, review exercises, specific updates for items in the chapter, and in "Destinations" many links to interesting related pathways on the Internet. *Geosystems Student Learning Center* is found at **http://www.prenhall.com/ christopherson/.**

Critical Thinking

A. Using Table 6.1 and Figure 6.5, on a day with wind, estimate wind speed and wind direction at least twice during the day, and record them in your notebook. If possible, check these against the reports from a local source of weather information. What changes in winds do you notice over several days? How did these changes relate to the weather you experienced?

B. Refer to the "Short Answer" section in Chapter 6 of the *Geosystems* Home Page. Questions 1 through 5 deal with satellite images of pressure systems, wind, and cloud patterns. Using information in this chapter, complete these five items. Can you identify any causes for these circulation patterns? Note the other nine short-answer items for further thought.

C. Go to **http://www.awea.org/,** the web site of the American Wind Energy Association, and to the European Wind Energy Association at **http://www.ewea.org/.** Sample the materials presented as you assess Focus Study 6.1 and the potential for wind-generated electricity. What are your thoughts concerning this resource—its potential, reasons for delays, and the competitive economics presented? What countries are leading the way? Propose a brief action plan for more rapid progress, or a justification for more delays, depending on your point of view.

PART II

The Water, Weather, and Climate Systems

The water, weather, and climate systems stretch to the horizon, across the Arctic Ocean. Distant precipitation makes its way to the surface in gray streaks of moisture. 2007 set a record for the lowest extent of sea ice since the 1970s and the satellite record. The polar regions are the scientific focus of the International Polar Year 2007–2009 that involves topics from throughout Geosystems.
[Photo by Bobbé Christopherson.]

Earth is the water planet, a topic at the heart of Part II. Chapter 7 describes the remarkable qualities and properties water possesses, where it occurs, and the way it is distributed. We see the daily dynamics of the atmosphere—the powerful interaction of moisture and energy, the resulting stability and instability, and the variety of cloud forms—as a preamble to understanding weather. Chapter 8 examines weather and its causes. Topics include the interaction of air masses, the daily weather map, and analysis of violent phenomena in thunderstorms, tornadoes, and hurricanes, and the recent trends for each of these.

Chapter 9 explains water circulation on Earth in the hydrologic cycle—an output of the water–weather system. We examine the water-budget concept, which is useful in understanding soil-moisture and water-resource relationships on global, regional, and local scales. Potable (drinking) water is emerging as the critical global political issue in this century. In Chapter 10, we see the spatial implications over time of the energy–atmosphere and water–weather systems and the generation of Earth's climatic patterns. In this way, Chapter 10 interconnects all the system elements from Chapters 2 through 9. Part Two closes with a discussion of global climate-change science, a look at present conditions, and a forecast of future climate trends.

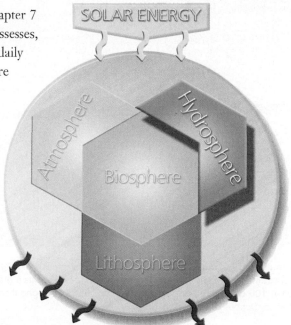

175

Cumulonimbus, cumulus, and cirrus clouds perfectly highlight Loch Tay in central Scotland, in the Grampian Mountains of the Highlands. The long loch (or lake) fills a glaciated valley. Annual precipitation averages about 100 cm (39 in.) and is evenly distributed throughout the year.

[Photo by Bobbé Christopherson.]

Water and Atmospheric Moisture

■ Key Learning Concepts

After reading the chapter, you should be able to:

■ **Describe** the origin of Earth's waters, **define** the quantity of water that exists today, and **list** the locations of Earth's freshwater supply.

■ **Describe** the heat properties of water, and **identify** the traits of its three phases: solid, liquid, and gas.

■ **Define** humidity and the expressions of the relative humidity concept, and **explain** dew-point temperature and saturated conditions in the atmosphere.

■ **Define** atmospheric stability, and **relate** it to a parcel of air that is ascending or descending.

■ **Illustrate** three atmospheric conditions—unstable, conditionally unstable, and stable—with a simple graph that relates the environmental lapse rate to the dry adiabatic rate (DAR) and moist adiabatic rate (MAR).

■ **Identify** the requirements for cloud formation, and **explain** the major cloud classes and types, including fog.

Walden is blue at one time and green at another, even from the same point of view. Lying between the earth and the heavens, it partakes of the color of both. . . . A lake is the landscape's most beautiful and expressive feature. It is earth's eye; looking into which the beholder measures the depth of his own nature. . . . Sky water. It needs no fence. Nations come and go without defiling it. It is a mirror which no stone can crack. . . . Nature continually

*Reprinted by permission of Merrill, an imprint of Macmillan Publishing Company, from *Walden* by Henry David Thoreau, pp. 192, 202, 204. Copyright © 1969 by Merrill Publishing. Originally published in 1854.

repairs . . . a mirror in which all impurity presented to it sinks, swept and dusted by the sun's hazy brush. A field of water . . . is continually receiving new life and motion from above. It is intermediate in its nature between land and sky.*

Thus did Thoreau speak of the water so dear to him— Walden Pond in Massachusetts, where he lived along the shore.

Water is critical to our daily lives and is an extraordinary compound in nature. Several satellite probes detected ice beneath the poles of the Moon. Three orbiting spacecraft and two robot landers imaged and detected flowing water early in Martian history. Scientists are finding evidence on Mars of existing subsurface water, polar water ice, and modern erosional features caused by water. Farther from the Sun, icy expanses can be seen on two of Jupiter's moons, Europa and Calisto—exciting discoveries about water in our Solar System. Yet, in the Solar System, water occurs in significant quantities only on our planet— water covers 71% of Earth by area.

Pure water is colorless, odorless, and tasteless; yet, because it is a solvent (dissolves solids), pure water rarely occurs in nature. Water weighs 1 g/cm³ (gram per cubic centimeter), or 1 kg/L (kilogram per liter). In the English system, water weighs 62.3 lb/ft³, or 8.337 lb/gal.

Water constitutes nearly 70% of our bodies by weight and is the major ingredient in plants, animals, and our food. A human can survive 50 to 60 days without food but only 2 or 3 days without water. The water we use must be adequate, both in quantity and quality, for its many tasks—everything from personal hygiene to vast national water projects. Water indeed occupies that place between land and sky, mediating energy and shaping both the lithosphere and atmosphere, as Thoreau revealed. Water is the medium of life.

In this chapter: We examine water on Earth and the dynamics of atmospheric moisture and stability—the essentials of weather. The key questions answered include: What is the origin of water? How much water is there? And where is water located? An irony exists with this most common compound because it possesses such uncommon physical characteristics. Water's unique heat properties and existence in all three states in nature are critical in powering Earth's weather systems. Condensation of water vapor and atmospheric conditions of stability and instability are key to cloud formation. We end the chapter with clouds, our beautiful indicators of atmospheric conditions.

Water on Earth

Earth's hydrosphere contains about 1.36 billion cubic kilometers of water (specifically, 1,359,208,000 km³, or 326,074,000 mi³). According to scientific evidence, much of Earth's water originated from icy comets and hydrogen- and oxygen-laden debris that were part of the planetesimals that coalesced to form the planet. In 2007, the orbiting Spitzer Space Telescope observed for the first time the presence of water vapor and ice as planets form in a system 1000 light-years from Earth. Such discoveries prove water to be abundant throughout the Universe. As a planet forms, water from within migrates to its surface and outgasses.

Outgassing is a continuing process by which water and water vapor emerge from layers deep within and below the crust, 25 km (15.5 mi) or more below Earth's surface. Figure 7.1 presents two such areas, among many sites worldwide, in New Zealand and Iceland. The Haukadalur Geysir geothermal area is in southeast Iceland, about 110 km from its capital, Reykjavík (Figure 7.1b). Written accounts of the Geysir activity date back to A.D. 1294 and it is the namesake of the term *geyser* we use today.

In the early atmosphere, massive quantities of outgassed water vapor condensed and then fell to Earth in

(a)

(b)

FIGURE 7.1 Water outgassing from the crust.
Outgassing of water from Earth's crust in geothermal areas: (a) near Wairakei on the North Island of New Zealand; and (b) the Smidur boiling spring in the Geysir area in Haukadalur, Iceland. [Photos by (a) Bill Bachman/Photo Researchers, Inc.; (b) Bobbé Christopherson.]

torrential rains. For water to remain on Earth's surface, land temperatures had to drop below the boiling point of 100°C (212°F), something that occurred about 3.8 billion years ago. The lowest places across the face of Earth then began to fill with water—first ponds, then lakes and seas, and eventually ocean-sized bodies of water. Massive flows of water washed over the landscape, carrying both dissolved and solid materials to these early seas and oceans. Outgassing of water has continued ever since and is visible in volcanic eruptions, geysers, and seepage to the surface.

Worldwide Equilibrium

Today, water is the most common compound on the surface of Earth, having attained the present volume approximately 2 billion years ago. This quantity has remained relatively constant, even though water is continuously being lost from the system. Water is lost when it dissociates into hydrogen and oxygen and the hydrogen escapes Earth's gravity to space, or when it breaks down and forms new compounds with other elements. Pristine water not previously at the surface, which emerges from within Earth's crust, replaces lost water in the system. The net result of these water inputs and outputs is that Earth's hydrosphere is in a steady-state equilibrium in terms of quantity.

Despite this overall net balance in water quantity, worldwide changes in sea level do occur. The concept **eustasy** describes the global sea-level condition. Eustatic changes relate to the water volume in the oceans and not changes in the overall quantity of planetary water. Some of these changes result when the amount of water stored in glaciers and ice sheets varies; these are **glacio-eustatic** factors (see Chapter 17). In cooler times, as more water is bound up in glaciers (on mountains worldwide and in high latitudes) and in ice sheets (Greenland and Antarctica), sea level lowers. In warmer times, less water is stored as ice, so sea level rises.

Some 18,000 years ago, during the most recent ice-age pulse, sea level was more than 100 m (330 ft) lower than it is today; 40,000 years ago it was 150 m (about 500 ft) lower. Over the past 100 years, mean sea level has risen by 20–40 cm (8–16 in.) and is still rising worldwide at an accelerating pace as higher temperatures melt more ice.

Isostasy refers to actual vertical physical movement in landmasses, such as continental uplift or subsidence. Such landscape changes cause apparent changes in sea level relative to coastal environments. Chapter 11 discusses such land elevation changes.

Distribution of Earth's Water Today

From a geographic point of view, ocean and land surfaces are distributed unevenly. If you examine a globe, it is obvious that most of Earth's continental land is in the Northern Hemisphere, whereas water dominates the Southern Hemisphere. In fact, when you look at Earth from certain angles, it appears to have an *oceanic hemisphere* and a *land hemisphere* (Figure 7.2).

An illustration of the present location of all of Earth's liquid and frozen water—whether fresh or saline, surface or underground—is in Figure 7.3. The oceans contain

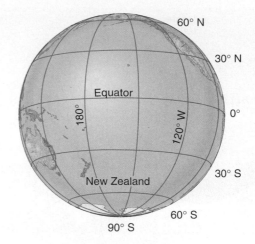

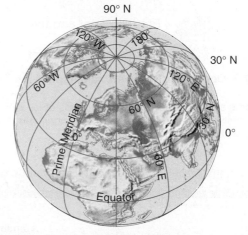

FIGURE 7.2 Land and water hemispheres. Two perspectives that roughly illustrate Earth's ocean hemisphere and land hemisphere.

97.22% of all water (Figure 7.3a). A table in the figure lists and details the four oceans—Pacific, Atlantic, Indian, and Arctic. The extreme southern portions of the Pacific, Atlantic, and Indian Oceans that surround the Antarctic continent are collectively the "Southern Ocean." Fourth in size among the five oceans, the Southern Ocean lacks precise definition and boundaries, so we meld it into its three parent oceans in the table.

Only 2.78% of all of Earth's water is freshwater (nonsaline and nonoceanic). The middle pie chart, along with Table 7.1, details this freshwater portion—surface water and subsurface water (Figure 7.3b). Ice sheets and glaciers are the greatest single repository of surface freshwater; they contain 77.14% of all of Earth's freshwater. Adding subsurface groundwater to frozen surface water accounts for 99.36% of all freshwater.

The remaining freshwater, which resides in lakes, rivers, and streams so familiar to us, actually represents less than 1% of all water (Figure 7.3c). All the world's freshwater lakes total only 125,000 km³ (30,000 mi³), with 80% of this volume in just 40 of the largest lakes and about 50% contained in just 7 lakes (listed with Table 7.1).

The greatest single volume of lakewater resides in 25-million-year-old Lake Baykal in Siberian Russia. It

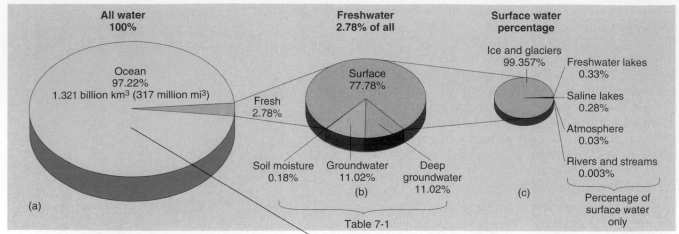

Earth's Water and the
Hydrologic Cycle

FIGURE 7.3 Ocean and freshwater distribution on Earth.
Pie diagrams show location and percentages of (a) all
water, (b) freshwater including subsurface water, and
(c) surface water.

Ocean	Earth's Ocean Area (%)	*Area (km² [mi²])	*Volume (km³ [mi³])	Mean Depth of Main Basin (m [ft])
Pacific	48	179,670 (69,370)	724,330 (173,700)	4280 (14,040)
Atlantic	28	106,450 (41,100)	355,280 (85,200)	3930 (12,890)
Indian	20	74,930 (28,930)	292,310 (70,100)	3960 (12,900)
Arctic	4	14,090 (5440)	17,100 (4100)	1205 (3950)

*Data in thousands (000): includes all marginal seas.

Table 7.1 Distribution of Freshwater on Earth

Location	Amount [km³ (mi³)]		Percentage of Freshwater	Percentage of Total Water
Surface Water				
Ice sheets and glaciers	29,180,000	(7,000,000)	77.14	2.146
Freshwater lakes*	125,000	(30,000)	0.33	0.009
Saline lakes and inland seas	104,000	(25,000)	0.28	0.008
Atmosphere	13,000	(3,100)	0.03	0.001
Rivers and streams	1,250	(300)	0.003	0.0001
Total surface water	29,423,250	(7,058,400)	77.78	2.164
Subsurface Water				
Groundwater—surface to 762 m (2500 ft) depth	4,170,000	(1,000,000)	11.02	0.306
Groundwater—762 to 3962 m (2500 to 13,000 ft) depth	4,170,000	(1,000,000)	11.02	0.306
Soil-moisture storage	67,000	(16,000)	0.18	0.005
Total subsurface water	8,407,000	(2,016,000)	22.22	0.617
Total Freshwater (rounded)	**37,800,000**	**(9,070,000)**	**100.00%**	2.78%

*Major Freshwater Lakes	Volume [km³ (mi³)]		Surface Area [km² (mi²)]		Depth (m [ft])	
Baykal (Russia)	22,000	(5280)	31,500	(12,160)	1620	(5315)
Tanganyika (Africa)	18,750	(4500)	39,900	(15,405)	1470	(4923)
Superior (U.S./Canada)	12,500	(3000)	83,290	(32,150)	397	(1301)
Michigan (U.S.)	4,920	(1180)	58,030	(22,400)	281	(922)
Huron (U.S./Canada)	3,545	(850)	60,620	(23,400)	229	(751)
Ontario (U.S./Canada)	1,640	(395)	19,570	(7,550)	237	(777)
Erie (U.S./Canada)	485	(115)	25,670	(9,910)	64	(210)

Not connected to the ocean are saline lakes and salty inland seas. They usually exist in regions of interior river drainage (no outlet to the ocean), which allows salts to become concentrated. They contain 104,000 km³ (25,000 mi³) of water. Examples of such lakes include Utah's Great Salt Lake, California's Mono Lake, Southwest Asia's Caspian and Aral Seas, and the Dead Sea between Israel and Jordan (Figure 7.4).

Think of all the moisture in the atmosphere and Earth's thousands of flowing rivers and streams. Combined, they amount to only 14,250 km³ (3400 mi³), or only 0.033% of freshwater, or 0.0011% of all water! Yet, this small amount is dynamic. A water molecule traveling through atmospheric and surface-water paths moves through the entire hydrologic cycle (ocean–atmosphere–precipitation–runoff) in less than two weeks. Contrast this to a water molecule in deep-ocean circulation, groundwater, or a glacier; moving slowly, it takes thousands of years to move through the system (Figure 7.5).

Unique Properties of Water

Earth's distance from the Sun places it within a most remarkable temperate zone when compared with the locations of the other planets. This temperate location allows all three states of water—ice, liquid, and vapor—to occur naturally on Earth.

Even though water is the most common compound on Earth's surface, it exhibits most uncommon properties. Two atoms of hydrogen and one of oxygen, which readily bond, comprise each water molecule (suggested in

FIGURE 7.4 Mono Lake, California, a saline lake. Mono Lake is in the western Great Basin, with the Sierra Nevada range as a backdrop (aerial view is to the southwest in February). Formed approximately 3 million years ago, this lake has no outlets other than evaporation. At its maximum, some 18,000 years ago, Lake Russell was about five times larger and six times deeper than the present Mono Lake—the basin marked by its ancient shorelines. [Photo by Bobbé Christopherson.]

contains almost as much water as all five U.S. Great Lakes combined. Africa's Lake Tanganyika contains the next-largest volume, followed by the five Great Lakes. Overall, 70% of lakewater is in North America, Africa, and Asia, with about a fourth of lakewater worldwide in small lakes too numerous to count. More than 3 million lakes exist in Alaska alone, and Canada has over 750 km² of lake surface.

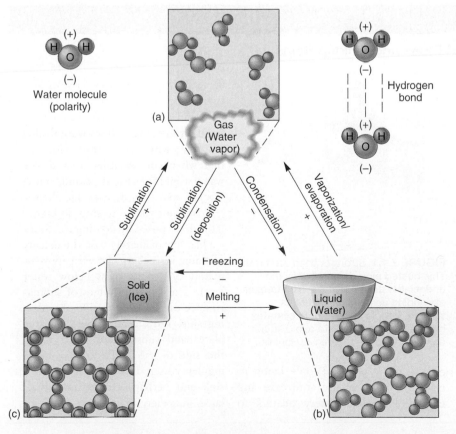
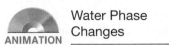

FIGURE 7.5 Three states of water and water's phase changes. The three physical states of water: (a) gas, or water vapor, (b) water, and (c) ice. Note the molecular arrangement in each state and the terms that describe the changes from one phase to another. Also note how the polarity of water molecules bonds them to one another, loosely in the liquid state and firmly in the solid state. The plus and minus symbols in the phase changes denote whether heat energy is absorbed (+) or liberated (released) (−).

Figure 7.5, upper left). Once hydrogen and oxygen atoms join in a covalent, or double, bond, they are difficult to separate, thereby producing a water molecule that remains stable in Earth's environment. This water molecule is a versatile solvent and possesses extraordinary heat characteristics.

The nature of the hydrogen–oxygen bond gives the hydrogen side of a water molecule a positive charge and the oxygen side a negative charge (see Figure 7.5, upper right). As a result of this *polarity*, water molecules attract each other: The positive (hydrogen) side of a water molecule attracts the negative (oxygen) side of another. This bonding between water molecules is *hydrogen bonding*.

The polarity of water molecules also explains why water "acts wet," sticking to things and dissolving many substances. Because of this solvent ability, pure water is rare in nature because something is usually dissolved in it.

The effects of hydrogen bonding in water are observable in everyday life. Hydrogen bonding creates the *surface tension* that allows you to float a steel needle on the surface of water, even though steel is much denser than water. This *surface tension* allows you to slightly overfill a glass with water; webs of millions of hydrogen bonds hold the water slightly above the rim.

Hydrogen bonding is the cause of *capillarity*, which you observe when you "dry" something with a paper towel. The towel draws water through its fibers because hydrogen bonds make each molecule pull on its neighbor. In chemistry laboratory classes, students observe the curved *meniscus*, or surface of the water, which forms in a cylinder

or a test tube because hydrogen bonding allows the water to slightly "climb" the glass sides. *Capillary action* is an important component of soil-moisture processes, discussed in Chapters 9 and 18. Without hydrogen bonding to hold molecules together in water and ice, water would be a gas at normal surface temperatures.

Heat Properties

For water to change from one state to another (solid, liquid, or vapor), *heat energy must be absorbed or liberated (released)*. To cause a change of state, the amount of heat energy must be sufficient to affect the hydrogen bonds between molecules. This relation between water and heat energy is important to atmospheric processes. In fact, the heat exchanged between physical states of water provides more than 30% of the energy that powers the general circulation of the atmosphere.

Figure 7.5 presents the three states of water and the terms describing a change from one state to another, a **phase change**. Along the bottom of the illustration, *melting* and *freezing* describe the familiar phase change between solid and liquid. At the right, the terms *condensation* and *evaporation* (or *vaporization* at boiling temperature) apply to the change between liquid and vapor. At the left, the term **sublimation** refers to the direct change of ice to water vapor or water vapor to ice; although when water vapor attaches directly to an ice crystal it is referred to as *deposition*. The deposition of water vapor to ice may form *frost* on surfaces.

News Report 7.1

Breaking Roads and Pipes and Sinking Ships

Road crews are busy in the summer repairing winter damage to streets and freeways in regions where winters are cold. A major contributor to this damage is the expansive phase change from water to ice. Rainwater seeps into roadway cracks and then expands as it freezes, thus breaking up the pavement. Perhaps you have noticed that bridges suffer the greatest damage. The reason is that cold air can circulate beneath a bridge and produces more freeze–thaw cycles on the bridge than in the roadbed on rock and soil.

The expansion of freezing water exerts a tremendous force—enough to crack plumbing or an automobile radiator or engine block (Figure 7.1.1). Wrapping water pipes with insulation to avoid damage is a common winter

FIGURE 7.1.1 Stronger than cast iron. This busted pipe demonstrates the undeniable power of water as it freezes, expanding as much as 9% of its volume. People living in cold climates must take precautions to avoid such damage. [Photo by Steven K. Huhtala.]

task in many places. People living in very cold climates use antifreeze and engine-block and battery heaters to

avoid damage to vehicles. Historically, this physical property of water was put to useful work in quarrying rock for building materials. Holes were drilled and filled with water before winter so that when cold weather arrived, the water would freeze and expand, cracking the rock into manageable shapes.

A major hazard to ships in higher latitudes is posed by floating ice. Since ice has approximately 0.86 the density of water, an iceberg sits with approximately 6/7 of its mass below water level. The irregular edges of submarine ice can impact the side of a passing ship. Hitting an iceberg buckled plates and substandard rivets along the side of the RMS *Titanic* on its maiden voyage in 1912, causing it to sink and perhaps triggering a brief lapse in society's faith in technology.

Ice, the Solid Phase As water cools, it behaves like most compounds and contracts in volume. However, it reaches its greatest density not as ice, but as water at 4°C (39°F). Below that temperature, water behaves differently from other compounds. It begins to expand as more hydrogen bonds form among the slower-moving molecules, creating the hexagonal (six-sided) structures shown in Figure 7.5c and Figure 7.6a. This six-sided preference applies to ice crystals of all shapes: plates, columns, needles, and dendrites (branching or treelike forms). Ice crystals demonstrate a unique interaction of chaos (all ice crystals are different) and the determinism of physical principles (all have a six-sided structure).

This expansion of water and ice that begins at 4°C continues to a temperature of –29°C (–20°F)—up to a 9% increase in volume is possible. This expansion is important in the weathering of rocks, highway and pavement damage, and burst water pipes. News Report 7.1 discusses some effects of ice. For more on ice crystals and snowflakes, see http://www.its.caltech.edu/~atomic/snowcrystals/ or see http://emu.arsusda.gov/snowsite/default.html.

The expansion in volume that accompanies the freezing process results in a decrease in density (the same number of molecules occupy a larger space). Specifically, pure ice has 0.91 times the density of water, so it floats. Without this change in density, much of Earth's freshwater would be bound in masses of ice on the ocean floor.

In nature, this density varies slightly given the age of the ice and air content. Therefore, icebergs float with slightly varying displacements that average about 1/7 (14%) of their mass exposed and about 6/7 (86%) for the submerged portion hidden beneath the ocean's surface (Figure 7.6d). With underwater portions melting faster than those above water, icebergs are inherently unstable and will overturn. You see in Figure 7.6b ruffled and fluted ice that was previously underwater.

(a) (b) (c) (d)

FIGURE 7.6 The uniqueness of ice forms.
(a) Computer-enhanced photo reveals ice-crystal patterns, which are dictated by the internal structure between water molecules. (b) Icebergs along the Greenland coast that overturned, exposing underwater portions that were sculpted by currents and melt channels. (c) Ice needles form in the freezing process of a meltwater stream in Greenland. (d) A bergy bit (small iceberg) off the coast of Antarctica illustrates the density–buoyancy state of floating ice. [(a) Photo enhancement © Scott Camazine/Photo Researchers, Inc., after W. A. Bentley; (b), (c), and (d) photos by Bobbé Christopherson.]

Chapter 13 discusses the freezing action of ice as an important physical weathering process. Chapter 17 discusses the freeze-and-thaw action of surface and subsurface water affecting approximately 30% of Earth's surface in periglacial landscapes, producing a variety of processes and landforms.

Water, the Liquid Phase As a liquid, water assumes the shape of its container and is a noncompressible fluid. For ice to change to water, heat energy must increase the motion of the water molecules to break some of the hydrogen bonds (Figure 7.5b). Despite the fact that there is no change in sensible temperature between ice at 0°C (32°F) and water at 0°C, 80 calories* of heat energy must be absorbed for the phase change of 1 g of ice to melt to 1 g of water (Figure 7.7, upper left). Heat energy involved in the phase change is **latent heat** and is hidden within the structure of water. It becomes liberated whenever the phase reverses and a gram of water freezes. The *latent heat of freezing* or the *latent heat of melting* each involves 80 calories.

To raise the temperature of 1 g of water at 0°C (32°F) to boiling at 100°C (212°F), we must add 100 cal, gaining an increase of 1 C° (1.8 F°) for each calorie added. No phase change is involved in this temperature gain.

Water Vapor, the Gas Phase Water vapor is an invisible and compressible gas in which each molecule moves independently of the others (Figure 7.5a). The phase change from liquid to vapor at boiling temperature, under normal sea-level pressure, requires the addition of 540 cal for each gram, the **latent heat of vaporization** (Figure 7.7). When water vapor condenses to a liquid,

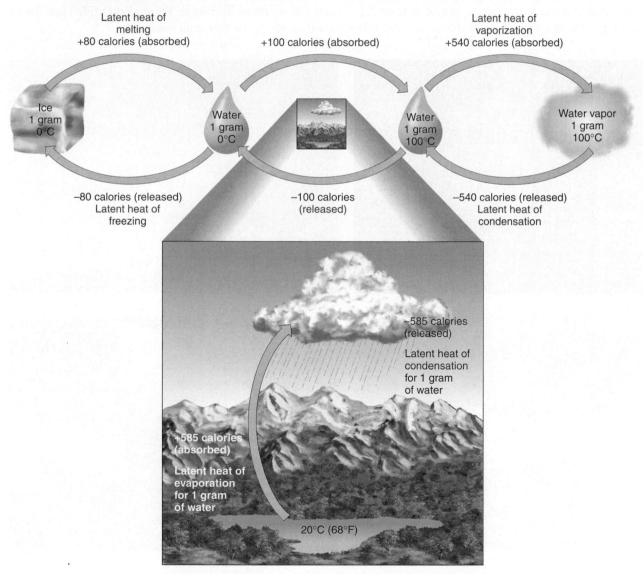

FIGURE 7.7 Water's heat-energy characteristics.
The phase changes of water absorb or release a lot of latent heat energy. To transform 1 g of ice at 0°C to 1 g of water vapor at 100°C requires 720 cal: 80 + 100 + 540. The landscape illustrates phase changes between water (lake at 20°C) and water vapor under typical conditions in the environment.

*Remember from Chapter 2: A calorie is the amount of energy required to raise the temperature of 1 g of water (at 15°C) 1 degree Celsius and is equal to 4.184 joules.

each gram gives up its hidden 540 cal as the **latent heat of condensation**. Perhaps you have felt the liberation of the latent heat of condensation on your skin from steam when you drained steamed vegetables, or pasta, or filled a hot teakettle.

In summary, taking 1 g of ice at 0°C and changing its phase to water, then to water vapor at 100°C—from a solid, to a liquid, to a gas—*absorbs* 720 cal (80 cal + 100 cal + 540 cal). Reversing the process, or changing the phase of 1 g of water vapor at 100°C to water and then to ice at 0°C, *liberates* 720 cal into the surrounding environment. Your *Student Animation CD* that accompanies this book has an excellent animation illustrating these concepts.

Heat Properties of Water in Nature

In a lake or stream or in soil water, at 20°C (68°F), every gram of water that breaks away from the surface through evaporation must absorb from the environment approximately 585 cal as the *latent heat of evaporation* (see the natural scene in Figure 7.7). This is slightly more energy than would be required if the water were at a higher temperature such as boiling (540 cal). You can feel this absorption of latent heat as evaporative cooling on your skin when it is wet. This latent heat exchange is the dominant cooling process in Earth's energy budget.

The process reverses when air cools and water vapor condenses back into the liquid state, forming moisture droplets and thus liberating 585 cal for every gram of water as the *latent heat of condensation*. When you realize that a small, puffy, fair-weather cumulus cloud holds 500–1000 tons of moisture droplets, think of the tremendous latent heat released when water vapor condensed to droplets.

Meteorologists estimated that the moisture in Hurricane Katrina (2005) weighed more than 30 trillion metric tons at its maximum power and mass. With 585 cal released for every gram as the latent heat of condensation, a weather event such as a hurricane involves a staggering amount of energy.

The **latent heat of sublimation** absorbs 680 cal as a gram of ice transforms into vapor. Water vapor freezing directly to ice releases a comparable amount of energy.

Humidity

Humidity refers to water vapor in the air. The capacity of air for water vapor is primarily a function of temperature—the temperatures of both the air and the water vapor, which are usually the same. There are several ways of expressing humidity, as discussed in this section.

We are all aware of humidity in the air, for its relationship to air temperature determines our sense of comfort. North Americans spend billions of dollars a year to adjust humidity, either with air-conditioning (extracting water vapor and cooling) or with air-humidifying (adding water vapor). We discussed the relation between humidity and temperature and the heat index in Chapter 5. To determine the energy available for powering weather,

one has to know the water vapor *content* of air and relate this in a ratio to the air's *saturation equilibrium* at a given temperature.

Relative Humidity

After air temperature and barometric pressure, the most common piece of information in local weather broadcasts is relative humidity. **Relative humidity** is a ratio (expressed as a percentage) of the amount of water vapor that is actually in the air compared to the maximum water vapor possible in the air at a given temperature.

Relative humidity varies because of water vapor or temperature changes in the air. The formula to calculate the relative humidity ratio and express it as a percentage places moisture as the numerator and the water vapor possible in the air as the denominator:

$$\text{Relative humidity} = \frac{\text{Actual water vapor in the air}}{\substack{\text{Maximum water vapor possible} \\ \text{in the air at that temperature}}} \times 100$$

Warmer air increases the evaporation rate from water surfaces, whereas cooler air tends to increase the condensation rate of water vapor to water surfaces. Because there is a maximum amount of water vapor that can exist in a volume of air at a given temperature, the rates of evaporation and condensation can reach equilibrium at some point; the air is then saturated with humidity. Relative humidity tells us how near the air is to saturation and is an expression of an ongoing process of water molecules moving between air and moist surfaces.

In Figure 7.8 at 5 P.M. the evaporation rate exceeds condensation at the higher temperatures of this time of day and relative humidity is at 20%. At 11 A.M. the evaporation rate still exceeds condensation, though not by as much since daytime temperatures are not as high, so the same volume of water vapor now occupies 50% of the maximum possible capacity. At 5 A.M., in the cooler morning air, saturation equilibrium exists and any further cooling or addition of water vapor produces net condensation. When the air is saturated with maximum water vapor for its temperature, the relative humidity percentage is 100%.

Saturation As stated, air is at **saturation**, at 100% relative humidity, when the rate of evaporation and the rate of condensation—the net transfer of water molecules—reach equilibrium. Saturation indicates that any further addition of water vapor or any decrease in temperature that reduces the evaporation rate results in active condensation (clouds, fog, or precipitation).

The temperature at which a given mass of air becomes saturated and net condensation begins to form water droplets is the **dew-point temperature**. *The air is saturated when the dew-point temperature and the air temperature are the same.* When temperatures are below freezing, the term *frost point* is sometimes used.

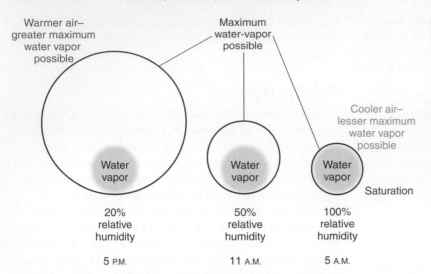

Warmer air–
greater maximum
water vapor
possible

Maximum
water-vapor
possible

Cooler air–
lesser maximum
water vapor
possible

Water
vapor

Water
vapor

Water
vapor

Saturation

20%
relative
humidity

50%
relative
humidity

100%
relative
humidity

5 P.M.

11 A.M.

5 A.M.

FIGURE 7.8 Water vapor, temperature, and relative humidity.
The maximum water vapor possible in warm air is greater (net evaporation) than that of cold air (net condensation), so relative humidity changes with temperature, even though in this example the actual water vapor present in the air stays the same during the day.

A cold drink in a glass provides a common example of these conditions. The water droplets that form on the outside of the glass condense from the air because the air layer next to the glass is chilled to below its dew-point temperature and thus becomes saturated (Figure 7.9). Figure 7.9 shows two additional examples of saturated air and active condensation above a rock surface and in a fog formation overlying a cool ocean surface. As you walk to classes on some mornings, you perhaps notice damp lawns, an indication of dew-point conditions in cool morning air.

Satellites using infrared sensors now routinely sense water vapor in the lower atmosphere. Water vapor

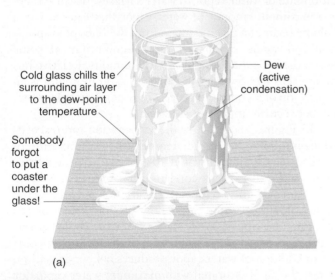

Cold glass chills the
surrounding air layer
to the dew-point
temperature

Dew
(active
condensation)

Somebody
forgot
to put a
coaster
under the
glass!

(a)

FIGURE 7.9 Dew-point temperature examples.
(a) The low temperature of the glass chills the surrounding air layer to the dew-point temperature and saturation. Thus, water vapor condenses out of the air and onto the glass as dew. (b) Cold air above the rain-soaked rocks is at the dew point and is saturated. Water evaporates from the rock into the air and condenses in a changing veil of clouds. (c) The cold ocean surface chills the moist air layer to the dew point and saturation; a dense fog forms. In the evening when temperatures lower over coastal lands, fog forms farther inland as dew-point temperatures move inland, giving the appearance that fog is moving ashore. [Photos by author.]

(b)

(c)

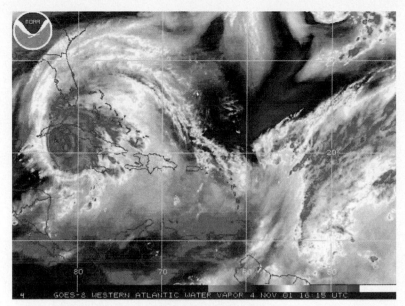

(a)

(b)

FIGURE 7.10 *GOES-8* infrared images of water vapor in the atmosphere.
(a) On the scale used for this image, color denotes high-altitude cloud tops that are cooler over the Gulf of Mexico and Caribbean. Hurricane Michelle and its strong vertical development is clearly visible heading toward western Cuba and near Florida, November 4, 2001. (b) Water vapor over the full Western Hemisphere with higher water-vapor content indicated by lighter-gray tones; note subpolar low-pressure circulation. [*GOES* images courtesy of NESDIS Satellite Services Division NOAA.]

absorbs long wavelengths (infrared), making it possible to distinguish areas of relatively high water vapor from areas of low water vapor. Figure 7.10 includes images of Hurricane Michelle and the Western Hemisphere showing water-vapor content recorded by sensors of the "water-vapor channel." This ability is important to forecasting because it shows the moisture available to weather systems and therefore the available latent heat energy and precipitation potential.

Daily and Seasonal Relative Humidity Patterns An inverse relation occurs during a typical day between air temperature and relative humidity—as temperature rises, relative humidity falls (Figure 7.11a). Relative humidity is highest at dawn, when air temperature is lowest. If you park outdoors, you know about the wetness of the dew that condenses on your car or bicycle overnight. In your own experience, you probably have noticed this pattern—the morning dew on windows, cars, and lawns evaporates by late morning as net evaporation increases with air temperature.

Relative humidity is lowest in the late afternoon, when higher temperatures increase the rate of evaporation. As shown in Figure 7.8, the actual water vapor present in the air may remain the same throughout the day. However, relative humidity changes because the temperature, and therefore the rate of evaporation, varies from morning to afternoon.

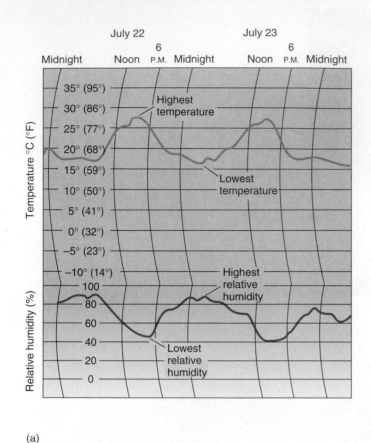

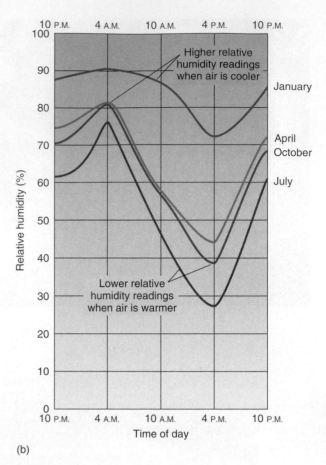

(a)

(b)

FIGURE 7.11 Daily and seasonal relative humidity patterns.
(a) Typical daily variations demonstrate temperature and relative humidity relations; (b) seasonal variations in daily relative humidity.

The seasonal variation in relative humidity by time of day confirms the relationship of temperature and relative humidity (Figure 7.11b). January readings are higher than July readings because air temperatures are lower overall in winter. Similar relative humidity records at most weather stations demonstrate the same relation among season, temperature, and relative humidity.

Expressions of Humidity

There are several ways to express humidity and relative humidity. Each has its own utility and application. Two measures involve vapor pressure and specific humidity.

Vapor Pressure As free water molecules evaporate from surfaces into the atmosphere, they become water vapor. Now part of the air, water-vapor molecules exert a portion of the air pressure along with nitrogen and oxygen molecules. The share of air pressure that is made up of water-vapor molecules is **vapor pressure**. Millibars (mb) express vapor pressure.

As explained earlier, *saturation* is reached when the movement of water molecules between surface and air is in equilibrium. Air that contains as much water vapor as

possible at a given temperature is at *saturation vapor pressure*. Any temperature increase or decrease will change the saturation vapor pressure.

Figure 7.12 graphs the saturation vapor pressure at various air temperatures. The graph illustrates that, for every temperature increase of 10 C° (18 F°), the saturation water vapor pressure in air nearly doubles. This relation explains why warm tropical air over the ocean can hold so much water vapor, thus providing much latent heat to power tropical storms. It also explains why cold air is "dry" and why cold air toward the poles does not produce a lot of precipitation (it contains too little water vapor, even though it is near the dew-point temperature).

As marked on the graph, air at 20°C (68°F) has a saturation vapor pressure of 24 mb; that is, the air is saturated if the water-vapor portion of the air pressure also is at 24 mb. Thus, if the water vapor actually present is exerting a vapor pressure of only 12 mb in 20°C air, the relative humidity is 50% (12 mb ÷ 24 mb = 0.50 × 100 = 50%). The inset in Figure 7.12 compares saturation vapor pressure over water and over ice surfaces at subfreezing temperatures. You can see that saturation vapor pressure is greater above a water surface than over an ice surface—that is, it takes more water-vapor molecules to saturate air

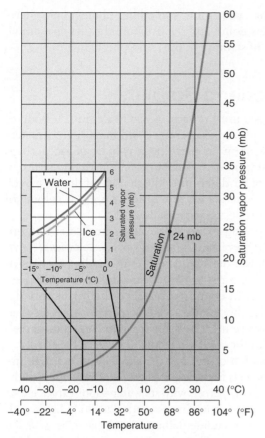

FIGURE 7.12 Saturation vapor pressure.
Saturation vapor pressure of air at various temperatures—the maximum possible water vapor, as measured by the pressure it exerts (mb). Inset compares saturation vapor pressures over water surfaces with those over surfaces at subfreezing temperatures. Note the 24-mb label for the text discussion.

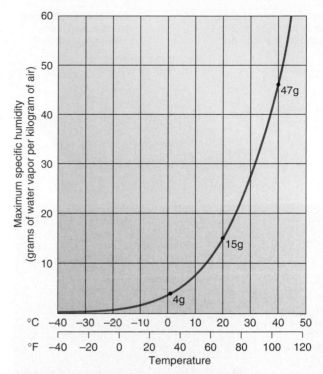

FIGURE 7.13 Maximum specific humidity.
Maximum specific humidity for a mass of air at various temperatures—the maximum possible water vapor in mass of water vapor per unit mass of air (g/kg). Note the 47-g, 15-g, and 4-g labels for the text discussion.

above water than it does above ice. This fact is important to condensation processes and rain-droplet formation, both of which are described later in this chapter.

Specific Humidity   A useful humidity measure is one that remains constant as temperature and pressure change. **Specific humidity** is the mass of water vapor (in grams) per mass of air (in kilograms) at any specified temperature. Because it is measured in mass, specific humidity is not affected by changes in temperature or pressure, such as when an air parcel rises to higher elevations. Specific humidity stays constant despite volume changes.*

The maximum mass of water vapor possible in a kilogram of air at any specified temperature is the *maximum specific humidity*, plotted in Figure 7.13. Noted on the graph is that a kilogram of air could hold a maximum specific

humidity of 47 g of water vapor at 40°C (104°F), 15 g at 20°C (68°F), and about 4 g at 0°C (32°F). Therefore, if a kilogram of air at 40°C has a specific humidity of 12 g, its relative humidity is 25.5% (12 g ÷ 47 g = 0.255 × 100 = 25.5%). Specific humidity is useful in describing the moisture content of large air masses that are interacting in a weather system and provides information necessary for weather forecasting.

Instruments for Measuring Humidity   Various instruments measure relative humidity. The **hair hygrometer** uses the principle that human hair changes as much as 4% in length between 0% and 100% relative humidity. The instrument connects a standardized bundle of human hair through a mechanism to a gauge. As the hair absorbs or loses water in the air, it changes length, indicating relative humidity (Figure 7.14a).

Another instrument used to measure relative humidity is a **sling psychrometer**. Figure 7.14b shows this device, which has two thermometers mounted side by side on a holder. One is the *dry-bulb thermometer*; it simply records the ambient (surrounding) air temperature. The other thermometer is the *wet-bulb thermometer*; it is set lower in the holder and is covered by a moistened cloth wick over the bulb. The psychrometer is then spun ("slung") by its handle or placed where a fan forces air over the wet bulb.

---

*Another similar measure used in relative humidity that approximates specific humidity is the *mixing ratio*, that is, the ratio of the mass of water vapor (grams) per mass of dry air (kilograms), as in g/kg.

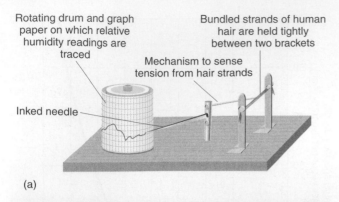

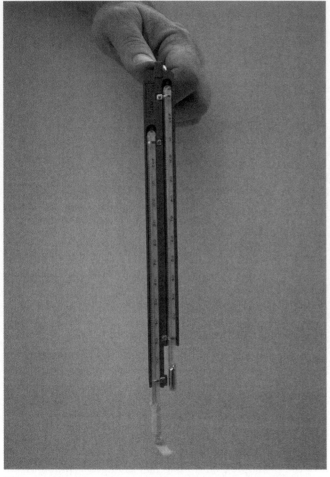

FIGURE 7.14 Instruments that measure relative humidity. (a) The principle of a hair hygrometer. (b) Sling psychrometer with wet and dry bulbs. [Photo by Bobbé Christopherson.]

The rate at which water evaporates from the wick depends on the relative saturation of the surrounding air. If the air is dry, water evaporates quickly, *absorbing specific heat as the latent heat of evaporation from the wet-bulb thermometer and its wick*, cooling the thermometer and causing its temperature to lower (the *wet-bulb depression*). In conditions of high humidity, little water evaporates from the wick; in low humidity, more water evaporates. After being spun a minute or two, a comparison of temperature on each bulb and a check on a relative humidity (psychrometric) chart give the relative humidity reading.

Now that you know something about atmospheric moisture, dew point, and relative humidity, let us examine the concept of stability in the atmosphere.

Atmospheric Stability

Meteorologists use the term *parcel* to describe a body of air that has specific temperature and humidity characteristics. Think of an air parcel as a volume of air, perhaps 300 m (1000 ft) in diameter or more. The temperature of the volume of air determines the density of the air parcel. Warm air produces a lower density in a given volume of air; cold air produces a higher density.

Two opposing forces work on a parcel of air: an upward *buoyant force* and a downward *gravitational force*. A parcel of lower density than the surrounding air rises (is more buoyant); a rising parcel expands as external pressure decreases. In contrast, a parcel of higher density descends (is less buoyant); a falling parcel compresses as external pressure increases. Figure 7.15 shows air parcels and illustrates these relationships.

An indication of weather conditions is the relative stability of air parcels in the atmosphere. **Stability** refers to the tendency of an air parcel, with its water-vapor cargo, either to remain in place or to change vertical position by ascending (rising) or descending (falling). An air parcel is *stable* if it resists displacement upward or, when disturbed, tends to return to its starting place. An air parcel is *unstable* if it continues to rise until it reaches an altitude where the surrounding air has a density and temperature similar to its own.

To visualize this, imagine a hot-air balloon launch. The air-filled balloon sits on the ground with the same air temperature inside as in the surrounding environment, like a stable air parcel. As the burner ignites, the balloon fills with hot (less-dense) air and rises buoyantly (Figure 7.16), like an unstable parcel of air. In the photo, why do you think these launches are taking place in the early morning? The concepts of stable and unstable air lead us to the specific temperature characteristics that produce these conditions.

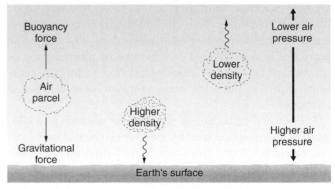

FIGURE 7.15 The forces acting on an air parcel. Buoyancy and gravitational forces work on an air parcel. Different densities produce rising or falling parcels in response to imbalance in these forces.

ANIMATION Atmospheric Stability

FIGURE 7.16 Principles of air stability and balloon launches.
Hot-air balloons being launched in southern Utah illustrate the principles of stability. As the temperature inside a balloon increases, the air in the balloon becomes less dense than the surrounding air and the buoyancy force causes the balloon to rise, acting like a warm-air parcel. [Photo by Steven K. Huhtala.]

Adiabatic Processes

Determining the degree of stability or instability requires measuring two temperatures: the temperature inside an air parcel and the temperature in the air surrounding the parcel. The contrast of these two temperatures determines stability. Such temperature measurements are made daily with *radiosonde* instrument packages carried aloft by helium-filled balloons at thousands of weather stations.

The *normal lapse rate*, introduced in Chapter 3, is the average decrease in temperature with increasing altitude, a value of 6.4 C°/1000 m (3.5 F°/1000 ft). This rate of temperature change is for still, calm air, and it can vary greatly under different weather conditions. Consequently, the *environmental lapse rate* is the actual lapse rate at a particular place and time. It can vary by several degrees per thousand meters.

Adiabatic describes the warming and cooling rates for a parcel of expanding or compressing air. *An ascending parcel of air tends to cool by expansion*, responding to the reduced pressure at higher altitudes. In contrast, *descending air tends to heat by compression*. *Diabatic* means occurring with an exchange of heat; *adiabatic* means occurring *without* a loss or gain of heat, *without any heat exchange between the surrounding environment and the vertically moving parcel of air*. Adiabatic temperatures are measured with

one of two specific rates, depending on moisture conditions in the parcel: dry adiabatic rate (DAR) and moist adiabatic rate (MAR).

Dry Adiabatic Rate   The **dry adiabatic rate (DAR)** is the rate at which "dry" air cools by expansion (if ascending) or heats by compression (if descending). "Dry" refers to air that is less than saturated (relative humidity is less than 100%). The average DAR is 10 C°/1000 m (5.5 F°/1000 ft).

The rising parcel at the left in Figure 7.17a illustrates the principle. To see how a specific example of dry air behaves, consider an unsaturated parcel of air at the surface with a temperature of 27°C (81°F). It rises, expands, and cools adiabatically at the DAR as it rises to 2500 m (approximately 8000 ft). What happens to the temperature of the parcel? Calculate the temperature change in the parcel, using the dry adiabatic rate:

$$(10 \text{ C°}/1000 \text{ m}) \times 2500 \text{ m} = 25 \text{ C° of total cooling}$$
$$(5.5 \text{ F°}/1000 \text{ ft}) \times 8000 \text{ ft} = 44 \text{ F° of total cooling}$$

Subtracting the 25 C° (44 F°) of adiabatic cooling from the starting temperature of 27°C (81°F) gives the temperature in the air parcel at 2500 m as 2°C (36°F).

In Figure 7.17b, assume that an unsaturated air parcel with a temperature of −20°C at 3000 m (−4°F at 9800 ft) descends to the surface, heating adiabatically. Using the dry adiabatic lapse rate, we determine the temperature of the air parcel when it arrives at the surface:

$$(10 \text{ C°}/1000 \text{ m}) \times 3000 \text{ m} = 30 \text{ C° of total warming}$$
$$(5.5 \text{ F°}/1000 \text{ ft}) \times 9800 \text{ ft} = 54 \text{ F° of total warming}$$

Adding the 30 C° of adiabatic warming from the starting temperature of −20°C gives the temperature in the air parcel at the surface as 10°C (50°F).

Moist Adiabatic Rate   The **moist adiabatic rate (MAR)** is the rate at which an ascending air parcel that is moist, or saturated, cools by expansion. The average MAR is 6 C°/1000 m (3.3 F°/1000 ft). This is roughly 4°C (2°F) less than the dry adiabatic rate. From this average, the MAR varies with moisture content and temperature and can range from 4 C° to 10 C° per 1000 m (2 F° to 5.5 F° per 1000 ft). (Note that a descending parcel of saturated air warms at the MAR as well, because the evaporation of liquid droplets, absorbing sensible heat, offsets the rate of compressional warming.)

The cause of this variability, and the reason that the MAR is lower than the DAR, is the latent heat of condensation. As water vapor condenses in the saturated air, latent heat is liberated, becoming sensible heat, thus decreasing the adiabatic rate. The release of latent heat may vary with temperature and water-vapor content. The

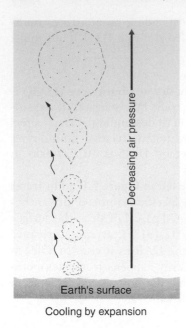

Cooling by expansion

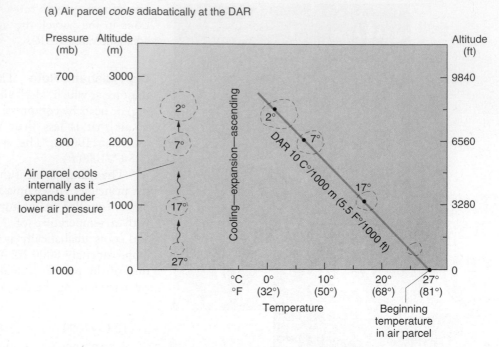

(a) Air parcel *cools* adiabatically at the DAR

Air parcel cools internally as it expands under lower air pressure

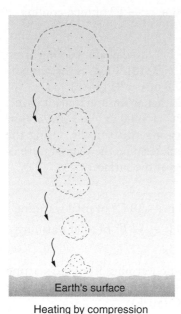

Heating by compression

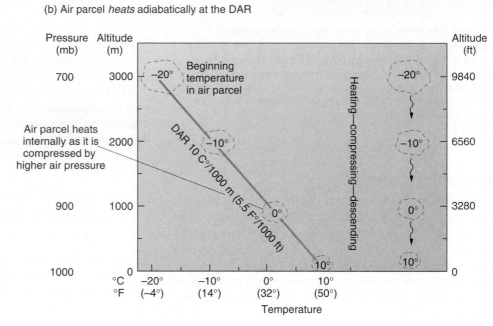

(b) Air parcel *heats* adiabatically at the DAR

Air parcel heats internally as it is compressed by higher air pressure

FIGURE 7.17 Vertically moving air experiences temperature changes—adiabatic cooling and heating.
(a) A rising air parcel that is less than saturated cools adiabatically at the dry adiabatic rate (DAR). (b) A descending air parcel that is less than saturated heats adiabatically by compression at the DAR.

Atmospheric Stability

MAR is much lower than the DAR in warm air, whereas the two rates are more similar in cold air.

Stable and Unstable Atmospheric Conditions

Now we bring this discussion together to determine atmospheric stability. The relation among the dry adiabatic rate (DAR), moist adiabatic rate (MAR), and the environmental lapse rate at a given time and place determines the stability of the atmosphere over an area. You see examples of three stability relationships in Figure 7.18.

Temperature relationships in the atmosphere produce the three different conditions: unstable, conditionally unstable, and stable. For the sake of illustration, the three examples in Figure 7.19 begin with an air parcel at the surface at 25°C (77°F). In each example, compare the temperatures of the air parcel and the surrounding

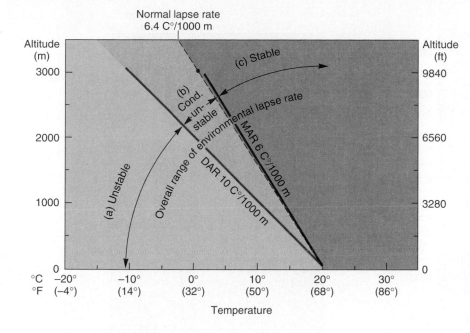

FIGURE 7.18 Temperature relationships and atmospheric stability. The relationship between dry and moist adiabatic rates and environmental lapse rates produces three atmospheric conditions: (a) unstable (environmental lapse rate exceeds the DAR), (b) conditionally unstable (environmental lapse rate is between the DAR and MAR), and (c) stable (environmental lapse rate is less than DAR and MAR).

environment. Assume that there is a lifting mechanism present, such as surface heating, a mountain range, or weather fronts, to get the parcel started (we examine lifting mechanisms in Chapter 8).

Given unstable conditions in Figure 7.19a, the air parcel continues to rise through the atmosphere because it is warmer (less dense and more buoyant) than the surrounding environment. Note the environmental lapse rate for this example is set at 12 C°/1000 m (6.6 F°/1000 ft). That is, the air surrounding the air parcel is cooler by 12 C° for every 1000-m increase in altitude. By 1000 m (about 3300 ft), the rising air parcel adiabatically cooled by expansion at the DAR from 25° to 15°C, while the surrounding air cooled from 25°C at the surface to 13°C. By comparing the temperature in the air parcel and the surrounding environment, you see that the temperature in the parcel is 2 C° (3.6 F°) warmer than the surrounding air at 1000 m. *Unstable* describes this condition because the less-dense air parcel will continue to lift.

Eventually, as the air parcel continues rising and cooling, it may achieve the dew-point temperature, saturation, and active condensation. This point of saturation forms the *lifting condensation level* that you see in the sky as the flat bottoms of clouds.

The example in Figure 7.19c shows stable conditions resulting when the environmental lapse rate is only 5 C°/1000 m (3 F°/1000 ft). An environmental lapse rate of 5 C°/1000 m is less than both the DAR and the MAR, a condition in which the air parcel has a lower temperature (is more dense and less buoyant) than the surrounding environment. The relatively cooler air parcel tends to settle back to its original position—it is *stable*. The denser air parcel resists lifting, unless forced by updrafts or a barrier, and the sky remains generally cloud-free. If clouds form, they tend to be stratiform (flat clouds) or cirroform (wispy), lacking vertical development. In regions experiencing air pollution, stable conditions in the atmosphere worsen the pollution by slowing exchanges in the surface air.

You may be wondering what stability condition exists if the environmental lapse rate is somewhere between the DAR and MAR, causing conditions to be neither unstable nor stable. In Figure 7.19b, the environmental lapse rate is measured at 7 C°/1000 m. Under these conditions, the air parcel resists upward movement, unless forced, if it is less than saturated. But, if the air parcel becomes saturated and cools at the MAR, it acts unstable and continues to rise.

One example of such conditionally unstable air occurs when stable air is forced to lift as it passes over a mountain range. As the air parcel lifts and cools to the dew point, the air becomes saturated and condensation begins. Now the MAR is in effect and the air parcel behaves in an unstable manner. The sky may be clear and without a cloud, yet huge clouds may develop over a nearby mountain range.

With these stability relationships in mind, let us look at the most visible expressions of stability and humidity in the atmosphere: clouds and fog.

Clouds and Fog

Clouds are more than whimsical, beautiful decorations in the sky; they are fundamental indicators of overall conditions, including stability, moisture content, and weather. They form as air becomes saturated with water. Clouds are the subject of much scientific inquiry, especially regarding their effect on net radiation patterns, as discussed in Chapters 4 and 5. With a little knowledge and practice, you can learn to "read" the atmosphere from its signature clouds.

A **cloud** is an aggregation of tiny moisture droplets and ice crystals that are suspended in air, great enough in

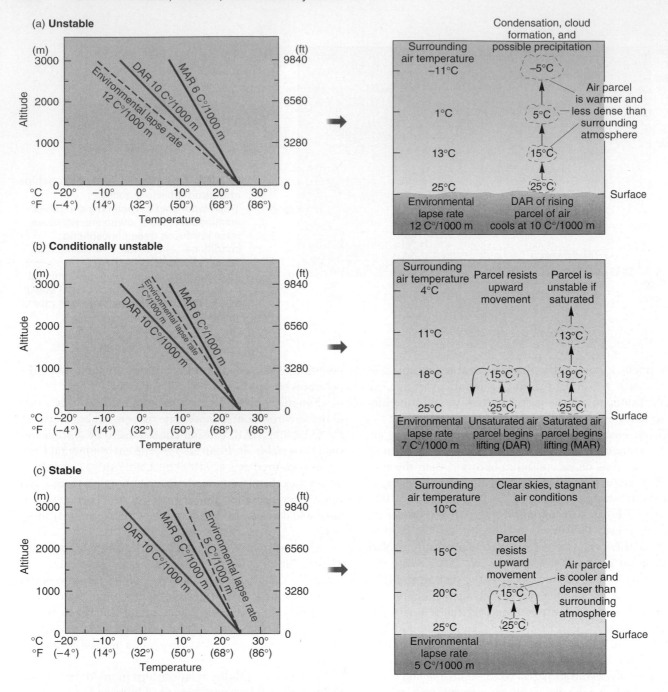

FIGURE 7.19 Stability—three examples.
Specific examples of (a) unstable, (b) conditionally unstable, and (c) stable conditions in the lower atmosphere. Note the response to these three conditions in the air parcel on the right side of each diagram.

volume and concentration to be visible. Fog is simply a cloud in contact with the ground. Cloud types are too numerous to fully describe here, so we present the most common examples in a simple classification scheme.

Cloud Formation Processes

Clouds may contain raindrops, but not initially. At the outset, clouds are a great mass of moisture droplets, each invisible without magnification. A **moisture droplet** is approximately 20 μm (micrometers) in diameter (0.002 cm, or 0.0008 in.). It takes a million or more such

droplets to form an average raindrop of 2000 μm in diameter (0.2 cm, or 0.078 in.), as shown in Figure 7.20.

As an air parcel rises it may cool to the dew-point temperature and 100% relative humidity. (Under certain conditions, condensation may occur at slightly less or more than 100% relative humidity.) More lifting of the air parcel cools it further, producing condensation of water vapor into water. Water does not just condense among the air molecules. Condensation requires **cloud-condensation nuclei**, microscopic particles that always are present in the atmosphere.

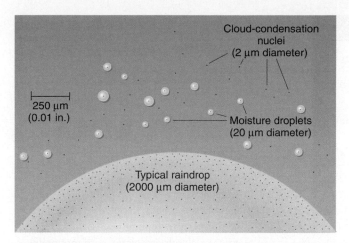

FIGURE 7.20 Moisture droplets and raindrops.
Cloud-condensation nuclei, moisture droplets, and a raindrop
enlarged many times—compared at roughly the same scale.

Continental air masses average 10 billion cloud-condensation nuclei per cubic meter. Ordinary dust, soot, and ash from volcanoes and forest fires and particles from burned fuel, such as sulfate aerosols, typically provide these nuclei. The air over cities contains great concentrations of such nuclei. In maritime air masses, nuclei average 1 billion per cubic meter, including sea salts derived from ocean sprays that supply some of the needed nuclei. The lower atmosphere never lacks cloud-condensation nuclei.

Given the presence of saturated air, the availability of cloud-condensation nuclei, and the presence of cooling (lifting) mechanisms in the atmosphere, condensation occurs. Two principal processes account for the majority of the world's raindrops and snowflakes: the *collision-coalescence process* and the *Bergeron ice-crystal process*. Figure 7.21 summarizes raindrop and snowflake formation.

Cloud Types and Identification

As noted, a cloud is a collection of moisture droplets and ice crystals suspended in air in sufficient volume and concentration to be visible. In 1803, English biologist and amateur meteorologist Luke Howard, in his article "On the Modification of Clouds," established a classification system for clouds and coined Latin names for them that we still use. About Howard's accomplishment his biographer stated,

Clouds were no longer exempt from human comprehension, and Howard, in contributing both a system of analysis and a full Latin nomenclature covering their

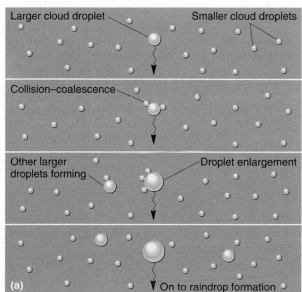
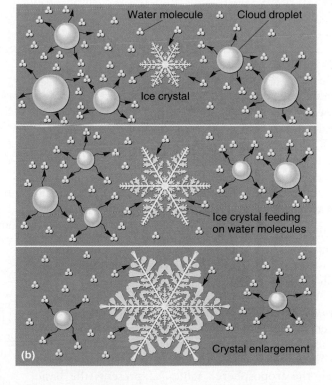

FIGURE 7.21 Raindrop and snowflake formation.
Principal processes for raindrop and snowflake formation: the *collision–coalescence process* and the *ice-crystal process*. (a) The collision–coalescence process predominates in clouds that form at above-freezing temperatures, principally in the warm clouds of the tropics. Initially, simple condensation takes place on small nuclei, some of which are larger and produce larger water droplets. As those larger droplets respond to gravity and fall through a cloud, they combine with smaller droplets, gradually coalescing into a raindrop. (b) The ice-crystal process occurs as supercooled water droplets (minute droplets of water that are below freezing and still in liquid form) evaporate rapidly near ice crystals, which then absorb the vapor. The ice crystals feed on the supercooled cloud droplets, grow in size, and eventually fall as snow or rain. Precipitation in middle and high latitudes can begin as ice and snow high in the clouds, then melt and gather moisture as it falls through the warmer portions of the cloud. [Art adapted from Lutgens and Tarbuck, *The Atmosphere: An Introduction to Meteorology*, 3rd ed., copyright © 1986, p. 127. Reprinted by permission of Prentice Hall, Inc.]

Table 7.2 Cloud Classes and Types

Class	Altitude/Composition at Midlatitudes	Type	Symbol	Description
Low clouds (C_L)	Up to 2000 m (6500 ft)	Stratus (St)		Uniform, featureless, gray, like high fog
	Water	Stratocumulus (Sc)		Soft, gray, globular masses in lines, groups, or waves, heavy rolls, irregular overcast patterns
		Nimbostratus (Ns)		Gray, dark, low, with drizzling rain
Middle clouds (C_M)	2000–6000 m (6500–20,000 ft)	Altostratus (As)		Thin to thick, no halos, Sun's outline just visible, gray day
	Ice and water	Altocumulus (Ac)		Patches of cotton balls, dappled, arranged in lines or groups, rippling waves, the lenticular clouds associated with mountains
High clouds (C_H)	6000–13,000 m (20,000–43,000 ft)	Cirrus (Ci)		Mares' tails, wispy, feathery, hairlike, delicate fibers, streaks, or plumes
	Ice	Cirrostratus (Cs)		Veil of fused sheets of ice crystals, milky, with Sun and Moon halos
		Cirrocumulus (Cc)		Dappled, "mackerel sky," small white flakes, tufts, in lines or groups, sometimes in ripples
Vertically developed clouds	Near surface to 13,000 m (43,000 ft)	Cumulus (Cu)		Sharply outlined, puffy, billowy, flat-based, swelling tops, fair weather
	Water below, ice above	Cumulonimbus (Cb)		Dense, heavy, massive, dark thunderstorms, hard showers, explosive top, great vertical development, towering, cirrus-topped plume blown into anvil-shaped head

families and genera, had contributed more than anyone to easing the path of understanding. . . . But the naming of clouds was a different kind of gesture for the hand of classification to have made. Here was the naming not of a solid, stable thing but of a series of self-canceling evanescences [disappearing entities]. Here was the naming of a fugitive presence that hastened to its onward dissolution. Here was the naming of clouds.*

Altitude and *shape* are key to cloud classification. Clouds occur in three basic forms—flat, puffy, and wispy—and in four primary altitude classes and ten basic cloud types. Horizontally developed clouds—flat and layered—are *stratiform* clouds. Vertically developed clouds—puffy and globular—are *cumuliform* clouds. Wispy clouds usually are quite high in altitude and are made of ice crystals; these are *cirroform*.

These three basic forms occur in four altitudinal classes: low, middle, high, and those vertically developed through the troposphere. Table 7.2 presents the basic cloud classes and types. The cloud symbols used today were of Luke Howard's invention. Figure 7.22 illustrates the general appearance of each type and includes representative photographs.

Low clouds, ranging from the surface up to 2000 m (6500 ft) in the middle latitudes, are simply *stratus* or

*R. Hamblyn, *The Invention of Clouds, How an Amateur Meteorologist Forged the Language of the Skies* (New York: Farrar, Straus, and Giroux, 2001), pp. 165, 171.

cumulus (Latin for "layer" and "heap," respectively). **Stratus** clouds appear dull, gray, and featureless. When they yield precipitation, they become **nimbostratus** (*nimbo-* denotes "stormy" or "rainy"), and their showers typically fall as drizzling rain (Figure 7.22e).

Cumulus clouds appear bright and puffy, like cotton balls. When they do not cover the sky, they float by in infinitely varied shapes. Vertically developed cumulus clouds are in a separate class in Table 7.2 because further vertical development can produce cumulus clouds that extend beyond low altitudes into middle and high altitudes (illustrated at the far right in Figure 7.22 and Figure 7.22d).

Sometimes near the end of the day **stratocumulus** may fill the sky in patches of lumpy, grayish, low-level clouds. Near sunset, these spreading puffy stratiform remnants may catch and filter the Sun's rays, sometimes indicating clearing weather.

The prefix *alto-* (meaning "high") denotes middle-level clouds. They are made of water droplets and, when cold enough, can be mixed with ice crystals. **Altocumulus** clouds, in particular, represent a broad category that occurs in many different styles: patchy rows, wave patterns, a "mackerel sky," or lens-shaped (lenticular) clouds.

Ice crystals in thin concentrations compose clouds occurring above 6000 m (20,000 ft). These wispy filaments, usually white except when colored by sunrise or sunset, are **cirrus** clouds (Latin for "curl of hair"), sometimes dubbed mares' tails. Cirrus clouds look as though an artist took a brush and added delicate feathery strokes high in the sky. Cirrus clouds can indicate an oncoming storm, especially if they thicken and lower in elevation. The prefix *cirro-*, as in *cirrostratus* and *cirrocumulus*, indicates other high clouds that form a thin veil or puffy appearance, respectively.

A cumulus cloud can develop into a towering giant called **cumulonimbus** (again, *-nimbus* in Latin denotes "rain storm" or "thundercloud"; Figure 7.23). Such clouds are *thunderheads* because of their shape and associated lightning and thunder. Note the surface wind gusts, updrafts and downdrafts, heavy rain, and the presence of ice crystals at the top of the rising cloud column. High-altitude winds may then shear the top of the cloud into the characteristic anvil shape of the mature thunderhead.

Fog

By international definition, **fog** is a cloud layer on the ground, with visibility restricted to less than 1 km (3300 ft). The presence of fog tells us that the air temperature and the dew-point temperature at ground level are nearly identical, indicating saturated conditions. A temperature-inversion layer generally caps a fog layer (warmer temperatures above and cooler temperatures below the inversion altitude), with as much as 22 C° (40 F°) difference in air temperature between the cool ground under the fog and the warmer, sunny skies above.

Almost all fog is warm—that is, its moisture droplets are above freezing. Supercooled fog, which occurs when the moisture droplets are below freezing, is special because it can be dispersed by means of artificial seeding with ice crystals or other crystals that mimic ice, following the principles of the ice-crystal formation process described earlier in Figure 7.21b. Let's briefly look at several types of fog.

Advection Fog   When air in one place *migrates* to another place where conditions are right for saturation, an **advection fog** forms. For example, when warm, moist air overlays cooler ocean currents, lake surfaces, or snow masses, the layer of migrating air directly above the surface becomes chilled to the dew point and fog develops. Off all subtropical west coasts in the world, summer fog forms in the manner just described (Figure 7.24). Some coastal desert communities actually extract usable water from such fog formations, as described in News Report 7.2.

 News Report 7.2

Harvesting Fog

Desert organisms have adapted remarkably to the presence of coastal fog along western coastlines in subtropical latitudes. For example, sand beetles in the Namib Desert in extreme southwestern Africa harvest water from the fog. They hold up their wings so condensation collects and runs down to their mouths. As the day's heat arrives, they burrow into the sand, only to emerge the next night or morning when the advection fog brings in more water for harvesting. Throughout history, people harvested water from fog. For centuries, coastal villages in the deserts of Oman collected water drips deposited on trees by coastal fogs.

In the Atacama Desert of Chile and Peru, residents stretch large nets to intercept the fog; moisture condenses on the netting and drips into trays, flowing through pipes to a 100,000-L (26,000-gal.) reservoir. Large sheets of plastic mesh along a ridge of the El Tofo Mountains harvest water from advection fog (Figure 7.2.1). Chungungo, Chile, receives 10,000 L (2,600 gal.) of water from 80 fog-harvesting

FIGURE 7.2.1 Fog harvesting.
In the mountains inland from Chungungo, Chile, polypropylene mesh stretched between two posts captures advection fog for local drinking-water supplies. [Photo by Robert S. Schemenauer.]

collectors in a project developed by Canadian (International Development Research Center) and Chilean interests and made operational in 1993. At least 30 countries across the globe experience conditions suitable for this water resource technology. (See **http://www.idrc.ca/en/ev-5077-201-1-DO_TOPIC.html, http://www.oas.org/dsd/publications/unit/oea59e/ch33.htm.**)

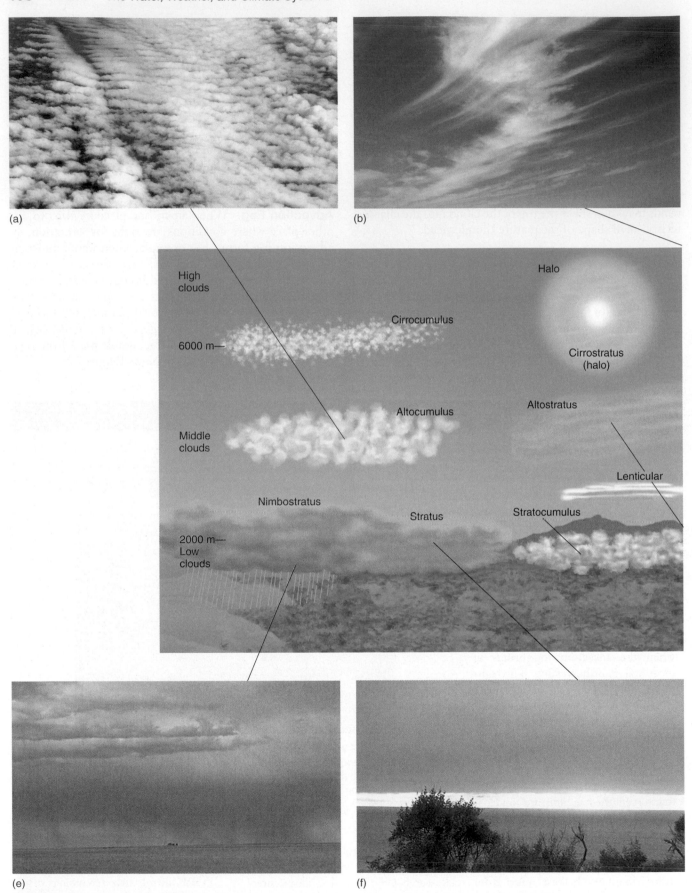

FIGURE 7.22 Principal cloud types.
Principal cloud types, classified by form (cirroform, stratiform, and cumuliform) and altitude (low, middle, high, and vertically developed across altitude): (a) altocumulus, (b) cirrus, (c) cirrostratus, (d) cumulonimbus, (e) nimbostratus, (f) stratus, (g) altostratus, and (h) cumulus. [Photos (a), (b), (d), (g), and (h) by Bobbé Christopherson; (c), (e), and (f) by author.]

(c)

(d)

Cirrus

Cirrostratus

Anvil-shaped head

Clouds with
vertical
development

Cumulonimbus

Cumulus
(fair weather)

(g)

(h)

FIGURE 7.23 Cumulonimbus thunderhead.
(a) Structure and form of a cumulonimbus cloud. Violent updrafts and downdrafts mark the circulation within the cloud. Blustery wind gusts occur along the ground. (b) Space Shuttle astronauts capture a dramatic cumulonimbus thunderhead as it moves over Galveston Bay, Texas. (c) Few acts of nature can match the sheer power released by an intense thunderstorm and related turbulence for air travelers.
(d) Towering cells in the upper Midwest with cloud tops at the same altitude as the airplane, 9000 m (29,500 ft). [Photos by (b) Space Shuttle photo from NASA; (c) and (d) by Bobbé Christopherson.]

Another type of advection fog forms when cold air lies over the warm water of a lake, ocean surface, or even a swimming pool. This wispy **evaporation fog**, or *steam fog*, may form as water molecules evaporate from the water surface into the cold overlying air, effectively humidifying the air to saturation followed by condensation to form fog (Figure 7.25). When evaporation fog happens at sea, it is a shipping hazard called *sea smoke*.

Advection fog also forms when moist air flows to higher elevations along a hill or mountain. This upslope lifting leads to adiabatic cooling by expansion as the air rises. The resulting **upslope fog** forms a stratus cloud at the condensation level of saturation. Along the Appalachians and the eastern slopes of the Rockies such fog is common in winter and spring. Another advection fog associated with topography is **valley fog**. Because cool air is denser than warm air, it settles in low-lying areas, producing a fog in the chilled, saturated layer near the ground in the valley (Figure 7.26).

Radiation Fog When radiative cooling of a surface chills the air layer directly above that surface to the dew-point temperature, creating saturated conditions, a **radiation fog** forms. This fog occurs over moist ground, especially on

FIGURE 7.26 Valley fog.
Cold air settles in the valleys of the Appalachian Mountains, chilling the air to the dew point and forming a valley fog. [Photo by author.]

FIGURE 7.24 Advection fog.
San Francisco's Golden Gate Bridge shrouded by an invading advection fog characteristic of summer conditions along a western coast. [Photo by author.]

clear nights; it does not occur over water because water does not cool appreciably overnight. Slight movements of air deliver even more moisture to the cooled area for more fog formation of greater depth (Figure 7.27).

In Figure 7.28, can you tell what two kinds of fog are pictured? The river water is warmer than the cold overlying air, producing an evaporation fog, especially beyond the bend in the river. The moist farmlands have radiatively cooled overnight, chilling the air along the surface to the dew point and active condensation. You see wisps of

radiation fog as light air movements flow from right to left in the photo.

Every year the media carry stories of multicar pileups on stretches of highway where vehicles drive at high speed in foggy conditions. These crash scenes can involve dozens of cars and trucks. Fog is a hazard to drivers, pilots, sailors, pedestrians, and cyclists, and its conditions of formation are quite predictable. The spatial aspects of fog occurrence should be a planning element for any proposed airport, harbor facility, or highway. The prevalence of fog throughout the United States and Canada is shown in Figure 7.29.

FIGURE 7.27 Radiation fog.
Satellite image of radiation fog in the southern Great Valley of California, November 20, 2002. This fog is locally known as a *tule fog* (pronounced "toolee") because of its association with the tule (bulrush) plants that line the low-elevation islands and marshes of the Sacramento River and San Joaquin River delta regions. [*Terra* MODIS image courtesy of MODIS Rapid Response Team, GSFC/NASA.]

FIGURE 7.25 Evaporation fog.
Evaporation fog, or sea smoke, highlights dawn on a cold morning at Donner Lake, California. Later that morning as air temperatures rose, what do you think happened to the evaporation fog? [Photo by Bobbé Christopherson.]

FIGURE 7.28 Two kinds of fog.
The cold air above the river creates evaporation fog along
portions of the channel, whereas radiation fog is forming over
the farmland. [Photo by Bobbé Christopherson.]

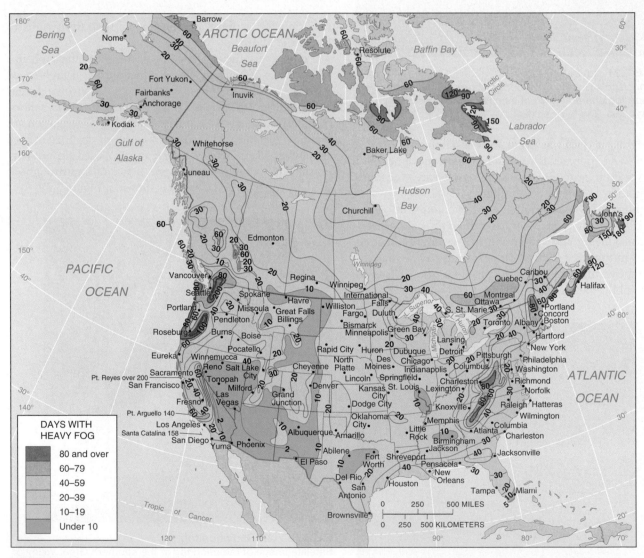

FIGURE 7.29 Fog incidence map.
Mean annual number of days with heavy fog in the United States and Canada. The foggiest spot in the United States is the
mouth of the Columbia River where it enters the Pacific Ocean at Cape Disappointment, Washington. One of the foggiest places
in the world is Newfoundland's Avalon Peninsula, specifically Argentia and Belle Isle, which regularly exceed 200 days of fog
each year. [Data courtesy of NWS; Map Series 3, *Climatic Atlas of Canada*, Atmospheric Environment Service Canada; and *The Climates
of Canada*, compiled by David Philips, Senior Climatologist, Environment Canada, 1990.]

Summary and Review—Water and Atmospheric Moisture

■ *Describe* the origin of Earth's waters, *define* the quantity of water that exists today, and *list* the locations of Earth's freshwater supply.

The next time it rains where you live, pause and reflect on the journey each of those water molecules made. Water molecules came from within Earth over a period of billions of years, in the **outgassing** process. Thus began endless cycling of water through the hydrologic system of evaporation–condensation–precipitation. Water covers about 71% of Earth. Approximately 97% of it is salty seawater, and the remaining 3% is freshwater—most of it frozen.

The present volume of water on Earth is estimated at 1.36 billion km³ (326 million mi³), an amount achieved roughly 2 billion years ago. This overall steady-state equilibrium might seem in conflict with the many changes in sea level that have occurred over Earth's history, but is not. **Eustasy** refers to worldwide changes in sea level and relates to the change in volume of water in the oceans. The amount of water stored in glaciers and ice sheets explains these changes as **glacio-eustatic** factors. At present, sea level is rising because of increases in the temperature of the oceans and the record melting of glacial ice.

outgassing (p. 178)
eustasy (p. 179)
glacio-eustatic (p. 179)

1. Approximately where and when did Earth's water originate?
2. If the quantity of water on Earth has been quite constant in volume for at least 2 billion years, how can sea level have fluctuated? Explain.
3. Describe the locations of Earth's water, both oceanic and fresh. What is the largest repository of freshwater at this time? In what ways is this distribution of water significant to modern society?
4. Why would climate change be a concern given this distribution of water?
5. Why might you describe Earth as the water planet? Explain.

■ *Describe* the heat properties of water, and *identify* the traits of its three phases: solid, liquid, and gas.

Water is the most common compound on the surface of Earth, and it possesses unusual solvent and heat characteristics. Part of Earth's uniqueness is that its water exists naturally in all three states—solid, liquid, and gas—owing to Earth's temperate position relative to the Sun. A change from one state to another is a **phase change**. The change from solid to vapor is **sublimation**; from liquid to solid, freezing; from solid to liquid, melting; from vapor to liquid, condensation; and from liquid to vapor, vaporization, or evaporation.

The heat energy required for water to change phase is **latent heat**, because, once absorbed, it is hidden within the structure of the water, ice, or water vapor. For 1 g of water to become 1 g of water vapor at boiling requires the addition of 540 cal, or the **latent heat of vaporization**. When this 1 g of water vapor condenses, the same amount of heat energy is liberated, as 540 cal of the **latent heat of condensation**. The **latent heat of sublimation** is the energy exchanged in the phase change from ice to vapor and vapor to ice. Weather is powered by the tremendous amount of latent heat energy involved in the phase changes among the three states of water.

phase change (p. 182)
sublimation (p. 182)
latent heat (p. 184)
latent heat of vaporization (p. 184)
latent heat of condensation (p. 185)
latent heat of sublimation (p. 185)

6. Describe the three states of matter as they apply to ice, water, and water vapor.
7. What happens to the physical structure of water as it cools below 4°C (39°F)? What are some visible indications of these physical changes?
8. What is latent heat? How is it involved in the phase changes of water?
9. Take 1 g of water at 0°C and follow it through to 1 g of water vapor at 100°C, describing what happens along the way. What amounts of energy are involved in the changes that take place?
10. Describe the physical structure of water and ice as it applies to Figure 7.6. Explain.

■ *Define* humidity and the expressions of the relative humidity concept, and *explain* dew-point temperature and saturated conditions in the atmosphere.

The amount of water vapor in the atmosphere is **humidity**. The maximum water vapor possible in air is principally a function of the temperature of the air and of the water vapor (usually these temperatures are the same). Warmer air produces higher net evaporation rates and maximum possible water vapor whereas cooler air can produce net condensation rates and lower possible water vapor.

Relative humidity is a ratio of the amount of water vapor actually in the air to the maximum possible at a given temperature. Relative humidity tells us how near the air is to saturation. Relatively dry air has a lower relative humidity value; relatively moist air has a higher relative humidity percentage. Air is said to be at **saturation** when the rate of evaporation and the rate of condensation reach equilibrium, and any further addition of water vapor or temperature lowering will result in active condensation (100% relative humidity). The temperature at which air achieves saturation is the **dew-point temperature**.

Among the various ways to express humidity and relative humidity are vapor pressure and specific humidity. **Vapor pressure** is that portion of the atmospheric pressure produced by the presence of water vapor. A comparison of vapor pressure with the saturation vapor pressure at any moment yields a relative humidity percentage. **Specific humidity** is the mass of water vapor (in grams) per mass of air (in kilograms) at any specified temperature. Because it is measured as a mass, specific humidity does not change as temperature or pressure changes, making it a valuable measurement in weather forecasting. A comparison of specific humidity with the maximum specific humidity at any moment produces a relative humidity percentage.

Two instruments measure relative humidity, and indirectly the actual humidity content of the air; they are the **hair hygrometer** and the **sling psychrometer**.

humidity (p. 185)
relative humidity (p. 185)
saturation (p. 185)

dew-point temperature (p. 185)
vapor pressure (p. 188)
specific humidity (p. 189)
hair hygrometer (p. 189)
sling psychrometer (p. 189)

11. What is humidity? How is it related to the energy present in the atmosphere? To our personal comfort and how we perceive apparent temperatures?
12. Define relative humidity. What does the concept represent? What is meant by the terms *saturation* and *dew-point temperature*?
13. Using different measures of humidity in the air than those used in the chapter, derive relative humidity values (vapor pressure/saturation vapor pressure; specific humidity/maximum specific humidity) in Figures 7.12 and 7.13.
14. How do the two instruments described in this chapter measure relative humidity?
15. How does the daily distribution of relative humidity compare with the daily trend in air temperature?

■ *Define* atmospheric stability, and *relate* it to a parcel of air that is ascending or descending.

Meteorologists use the term *parcel* to describe a body of air that has specific temperature and humidity characteristics. Think of a parcel of air as a volume of air, perhaps 300 m (1000 ft) in diameter. The temperature of the volume of air determines the density of the air parcel. Warm air has a lower density in a given volume of air; cold air has a higher density.

Stability refers to the tendency of an air parcel, with its water-vapor cargo, either to remain in place or to change vertical position by ascending (rising) or descending (falling). An air parcel is *stable* if it resists displacement upward or, when disturbed, it tends to return to its starting place. An air parcel is *unstable* if it continues to rise until it reaches an altitude where the surrounding air has a density (air temperature) similar to its own.

stability (p. 190)

16. Differentiate between stability and instability relative to a parcel of air rising vertically in the atmosphere.
17. What are the forces acting on a vertically moving parcel of air? How are they affected by the density of the air parcel?

■ *Illustrate* three atmospheric conditions—unstable, conditionally unstable, and stable—with a simple graph that relates the environmental lapse rate to the dry adiabatic rate (DAR) and moist adiabatic rate (MAR).

An ascending (rising) parcel of air cools by expansion, responding to the reduced air pressure at higher altitudes. A descending (falling) parcel heats by compression. These temperature changes internal to a moving air parcel are explained by physical laws that govern the behavior of gases. Temperature changes in both ascending and descending air parcels occur without any significant heat exchange between the surrounding environment and the vertically moving parcel of air. The warming and cooling rates for a parcel of expanding or compressing air are termed **adiabatic**.

The **dry adiabatic rate (DAR)** is the rate at which "dry" air cools by expansion (if ascending) or heats by compression (if descending). The term *dry* is used when air is less than saturated (relative humidity less than 100%). The DAR is 10 C°/1000 m (5.5 F°/1000 ft). The **moist adiabatic rate (MAR)** is the average rate at which ascending air that is moist (saturated) cools by expansion or descending air warms by compression. The average MAR is 6 C°/1000 m (3.3 F°/1000 ft). This is roughly

4 C° (2 F°) less than the dry rate. The MAR, however, varies with moisture content and temperature and can range from 4 to 10 C° per 1000 m (2 to 5.5 F° per 1000 ft).

A simple comparison of the dry adiabatic rate (DAR) and moist adiabatic rate (MAR) in a vertically moving parcel of air with that of the environmental lapse rate in the surrounding air determines the atmosphere's stability—whether it is unstable (lifting of air parcel continues), stable (air parcel resists vertical displacement), or conditionally unstable (air parcel behaves as though unstable if the MAR is in operation, and stable otherwise).

adiabatic (p. 191)
dry adiabatic rate (DAR) (p. 191)
moist adiabatic rate (MAR) (p. 191)

18. How do the adiabatic rates of heating or cooling in a vertically displaced air parcel differ from the normal lapse rate and environmental lapse rate?
19. Why is there a difference between the dry adiabatic rate (DAR) and the moist adiabatic rate (MAR)?
20. What atmospheric temperature and moisture conditions would you expect on a day when the weather is unstable? When it is stable? Relate in your answer what you would experience if you were outside watching.
21. Use the "Atmospheric Stability" animation on your *Student Animation CD*. Try different temperature settings on the sliders to produce stable and unstable conditions.

■ *Identify* the requirements for cloud formation, and *explain* the major cloud classes and types, including fog.

A **cloud** is an aggregation of tiny moisture droplets and ice crystals suspended in the air. Clouds are a constant reminder of the powerful heat-exchange system in the environment. **Moisture droplets** in a cloud form when saturated air and the presence of **cloud-condensation nuclei** lead to *condensation*. Raindrops are formed from moisture droplets through either the *collision-coalescence process* or the *Bergeron ice-crystal process*.

Low clouds, ranging from the surface up to 2000 m (6500 ft) in the middle latitudes, are **stratus** (flat clouds, in layers) or **cumulus** (puffy clouds, in heaps). When stratus clouds yield precipitation, they are **nimbostratus**. Sometimes near the end of the day, lumpy, grayish, low-level clouds called **stratocumulus** may fill the sky in patches. Middle-level clouds are denoted by the prefix *alto-*. **Altocumulus** clouds, in particular, represent a broad category that occurs in many different styles. Clouds at high altitude, principally composed of ice crystals, are called **cirrus**. A cumulus cloud can develop into a towering giant **cumulonimbus** cloud (-*nimbus* in Latin denotes "rain storm" or "thundercloud"). Such clouds are called *thunderheads* because of their shape and their associated lightning, thunder, surface wind gusts, updrafts and downdrafts, heavy rain, and hail.

Fog is a cloud that occurs at ground level. **Advection fog** forms when air in one place migrates to another place where conditions exist that can cause saturation—for example, when warm, moist air moves over cooler ocean currents. Another type of advection fog forms when cold air flows over the warm water of a lake, ocean surface, or swimming pool. This **evaporation fog**, or steam fog, may form as the water molecules evaporate from the water surface into the cold overlying air. **Upslope fog** is produced when moist air is forced to higher elevations along a hill or mountain. Another fog caused by topography is **valley fog**, formed because cool, denser air settles in low-lying areas, producing fog in the chilled, saturated layer near the ground. Radiative cooling of a surface that chills the air layer directly above the surface to the dew-point temperature creates saturated conditions and a **radiation fog**.

cloud (p. 193)
moisture droplet (p. 194)
cloud-condensation nuclei (p. 194)
stratus (p. 196)
nimbostratus (p. 196)
cumulus (p. 196)
stratocumulus (p. 196)
altocumulus (p. 196)
cirrus (p. 196)
cumulonimbus (p. 197)
fog (p. 197)
advection fog (p. 197)
evaporation fog (p. 200)
upslope fog (p. 200)

valley fog (p. 200)
radiation fog (p. 200)

22. Specifically, what is a cloud? Describe the droplets that form a cloud.
23. Explain the condensation process: What are the requirements? What two principal processes are discussed in this chapter?
24. What are the basic forms of clouds? Using Table 7.2, describe how the basic cloud forms vary with altitude.
25. Explain how clouds might be used as indicators of the conditions of the atmosphere and of expected weather.
26. What type of cloud is fog? List and define the principal types of fog.
27. Describe the occurrence of fog in the United States and Canada. Where are the regions of highest incidence?

 NetWork

The *Geosystems* Student Learning Center provides online resources for this chapter on the World Wide Web. To begin: Once at the Center, click on the cover of this textbook, scroll the Table of Contents menu, and select this chapter. You will find self-tests that are graded, review exercises, specific updates for items in the chapter, and in "Destinations" many links to interesting related pathways on the Internet. *Geosystems* Student Learning Center is found at **http://www.prenhall.com/christopherson/**.

Critical Thinking

A. Examine the iceberg photographs in Figure 7.6b and Figure 7.6d. These photos show Greenland and Antarctic waters. Determine what caused the "shoreline" thermal notch above the ocean surface around the iceberg in 7.6b. Why do you think the iceberg appears to be riding higher in the water than several weeks earlier? In rough terms, how much ice (in percent) do you think is beneath the surface in comparison to the amount you see above sea level in the iceberg photos? Explain why this physical trait in larger icebergs can be a hazard to shipping.

B. Using Figure 7.22, begin to observe clouds on a regular basis. See if you can relate the cloud type to particular weather conditions at the time of observation. You may want to keep a log in your notebook during this physical geography course. Seeing clouds in this way will help you to understand weather and make studying Chapter 8 easier.

C. Given our discussion of relative humidity, air temperature, dew-point temperature, and saturation, please refer to the "Short Answer" section of Chapter 7, items 1 and 2, in the *Geosystems* Student Learning Center. View the dew-point temperature map in relation to the map of air temperatures. As you compare and contrast the maps, describe in general terms what relative humidity conditions you find in the Northwest and along the Gulf Coast in the Southeast.

Such catastrophes as Hurricane Katrina link weather events to the failure of floodplain and levee management, discussed in Chapter 14. Damage to houses and cars in the Lakeview neighborhood of New Orleans (upper-left). New Basin Canal Park, a city park, serves as a temporary dump for life's debris (upper-right). A 91-m (300-ft) break in the London Avenue Canal levee that leads from Lake Pontchartrain partially into New Orleans; inset photo shows the dried, toxic "muck" that covered the region after floodwaters retreated (lower-left). Houses missing from their foundations, debris, and broken trees and lives in Long Beach, Mississippi (lower-right). Sadly, after years, the tragedy continues for the region as a result of the failure of government institutions. [All photos by Randall Christopherson. All rights reserved.]

Weather

■ Key Learning Concepts

After reading the chapter, you should be able to:

- ■ *Describe* air masses that affect North America, and *relate* their qualities to source regions.

- ■ *Identify* four types of atmospheric lifting mechanisms, and *describe* four principal examples.

- ■ *Analyze* the pattern of orographic precipitation, and *describe* the link between this pattern and global topography.

- ■ *Describe* the life cycle of a midlatitude cyclonic storm system, and *relate* this to its portrayal on weather maps.

- ■ *List* the measurable elements that contribute to modern weather forecasting, and *describe* the technology and methods employed.

- ■ *Analyze* various forms of violent weather and the characteristics of each, and *review* several examples of each from the text.

W ater has a leading role in the vast drama played out daily on Earth's stage. It affects the stability of air masses and their interactions and produces powerful and beautiful special effects in the lower atmosphere. Air masses come into conflict; they move and shift, dominating now one region then another, varying in strength and characteristics. Think of the weather as a play, North America the stage, and the air masses as actors of varying ability.

In this chapter: We follow huge air masses across North America, observe powerful lifting mechanisms in the atmosphere, revisit the concepts of stable and unstable conditions, examine migrating cyclonic systems with attendant cold and warm fronts, and conclude with a portrait of violent and dramatic weather so often in the news in recent years. Water, with its ability to absorb and release vast quantities of heat energy, drives this daily drama in the atmosphere. Also, modern weather forecasting is introduced. The spatial implications of these weather phenomena and their relationship to human activities strongly link meteorology to physical geography and this chapter. We begin with a look at some weather essentials.

Weather Essentials

Weather is the short-term, day-to-day condition of the atmosphere, contrasted with *climate*, which is the long-term average (over decades) of weather conditions and extremes in a region. Weather is, at the same time, both a "snapshot" of atmospheric conditions and a technical status report of the Earth–atmosphere heat-energy budget. Important elements that contribute to weather are temperature, air pressure, relative humidity, wind speed and direction, and seasonal factors such as insolation receipt related to daylength and Sun angle.

We turn for weather forecasts to the National Weather Service (NWS) in the United States (**http://www.nws. noaa.gov/**) or the Canadian Meteorological Centre, a branch of the Meteorological Service of Canada (MSC) (**http://www.msc-smc.ec.gc.ca/cmc/index_e.html**) to see current satellite images and to hear weather analysis. Internationally, the World Meteorological Organization coordinates weather information (see **http://www. wmo.ch/**). Many sources of weather information and related topics are found in the "Destinations" section for this chapter on the *Geosystems* Student Learning Center.

Meteorology is the scientific study of the atmosphere. (*Meteor* means "heavenly" or "of the atmosphere.") Meteorologists study the atmosphere's physical characteristics and motions, related chemical, physical, and geologic processes, the complex linkages of atmospheric systems, and weather forecasting. Computers handle volumes of data for accurate forecasting of near-term weather and for studying trends in long-term weather, climatology, and climatic change. New developments in supercomputing, an Earth-bound instrument network in the Automated Surface Observing System (ASOS) arrays, orbiting observation systems, and Doppler radar installations are rapidly advancing the science of the atmosphere.

An important tool in each NWS Weather Forecast Office is the Advanced Weather Interactive Processing System (AWIPS, see Figure 8.17). AWIPS allows forecasters to integrate meteorological, hydrological, satellite, and radar data at a computer workstation. The forecaster and scientist have an interactive ability to observe, analyze, and manipulate geospatial weather data. The network of AWIPS stations, including 13 River Forecast Centers, covers the country.

An essential part of understanding weather is Doppler radar. Using backscatter from two radar pulses, it detects the direction of moisture droplets toward or away from the radar, indicating wind direction and speed. This information is critical to making accurate severe storm warnings. As part of the NEXRAD (Next Generation Weather Radar) program, 158 WSR-88D (*W*eather *S*urveillance *R*adar) Doppler radar systems are operational through the National Weather Service, Federal Aviation Administration, and the Department of Defense (Figure 8.1). In Canada, the

(a)

(b)

FIGURE 8.1 Weather installation.
(a) Doppler radar installation at the Indianapolis International Airport operated by the National Weather Service. The radar antenna is sheltered within the dome structure. (b) WSR-88D coverage in the United States. Other installations are maintained in Okinawa, Guam, South Korea, and the Azores.
[(a) Photo by Bobbé Christopherson; (b) adapted map courtesy of Radar Operations Center, Norman, Oklahoma.]

National Radar Project placed 25 CWSR-98 radars in service out of an eventual 30 units.

The cost of weather-related destruction has risen more than 500% between 1975 and 1995, from an average of $2 billion to $10 billion annually. In 1998 alone, damage topped $90 billion worldwide, which exceeded the total for all the 1980s, even when adjusted for inflation. Damage costs are rising. Hurricane Katrina alone produced $125 billion in losses. A 2005 U.N. Environment Programme report estimated weather-related damage losses annually (drought, floods, hail, tornadoes, derechos, tropical systems, storm surge, blizzard and ice storms, and wildfires) will exceed $1 trillion by A.D. 2040 (adjusted to current dollars).

Air Masses

Each area of Earth's surface imparts its temperature and moisture characteristics to overlying air. The effect of the surface on the air creates regional air masses with a homogenous mix of temperature, humidity, and stability. These masses of air interact to produce weather patterns—in essence, air masses are the actors in our weather drama. Such a distinctive body of air is an **air mass**, and it initially reflects the characteristics of its *source region* and sometimes extends through the lower half of the troposphere. For example, weather forecasters speak of a "cold Canadian air mass" and "moist tropical air mass."

Air Masses Affecting North America

We classify air masses generally according to the moisture and temperature characteristics of their source regions:

1. *Moisture*—designated **m** for maritime (wet) and **c** for continental (dry).
2. *Temperature* (latitude factor)—designated **A** (arctic), **P** (polar), **T** (tropical), **E** (equatorial), and **AA** (Antarctic).

The principal air masses that affect North America in winter and summer are mapped in Figure 8.2.

Continental polar (cP) air masses form only in the Northern Hemisphere and are most developed in winter and cold-weather conditions. These cP air masses are major players in middle- and high-latitude weather. The cold, dense, cP air displaces moist, warm air in its path, producing lifting, cooling, and condensation. An area covered by cP air in winter experiences cold, stable air, clear skies, high pressure, and anticyclonic wind flow, all visible on the weather map in Figure 8.3. The Southern Hemisphere lacks the necessary continental landmasses at high latitudes to create such a cP air mass.

Maritime polar (mP) air masses in the Northern Hemisphere exist northwest and northeast of the North American continent over the northern oceans. Within

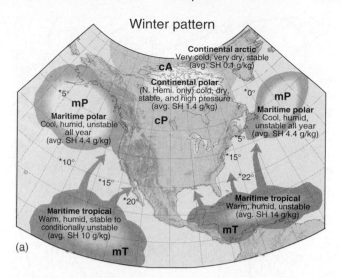

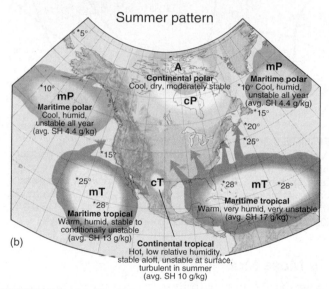

FIGURE 8.2 Principal air masses.
Air masses and their source regions influence North America during winter (a) and summer (b). (*Sea-surface temperature in °C; SH = specific humidity.)

them, cool, moist, unstable conditions prevail throughout the year. The Aleutian and Icelandic subpolar low-pressure cells reside within these mP air masses, especially in their well-developed winter pattern (see the January isobar pressure map in Figure 6.10a).

Two maritime tropical (mT) air masses—the *mT Gulf/Atlantic* and the *mT Pacific*—influence North America. The humidity experienced in the East and Midwest is created by the mT Gulf/Atlantic air mass, which is particularly unstable and active from late spring to early fall (Figure 8.4). In contrast, the mT Pacific is stable to conditionally unstable and generally lower in moisture content and available energy. As a result, the western United States, influenced by this weaker Pacific air mass, receives lower average precipitation than the rest of the country. Please review Figure 6.13 and the discussion of subtropical high-pressure cells.

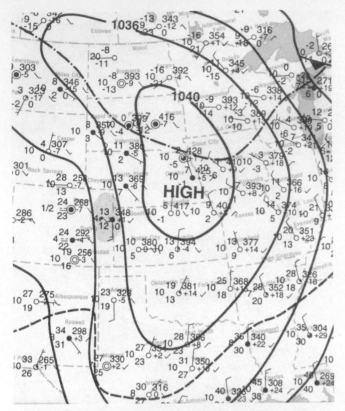

FIGURE 8.3 Winter high-pressure system.
A cP air mass dominates the Midwest with a central pressure of 1042.8 mb (30.76 in.), air temperature of −17°C, dew-point temperature of −21°C (2° and −5°F, respectively), with clear, calm, stable conditions. Note the pattern of isobars portraying the cP air mass. The dotted lines are the −18°C (0°F) and 0°C (32°F) isotherms. [Weather map courtesy of NWS, NOAA.]

Air Mass Modification

As air masses migrate from source regions, their temperature and moisture characteristics modify and slowly take on the characteristics of the land over which they pass. For example, an mT Gulf/Atlantic air mass may carry humidity to Chicago and on to Winnipeg but gradually loses its initial characteristics of high humidity and warmth with each day's passage northward.

Similarly, below-freezing temperatures occasionally reach into southern Texas and Florida, brought by an invading winter cP air mass from the north. However, that air mass warms from the −50°C (−58°F) of its winter source region in central Canada, warming especially after it leaves areas covered by snow.

Modification of cP air as it moves south and east produces snowbelts that lie to the east of each of the Great Lakes. As below-freezing cP air passes over the warmer Great Lakes, it absorbs heat energy and moisture from the lake surfaces and becomes *humidified*. This enhancement produces heavy lake-effect snowfall downwind into Ontario, Québec, Michigan, Pennsylvania, and New York—some areas receiving in excess of 250 cm (100 in.) in average snowfall a year (Figure 8.5).

A couple of such storms in western New York serve as examples. In October 2006, a lake-effect storm dubbed the "surprise storm" dropped 0.6 m (2 ft) of wet snow. This was the earliest date for such a storm in 137 years of records and remarkable because the ratio of snow to water content was 6:1. An "avalanche of a storm" hit in February 2007 when 2 m (6.5 ft) of snow fell in 1 day, over 3.7 m (12 ft) over 7 days. Severity correlates to a low-pressure system positioned north of the Great Lakes, with counterclockwise winds pushing air across the lakes. With global warming, regional temperatures are increasing, which leads forecasters to predict enhanced lake-effect snowfall over the next several decades, since warmer air can absorb more water vapor. Later in the century, snowfall will decrease as temperatures rise, although rainfall totals will continue to increase over the leeside region.

Atmospheric Lifting Mechanisms

For air masses to cool adiabatically (by expansion) and to reach the dew-point temperature and saturate, condense, and form clouds and perhaps precipitation, they must lift and rise in altitude. Four principal lifting mechanisms operate in the atmosphere:

- Convergent lifting—air flows toward an area of low pressure
- Convectional lifting—stimulated by local surface heating

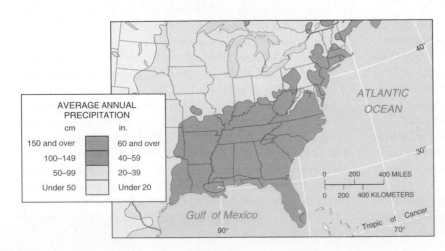

FIGURE 8.4 Influence of mT Gulf/Atlantic air mass.
The pattern of precipitation over the southeastern United States shows the influence of the warm, moist, and generally unstable mT Gulf/Atlantic air mass. Precipitation decreases with distance inland from the source region over the Gulf of Mexico and Atlantic Ocean.

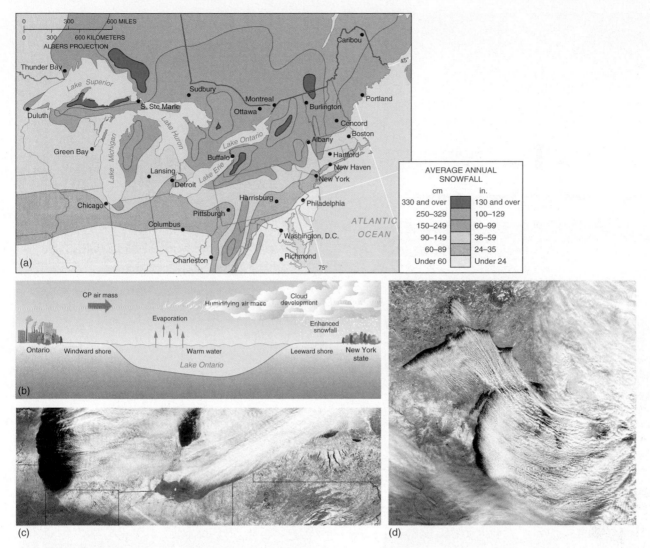

FIGURE 8.5 Lake-effect snowbelts of the Great Lakes.
(a) Locally, heavy areas of snowfall are associated with the lee side of each of the Great Lakes. In winter, cold cP and cA air masses pass from colder land surfaces across the relatively warmer water of these lakes. (b) The air masses are warmed and humidified (water vapor is added) from the lake water. The humid, now unstable air yields heavy snowfall as it moves onshore and becomes chilled. The strongest effect is generally limited to about 50 km (30 mi) inland up to 100 km (60 mi). (c) and (d) Satellite images show the lake-effect weather. [(a) Snowfall data from the *Climatic Atlas of the United States* (Washington, DC: Department of Commerce, NOAA, 1983), p. 53. (c) MODIS *Terra* image, December 7, 2002, courtesy of the Rapid Response Team; and (d) *OrbView-2* image, December 5, 2000, courtesy *SeaWiFS* Project, GeoEye; both NASA/GSFC]

- Orographic lifting—air is forced over a barrier such as a mountain range
- Frontal lifting—along the leading edges of contrasting air masses

Descriptions of all four mechanisms follow and are in Figure 8.6.

Convergent Lifting

Air flowing from different directions into the same low-pressure area is converging, displacing air upward in **convergent lifting**. All along the equatorial region, the southeast and northeast trade winds converge, forming the intertropical convergence zone (ITCZ) and areas of extensive uplift, towering cumulonimbus cloud development, and high average annual precipitation (see Figures 6.11 and 6.12).

Convectional Lifting

When an air mass passes from a maritime source region to a warmer continental region, heating from the warmer land causes lifting and convection in the air mass. Other sources of surface heating might include an urbanized area heat island or an area of dark soil in a plowed field—the warmer surfaces produce **convectional lifting**. If conditions are unstable, initial lifting continues and clouds develop. Figure 8.7 illustrates convectional action

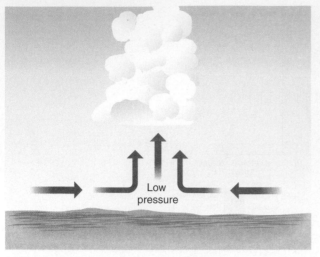

(a) Convergent

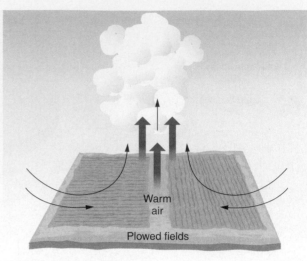

(b) Convectional (local heating)

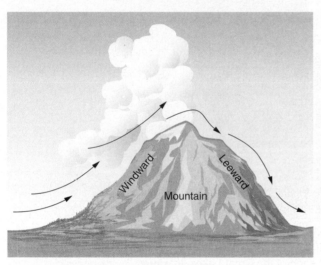

(c) Orographic (barrier)

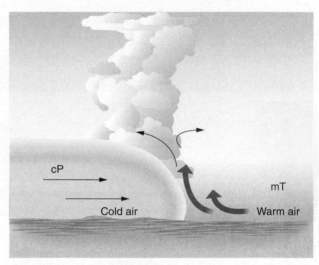

(d) Frontal (e.g., cold front)

FIGURE 8.6 Four atmospheric lifting mechanisms.
(a) Convergent lifting; (b) convectional lifting; (c) orographic lifting; and (d) frontal lifting.

ANIMATION

Atmospheric Stability

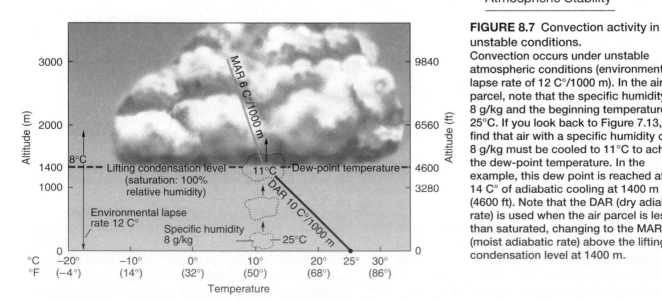

FIGURE 8.7 Convection activity in unstable conditions.
Convection occurs under unstable atmospheric conditions (environmental lapse rate of 12 C°/1000 m). In the air parcel, note that the specific humidity is 8 g/kg and the beginning temperature is 25°C. If you look back to Figure 7.13, you find that air with a specific humidity of 8 g/kg must be cooled to 11°C to achieve the dew-point temperature. In the example, this dew point is reached after 14 C° of adiabatic cooling at 1400 m (4600 ft). Note that the DAR (dry adiabatic rate) is used when the air parcel is less than saturated, changing to the MAR (moist adiabatic rate) above the lifting-condensation level at 1400 m.

SATELLITE

Convectional Heating
and Tornado in Florida

FIGURE 8.8 Convectional activity over the Florida peninsula.
Cumulus clouds cover the land, with several cells developing into cumulonimbus thunderheads, demonstrating convectional lifting. Local heating lifts warm, moist air from the Gulf of Mexico and Atlantic as it passes over the relatively warmer land. [MODIS *Terra* image courtesy of the Rapid Response Team, NASA/GSFC.]

stimulated by local heating, with unstable conditions present in the atmosphere. The rising parcel of air continues its ascent because it is warmer and therefore less dense than the surrounding environment (Figure 8.7).

Florida's precipitation generally illustrates both these lifting mechanisms: convergence and convection. Heating of the land produces convergence of onshore winds from the Atlantic and the Gulf of Mexico. As an example of local heating and convectional lifting, Figure 8.8 depicts a day on which the landmass of Florida was warmer than the surrounding Gulf of Mexico and Atlantic Ocean. Because the Sun's radiation gradually heats the land throughout the day and warms the air above it, convectional showers tend to form in the afternoon and early evening, causing the highest frequency of days with thunderstorms in the United States. Florida appears highlighted with clouds.

Orographic Lifting

The physical presence of a mountain acts as a topographic barrier to migrating air masses. **Orographic lifting** (*oro* means "mountain") occurs when air is forcibly lifted upslope as it is pushed against a mountain. The lifting air cools adiabatically. Stable air forced upward in this manner may produce stratiform clouds, whereas unstable or conditionally unstable air usually forms a line of cumulus and cumulonimbus clouds. An orographic barrier enhances convectional activity and causes additional lifting during the passage of weather fronts and cyclonic systems, thereby extracting more moisture from passing air masses.

Figure 8.9a illustrates the operation of orographic lifting under unstable conditions. The wetter intercepting slope is the *windward slope*, as opposed to the drier far-side slope, known as the *leeward slope*. Moisture condenses from the lifting air mass on the windward side of the mountain; on the leeward side, the descending air mass heats by compression, and any remaining water in the air evaporates. Thus, air beginning its ascent up a mountain can be warm and moist, but finishing its descent on the leeward slope it becomes hot and dry.

The state of Washington provides an excellent example of this concept, as shown in Figure 8.10. The Olympic Mountains and Cascade Mountains orographically lift invading mP air masses from the North Pacific Ocean, squeezing out annual precipitation of more than 500 cm and 400 cm, respectively (200 in. and 160 in.).

The Quinault Ranger and Rainier Paradise weather stations demonstrate windward-slope rainfall. The cities of Sequim, in the Puget Trough, and Yakima, in the Columbia Basin, are in the rain shadow on the leeward side of these mountain ranges and are characteristically low in annual rainfall. Find these stations on the landscape profile and on the precipitation map.

In North America, **chinook winds** (called *föhn* or *foehn* winds in Europe) are the warm, downslope air flows characteristic of the leeward side of mountains. Such winds can bring a 20 C° (36 F°) jump in temperature and greatly reduced relative humidity on the lee side of the mountains.

The term **rain shadow** is applied to dry regions leeward of mountains. East of the Cascade Range, Sierra Nevada, and Rocky Mountains, such rain-shadow

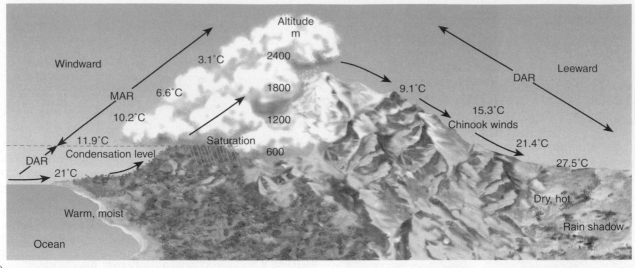

(a)

(b)

FIGURE 8.9 Orographic precipitation.
(a) Orographic barrier and precipitation patterns—unstable conditions assumed. Prevailing winds force warm, moist air upward against a mountain range, producing adiabatic cooling, eventual saturation and net condensation, cloud formation, and precipitation. On the leeward slope, as the "dried" air descends, compressional heating warms it and net evaporation dominates, creating the hot, relatively dry rain shadow of the mountain. (b) Rain shadow produced by descending, warming air contrasts with the clouds of the windward side. Dust is stirred up by leeward downslope winds. [Photo by author.]

patterns predominate. In fact, the precipitation pattern of windward and leeward slopes persists worldwide, as confirmed by the precipitation maps for North America (Figure 9.5) and the world (Figure 10.2). See News Report 8.1 for more about the role of orographic barriers in setting precipitation records.

Frontal Lifting (Cold and Warm Fronts)

The leading edge of an advancing air mass is its *front*. Vilhelm Bjerknes (1862–1951) first applied the term while working with a team of meteorologists in Norway during World War I. Weather systems seemed to them to be migrating air mass "armies" doing battle

along fronts. A front is a place of atmospheric discontinuity, a narrow zone forming a line of conflict between two air masses of different temperature, pressure, humidity, wind direction and speed, and cloud development. The leading edge of a cold air mass is a **cold front**, whereas the leading edge of a warm air mass is a **warm front**.

Cold Front On weather maps, such as the examples in Figures 8.14 and 8.15 and in News Report 8.2, a cold front is a line with triangular spikes that point in the direction of frontal movement along an advancing cP or mP air mass. The steep face of the cold air mass suggests

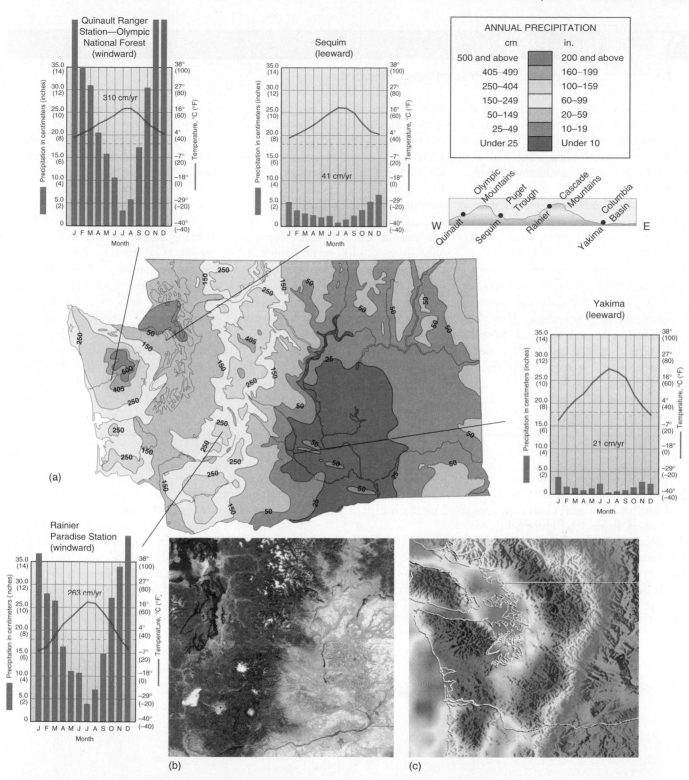

FIGURE 8.10 Orographic patterns in Washington State.
(a) Four stations in Washington provide examples of orographic effects: windward precipitation and leeward rain shadows. Isohyets (isolines of equal precipitation amounts) on the map indicate rainfall (in inches). (b) Vegetation and landscapes in this true-color MODIS image from *Terra* demonstrate these precipitation patterns. (c) Rainfall totals for October 31 to November 8, 2006; data from the *TRMM* satellite show orographic influence: darker red areas exceeded 250 mm (10 in.); greens and blues denote less precipitation. [(a) Data from J. W. Scott and others, *Washington: A Centennial Atlas* (Bellingham, WA: Center for Pacific Northwest Studies, Western Washington University, 1989), p. 3; (b) MODIS *Terra* image courtesy of Land Response Team, NASA/GSFC; (c) *TRMM* precipitation analysis courtesy of Hal Pierce, SSAI/NASA/GSFC.]

News Report 8.1

Mountains Set Precipitation Records

Because orographic lifting is limited in areal extent to locations where a topographic barrier exists, it is the least dominant lifting mechanism worldwide. But mountain ranges are the most consistent of all the precipitation-inducing mechanisms. Both the greatest average annual precipitation and the greatest maximum annual precipitation on Earth occur on the windward slopes of mountains that intercept moist tropical trade winds. During the days captured in Figure 8.10c, Stampede Pass in western Washington received 20.9 cm (8.22 in.) on November 8, a new record, swamping six local rivers, all rising to flood stage.

The world's greatest average annual precipitation occurs in the United States on Mount Waialeale, on the island of Kaua'i, Hawai'i. This mountain rises 1569 m (5147 ft) above sea level. On its windward slope, rainfall averaged 1234 cm (486 in., or 40.5 ft) a year for the years 1941–1992 (topping the mountain's previous average of 460 in.). In contrast, the rain-shadow side of Kaua'i receives only 50 cm (20 in.) of rain annually. If no islands existed at this location, this portion of the Pacific Ocean would receive only an average 63.5 cm (25 in.) of precipitation a year. (Note the importance of using established weather stations with a consistent record of weather data reporting when making such assessments. There are several stations that claim higher values but fall short of dependable measurement criteria.)

Another place receiving world-record precipitation is Cherrapunji, India, 1313 m (4309 ft) above sea level at 25° N latitude, in the Assam Hills south of the Himalayas. Because of the summer monsoons that pour in from the Indian Ocean and the Bay of Bengal, Cherrapunji has received 930 cm (366 in., or 30.5 ft) of rainfall in 1 month and a total of 2647 cm (1042 in., or 86.8 ft) in 1 year! Not surprisingly, Cherrapunji is the all-time precipitation record holder for a single year and for every other time interval from 15 days to 2 years. The average annual precipitation there is 1143 cm (450 in., 37.5 ft), placing it second only to Mount Waialeale.

its ground-hugging nature, caused by its greater density and uniform physical character compared to the warmer air mass it displaces (Figure 8.11).

Warm, moist air in advance of the cold front lifts upward abruptly and experiences the same adiabatic rates of cooling and factors of stability or instability that pertain to all lifting air parcels. A day or two ahead of the cold front's passage, high cirrus clouds appear, telling observers that a lifting mechanism is on the way.

A wind shift, temperature drop, and lowering barometric pressure mark a cold front's advance due to lifting along the front. As the line of most intense lifting passes, usually just ahead of the front itself, air pressure drops to a local low. Clouds may build along the cold front into

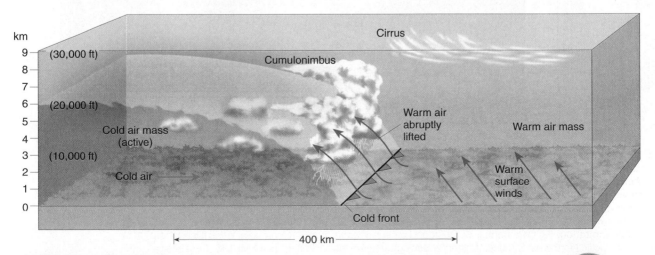

FIGURE 8.11 A typical cold front.
Denser, advancing cold air forces warm, moist air to lift abruptly. As the air is lifted, it cools by expansion at the DAR, cooling to the dew-point temperature as it rises to a level of condensation and cloud formation. Cumulonimbus clouds may produce large raindrops, heavy showers, lightning and thunder, and hail.

ANIMATION

Cold and Warm Fronts

characteristic cumulonimbus form and may appear as an advancing wall of clouds. Precipitation usually is heavy, containing large droplets, and can be accompanied by hail, lightning, and thunder.

The aftermath of a cold front passage usually brings northerly winds in the Northern Hemisphere and southerly winds in the Southern Hemisphere as anticyclonic high-pressure advances; lower temperatures; increasing air pressure produced by the cooler, more dense air; and broken cloud cover.

The particular shape and size of the North American landmass and its latitudinal position present conditions where cP and mT air masses are best developed and have the most direct access to each other. The resulting contrast can lead to dramatic weather, particularly in late spring, with sizable temperature differences from one side of a cold front to the other.

A fast-advancing cold front can cause violent lifting and create a zone right along or slightly ahead of the front called a **squall line**. Along a squall line, such as the one in the Gulf of Mexico shown in Figure 8.12, wind patterns are turbulent and wildly changing and precipitation is intense. The well-defined frontal clouds in the photograph rise abruptly, with new thunderstorms forming along the front. Tornadoes also may develop along such a squall line.

Warm Front   A line with semicircles facing in the direction of frontal movement denotes a warm front on weather maps (see Figure 8.14 map). The leading edge of an advancing warm air mass is unable to displace cooler, passive air, which is denser along the surface. Instead, the warm air tends to push the cooler, underlying air into a characteristic wedge shape, with the warmer air sliding up over the cooler air. Thus, in the cooler-air region a temperature inversion is present, sometimes causing poor air drainage and stagnation.

FIGURE 8.12 Cold front and squall line.
A sharp line of cumulonimbus clouds near the Texas coast and Gulf of Mexico mark a cold front and squall line. The cloud formation rises to 17,000 m (56,000 ft). The passage of such a frontal system over land often produces strong winds and possibly tornadoes. [Space Shuttle astronaut photo from NASA.]

Figure 8.13 illustrates a typical warm front in which mT air is gently lifted, leading to stratiform cloud development and characteristic nimbostratus clouds and drizzly precipitation. A warm front presents a progression of cloud development to an observer: High cirrus and cirrostratus clouds announce the advancing frontal system; then clouds lower and thicken to altostratus; and finally the clouds lower and thicken to stratus within several hundred kilometers of the front.

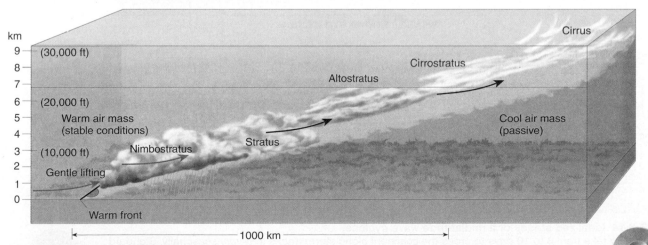

FIGURE 8.13 A typical warm front.
Note the sequence of cloud development as the warm front approaches. Warm air slides upward over a wedge of cooler, passive air near the ground. Gentle lifting of the warm, moist air produces nimbostratus and stratus clouds and drizzly rain showers, in contrast to the more dramatic precipitation associated with the passage of a cold front.

ANIMATION

Cold and
Warm Fronts

Midlatitude Cyclonic Systems

The conflict between contrasting air masses can develop a **midlatitude cyclone**, or **wave cyclone**. This migrating low-pressure center with converging, ascending air spirals inward counterclockwise in the Northern Hemisphere, or inward clockwise in the Southern Hemisphere. Because of the undulating nature of frontal boundaries and the steering flow of the jet streams, the term *wave* is appropriate. The combination of the *pressure gradient force*, *Coriolis force*, and *surface friction* generates this cyclonic motion (see discussion in Chapter 6).

Before World War I, weather maps displayed only pressure and wind patterns. V. Bjerknes added the concept of fronts, and his son Jacob Bjerknes contributed the concept of migrating centers of cyclonic low-pressure systems.

Wave cyclones dominate weather patterns in the middle and higher latitudes of both the Northern and Southern Hemispheres and act as a catalyst for air mass conflict. Such a midlatitude cyclone can initiate along the polar front, particularly in the region of the Icelandic and Aleutian subpolar low-pressure cells in the Northern Hemisphere. The intense high-speed winds of the jet streams guide cyclonic systems along their tracks (see Figures 6.16 and 6.17).

Life Cycle of a Midlatitude Cyclone

Figure 8.14 shows the birth, maturity, and death of a typical midlatitude cyclone in several stages, along with an idealized weather map. On the average, a midlatitude cyclonic system takes 3–10 days to progress through these

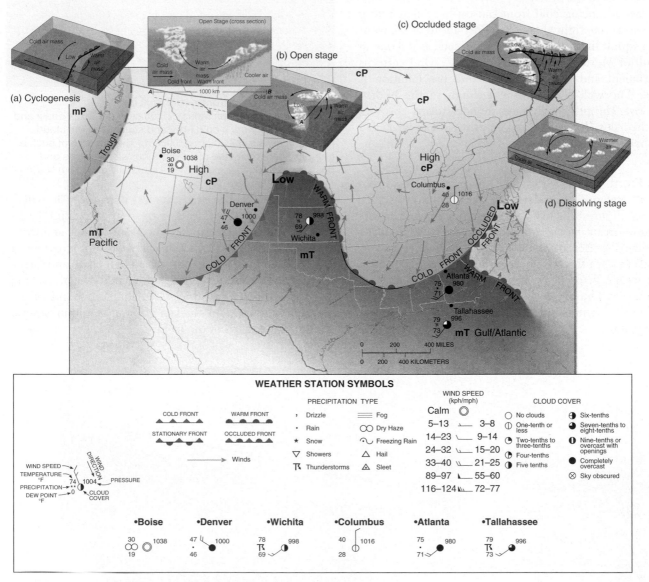

FIGURE 8.14 Idealized stages of a midlatitude wave cyclone.
A system may take from 3 to 10 days to cross the continent along storm tracks that vary seasonally. Standard weather station symbols and current conditions for six cities appear below the map. The block diagrams show (a) cyclogenesis where surface convergence and lifting begin, (b) the open stage, (c) the occluded stage, and (d) the dissolving stage, reached at the end of the storm track as the cyclone spins down, no longer energized by the latent heat from condensing moisture. After studying the text, can you describe conditions in Boise, Denver, Wichita, Columbus, Atlanta, and Tallahassee as depicted on this weather map?

ANIMATION

Midlatitude Cyclones

stages from the area where it develops to the area where it finally dissolves. However, chaos rules and every day's weather map departs from the ideal in some manner.

Cyclogenesis The atmospheric process in which low-pressure wave cyclones develop and strengthen is **cyclogenesis**. This process usually begins along the polar front, where cold and warm air masses converge and are drawn into conflict.

The polar front is a discontinuity of temperature, moisture, and winds that establishes potentially unstable conditions. For a wave cyclone to form along the polar front, a compensating area of divergence aloft matches a surface point of air convergence. Even a slight disturbance along the polar front, perhaps a small change in the path of the jet stream, can initiate the converging, ascending flow of air and thus a surface low-pressure system (illustrated in Figure 8.14a).

In addition to the polar front, certain other areas are associated with wave cyclone development and intensification: the eastern slope of the Rockies and other north–south mountain barriers, the Gulf Coast, and along the east coasts of North America and Asia.

Open Stage To the east of the developing low-pressure center, warm air begins to move northward along an advancing front, while cold air advances southward to the west of the center. See this organization in Figure 8.14b and around the center of low pressure, located over western Nebraska. The growing circulation system then vents into upper-level winds. As the midlatitude cyclone matures, the counterclockwise flow draws the cold air mass from the north and west and the warm air mass from the south. In the cross section, you can see the profiles of both a cold front and a warm front and each air mass segment. Compare with the illustrations in Figures 8.11 and 8.13.

On the map, Denver has just experienced the passage of a cold front. Before the front passed, you see on the map that winds were from the southwest, but now the winds have shifted and are from the northwest. Temperature and humidity went from the warm, moist mT air mass to colder conditions as the cP air mass moves over Denver. Meanwhile, Wichita, Kansas, experienced the passage of a warm front and now is in the midst of the warm-air segment of the cyclone. A weather map showing a classic open stage from March 31, 2007 is in the "Critical Thinking" section at the end of this chapter.

Another example of an open stage of a midlatitude cyclone occurred April 20, 2000, with low pressure centered over the Iowa–Missouri border. Figure 8.15a and b shows you a satellite image and the daily weather map for that day. Note how the isobars portray the cyclone and the low of 997.6 mb (29.45 in.) on the map. And in another example, note the cloud pattern stretching along the cold front and the swirling clouds around the low. Figure 8.15c and d is a satellite image and weather map of a frontal storm system hitting the Northwest on February 1, 2000.

Occluded Stage Remember the relation between air temperature and the density of an air mass. The colder cP air mass is denser than the warmer mT air mass. This cooler, more unified air mass acts like a bulldozer blade and therefore moves faster than the warm front. Cold fronts can travel at an average 40 kmph (25 mph), whereas warm fronts average roughly half that at 16–24 kmph (10–15 mph). Thus, a cold front often overtakes the cyclonic warm front, wedging beneath it, producing an **occluded front** (*occlude* means "to close") in Figure 8.14c.

On the idealized weather map an occluded front stretches south from the center of low pressure in Virginia to the border between the Carolinas. Precipitation may be moderate to heavy initially and then taper off as the warmer air wedge is lifted higher by the advancing cold air mass. Note the still active warm front in the extreme Southeast and the flow of mT air. What conditions do you observe in Tallahassee, Florida? What might it be like there in the next 12 hours if the cold front passes south of the city?

When there is a stalemate between cooler and warmer air masses where air flow on either side is almost parallel to the front, although in opposite directions, a **stationary front** results. Some gentle lifting might produce light to moderate precipitation. Eventually, the stationary front will begin to move, as one of the air masses assumes dominance, evolving into a warm or a cold front.

Dissolving Stage The final, dissolving, stage of the midlatitude cyclone occurs when its lifting mechanism is completely cut off from the warm air mass, which was its source of energy and moisture. Remnants of the cyclonic system then dissipate in the atmosphere, perhaps after passage across the country (Figure 8.14d).

Storm Tracks Cyclonic storms—1600 km (1000 mi) wide—and their attendant air masses move across the continent along **storm tracks**, which shift in latitude with the Sun and the seasons. Typical storm tracks that cross North America are farther northward in summer and farther southward in winter (Figure 8.16). Noted on the map are some of the regional names for cyclonic source regions. As the storm tracks begin to shift northward in the spring, cP and mT air masses are in their clearest conflict. This is the time of strongest frontal activity, featuring thunderstorms and tornadoes. Storm tracks follow the path of upper-air winds, which direct storm systems across the continent.

A map of actual storm tracks for March 1991 in Figure 8.16b demonstrates several areas of cyclogenesis: the northwest over the Pacific Ocean, the Gulf of Mexico (Gulf Lows), the eastern seaboard (Nor'easter, Hatteras Low), and the Arctic. Cyclonic circulation also frequently develops on the lee side of mountain ranges, as along the Rockies from Alberta (Alberta Clipper) south to Colorado (Colorado Low). By crossing the mountains, such systems gain access to the moisture-laden, energy-rich mT air masses from the Gulf of Mexico and are strengthened.

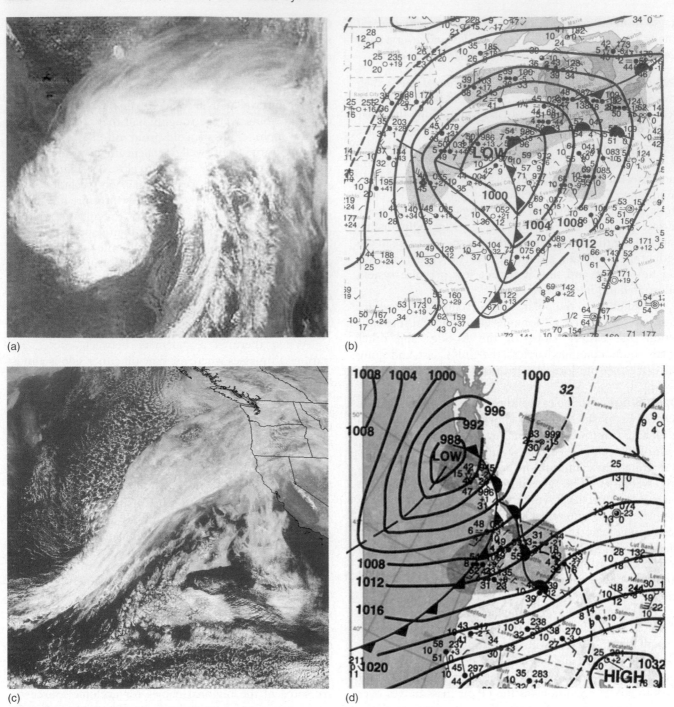

(a)

(b)

(c)

(d)

FIGURE 8.15 Open stage of a midlatitude cyclone.
(a) *SeaWiFS* image of a cyclonic system over the Midwest. The cloud patterns are areas of
precipitation; clear skies are in a wedge behind the cold front as a cP air mass moves over the landscape.
(b) A segment from the April 20, 2000, daily weather map, about 8 hours earlier than the image, showing
the low-pressure system centered on 997.6 mb (29.45 in.). (c) Parts of Washington and Oregon receive
heavy snow in the mountains and rain elsewhere from a jet-stream-guided storm system, February 1,
2000. Wind gusts reach 110 kmph (70 mph) along the Oregon Coast. (d) Segment of the daily weather
map. [(a) *SeaWiFS* image used with permission of GeoEye/NASA. All rights reserved; (b) and (d) Segments of
"Daily Weather Map," courtesy of NWS, NOAA; (c) *GOES-10* image courtesy of NESDIS/NOAA.]

Analysis of Daily Weather Maps— Forecasting

Synoptic analysis is the evaluation of weather data collected
at a specific time. Building a database of wind, pressure,
temperature, and moisture conditions is key to *numerical*

(computer-based) *weather prediction* and the development
of weather-forecasting models. Development of numerical models is a great challenge because the atmosphere
operates as a nonlinear system, tending toward chaotic
behavior. Slight variations in input data or slight changes

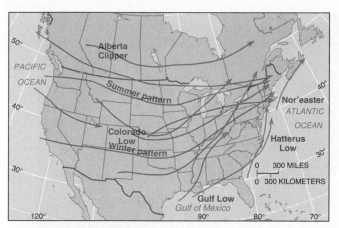

(a) Average storm tracks

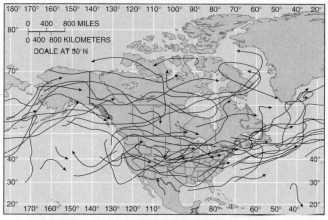

(b) Actual storm tracks in March 1991

FIGURE 8.16 Typical and actual storm tracks.
(a) Cyclonic storm tracks over North America vary seasonally. The tracks indicate several locations of cyclogenesis; note regional names. (b) Actual cyclonic tracks during March 1991 over North America. [(b) From *Storm Data*, vol. 33, no. 3 (March 1991), Asheville, NC: NOAA/NESDIS, National Climatic Data Center.]

in the basic assumptions of the model's behavior can produce widely varying forecasts. As our knowledge of the interactions that produce weather and as instruments and software improve, so too will the accuracy of our forecasts.

Weather data necessary for the preparation of a synoptic map and forecast include:

- Barometric pressure (sea level and altimeter setting)
- Pressure tendency (steady, rising, falling)
- Surface-air temperature
- Dew-point temperature
- Wind speed, direction, and character (gusts, squalls)
- Type and movement of clouds
- Current weather
- State of the sky (current sky conditions)
- Visibility; vision obstruction (fog, haze)
- Precipitation since last observation

For links to weather maps, current forecasts, satellite images, and the latest radar, please go to the *Geosystems* Student Learning Center, to Chapter 8, and click "Destinations," where you find many related Internet links. With the Internet, there is no need to wait for television or the newspaper to bring you the latest satellite images, for they are at your fingertips.

NOAA's Earth System Research Laboratory (ESRL), formerly the Forecast Systems Laboratory (FSL), in Boulder, Colorado, is developing new forecasting tools (see **http://www.esrl.noaa.gov/**). A few innovations include: wind profilers using radar to profile winds from the surface to high altitudes; a High Performance Computing System (hundreds of CPUs networked to act as one) to model software for 3-D weather models and other computations; improved international cooperation and weather data sharing; and continuous software upgrades of the Advanced Weather Interactive Processing System (AWIPS) and the Automated Surface Observing System, or ASOS (Figure 8.17).

ASOS sensor instrument arrays are a primary surface weather-observing network (Figure 8.17a). An ASOS installation includes: rain gauge (tipping bucket type), temperature/dew-point sensor, barometer, present weather identifier, wind-speed indicator, direction sensor, cloud-height indicator, freezing-rain sensor, thunderstorm sensor, and visibility sensor, among other items.

Preparing a weather report and forecast requires analysis of daily weather maps and the use of standard weather symbols, such as those in Figure 8.14. Although the actual pattern of cyclonic passage over North America is widely varied in shape and duration, you can apply this general model, along with your understanding of warm and cold fronts, to reading the daily weather map. A weather map is a useful tool, especially when weather turns dramatic in violent episodes.

Violent Weather

Weather provides a continuous reminder of the flow of energy across the latitudes that at times can set into motion destructive, violent weather conditions. In dollar value, weather-related damage is increasing each year as population increases in areas prone to violent weather and as climate change intensifies weather anomalies. We focus on thunderstorms, tornadoes, and hurricanes.

However, we must include mention of such things as *ice storms* of **sleet** (freezing rain, ice glaze, and ice pellets), snow blizzards, and low temperatures. Sleet is caused when precipitation falls through a below-freezing layer of air near the ground. During the ice storm of January 1998, a large region in Canada and the United States with 700,000 residents was without power for weeks (Figure 8.18). Imagine a coating of ice on everything, weighing on power lines and tree limbs. More than 80 hours of freezing rain and drizzle hit the region; this is more than double typical ice-storm duration. In Montreal, over 100 mm (4 in.) of ice accumulated. Hypothermia claimed 25 human lives.

(a)

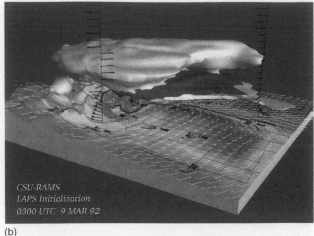

CSU-RAMS
LAPS Initialization
0300 UTC 9 MAR 92

(b)

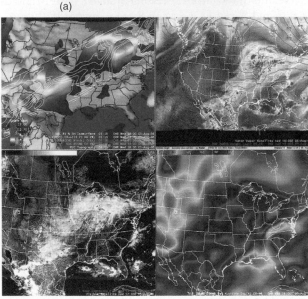

(c)

(d)

(e)

FIGURE 8.17 ASOS weather instruments and AWIPS workstation display.
(a) Automated Surface Observing System (ASOS) installation. (b) Forecast Systems Laboratory (FSL) scientists develop 3-D modeling programs to portray weather and simulate future weather for more accurate forecasting. On the image note the gray area that depicts significant cloud water content and the red layer within the cloud where aircraft icing is forecast. The black wind barbs off the two vertical axes denote wind profiles through the atmosphere. (c) Powerful computers are needed to process the data and produce these 3-D virtual images on the Advanced Weather Interactive Processing System (AWIPS), employing screens with multiple frames. Of the many items that can be displayed, we see in Figure 8.17c a simulated 3-D pressure analysis, water vapor, and visible satellite image—including lightning strikes—and humidity. (d) An AWIPS workstation. (e) A portion of the High Performance Computing System, known as "Jet" at ESRL. [Photo (a) and (e) by Bobbé Christopherson; (b–d) images courtesy of the FSL, NOAA, Boulder, Colorado.]

(a)

(b)

FIGURE 8.18 Aftermath of record ice storm. Record blizzards and ice from a nor'easter storm blanketed New England and eastern Canada, causing almost a billion dollars in damage, January 1998. (a) Broken birch trees in Maine, snapped under the weight of an ice coating as thick as 8 cm (3 in.), remain as reminders of this weather disaster. (b) Collapsed Hydro-Québec high-voltage towers south of Montreal. Power to more than 1 million homes was out for weeks. [Photos by (a) Bobbé Christopherson; (b) courtesy of Environment Canada. Used by Permission, Minister of Public Works, Government Services Canada.]

Government research and monitoring of violent weather is centered at NOAA's National Severe Storms Laboratory and Storm Prediction Center—see http://www.nssl.noaa.gov/, http://www.spc.noaa.gov/, and http://www.esrl.noaa.gov/; consult these sites for each of the topics that follow.

Thunderstorms

The condensation of large quantities of water vapor in clouds liberates tremendous amounts of energy. This process locally heats the air, rapidly changing its density and buoyancy and causing violent updrafts. As rising air parcels pull surrounding air into the cumulonimbus cloud system, updrafts become stronger. Raindrops form and descend through the cloud, and their frictional drag pulls air toward the ground, causing violent downdrafts. The resulting giant cumulonimbus clouds signal dramatic weather moments—heavy precipitation, lightning, thunder, hail, blustery winds, and maybe tornadoes.

Such thunderstorms may develop under three conditions: within an air mass (particularly in warm, moist air), in a line along a cold front or other convergent boundary, or where mountain slopes cause orographic lifting. With a slowly rotating updraft, a dangerous convective supercell thunderstorm might develop, featuring strong winds, hail, and persistent tornadoes of great strength.

Thousands of thunderstorms occur on Earth at any given moment. Equatorial regions and the ITCZ experience many of them, exemplified by the city of Kampala in Uganda, East Africa (north of Lake Victoria), which sits virtually on the equator and averages a record 242 days a year of thunderstorms. Figure 8.19 shows the annual distribution of days with thunderstorms across the United States and Canada. You can see that, in North America, most thunderstorms occur in areas dominated by mT air masses.

Atmospheric Turbulence Most airplane flights experience at least some turbulence—the encountering of air of different densities or air layers moving at different speeds and directions. This is a natural state of the atmosphere and passengers are asked to keep seat belts fastened when in their seats to avoid injury, even if the seat-belt light is turned off.

Thunderstorms can produce severe turbulence in the form of downbursts, which are exceptionally strong downdrafts. Downbursts are classified by size: a *macroburst* is at least 4.0 km (2.5 mi) wide and in excess of 210 kmph (130 mph); a *microburst* is smaller in size and speed. A microburst causes rapid changes in wind speed and direction and it causes the dreaded *wind shear* that can bring down aircraft. Such turbulence events are short-lived and hard to detect, although the Forecast Systems Laboratory, among others, is making progress in developing forecasting methods.

Lightning and Thunder An estimated 8 million lightning strikes occur each day on Earth. **Lightning** refers to flashes of light caused by enormous electrical discharges—tens of millions to hundreds of millions of volts—that briefly superheat the air to temperatures of 15,000°–30,000°C (27,000°–54,000°F). A buildup of electrical energy polarity between areas within a cumulonimbus cloud or between the cloud and the ground

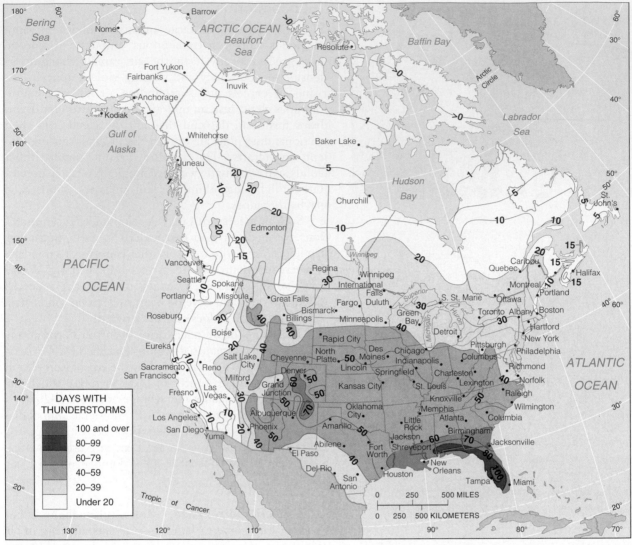

FIGURE 8.19 Thunderstorm occurrence.
Average annual number of days experiencing thunderstorms. [Data courtesy of U.S. National Weather Service; Map Series 3; *Climatic Atlas of Canada*, Atmospheric Environment Service, Canada.]

creates lightning (Figure 8.20a). The violent expansion of this abruptly heated air sends shock waves through the atmosphere as the sonic bang of **thunder**.

NASA's Lightning Imaging Sensor (LIS) aboard the *Tropical Rainfall Measuring Mission* (*TRMM*) satellite monitors lightning and other phenomena. The LIS can image lightning strikes day or night, within clouds or between cloud and ground. The sensor's data show that about 90% of all strikes occur over land in response to increased convection over relatively warmer continental surfaces, with expected seasonal shifts with the high Sun, as shown in Figure 8.20b and c (see **http://thunder. msfc.nasa.gov/lis/**).

Lightning poses a hazard to aircraft, people, animals, trees, and structures. Certain precautions are mandatory when a lightning discharge threatens, because lightning causes nearly 200 deaths and thousands of injuries each year in the United States and Canada. When lightning is imminent, the National Weather Service issues *severe storm warnings* and cautions people to remain indoors. If

the hair on your head or neck begins to stand on end, get indoors, or if in the open get low to the ground, crouched on both feet and not laying down, in a low spot, if possible. Your hair is telling you that a charge is building in that area. The place *not* to seek shelter is beneath a tree, for trees are good conductors of electricity and often are hit by lightning.

Hail Ice pellets called **hail** generally form within a cumulonimbus cloud. Raindrops circulate repeatedly above and below the freezing level in the cloud, adding layers of ice until the circulation in the cloud can no longer support their weight (Figure 8.21a). Hail may also grow from the addition of moisture on a snow pellet.

Pea-sized and marble-sized hail is common, although hail the size of golf balls and baseballs happens at least once or twice a year somewhere in North America. For larger hail to form, the frozen pellets must stay aloft for longer periods. The largest authenticated hailstone in the

(a)

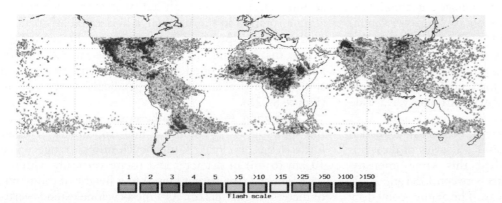

(b)   Winter (Dec. 1999, Jan. and Feb. 2000)

1	2	3	4	5	>5	>10	>15	>25	>50	>100	>150

Flash scale

(c)   Summer (June, July, August 2000)

FIGURE 8.20 Seasonal images show global lightning.
(a) Multiple cloud-to-ground lightning strikes captured in a time-lapse photo. (b) A composite of 3 months' data derived from NASA's Lightning Imaging Sensor (LIS) records all lightning strikes between 35° N and 35° S latitudes during (b) winter (December 1999–February 2000) and (c) summer (June, July, August 2000). The LIS is aboard the *Tropical Rainfall Measuring Mission* satellite launched in 1997. [(a) Photo from C. Clark, NOAA Photo Library, NOAA Central Library, NSSL; (b) and (c) *TRMM* images courtesy of NASA's Global Hydrology and Climate Center, Marshall Spaceflight Center, Huntsville, Alabama.]

(a)   (b)

FIGURE 8.21 Hailstones and hailstone damage.
(a) Irregular layers give evidence of numerous convective trips through a cumulonimbus cloud. (b) Hail
can severely damage windows, automobiles, crops, livestock, and people. [(a) © Arthur C. Smith III/Grant
Heilman Photography, Inc.; (b) photo by *The Denver Post*, 1990.]

world landed in Kansas in 1970; it measured 14 cm
(5.5 in.) in diameter and weighed 758 g (1.67 lb)!

Hail is common in the United States and Canada, although somewhat infrequent at any given place. Hail occurs perhaps every 1 or 2 years in the highest-frequency
areas. Annual hail damage in the United States tops $800
million (Figure 8.21b). The pattern of hail occurrence
across the United States and Canada is similar to that of
thunderstorms shown in Figure 8.19.

Derechos

Although tornadoes and hurricanes grab headlines,
straight-line winds associated with thunderstorms and
bands of showers cause significant damage and crop losses.
These strong linear winds in excess of 26 m/s (58 mph)
are known as **derechos**, or in Canada sometimes called
"plow winds." Wind outbursts from convective storms
can produce groups of downburst clusters from a thunderstorm system. These derechos tend to blast in linear
paths fanning out along curved-wind fronts over a wide
swath of land.

A derecho in 1998 in eastern Wisconsin exceeded
57 m/s (128 mph). In August 2007, a series of derechos
across northern Illinois reached this same intensity.
Researchers identified 377 events between 1986 and 2003,
an average of about 21 per year. The name, coined by
physicist G. Hinrichs in 1888, derives from a Spanish
word meaning "direct" or "straight ahead."

Derechos pose distinct hazards to summer outdoor
activities by overturning boats, hurling flying objects, and
causing broken trees and limbs. Their highest frequency
(about 70%) is from May to August in the region
stretching from Iowa, across Illinois, and into the Ohio

River Valley in the upper Midwest. By September and on
through to April, although less than in summer, areas of
activity migrate southward to eastern Texas through
Alabama. For more information and a description of
numerous derechos, see **http://www.spc.noaa.gov/misc/
AbtDerechos/derechofacts.htm**.

Tornadoes

The updrafts associated with cumulonimbus development appear on satellite images as pulsing bubbles of
clouds, sometimes as supercells of rotating updrafts. Friction with the ground slows surface winds, but higher
in the troposphere, winds blow faster. Thus, a body of
air moves faster at higher altitude than at the surface, creating a rotation in the air along a horizontal axis parallel
to the ground (picture a rolling pin; see Figure 8.22a).
When that rotating air encounters the strong updrafts
associated with frontal activity, the axis of rotation
shifts to a vertical alignment perpendicular to the ground
(Figure 8.22b).

According to one hypothesis for *supercell tornadoes*,
this spinning, rising column of mid-troposphere-level air
forms a mesocyclone. A **mesocyclone** can range up to
10 km (6 mi) in diameter and rotate vertically within a
supercell cloud (the parent cloud) to a height of thousands
of meters (Figure 8.22c). As a mesocyclone extends vertically and contracts horizontally, wind speeds accelerate in
an inward vortex (as ice skaters draw their arms in to
accelerate while spinning or as water speeds up when it
spirals down a sink drain).

As more moisture-laden air draws up into the circulation of a mesocyclone, more energy is liberated and the
rotation of air becomes more rapid as a *wall cloud* forms

ANIMATION

Tornado Wind Patterns

FIGURE 8.22 Mesocyclone and tornado formation.
(a) Strong wind aloft establishes spinning along a horizontal axis. (b) Updraft from thunderstorm development tilts the rotating air along a vertical axis. (c) Mesocyclone forms as a rotating updraft within the thunderstorm. If one forms, a tornado will descend from the lower portion of the mesocyclone. (d) A tornado leaves its mark, as this *Landsat-7* image captures. The damage path tracks for 39 km (24 mi) where an EF-4 tornado tore through La Plata, Maryland; more than 900 homes and 200 businesses were destroyed. (e) EF-5 tornado damage in May 1999 in Oklahoma. The church visible in the background was used for an emergency shelter. (f) A fully developed tornado swirls dust and debris as it locks onto the ground. (g) A tornado emerges from the base of a wall cloud near Oakfield, Wisconsin. [(d) *Landsat-7* image courtesy of Satellite Systems Branch, USGS EROS Data Center. Photos by (e) Todd Lindlay; (f) C. Lavoie/First Light; (g) Don Lloyd/AP/Wide World Photos.]

FIGURE 8.23 Supercell tornado.
A supercell tornado descends from the cloud base near Spearman, Texas. Strong hail is falling to the left of the tornado. [Photo by Howard Bluestein, all rights reserved.]

Tornado Measurement and Science A tornado's diameter can range from a few meters to a few hundred meters or larger, as in Oklahoma City and Greensburg. It can last from a few seconds to tens of minutes. Air pressure inside a tornado funnel usually is about 10% less than in the surrounding air. This disparity is similar to the pressure difference between sea level and an altitude of 1000 m (3300 ft). Such a sharp horizontal-pressure gradient causes an in-rushing convergence and severe wind speeds at the funnel.

The late Theodore Fujita, a noted meteorologist from the University of Chicago, designed the Fujita Scale (1971), which classifies tornadoes according to wind speed as indicated by related property damage. A refinement in the scale, adopted in February 2007, led to the *Enhanced Fujita Scale*, or *EF Scale* (Table 8.1). The revision filled the need to better assess damage, correlate wind speed to damage caused, and account for construction quality of structures. To assist with wind estimates, the EF Scale contains 28 Damage Indicators (DI) relating to types of structures and vegetation affected along with Degrees of Damage (DOD) ratings, both of which you can view at the URL listed below the table.

(Figure 8.23). The swirl of the mesocyclone itself is visible, as are dark gray **funnel clouds** that pulse from the bottom side of the parent cloud. The terror of this stage of development is the lowering of such a funnel to Earth. This tightly rotating vortex is the dreaded **tornado**, from the Latin *tornare*, "to make round by turning" (Figures 8.22f and g and 8.23). When tornado circulation occurs over water, a **waterspout** forms, and the sea or lake surface water is drawn up into the funnel.

A tornado's scoured path is visible on the ground in a series of semicircular "suction" prints and, of course, in the trail of damage, such as in the image of a tornado's path through Maryland (Figure 8.22d). Figure 8.22e shows EF-5 destruction from a tornado in Oklahoma City. This extreme tornado was 1.6-km (1-mi) wide at its base and lasted an hour, killing 36 people. Resulting property damage exceeded $1 billion.

Greensburg, Kansas, was essentially erased from the map by an EF-5 tornado in May 2007. The statistics are unprecedented: Winds exceeded 330 kmph (205 mph) in the 2.7-km (1.7-mi) wide funnel that was on the ground along a 35-km (22-mi) track. See the chapter-opening photo and satellite images that begin Chapter 1. Improved warning systems held the loss of life to 10 people.

Table 8.1	The Enhanced Fujita Scale
EF-number	**3-Second Gust Wind Speed; Damage Specs**
F0 Gale	105–137 kmph; 65–85 mph: *light damage*: break branches, damage chimneys.
F1 Weak	138–177 kmph; 86–110 mph: *moderate damage*: beginning of hurricane wind-speed designation, roof coverings peel off, mobile homes pushed off foundations.
F2 Strong	178–217 kmph; 111–135 mph: *considerable damage*: roofs torn off frame houses, large trees uprooted or snapped, box cars pushed over, small missiles generated.
F3 Severe	218–266 kmph; 136–165 mph: *severe damage*: roofs torn off well-constructed houses, trains overturned, trees uprooted, cars thrown.
F4 Devastating	267–322 kmph; 166–200 mph: *devastating damage*: well-built houses leveled, cars thrown, large missiles generated.
F5 Incredible	More than 322 kmph; >200 mph: *incredible damage*: houses lifted and carried distance to disintegration, car-sized missiles fly farther than 100 m, bark removed from trees.

Note: See the EF-Scale pdf file at **http://www.wind.ttu.edu/ General/Research.php**.

North America receives more tornadoes than anywhere on Earth because its latitudinal position, topography, and the nature of the landscape permit contrasting air masses to confront one another. Tornadoes have struck all 50 states and all the Canadian provinces and territories. May and June are the peak months, as you see in Figure 8.24, based on 51 years of records. Note the comparison with statistics for the 2003–2005 period and the annual trend for tornado occurrence on the graph.

Canada experiences an average of 80 observed tornadoes per year, although unpopulated rural areas go unreported. University of Leeds researchers report that in the United Kingdom as many as 100 tornadoes may hit in a year, although observers report 60 to 80 a year. Other continents report a small number of tornadoes each year.

The Storm Prediction Center in Kansas City, Missouri, provides short-term forecasting for thunderstorms and tornadoes to the public and to NWS field offices.

Warning times of from 12 to 30 minutes are possible with current technology (see the Storm Prediction Center at http://www.spc.noaa.gov/). News Report 8.2 details a tornado outbreak in May 2003 right through the heart of the infamous "Tornado Alley"—a corridor that experiences the highest average number of tornadoes on Earth.

Recent Tornado Records In the United States, 48,821 tornadoes were recorded in the 57 years from 1950 through 2006, causing 4793 deaths, or about 84 deaths per year. Of note is the fact that these tornadoes resulted in more than 80,000 injuries and property damage of over $26 billion for this period of record. The yearly average of $500 million in damage is rising each year.

The annual average number of tornadoes before 1990 was 787. Interestingly, since 1990 the average per year has risen to over 1000, with 2004 experiencing 1819 tornadoes, 2005 with 1264, and 2006 with 1105 tornadoes. Importantly, the number of EF-4 and EF-5 tornadoes is on the increase, reaching more than a dozen each season.

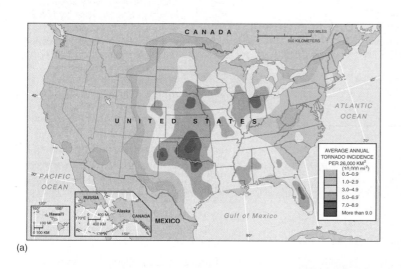

(a)

Convectional Heating and Tornado in Florida

NOTEBOOK

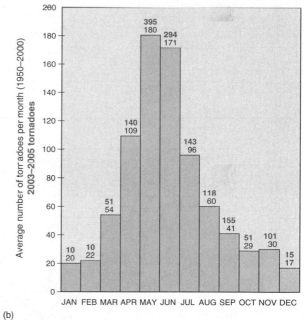

(b)

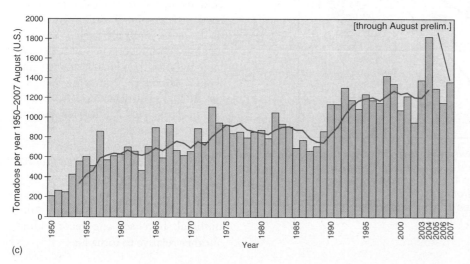

(c)

FIGURE 8.24 Tornado occurrence in the United States.
(a) Average number of tornadoes per 26,000 km² (10,000 mi²) for 1950 to 2003.
(b) Average number of tornadoes per month 1950–2000. Remarkably, the 2003–2005 average tornadoes by month shows a significant increase in numbers.
(c) Tornadoes per year 1950 through August 2007, including 5-year mean.
[Data courtesy of the Storm Prediction Center, NWS, Kansas City, Missouri, and NOAA sources.]

News Report 8.2

May 2003 Tornado Outbreak in Tornado Alley

During May 2003, 543 tornadoes struck the United States. The first 12 days of the month featured 354 tornadoes. Figure 8.2.1 maps the tracks for these record-setting May tornadoes across some 20 states. This is 366 beyond the average for May, and more tornadoes for any single month since 1950. The results caused more than 40 deaths and hundreds of millions of dollars in property damage.

Eighty-four tornadoes hit on Sunday, May 4, in a path across eight states. Tornado damage ratings of four EF-4s and eight EF-3s struck in the Midwest. Five tornadoes struck in or near Kansas City. One of the Kansas City EF-4 tornadoes achieved a funnel-base width of 450 m (1500 ft). Some damage appeared to be from downbursts and derecho linear-wind effects. (For an

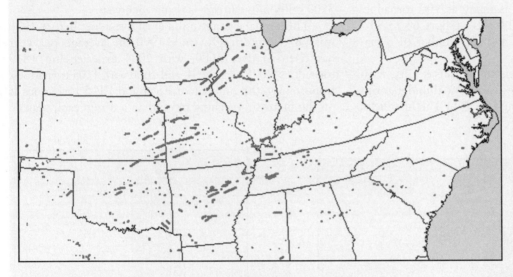

FIGURE 8.2.1 May 2003 tornado outbreak.
Map of 543 tornado tracks that hit during May 2003 as listed in *Storm Data* by NWS Forecast Offices. Tornado paths may blend, or cross, or fall on top of one another on a map of this small scale. This was a record for the month of May. [NWS Forecast Offices in *Storm Data*, Storm Prediction Center/NOAA.]

animation of satellite images, see **http://www.osei.noaa.gov/Events/Severe/US_Midwest/ 2003/SVRusMW125_G12.avi**.)

The weather map for May 5, 7:00 A.M., shows the system that produced these storms the day before (Figure 8.2.2). Local NWS forecasters broadcast in advance that conditions were perfect for a severe weather outbreak. Several supercell storms resulted from the passage of a cold front, where cold, dry air from the Rockies collided with warm, moist air from the Gulf of Mexico. Upper-air flows with divergence aloft helped trigger giant supercells along the frontal boundary. Following numerous hits, the entire state of Oklahoma was declared a federal disaster area on May 10 (Figure 8.2.3).

For those of you who live in this region of the country, tornado alerts and tornado warnings are something with which you are familiar. Unfortunately, the years that followed this outbreak have set one record after another relative to tornadoes.

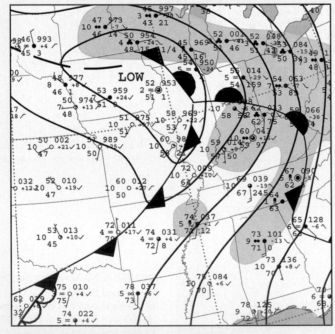

FIGURE 8.2.2 May 5, 2003, 7:00 A.M. weather map segment.
The cold front, now partially occluded and still potent, that swept across Oklahoma, Kansas, and Missouri on May 4. The conflict between cold-dry and warm-moist air masses sets up perfect frontal conditions for the formation of supercells, spawning many mesocyclones and tornadoes. [Weather map segment courtesy of "Daily Weather Maps," NWS, NOAA.]

FIGURE 8.2.3 Tornado damage near Oklahoma City. A chaplain inspects what is left of his church in Bethany, Oklahoma, following a May 10, 2003, outbreak. The entire state was declared a disaster area on this day following a week of severe weather. [Photo by Richard Michael Pruitt/*Dallas Morning News*.]

Reasons for the annual increase in tornado frequency in North America range from global climate change and more intense thunderstorm activity to better reporting through Doppler radar and a larger population with videotape cameras. Researchers are pursuing many questions with an enthusiasm for the science and an awe of the subject, as summarized by Howard Bluestein, foremost tornado researcher and meteorologist:

The source of rotation in some tornadoes appears to be preexisting vortices near the ground above where convective storms are growing; in others, the source seems to be linked to the mesocyclone. Does the mesocyclone itself descend to the ground and intensify to become the tornado, or does it trigger events near the ground that create tornadoes from other sources of rotation? Can it do both? We hope the answers to these questions are soon forthcoming. Our quest for discovery has not taken away from the respect we have for the awesome power the tornado harbors, nor the thrill for viewing the violent motions in the tornado or the beauty of the storm. We eagerly await the next act in the atmospheric play starring the tornado.*

Tropical Cyclones

A powerful manifestation of the Earth–atmosphere energy budget is the **tropical cyclone**, which originates entirely within tropical air masses. The tropics extend from the Tropic of Cancer at 23.5° N to the Tropic of Capricorn at 23.5° S, containing the equatorial zone between 10° N and 10° S. Approximately 80 tropical cyclones occur annually worldwide. Some 45 per year are powerful enough to be classified as hurricanes, typhoons, and cyclones (different regional names for the same type of tropical storm)—30% of these occur in the western North Pacific.

Cyclonic systems forming in the tropics are quite different from midlatitude cyclones because the air of the tropics is essentially homogeneous, with no fronts or conflicting

*H. Bluestein, *Tornado Alley, Monster Storms of the Great Plains* (New York: Oxford University Press, 1999), p. 162.

air masses of differing temperatures. In addition, the warm air and warm seas ensure abundant water vapor and thus the necessary latent heat to fuel these storms. Tropical cyclones convert heat energy from the ocean into mechanical energy in the wind—the warmer the ocean and atmosphere, the more intense the conversion and powerful the storm.

What mechanism triggers the start of a tropical cyclone? Meteorologists now think that cyclonic motion begins with slow-moving easterly waves of low pressure in the trade-wind belt of the tropics, such as the Caribbean area (Figure 8.25). For these processes, sea-surface temper-

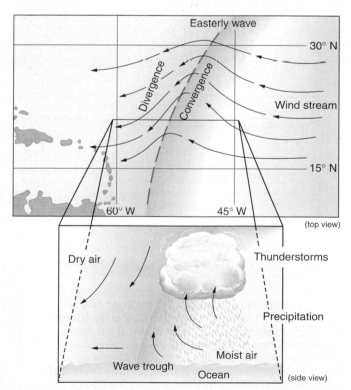

FIGURE 8.25 Easterly wave in the tropics. A low-pressure center develops along an easterly (westward-moving) wave. Moist air rises in an area of convergence at the surface to the east of the wave trough. Wind flows bend and converge before the trough and diverge downwind from the trough.

atures must exceed approximately 26°C (79°F). Tropical cyclones form along the eastern (leeward) side of these migrating troughs of low pressure, a place of convergence and rainfall. Surface air flow then converges into the low-pressure area, ascends, and flows outward aloft. This important divergence aloft acts as a chimney, pulling more moisture-laden air into the developing system. To maintain and strengthen this vertical convective circulation, there must be little or no wind shear to interrupt or block the air flow.

Hurricanes, Typhoons, and Cyclones   Tropical cyclones are potentially the most destructive storms experienced by humans, claiming thousands of lives each year worldwide. This is especially true when they attain wind speeds and low-pressure readings that upgrade their status to a full-fledged **hurricane, typhoon,** or *cyclone* (>65 knots, >74 mph, >119 kmph). Across the globe, such storms bear these three different names, among others: *hurricane* around North America, *typhoon* in the western Pacific (Japan, Philippines), and *cyclone* in Indonesia, Bangladesh, and India. By any name they can be destructive killers. Worldwide, about 10% of all tropical disturbances have the right ingredients to become hurricanes or typhoons. For coverage and reporting, see the National Hurricane Center at **http://www.nhc.noaa.gov/** or the Joint Typhoon Warning Center at **https://metocph. nmci.navy.mil/jtwc.php.**

The map in Figure 8.26 shows areas in which tropical cyclones form and some of the characteristic storm tracks traveled. The map indicates the range of months during

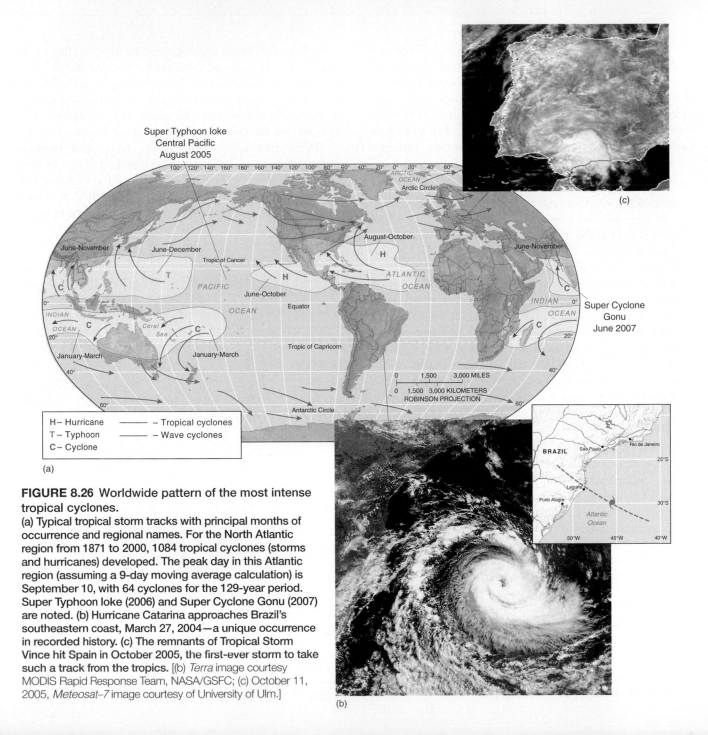

FIGURE 8.26 Worldwide pattern of the most intense tropical cyclones.

(a) Typical tropical storm tracks with principal months of occurrence and regional names. For the North Atlantic region from 1871 to 2000, 1084 tropical cyclones (storms and hurricanes) developed. The peak day in this Atlantic region (assuming a 9-day moving average calculation) is September 10, with 64 cyclones for the 129-year period. Super Typhoon Ioke (2006) and Super Cyclone Gonu (2007) are noted. (b) Hurricane Catarina approaches Brazil's southeastern coast, March 27, 2004—a unique occurrence in recorded history. (c) The remnants of Tropical Storm Vince hit Spain in October 2005, the first-ever storm to take such a track from the tropics. [(b) *Terra* image courtesy MODIS Rapid Response Team, NASA/GSFC; (c) October 11, 2005, *Meteosat–7* image courtesy of University of Ulm.]

which tropical cyclones are most likely to appear. For example, storms that strike the southeastern United States do so mostly between August and October. Officially, tropical cyclone season for the Atlantic is from June 1 to November 30 each year. (For comparison, the average midlatitude cyclonic storm tracks are in blue on the map.)

Relative to North and Central America, tropical depressions (low-pressure areas) intensify into tropical storms as they cross the Atlantic. A rule of thumb has emerged: If tropical storms mature early along their track, they tend to curve northward toward the North Atlantic and miss the United States. The critical position is approximately 40° W longitude. If a tropical storm matures after it reaches the longitude of the Dominican Republic (70° W), then it has a higher probability of hitting the United States.

In the Southern Hemisphere, no hurricane was ever observed turning from the equator into the south Atlantic (note on the map). However, in March 2004 Hurricane Catarina made landfall in Brazil south of the resorts at Laguna, state of Santa Catarina. NASA's *Terra* satellite in Figure 8.26b clearly shows an organized hurricane, with characteristic central eye and rain bands. Compare the circulation patterns of Hurricane Gilbert (see Figure 8.27a) and Hurricane Catarina (8.27d). What do you see? Note the Coriolis force in action, counterclockwise in the Northern Hemisphere and clockwise in the Southern Hemisphere.

If this Brazilian hurricane wasn't strange enough, in October 2005, the remnants of Tropical Storm Vince became the first-ever Atlantic tropical cyclone to strike Spain (8.26c)! Then in 2006, the first tropical storm to become a category 5 super typhoon developed in the Central Pacific, some 1285 km (800 mi) south of Hawai'i. Super Typhoon Ioke continued for almost three weeks, with final remnants causing large-scale erosion along Alaskan shores. And in 2007, Super Cyclone Gonu became the strongest tropical cyclone on record occurring in the Arabian Sea, eventually hitting Oman and the Arabian Peninsula. These four storms apparently are signaling changes underway in tropical meteorology relative to the increased occurrence and intensity of cyclonic systems, as related to higher oceanic and atmospheric temperatures.

Table 8.2 presents the criteria for classification of tropical cyclones on the basis of wind speed and lists

Table 8.2 Tropical Cyclone Classification

Designation	Winds	Features
Tropical disturbance	Variable, low	Definite area of surface low pressure; patches of clouds
Tropical depression	Up to 34 knots 63 kmph (39 mph)	Gale force, organizing circulation; light to moderate rain
Tropical storm	35–63 knots 63–118 kmph (39–73 mph)	Closed isobars; definite circular organization; heavy rain; assigned a name
Hurricane (Atlantic and E. Pacific) Typhoon (W. Pacific) Cyclone (Indian Ocean, Australia)	Greater than 65 knots 119 kmph (74 mph)	Circular, closed isobars; heavy rain, storm surges; tornadoes in right-front quadrant

Saffir–Simpson Hurricane Damage Potential Scale

Category	Wind Speed Central Pressure (mb)	Notable Atlantic Examples (rating at landfall, No. and Cent. Americas)
1	65–82 knots 119–154 kmph (74–95 mph) > 980 mb	—
2	83–95 knots 155–178 kmph (96–110 mph) 965–979 mb	1954 Hazel; 1999 Floyd; 2003 Isabel (was a cat. 5), Juan; 2004 Francis
3	96–113 knots 179–210 kmph (111–130 mph) 945–964 mb	1985 Elena; 1991 Bob; 1995 Roxanne, Marilyn; 1998 Bonnie; 2003 Kate; 2004 Ivan (was a cat. 5), Jeanne; 2005 Dennis; Rita and Wilma (were cat. 5); 2007 Henrietta
4	114–135 knots 211–250 kmph (131–155 mph) 920–944 mb	1979 Frederic; 1985 Gloria; 1995 Felix, Luis, Opal; 1998 Georges; 2004 Charley; 2005 Emily, Katrina (were cat. 5)
5	> 135 knots > 250 kmph (> 155 mph) < 920 mb	1935 No. 2; 1938 No. 4; 1960 Donna; 1961 Carla; 1969 Camille; 1971 Edith; 1977 Anita; 1979 David; 1980 Allen; 1988 Gilbert, Mitch; 1989 Hugo; 1992 Andrew; 2004 Ivan; 2007 Dean, Felix

features of each designation. When you hear meteorologists speak of a "category 4" hurricane, they are using the *Saffir–Simpson Hurricane Damage Potential Scale* to estimate possible damage from hurricane-force winds. Using wind speeds and central pressure criteria, the scale ranks hurricanes and typhoons in five categories, from smaller, category 1 storms to extremely dangerous category 5. Damage depends on the degree of property development at a storm's landfall site, how prepared citizens are for the blow, the height of storm surge, the presence of embedded tornadoes, and the overall intensity of the storm.

Physical Structure Fully organized tropical cyclones have an intriguing physical appearance (Figure 8.27). They range in diameter from a compact 160 km (100 mi), to 1000 km (600 mi), to some western Pacific super typhoons that attain 1300–1600 km (800–1000 mi). Vertically, these storms dominate the full height of the troposphere.

(a)

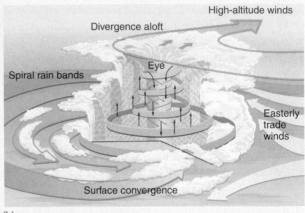

(b)

(c)

FIGURE 8.27 Profile of a hurricane.
For the Western Hemisphere, Hurricane Gilbert attained the record size (1600 km, or 1000 mi, in diameter) and second lowest barometric pressure (888 mb, or 26.22 in.). Gilbert's sustained winds reached 298 kmph (185 mph) with peaks exceeding 320 kmph (200 mph). Hurricane Gilbert, September 13, 1988: (a) *GOES-7* satellite image; (b) a stylized portrait of a mature hurricane, drawn from an oblique perspective (cutaway view shows the eye, rain bands, and wind-flow patterns); (c) *SLAR* (side-looking airborne radar) image from an aircraft flying through the center of the storm. Rain bands of greater cloud density are false-colored in yellows and reds. The clear sky in the central eye is dramatically portrayed. (d) Catarina demonstrates divergence aloft in a Southern Hemisphere hurricane; see the outflowing high cirrus clouds. [(a), (b), (c) NOAA and the NHC, Miami; (d) ISS astronaut photo March 26, 2004, JSC/NASA.]

(d)

ANIMATION

Hurricane
Wind Patterns

SATELLITE

Hurricane Georges

2004 Florida
Hurricanes
Satelite Coverage

Hurricane Isabel
9/6—9/19/04

The inward-spiraling clouds form dense rain bands, with a central area designated the *eye*. Around the eye swirls a thunderstorm cloud called the *eyewall*, which is the area of most intense precipitation. The eye remains an enigma, for in the midst of devastating winds and torrential rains the eye is quiet and warm with even a glimpse of blue sky or stars possible. The structure of the rain bands, central eye, and eyewall is clearly visible for Hurricane Gilbert in Figure 8.27a (top view), 8.27b (oblique view from an artist's perspective), and 8.27c (side-radar view). An astronaut photo from the International Space Station (8.27d) shows Hurricane Catarina before it hit Brazil. Note the divergence aloft marked by outward-flowing cirrus clouds. And, compare the Southern Hemisphere-clockwise winds to those of Gilbert.

The strongest winds of a tropical cyclone are usually recorded in its right-front quadrant (relative to the storm's directional path), where dozens of fully developed tornadoes may be embedded at the time of landfall. For example, Hurricane Camille in 1969 had up to 100 tornadoes embedded in that quadrant.

A tropical cyclone moves along at 16–40 kmph (10–25 mph). When it makes **landfall**, or moves ashore, the storm pushes seawater inland, causing dangerous **storm surges** often several meters in depth. Storm surges often catch people by surprise and cause the majority of hurricane drownings. In 1998, hurricanes Bonnie, Georges, and Mitch moved slowly onshore, producing damaging storm surges along North Carolina, Mississippi, and Central America, respectively. The slow progress of these storms produced extensive flooding from the sustained rains.

The storm surge from Hurricane Katrina in 2005 (see Focus Study Figure 8.1.1) caused failure in numerous poorly conceived and constructed levees and canals; therefore we discuss this in Chapter 14 under floodplain management. The case is properly made that Katrina, actually downgraded to a category 3 hurricane when it made landfall, was the storm that hit but the flood disaster that befell New Orleans was human-caused!

Sample Devastating Tropical Cyclones   The tropical cyclone that struck Bangladesh in 1970 killed an estimated 300,000 people, and the one in 1991 claimed over 200,000. In the United States, death tolls are much lower, but still significant. The Galveston, Texas, hurricane of 1900 killed 6000; Hurricane Katrina and engineering failures killed more than 1000 in Louisiana, Mississippi, and Alabama in 2005; Hurricane Audry (1957), 400; Hurricane Gilbert (1988), 318; Hurricane Camille (1969), 256; Hurricane Agnes (1972), 117. In 1998, hurricanes Bonnie and Georges damaged property and took lives. Hurricane Mitch (October 26–November 4, 1998) was the deadliest Atlantic Hurricane in two centuries, killing more than 12,000 people in Central America. In 2007, for the first time in history in a single Atlantic season, two category 5 hurricanes simultaneously made

landfall, Hurricane Dean (Yucatán) and Hurricane Felix (Honduras).

In September 2003, Hurricane Isabel struck the Outer Banks and Cape Hatteras of North Carolina, causing extensive flooding and wind damage to these fragile barrier islands as well as 36 deaths and about $2 billion in damages (Figure 8.28). Hurricane Charley brought category 4 winds to Florida's Gulf Coast in 2004, causing approximately $12 billion in damage and taking about two dozen lives. The Ft. Meyers region will take years to recover from Charley (Figure 8.29), along with Florida's Atlantic coast in the aftermath of Hurricane Andrew.

The tragedy of Hurricane Andrew in 1992 is that the storm destroyed or seriously damaged 70,000 homes

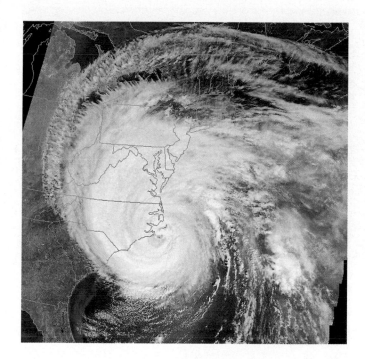

FIGURE 8.28 Cape Hatteras and the Outer Banks, hit again.
At one point in its path when far out at sea, Hurricane Isabel reached a category 5 rating. At the time of landfall during the afternoon of September 18, 2003, it had slowed to a category 2—still a damaging storm. Rainfall exceeded 12.7 cm (5 in.) over eastern North Carolina. Its remnants reached Lake Erie 24 hours later. [MODIS *Terra* image, September 18, 2003, courtesy of Rapid Response Team, NASA/GSFC.]

(a)

(b)

FIGURE 8.29 Damage from Hurricane Charley.
Category 4 Hurricane Charley in 2004 devastated Punta Gorda, near Port Charlotte (a) and Sanibel Island, near Fort Myers (b), in southwest Florida. [Photos by Bobbé Christopherson.]

and left 200,000 people homeless between Miami and the Florida Keys. The Everglades, the coral reefs north of Key Largo, some 10,000 acres of mangrove wetlands, and coastal southern pine forests all had significant damage from Andrew. Natural recovery will take years, which offers scientists a research opportunity. Scientific assessments judge the Everglades to be quite resilient because the region's ecosystems evolved naturally with periodic hurricanes over the millennia. An important fact is that storms damage human structures more than they damage natural systems. Remember this perspective: *Urbanization, agriculture, pollution, and water diversion pose a greater ongoing threat to the Everglades than do hurricanes!*

The Future We appear to be in a higher frequency hurricane cycle similar to the 1941 to 1970 pattern when there were 24 major hurricanes, as compared to 14 between 1971 and 1994. In addition, warmer Atlantic

waters along the 20th parallel between 20° W and 90° W and the Caribbean and Gulf of Mexico are fueling a greater intensity than previously experienced. A study published in *Science* stated,

> ...the trend of increasing numbers of category 4 and 5 hurricanes for the period 1970–2004 is directly linked to the trend in sea-surface temperature.... Higher SST [sea-surface temperatures] was the only statistically significant controlling variable related to the upward trend in global hurricane strength since 1970.*

Since record keeping began in 1870, the year for the greatest number of Atlantic hurricanes was 2005. The remarkable 2005 Atlantic season and forecasting Atlantic hurricanes are discussed in Focus Study 8.1.

Ironically, the relatively mild hurricane seasons from 1971 to 1994 encouraged weak zoning and rapid development of vulnerable coastal lowlands. New buildings, apartments, and government offices are opening right next to still visible rubble and bare foundation pads. Unfortunately, public or private decision makers practice thoughtful hazard planning infrequently, whether along coastal lowlands, river floodplains, or earthquake fault zones. The consequence is that our entire society bears the financial cost of planning failure, whether in Florida, North Carolina, California, or the Midwest, not to mention those victims who directly shoulder the physical, emotional, and economic hardship of the event.

This recurrent, yet avoidable, cycle—construction, devastation, reconstruction, devastation—was reinforced ironically in *Fortune* magazine more than 35 years ago after the destruction caused by Hurricane Camille:

> Before long the beachfront is expected to bristle with new motels, apartments, houses, condominiums, and office buildings. Gulf Coast businessmen, incurably optimistic, doubt there will ever be another hurricane like Camille, and even if there is, they vow, the Gulf Coast will rebuild bigger and better after that.**

Sadly, Hurricane Katrina obliterated the same Gulf Coast towns in 2005—Waveland, Bay Saint Louis, Pass Christian, Long Beach, Gulfport, among others—that Camille washed away in 1969. The below-sea-level sections of New Orleans may never recover their former status; however we are told that "...the Gulf Coast will rebuild bigger and better after that."

*C. D. Hoyos, P. A. Agudelo, P. J. Webster, and J. A. Curry, "Deconvolution of Factors Contributing to the Increase in Global Hurricane Intensity," *Science* (April 7, 2006), v. 312, no. 5770: 94–97.

**Fortune, October 1969, p. 62.

Focus Study 8.1

Atlantic Hurricanes and the 2005 Season

Forecasters at the National Hurricane Center (NHC) in Miami were busy throughout the 1995–2006 hurricane seasons. This was the most active 12-year period in the history of the NHC, with 167 named tropical storms, including 94 hurricanes (40 intense, meaning category 3 or higher). This was a record level of activity and individual storm intensity despite the reduced number of tropical cyclones during the 1997 El Niño season.

Statistically, the damage caused by tropical cyclones is increasing substantially as more and more development occurs along susceptible coastlines, whereas loss of life is decreasing in most parts of the world owing to better forecasting of these storms. The journal *Science* offered this "Perspective":

The risk of human losses is likely to remain low, however, because of a well-established warning and rescue system and ongoing improvements in hurricane prediction. A main concern is the risks of high damage costs (up to $100 billion in a single event) because of ongoing population increases in coastal areas and increasing investment in buildings and extensive infrastructure in general.*

As Atlantic hurricane seasons go, 1995 was the third most active since record keeping began in 1870. Of 19 tropical storms, 11 became hurricanes, and 5 of these achieved category 3 status or higher. Only 1933, with 21 tropical storms, and 1969, with 12 hurricanes, exceeded these two totals, until 2005. A map of the 1995 season and information on each storm is on our *Geosystems* Student Learning Center web site under this chapter.

The 2005 Atlantic hurricane season broke several records, including the most named tropical storms in a

single year, totaling 27 (the average is 10); most hurricanes, 15 (the average is 5); and 7 of these were rated intense, or greater than category 3 (the average is 2). The 2005 season was the only time two category 5 storms were in the Gulf of Mexico (Katrina and Rita), and the first time that three category 3 or more hurricanes made U.S. landfall in subsequent years (2004 and 2005). The 2005 season also recorded the greatest damage total in 1 year, at $130 billion. Hurricane Wilma is now the most intense storm (lowest central pressure) in recorded history for the Atlantic. Figure 8.1.1 presents a map and a detailed table of all the storms in this remarkable 2005 season of tropical cyclones and hurricanes, with five of the storm tracks highlighted.

Each storm went through the stages of tropical depression, tropical storm, hurricane, and extratropical depression, based on wind speed (depression, storm, hurricane) and location of the low-pressure center (tropical or extratropical). Top wind speed and lowest central pressure are noted in the table for each hurricane.

The Gulf Coast and Atlantic Seaboard sustained hits and near misses in 2005 from Hurricanes Cindy, Dennis, Katrina, Ophelia, Rita, Wilma, and Gamma (Figure 8.1.1); and, when added to Charley, Francis, Ivan, and Jeanne in 2004, the United States sustained more than $200 billion in damages and more than 2500 lives lost.

Given what we know, how do we cope with the televised images of the Gulf Coast tragedy or Katrina's power, as sampled in Figure 8.1.2? As we learn through hearings and investigations of the poststorm failure to respond and assist the region and as we see continuing mistakes in planning and engineering, the case for an integrated scientific approach is made.

How do we grapple with the "before-and-after" shock of such

events? In Figure 8.1.3a, we see a gambling casino, built as a mock pirate ship (2003 photo), which was destroyed by Hurricane Katrina in 2005. The state of Mississippi required that casinos be floating and moored offshore. This law led to strange scenes, such as Figure 8.1.2e, where Katrina's storm surge forced one casino onshore and parked it on top of a Holiday Inn hotel. Or in Figure 8.1.3b, we see an old boat-turned-gift shop selling hurricane souvenirs (far right) parked where it was blown ashore by Hurricane Camille in 1969 (dubbed the "SS Hurricane Camille") yet destroyed again by Katrina—the gift shop is gone and the surrounding landscape changed. We discuss the engineering/planning, floodplain management, levee considerations, and other mitigation strategy failures in Chapter 14.

Prediction and the Future

The Tropical Prediction Center is at the National Hurricane Center (NHC) in Miami, Florida. Prediction forecasts are posted on the NHC website (http://www.nhc.noaa.gov/). Their analysis of weather records for the period 1900–2006 disclosed a significant causal relationship between Atlantic tropical storms, including hurricanes that make landfall along the U.S. Gulf and East coasts, and several meteorological variables, such as sea-surface temperatures, the presence of upper tropospheric cross winds that cause shearing of vertical tropical storm circulation, and conditions in the Pacific Ocean, among other factors.

Research established a new "total dissipation index" to rate the potential destructiveness of tropical cyclones, integrated over their lifetime. The new research relates the marked increase in tropical cyclone intensity. MIT scientist Kerry Emanuel defined an "index of potential destructiveness" based on a hurricane's total

*L. Bengtsson, "Hurricane threats," *Science* 293 (July 20, 2001): 441.

(continued)

Focus Study 8.1 (continued)

FIGURE 8.1.1 2005 Atlantic hurricane season breaks records.
(a) The remarkable tropical cyclones of 2005 set several records—a total of 27 named storms, including 7 intense hurricanes (category 3 or higher). (b) NOAA aerial photo shows the power of wind and storm surge as entire Mississippi Gulf Coast neighborhoods are obliterated, such as here in Long Beach, with the debris carried by storm surge a kilometer from shore. [(a) Data for map from NOAA and the National Hurricane Center; photo (b) courtesy of NOAA aerial photography program.]

T-Tropical Storm, or H (cat. no.)-Hurricane (affected US)	2005 Dates (inclusive)	Top average wind speed (mph/kmph)	Lowest Central Pressure (mb)	Days as: H or T	
				H	T
T-Arlene (US)	6/8 - 6/12	70/113	989	–	3
T-Bret	6/28 - 6/30	40/64	1004	–	1
T-Cindy (US)	7/03 - 7/06	75/121	991	2	3
H4-Dennis (US)	7/05 - 7/11	150/241	930	4	3
H4-Emily	7/11 - 7/21	155/249	930	7	4
T-Franklin	7/21 - 7/29	70/113	997	–	8
T-Gert	7/23 - 7/25	45/72	1005	–	2
T-Harvey	8/02 - 8/08	65/105	994	–	6
H2-Irene	8/04 - 8/18	100/161	975	3	8
T-Jose	8/22 - 8/23	50/80	1001	–	1
H5-Katrina (US)	8/23 - 8/30	175/282	902	4	3
T-Lee	8/28 - 9/02	40/64	1007	–	1
H3-Maria	9/01 - 9/10	115/185	960	5	3
H1-Nate	9/05 - 9/10	90/145	979	3	4
H1-Ophilia (US	9/06 - 9/18	85/137	976	7	9
H1-Philippe	9/17 - 9/24	80/129	985	2	5
H5-Rita (US)	9/18 - 9/25	175/282	987	5	4
H1-Stan	10/01 - 10/05	80/129	979	1	3
T-Tammy (US)	10/05 - 10/06	50/80	1001	–	2
H1-Vince	10/09 - 10/11	75/121	987	2	2
H5-Wilma (US)	10/15 - 10/25	175/282	882*	7	2
T-Alpha	10/22 - 10/24	50/80	998	–	2
H3-Beta	10/27 - 10/31	115/185	960	2	4
T-Gamma	11/14 - 11/21	45/72	1004	–	3
T-Delta	11/23 - 11/28	70/113	980	–	6
H1-Epsilon	11/29 - 12/08	80/129	987	3	7
T-Zeta	12/30 - 1/5/06	65/105	992	–	7

* Record low pressure for US.

2005: 27 Storms, Arlene to Zeta

SATELLITE

(b)

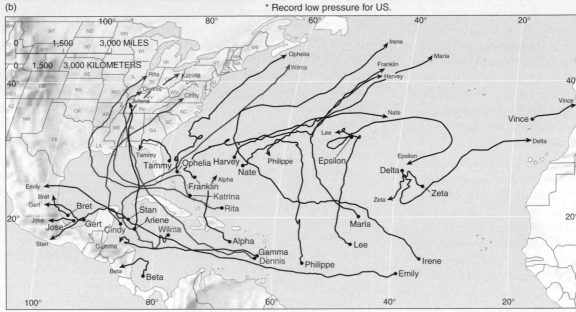

(a)

dissipation of power over its lifetime, and stated,

...this index has increased markedly since the mid-1970s. This trend is due to both longer storm lifetimes and greater storm intensities .... highly correlated with tropical sea surface temperature, reflecting well-documented climate signals, including multidecadal oscillations ... and global warming .... the near doubling of power dissipation over the period of record should be a matter of some concern.[†]

[†]K. Emanuel, "Increasing destructiveness of tropical cyclones over the past 30 years," *Nature* 436 (August 4, 2005): 686–688.

(a)

(b)

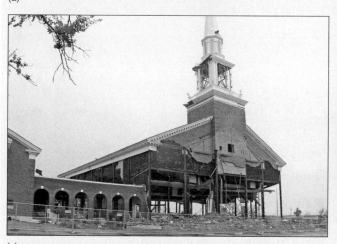

(c)

(d)

(e)

FIGURE 8.1.2 Gulf Coast scenes in the aftermath of Hurricane Katrina.
(a) Former New Orleans neighborhood next to the 17th-Street Canal where a 153-m (500-ft) section of levee failed; inset photo shows the changing water marks on abandoned homes left by floodwaters in the Lakeview area. The red "X" on houses tells recovery forces the date, search group involved, part of house searched, and victims found. (b) Chaos and jumble at the Orleans Marina on Lake Pontchartrain. (c) A church in ruins in Gulfport, Mississippi. (d) Remnants of the U.S. 90 bridge and causeway out of Bay St. Louis, Mississippi. (e) The storm surge picked up and dropped a casino on a portion of a Holiday Inn hotel.

Dr. Emanuel and others suggest that future warming with climate change, when considered alongside increased coastal population settlement, will produce unwanted, yet substantial, hurricane-related damage losses. Katrina, Rita, and Wilma in 2005, and Dean and Felix in 2007, all recorded remarkable central pressure drops as they passed over record sea-surface temperatures in the Gulf of Mexico and the Caribbean. Wilma experienced a 100-mb central pressure drop in a little over 24 hours, October 19, 2005. The record sea-surface temperature produces warmer water to depth, and the natural mixing that such storms trigger only brings up more heat to fuel the circulation. Wilma became the record holder for lowest air pressure in the United States, followed by Gilbert, Rita, Katrina, and Dean.

No matter how accurate storm forecasts become, coastal and lowland property damage will continue to increase until better hazard zoning and development restrictions are in place. The property insurance industry appears to be taking action to promote

(continued)

Focus Study 8.1 (continued)

FIGURE 8.1.3 The "before-and-after" shock along the Gulf Coast.
(a) Treasure Bay Casino, Biloxi, MS, built as a mock pirate ship in 2003 and after Katrina tore it and the attached hotel apart and moved it in 2005. (b) *SS Hurricane Camille* relic boat and gift shop in 2003 and what is left after Katrina. The Gulfport, Mississippi, gift shop to the right is no more, the souvenirs were all swept away; note the anchor bent over. [Before photos by Bobbé Christopherson; after photos by Randall Christopherson. All rights reserved.]

these improvements. It is requiring tougher building standards to obtain coverage—or, in some cases, it is refusing to insure property along vulnerable coastal lowlands. Given the increased intensity and "power dissipation" of these storms and a rise in sea level, the public, their politicians, and business interests must respond to somehow mitigate this hazardous predicament. Given what we have experienced and what we know of Earth systems, did society learn anything? The Atlantic hurricane season begins each June 1.

Summary and Review—Weather

Weather is the short-term condition of the atmosphere; **meteorology** is the scientific study of the atmosphere. The spatial implications of atmospheric phenomena and their relationship to human activities strongly link meteorology to physical geography. Analyzing and understanding patterns of wind, air pressure, temperature, and moisture conditions portrayed on daily weather maps is key to numerical (computer) weather forecasting.

weather (p. 208)
meteorology (p. 208)

■ *Describe* air masses that affect North America, and *relate* their qualities to source regions.

Specific conditions of humidity, stability, and cloud coverage occur in a regional, homogenous **air mass**. The longer an air mass remains stationary over a region, the more definite its physical attributes become. The homogeneity of temperature and humidity in an air mass sometimes extends through the lower half of the troposphere. Air masses are categorized by their moisture content—**m** for maritime (wetter) and **c** for continental (drier)—and their temperature, a function of latitude—designated **A** (arctic), **P** (polar), **T** (tropical), **E** (equatorial), and **AA** (Antarctic).

air mass (p. 209)

1. How does a source region influence the type of air mass that forms over it? Give specific examples of each basic classification.
2. Of all the air masses, which are of greatest significance to the United States and Canada? What happens to them as they migrate to locations different from their source regions? Give an example of air mass modification.

■ *Identify* four types of atmospheric lifting mechanisms, and *describe* four principal examples.

Air masses can rise through **convergent lifting** (air flows conflict, forcing some of the air to lift); **convectional lifting** (air passing over warm surfaces gains buoyancy); **orographic lifting** (passage over a topographic barrier); and *frontal lifting*. In North America, **chinook winds** (called *föhn* or *foehn* winds in Europe) are the warm, downslope air flows characteristic of the leeward side of mountains. Orographic lifting creates wetter windward slopes and drier leeward slopes situated in the **rain shadow** of the mountain. Conflicting air masses at a front produce a **cold front** (and sometimes a zone of strong wind and rain) or a **warm front**. A zone right along or slightly ahead of the front, called a **squall line**, is characterized by turbulent and wildly changing wind patterns and intense precipitation.

convergent lifting (p. 211)
convectional lifting (p. 211)
orographic lifting (p. 213)
chinook winds (p. 213)
rain shadow (p. 213)
cold front (p. 214)
warm front (p. 214)
squall line (p. 217)

3. Explain why it is necessary for an air mass to be lifted if there is to be saturation, condensation, and precipitation.
4. What are the four principal lifting mechanisms that cause air masses to ascend, cool, condense, form clouds, and perhaps produce precipitation? Briefly describe each.
5. Differentiate between the structure of a cold front and a warm front.

■ *Analyze* the pattern of orographic precipitation, and *describe* the link between this pattern and global topography.

The physical presence of a mountain acts as a topographic barrier to migrating air masses. *Orographic lifting* (oro- means "mountain") occurs when air is forcibly lifted upslope as it is pushed against a mountain. It cools adiabatically. An orographic barrier enhances convectional activity and causes additional lifting during the passage of weather fronts and cyclonic systems, thereby extracting more moisture from passing air masses. The precipitation pattern of windward and leeward slopes persists worldwide. The state of Washington is presented in the chapter as an excellent example of this concept.

6. When an air mass passes across a mountain range, many things happen to it. Describe each aspect of a moist air mass crossing a mountain. What is the pattern of precipitation that results?
7. Explain how the distribution of precipitation in the state of Washington is influenced by the principles of orographic lifting.

■ *Describe* the life cycle of a midlatitude cyclonic storm system, and *relate* this to its portrayal on weather maps.

A **midlatitude cyclone**, or **wave cyclone**, is a vast low-pressure system that migrates across the continent, pulling air masses into conflict along fronts. **Cyclogenesis**, the birth of the low-pressure circulation, can occur off the west coast of North America, along the polar front, along the lee slopes of the Rockies, in the Gulf of Mexico, and along the East Coast. A midlatitude cyclone can be thought of as having a life cycle of birth, maturity, old age, and dissolution. An **occluded front** is produced when a cold front overtakes a warm front in the maturing cyclone. Sometimes a **stationary front** develops between conflicting air masses, where air flow is parallel to the front on both sides. These systems are guided by the jet streams of the upper troposphere along seasonally shifting **storm tracks**.

midlatitude cyclone (p. 218)
wave cyclone (p. 218)
cyclogenesis (p. 219)
occluded front (p. 219)
stationary front (p. 219)
storm tracks (p. 219)

8. Differentiate between a cold front and a warm front as types of frontal lifting and describe what you would experience with each one.
9. How does a midlatitude cyclone act as a catalyst for conflict between air masses?
10. What is meant by cyclogenesis? In what areas does it occur and why? What is the role of upper-tropospheric circulation in the formation of a surface low?
11. Diagram a midlatitude cyclonic storm during its open stage. Label each of the components in your illustration and add arrows to indicate wind patterns in the system.

■ *List* the measurable elements that contribute to modern weather forecasting, and *describe* the technology and methods employed.

Synoptic analysis involves the collection of weather data at a specific time, as shown on the synoptic weather maps inserted in Figures 8.3, 8.15b, and 8.15d, and in News Report 8.2. The standard weather station symbols used on these maps are in the legend to Figure 8.14. Building a database of wind, pressure, temperature, and moisture conditions is key to computer-based weather prediction and the development of weather forecasting models. Preparing a weather report and forecast requires analysis of a daily weather map and satellite images that depict atmospheric conditions. Weather data from these sources include the following:

- Barometric pressure (sea level and altimeter setting)
- Pressure tendency (steady, rising, falling)
- Surface air temperature
- Dew-point temperature
- Wind speed, direction, and character (gusts, squalls)
- Type and movement of clouds
- Current weather
- State of the sky (current sky conditions)
- Visibility; vision obstruction (fog, haze)
- Precipitation since last observation

Technology employed includes Doppler radar, wind profilers, AWIPS consoles, ASOS instrument arrays, various satellite platforms, and high-speed computers.

12. What is your principal source of weather data, information, and forecasts? Where does your source obtain its data? Have you used the Internet and World Wide Web to obtain weather information? In what ways will you personally apply this knowledge in the future? What benefits do you see?

■ *Analyze* various forms of violent weather and the characteristics of each, and *review* several examples of each from the text.

The violent power of some weather phenomena poses a hazard to society. Severe ice storms involve **sleet** (freezing rain, ice glaze, and ice pellets), snow blizzards, and crippling ice coatings on roads, power lines, and crops. Thunderstorms produce **lightning** (electrical discharges in the atmosphere), **thunder** (sonic bangs produced by the rapid expansion of air after intense heating by lightning), and **hail** (ice pellets formed within cumulonimbus clouds).

Strong linear winds in excess of 26 m/s (58 mph) are known as **derechos**, associated with thunderstorms and bands of showers crossing a region. Straight-line winds can cause significant damage and crop losses. A spinning, cyclonic column rising to mid-troposphere level—a **mesocyclone**—is sometimes visible as the swirling mass of a cumulonimbus cloud, especially in a *supercell* system. Dark gray **funnel clouds** pulse from the bottom side of the parent cloud. A **tornado** is formed when the funnel connects with Earth's surface. A **waterspout** forms when a tornado circulation occurs over water.

Within tropical air masses, large low-pressure centers can form along easterly wave troughs. Under the right conditions, a tropical cyclone is produced. Depending on wind speeds and central pressure, a **tropical cyclone** can become a **hurricane**, **typhoon**, or cyclone, when winds exceed 65 knots (74 mph, 119 kmph). As forecasting and the public's perception of weather-related hazards improve, loss of life decreases, although property damage continues to increase. Great damage occurs to occupied coastal lands when hurricanes make **landfall** and when winds drive ocean water inland in **storm surges**.

sleet (p. 221)
lightning (p. 223)

thunder (p. 224)
hail (p. 224)
derechos (p. 226)
mesocyclone (p. 226)
funnel clouds (p. 228)
tornado (p. 228)
waterspout (p. 228)
tropical cyclone (p. 231)
hurricane (p. 232)
typhoon (p. 232)
landfall (p. 235)
storm surges (p. 235)

13. What constitutes a thunderstorm? What type of cloud is involved? What type of air mass would you expect in an area of thunderstorms in North America?

14. Lightning and thunder are powerful phenomena in nature. Briefly describe how they develop.

15. Describe the formation process of a mesocyclone. How is this development associated with that of a tornado?

16. Evaluate the pattern of tornado activity in the United States. Where is Tornado Alley? What generalizations can you make about the distribution and timing of tornadoes? Do you perceive a trend in tornado occurrences in the United States? Explain.

17. What are the different classifications for tropical cyclones? List the various names used worldwide for hurricanes. Have any hurricanes ever occurred in the south Atlantic?

18. What factors contributed to the incredible damage cost of Hurricane Andrew? Why have such damage figures increased even though loss of life has decreased over the past 30 years?

19. What have scientists said to explain the intensity of tropical cyclones since 1970? Put into your own words the quotations in this chapter from scientific journals related to changes in storm intensity.

20. Relative to improving weather forecasting, what are some of the technological innovations discussed in this chapter?

 NetWork

The *Geosystems* Student Learning Center provides online resources for this chapter on the World Wide Web. To begin: Once at the Center, click on the cover of this textbook, scroll the Table of Contents menu, and select this chapter. You will find self-tests that are graded, review exercises, specific updates for items in the chapter, and in "Destinations" many links to interesting related pathways on the Internet. *Geosystems* Student Learning Center is found at **http://www.prenhall.com/ christopherson/**.

Critical Thinking

A. In the following figure you see a weather map of conditions on March 31, 2007, 7 A.M. EST, along with a *GOES-12* satellite image of water vapor at the time of the map. Using the legend for weather map symbols in Figure 8.14, briefly analyze the map: Find the center of low pressure, note the counterclockwise winds, compare temperatures on either side of the cold front, note air temperatures and dew-point temperatures, determine why the fronts are placed where they are on the map, and locate the center of high pressure.

You can see that the air and dew-point temperatures in New Orleans are 68°F and 63°F, respectively; yet, in St. George, Utah (southwest corner of state), you see 27°F and 22°F, respectively. To become saturated, the warm, moist air in New Orleans needs to cool to 63°F, whereas in St. George the dry, cold air needs to cool to 22°F for saturation. In the far north, find Churchill on the western shore of Hudson Bay. The air and dew-point temperatures are 6°F and 3°F, respectively, and the state of the sky is clear. If you were there, what would you experience at this time? Why would you be reaching for the lip balm?

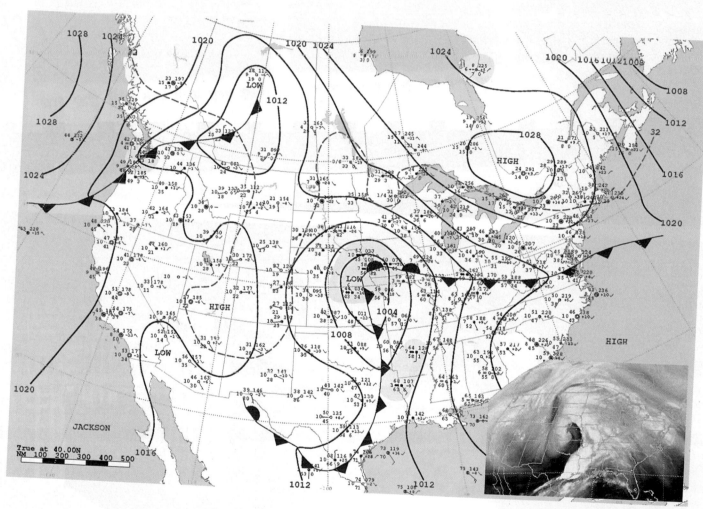

B. Relative to coastal devastation from tropical weather, the following statement appears in the text: "This recurrent, yet avoidable cycle—construction, devastation, reconstruction, devastation...." This describes the ever-increasing dollar losses to property from tropical storms and hurricanes, yet at a time when improved forecasts have resulted in a significant reduction in loss of life. (These issues are discussed further in Chapter 16's Focus Study 16.1.) In your opinion, what is the solution to halting these increasing losses, this cycle of destruction? How would you implement your plan?

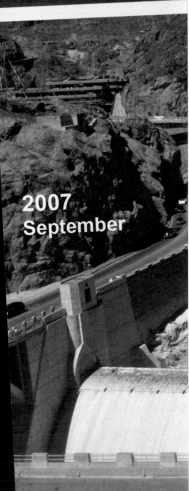

1983
July

2007
September

In the last two chapters, we saw how the exchanges of energy between water and the atmosphere drive Earth's weather systems. The flow of water links the atmosphere, ocean, land, and living things through exchanges of energy and matter. In particular, the energy and moisture exchange between plants and the atmosphere is important to the status of the climate system and the range of responses from plant communities to climate change.

Water is not always naturally available where and when we want it. Consequently, we rearrange surface-water resources to suit our needs. We drill wells, build cisterns and reservoirs, and dam and divert streams to redirect water either spatially (geographically, from one area to another) or temporally (over time, from one part of the calendar to another). All of this activity constitutes water-resource management.

Fortunately, water is a renewable resource, constantly cycling through the environment, endlessly renewed. Even so, some 1.1 billion people lack safe drinking water. People in 80 countries face impending water shortages, either in quantity or quality, or both. Approximately 2.4 billion people lack adequate sanitary facilities—80% of these in Africa, 13% in Asia. This translates to approximately 2 million deaths a year due to lack of water and 5 million deaths a year from waterborne infections and disease. An investment in safe drinking water, sanitation, and hygiene could decrease these numbers.

During the first half of this century, water availability per person will drop by 74%, as population increases and adequate water decreases. Author Peter Gleick's *The World's Water*, a biennial report published since 1998 (Washington, DC: Island Press; http://www.worldwater. org/), summarized:

> The overall economic and social benefits of meeting basic water requirements far outweigh any reasonable assessment of the costs of providing those needs.... Water-related diseases cost society on the order of $125 billion per year. . . . Yet the cost of providing new infrastructure needs for all major urban water sectors has been estimated at around $25 to $50 billion per year. While these costs are far below the costs of failing to meet these needs, they are two to three times the average rate of spending for water during the 1980s and 1990s.

Gordon Young, coordinator of the World Water Assessment Program, UNESCO (http://www.unesco.org/ water/wwap/), called for an integrated approach:

> As we enter the twenty-first century a global water crisis is threatening the security, stability, and environmental sustainability of all nations, particularly those in the developing world. Millions die each year from water-related diseases, while water pollution and ecosystem destruction grow. . . . Current thinking accepts that the management of water resources must be undertaken using an integrated approach.

Thus the role for geographic science with its integrative, spatial approach is front and center in analyzing a systematic, comprehensive global portrait of the changing status of water resources.

In this chapter: The hydrologic cycle and global water balance give us a model for understanding the global plumbing system, so this important cycle begins the chapter. Water spends time in the ocean, in the air, on the surface, and underground as groundwater. Water availability to plants from precipitation and from the soil is critical to water-resource issues.

We look at the water resource using a water-budget approach—similar in many ways to a money budget—in which we examine water "receipts" and "expenses" at specific locations. Precipitation provides the principal receipt of moisture, whereas evaporation and plant transpiration are the principal expenditures. This budget approach can be applied at any scale, from a small garden, to a farm, to a regional landscape.

About half the U.S. population draws freshwater from the groundwater resource, yet groundwater is tied to surface-water supplies for recharge. There are finite limits to groundwater. We discuss groundwater resources, overuse through groundwater mining, the irreversibility of groundwater pollution, and future prospects.

This chapter concludes by considering the quantity and quality of the water we withdraw and consume for irrigation, industrial, and municipal uses—our specific water supply. Adequate water supplies in terms of quantity and quality loom as *the resource issue* for many parts of the world in this century. Ismail Serageldin, chair of the World Water Commission, bluntly declares about water issues: " . . . the wars of the twenty-first century will be fought over water. . . . Water is the most critical issue facing human development."*

The Hydrologic Cycle

Vast currents of water, water vapor, ice, and energy are flowing about us continuously in an elaborate, open, global plumbing system. Together they form the **hydrologic cycle**, which has operated for billions of years, from the lower atmosphere to several kilometers beneath Earth's surface. The cycle involves the circulation and transformation of water throughout Earth's atmosphere, hydrosphere, lithosphere, and biosphere. (See NASA's Global Hydrology and Climate Center site at http://weather. msfc.nasa.gov/surface_hydrology/ for a combined government and academic effort to study the global hydrologic cycle and related climatic effects.)

A Hydrologic Cycle Model

Figure 9.1a is a simplified model of this complex system. Let's use the ocean as a starting point for our discussion, although we could jump into the model at any point. More than 97% of Earth's water is in the ocean, and here most evaporation and precipitation occur. We can trace 86% of all evaporation to the ocean. The other 14% is from the land, including water moving from the soil into plant roots and passing through their leaves (a process called *transpiration*, described later in this chapter).

*M. De Villiers, *Water, The Fate of Our Most Precious Resource* (New York: Houghton Mifflin Co., 2000), p. 13–14.

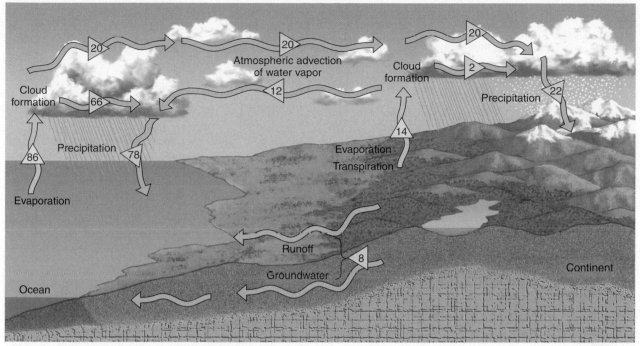

(a)

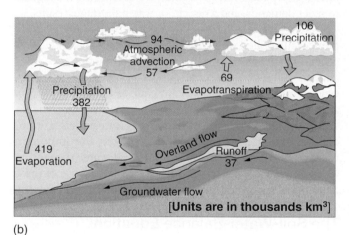

(b)

ANIMATION

Earth's Water and the Hydrologic Cycle

FIGURE 9.1 The hydrologic cycle model.
The model shows how water travels endlessly through the hydrosphere, atmosphere, lithosphere, and biosphere. (a) The triangles show global average values as percentages. Note that all evaporation (86% + 14% = 100%) equals all precipitation (78% + 22% = 100%) and advection in the atmosphere is balanced by surface and subsurface water runoff when all of Earth is considered. (b) The annual volume of water in all parts of the hydrologic cycle as measured in thousands of cubic kilometers (1 km³ × 0.24 = 1 mi³).

In the figure, you can see that of the 86% of evaporation rising from the ocean, 66% combines with 12% advected (moving horizontally) from the land to produce the 78% of all precipitation that falls back into the ocean. The remaining 20% of moisture evaporated from the ocean, plus 2% of land-derived moisture, produces the 22% of all precipitation that falls over land. Clearly, the bulk of continental precipitation comes from the oceanic portion of the cycle. Regionally, various parts of the cycle will vary, creating imbalances and, depending on climate, surpluses in one region and shortages in another.

Figure 9.1b presents the volume (in 1000 km³) of water flowing along these pathways in the hydrologic cycle. In this global water balance you see precipitation totaling 488,000 km³ (117,120 mi³) is equaled by evaporation and transpiration of the same volume.

Surface Water

Precipitation that reaches Earth's surface follows two basic pathways: It either flows overland or soaks into the soil.

Along the way **interception** occurs when precipitation strikes vegetation or other ground cover. Intercepted water that drains across plant leaves and down their stems to the ground is *stem flow* and can be an important moisture route to the ground. Precipitation that falls directly to the ground, coupled with drips from vegetation (excluding stem flow), constitutes *throughfall*. Water soaks into the subsurface through **infiltration**, or penetration of the soil surface. It further permeates soil or rock through the downward movement of **percolation**. These concepts are shown in Figure 9.2.

The atmospheric advection of water vapor from sea to land and land to sea at the top of Figure 9.1 appears to be unbalanced: 20% (94,000 km³) moving inland but only 12% (57,000 km³) moving out to sea. However, this exchange is balanced by the 8% (37,000 km³) land runoff that flows to the sea. Most of this runoff—about 95%—comes from surface waters that wash across land as overland flow and streamflow. Only 5% of runoff is slow-moving subsurface groundwater. These percentages indicate that the

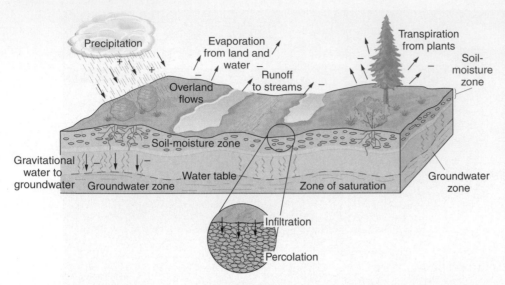

FIGURE 9.2 The soil-moisture environment.
Precipitation supplies the soil-moisture environment. The principal pathways for water include interception by plants; throughfall to the ground; collection on the surface, forming overland flow to streams; transpiration and evaporation from plants; evaporation from land and water; and gravitational water moving to subsurface groundwater.

small amount of water in rivers and streams is dynamic; in contrast, the large quantity of subsurface water is sluggish and represents only a small portion of total runoff.

The residence time for a water molecule in any part of the hydrologic cycle determines its relative importance in affecting Earth's climates. The short time spent by water in transit through the atmosphere (an average of 10 days) plays a role in temporary fluctuations in regional weather patterns. Long residence times, such as the 3000–10,000 years in deep-ocean circulation, groundwater aquifers, and glacial ice, act to moderate temperature and climatic changes. These slower parts of the cycle work as a "system memory"; the long periods over which heat energy is stored and released can buffer the effects of system changes.

To observe and describe the hydrologic cycle and its related energy budgets, scientists established in 1988 the Global Energy and Water Cycle Experiment (GEWEX); this is part of the World Climate Research Program. As part of this effort, the GEWEX Americas Prediction Project (GAPP), a multiscale hydrometeorological investigation of the hydrologic cycle, is underway (see **http://www.ogp.noaa.gov/mpe/gapp/**). The goal is to correlate land surface, general circulation, regional climate, and hydrologic models to better predict seasonal and annual water-resource changes. Now that you are acquainted with the hydrologic cycle, let's examine the concept of the soil-water budget as a method of assessing water resources.

Soil-Water-Budget Concept

A **soil-water budget** can be established for any area of Earth's surface—a continent, country, region, field, or front yard. Key is measuring the precipitation "supply" input and its distribution to satisfy the "demand" outputs of plants, evaporation, and soil-moisture storage in the area considered. Such a budget can examine any time frame, from minutes to years.

Think of a soil-water budget as a money budget: Precipitation income must be balanced against expenditures of evaporation, transpiration, and runoff. Soil-moisture

storage acts as a savings account, accepting deposits and withdrawals of water. Sometimes all expenditure demands are met, and any extra water results in a surplus. At other times, precipitation and soil-moisture income are inadequate to meet demands, and a deficit, or water shortage, results.

Geographer Charles W. Thornthwaite (1899–1963) pioneered applied water-resource analysis and worked with others to develop a water-balance methodology. They applied water-balance concepts to geographic problems, especially to irrigation, which requires accurate quantity and timing of water application to maximize crop yields. Thornthwaite also developed methods for estimating evaporation and transpiration. He recognized the important relation between water supply and local water demand as an essential climatic element. In fact, his initial use of these techniques was to develop a climatic classification system.

The Soil-Water-Balance Equation

To understand Thornthwaite's water-balance methodology and "accounting" or "bookkeeping" procedures, we must first understand some terms and concepts. Figure 9.2 illustrates the essential aspects of a soil-water budget. Precipitation (mostly rain and snow) provides the moisture input. The object, as with a money budget, is to account for the ways in which this supply is distributed: actual water taken by evaporation and plant transpiration, extra water that exits in streams and subsurface groundwater, and recharge or utilization of soil-moisture storage.

Figure 9.3 organizes the water-balance components into an equation. As in all equations, the two sides must balance; that is, the precipitation receipt (left side) must be fully accounted for by expenditures (right side). Follow this water-balance equation as you read the following paragraphs. To help you learn these concepts, we use letter abbreviations (such as PRECIP) for the components.

Precipitation (PRECIP) Input The moisture supply to Earth's surface is **precipitation** (PRECIP, or P). It arrives in several different forms depending upon temperature and moisture supply. Table 9.1 summarizes the different

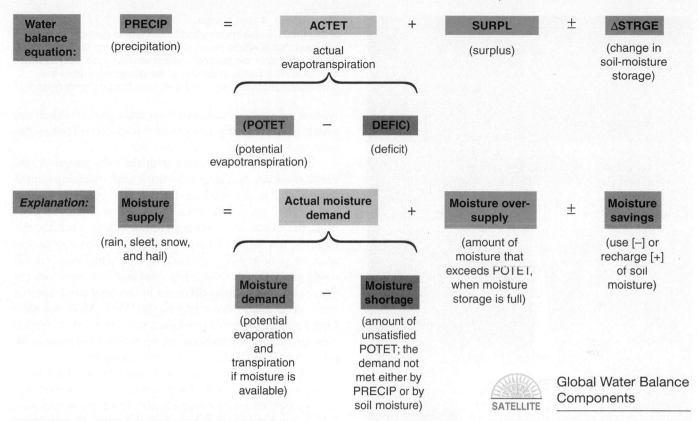

FIGURE 9.3 The water-balance equation explained.
The outputs (components to the right of the equal sign) are an accounting of expenditures from the precipitation moisture-supply input (to the left).

Table 9.1	Types of Precipitation	
Type	**Characteristic**	**Typical Amounts**
Dew	Condensation on surfaces, especially vegetation; hoar frost when frozen, formed by direct deposition	0.1 to 1.0 mm per night, small input to supply, may be locally significant
Fog-drip	Deposited on vegetation and other surfaces from fog; rime when frozen	A small input to supply; may be significant in foggy climates
Drizzle (mist)	Numerous, uniform water drops	Light 0.3 mm/hr; moderate 0.3–0.5 mm/hr; heavy > 0.5 mm/hr; if over 1 mm/h, it's rain
Hail	Roughly spherical, irregular lumps of ice > 5 mm in diameter, with a layered structure of opaque and clear ice in cross section; produced in convective clouds, almost always cumulonimbus	Highly variable, can cover the ground; hailstones can go to more than 15 cm in diameter
Rain	Water drops > 0.5 mm in diameter, or widely scattered drops might be smaller	Light < 0.25 cm/hr Moderate 0.26–0.76 cm/hr Heavy > 0.76 cm/hr
Snowflakes	Cluster of ice crystals, or a single crystal, that fall from a cloud; an aggregate can exceed 6 cm in diameter	Highly variable
Snow Grains (granular snow)	Quite small, flat, opaque grains of ice; solid equivalent of drizzle	Highly variable
Snow Pellets (graupel or soft hail)	Opaque, approximately round pellets of ice, 2 to 5 mm in diameter falling in showers; often before or with snow	Highly variable
Ice pellets	Clear ice encasing a snowflake or snow; transparent pellets of ice < 5 mm in diameter	Highly variable
Sleet (U.S., UK)	Mixture of rain and partially melted snow; *freezing precipitation* in Canada; can form layer on below-freezing ground	Highly variable

FIGURE 9.4 A rain gauge.
A standard rain gauge is cylindrical. A funnel guides water into a bucket that is sitting on an electronic weighing device. The gauge minimizes evaporation, which would cause low readings. The wind shield around the top of the gauge minimizes the undercatch produced by wind. [Photo by Bobbé Christopherson.]

types of precipitation. As you read through this table recall which types of precipitation you have experienced—rain, sleet, snow, and hail being most common.

Precipitation is measured with the **rain gauge**. A rain gauge is essentially a large measuring cup, collecting rainfall and snowfall so the water can be measured by depth, weight, or volume (Figure 9.4). Wind can cause an undercatch because the drops or snowflakes are not falling vertically. For example, a wind of 37 kmph (23 mph) produces an undercatch as great as 40%, meaning that an actual 1-in. rainfall might gauge at only 0.6 in. The *wind shield* you see above the gauge's opening reduces this error by catching raindrops that arrive at an angle. According to the World Meteorological Organization (**http://www.wmo.ch/**), more than 40,000 weather-monitoring stations are operating worldwide, with more than 100,000 places measuring precipitation.

Figure 9.5 maps precipitation patterns for the United States and Canada. Note the patterns of wetness and dryness as you remember our discussion of air masses and lifting mechanisms in Chapter 8. (Chapter 10 presents a world precipitation map in Figure 10.2.) PRECIP is the principal input to the water-balance equation (Figure 9.3).

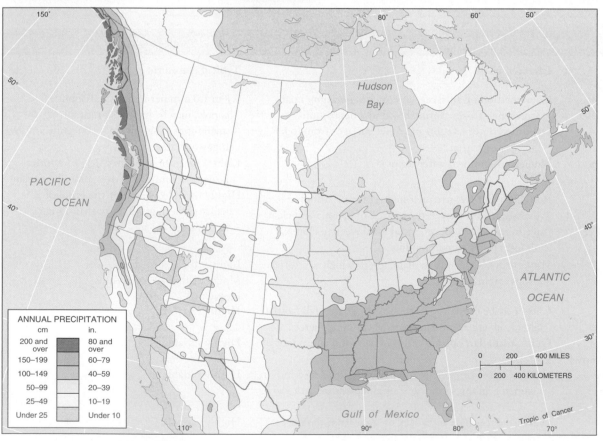

ANNUAL PRECIPITATION	
cm	in.
200 and over	80 and over
150–199	60–79
100–149	40–59
50–99	20–39
25–49	10–19
Under 25	Under 10

FIGURE 9.5 Precipitation (PRECIP) in North America—the supply.
Annual precipitation (water supply, PRECIP) in the United States, Canada, and northern Mexico.
[Adapted from NWS, U.S. Department of Agriculture, and Environment Canada.]

SATELLITE

Global Water Balance Components

All the components discussed next are outputs or expenditures of this water.

Actual Evapotranspiration (ACTET) Evaporation is the net movement of free water molecules away from a wet surface into air that is less than saturated. **Transpiration** is a cooling mechanism in plants. When a plant transpires, it moves water through small openings (stomata) in the underside of its leaves. The water evaporates, cooling the plant, much as perspiration cools humans. Transpiration is partially regulated by the plants themselves. Control cells around the stomata conserve or release water. Transpired quantities can be significant: On a hot day, a single tree can transpire hundreds of liters of water; a forest, millions of liters.

Evaporation and transpiration are important water-budget expenditures, and both respond directly to air temperature and humidity. Rates decrease when air is cold (can absorb less moisture) or when there is high humidity (at or near saturation); both rates increase when air is hot (can absorb more moisture) and dry (less than saturation). Evaporation and transpiration are combined into one term—**evapotranspiration**. In the hydrologic cycle, Figure 9.1, 14% of evaporation and transpiration occurs from land and plants. Now, let's examine ways to estimate evapotranspiration rates.

Potential Evapotranspiration (POTET) Evapotranspiration is an actual expenditure of water. In contrast, **potential evapotranspiration** (POTET, or PE) is the amount of water that would evaporate and transpire under optimum moisture conditions when adequate precipitation and adequate soil-moisture supply are present.

Filling a bowl with water and letting the water evaporate illustrates this concept: When the bowl becomes dry, is there still an evaporation demand? The demand remains, of course, regardless of whether the bowl is dry. If you always kept water in the bowl, the amount of water that would evaporate or transpire is the POTET, or the optimum demand given a constant supply of water. If you let the bowl dry out, the amount of POTET demand that went unmet is the shortage, or deficit (DEFIC). Note that in the water-balance equation, when we subtract the deficit from the potential evapotranspiration, we derive what actually happened—ACTET.

Figure 9.6 presents POTET values derived using Thornthwaite's approach for the United States and Canada. Note that higher values occur in the South, with highest readings in the Southwest where higher average air temperature and lower relative humidity exist. Lower POTET values are found at higher latitudes and elevations, which have lower average temperatures.

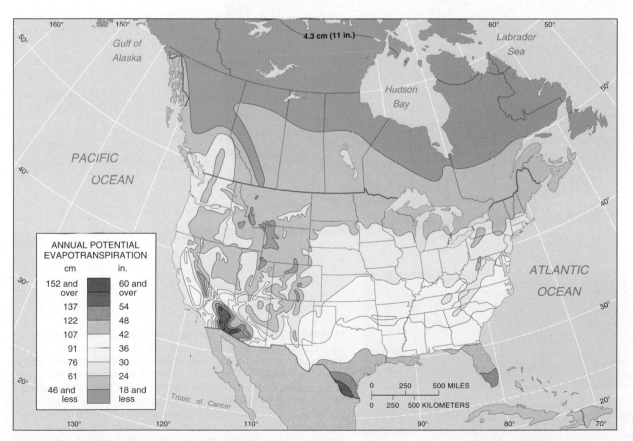

FIGURE 9.6 Potential evapotranspiration (POTET) for the United States and Canada—the demand. [From C. W. Thornthwaite, "An approach toward a rational classification of climate," *Geographical Review* 38 (1948): 64. Adapted by permission from the American Geographical Society. Canadian data adapted from M. Sanderson, "The climates of Canada according to the new Thornthwaite classification," *Scientific Agriculture* 28 (1948): 501–17.]

SATELLITE

Global Water Balance Components

Compare this POTET (demand) map with the PRE-CIP (supply) map in Figure 9.5. The relation between the two determines the remaining components of the water-balance equation in Figure 9.3. From the two maps, can you identify regions where PRECIP is greater than POTET (for example, the eastern United States)? Or where POTET is greater than PRECIP (for example, the southwestern United States)? Where you live, is the water demand usually met by the precipitation supply? Or does your area experience a natural shortage? How might you find out?

Determining POTET   Although exact measurement is difficult, one method of measuring POTET employs an **evaporation pan**, or *evaporimeter*. As evaporation occurs, water in measured amounts is automatically replaced in the pan, equaling the amount that evaporated. Mesh screens over the pan protect against overmeasurement due to wind, which accelerates evaporation.

A more elaborate measurement device is a **lysimeter**. A tank, approximately a cubic meter in size or larger, is buried in a field with its upper surface left open. The lysimeter isolates a representative volume of soil, subsoil, and plant cover, thus allowing measurement of the moisture moving through the sampled area. A weighing lysimeter has this embedded tank resting on a weighing scale (Figure 9.7a). A rain gauge next to the lysimeter measures the precipitation input. (See the Agricultural Research Service at **http://www.ars.usda.gov/**.)

The number of lysimeters and evaporation pans is somewhat limited across North America, but they still provide a database from which to develop ways of estimating POTET. Other remote sensors measure POTET and water-balance components, such as the instrumentation in Death Valley, shown in Figure 9.7b. Scientists are attempting to estimate the amount of groundwater that is lost to evaporation from this hot, dry valley floor.

Several methods of estimating POTET on the basis of meteorological data are used widely and are easily implemented for regional applications. Thornthwaite developed one of these methods. He discovered that if

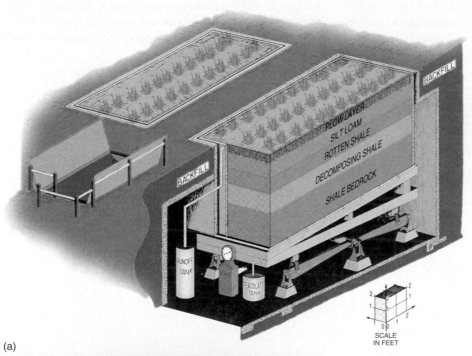

(a)

(b)

FIGURE 9.7 Lysimeter and remote weather instruments.
(a) A weighing lysimeter for measuring evaporation and transpiration. The various pathways of water are tracked: Some water remains as soil moisture, some is incorporated into plant tissues, some drains from the bottom of the lysimeter, and the remainder is credited to evapotranspiration. Given natural conditions, the lysimeter measures actual evapotranspiration. (b) Weather instruments placed near Badwater, the lowest elevation in the Western Hemisphere (–86 m, –282 ft, below sea level), record the harsh water-budget conditions. The USGS is studying temperatures, wind, and evaporation–transpiration rates as part of regional groundwater analysis. [(a) Adapted from illustration courtesy of Lloyd Owens, Agricultural Research Service, USDA, Coshocton, Ohio; (b) Photo by Bobbé Christopherson.]

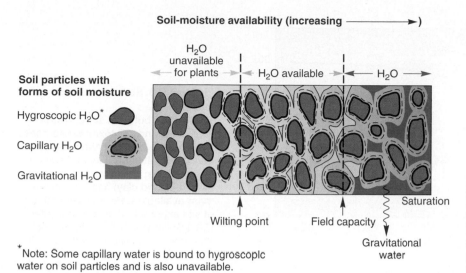

Soil-moisture availability (increasing ————▶)

H₂O unavailable for plants ◀——▶ H₂O available ◀——▶ H₂O

Soil particles with forms of soil moisture

Hygroscopic H₂O*

Capillary H₂O

Gravitational H₂O

Wilting point   Field capacity

Saturation

Gravitational water

FIGURE 9.8 Types of soil moisture. Hygroscopic and gravitational water are unavailable to plants; only capillary water is available. [After D. Steila, *The Geography of Soils*, © 1976, p. 45. Reprinted by permission of Prentice Hall, Inc., Upper Saddle River, NJ.]

*Note: Some capillary water is bound to hygroscopic water on soil particles and is also unavailable.

you know monthly mean air temperature and daylength, you could approximate POTET. These data are readily available, so calculating POTET is easy and fairly accurate for most midlatitude locations. (Recall that daylength is a function of a station's latitude.) His method can work with data from hourly, daily, or annual time frames, or with historical data to re-create water conditions in past environments.

Thornthwaite's method works better in some climates than in others. It ignores warm or cool air movements across a surface, sublimation and surface-water retention at subfreezing temperatures, poor drainage, and frozen soil (*permafrost*, Chapter 17), although there is allowance for snow accumulations. Nevertheless, we use his water-balance method here because of its great utility as a teaching tool and its overall ease of application and acceptable results for geographic studies.

Deficit (DEFIC)   The POTET demand can be satisfied in three ways: by PRECIP, by moisture stored in the soil, or through artificial irrigation. If these three sources are inadequate to meet moisture demands, the location experiences a moisture shortage. This unsatisfied POTET is **deficit** (DEFIC).

By subtracting DEFIC from POTET, we determine the **actual evapotranspiration**, or ACTET, that takes place (see Figure 9.3). Under ideal conditions for plants, potential and actual amounts of evapotranspiration are about the same, so plants do not experience a water shortage; droughts result from deficit conditions.

Surplus (SURPL)   If POTET is satisfied and the soil is full of moisture, then additional water input becomes **surplus** (SURPL). This excess water might sit on the surface in puddles, ponds, and lakes, or it might flow across the surface toward stream channels or percolate through the soil underground. The **overland flow** combines with precipitation and groundwater flows into river channels to make up the **total runoff**. Because surplus water generates most streamflow, or runoff, the water-balance approach is useful for indirectly estimating streamflow.

Soil-Moisture Storage (Δ STRGE)   A "savings account" of water that receives recharge "deposits" and provides for utilization "withdrawals" is **soil-moisture storage** (Δ STRGE). This is the volume of water stored in the soil that is accessible to plant roots. The delta symbol, Δ, means that this component includes both *recharge* and *utilization* (use) of soil moisture; the Δ in math means "change." Soil moisture comprises two categories of water—hygroscopic and capillary—but only capillary water is accessible to plants (Figure 9.8).

Hygroscopic water is inaccessible to plants because it is a molecule-thin layer that is tightly bound to each soil particle by the hydrogen bonding of water molecules (Figure 9.8, left). Hygroscopic water exists even in the desert, but it is unavailable to meet POTET demands. Relative to plants, soil is at the **wilting point** when all that remains is this inaccessible water; plants wilt and eventually die after a prolonged period of such moisture stress.

Capillary water is generally accessible to plant roots because it is held against the pull of gravity in the soil by hydrogen bonds between water molecules, which causes surface tension, and by hydrogen bonding between water molecules and the soil. Most capillary water that remains in the soil is **available water** in soil-moisture storage. After some water drains from the larger pore spaces, the amount of available water remaining for plants is termed **field capacity**, or storage capacity. It is removable to meet POTET demands through the action of plant roots and surface evaporation (Figure 9.8, center). Field capacity is specific to each soil type, and the amount can be determined by soil surveys.

When soil becomes saturated after a precipitation event, any water surplus in the soil body becomes **gravitational water**. It percolates from the shallower capillary zone to the deeper groundwater zone (Figure 9.8, right).

Figure 9.9 shows the relation of soil texture to soil-moisture content. Different plant species send roots to different depths and therefore reach different amounts of soil moisture. A soil blend that maximizes available water

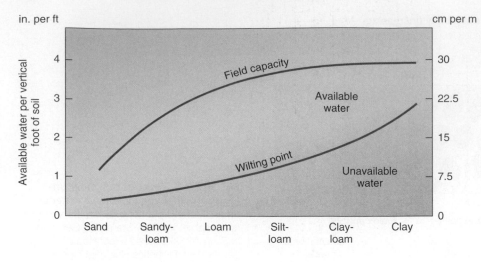

FIGURE 9.9 Soil-moisture availability.
The relation between soil-moisture availability and soil texture determines the distance between the two curves that show field capacity and wilting point. A loam soil (one-third each of sand, silt, and clay) has roughly the most available water per vertical foot of soil exposed to plant roots. [After U.S. Department of Agriculture, *1955 Yearbook of Agriculture—Water*, p. 120.]

is best for plants. On the basis of Figure 9.9, can you determine the soil texture with the greatest quantity of available water?

As **soil-moisture utilization** removes soil water, the plants must exert greater effort to extract the amount of moisture they need. As a result, even though a small amount of water may remain in the soil, plants may be unable to exert enough pressure to use it. The resulting *unsatisfied demand* is a *deficit* (DEFIC). Avoiding a deficit and reducing plant growth inefficiencies are the goals of irrigation, for the harder plants must work to get water, the less their yield and growth will be.

Whether from natural precipitation or artificial irrigation, water infiltrates the soil and replenishes available water, the process of **soil-moisture recharge**. The texture and the structure of the soil dictate available pore spaces, or **porosity**. The property of the soil that determines the rate of soil-moisture recharge is its **permeability**. Permeability depends on particle sizes and the shape and packing of soil grains.

Water infiltration is rapid in the first minutes of precipitation and slows as the upper soil layers become saturated, even though the deeper soil is still dry. Agricultural practices, such as plowing and adding sand or manure to loosen soil structure, can improve both soil permeability and the depth to which moisture can efficiently penetrate to recharge soil-moisture storage. You may have found yourself working to improve soil permeability with a houseplant or garden, working the soil to increase the rate of soil-moisture recharge.

Drought

Drought might seem a simple thing to define: less precipitation and higher temperatures make for drier conditions. However, there is no precise definition of **drought**. The key is to know the water-resource demand as it relates to the lack of water. There must be a consideration of not just precipitation, temperature, and soil-moisture content, but whether crops are in the growing season, or if low precipitation is during winter and the season for

snowpack accumulation, or the overall relation of precipitation patterns to the climate of a region.

A *meteorological drought* relates to dry conditions of lower precipitation and higher temperatures and losses in soil moisture. An *agricultural drought* occurs when changes in soil moisture and weather elements affect crop yields. Losses can be significant and are running in the tens of billions of dollars each year in the United States, although drought disasters seem to slowly evolve, getting little coverage in the media, and lack the shock of a hurricane or earthquake.

When reservoir levels drop, or mountain snowpack declines, streamflow decreases, and groundwater mining increases, we have a *hydrologic drought*. A more comprehensive measure considers loss of life, water rationing, wildfire events, and other impacts as they become widespread in *socioeconomic drought*.

The National Drought Mitigation Center, University of Nebraska–Lincoln (NDMC, **http://www.drought.unl.edu/**) publishes a weekly "Drought Monitor" map, released each Thursday, and a newsletter *Drought Scape*. Figure 9.10a presents the map for September 4, 2007, on which you can see the drought conditions in the West mentioned in the chapter-opening photo caption as well as the impacted Southeast and northern Midwest. Remnants of Hurricane Humberto in 2007 helped lessen some of the southeastern drought. Figure 9.10b from the NDMC presents the approximate changes in drought intensity between late September 2006 and late August 2007.

Droughts of various intensities are occurring on every continent through 2007. Australia in 2008 was in the seventh year of its worst drought in 100 years. Regarding the U.S. Southwest drought, underway since 2000, scientists reported in May 2007 this remarkable situation linked directly to global climate change:

The projected future climate of intensified aridity in the Southwest is caused by . . . a poleward expansion of the subtropical dry zones. The drying of subtropical land areas . . . imminent or already underway is unlike any climate state we have seen in the instrumental record. It is also distinct from the multi-decadal mega-droughts

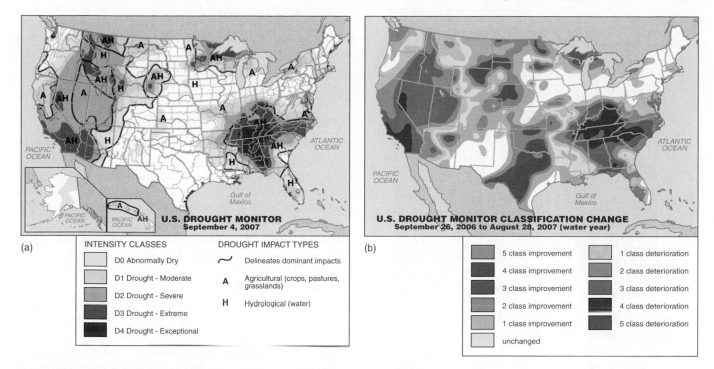

(a)

INTENSITY CLASSES

- D0 Abnormally Dry
- D1 Drought - Moderate
- D2 Drought - Severe
- D3 Drought - Extreme
- D4 Drought - Exceptional

DROUGHT IMPACT TYPES

~ Delineates dominant impacts

A Agricultural (crops, pastures, grasslands)

H Hydrological (water)

(b)

- 5 class improvement
- 4 class improvement
- 3 class improvement
- 2 class improvement
- 1 class improvement
- unchanged
- 1 class deterioration
- 2 class deterioration
- 3 class deterioration
- 4 class deterioration
- 5 class deterioration

FIGURE 9.10 U.S. Drought Monitor.
(a) The Western, Southwestern, and Southeastern regions of the United States stand out on this map from the National Drought Mitigation Center as they suffer a multiyear drought. (b) Approximate changes in drought intensity between late September 2006 and late August 2007. The colors in the legend denote intensity-class improvements or deterioration of drought conditions. [Maps courtesy of Thomas Heddinghaus, CPC/NOAA, and Michael Hayes, NDMC, University of Nebraska–Lincoln.]

that afflicted the American Southwest during Medieval times. . . . The most severe future droughts will still occur during persistent La Niña events, but they will be worse than any since the Medieval period, because the La Niña conditions will be perturbing a base state that is drier than any state experienced recently.[*]

In other words, previous droughts in the region related to sea-surface temperature changes in the distant tropical Pacific Ocean will still occur, but they will be worsened by climate change and the unwanted expansion farther into the region of the hot, dry, subtropical high-pressure system and the summertime continental tropical (cT) air mass. The spatial implications of such semipermanent drought in a region of explosive population growth and urbanization are serious and tie in directly to the need for water-resource planning.

Sample Water Budgets

Using all of these concepts, we graph the water-balance components for several representative cities. Let's begin by looking at Kingsport, in the extreme northeastern corner of Tennessee at 36.6° N 82.5° W, elevation 390 m (1280 ft).

Figure 9.11 graphs the water-balance components for long-term supply and demand, using monthly averages.

The monthly values for PRECIP and POTET smooth the actual daily and hourly variability. On the graph, a comparison of PRECIP and POTET by month determines whether there is a *net supply* or a *net demand* for water. There is a net supply (blue line) from October to May, but the warm days from June to September create a net water demand. If we assume a soil-moisture storage capacity of 100 mm (4.0 in.), typical of shallow-rooted plants, the net water-demand months are satisfied through soil-moisture utilization (green area), with a small summer soil-moisture deficit.

Table 9.2 shows the actual data for Kingsport's water-balance equations for the year and for the months of March and September. Here, you can see the interaction of these various components. Check these three equations and perform the functions indicated to see whether we accounted for all the PRECIP received. Compare March and September in the equations with the same months in the graph in Figure 9.11.

Obviously, not all stations experience the surplus moisture patterns of this humid continental region. Different climatic regimes experience different relations among water-balance components. Figure 9.12 presents water-balance graphs for several other cities in North America and the Caribbean Sea region. Among these examples, compare the summer minimum precipitation of Berkeley, California, with the summer maximum of Seabrook, New Jersey; the drier prairies of Saskatchewan with the more humid conditions in Ottawa; and the

[*]Richard Seager et al., "Model projections of an imminent transition to a more arid climate in southwestern North America," *Science* 316, no. 5828 (May 25, 2007): 1184.

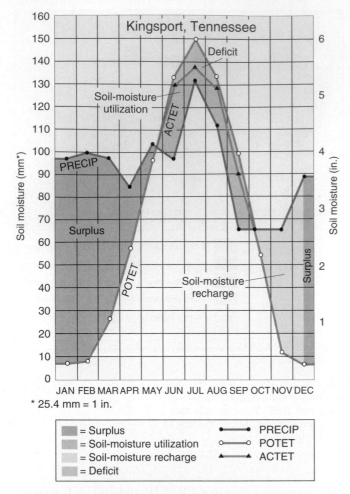

FIGURE 9.11 Sample water budget for Kingsport, Tennessee.
Compare the average plots for precipitation inputs and potential evapotranspiration outputs to determine the condition of the soil-moisture environment. A typical pattern of spring surplus, summer soil-moisture utilization, a small summer deficit, autumn soil-moisture recharge, and ending surplus highlights the year.

the Dust Bowl (see News Report 15.1), an urban flood, or a hurricane. As an example, Focus Study 9.1 presents Hurricane Camille (1969) and its generally positive effects on regional water resources.

Water Budget and Water Resources

Water distribution is uneven over space and time. Because we require a steady supply, we build large-scale management projects intended to redistribute the water resource either geographically, by moving water from one place to another, or over time, by storing water from time of receipt until it is needed. In this way, deficits are reduced, surpluses are held for later release, and water availability is improved to satisfy evapotranspiration and society's demands. The water balance permits analysis of the contribution streams make to water resources.

Streams may be *perennial* (constantly flowing; *perennial* is Latin for "through the year") or *intermittent*. In either case, the total runoff that moves through them comes from surplus surface-water runoff, subsurface throughflow, and groundwater. Figure 9.13 maps annual global river runoff for the world. Highest runoff amounts are along the equator within the tropics, reflecting the continual rainfall along the Intertropical Convergence Zone (ITCZ). Southeast Asia also experiences high runoff, as do northwest coastal mountains in the Northern Hemisphere. In countries having great seasonal fluctuations in runoff, groundwater becomes an important reserve water supply. Regions of lower runoff coincide with Earth's subtropical deserts, rain-shadow areas, and continental interiors, particularly in Asia.

Here are a few examples of water redistribution projects. The new Three Gorges Dam controls the treacherous Yangtze River in China (Figure 9.14). At a length of 2.3 km (1.5 mi) and a height of 185 m (607 ft), it is the largest dam and related construction in the world in overall size. Forced relocation of entire cities and more than 1 million people made room for the 600-km- (370-mi-) long reservoir. Controversies surrounded the immense scale of environmental, historical, and cultural losses. Gained is flood control, water resources for redistribution,

precipitation receipts of San Juan, Puerto Rico, with the arid desert of Phoenix, Arizona.

An interesting application of the water-balance approach to water resources is analyzing an event, such as

Table 9.2 Annual, March, and September Water Balances for Kingsport, Tennessee

	PRECIP	=	(POTET	−	DEFIC)	+	SURPL	±	ΔSTRGE
Annual	1119.0	=	(781.0	−	16.0)	+	354.0	±	0
	(44.1)	=	(30.7	−	0.6)	+	(13.9)	±	0
March	97.0	=	(24.0	−	0)	+	73.0	±	0
	(3.8)	=	(0.9	−	0)	+	(2.9)	±	0
September	66.0	=	(99.0	−	8.0)	+	0	−	25.0
	(2.6)	=	(3.9	−	0.3)	+	0	−	(1.0)

Note: All quantities in millimeters (inches).

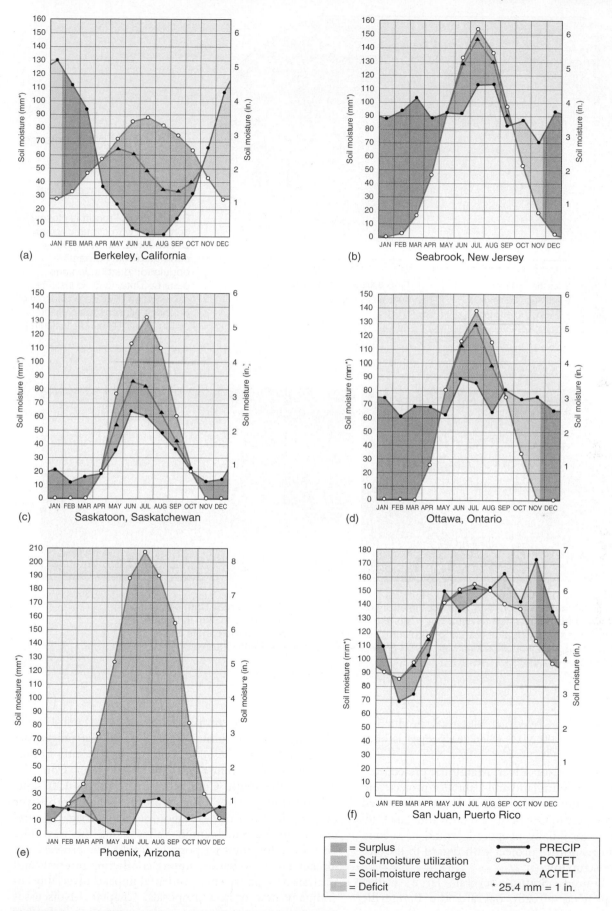

FIGURE 9.12 Sample water budgets for selected stations.
Sample water-balance regimes for selected stations in North America and the Caribbean region.

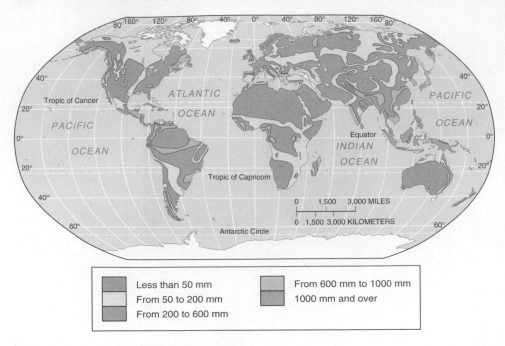

Global Water Balance
Components

FIGURE 9.13 Annual global river runoff.
Distribution of runoff is closely correlated with climatic region, as expected. But it is poorly correlated with human population distribution and density. [Data from the Institute of Geography, Russian Academy of Sciences, Moscow, as presented by the World Resources Institute.]

Map legend:
- Less than 50 mm
- From 50 to 200 mm
- From 200 to 600 mm
- From 600 mm to 1000 mm
- 1000 mm and over

FIGURE 9.14 Major dam in China.
Satellite views in 2000 (construction) and 2006 (completion) of the $25-billion Three Gorges Dam on the Yangtze River, flowing from upper left to right. For scale, the main portion of the dam is 2.3 km in length. This is the world's largest individual construction project. The channel to the north of the dam is a system of locks for shipping. [Images courtesy of ASTER Science Team, Jesse Allen Earth Observatory, NASA/GSFC/MITI/ERSDAC/JAROS.]

and power production. Eventual electrical production capacity will reach 22,000 MW when the project is built out in 2010.

In Canada, the Nelson River, Manitoba, project, the Churchill Falls, Newfoundland and Labrador project, and the proposed $12 billion Gull Island Dam and related hydroelectric projects in Québec are significant. The Gull Island project, as originally proposed, would be second only to the Three Gorges project with regard to the size of proposed construction.

In the United States, several large water-management projects already operate: the Bonneville Power Authority in Washington State, the Tennessee Valley Authority in the Southeast, the California Water Project, and the Central Arizona Project (Figure 9.15). Literally, the California Water Project rearranges the water budget in the state, holding back winter runoff for release in summer and pumping water from the north to the south. Completed in 1971, this 1207-km- (750-mi-) long "river" flows from the Sacramento River delta to the Los Angeles region, servicing irrigated agriculture in the San Joaquin Valley along the way. The best sites for multipurpose hydroelectric projects are already taken, so battles among conflicting interests and analysis of negative environmental impacts invariably accompany new project proposals. Chapter 14 discusses river management through such construction in further detail.

(a)

(b)

FIGURE 9.15 Two major water projects in the United States.
(a) The California Aqueduct transports water from northern to southern California, as part of the California Water Project. Near Palmdale from the Lamont–Odett vista point, the aqueduct flows parallel to the San Andreas fault zone. The Palmdale Reservoir beyond the aqueduct is in a depression in the fault zone. (b) Grand Coulee Dam holds back the Columbia River, as part of the federal Bonneville Power Administration, in Washington State. Comprising 31 hydroelectric dams and a transmission grid, BPA began in 1937. [Photos by (a) Bobbé Christopherson; (b) author.]

Focus Study 9.1

Hurricane Camille, 1969: Water-Balance Analysis Shows Moisture Benefits

Hurricane Camille was one of the most devastating hurricanes of the twentieth century. Hurricanes Ivan in 2004, Rita, Katrina, and Wilma in 2005, and Humberto in 2007 brought back memories of Camille to Gulf Coast residents. Ironically, Camille was significant not only for the disaster it brought (256 dead, $1.5 billion in damage) but also for the drought it ended.

Figure 9.1.1 shows Camille inland from the Gulf Coast north of Biloxi, Mississippi. Hurricane-force winds sharply diminished after the hurricane made landfall, leaving a vast rainstorm that traveled from the Gulf Coast through Mississippi, western Tennessee, Kentucky, and into central Virginia (Figure 9.1.2). Severe flooding drowned the Gulf Coast near landfall and the James River basin of Virginia, where torrential rains produced record floods. But Camille actually had beneficial aspects; it

FIGURE 9.1.1 *ESSA-9* satellite image of Hurricane Camille.
Hurricane Camille made landfall on the night of August 17, 1969, and continued inland on August 18. The storm is progressing northward through Mississippi in the image.
[Image from NOAA.]

ended a year-long drought along major portions of its storm track.

Figure 9.1.2a maps the precipitation from Camille and portrays the storm's track from landfall in Mississippi to the coast of Virginia and Delaware. By comparing actual water budgets along Camille's track with water budgets for the same three days with Camille's rainfall totals removed artificially, the moisture impact of the storm on the region's water budget could be analyzed. Figure 9.1.2b maps the moisture

shortages that were avoided because of Camille's rains—termed "deficit abatement." Think of this as drought that did not continue because Camille's rains occurred. Over vast portions of the affected area, Camille reduced dry-soil conditions, restored pastures, and filled low reservoirs.

Thus, Camille's monetary benefits inland outweighed its damage by an estimated 2-to-1 ratio. (Of course, the tragic loss of life does not fit into a financial equation.) In 2007, Hurricane

Humberto, a category 1 storm, brought some needed relief across the Southeast once it moved inland. Hurricanes should be viewed as normal and natural meteorological events that not only have terrible destructive potential, mainly to coastal lowlands, but also contribute to the precipitation regimes of the southern and eastern United States. According to weather records, about one-third of all hurricanes making landfall in the United States provide beneficial precipitation to local water budgets.

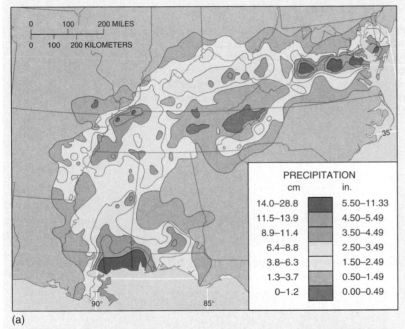

(a)

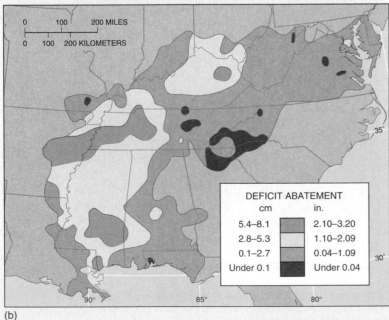

(b)

FIGURE 9.1.2 Camille affects water budgets in a beneficial way.
A water-resource view of Hurricane Camille's impact on local water budgets: (a) precipitation attributable to the storm (moisture supply), (b) resulting deficit abatement (avoided moisture shortages) attributable to Camille. [Data and maps by R. Christopherson. All rights reserved.]

Groundwater Resources

Groundwater is an important part of the hydrologic cycle, although it lies beneath the surface, beyond the soil-moisture root zone. It is tied to surface supplies through pores in soil and rock. Groundwater is the largest potential freshwater source in the hydrologic cycle—larger than all surface lakes and streams combined. Between Earth's land surface and a depth of 4 km (13,000 ft) worldwide, some 8,340,000 km³ (2,000,000 mi³) of water resides, a volume comparable to 70 times all the freshwater lakes in the world. Despite this volume and its obvious importance, pollution threatens groundwater quality and overconsumption depletes groundwater volume in quantities beyond natural replenishment rates.

Remember: *Groundwater is not an independent source of water, for it is tied to surface supplies for recharge.* An important consideration in many regions is that groundwater accumulation occurred over millions of years, so care must be taken not to exceed this long-term buildup with excessive short-term demands.

Globally, groundwater is the raw material extracted most from nature by people, totaling more than 700 km³/year in volume, and is the basis for agriculture's "green revolution," used for irrigation more widely than surface water in many countries. For many uses, quality groundwater is a better source than surface water. It is available in many parts of the world that lack dependable surface runoff. Whereas surface supplies are affected by short-term drought, groundwater is not (although long-term drought affects both). Except in severely polluted areas, groundwater is generally free of sediment, color, and disease organisms, although polluted groundwater conditions are considered irreversible.

About 50% of the U.S. population derives a portion of its freshwater from groundwater sources. In some states, such as Nebraska, groundwater supplies 85% of water needs and as high as 100% in rural areas. In Canada, about 6 million people (two-thirds of them live in rural areas) rely on groundwater for domestic needs. Between 1950 and 2000, annual groundwater withdrawal in the United States and Canada increased more than 150%. Figure 9.16 shows potential groundwater resources in both countries.

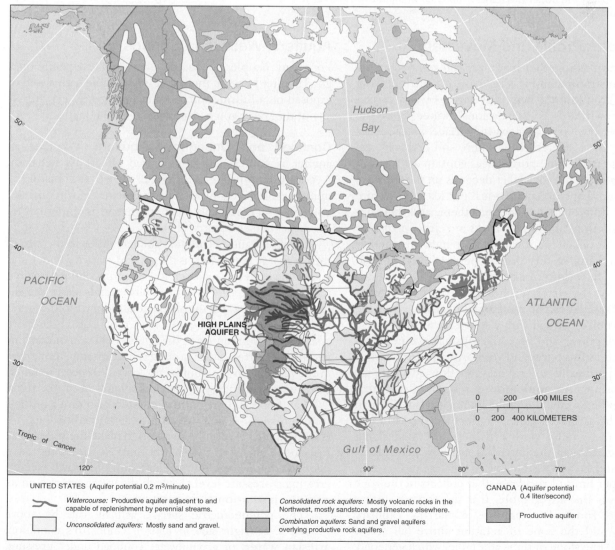

FIGURE 9.16 Groundwater resource potential for the United States and Canada.
Highlighted areas of the United States are underlain by productive aquifers capable of yielding freshwater to wells at 0.2 m³/min or more (for Canada, 0.4 L/s). [Courtesy of Water Resources Council for the United States and the *Inquiry on Federal Water Policy* for Canada.]

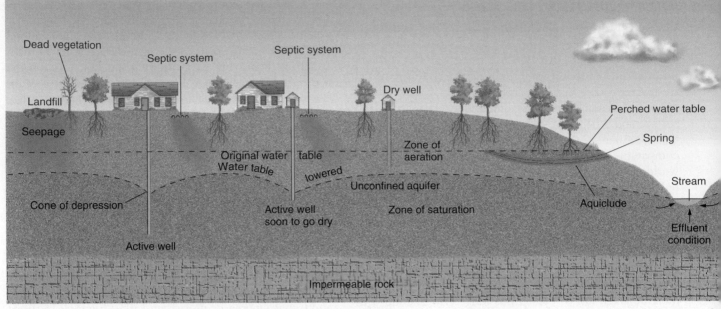

FIGURE 9.17 Groundwater characteristics.
Subsurface groundwater characteristics, processes, water flows, and human interactions. Follow the concepts across the illustration left-to-right as you read along in the text.

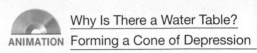

Why Is There a Water Table?
Forming a Cone of Depression

Groundwater Profile and Movement

Figure 9.17 brings together many groundwater elements in a single illustration, and it is the basis for the following discussion. Groundwater begins as surplus water, which percolates downward as gravitational water from the zone of capillary water. This excess surface water moves through the **zone of aeration**, where soil and rock are less than saturated (some pore spaces contain air).

The *porosity* of any rock layer depends on the arrangement, size, and shape of its individual particles, the nature of any cementing elements between them, and their degree of compaction. Subsurface rocks are either *permeable* or *impermeable*. Their permeability depends on whether they conduct water readily (higher permeability) or tend to obstruct its flow (lower permeability).

Eventually, the water reaches an area where subsurface water accumulates, the **zone of saturation**. Here, the pores are completely filled with water. Like a hard sponge made of sand, gravel, and rock, the saturation zone stores water in its countless pores and voids. An **aquifer** is a rock layer that is permeable to groundwater flow adequate for wells and springs. An **aquiclude** is a body of rock that does not conduct water in usable amounts (also called an *aquitard*). The saturated zone may include the saturated portion of the aquifer and a part of the underlying *aquiclude*, even though the latter has such low permeability that its water is inaccessible.

The upper limit of the water that collects in the zone of saturation is the **water table**. It is the contact surface between the zone of saturation, a narrow capillary fringe layer, and the zone of aeration above (all across Figure 9.17). The slope of the water table, which generally follows the contours of the land surface, controls groundwater movement.

Aquifers, Wells, and Springs

For many people, adequate water supply depends on a good aquifer beneath them that is accessible with a well or exposed on a hillside where water emerges as a spring. Is any of the water you consume from a well or spring?

Confined and Unconfined Aquifers Whether an aquifer is confined or unconfined affects its behavior. A **confined aquifer** is bounded above and below by impermeable layers of rock or sediment. An **unconfined aquifer** has a permeable layer on top and an impermeable one beneath (see Figure 9.17).

Confined and unconfined aquifers differ in the size of their recharge area, which is the ground surface where water enters an aquifer to recharge it. For an unconfined aquifer, the **aquifer recharge area** generally extends above the entire aquifer; the water simply percolates down to the water table. But in a confined aquifer, the recharge area is far more restricted, as you can see in the figure. Pollution of this limited recharge area causes groundwater contamination; note in the illustration the improperly located disposal pond on the aquifer recharge area, contaminating wells to the right.

Confined and unconfined aquifers also differ in their water pressure. A well drilled into an unconfined aquifer (see Figure 9.17) must be pumped to make the water level rise above the water table. In contrast, the water in a confined aquifer is under the pressure of its own weight, creating a pressure level to which the water can rise on its own, the **potentiometric surface**.

The potentiometric surface actually can be above ground level (right side in figure). Under this condition, **artesian water**, or groundwater confined under pressure, may rise in a well and even flow at the surface without pumping if the top of the well is lower than the potentiometric sur-

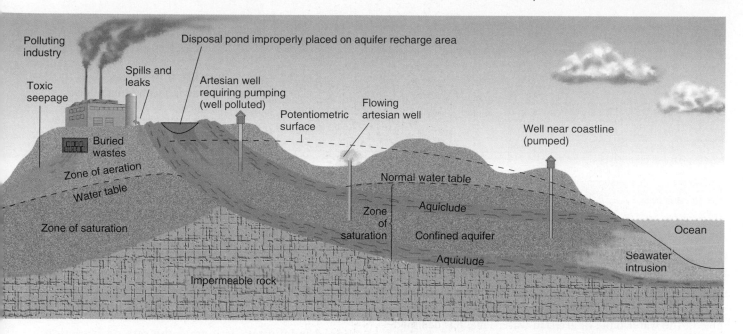

face. (These wells are called *artesian*, for the Artois area in France where they are common.) In other wells, however, pressure may be inadequate, and the artesian water must be pumped the remaining distance to the surface.

Wells, Springs, and Streamflows   The slope of the water table, which broadly follows the contour of the land surface, controls groundwater movement (see Figure 9.17). Groundwater tends to move toward areas of lower pressure and elevation.

Water wells can work only if they penetrate the water table. Too shallow a well will be a "dry well"; too deep a well will punch through the aquifer and into the impermeable layer below, also yielding little water. This is why water wells should be drilled in consultation with a *hydrogeologist*.

Where the water table intersects the surface, it creates springs (Figure 9.18). Such an intersection also occurs in lakes and riverbeds. Ultimately, groundwater may enter stream channels to flow as surface water (stream near center in Figure 9.17). In fact, during dry periods, the water table may sustain river flows.

Figure 9.19 illustrates the relation between the water table and surface streams in two different climatic settings. In humid climates, where the water table generally supplies a continuous base flow to a stream and is higher than the stream channel, the stream is *effluent* because it receives the water flowing out (effluent) from the surrounding ground. The Mississippi River and countless other streams are examples. In drier climates, with lower water tables, water from *influent* streams flows into the adjacent ground, sustaining vegetation along the stream. The Colorado River and the Rio Grande of the American West are examples of influent streams.

Overuse of Groundwater

As water is pumped from a well, the surrounding water table within an unconfined aquifer may experience

FIGURE 9.18 An active spring.
Springs provide evidence of groundwater emerging at the surface. This spring flows into Crystal Lake, Salt River Range, Wyoming. [Photo by Bobbé Christopherson.]

drawdown, or become lowered. Drawdown occurs if the pumping rate exceeds the replenishment flow of water into the aquifer or the horizontal flow around the well. The resultant lowering of the water table around the well is a **cone of depression** (see Figure 9.17, left).

Overpumping   Aquifers frequently are pumped beyond their flow and recharge capacities, a condition known as

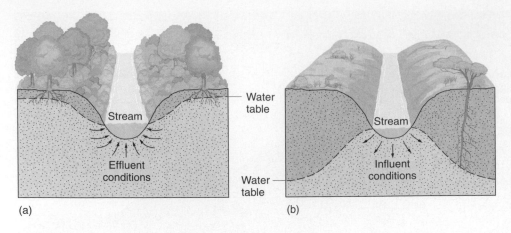

FIGURE 9.19 Groundwater interaction with streamflow. (a) Effluent stream base flow is partially supplied by a high water table, characteristic of humid regions. (b) Influent stream supplies a lower water table, characteristic of drier regions. The water table may drop below the stream channel, effectively drying it out.

groundwater mining. Today, large tracts experience chronic groundwater overdrafts in the Midwest, West, lower Mississippi Valley, Florida, and the intensely farmed Palouse region of eastern Washington State. In many places, the water table or artesian water level has declined more than 12 m (40 ft). In the United States, groundwater mining is of special concern in the great High Plains Aquifer, which is the topic of Focus Study 9.2.

About half of India's irrigation water and half of industrial and urban water needs are met by the groundwater reserve. In rural areas, groundwater supplies 80% of domestic water from some 3 million hand-pumped bore holes. And, in approximately 20% of India's agricultural districts, groundwater mining through more than 17 million wells (motorized dug wells and tube wells) is beyond recharge rates.

In the Middle East, conditions are even more severe, as detailed in News Report 9.1. The groundwater resource beneath Saudi Arabia accumulated over tens of thousands of years, forming "fossil aquifers," but the increasing withdrawals are not being naturally recharged to any appreciable degree at present due to the desert climate—in essence, it is now a nonrenewable groundwater resource. Some

researchers suggest that groundwater in the region will be depleted in a decade, although worsening water-quality problems will no doubt arise before this date.

Like any renewable resource, groundwater can be tapped indefinitely as long as the rate of extraction does not exceed the rate of replenishment. But just like a bank account, a groundwater reserve will dwindle if withdrawals exceed deposits. Few governments have established and enforced rules and regulations to insure that groundwater sources are exploited at a sustainable rate. . . . No government has yet adequately tackled the issue of groundwater depletion, but it is at least getting more attention.*

Collapsing Aquifers A possible effect of water removal from an aquifer is that the aquifer, which is a layer of rock or sediment, will lose its internal support. Water in the pore spaces between rock grains is not compressible, so it

*L. Brown et al. and the Worldwatch Institute, *Vital Signs, The Environmental Trends That Are Shaping Our Future* (New York: W.W. Norton & Co., 2000), pp. 122, 123.

Focus Study 9.2

High Plains Aquifer Overdraft

North America's largest known aquifer is the High Plains aquifer. It lies beneath the American High Plains, an eight-state, 450,600-km² (174,000-mi²) area from southern South Dakota to Texas (Figure 9.2.1a). Precipitation over the region varies from 30 cm in the southwest to 60 cm in the northeast (12 to 24 in.).

For several hundred thousand years, the aquifer's sand and gravel were charged with meltwaters from retreating glaciers. However, heavy

mining of High Plains groundwater has occurred for the past 100 years, and mining intensified after World War II with the introduction of center-pivot irrigation (Figure 9.2.2). These large circular devices provide vital water to wheat, sorghums, cotton, corn, and about 40% of the grain fed to cattle in the United States.

The High Plains aquifer irrigates about one-fifth of all U.S. cropland: 120,000 wells provide water for 5.7 million hectares (14 million acres), or

about 30% of all irrigation water for crops in the United States. This is down from the peak of 170,000 wells in 1978. In 1980, water was pumped from the aquifer at the rate of 26 billion cubic meters (21 million acre-feet) a year, an increase of more than 300% since 1950. By the turn of this century, withdrawals had decreased 10% due to declining well yields and increasing pumping costs.

During the past five decades, the water table in the aquifer dropped

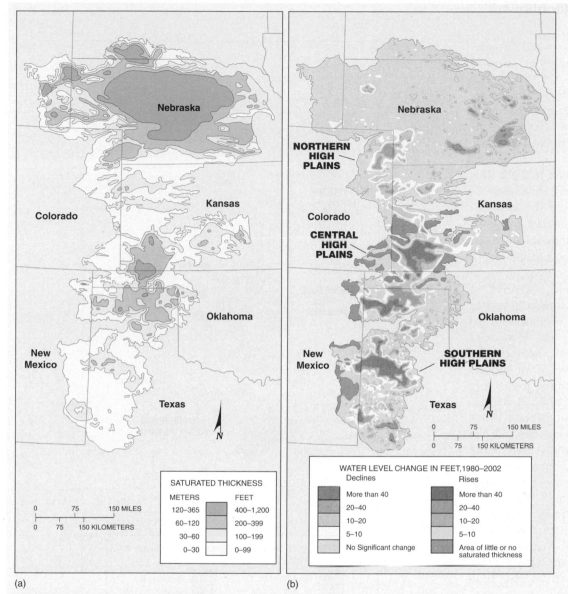

(a)

(b)

FIGURE 9.2.1 High Plains aquifer.
(a) The High Plains is the largest known aquifer in North America, mapped here showing its average saturated thickness. (b) Water-level changes in the aquifer from 1980 to 2002, given in feet. [Maps: (a) After D. E. Kromm and S. E. White, "Interstate groundwater management preference differences: The High Plains region," *Journal of Geography* 86, no. 1 (January–February 1987): 5 and (b) adapted from "Water-Level Changes in the High Plains Aquifer, Predevelopment to 2002, 1980 to 2002, and 2001 to 2002," by V. L. McGuire, USGS Fact Sheet 2004–3026, Fig. 2.]

FIGURE 9.2.2 Center-pivot irrigation.
Myriad center-pivot irrigation systems water crops in the western Great Plains. A growing season for corn requires from 10 to 20 revolutions of the sprinkler arm, depending on the weather. The arm delivers about 3 cm (1.18 in.) of water per revolution (rainfall equivalent). [Photo by Bobbé Christopherson.]

(continued)

Focus Study 9.2 (continued)

more than 30 m (100 ft), and through-out the 1980s it averaged a 2-m (6-ft) drop each year. The USGS estimates that recovery of the High Plains aquifer (those portions that have not collapsed) would take at least 1000 years if groundwater mining stopped today!

Figure 9.2.1b maps changes in water levels from 1980 to 2002. De-clining water levels are most severe in northern Texas, where the saturated thickness of the aquifer is least, through the Oklahoma panhandle and into western Kansas. Rising water levels are noted in portions of south-central Nebraska and a portion of Texas owing to recharge from surface irrigation, a period of above-normal precipitation years, and downward percolation from canals and reservoirs.

The USGS began monitoring this groundwater mining from more than 7000 wells in 1988. A six-year High Plains Regional Ground Water Study began in 1999 and concluded at the end of 2005. The goal is an overall assessment of the resource, including groundwater quality and contaminant sources. A 2004 USGS map of the water-level changes from predevelop-ment (about 1950) to 2002 is in Figure 9.2.3. Wells were measured in the winter or early spring when water levels were recovered from the previous irrigation season. (See **http://ne.water.usgs.gov/** for links or **http://co.water.usgs.gov/nawqa/hpgw/HPGW_home.html**.)

Obviously, billions of dollars of agricultural activity cannot be abrupt-ly halted, but neither can profligate water mining continue. This issue raises tough questions: How best to manage cropland? Can extensive irri-gation continue? Can the region con-tinue to meet the demand to produce commodities for export? Should we continue high-volume farming of cer-tain crops that are in chronic oversup-ply? Should we rethink federal policy on crop subsidies and price supports? What changes will biofuels produc-tion force? What would be the impact on farmers and rural communities of any changes to the existing system?

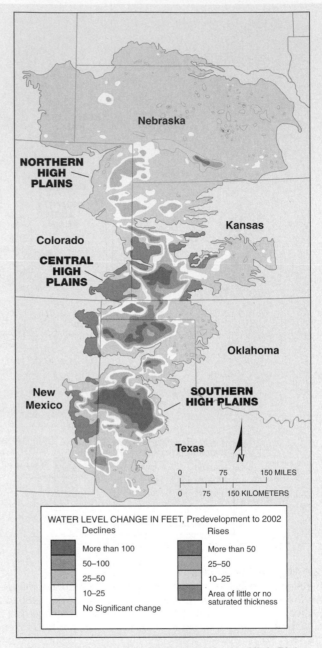

FIGURE 9.2.3 Water-level changes in the High Plains aquifer, predevelopment to 2002.
Using 1950 to represent "predevelopment," this map of changes in groundwater levels was derived from sampling some 7000 wells. The color scale in the legend indicates widespread declines and a few areas of rises in water levels. [Map adapted from "Water-Level Changes in the High Plains Aquifer, Predevelopment to 2002, 1980 to 2002, and 2001 to 2002," by V. L. McGuire, USGS Fact Sheet 2004–3026, Fig. 1.]

Present irrigation practices, if continued, will destroy about half of the High Plains aquifer resource (and two-thirds of the Texas portion) by the year 2020. Add to this the approx-imate 10% loss of soil moisture due to increased evapotranspiration demand caused by climatic warming for this region by 2050, as forecast by com-puter models, and we have a portrait of a major regional water problem and a challenge for society.

News Report 9.1

Middle East Water Crisis: Running on Empty

The Persian Gulf states soon may run out of freshwater. Their vast groundwater resource is overpumped to such an extent that salty seawater is encroaching into aquifers tens of kilometers inland. In this decade, groundwater on the Arabian Peninsula may become undrinkable. Imagine having the hottest issue in the Middle East become water, not oil!

Remedies for groundwater overuse are neither easy nor cheap. In the Persian Gulf area, additional freshwater is obtained by desalination of seawater, using desalination plants along the coasts. These processing plants remove salt from seawater by distillation and evaporation processes. In fact, approximately 50% of all desalination plants are in the Middle East. In Saudi Arabia, 30 desalination plants supply 70% of drinking water needs (see Figure 9.22).

One water pipeline proposal brings water overland from Turkey to the Middle East through two branches. The Gulf water line would run southeast through Jordan and Saudi Arabia, with extensions into Kuwait, Abu Dhabi, and Oman. The other branch would run south through Syria, Jordan, and Saudi Arabia to the cities of Makkah (Mecca) and Jeddah. This pipeline would cover some 1500 km (930 mi), the distance from New York City to St. Louis. (See the links listed under Middle East Water Information Network at **http://www.columbia.edu/cu/lweb/indiv/mideast/cuvlm/water.html**.)

Other remedies are possible. Traditional agricultural practices could be modernized to use less water. Urban water use could be made more efficient to reduce groundwater demand. Aquifers and rivers could be shared, although at present no negotiated accords exist for this purpose in the Middle East. Conflicts at various scales pose a grave threat to water pipelines and desalination facilities throughout the region.

> There are severe water shortages in the Middle East. . . . The resources of the Nile, the Tigris–Euphrates, and the Jordan are overextended owing both to natural causes and to those deriving from human behavior. . . . In addition to a severe shortage in the quantity of water, there is a growing concern over water quality.*

* N. Kliot, *Water Resources and Conflict in the Middle East* (London: Routledge, 1994), p. 1.

---

adds structural strength to the rock. If the water is removed through overpumping, air infiltrates the pores. Air is readily compressible, and the tremendous weight of overlying rock may crush the aquifer. On the surface, the visible result may be land subsidence, cracked house foundations, and changes in drainage.

Houston, Texas, provides an example. The removal of groundwater and crude oil caused land within an 80-km (50-mi) radius of Houston to subside more than 3 m (10 ft) over the years. Yet another example is the Fresno area of California's San Joaquin Valley. After years of intensive pumping of groundwater for irrigation, land levels have dropped almost 10 m (33 ft) because of a combination of water removal and soil compaction from agricultural activity.

Surface mining ("strip mining") for coal in eastern Texas, northeastern Louisiana, Wyoming, Arizona, and other locales also destroys aquifers. Aquifer collapse is an increasing problem in West Virginia, Kentucky, and Virginia, where low-sulfur coal is mined to meet current air pollution requirements for coal-burning electric power plants.

Unfortunately, collapsed aquifers may not be rechargeable, even if surplus gravitational water becomes available, because pore spaces may be permanently collapsed. The water-well fields that serve the Tampa Bay–St. Petersburg, Florida, area are a case in point. Pond and lake levels, swamps, and wetlands in the area are declining and land surfaces subsiding as the groundwater drawdown for export to the cities increases.

Saltwater Encroachment When aquifers are overpumped near the ocean, another problem arises. Along a coastline, fresh groundwater and salty seawater establish a natural interface, or *contact surface*, with the less-dense freshwater flowing on top of the saline groundwater. But excessive withdrawal of freshwater can cause this interface to migrate inland. As a result, wells near the shore may become contaminated with saltwater and the aquifer may become useless as a freshwater source as local drawdowns allow underlying saltwater to *upcone*, or rise. Figure 9.17 illustrates this seawater intrusion (far-right side). Pumping freshwater back into the aquifer may halt contamination by seawater, but once contaminated, the aquifer is difficult to reclaim.

Pollution of Groundwater

When surface water is polluted, groundwater inevitably becomes contaminated because it is recharged from surface-water supplies. Surface water flows rapidly and flushes pollution downstream, but slow-moving groundwater, once contaminated, remains polluted virtually forever.

Pollution can enter groundwater from many sources: industrial injection wells (wastes pumped into the ground), septic tank outflows, seepage from hazardous-waste disposal sites, industrial toxic waste, agricultural residues (pesticides, herbicides, fertilizers), and urban solid-waste landfills. As an example, leakage from some 10,000 underground gasoline storage tanks at U.S. gasoline stations is suspected. Some cancer-causing additives to gasoline, such as the oxygenate MTBE (methyl tertiary butyl ether), are contaminating thousands of local water supplies.

For management purposes, about 35% of pollution is *point source* (such as a gasoline tank or septic tank); 65% is *nonpoint source* (from a broad area, such as an agricultural field or urban runoff). Regardless of the spatial nature of the source, pollution can spread over a great distance, as illustrated in Figure 9.17. Since the nature of aquifers makes them inaccessible, the extent of groundwater pollution is underestimated.

Government zoning to prohibit pollution can be effective in the case of confined and unconfined aquifers. For example, once recharge areas are identified, an informed population should insist on a ban on pollution discharges, septic and sewage-system installations, or hazardous-material dumping.

In the face of government inaction, and even attempts to reverse some protection laws, serious groundwater contamination continues nationwide, adding validity to a quote from more than 20 years ago:

> One characteristic is the practical irreversibility of groundwater pollution, causing the cost of clean-up to be prohibitively high. . . . It is a questionable ethical practice to impose the potential risks associated with groundwater contamination on future generations when steps can be taken today to prevent further contamination.*

*J. Tripp, "Groundwater Protection Strategies," in *Groundwater Pollution, Environmental and Legal Problems* (Washington, DC: American Association for the Advancement of Science, 1984), p. 137. Reprinted by permission.

Our Water Supply

Human thirst for adequate water supplies, both in quantity and quality, will be a major issue in this century. Internationally, increases in per capita water use are double the rate of population growth. Since we are so dependent on water, it seems that humans should cluster where good water is plentiful. But accessible water supplies are not well correlated with population distribution or the regions where population growth is greatest.

Table 9.3 shows estimated world water supplies, world population in 2007, estimated population figures for 2025, a population-change forecast between 2007 and 2050 at present growth rates, and a comparison of population and global runoff percentages for each continent. Also note estimates for urban and rural access to improved sanitation for each region. The table gives an idea of the unevenness of Earth's water supply, which is tied to climatic variability, and water demand, which is tied to level of development, affluence, and per capita consumption.

For example, North America's mean annual discharge is 5960 km³ and Asia's is 13,200 km³. However, North America has only 6.7% of the world's population, whereas Asia has 60.5%, with a population doubling time less than half of North America's. In northern China, 550 million people living in approximately 500 cities lack adequate water supplies. For comparison, note that the 1990 floods cost China $10 billion, whereas water shortages are running at more than $35 billion a year in costs to the Chinese economy. The Yellow River, a water resource for many Chinese, has run dry every year since the early 1970s. In 1997, it failed to reach the sea for almost 230 days! An analyst for the World

Table 9.3	Estimate of Available Global Water Supply Compared by Region and Population								
Region (2007 population in millions)	**Land Area in Thousands of**		**Share of Mean Annual Discharge in**		**Global Stream Runoff (%)**	**Projected Global Population, 2025 (%)**	**Projected Population, 2025 (millions)**	**Projected Population Change (+) 2007–2050 (%)**	**Population with Access to Improved Sanitation (%) For 2006 Urban / Rural**
	km²	**(mi²)**	**km³/yr**	**(BGD)**					
Africa (944)	30,600	(11,800)	4,220	(3,060)	10.8	17.1	1,359	107	62 / 30
Asia (4010)	44,600	(17,200)	13,200	(9,540)	34.0	59.9	4,768	34	74 / 31
Australia–Oceania (35)	8,420	(3,250)	1,960	(1,420)	5.1	0.53	42	41	98 / 58
Europe (733)	9,770	(3,770)	3,150	(2,280)	8.1	9.0	719	-9	100 / 100
North America (442) (Canada, Mexico, U.S.)	22,100	(8,510)	5,960	(4,310)	15.3	6.4	512	38	100 / 100 Mexico: 90 / 39
Central and South America (463)	17,800	(6,880)	10,400	(7,510)	26.7	7.1	566	41	84 / 43
Global (6625) (excluding Antarctica)	134,000	(51,600)	38,900	(28,100)	—	100.00	7,965	40	More developed: 100 / 92 Less developed: 73 / 31

Note: BGD, billion gallons per day. Population data from *2007 World Population Data Sheet* (Washington, D.C. Population Reference Bureau, 2007). Numbers are rounded.

News Report 9.2

Personal Water Use and Water Measurements

On an individual level, statistics indicate that urban dwellers directly use an average of 680 L (180 gal) of water per person per day, whereas rural populations average only 265 L (70 gal) or less per person per day. However, each of us indirectly accounts for a much greater water demand because we all consume food and other products produced with water. For the United States, dividing the total water withdrawal (408 billion gallons a day, BGD, in 2000, the latest year data are available) by a population of 285 million people yields a per capita direct and indirect use of 5419 L (1431 gal) of water per day! As population and per capita affluence increase, greater demands are placed on the water-resource base, which is essentially fixed.

Simply providing the variety of food we enjoy requires voluminous water. For example: 77 g (2.7 oz) of broccoli requires 42 L (11 gal) of water to grow and process; producing 250 ml (8 oz) of milk requires 182 L (48 gal) of water; producing 28 g (1 oz) of cheese requires 212 L (56 gal); producing 1 egg requires 238 L (63 gal); and a 113-g (4-oz) beef patty requires 2314 L (616 gal).

And then there are the toilets, the majority of which still flush approximately 4 gal of water. Imagine the spatial complexity of servicing the desert city of Las Vegas, with 135,000 hotel rooms times the number of toilet flushes per day in those rooms; of providing all support services for 38 million visitors per year, plus 38 golf courses. One hotel confirms that it washes 14,000 pillowcases a day. In addition, there is the usage of a resident population of 1.5 million people—together, both tourists' and residents' water demand exceeds Nevada's annual allotment from the Colorado River.

Clearly, the average American lifestyle depends on enormous quantities of water and thus is vulnerable to shortfalls or quality problems. Demand-side water conservation and efficiency remain as potentially our best water resource.

How is all this water measured? In most of the United States, hydrologists measure streamflow in cubic feet per second (ft³/s); Canadians use cubic meters per second (m³/s). In the eastern United States, or for large-scale assessments, water managers use millions of gallons a day (MGD), billions of gallons a day (BGD), or billions of liters per day (BLD).

In the western United States, where irrigated agriculture is so important, the measure frequently used is acre-feet per year. One acre-foot is an acre of water, 1 ft deep, equivalent to 325,872 gal (43,560 ft³, or 1234 m³, or 1,233,429 L). An acre is an area that is about 208 ft on a side and is 0.4047 hectares. For global measurements, 1 km³ = 1 billion cubic meters = 810 million acre-feet; 1000 m³ = 264,200 gal = 0.81 acre-feet. For smaller measures, 1 m³ = 1000 L = 264.2 gal.

Bank stated that China's water shortages pose a more serious threat than floods during this century, yet in news coverage we hear only of floods when they happen.

Water is critical for survival. Just to produce our food requires enormous quantities of water for growing, cleaning, processing, and waste disposal. See News Report 9.2 for personal water use and water measurements. We sometimes take for granted the quality and quantity of our water supply because it always seems to be there when we need it.

In Africa, 56 countries draw from a varied water-resource base; these countries share more than 50 river and lake watersheds. Climate change is affecting the sustainability of water resources, a condition of expanding aridity and desertification discussed in Figure 15.25. Increasing population growth and percentage of urban dwellers cause water demands to rise and lead to an overexploitation of the resource. Twelve African countries experience water stress (less than 1700 m³ per person/year) and 14 have water scarcity (1000 m³ per person/year).

At present, the world's people are withdrawing 30% of the runoff that is accessible. This idea of "accessibility" is important, because about 20% of global runoff is remote and not readily available to meet water demand. Timing of flows is also important, for approximately 50% of total runoff remains beyond use as uncaptured floodwater. For example, the greatest river in terms of runoff discharge on Earth—the Amazon—is about 95% remote from population centers, that is, not readily available.

Water resources differ from other resources in that there is no alternative for water. Water shortages increase the probability of international conflict, endanger public health, reduce agricultural productivity, and damage life-supporting ecological systems. New dams and river-management schemes might increase runoff accessibility by 10% over the next 30 years, but population is projected to grow by 32% in the same time period. However, for the twenty-first century a transition to a new approach is required, away from the large, centralized structures toward decentralized, community-based strategies, focused on demand management and efficient technologies.

Water Supply in the United States

The U.S. water supply derives from surface and groundwater sources that are fed by an average daily precipitation of 4200 BGD (billion gallons a day). That sum is based on an average annual precipitation value of 76.2 cm (30 in.) divided evenly among the 48 contiguous states (excluding Alaska and Hawai'i).

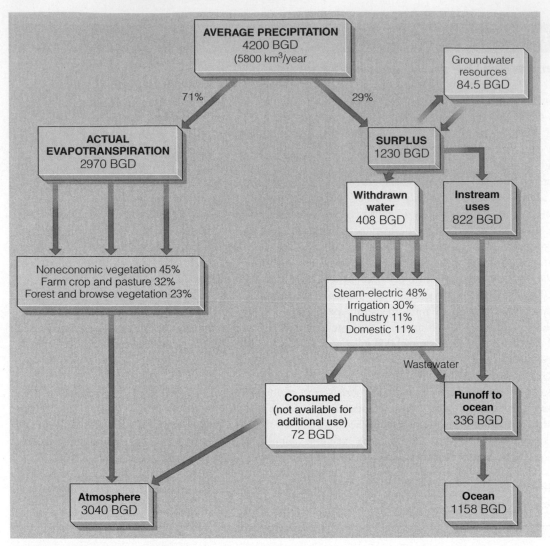

FIGURE 9.20 U.S. water budget.
Daily water budget for the contiguous 48 states in billions of gallons a day (BGD). As of 2000, approximately 33% of available surplus was withdrawn for agricultural (irrigation–livestock), municipal (domestic–commercial), and industrial (industry–mining–thermoelectric) use. [Data from Susan Hutson, Nancy Barber, Joan Kenny, Kristin Linsey, Deborah Lumia, and Molly Maupin, *Estimated Use of Water in the United States in 2000* (Denver, CO: USGS Circular 1268, 2000)—the latest year data available.]

The 4200 billion gallons of average daily precipitation mentioned is unevenly distributed across the country and unevenly distributed throughout the year. For example, New England's water supply is so abundant that only about 1% of available water is consumed each year. (The same is true in Canada, where the resource greatly exceeds that in the United States.) But in the dry Colorado River Basin mentioned earlier the discharge is completely consumed. In fact, by treaty and compact agreements, the Colorado actually is budgeted beyond its average flow. This paradox results from political misunderstanding of the water resource.

Figure 9.20 shows how the U.S. supply of 4200 BGD is distributed. Viewed daily, the national water budget has two general outputs: 71% actual evapotranspiration (ACTET) and 29% surplus (SURPL). The 71% actual evapotranspiration involves 2970 BGD of the daily supply. It passes through nonirrigated land, including farm crops and pasture, forest and browse (young leaves, twigs,

and shoots), and nonagricultural vegetation. Eventually, it returns to the atmosphere to continue its journey through the hydrologic cycle. The remaining 29% surplus is what we directly use.

Instream, Nonconsumptive, and Consumptive Uses

The surplus 1230 BGD is runoff, available for withdrawal, consumption, and various instream uses.

- *Instream* uses are those that use stream water in place: navigation, wildlife and ecosystem preservation, waste dilution and removal, hydroelectric power production, fishing resources, and recreation.
- *Nonconsumptive uses*, sometimes called **withdrawal,** or *offstream use*, remove water from the supply, use it, and then return it to the same supply. Non-consumptive water is used by industry, agriculture, municipalities,

and in steam–electric power generation. A portion of the water withdrawn is consumed.

- **Consumptive uses** remove water from a stream but do not return it, so it is not available for a second or third use. Some consumptive examples include water that evaporates or is vaporized in steam–electric plants.

When water returns to the system, water quality usually is altered—water is contaminated chemically with pollutants, or waste, or thermally with heat energy. In Figure 9.20, this portion of the budget is returned to runoff and eventually the ocean. This wastewater represents an opportunity to extend the resource through reuse.

Contaminated or not, returned water becomes a part of all water systems downstream. An example is New Orleans, the last city to withdraw municipal water from the Mississippi River. New Orleans receives diluted and mixed contaminants added throughout the entire Missouri–Ohio–Mississippi River drainage. This includes: the effluent from chemical plants, runoff from millions of acres of farm fields treated with fertilizer and pesticides, treated and untreated sewage, oil spills, gasoline leaks, wastewater from thousands of industries, runoff from urban streets and storm drains, and turbidity from countless construction sites and from mining, farming, and logging activities. Abnormally high cancer rates among citizens living along the Mississippi River between Baton Rouge and New Orleans have led to the ominous label "Cancer Alley" for the region. The infamous "muck" that covered the region when the levees broke in 2005 contained some of this in its brew.

The estimated U.S. withdrawal of water for 2000 was 408 BGD, almost 3 times the usage rate in 1940, but a decrease of 10% since the peak year of 1980 and similar to 1990. Total per capita use dropped from 1340 gal/day in 1990 to 1280 gal/day in 2000. A shift in emphasis from supply-side to demand-side planning, pricing, and management produced these efficiencies. The 2000 study is the most recent water-withdrawal study completed by the USGS; a 2005 update was still underway at the end of 2007.

The four main uses of withdrawn water in the United States in 2000 were steam–electric power (48%), industry–mining (11%), irrigation–livestock (30%), and domestic–commercial (11%). In contrast, Canada uses only 12% of its withdrawn water for irrigation and 70% for industry. Figure 9.21 compares regions by their use of withdrawn water during 1998. It graphically illustrates the differences between more-developed and less-developed parts of the world. (For studies of water use in the United States, see http://water.usgs.gov/public/watuse/.)

Desalination

Desalination of seawater to augment diminishing groundwater supplies is becoming increasingly important as a freshwater source. Plant construction is expected to increase 140% between 2007 and 2017, some $60 billion in investments; somewhere between 1500 and 2000 plants are now in operation worldwide. Australia has $4.9 billion in construction underway. In the Middle East, desalination is presently in widespread use. Desalination is an alternative to further groundwater mining and saltwater intrusion under Saudi Arabia (Figure 9.22a). Also, along the coast of southern California and in Florida desalination is slowly increasing in use.

The Tampa–St. Petersburg area in Florida took a giant step forward in 2003 with the opening of its own Tampa Bay desalination plant producing 25 million gallons a day (MGD), making it the largest in the United States. The seawater intake is from the Big Bend Power Station's cooling water discharge. The plant produces drinking water through a series of filters to remove algae

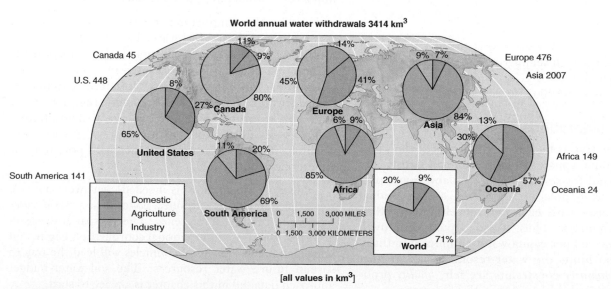

[all values in km³]

FIGURE 9.21 Water withdrawal by sector.
The world annual water withdrawal estimated total is 3414 km³ (819.4 mi³). Compare industrial water use among the geographic areas, as well as agricultural and municipal uses. [After World Resources Institute, *World Resources 2000–2001*, Data Table FW.1, pp. 276–77.]

(a)

(b)

FIGURE 9.22 Water desalination.
(a) Freshwater is supplied to Saudi Arabia from the Jabal water desalination plant along the Red Sea. Saudi Arabia obtains a significant amount of its water from desalination of seawater. (b) The Tampa Bay desalination plant provides 25 MGD to the region's water resources through filtration and reverse osmosis processes. [Photos by (a) Liaison Agency; (b) Tampa Bay Water.]

and particles and a reverse osmosis process where the water is forced through semipermeable membranes. The plant is to expand to 35 MGD, or about 10% of the regional water supply, by 2008 (Figure 9.22b).

Future Considerations

When precipitation is budgeted, the limits of the water resource become apparent. How can we satisfy the growing demand for water? Water availability per person declines as population increases, and individual demand increases with economic development, affluence, and technology. Thus, world population growth since 1970 reduced per capita water supplies by a third. Also, pollution limits the water-resource base, so that even before *quantity* constraints are felt, *quality* problems may limit the health and growth of a region.

The international aspect of the water-resource problem is illustrated by the 200 major river drainage basins in the world. One hundred and forty-five countries possess territory within an international river drainage basin—truly a global commons.

Working Group II, *Climate Change Impacts, Adaptation, and Vulnerability*, in their "Summary for Policy Makers" (Intergovernmental Panel on Climate Change, *Fourth Assessment Report*, April 2007) concluded that global climate change and greenhouse warming will cause added water problems:

> By mid-century, annual river runoff and water availability are projected to increase 10–40% at high latitudes and in some wet tropical areas, and decrease by 10–30% over some dry regions at mid-latitude and in dry tropics, some of which are presently water stressed areas. . . . Drought-affected areas will likely increase in extent. . . . In the course of the century, water supplies stored in glaciers and snow cover are projected to decline, reducing water availability in regions supplied by meltwater from mountain ranges, where more than one-sixth of the world population currently lives.

The IPCC concluded that given higher temperatures, water demand increases to meet potential evapotranspiration needs and algal blooms arise in the warmer runoff waters. Those areas that experience increased intensity of heavy precipitation events will experience surface and groundwater contamination episodes. Also, where groundwater mining occurs to meet rising demands, the drawdown of the water table and saltwater intrusion near coastlines will worsen.

Many ongoing studies of available water resources, withdrawal and consumption patterns, essential water needs of natural systems that should not be available for exploitation, population growth, and national income are revealing regional scarcities and water stress in Africa, the Middle East, Asia, southeastern Australia, Peru, Mexico, and the American Southwest and Great Plains. An important consideration to remember is this:

> Unlike other important commodities such as oil, copper, or wheat, freshwater has no substitutes for most of its uses. It is also impractical to transport the large quantities of water needed in agriculture and industry more than several hundred kilometers. Freshwater is now scarce in many regions of the world, resulting in severe ecological degradation, limits on agriculture and industrial production, threats to human health, and increased potential for international conflict.*

Clearly, cooperation is needed, yet we continue toward a water crisis without a concept of a *world water economy* as a frame of reference. The geospatial question is: When will more international coordination begin, and which country or group of countries will lead the way to sustain future water resources? The soil-water-budget approach detailed in this chapter is a place to start.

*S. L. Postal et al., "Human appropriation of renewable fresh water," *Science* 271, no. 5250 (February 9, 1996): 785.

Summary and Review—Water Resources

■ *Illustrate* the hydrologic cycle with a simple sketch, and *label* it with definitions for each water pathway.

The flow of water links the atmosphere, ocean, and land through energy and matter exchanges. The **hydrologic cycle** is a model of Earth's water system, which has operated for billions of years from the lower atmosphere to several kilometers beneath Earth's surface. **Interception** occurs when precipitation strikes vegetation or other ground cover. Water soaks into the subsurface through **infiltration**, or penetration of the soil surface. It further permeates soil or rock through vertical movement called **percolation**.

hydrologic cycle (p. 246)
interception (p. 247)
infiltration (p. 247)
percolation (p. 247)

1. Sketch and explain a simplified model of the complex flows of water on Earth—the hydrologic cycle.
2. What are the possible routes that a raindrop may take on its way to and into the soil surface?
3. Compare precipitation and evaporation volumes from the ocean with those over land. Describe advection flows of moisture and countering surface and subsurface runoff.

■ *Relate* the importance of the water-budget concept to your understanding of the hydrologic cycle, water resources, and soil moisture for a specific location.

A **soil-water budget** can be established for any area of Earth's surface by measuring the precipitation input and the output of various water demands in the area considered. Understanding both the supply of the water resource and the natural demands on the resource is essential to sustainable human interaction with the hydrologic cycle. Water resources represent the ultimate output of Earth's water system.

Groundwater is the largest potential freshwater source in the hydrologic cycle and is tied to surface supplies. Streams represent only a tiny fraction of all water (1250 km³, or 300 mi³), the smallest volume of any of the freshwater categories. Yet streams represent four-fifths of all the water we use.

soil-water budget (p. 248)

4. How might an understanding of the hydrologic cycle in a particular locale or a soil-moisture budget of a site assist you in assessing water resources? Give some specific examples.

■ *Construct* the water-balance equation as a way of accounting for the expenditures of water supply, and *define* each of the components in the equation and their specific operation.

The moisture supply to Earth's surface is **precipitation** (PRECIP, or P), arriving as rain, sleet, snow, and hail. Precipitation is measured with the **rain gauge**. **Evaporation** is the net movement of free water molecules away from a wet surface into air. **Transpiration** is the movement of water through plants and back into the atmosphere; it is a cooling mechanism for plants. Evaporation and transpiration are combined into one term—**evapotranspiration**. The ultimate demand for moisture is **potential evapotranspiration** (POTET, or PE), the amount of water that *would* evaporate and transpire under optimum mois-

ture conditions (adequate precipitation and adequate soil moisture). Evapotranspiration is measured with an **evaporation pan** (*evaporimeter*) or the more elaborate **lysimeter**.

Unsatisfied POTET is a **deficit** (DEFIC). By subtracting DEFIC from POTET, we determine **actual evapotranspiration**, or ACTET. Ideally, POTET and ACTET are about the same, so that plants have sufficient water. If POTET is satisfied and the soil is full of moisture, then additional water input becomes **surplus** (SURPL), which may puddle on the surface, flow across the surface toward stream channels, or percolate underground through the soil. The **overland flow** to streams includes precipitation and groundwater flows into river channels to make up the **total runoff** from the area.

A "savings account" of water that receives deposits and provides withdrawals as water-balance conditions change is the **soil-moisture storage**, ΔSTRGE. This is the volume of water stored in the soil that is accessible to plant roots. In soil, **hygroscopic water** is inaccessible because it is a molecule-thin layer that is tightly bound to each soil particle by hydrogen bonding. As available water is utilized, soil reaches the **wilting point** (all that remains is unextractable water). **Capillary water** is generally accessible to plant roots because it is held in the soil by surface tension and hydrogen bonding between water and soil. Almost all capillary water that remains in the soil is **available water** in soil-moisture storage. After water drains from the larger pore spaces, the available water remaining for plants is termed **field capacity**, or storage capacity. When soil is saturated after a precipitation event, surplus water in the soil becomes **gravitational water** and percolates to groundwater. As **soil-moisture utilization** removes soil water, the plants work harder to extract the same amount of moisture, whereas **soil-moisture recharge** is the rate at which needed moisture enters the soil.

The texture and the structure of the soil dictate available pore spaces, or **porosity**. The soil's **permeability** is the degree to which water can flow through it. Permeability depends on particle sizes and the shape and packing of soil grains.

Drought does not have a simple water-budget definition; rather, it can occur in at least four forms: *meteorological drought, agricultural drought, hydrologic drought,* and/or a *socioeconomic drought.*

precipitation (p. 248)
rain gauge (p. 250)
evaporation (p. 251)
transpiration (p. 251)
evapotranspiration (p. 251)
potential evapotranspiration (p. 251)
evaporation pan (p. 252)
lysimeter (p. 252)
deficit (p. 253)
actual evapotranspiration (p. 253)
surplus (p. 253)
overland flow (p. 253)
total runoff (p. 253)
soil-moisture storage (p. 253)
hygroscopic water (p. 253)
wilting point (p. 253)
capillary water (p. 253)
available water (p. 253)

field capacity (p. 253)
gravitational water (p. 253)
soil-moisture utilization (p. 254)
soil-moisture recharge (p. 254)
porosity (p. 254)
permeability (p. 254)
drought (p. 254)

5. What does this statement mean? "The soil-water budget is an assessment of the hydrologic cycle at a specific site."

6. What are the components of the water-balance equation? Construct the equation and place each term's definition below its abbreviation in the equation.

7. Using the annual water-balance data for Kingsport, Tennessee, in Table 9.2, work the values through the water-balance "bookkeeping" method. Does the equation balance?

8. Explain how to derive actual evapotranspiration (ACTET) in the water-balance equation.

9. What is potential evapotranspiration (POTET)? How do we go about estimating this potential rate? What factors did Thornthwaite use to determine this value?

10. Explain the operation of soil-moisture storage, soil-moisture utilization, and soil-moisture recharge. Include discussion of the field capacity, capillary water, and wilting point concepts.

11. In the case of silt-loam soil from Figure 9.9, roughly what is the available water capacity? How is this value derived?

12. Describe the four types of drought. Why do you think media coverage of drought seems lacking? If wars in the twenty-first century are predicted to be about water availability in quantity and quality, what action could we take to understand the issues and avoid the conflicts?

■ *Describe* the nature of groundwater, and *define* the elements of the groundwater environment.

Groundwater is a part of the hydrologic cycle, but it lies beneath the surface beyond the soil-moisture root zone. Groundwater does not exist independently because its replenishment is tied to surface surpluses. Excess surface water moves through the **zone of aeration**, where soil and rock are less than saturated. Eventually, the water reaches the **zone of saturation**, where the pores are completely filled with water.

The *permeability* of subsurface rocks depends on whether they conduct water readily (higher permeability) or tend to obstruct its flow (lower permeability). They can even be impermeable. An **aquifer** is a rock layer that is permeable to groundwater flow in usable amounts. An **aquiclude** (aquitard) is a body of rock that does not conduct water in usable amounts.

The upper limit of the water that collects in the zone of saturation is the **water table**; it is the contact surface between the zones of saturation and aeration. A **confined aquifer** is bounded above and below by impermeable layers of rock or sediment. An **unconfined aquifer** has a permeable layer on top and an impermeable one beneath. The **aquifer recharge area** extends over an entire unconfined aquifer. Water in a confined aquifer is under the pressure of its own weight, creating a pressure level to which the water can rise on its own, the **potentiometric surface**, which can be above ground level. Groundwater confined under pressure is **artesian water**; it may rise up in wells and even flow out at the surface without pumping if the head of the well is below the potentiometric surface.

As water is pumped from a well, the surrounding water table within an unconfined aquifer will experience **drawdown**,

or become lower, if the rate of pumping exceeds the horizontal flow of water in the aquifer around the well. This excessive pumping causes a **cone of depression**. Aquifers frequently are pumped beyond their flow and recharge capacities, a condition known as **groundwater mining**.

groundwater (p. 260)
zone of aeration (p. 262)
zone of saturation (p. 262)
aquifer (p. 262)
aquiclude (p. 262)
water table (p. 262)
confined aquifer (p. 262)
unconfined aquifer (p. 262)
aquifer recharge area (p. 262)
potentiometric surface (p. 262)
artesian water (p. 262)
drawdown (p. 263)
cone of depression (p. 263)
groundwater mining (p. 264)

13. Are groundwater resources independent of surface supplies, or are the two interrelated? Explain your answer.

14. Make a simple sketch of the subsurface environment, labeling zones of aeration and saturation and the water table in an unconfined aquifer. Then add a confined aquifer to the sketch.

15. At what point does groundwater utilization become groundwater mining? Use the High Plains aquifer example to explain your answer.

16. What is the nature of groundwater pollution? Can contaminated groundwater be cleaned up easily? Explain.

■ *Identify* critical aspects of freshwater supplies for the future, and *cite* specific issues related to sectors of use, regions and countries, and potential remedies for any shortfalls.

Nonconsumptive uses, or water **withdrawal**, or *offstream use*, remove water from the supply, use it, and then return it to the stream. **Consumptive uses** remove water from a stream but do not return it, so the water is not available for a second or third use. Americans in the 48 contiguous states withdraw approximately one-third of the available surplus runoff for irrigation, industry, and municipal uses. Regional and global water-resource planning, using water-budget principles, is essential if Earth's societies are to have enough water of adequate quality. **Desalination** as a water resource involves the removal of organics, debris, and salinity from seawater through distillation or reverse osmosis. This processing yields potable water for domestic uses.

withdrawal (p. 270)
consumptive uses (p. 271)
desalination (p. 271)

17. Describe the principal pathways involved in the water budget of the contiguous 48 states. What is the difference between withdrawal and consumptive use of water resources? Compare these with instream uses.

18. Characterize each of the sectors withdrawing water: irrigation, industry, and municipalities. What are the present usage trends in more-developed and less-developed nations?

19. Briefly assess the status of world water resources. What challenges exist in meeting the future needs of an expanding population and growing economies?

NetWork

The *Geosystems* Student Learning Center provides online resources for this chapter on the World Wide Web. To begin: Once at the Center, click on the cover of this textbook, scroll the Table of Contents menu, and select this chapter. You will find self-tests that are graded, review exercises, specific updates for items in the chapter, and in "Destinations" many links to interesting related pathways on the Internet. *Geosystems* Student Learning Center is found at **http://www.prenhall.com/christopherson/**.

Critical Thinking

A. Select your campus, yard, or perhaps a house plant and apply the water-balance concepts. What is the supply of water? Estimate the ultimate water supply and demand for the area you selected. Estimate water needs and how they vary with season as components of the water-budget change.

B. You have no doubt had several glasses of water between the time you awoke this morning and the time you are reading these words. Where did this water originate? Obtain the name of the water company or agency, determine whether it is using surface or groundwater to meet demands, and check how the water is metered and billed. If your state or province requires water-quality reporting, obtain a copy of the analysis of your tap water.

C. What about the water on your campus? If the campus has its own wells, how is the quality tested? Who on campus is in charge of supervising these wells? Lastly, what is your subjective assessment of the water you use: taste, smell, hardness, clarity? Compare these perceptions with others in your class.

Palm–oak forest hammock ("shady place"), featuring a mixed hardwood forest. There are bromeliads (epiphytes) growing on the live oak trees and a few on the cabbage, or sabal, palms. This palm is Florida's state tree. A hammock is slightly higher ground, surrounded by wet prairies, mixed swamp, and cypress forest marshes. Such "islands" of vegetation result from having few fires, protected as they are by surrounding marshes. These two photos are in Myakka River State Park, Sarasota County, about 15 miles inland from the central Gulf Coast. [Overhead and forest-floor photos by Bobbé Christopherson.]

10

Global Climate Systems

■ Key Learning Concepts_____

After reading the chapter, you should be able to:

- ■ *Define* climate and climatology, and *explain* the difference between climate and weather.

- ■ *Review* the role of temperature, precipitation, air pressure, and air mass patterns used to establish climatic regions.

- ■ *Review* the development of climate classification systems, and *compare* genetic and empirical systems as ways of classifying climate.

- ■ *Describe* the principal climate classification categories other than deserts, and *locate* these regions on a world map.

- ■ *Explain* the precipitation and moisture efficiency criteria used to determine the arid and semiarid climates, and *locate* them on a world map.

- ■ *Outline* future climate patterns from forecasts presented, and *explain* the causes and potential consequences of climate change.

Earth experiences an almost infinite variety of *weather*—conditions of the atmosphere—at any given time and place. But if we consider the weather over many years, including its variability and extremes, a pattern emerges that constitutes **climate**. Think of climatic patterns as dynamic rather than static, owing to the fact that we are witnessing climate change. Climate is more than a consideration of simple averages of temperature and precipitation.

Today, climatologists know that intriguing global-scale linkages exist in the Earth–atmosphere–ocean system. For instance, strong monsoonal rains in West Africa are correlated with the development of intense Atlantic hurricanes; or, one year an El Niño in the Pacific is tied to rains in the American West, floods in Louisiana and Northern Europe, and a weak Atlantic hurricane season. Yet, the persistent La Niña in 2007 strengthened drought's six-year hold on the West. The El Niño/La Niña phenomenon is the subject of Focus Study 10.1.

Climatologists, among other scientists, are analyzing global climate change—record-breaking global average temperatures, glacial ice melt, drying soil-moisture conditions, changing crop yields, spreading of infectious disease, changing distributions of plants and animals, declining coral reef health and fisheries, and the thawing of high-latitude lands and seas. Climatologists are concerned about observed changes occurring in the global climate, as these are at a pace not evidenced in the records of the past millennia. Climate and natural vegetation shifts during the next 50 years could exceed the total of all changes since the peak of the last ice-age episode, some 18,000 years ago. In Chapter 1, *Geosystems* began with these words from scientist Jack Williams,

> By the end of the 21st century, large portions of the Earth's surface may experience climates not found at present, and some 20th-century climates may disappear. . . . Novel climates are projected to develop primarily in the tropics and subtropics. . . . Disappearing climates increase the likelihood of species extinctions and community disruption for species endemic to particular climatic regimes, with the largest impacts projected for poleward and tropical montane regions.*

We need to realize that the climate map and climate designations we study in this chapter are not fixed but are on the move as temperature and precipitation relationships alter.

In this chapter: Climates are so diverse that no two places on Earth's surface experience exactly the same climatic conditions; in fact, Earth is a vast collection of microclimates. However, broad similarities among local climates permit their grouping into climatic regions.

Many of the physical elements of the environment, studied in the first nine chapters of this text, link together to explain climates. Here we survey the patterns of climate using a series of sample cities and towns. *Geosystems* uses a simplified classification system based on physical factors that help uncover the "why" question—why climates are in certain locations. Though imperfect, this method is easily understood and is based on a widely used classification system devised by climatologist Wladimir Köppen (pronounced KUR-pen). For reference, Appendix B details the Köppen climate classification system and all its criteria.

Climatologists use powerful computer models to simulate changing complex interactions in the atmosphere, hydrosphere, lithosphere, and biosphere. This chapter concludes with a discussion of climate change and its vital implications for society. Climate patterns are changing at an unprecedented rate. Especially significant are changes occurring in the polar regions of the Arctic and Antarctic.

Earth's Climate System and Its Classification

Climatology, the study of climate and its variability, analyzes long-term weather patterns over time and space and the controls that produce Earth's diverse climatic conditions. One type of climatic analysis locates areas of similar weather statistics and groups them into **climatic regions**. Observed patterns grouped into regions are at the core of climate classification.

The climate where you live may be humid with distinct seasons, or dry with consistent warmth, or moist and cool—almost any combination is possible. There are places where it rains more than 20 cm (8 in.) each month, with monthly average temperatures remaining above 27°C (80°F) year-round. Other places may be rainless for a decade at a time. A climate may have temperatures that average above freezing every month yet still threaten severe frost problems for agriculture. Students reading *Geosystems* in Singapore experience precipitation every month, ranging from 13.1 to 30.6 cm (5.1 to 12.0 in.), or 228.1 cm (89.8 in.) during an average year, whereas students at the university in Karachi, Pakistan, measure only 20.4 cm (8 in.) of rain over an entire year.

Climates greatly influence *ecosystems*, the natural, self-regulating communities formed by plants and animals in their nonliving environment. On land, the basic climatic regions determine to a large extent the location of the world's major ecosystems. These regions, called *biomes*, include forest, grassland, savanna, tundra, and desert. Plant, soil, and animal communities are associated with these biomes. Because climate cycles through periodic change, it is never really stable; therefore, ecosystems should be thought of as being in a constant state of adaptation and response.

The present global climatic warming trend is producing changes in plant and animal distributions. Figure 10.1 presents a schematic view of Earth's climate system, showing both internal and external processes and linkages that influence climate and thus regulate such changes.

Climate Components: Insolation, Temperature, Pressure, Air Masses, and Precipitation

The principal elements of climate are insolation, temperature, pressure, air masses, and precipitation. The first nine chapters discussed each of these elements. We review them briefly here. Insolation is the energy input for the

---

*Jack Williams et al., "Projected distributions of novel and disappearing climates by A.D. 2100," *Proceedings of the National Academy of Sciences* (April 3, 2007): 5739.

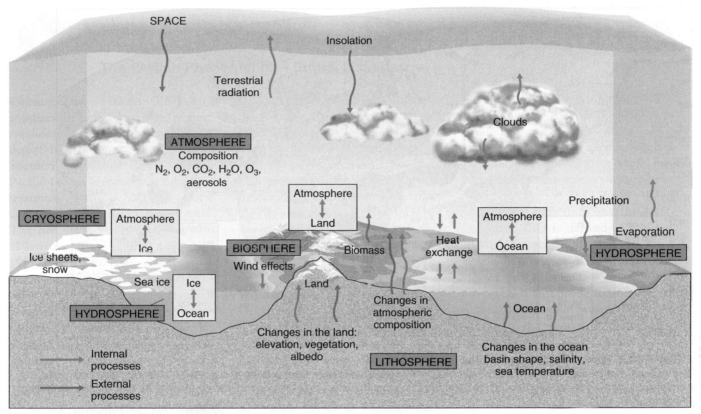

FIGURE 10.1 A schematic of Earth's climate system.
Imagine you are hired to write a computer program that simulates Earth's climates. *Internal processes* that influence climate involve the atmosphere, hydrosphere (streams and oceans), cryosphere (polar ice masses and glaciers), biosphere, and lithosphere (land)—all energized by insolation. *External processes*, principally from human activity, affect this climatic balance and force climate change. [After J. Houghton, *The Global Climate* (Cambridge, UK: Cambridge University Press, 1984); and the Global Atmospheric Research Program.]

climate system, but it varies widely over Earth's surface by latitude (see Chapter 2 and Figures 2.9, 2.10, and 2.11). Daylength and temperature patterns vary diurnally (daily) and seasonally. The principal controls of temperature are latitude, altitude, land-water heating differences, and cloud cover. The pattern of world temperatures and their annual ranges are in Chapter 5 (see Figures 5.14, 5.15, 5.17, and 5.18).

Temperature variations result from a coupling of dynamic forces in the atmosphere to Earth's pattern of atmospheric pressure and resulting global wind systems (see Figures 6.10 and 6.12). Important, too, are the location and physical characteristics of air masses, those vast bodies of homogeneous air that form over oceanic and continental source regions.

Moisture is the remaining input to climate. The hydrologic cycle transfers moisture, with its tremendous latent heat energy, through Earth's climate system (see Figure 9.1). The moisture input to climate is precipitation in all its forms. Figure 10.2 shows the worldwide distribution of precipitation, our moisture supply. Its patterns are important, for it is a key climate control factor. Average temperatures and daylength help us approximate POTET (potential evapotranspiration), a measure of natural moisture demand.

Most of Earth's desert regions, areas of permanent water deficit, are in lands dominated by subtropical high-pressure cells, with bordering lands grading to grasslands and to forests as precipitation increases. The most consistently wet climates on Earth straddle the equator in the Amazon region of South America, the Congo region of Africa, and Indonesia and Southeast Asia, all of which are influenced by equatorial low pressure and the intertropical convergence zone (ITCZ, see Figure 6.11).

Simply relating the two principal climatic components—temperature and precipitation—reveals general climate types (Figure 10.3). Temperature and precipitation patterns, plus other weather factors, provide the key to climate classification.

Classification of Climatic Regions

The ancient Greeks simplified their view of world climates into three zones: The "torrid zone" referred to warmer areas south of the Mediterranean; the "frigid zone" was to the north; and the area where they lived was labeled the "temperate zone," which they considered the optimum climate. They believed that travel too close to the equator or too far north would surely end in death. But the world is a diverse place and Earth's myriad climatic variations are more complex than these simple views.

News Report 10.1

What's in a Boundary?

The boundary between mesothermal and microthermal climates is sometimes placed along the isotherm where the coldest month is −3°C (26.6°F) or lower. That might be an accurate criterion for Europe, but for conditions in North America, the 0°C (32°F) isotherm is considered more appropriate. The difference between the 0 and −3°C isotherms covers an area about the width of the state of Ohio. Remember, these isotherm lines are really transition zones and do not mean abrupt change from one temperature to another.

A line denoting at least one month below freezing runs from New York City roughly along the Ohio River, trending westward until it meets the dry climates in the southeastern corner of Colorado. In Figure 10.5 you see this boundary used. A line marking −3°C as the coldest month would run farther north along Lake Erie and the southern tip of Lake Michigan. In addition, remember that from year to year, the position of the 0°C isotherm for January can shift several hundred kilometers as weather conditions vary.

Climate change adds another dimension to this question of accurate placement of statistical boundaries—

for they are shifting. The Inter-governmental Panel on Climate Change (IPCC, discussed later in this chapter) predicts a 150- to 550-km (90- to 350-mi) range of possible poleward shift of climatic patterns in the midlatitudes during this century. Such change in climate would place Ohio, Indiana, and Illinois within climate regimes now experienced in Arkansas and Oklahoma. As you examine North America in Figure 10.5, use the graphic scale to get an idea of the magnitude of these potential shifts. Boundaries are indeed dynamic.

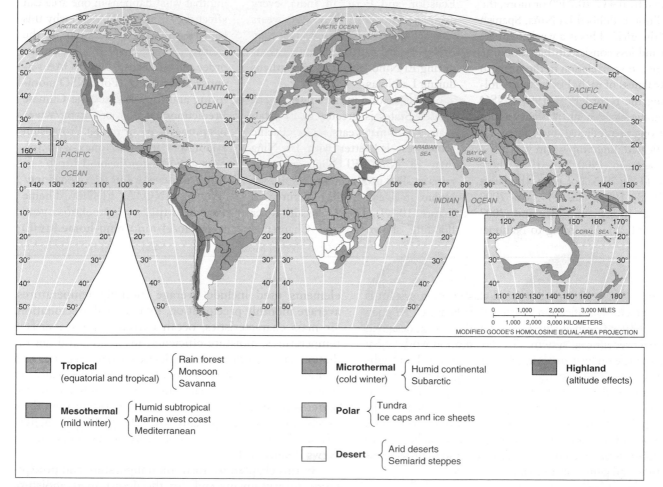

FIGURE 10.4 Climate regions generalized.
Six general climate categories, keyed to the legend and coloration in the Figure 10.5 map. Relative to the question asked about your campus and birthplace in the caption to Figure 10.3, locate these two places on this climate map.

Global Climate Maps, World Map References

and their regional types that provide us with a structure for our discussion in this chapter.

- Tropical (equatorial and tropical latitudes)
    rain forest (rainy all year)
    monsoon (6 to 12 months rainy)
    savanna (less than 6 months rainy)
- Mesothermal (midlatitudes, mild winter)
    humid subtropical (hot summers)
    marine west coast (warm to cool summers)
    Mediterranean (dry summers)
- Microthermal (mid and high latitudes, cold winter)
    humid continental (hot to warm summers)
    subarctic (cool summers to very cold winters)
- Polar (high latitudes and polar regions)
    tundra (high latitude or high altitude)
    ice caps and ice sheets (perpetually frozen)
    polar marine
- Highland (compared to lowlands at the same latitude, highlands have lower temperatures—recall the normal lapse rate)

Only one climate category is based on moisture efficiency as well as temperature:

- Desert (permanent moisture deficits)
    arid deserts (tropical, subtropical hot and midlatitude cold)

semiarid steppes (tropical, subtropical hot and midlatitude cold)

Global Climate Patterns Adding detail to Figure 10.4, we develop the world climate map presented in Figure 10.5. The following sections describe specific climates, organized around each of the main climate categories listed previously. An opening box at the beginning of each climate section gives a simple description of the climate category and causal elements that are in operation. A world map showing distribution and the featured representative cities also is in the introductory box for each climate. The names of the climates appear in italics in the chapter.

Climographs exemplify particular climates for selected cities. A **climograph** is a graph that shows monthly temperature and precipitation, location coordinates, average annual temperature, total annual precipitation, elevation, the local population, annual temperature range, annual hours of sunshine (if available, as an indication of cloudiness), and a location map. Along the top of each climograph are the dominant weather features that are influential in that climate.

Discussions of soils, vegetation, and major terrestrial biomes that fully integrate these global climate patterns are in Part IV. Table 20.1 synthesizes all this information and enhances your understanding of this chapter, so please place a tab on that page and refer to it as you read.

Tropical Climates (equatorial and tropical latitudes)

Tropical climates occupy about 36% of Earth's surface, including both ocean and land areas—Earth's most extensive climate category. The tropical climates straddle the equator from about 20° N to 20° S, roughly between the Tropics of Cancer and Capricorn, thus the designation *tropical*. Tropical climates stretch northward to the tip of Florida and south-central Mexico, central India, and Southeast Asia. These climates are truly winterless. Consistent daylength and almost perpendicular Sun angle throughout the year generate this warmth. Important causal elements include:

- Consistent daylength and insolation input, which produce consistently warm temperatures;
- Intertropical convergence zone (ITCZ), which brings rain as it shifts seasonally with the high Sun;
- Warm ocean temperatures and unstable maritime air masses.

Tropical climates have three distinct regimes: *tropical rain forest* (ITCZ present all year), *tropical monsoon* (ITCZ present for

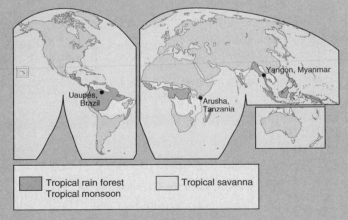

6 to 12 months), and *tropical savanna* (ITCZ present for less than 6 months).

Tropical Rain Forest Climates

The *tropical rain forest* climate is constantly moist and warm. Convectional thunderstorms, triggered by local heating and trade-wind convergence, peak each day from

mid-afternoon to late evening inland and earlier in the day where marine influence is strong along coastlines. Precipitation follows the migrating intertropical convergence zone (ITCZ, Chapter 6). The ITCZ shifts northward and southward with the summer Sun throughout the

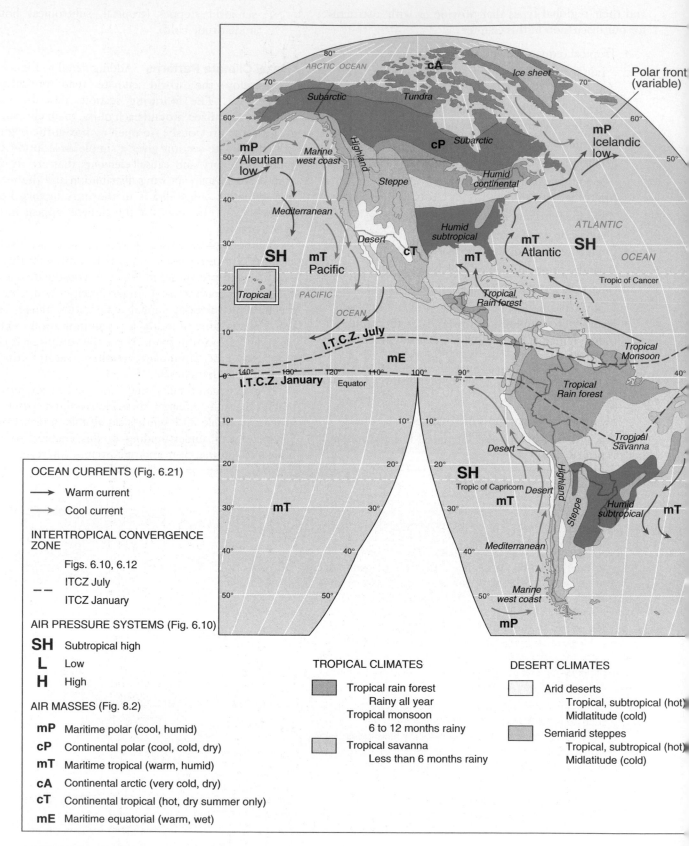

FIGURE 10.5 World climate classification.
Annotated on this map are selected air masses, near-shore ocean currents, pressure systems, and the January and July locations of the ITCZ. Use the colors in the legend to locate various climate types; some labels of the climate names appear in italics on the map to guide you.

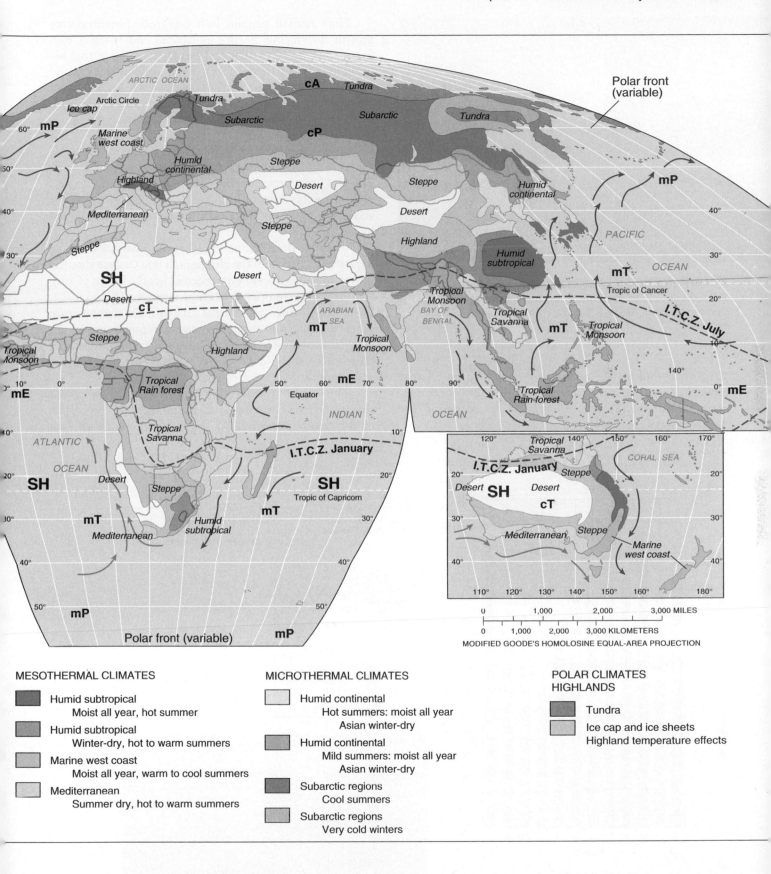

MESOTHERMAL CLIMATES

Humid subtropical
Moist all year, hot summer

Humid subtropical
Winter-dry, hot to warm summers

Marine west coast
Moist all year, warm to cool summers

Mediterranean
Summer dry, hot to warm summers

MICROTHERMAL CLIMATES

Humid continental
Hot summers: moist all year
Asian winter-dry

Humid continental
Mild summers: moist all year
Asian winter-dry

Subarctic regions
Cool summers

Subarctic regions
Very cold winters

POLAR CLIMATES
HIGHLANDS

Tundra

Ice cap and ice sheets
Highland temperature effects

FIGURE 10.6 The tropical rain forest.
The lush equatorial rain forest and Sangha River, near Ouesso, Congo, Africa. [Photo by BIOS M. Gunther/Peter Arnold, Inc.]

High rainfall sustains lush evergreen broadleaf tree growth, producing Earth's equatorial and tropical rain forests (Figure 10.6). The leaf canopy is so dense that little light diffuses to the forest floor, leaving the ground surface dim and sparse in plant cover. Dense surface vegetation occurs along riverbanks, where light is abundant. (Widespread deforestation of Earth's rain forest is detailed in Chapter 20.)

High temperature promotes energetic bacterial action in the soil so that organic material is quickly consumed. Heavy precipitation washes away certain minerals and nutrients. The resulting soils are somewhat sterile and can support intensive agriculture only if supplemented by fertilizer.

Uaupés, Brazil (Figure 10.7), is characteristic of *tropical rain forest*. On the climograph you can see that the lowest-precipitation month receives nearly 15 cm (6 in.), and the annual temperature range is barely 2 C° (3.6 F°). In all such climates, the diurnal (day-to-night) temperature range exceeds the annual average minimum–maximum (coolest to warmest) range: Day–night temperatures can range more than 11 C° (20 F°), more than 5 times the annual monthly average range.

The only interruption of *tropical rain forest* climates across the equatorial region is in the highlands of the South American Andes and in East Africa (see

year, but it influences *tropical rain forest* regions all year long. Not surprisingly, water surpluses are enormous, creating the world's greatest stream discharges in the Amazon and Congo rivers.

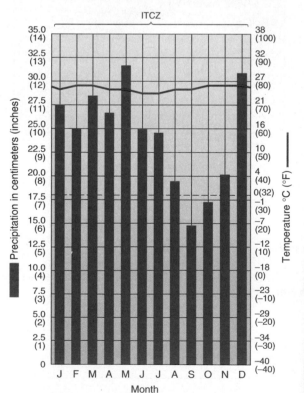

Station: Uaupés, Brazil   **Elevation:** 86 m (282.2 ft)
Lat/long: 0°08' S 67°05' W   **Population:** 10,000
Avg. Ann. Temp.: 25°C (77°F) **Ann. Temp. Range:**
Total Ann. Precip.:   2 C° (3.6 F°)
   291.7 cm (114.8 in.) **Ann. Hr of Sunshine:** 2018

FIGURE 10.7 Tropical rain forest climate.
(a) Climograph for Uaupés, Brazil (*tropical rain forest*).
(b) The rain forest near Uaupés along a tributary of the Rio Negro. [Photo by Will and Deni McIntyre/Photo Researchers, Inc.]

Figure 10.5). There, higher elevations produce lower temperatures; Mount Kilimanjaro is less than 4° south of the equator, but at 5895 m (19,340 ft) it has permanent glacial ice on its summit (although this ice is shrinking due to higher air temperatures). Such mountainous sites fall within the highland climate category.

Tropical Monsoon Climates

The *tropical monsoon* climates feature a dry season that lasts 1 or more months. Rainfall brought by the ITCZ falls in these areas from 6 to 12 months of the year. (Remember, the ITCZ affects the *tropical rain forest* climate region throughout the year.) The dry season occurs when the convergence zone is not overhead. Yangon, Myanmar (formerly Rangoon, Burma), is an example of

this climate type, as illustrated by the climograph and photograph in Figure 10.8. Mountains prevent cold air masses from central Asia getting into Yangon, resulting in its high average annual temperatures.

About 480 km (300 mi) north in another coastal city, Sittwe (Akyab), Myanmar, on the bay of Bengal, annual precipitation rises to 515 cm (203 in.) compared to Yangon's 269 cm (106 in.). Therefore, Yangon is a drier tropical monsoon area than farther north along the coast but still receives more than the 250-cm criteria.

Tropical monsoon climates lie principally along coastal areas within the tropical rain forest climatic realm and experience seasonal variation of wind and precipitation. Evergreen trees grade into thorn forests on the drier margins near the adjoining savanna climates.

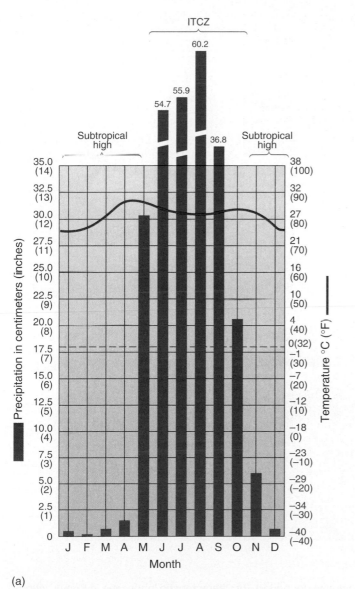

Station: Yangon, Myanmar*
Lat/long: 16°47' N 96°10' E
Avg. Ann. Temp.: 27.3°C (81.1°F)
Total Ann. Precip.:
    268.8 cm (105.8 in.)

*(Formerly Rangoon, Burma)

Elevation: 23 m (76 ft)
Population: 6,000,000
Ann. Temp. Range:
    5.5 C° (9.9 F°)

FIGURE 10.8 Tropical monsoon climate.
(a) Climograph for Yangon, Myanmar (formerly Rangoon, Burma) (*tropical monsoon*); city of Sittwe also noted on map. (b) The monsoonal forest near Malang, Java, at the Purwodadi Botanical Gardens.
[Photo by Tom McHugh/Photo Researchers, Inc.]

Tropical Savanna Climates

Tropical savanna climates exist poleward of the *tropical rain forest* climates. The ITCZ reaches these climate regions for about six months or less of the year as it migrates with the summer Sun. Summers are wetter than winters because convectional rains accompany the shifting ITCZ when it is overhead. This produces a notable dry condition when the ITCZ is farthest away and high pressure dominates. Thus, POTET (natural moisture demand) exceeds PRECIP (natural moisture supply) in winter, causing water-budget deficits.

Temperatures vary more in tropical savanna climates than in tropical rain forest regions. The tropical savanna regime can have two temperature maximums during the year because the Sun's direct rays are overhead twice—before and after the summer solstice in each hemisphere as the Sun moves between the equator and the tropic. Dominant grasslands with scattered trees, drought resistant to cope with the highly variable precipitation, characterize the *tropical savanna* regions.

Arusha, Tanzania, is a characteristic *tropical savanna* city (Figure 10.9). This metropolitan area of more

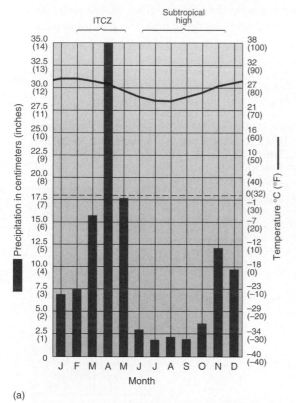

Station: Arusha, Tanzania
Lat/long: 3°24' S 36°42' E
Avg. Ann. Temp.: 26.5°C (79.7°F)
Total Ann. Precip.: 119 cm (46.9 in.)

Elevation: 1387 m (4550 ft)
Population: 1,368,000
Ann. Temp. Range: 4.1 C° (7.4 F°)
Ann. Hr of Sunshine: 2600

(a)

(b)

(c)

FIGURE 10.9 Tropical savanna climate.
(a) Climograph for Arusha, Tanzania (*tropical savanna*); note the intense dry period. (b) Characteristic landscape in Kenya, with plants adapted to seasonally dry water budgets. (c) Extreme northern Australia on the Cape York Peninsula; the tropical savanna features eucalyptus trees, grasses, and conical termite mounds, in an open savanna in Mungkan Kandju National Park. [Photos by (b) Stephen J. Krasemann/DRK Photo; (c) B. G. Thomson, courtesy of The Wilderness Society.]

than 1,368,000 people is east of the famous Serengeti Plains savanna and Olduvai Gorge, site of human origins, and north of Tarangire National Park. Temperatures are consistent with tropical climates, despite the elevation (1387 m) of the station. Note the marked

dryness from June to October, which defines changing dominant pressure systems rather than annual changes in temperature. This region is near the transition to the dryer desert hot-steppe climates to the northeast.

Mesothermal Climates (midlatitudes, mild winters)

Mesothermal, meaning "middle temperature," describes these warm and temperate climates where true seasonality begins and seasonal contrasts in vegetation, soil, and human lifestyle adaptations are evident. Mesothermal climates occupy the second-largest percentage of Earth's land and sea surface—more than half of Earth's oceans and about one-third of its land area. Approximately 55% of the world's population resides in these climates.

The mesothermal climates, and nearby portions of the microthermal (cold winter) climates, are regions of great weather variability, for these are the latitudes of greatest air mass interaction. Causal elements include:

- Shifting air masses of maritime and continental origin are guided by upper-air westerly winds and undulating Rossby waves and jet streams.
- Migrating cyclonic (low-pressure) and anticyclonic (high-pressure) systems bring changeable weather conditions and air mass conflicts.
- Sea-surface temperatures of offshore ocean currents influence air mass strength: cooler water temperatures along west coasts (weaken) and warmer water along east coasts (strengthen).
- Summers transition from hot to warm to cool as you move away from the tropics. Climates are humid, except where subtropical high pressure produces dry summer conditions.

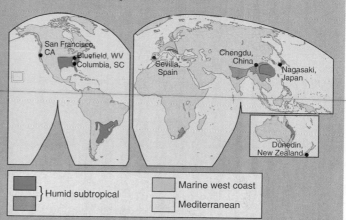

Mesothermal climates have four distinct regimes based on precipitation variability: *humid subtropical hot-summer* (moist all year), *humid subtropical winter-dry* (hot to warm summers, in Asia), *marine west coast* (warm to cool summers, moist all year), and *Mediterranean* (warm to hot summers, dry summers).

Humid Subtropical Climates

The *humid subtropical hot-summer climates* are either moist all year or have a pronounced winter-dry period, as occurs in eastern and southern Asia. Maritime tropical air masses generated over warm waters off eastern coasts influence *humid subtropical hot-summer* climates during summer. This warm, moist, unstable air produces convectional showers over land. In fall, winter, and spring, maritime tropical and continental polar air masses interact, generating frontal activity and frequent midlatitude cyclonic storms. These two mechanisms produce year-round precipitation. Overall, precipitation averages 100–200 cm (40–80 in.) a year.

Nagasaki, Japan (Figure 10.10), is characteristic of an Asian *humid subtropical hot-summer* station, whereas Columbia, South Carolina, is characteristic of the North American climate region (Figure 10.11). Unlike the precipitation of humid subtropical hot-summer cities in the United States (Atlanta, Memphis, Norfolk, New

Orleans, and Columbia), Nagasaki's winter precipitation is a bit less because of the effects of the Asian monsoon. However, the lower precipitation of winter is not quite dry enough to change its category to a winter-dry. In comparison to higher rainfall amounts in Nagasaki (196 cm, 77 in.), Columbia's precipitation totals 126.5 cm (49.8 in.); compare to Atlanta's 122 cm (48 in.) annually (Figure 10.11b).

Humid subtropical winter-dry climates are related to the winter-dry, seasonal pulse of the monsoons. They extend poleward from tropical savanna climates and have a summer month that receives 10 times more precipitation than their driest winter month. A representative station is Chengdu, China. Figure 10.12 demonstrates the strong correlation between precipitation and the high-summer Sun.

The habitability of the humid *subtropical hot-summer* and *humid subtropical winter-dry* climates and their ability to sustain populations are borne out by the concentration

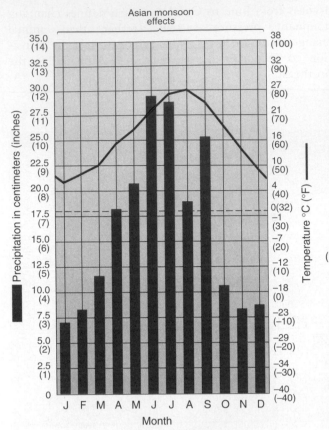

(b)

[(b) Photo by Ken Straiton/First Light.]

Station: Nagasaki, Japan
Lat/long: 32°44' N 129°52' E
Avg. Ann. Temp.: 16°C (60.8°F)
Total Ann. Precip.:
　　　　195.7 cm (77 in.)
Elevation: 27 m (88.6 ft)
Population: 1,585,000
Ann. Temp. Range: 21 C° (37.8 F°)
Ann. Hr of Sunshine: 2131

(a)

FIGURE 10.10 Humid subtropical hot-summer climate, rainy all year, Asian region.
(a) Climograph for Nagasaki, Japan (*humid subtropical*). (b) Landscape near Nagasaki.

of people in north-central India, the bulk of China's 1.3 billion people, and the many who live in climatically similar portions of the United States.

The intense summer rains of the Asian monsoon can cause problems, as they did each year between 2004 and 2007, producing floods in India and Bangladesh. Occasional dramatic thunderstorms and tornadoes are notable in the southeastern United States, with tornado occurrences seemingly breaking previous records each year.

The monsoonal winter-dry climates hold several precipitation records. Cherrapunji, India, in the Assam Hills south of the Himalayas, is the all-time precipitation record holder for a single year and for every other time interval from 15 days to 2 years. Because of the summer monsoons that pour in from the Indian Ocean and the Bay of Bengal, Cherrapunji has received 930 cm (30.5 ft) of rainfall in 1 month and 2647 cm (86.8 ft) in 1 year—both records.

Marine West Coast Climates

Marine west coast climates, featuring mild winters and cool summers, dominate Europe and other middle-to-high-latitude west coasts (see Figure 10.5). In the United States, these climates with their cooler summers are in contrast to the hot-summer humid climate of the southeastern United States.

Maritime polar air masses—cool, moist, unstable—dominate *marine west coast* climates. Weather systems forming along the polar front and maritime polar air masses move into these regions throughout the year, making weather quite unpredictable. Coastal fog, annually totaling 30 to 60 days, is a part of the moderating marine influence. Frosts are possible and tend to shorten the growing season.

Marine west coast climates are unusually mild for their latitude. They extend along the coastal margins of the Aleutian Islands in the North Pacific, cover the southern third of Iceland in the North Atlantic and coastal Scandinavia, and dominate the British Isles. It is hard to imagine that such high-latitude locations can have average monthly temperatures above freezing throughout the year.

Unlike the extensive influence of marine west coast regions in Europe, mountains restrict this climate to coastal environs in Canada, Alaska, Chile, and Australia. The temperate rain forest of Vancouver Island is representative of these moist and cool conditions (Figure 10.13). (A temperature graph for Vancouver, British Columbia, appears in Figure 5.12 with a photo of this *marine west coast* city.)

The climograph for Dunedin, New Zealand, demonstrates the moderate temperature patterns and the annual temperature range for a *marine west coast* city in the Southern Hemisphere (Figure 10.14).

Station: Columbia, South Carolina
Lat/long: 34° N 81° W
Avg. Ann. Temp.: 17.3°C (63.1°F)
Total Ann. Precip.:
126.5 cm (49.8 in.)

Elevation: 96 m (315 ft)
Population: 116,000
Ann. Temp. Range:
20.7 C° (37.3 F°)
Ann. Hr of Sunshine: 2800

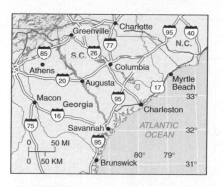

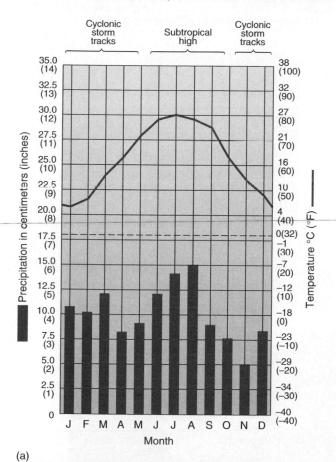

(a)

(b)

(c)

FIGURE 10.11 Humid subtropical hot-summer climate, rainy all year, American region. (a) Climograph for Columbia, South Carolina (*humid subtropical*). Note the more consistent precipitation pattern compared to Nagasaki, as Columbia receives seasonal cyclonic storm activity and summer convection showers within maritime tropical air. (b) The mixed deciduous and evergreen forest in southern Georgia, typical of the humid subtropical southeastern United States. (c) This scene is from the humid subtropical portion of Argentina along the Paraná River, northwest of **Buenos Aires.** [(b) and (c) Photos by Bobbé Christopherson.]

An interesting anomaly occurs in the eastern United States. In portions of the Appalachian highlands, increased elevation lowers summer temperatures in the surrounding *humid subtropical hot-summer* climate, producing a *marine west coast* cooler summer. The climograph for Bluefield, West Virginia (Figure 10.15), reveals marine west coast temperature and precipitation patterns, despite its location in the east. Vegetation similarities between the Appalachians and the Pacific Northwest have enticed many emigrants from the East to settle in these climatically familiar environments in the Northwest.

Mediterranean Dry-Summer Climates

Across the planet during summer months, shifting cells of subtropical high pressure block moisture-bearing winds from adjacent regions. This shifting of stable, warm to

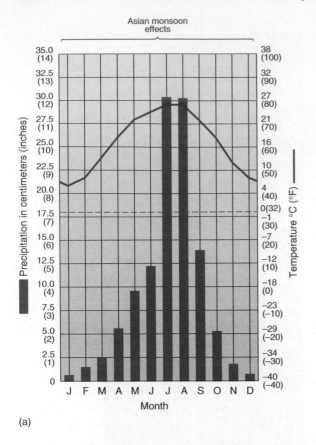

Asian monsoon effects

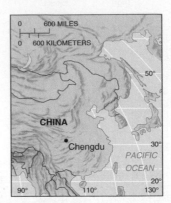

Station: Chengdu, China   **Elevation:** 498 m (1633.9 ft)
Lat/long: 30°40' N 104°04' E   **Population:** 2,500,000
Avg. Ann. Temp.: 17°C (62.6°F)   **Ann. Temp. Range:**
Total Ann. Precip.:   20 C° (36 F°)
        114.6 cm (45.1 in.)   **Ann. Hr of Sunshine:** 1058

(a)

(b)

FIGURE 10.12 Humid subtropical winter-dry climate.
(a) Climograph for Chengdu, China (*humid subtropical*).
Note the summer-wet monsoonal precipitation. (b)
Landscape of southern interior China characteristic of
this winter-dry climate. This valley is near Mount Daliang
in Sichuan Province. [Photo by Jin Zuqi/Sovfoto/Eastfoto.]

FIGURE 10.13 A marine west coast climate.
Temperate rain forest in the Macmillan Provincial Park, central
Vancouver Island, British Columbia, Canada (*marine west
coast*), features a Douglas fir forest, with western red cedar
and hemlock. [Photo by Bobbé Christopherson.]

hot, dry air over an area in summer and away from these
regions in the winter creates a pronounced dry-summer
and wet-winter pattern. For example, the continental
tropical air mass over the Sahara in Africa shifts northward in summer over the Mediterranean region and
blocks maritime air masses and cyclonic storm tracks. The
Mediterranean climate designation specifies that at least
70% of annual precipitation occurs during the winter
months. This is in contrast to the majority of the world
that experiences summer-maximum precipitation.

Worldwide, cool offshore ocean currents (the California current, Canary current, Peru current, Benguela
current, and West Australian current) produce stability in
overlying air masses along west coasts, poleward of
subtropical high pressure. The world climate map (see
Figure 10.5) shows Mediterranean dry-summer climates
along the western margins of North America, central
Chile, and the southwestern tip of Africa, as well as across
southern Australia and the Mediterranean Basin—the climate's namesake region. Examine the offshore currents
along each of these regions on the world climate map.

Figure 10.16 compares the climographs of the
Mediterranean dry-summer cities of San Francisco and

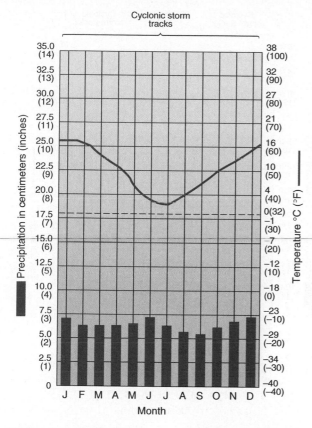

Cyclonic storm tracks

Station: Dunedin, New Zealand
Lat/long: 45°54' S 170°31' E
Avg. Ann. Temp.: 10.2°C (50.3°F)
Total Ann. Precip.:
78.7 cm (31.0 in.)

Elevation: 1.5 m (5 ft)
Population: 120,000
Ann. Temp. Range:
14.2 C° (25.5 F°)

(a)

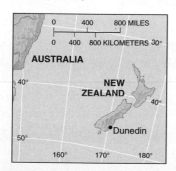

(b)

FIGURE 10.14 A Southern Hemisphere marine west coast climate.
(a) Climograph for Dunedin, New Zealand (*marine west coast*).
(b) Meadow, forest, and mountains on South Island, New Zealand. [Photo by Brian Enting/Photo Researchers, Inc.]

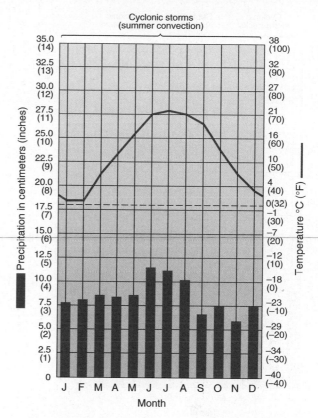

Cyclonic storms (summer convection)

Station: Bluefield, West Virginia
Lat/long: 37°16' N 81°13' W
Avg. Ann. Temp.:
12°C (53.6°F)
Total Ann. Precip.:
101.9 cm (40.1 in.)

Elevation: 780 m (2559 ft)
Population: 11,000
Ann. Temp. Range:
21 C° (37.8 F°)

(a)

(b)

FIGURE 10.15 Marine west coast climate in the Appalachians of the East.
(a) Climograph for Bluefield, West Virginia (*marine west coast*). (b) Characteristic mixed forest of Dolly Sods Wilderness in the Appalachian highlands. [Photo by David Muench Photography, Inc.]

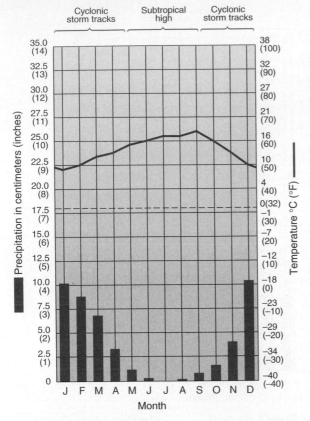

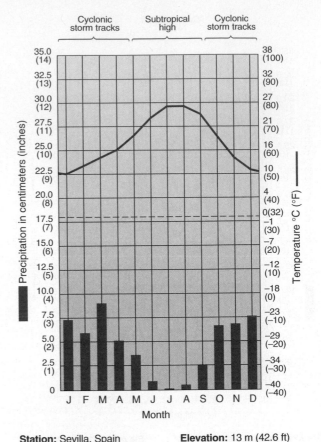

Station: San Francisco, California
Lat/long: 37°37' N 122°23' W
Avg. Ann. Temp.:
    14°C (57.2°F)
Total Ann. Precip.:
    47.5 cm (18.7 in.)

Elevation: 5 m (16.4 ft)
Population: 747,000
Ann. Temp. Range:
    9 C° (16.2 F°)
Ann. Hr of Sunshine:
    2975

(a)

Station: Sevilla, Spain
Lat/long: 37°22' N 6°00' W
Avg. Ann. Temp.:
    18°C (64.4°F)
Total Ann. Precip.:
    55.9 cm (22 in.)

Elevation: 13 m (42.6 ft)
Population: 1,764,000
Ann. Temp. Range:
    16 C° (28.8 F°)
Ann. Hr of Sunshine:
    2862

(b)

(c)

(d)

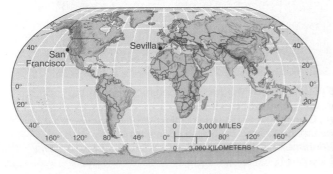

FIGURE 10.16 Mediterranean climates, California and Spain.
Climographs for (a) San Francisco, California, with its cooler dry summer and (b) Sevilla, Spain, with its hotter dry summer. (c) Central California Mediterranean landscape of oak savanna. (d) The countryside around Olvera, Andalusia, Spain. [Photos by (c) Bobbé Christopherson; (d) Kaz Chiba/Liaison Agency, Inc.]

Sevilla (Seville), Spain. Coastal maritime effects moderate San Francisco's climate, producing a cooler summer. The transition to a hot summer occurs no more than 24–32 km (15–20 mi) inland from San Francisco. The photos in Figure 10.16 show oak-savanna landscapes near Olvera, Spain, and in central California.

The *Mediterranean dry-summer* climate brings summer water-balance deficits. Winter precipitation recharges soil moisture, but water use usually exhausts soil moisture by late spring. Large-scale agriculture requires irrigation, although some subtropical fruits, nuts, and vegetables are uniquely suited to these conditions. Natural vegetation features a hard-leafed, drought-resistant variety known locally as *chaparral* in the western United States. (Chapter 20 discusses local names for this type of vegetation in other parts of the world.)

Microthermal Climates (mid- and high-latitudes, cold winters)

Humid microthermal climates have longer winters than mesothermal climates, with summer warmth. Here the term *microthermal* means cool temperate to cold. Approximately 21% of Earth's land surface is influenced by these climates, equaling about 7% of Earth's total surface. These climates occur poleward of the mesothermal climates and experience great temperature ranges related to continentality and air mass conflicts.

Temperatures decrease with increasing latitude and toward the interior of continental landmasses and result in intensely cold winters. Precipitation varies between moist-all-year regions (the northern tier across the United States and Canada, eastern Europe through the Ural Mountains) and winter-dry regions associated with the Asian monsoon.

In Figure 10.5, note the absence of microthermal climates in the Southern Hemisphere. Because the Southern Hemisphere lacks substantial landmasses, microthermal climates develop there only in highlands. Important causal elements include:

- Increasing seasonality (daylength and Sun altitude) and greater temperature ranges (daily and annually).
- Upper-air westerly winds and undulating Rossby waves, which bring warmer air northward and colder air southward for cyclonic activity, and convectional thunderstorms from mT air masses in summer.
- Asian winter-dry pattern for the microthermal climates, increasing east of the Ural Mountains to the Pacific Ocean and eastern Asia.

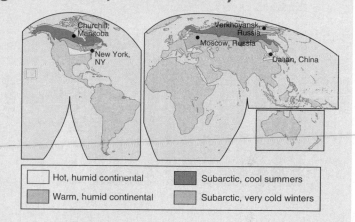

Hot, humid continental Subarctic, cool summers
Warm, humid continental Subarctic, very cold winters

- Hot summers cooling northward from the mesothermal climates, and short spring and fall seasons surrounding winters that are cold to very cold.
- Continental interiors serving as source regions for intense continental polar (cP) air masses that dominate winter, blocking cyclonic storms.

Microthermal climates have four distinct regimes based on increasing cold with latitude and precipitation variability: *humid continental hot-summer* (Chicago, New York); *humid continental mild-summer* (Duluth, Toronto, Moscow); and *subarctic* climates featuring cool summers, such as Churchill, Manitoba, and the formidable extremes of frigid, very cold winters in Verkhoyansk and northern Siberia.

Humid Continental Hot-Summer Climates

Humid continental hot-summer climates are differentiated by their annual precipitation distribution. In the summer, maritime tropical air masses influence both humid continental moist-all-year and winter-dry climates. In North America, frequent weather activity is possible between conflicting air masses—maritime tropical and continental polar—especially in winter. The climograph for New York City and photo (Figure 10.17a, c) and Dalian, China (10.17b,d), illustrate these two hot-summer microthermal climates.

Before European settlement, forests covered the *humid continental hot-summer* climatic region of the United States as far west as the Indiana–Illinois border. Beyond that approximate line, tall-grass prairies extended westward to about the 98th meridian (98° W in central Kansas) and the approximate location of the 51-cm (20-in.) isohyet (line of equal precipitation). Further west, the short-grass prairies reflected lower precipitation receipts.

The dry winter associated with the vast Asian landmass, specifically Siberia, results from a dry-winter high-pressure anticyclone. The dry monsoons of southern and eastern Asia are produced in the winter months by this system, as winds blow out of Siberia toward the Pacific and Indian oceans. The Dalian, China, climograph demonstrates this dry-winter tendency. The intruding cold of continental air is a significant winter feature.

Deep sod made farming difficult for the first settlers of the American prairies, as did the climate. However, native grasses soon were replaced with domesticated

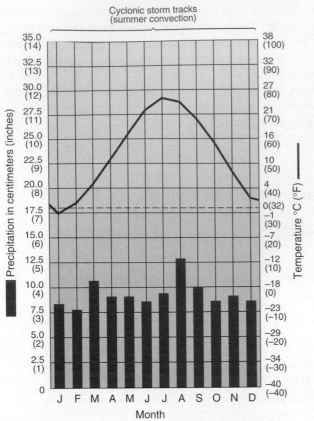

Cyclonic storm tracks
(summer convection)

Station: New York, New York
Lat/long: 40°46' N 74°01' W
Avg. Ann. Temp.:
        13°C (55.4°F)
Total Ann. Precip.:
        112.3 cm (44.2 in.)

Elevation: 16 m (52.5 ft)
Population: 8,092,000
Ann. Temp. Range:
        24 C° (43.2 F°)
Ann. Hr of Sunshine:
        2564

(a)

Asian monsoon
effects

Station: Dalian, China
Lat/long: 38°54' N 121°54' E
Avg. Ann. Temp.:
        10°C (50°F)
Total Ann. Precip.:
        57.8 cm (22.8 in.)

Elevation: 96 m (314.9 ft)
Population: 5,550,000
Ann. Temp. Range:
        29 C° (52.2 F°)
Ann. Hr of Sunshine:
        2762

(b)

(c)

(d)

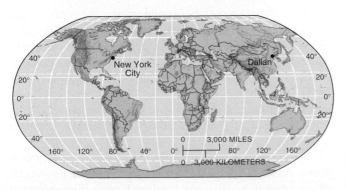

FIGURE 10.17 Humid continental hot-summer climates, New York and China.
Climographs for (a) New York City (*humid continental hot summer* that is moist all year) and (b) Dalian, China (*humid continental hot summer* that is dry in winter). (c) The "Literary Walk" with a canopy of American elms in New York City's Central Park emerging from winter just as spring and the return of leaves and warmth begin. (d) Dalian, China, cityscape and park in summer. [Photos (c) by Bobbé Christopherson; and (d) courtesy of the Paul Louis collection.]

(a)

(b)

FIGURE 10.18 Central Indiana and Buffalo, NY, landscapes.
(a) Typical humid continental deciduous forest and ready-to-harvest soybean field in central Indiana near Zelma in the American Midwest. (b) This forest near Buffalo, NY, with its characteristic mix of maple–beech–birch, with some elm and ash, is in transition as temperatures increase this century.
[Photos by Bobbé Christopherson.]

wheat and barley. Various inventions (barbed wire, the self-scouring steel plow, well-drilling techniques, windmills, railroads, and the six-shooter) aided nonnative peoples' expansion into the region. In the United States today, the *humid continental hot-summer* region is the location of corn, soybean, hog, feed crop, dairy, and cattle production (Figure 10.18a).

The region around Buffalo, New York, is in a climatic transition from milder to hotter summers. As the average July temperature increases further above 22°C (71.6°F), Buffalo's climatic designation will become more like that of New York and Chicago. The surrounding forest in the latter half of this century will shift from the present maple–beech–birch, with some elm and ash, to a dominant oak–hickory–hemlock mix (Figure 10.18b). Such shifting of climatic boundaries and their related ecosystems is occurring worldwide.

Humid Continental Mild-Summer Climates

Soils are thinner and less fertile in the cooler microthermal climates, yet agricultural activity is important and includes dairy cattle, poultry, flax, sunflowers, sugar beets, wheat, and potatoes. Frost-free periods range from fewer than 90 days in the north to as many as 225 days in the south. Overall, precipitation is less than in the hot-summer regions to the south; however, notably heavier snowfall is important to soil-moisture recharge when it melts. Various snow-capturing strategies are in use, including fences and tall stubble left standing after harvest

in fields to create snowdrifts and thus more moisture retention on the soil.

Characteristic cities are Duluth, Minnesota, and Saint Petersburg, Russia. Figure 10.19 presents a climograph for Moscow, which is at 55° N, or about the same latitude as the southern shore of Hudson Bay in Canada. The photos of landscapes near Moscow and Sebago Lake, inland from Portland, Maine, show summer and late winter scenes, respectively.

The dry-winter aspect of the mild-summer climate occurs only in Asia, in a far-eastern area poleward of the winter-dry mesothermal climates. A representative *humid continental mild-summer* climate along Russia's east coast is Vladivostok, usually one of only two ice-free ports in that country.

Subarctic Climates

Farther poleward, seasonal change becomes greater. The short growing season is more intense during long summer days. The subarctic climates include vast stretches of Alaska, Canada, northern Scandinavia with their cool summers, and Siberian Russia with its very cold winters. Discoveries of minerals and petroleum reserves and the Arctic Ocean sea-ice losses have led to new interest in portions of these regions.

Areas that receive 25 cm (10 in.) or more of precipitation a year on the northern continental margins and are covered by the so-called snow forests of fir, spruce, larch, and birch are the *boreal forests* of Canada and the *taiga* of Russia. These forests are in transition to the more open northern woodlands and to the tundra region of the far north. Forests thin out to the north when the warmest

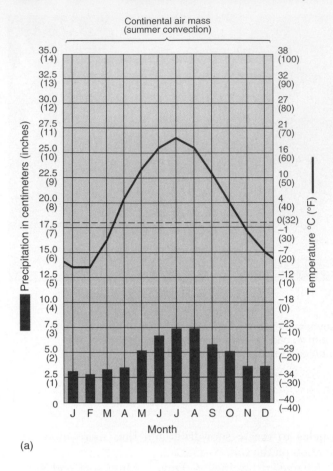

Continental air mass (summer convection)

(a)

Station: Moscow, Russia
Lat/long: 55°45′ N 37°34′ E
Avg. Ann. Temp.: 4°C (39.2°F)
Total Ann. Precip.:
57.5 cm (22.6 in.)

Elevation: 156 m (511.8 ft)
Population: 11,460,000
Ann. Temp. Range:
29 C° (52.2 F°)
Ann. Hr of Sunshine:
1597

(b)

(c)

FIGURE 10.19 Humid continental mild-summer climate.
(a) Climograph for Moscow, Russia (*humid continental mild summer*). (b) Fields near Saratov, Russia, during the short summer season. (c) Late-winter scene of Sebago Lake and forests inland from Portland, Maine. [(b) Photo by Wolfgang Kaehler/Liaison Agency, Inc.; (c) photo by Bobbé Christopherson.]

month drops below an average temperature of 10°C (50°F). During the decades ahead, the boreal forests are shifting northward into the tundra in response to climate change with higher temperatures (Figure 10.20b).

Soils are thin in these lands once scoured by glaciers. Precipitation and potential evapotranspiration both are low, so soils are generally moist and either partially or totally frozen beneath the surface, a phenomenon known as *permafrost* (discussed in Chapter 17).

The Churchill, Manitoba, climograph (Figure 10.20) shows average monthly temperatures below freezing for 7 months of the year, during which time light snow cover and frozen ground persist. High pressure dominates Churchill during its cold winter—this is the source region for the continental polar air mass. Churchill is representative of the *subarctic* climate, with a cool summer: annual temperature range of 40 C° (72 F°) and low precipitation of 44.3 cm (17.4 in.).

The *subarctic* climates that feature a dry and very cold winter occur only within Russia. The intense cold of Siberia and north-central and eastern Asia is difficult to comprehend, for these areas experience an average temperature lower than freezing for 7 months; minimum temperatures of below −68°C (−90°F) were recorded there, as described in Chapter 5. Yet summer maximum temperatures in these same areas can exceed +37°C (+98°F).

An example of this extreme subarctic climate with very cold winters is Verkhoyansk, Siberia (Figure 10.21). For 4 months of the year, average temperatures fall below −34°C (−30°F). Verkhoyansk has probably the world's greatest annual temperature range from winter to summer: a remarkable 63 C° (113.4 F°). Winters feature brittle metals and plastics, triple-thick windowpanes, and temperatures that render straight antifreeze a solid.

Continental air mass

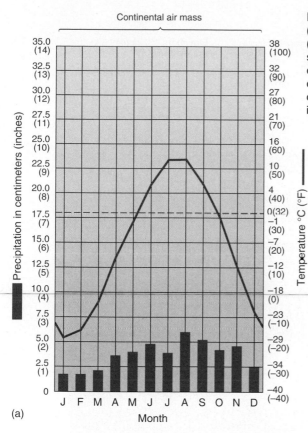

(a)

FIGURE 10.20 Subarctic cool summer climate.
(a) Climograph for Churchill, Manitoba (*subarctic*, cool summer).
(b) Winter scene on the edge of the boreal forest; young pioneer spruce trees move into the tundra. (c) Lone polar bear hunkered down in a protective dugout next to a frozen pond. (d) Mom and two cubs-of-the-year in the tundra near Hudson Bay, west of Churchill; characteristic willows in the background. (e) November street scene in Churchill. [All photos by Bobbé Christopherson.]

Station: Churchill, Manitoba
Lat/long: 58°45' N 94°04' W
Avg. Ann. Temp.: −7°C (19.4°F)
Total Ann. Precip.:
    44.3 cm (17.4 in.)

Elevation: 35 m (114.8 ft)
Population: 1400
Ann. Temp. Range:
    40 C° (72 F°)
Ann. Hr of Sunshine:
    1732

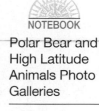

NOTEBOOK

Polar Bear and High Latitude Animals Photo Galleries

(b)

(c)

(d)

(e)

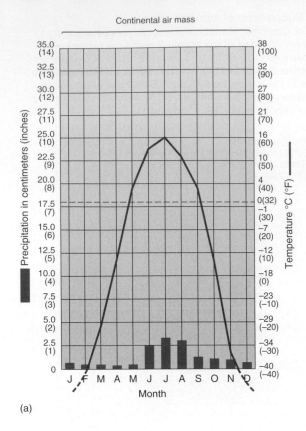

Continental air mass

(a)

Station: Verkhoyansk, Russia
Lat/long: 67°35' N 133°27' E
Avg. Ann. Temp.: −15°C (5°F)
Total Ann. Precip.:
    15.5 cm (6.1 in.)

Elevation: 137 m (449.5 ft)
Population: 1500
Ann. Temp. Range:
    63 C° (113.4 F°)

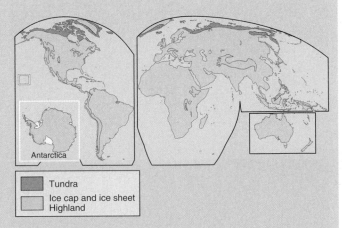

(b)

FIGURE 10.21 Extreme subarctic cold winter climate.
(a) Climograph for Verkhoyansk, Russia (*subarctic*, very cold winter). (b) Scene in the town of
Verkhoyansk during the short summer. [Photo by Dean Conger/National Geographic Society.]

Polar and Highland Climates

The polar climates cover about 19% of Earth's total surface and about 17% of its land area. These climates have no true summer like that in lower latitudes. Poleward of the Arctic and Antarctic Circles, daylength in summer becomes continuous, yet average monthly temperatures never rise above 10°C (50°F). These temperature conditions are intolerant to tree growth. Daylength, which in part determines the amount of insolation received, and low Sun altitude in the sky are the principal climatic factors in these frozen and barren regions. Yet, in winter, the Sun drops below the horizon poleward of 66.5° latitude, producing continuous night. An extended dawn and twilight period eases this seasonal shock a little and reduces the time of true night.

Principal climatic factors in these frozen and barren regions are the following:

- Extremes of daylength between winter and summer determine the amount of insolation received.
- Low Sun altitude even during the long summer days is the principal climatic factor.
- Extremely low humidity produces low precipitation amounts—these regions are Earth's frozen deserts.
- Light-colored surfaces of ice and snow reflect substantial energy away from the ground surface, thus reducing net radiation.

Polar climates have three regimes: *tundra* (high latitude, or elevation); *ice caps* and *ice sheet* (perpetually frozen); and *polar marine* (oceanic association, slight moderation of extreme cold).

Also in this climate category we include highland climates, for even at low latitudes the effects of elevation can produce tundra and polar conditions. Glaciers on tropical mountain summits attest to the cooling effects of elevation. Highland climates on the map follow the pattern of Earth's mountain ranges.

Tundra Climate

In a *tundra* climate, land is under continuous snow cover for 8–10 months, with the warmest month above 0°C yet never warming above 10°C (50°F). Because of elevation, the summit of Mount Washington in New Hampshire (1914 m, or 6280 ft) statistically qualifies as a highland *tundra* climate on a small scale.

In spring when the snow melts, numerous plants appear—stunted sedges, mosses, flowering plants, and lichens. Some of the little (7.5-cm-, 3-in.-tall) willows can exceed 300 years in age. The September photo shows the emerging fall colors of these small plants (Figure 10.22a). Much of the area experiences permafrost and ground ice conditions; these are Earth's periglacial regions (see Chapter 17).

Approximately 410,500 km² (158,475 mi²) of Greenland are ice-free, an area of tundra and rock about the size of California. The rest of Greenland is ice sheet, covering 1,756,00 km² (677,900 mi²). Despite the severe climate, a permanent population of 56,500 lives in this province of Denmark. There are only a couple of towns along Greenland's east coast. Ittoqqortoormiit, or Scoresby Sund, has 850 permanent residents (Figure 10.22b). Here is a village on the tundra.

Global warming is bringing dramatic changes to the tundra and its plants, animals, and permafrost ground conditions. In parts of Canada and Alaska, registered temperatures as much as 5 C° above average are a regular occurrence, setting many records. As organic peat deposits in the tundra thaw, vast stores of carbon are released to the atmosphere, further adding to the greenhouse gas problem. Temperatures in the Arctic are warming at a rate twice that of the global average increase.

Tundra climates are strictly in the Northern Hemisphere, except for elevated mountain locations in the Southern Hemisphere and a portion of the Antarctic Peninsula.

Ice-Cap and Ice-Sheet Climate

Most of Antarctica and central Greenland fall within the *ice-sheet* climate, as does the North Pole, with all months averaging below freezing. Both regions are dominated by dry, frigid air masses, with vast expanses that never warm above freezing. The area of the North Pole is actually a sea covered by ice, whereas Antarctica is a substantial continental landmass covered by Earth's greatest ice sheet. For comparison, winter minimums in central Antarctica (July) frequently drop below the temperature of solid carbon dioxide or "dry ice" (–78°C, or –109°F). Ice caps are smaller in extent than ice sheets, roughly less than 50,000 km² (19,300 mi²), yet they completely bury the landscape like an ice sheet. The Vatnajökull Ice Cap in southeastern Iceland is an example.

Antarctica is constantly snow-covered but receives less than 8 cm (3 in.) of precipitation each year. However, Antarctic ice has accumulated to several kilometers deep and is the largest repository of freshwater on Earth. Earth's two ice sheets cover the Antarctic continent and most of the island of Greenland. Figure 10.23 shows two scenes of these repositories of multiyear ice. The status of this ice is the focus of much attention during the present International Polar Year of scientific research.

This ice contains a vast historical record of Earth's atmosphere. Within it, evidence of thousands of past volcanic eruptions from all over the world and ancient combinations of atmospheric gases lie trapped in frozen bubbles.

(a)

(b)

East Greenland Photos
NOTEBOOK

FIGURE 10.22 Greenland tundra and a small town.
(a) Tundra is marked by an uneven, hummocky surface of mounds resulting from an active layer that freezes and thaws with the seasons, as it is here in east Greenland. Large trees are absent in the tundra; however, relatively lush vegetation for the harsh conditions includes willow, dwarf birch and shrubs, sedges, moss, lichen, and cotton grass (white tufts). (b) A town in the tundra, Scoresby Sund fjord in the distance. In the foreground, sledge dogs rest to get ready for the winter's work ahead. [Photos by Bobbé Christopherson.]

(a)

(b)

FIGURE 10.23 Earth's ice sheets—Antarctica and Greenland.
These are Earth's frozen freshwater reservoirs. (a) On the Antarctic Peninsula, basaltic mountains
and glacial multiyear ice along the Graham Coast and Flandres Bay, Cape Renard in the distance.
(b) Three outlet glaciers drain the Greenland ice sheet into the North Atlantic from southeastern
Greenland; the glacial front is in retreat. [Photos by Bobbé Christopherson.]

High Latitude
Connection
Videos

Chapter 17 presents analysis of ice cores taken from Greenland and the latest one from Antarctica, which pushed the climate record to 800,000 years before the present.

Polar Marine Climate

Polar marine stations are more moderate than other polar climates in winter, with no month below −7°C (20°F), yet they are not as warm as *tundra* climates.

Because of marine influences, annual temperature ranges are low. This climate exists along the Bering Sea, the southern tip of Greenland, northern Iceland, Norway, and in the Southern Hemisphere, generally over oceans between 50° S and 60° S. Macquarie Island at 54° S in the Southern Ocean, south of New Zealand, is polar marine. Precipitation, which frequently falls as sleet (ice pellets), is greater in these regions than in continental polar climates.

Arid and Semiarid Climates (permanent moisture deficits)

Dry climates are the world's arid deserts and semiarid regions, where we consider moisture efficiency along with temperature for understanding the climate. These regions have unique plants, animals, and physical features. Arid and semiarid regions occupy more than 35% of Earth's land area and clearly are the most extensive climate over land.

The mountains, long vistas, and resilient struggle for life are all magnified by the dryness. Sparse vegetation leaves the landscape exposed; moisture demand exceeds moisture supply throughout, creating permanent water deficits (water balance is discussed in Chapter 9). The extent of this dryness distinguishes desert and steppe climatic regions. (In addition, refer to specific annual and daily desert temperature regimes, including the highest recorded temperatures, discussed in Chapter 5; surface energy budgets covered in Chapter 4; desert landscapes in Chapter 15; and desert environments in Chapter 20.)

Important causal elements in these drylands include

- Dry, subsiding air in subtropical high-pressure systems dominates.
- Midlatitude deserts and steppes form in the rain shadow of mountains, those regions to the lee of precipitation-intercepting mountains.

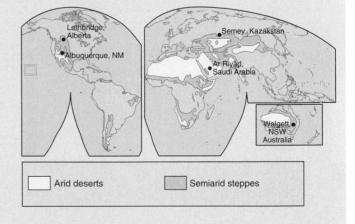

| Arid deserts | Semiarid steppes |

- Continental interiors, particularly central Asia, are far from moisture-bearing air masses.
- Shifting subtropical high-pressure systems produce semiarid steppe lands around the periphery of arid deserts.

Dry climates are distributed by latitude and the amount of moisture deficits in four distinct regimes: *arid deserts* (tropical, subtropical hot, midlatitude cold) and *semiarid steppes* (tropical, subtropical hot, midlatitude cold).

Desert Characteristics

The world climate map in Figure 10.5 reveals the pattern of Earth's dry climates, which cover broad regions between 15° and 30° N and S latitudes. In these areas, subtropical high-pressure cells predominate, with subsiding, stable air and low relative humidity. Under generally cloudless skies, these subtropical deserts extend to western continental margins, where cool, stabilizing ocean currents operate offshore and summer advection fog forms. The Atacama Desert of Chile, the Namib Desert of Namibia, the Western Sahara of Morocco, and the Australian Desert each lie adjacent to such a coastline.

Orographic lifting intercepts moisture-bearing weather systems to create rain shadows along mountain ranges that extend these dry regions into higher latitudes (Figure 10.24). Note these rain shadows in North and South America on the climate map. The isolated interior of Asia, far distant from any moisture-bearing air masses, falls within the *dry arid and semiarid* climates as well.

Major subdivisions include: *deserts* (precipitation supply roughly less than one-half of the natural moisture demand) and *semiarid steppes* (precipitation supply roughly more than one-half of natural moisture demand). Important is whether precipitation falls principally in the winter with a dry summer, in the summer with a dry winter, or is evenly distributed. Winter rains are most effective because they fall at a time of lower moisture demand. Relative to temperature, the lower-latitude deserts and steppes tend to be hotter with less seasonal change than the midlatitude deserts and steppes, where mean annual temperatures are below 18°C (64.4°F) and freezing winter temperatures are possible.

Tropical, Subtropical Hot Desert Climates

Tropical, subtropical hot desert climates are Earth's true tropical and subtropical deserts and feature annual average temperatures above 18°C (64.4°F). They generally reside on the western sides of continents, although Egypt, Somalia, and Saudi Arabia also fall within this classification. Rainfall is from local summer convectional showers. Some regions receive almost no rainfall, whereas others may receive up to 35 cm (14 in.) of precipitation a year. A representative *subtropical hot desert* city is Ar Riyāḍ (Riyadh), Saudi Arabia (Figure 10.25).

Along the Sahara's southern margin is a drought tortured region. Human populations suffered great hardship as desert conditions gradually expanded over their homelands. The sparse environment sets the stage for a rugged lifestyle and subsistence economies, pictured here near Timbuktu, Mali (Figure 10.25c). Chapter 15 presents the process of desertification (expanding desert conditions).

Death Valley, California, features such a hot desert climate with an average annual temperature of 24.4°C (76°F). July and August average temperatures are 46°C and 45°C (115°F, 113°F), respectively. Temperatures over 50°C (122°F) are not uncommon.

(a)

(b)

FIGURE 10.24 Desert landscapes.
Desert vegetation is typically *xerophytic*: drought-resistant, waxy, hard-leafed, and adapted to aridity and low transpiration loss. (a) Ocotillo (to left) and creosote (to right) in the Anza-Borrego Desert, southern California; such plants are particularly well adapted to the harsh environment. (b) Colorful rock formations and view along a winding desert highway in Valley of Fire State Park, southern Nevada.
[Photos by Bobbé Christopherson.]

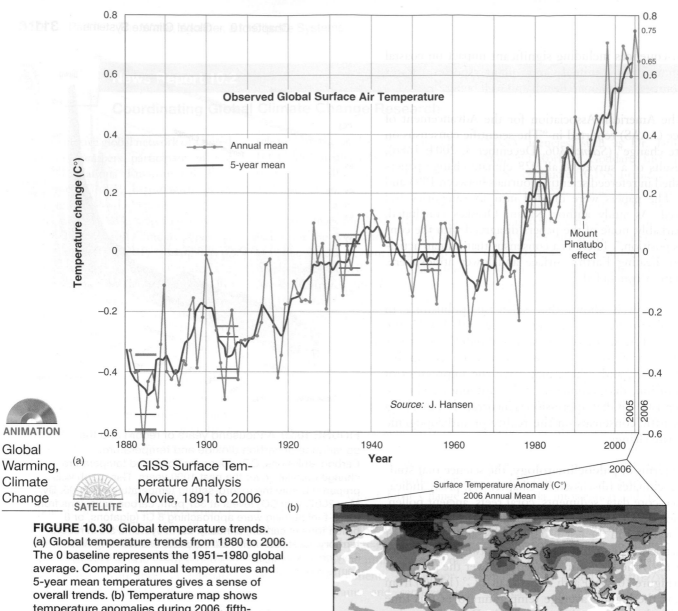

ANIMATION

Global
Warming,
Climate
Change

SATELLITE

(a) GISS Surface Tem-
perature Analysis
Movie, 1891 to 2006

(b)

FIGURE 10.30 Global temperature trends.
(a) Global temperature trends from 1880 to 2006.
The 0 baseline represents the 1951–1980 global
average. Comparing annual temperatures and
5-year mean temperatures gives a sense of
overall trends. (b) Temperature map shows
temperature anomalies during 2006, fifth-
warmest year on record. The coloration
represents C° departures from the base period
1951–1980. On the CD-ROM that accompanies
this text there is a movie of temperature
anomalies from 1881 to 2006 in which you can
see the warming patterns over this time span.
[(a) and (b) Data courtesy of Dr. James Hansen,
GISS/NASA, and NCDC/NOAA.]

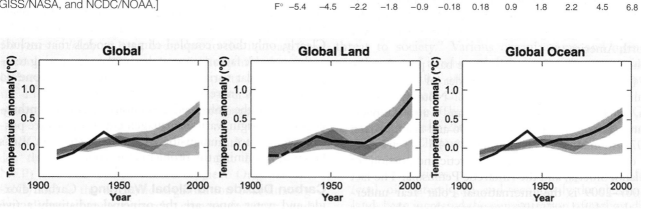

FIGURE 10.31 Explaining global temperature changes.
Computer models accurately track observed temperature change (black line) when they factor in human-forced influences on
climate (red shading). Solar activity and volcanoes do not explain the increases (blue shading). [Graphs from *Climate Change
2007: The Physical Science Basis,* Working Group I, IPCC *Fourth Assessment Report*, February 2007: Fig. SPM-4, p. 11.]

absorb and radiate longwave energy. Figure 10.32 plots changes in three of these greenhouse gases over the past 10,000 years.

These gases are transparent to light but opaque to the longer wavelengths radiated by Earth. Thus, they transmit light from the Sun to Earth but delay heat-energy loss to space. While detained, this heat energy is absorbed and emitted over and over, warming the lower atmosphere. As concentrations of these greenhouse gases increase, more heat energy remains in the atmosphere and temperatures increase.

Changes in Greenhouse Gases from Ice-Core and Modern Data

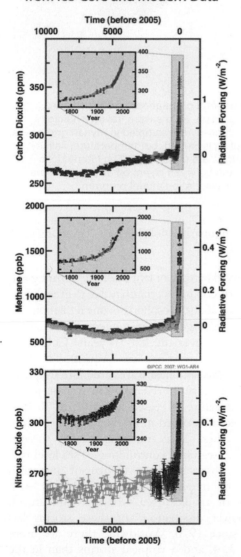

FIGURE 10.32 Greenhouse gas changes over the last 10,000 years.
Ice-core and modern data show the trends in carbon dioxide, methane, and nitrous oxide over the last 10,000 years. It is immediately obvious that we are living in unique territory on these graphs. During May 2007, CO_2 passed the 387 ppm level, CH_4 was at 1780 ppb, and N_2O approached 330 ppb.
[Graphs from *Climate Change 2007: The Physical Science Basis*, Working Group I, IPCC *Fourth Assessment Report*, February 2007: Fig. SPM-1, p. 3.]

The present CO_2 concentration tops anything over the last 800,000 years in the Dome C ice core. In fact, over this ice-core record, changes of as much as 30 ppm in CO_2 took at least 1000 years, yet, the concentration has changed 30 ppm in just the last 17 years. With 4.5% of the world's population, the United States continues to produce 24% of global CO_2 emissions. China, with 20% of global population, is responsible for 18% of CO_2 emissions and is increasing its output. Clearly, per capita CO_2 emissions in the United States are far above what an average person in China produces.

Methane and Global Warming   Another radiatively active gas contributing to the overall greenhouse effect is methane (CH_4), which, at more than 1% per year, is increasing in concentration even faster than carbon dioxide. In ice cores, methane levels never topped 750 ppb in the past 800,000 years, yet in Figure 10.32 we see present levels at 1780 ppb. We are at an atmospheric concentration of methane that is higher than at any time in the past 800 millennia.

Methane is generated by such organic processes as digestion and rotting in the absence of oxygen (anaerobic processes). About 50% of the excess methane comes from bacterial action in the intestinal tracts of livestock and from organic activity in flooded rice fields. Burning of vegetation causes another 20% of the excess, and bacterial action inside the digestive systems of termite populations also is a significant source. Methane is thought responsible for at least 19% of the total atmospheric warming.

Other Greenhouse Gases   Nitrous oxide (N_2O) is the third most important greenhouse gas that is forced by human activity—up 17% in atmospheric concentration since 1750, higher than at any time in the past 10,000 years (Figure 10.32). Fertilizer use increases the processes in soil that emit nitrous oxide, although more research is needed to fully understand the relationships. Chlorofluorocarbons (CFCs) and other halocarbons also contribute to global warming. CFCs absorb longwave energy missed by carbon dioxide and water vapor in the lower troposphere. As radiatively active gases, CFCs enhance the greenhouse effect in the troposphere and are a cause of ozone depletion and slight cooling in the stratosphere.

Climate Models and Future Temperatures

The scientific challenge in understanding climate change is to sense climatic trends in what is essentially a nonlinear, chaotic natural system. Imagine the tremendous task of building a computer model of all climatic components and programming these linkages (shown in Figure 10.1) over different time frames and at various scales.

Using mathematical models originally established for forecasting weather, scientists developed a complex computer climate model known as a **general circulation model (GCM)**. There are at least a dozen established GCMs now operating around the world. Submodel programs for the atmosphere, ocean, land surface, cryosphere, and biosphere operate within the GCM. The most sophisticated models couple atmosphere and ocean submodels and are known as *Atmosphere-Ocean General Circulation Models* (*AOGCMs*).

The first step in describing a climate is defining a manageable portion of Earth's climatic system for study. Climatologists create dimensional "grid boxes" that extend from beneath the ocean to the tropopause, in multiple layers (Figure 10.33). Resolution of these boxes in the atmosphere is about 250 km (155 mi) in the horizontal and 1 km (0.6 mi) in the vertical; in the ocean the boxes use the same horizontal resolution and a vertical resolution of about 200 to 400 m (650–1300 ft). Analysts deal not only with the climatic components within each grid layer but also with the interaction among the layers on all sides.

A comparative benchmark among the operational GCMs is *climatic sensitivity* to doubling of carbon dioxide levels in the atmosphere. GCMs do not predict specific temperatures, but they do offer various scenarios of global warming. GCM-generated maps correlate well with the observed global warming patterns experienced since 1990.

The 2007 IPCC *Fourth Assessment Report*, using a variety of GCM forecast scenarios, predicted a range of average surface warming for this century. Figure 10.34 illustrates six of these scenarios, each with its own assumptions of economics, population, degree of global cooperation, and greenhouse gas emission levels. The orange line is the simulation experiment where greenhouse gas emissions are held at 2000 values with no increases. The gray bars give you the best estimate and likely ranges of outcomes; for example, from a "low forecast" in B1 to a "high forecast" in A1Fl. Even the "B" scenario represents a significant increase in global land and ocean temperatures and will produce consequences. Although regionally variable and subject to revision, the IPCC temperature change forecasts for the twenty-first century are:

- High forecast: 6.4 C° (11.5 F°)
- Middle forecast: 1.8 C°–4.0 C° (3.1 F°–7.2 F°)
- Low forecast: 1.1 C° (2.0 F°)

Figure 10.35 offers us a look at the world of 2020–2029 and 2090–2099 using three scenarios from several different AOGCMs. Find the three scenarios used for these three pairs of maps on the graph in Figure 10.34. You can see from the maps why scientists are concerned about temperature trends in the higher latitudes.

Consequences of Global Warming

The consequences of uncontrolled atmospheric warming are complex. Regional climate responses are expected as temperature, precipitation, soil-moisture, and air mass characteristics change. Although the ability to accurately forecast such regional changes is still evolving, some consequences of warming have been forecasted and in several regions are already underway. The challenge for science is to analyze such effects on a global scale.

The following list is a brief overview from the IPCC *Fourth Assessment Report* "Summary for Policy Makers" and

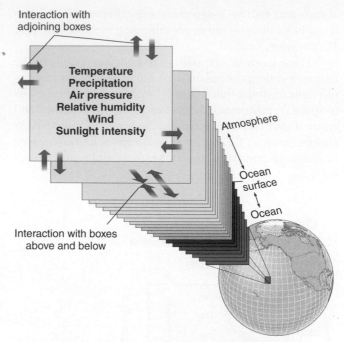

FIGURE 10.33 A general circulation model scheme. Temperature, precipitation, air pressure, relative humidity, wind, and sunlight intensity are sampled in myriad grid boxes. In the ocean, sampling is limited, but temperature, salinity, and ocean current data are considered. The interactions within a grid layer, and between layers on all six sides, are modeled in a general circulation model program.

other sources that summarize global impacts emerging from climate change.

- The observed pattern of tropospheric warming and stratospheric cooling is *very likely* due to greenhouse gas increases and stratospheric ozone depletion.
- Widespread changes in extreme temperatures have been observed over the last 50 years. Cold days, cold nights, and frost are less frequent, while hot days, hot nights, and heat waves are more frequent.
- Observations since 1961 show the average global ocean temperature increased to depths of 3000 m and the ocean absorbed more than 80% of climate system heating. Such warming causes thermal expansion of seawater, contributing to sea level rise.
- There is observational evidence of increased intensity of tropical cyclones correlated with increases of tropical sea-surface temperatures. Total "power dissipation" of these storms has doubled since 1970. Worldwide there are more category 3, 4, and 5 tropical storms than in the previous record.
- Mountain glaciers and snow cover declined on average in both hemispheres, contributing to sea-level rise.
- Mount Kilimanjaro in Africa, portions of the South American Andes, and the Himalayas will very likely lose most of their glacial ice within the next two decades, affecting local water resources. Glacial ice continues its retreat in Alaska.

Multi-model Averages and Assessed Ranges for Surface Warming

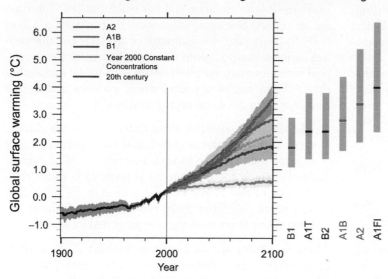

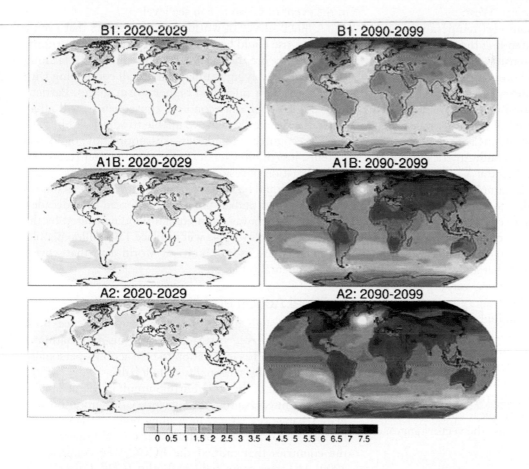

FIGURE 10.34 Model scenarios for surface warming.
Ranging from business as usual, lack of international cooperation, and continuation of the use of fossil fuels ("A2FI") to a low-emission scenario ("B1"), each scenario spells climatic catastrophe of some degree. Holding emissions at the 2000 concentration (orange line plot) results in the least temperature impacts. [Graph from *Climate Change 2007: The Physical Science Basis,* Working Group I, IPCC *Fourth Assessment Report*, February 2007: Fig. SPM-5, p. 14.]

FIGURE 10.35 Model projections of surface temperature for three scenarios.
Three pairs of model simulations for 2020–2029 and 2090–2099 for the B1, A1B, and A2 scenarios. The colors are defined in the temperature scale along the bottom. Note the severity of warming in the worst case. Correlate these three cases with the graph in Figure 10.34. [Maps from *Climate Change 2007: The Physical Science Basis*, Working Group I, IPCC *Fourth Assessment Report*, February 2007: Fig. SPM-6, p. 15.]

- The average atmospheric water vapor content has increased over land and ocean as well as in the upper troposphere. The increase is consistent with the fact that warmer air can absorb more water vapor.
- Flow speed accelerated for some Greenland and Antarctic outlet glaciers as they drain ice from the interior of the ice sheets. In Greenland this rate of loss exceeds snowfall accumulation, with losses per year more than doubling between 1996 and 2005 (mass loses of 91 km³ compared to 224 km³).

- Average Arctic temperatures increased at almost twice the global average rate in the past 100 years. Wintertime lower atmosphere temperatures over Antarctica are warming at nearly three times the global average, first reported in 2006.
- Temperatures in the permafrost active layer increased overall since the 1980s in the Arctic up to 3 C°. The maximum area of seasonally frozen ground decreased 7% in the Northern Hemisphere, with a decrease in spring of up to 15%.

- Since 1978, annual average Arctic sea ice extent shrunk in summer by 7.4% per decade, according to satellite measurements. September 2007 Arctic sea ice extent retreated to its lowest area coverage in the record.
- Changes in precipitation and evaporation over the oceans and the melting of ice are freshening mid- and high-latitude oceans and seas, together with increased salinity in low-latitudes. Oceans are acidifying (lower pH) in response to absorption of increasing atmospheric CO_2.
- Mid-latitude westerly winds strengthened in both hemispheres since the 1960s.
- More intense and longer droughts have been observed over wider areas since the 1970s, particularly in the tropics and subtropics. Increased drying linked to higher temperatures and decreased precipitation observed in the Sahel, the Mediterranean, southern Africa, parts of southern Asia, Australia, and the American West.
- Higher spring and summer temperatures and earlier snowmelt are extending the wildfire season and increasing the intensity of wildfires in the western U.S. and elsewhere.
- The frequency of heavy precipitation events increased over most land areas, consistent with warming and observed increases of atmospheric water vapor. Significantly increased precipitation has been observed in eastern parts of North and South America, northern Europe, and northern and central Asia.
- Crop patterns, as well as natural habitats of plants and animals, will shift to maintain preferred temperatures. According to climate models, climatic regions in the midlatitudes could shift poleward by 150 to 550 km (90 to 350 mi) during this century.
- Biosphere models predict that a global average of 30% of the present forest cover (varying regionally from 15% to 65%) will undergo major species redistribution—greatest at high latitudes and least in the tropics.
- Populations previously unaffected by malaria, dengue fever, lymphatic filariasis, and yellow fever (all mosquito vector), schistosomiasis (water snail vector), and sleeping sickness (tsetse fly vector) will be at greater risk in subtropical and midlatitude areas as temperatures increase the vector ranges.

Changes in Sea Level Sea-level rise must be expressed as a range of values that are under constant reassessment. During the last century, sea level rose 10–20 cm (4–8 in.), a rate 10 times higher than the average rate during the last 3000 years.

The 2007 IPCC forecast scenarios for global mean sea-level rise this century, given regional variations, are:

- Low forecast: 0.18 m (7.1 in.)
- Middle forecast: 0.39 m (15.4 in.)
- High forecast: 0.59 m (23.2 in.)

Unfortunately the new 2006–2007 measurements of Greenland's ice-loss acceleration did not reach the IPCC in time for its report. Scientists are considering at least a 1.2-m (3.94-ft) "high case" for estimates of sea-level rise this century as more realistic given Greenland's present losses coupled with mountain glacial ice losses worldwide. Remember that a 0.3-m rise in sea level would produce a shoreline retreat of 30 m (98 ft) on average. Here is the concern:

> The data now available raise concerns that the climate system, in particular sea level, may be responding more quickly than climate models indicate. . . . The rate of sea-level rise for the past 20 years is 25% faster than the rate of rise in any 20-year period in the preceding 115 years. . . . Since 1990 the observed sea-level has been rising faster than the rise projected by models.[*]

These increases would continue beyond 2100 even if greenhouse gas concentrations were stabilized. In Chapter 16, maps present what coastlines will experience in the event of a 1-m rise in sea level.

A quick survey of world coastlines shows that even a moderate rise could bring change of unparalleled proportions. At stake are the river deltas, lowland coastal farming valleys, and low-lying mainland areas, all contending with high water, high tides, and higher storm surges. Particularly tragic social and economic consequences will affect small island states, which are unable to adjust within their present country boundaries—disruption of biological systems, loss of biodiversity, reduction in water resources, and evacuation of residents are among the impacts.

There could be both internal and international migration of affected human populations, spread over decades, as people move away from coastal flooding caused by the sea-level rise—a 1-m rise will displace 130 million people. Presently, there is no body of world law that covers "environmental refugees."

Political Action and "No Regrets"

Reading through all this climate-change science must seem pretty "heavy." Instead, think of this information as empowering and as motivation to take action—personally, locally, regionally, nationally, and globally.

A product of the 1992 Earth Summit in Rio de Janeiro was the United Nations Framework Convention on Climate Change (FCCC). The leading body of the Convention is the Conference of the Parties (COP) operated by the countries that ratified the FCCC, 186 countries by 2000. Meetings were held in Berlin (*COP-1*, 1995) and Geneva (*COP-2*, 1996). These meetings set the stage for *COP-3* in Kyoto, Japan, in December 1997, where 10,000 participants adopted the *Kyoto Protocol* by consensus. Seventeen national academies of science endorsed the Kyoto Protocol. (For updates on the status of the Kyoto Protocol, see **http://unfccc.int/2860.php**.) The latest 2007 gathering, COP-13, was in Bali, Indonesia.

[*]S. Rahmstorf, et al., "Recent climate observations compared to projections," AAAS *Science* 316 (May 4, 2007): 709.

The Kyoto Protocol binds more-developed countries to a collective 5.2% reduction in greenhouse gas emissions as measured at 1990 levels for the period 2008 to 2012. Within this group goal, various countries promised cuts: Canada is to cut 6%, the European Union 8%, and Australia 8%, among many others. The *Group of 77* countries plus China favor a 15% reduction by 2010. With Russian ratification in November 2004, the Kyoto Protocol is now international law, with nearly 140 national signatories; this is without United States participation.

The Intergovernmental Panel on Climate Change (IPCC) declared that "no regrets" opportunities to reduce carbon dioxide emissions are available in most countries. The IPCC Working Group III defines this as follows:

> No regrets options are by definition greenhouse gas emissions reduction options that have negative net costs, because they generate direct and indirect benefits that are large enough to offset the costs of implementing the options.

Benefits that equal or exceed their cost to society include reduced energy cost, improved air quality and health, reduction in tanker spills and oil imports, and deployment of renewable and sustainable energy sources, among others. This holds true without even considering the benefits of slowing the rate of climate change. For Europe, scientists determined that carbon emissions could be reduced to less than half the 1990 level by 2030, at a negative cost. One key to "no regrets" is the untapped energy-efficiency potential.

In the United States, five Department of Energy national laboratories (Oak Ridge, Lawrence Berkeley, Pacific Northwest, National Renewable Energy, and Argonne) reported that the United States can meet the Kyoto carbon emission reduction targets with negative overall costs (cash benefit savings) ranging from −$7 to −$34 billion. (For more, see Working Group III, *Climate Change 2001, Mitigation,* London: Cambridge University Press, 2001, pp. 21, 474–76 and 506–507.)

Summary and Review—Global Climate Systems

■ *Define* climate and climatology, and *explain* the difference between climate and weather.

Climate is dynamic, not static. **Climate** is a synthesis of weather phenomena at many scales, from planetary to local, in contrast to weather, which is the condition of the atmosphere at any given time and place. Earth experiences a wide variety of climatic conditions that can be grouped by general similarities into climatic regions. **Climatology** is the study of climate and attempts to discern similar weather statistics and identify **climatic regions**.

climate (p. 277)
climatology (p. 278)
climatic regions (p. 278)

1. Define climate and compare it with weather. What is climatology?
2. Explain how a climatic region synthesizes climate statistics.
3. How does the El Niño phenomenon produce the largest interannual variability in climate? What are some of the changes and effects that occur worldwide?

■ *Review* the role of temperature, precipitation, air pressure, and air mass patterns used to establish climatic regions.

Climatic inputs include insolation (pattern of solar energy in the Earth–atmosphere environment), temperature (sensible heat energy content of the air), precipitation (rain, sleet, snow, and hail; the supply of moisture), air pressure (varying patterns of atmospheric density), and air masses (regional-sized homogeneous units of air). Climate is the basic element in ecosystems, the natural, self-regulating communities of plants and animals that thrive in specific environments.

4. How do radiation receipts, temperature, air-pressure inputs, and precipitation patterns interact to produce climate types? Give an example from a humid environment and one from an arid environment.
5. Evaluate the relationships among a climatic region, ecosystem, and biome.

■ *Review* the development of climate classification systems, and *compare* genetic and empirical systems as ways of classifying climate.

Classification is the process of ordering or grouping data in related categories. A **genetic classification** is based on causative factors, such as the interaction of air masses. An **empirical classification** is one based on statistical data, such as temperature or precipitation. This text analyzes climate using aspects of both approaches, with a map based on climatological elements.

classification (p. 280)
genetic classification (p. 280)
empirical classification (p. 280)

6. What are the differences between a genetic and an empirical classification system?
7. What are some of the climatological elements used in classifying climates? Why each of these?

■ *Describe* the principal climate classification categories other than deserts, and *locate* these regions on a world map.

Here we focus on temperature and precipitation measures. Keep in mind these are measurable results produced by interacting elements of weather and climate. These data are plotted on a **climograph** to display the characteristics of the climate.

There are six basic climate categories. Temperature and precipitation considerations form the basis of five climate categories and their regional types:

- Tropical (equatorial and tropical latitudes)
  rain forest (rainy all year)
  monsoon (6 to 12 months rainy)
  savanna (less than 6 months rainy)
- Mesothermal (midlatitudes, mild winters)
  humid subtropical (hot summers)

marine west coast (warm to cool summers)

Mediterranean (dry summers)

- Microthermal (mid- and high latitudes, cold winters)

humid continental (hot to warm summers)

subarctic (cool summers to very cold winters)

- Polar (high latitudes and polar regions)

tundra (high latitude or high altitude)

ice caps and ice sheets (perpetually frozen)

polar marine

- Highland (compared to lowlands at the same latitude, highlands have lower temperatures—recall the normal lapse rate)

Only one climate category is based on moisture efficiency as well as temperature:

- Desert (permanent moisture deficits)

arid deserts (tropical, subtropical hot and midlatitude cold)

semiarid steppes (tropical, subtropical hot and midlatitude cold)

climograph (p. 285)

8. List and discuss each of the principal climate categories. In which one of these general types do you live? Which category is the only type associated with the annual distribution and amount of precipitation?

9. What is a climograph, and how is it used to display climatic information?

10. Which of the major climate types occupies the most land and ocean area on Earth?

11. Characterize the tropical climates in terms of temperature, moisture, and location.

12. Using Africa's tropical climates as an example, characterize the climates produced by the seasonal shifting of the ITCZ with the high Sun.

13. Mesothermal (subtropical and midlatitude, mild winter) climates occupy the second-largest portion of Earth's entire surface. Describe their temperature, moisture, and precipitation characteristics.

14. Explain the distribution of the *humid subtropical hot-summer* and *Mediterranean dry-summer* climates at similar latitudes and the difference in precipitation patterns between the two types. Describe the difference in vegetation associated with these two climate types.

15. Which climates are characteristic of the Asian monsoon region?

16. Explain how a *marine west coast* climate type can occur in the Appalachian region of the eastern United States.

17. What role do offshore ocean currents play in the distribution of the *marine west coast* climates? What type of fog is formed in these regions?

18. Discuss the climatic conditions for the coldest places on Earth outside the poles.

■ *Explain* the precipitation and moisture efficiency criteria used to determine the arid and semiarid climates, and *locate* them on a world map.

The dry and semiarid climates are described by precipitation rather than temperature. Dry climates are the world's arid deserts and

semiarid regions, with their unique plants, animals, and physical features. The arid and semiarid climates occupy more than 35% of Earth's land area, clearly the most extensive climate over land.

Major subdivisions are *arid deserts* in tropical and midlatitude areas (precipitation—natural water supply—less than one-half of natural water demand) and *semiarid steppes* in tropical and midlatitude areas (precipitation more than one-half of natural water demand).

19. In general terms, what are the differences among the four desert classifications? How are moisture and temperature distributions used to differentiate these subtypes?

20. Relative to the distribution of arid and semiarid climates, describe at least three locations where they occur across the globe and the reasons for their presence in these locations.

■ *Outline* future climate patterns from forecasts presented, and *explain* the causes and potential consequences of climate change.

Various activities of present-day society are producing climatic changes, particularly a global warming trend. The highest average annual temperatures experienced since the advent of instrumental measurements have dominated the last 25 years. There is a scientific consensus building that global warming is related to the anthropogenic impacts on the natural greenhouse effect.

The 2007 *Fourth Assessment Report* from the Intergovernmental Panel on Climate Change affirms this consensus. The IPCC has predicted surface temperature response to a doubling of carbon dioxide ranging from an increase of 1.1 C° (2.0 F°) to 6.4 C° (11.5 F°) between the present and 2100. Natural climatic variability over the span of Earth's history is the subject of **paleoclimatology**. A **general circulation model (GCM)** forecasts climate patterns and is evolving to greater capability and accuracy than in the past. People and their political institutions can use GCM forecasts to form policies aimed at reducing unwanted climate change.

paleoclimatology (p. 311)

general circulation model (GCM) (p. 314)

21. Explain climate forecasts. How do general circulation models (GCMs) produce such forecasts?

22. Describe the potential climatic effects of global warming on polar and high-latitude regions. What are the implications of these climatic changes for persons living at lower latitudes?

23. How is climatic change affecting agricultural and food production? Natural environments? Forests? The possible spread of disease?

24. What are the actions being taken at present to delay the effects of global climate change? What is the Kyoto Protocol? What is the current status of U.S. and Canadian government action on the protocol?

NetWork

The *Geosystems* Student Learning Center provides on-line resources for this chapter on the World Wide Web. To begin: Once at the Center, click on the cover of this textbook, scroll the Table of Contents menu, and select this chapter. You will find self-tests that are graded, review exercises, specific updates for items in the chapter, and in "Destinations" many links to interesting related pathways on the Internet. *Geosystems* Student Learning Center is found at **http://www.prenhall.com/christopherson/**.

Critical Thinking

A. The text asked that you find the climate conditions for your campus and your birthplace and locate these two places on Figures 10.3, 10.4, and 10.5. Briefly describe the information sources you used: library, Internet, teacher, and phone calls to state and provincial climatologists. Now, refer to Appendix B to refine your assessment of climate for the two locations. Briefly show how you worked through the Köppen climate criteria given in the appendix that established the climate classification for your two cities.

B. Many external factors force climate. The chart "Global and annual mean radiative forcing for the year 2000, relative to 1750" is presented here (from IPCC *Climate Change 2001, The Scientific Basis*, Washington: Cambridge University Press, 2001, Figure 3, p. 8, and Figure 6.6, p. 392).

The estimates of radiative forcing in Watts per square meter units are given on the *y*-axis (vertical axis). The level of scientific understanding is noted along the *x*-axis (horizontal axis), arranged from "high" to "very low." Those columns above the "0" value (in red) indicate *positive forcing*, such as the greenhouse gases grouped in the far-left column. Columns that fall below the "0" value (in blue) indicate *negative forcing*, such as the haze from sulfate aerosols, fourth column from the left. The vertical line between the markers on each column is an estimate of the uncertainty range. Where no column appears but there is instead a line denoting a range, there is no central estimate given present uncertainties, such as for mineral dust.

Assume you are a policymaker with a goal of reducing the rate of global warming, that is, reducing positive radiative forcing of the climate system. What strategies do you suggest to alter the height of the columns and adjust the mix of elements that cause warming? Assign priorities to each suggested strategy to denote most-to-least effective in moderating climate change. Brainstorm and discuss your strategies with others.

**The Global Mean Radiative Forcing of the Climate System
for the Year 2000, Relative to 1750**

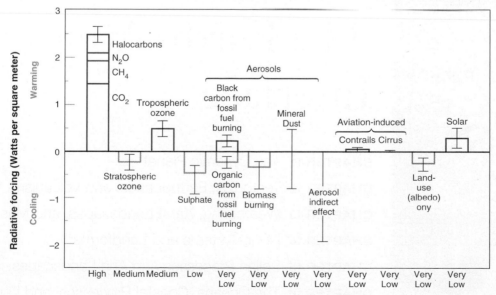

PART III

The Earth–Atmosphere Interface

Beds of fine shales turned on edge rise along Duvefjord, Nordaustlandet, Spitsbergen (Svalbard) Archipelago, in the Arctic Ocean. The frost action works the layers apart in a delicate physical weathering. [Photo by Bobbé Christopherson.]

SOLAR ENERGY

Atmosphere

Hydrosphere

Biosphere

Lithosphere

Earth is a dynamic planet whose surface is shaped by active physical agents of change. Two broad systems organize these agents in Part III—endogenic and exogenic. The **endogenic system** (Chapters 11 and 12) encompasses internal processes that produce flows of heat and material from deep below Earth's crust. Radioactive decay principally powers these processes. The materials involved constitute the *solid realm* of Earth. Earth's surface responds by moving, warping, and breaking, sometimes in dramatic episodes of earthquakes and volcanic eruptions, constructing the crust.

At the same time, the **exogenic system** (Chapters 13 through 17) involves external processes that set into motion air, water, and ice, all powered by solar energy—this is the *fluid realm* of Earth's environment. These media carve, shape, and reduce the landscape. One such process, weathering, breaks up and dissolves the crust. Erosion picks up these materials; transports them in rivers, winds, coastal waves, and flowing glaciers; and deposits them along the way. Thus, Earth's surface is the interface between two vast open systems: one that builds the landscape and topographic relief and one that tears it down into sedimentary plains.

NOTEBOOK
Earth's Varied Landscapes

A weathered mountain rises above calm waters in Hornsund in southwest Spitsbergen Island. Ice drifting by is from several calving glaciers at the head of the fjords. Quiet reflections as we contemplate The Dynamic Planet, Earth. [Photo by Bobbé Christopherson.]

The Dynamic Planet

■ Key Learning Concepts

After reading the chapter, you should be able to:

- ■ *Distinguish* between the endogenic and exogenic systems, *determine* the driving force for each, and *explain* the pace at which these systems operate.

- ■ *Diagram* Earth's interior in cross section, and *describe* each distinct layer.

- ■ *Illustrate* the geologic cycle, and *relate* the rock cycle and rock types to endogenic and exogenic processes.

- ■ *Describe* Pangaea and its breakup, and *relate* several physical proofs that crustal drifting is continuing today.

- ■ *Portray* the pattern of Earth's major plates, and *relate* this pattern to the occurrence of earthquakes, volcanic activity, and hot spots.

The twentieth century was a time of great discovery about Earth's internal structure and dynamic crust, yet much remains on the scientific frontier of Earth systems science. Discoveries have revolutionized our understanding of how continents and oceans came to be arranged as they are. A new era of understanding is emerging, combining various sciences within the study of physical geography. One task of physical geography is to explain the *spatial implications* of all this new information.

In this chapter: Earth's interior is organized as a core surrounded by roughly concentric shells of material. It is unevenly heated by the radioactive decay of unstable elements. A rock cycle produces three classes of rocks through igneous, sedimentary, and metamorphic processes. Internal processes coupled with the rock cycle, hydrologic cycle, and tectonic cycle result in a varied crustal surface, featuring irregular fractures, extensive mountain ranges both on land and the ocean floor, drifting continental and oceanic crust, and frequent earthquakes and volcanic events. All of this movement of materials results from *endogenic* forces within Earth—the subject of this chapter.

The Pace of Change

The **geologic time scale** is a summary timeline of all Earth history, shown in Figure 11.1. It reflects currently accepted names of time intervals for each segment of Earth's history, from vast *eons* through briefer *eras, periods,* and *epochs*. Present thinking places Earth's age at 4.567 billion years, with the Moon about 30 million years younger, formed when a Mars-sized object struck the early Earth.

The time scale depicts two important kinds of time: *relative* (what happened in what order) and *absolute* (actual number of years before the present). Also on the geologic time scale, see the labels denoting the six major extinctions of life forms in Earth history. Rather than being complete extinctions, think of some of these spasms in planetary life as significant *depletions*. These range from 440 million years ago (m.y.a.) to the ongoing present-day episode caused by modern civilization.

Relative time is the *sequence* of events, based on the relative positions of rock strata above or below each other. Relative time is based on the important general principle of *superposition,* which states that *rock and sediment always are arranged with the youngest beds "superposed" toward the top of a rock formation and the oldest at the base, if they have not been disturbed.* The study of these sequences is called *stratigraphy.* Thus, relative time places the Precambrian at the bottom (beginning) of the time scale and the Holocene (today) at the top. Important time clues—namely, *fossils,* the remains of ancient plants and animals—lie embedded within these strata. Since approximately 4.0 billion years ago, life has left its evolving imprint in the rocks.

Scientific methods such as radiometric dating determine *absolute time,* the actual "millions of years ago" shown on the time scale. These absolute ages permit scientists to actively update geologic time, refining the time-scale sequence and lending greater accuracy to relative dating sequences. See News Report 11.1 for more information on radioactivity and Earth's time clock. For more on the geologic time scale, see **http://www.ucmp. berkeley.edu/exhibit/geology.html**.

Holocene is the name given to the youngest epoch in the geologic time scale, characteristic of postglacial conditions since the retreat of the continental glaciers, approximately the last 11,500 years. Because of the impact of human society on planetary systems, discussion is

News Report 11.1

Radioactivity: Earth's Time Clock

The age of Earth and the age of the earliest known crustal rock are astounding, for we think in terms of Earth's trips around the Sun and the pace of our own lives. We need something greater than human time to measure the vastness of geologic time. Nature has provided a way: *radiometric dating.* It is based on the steady decay of certain atoms.

An atom contains protons and neutrons in its nucleus. Certain forms of atoms called isotopes have unstable nuclei; that is, the protons and neutrons do not remain together indefinitely. As particles break away and the nucleus disintegrates, radiation is emitted and the atom decays into a different element—this process is *radioactivity.*

Radioactivity provides the steady time clock needed to measure the age of ancient rocks. It works because the decay rates for different isotopes are

determined precisely, and they do not vary beyond established uncertainties. Further refining of these geochronologies, the chronologic sequence of geologic time, is an ongoing process.

The decay rate is expressed as *half-life,* the time required for one-half of the unstable atoms in a sample to decay into "daughter" isotopes. Some examples of unstable elements that become stable elements and their half-lives include uranium-238 to lead-206 (4.5 billion years); thorium-232 to lead-208 (14.1 billion years); potassium-40 to argon-40 (1.3 billion years); and, in organic materials, carbon-14 to nitrogen-14 (5730 years), although with correlation from ice-core data, carbon-dating accuracy is being pushed back to more than 24,000 years.

The presence of these decaying elements and stable end products in

sediment or rock allows scientists to read the radiometric "clock." They compare the amount of original isotope in the sample with the amount of decayed end product in the sample. If the two are in a ratio of 1:1 (equal parts), one half-life has passed. Errors can occur if the sample was disturbed or subjected to natural weathering processes that might alter its radioactivity.

To increase accuracy, investigators may check a sample using more than one radiometric measurement. Calibration of the past 10,000 years correlates through tree-ring analysis; the past 45,000 years is calibrated using fossils found in lake sediments, among several other cross-checks available. The dates for Earth's oldest known rocks from Canada and Greenland were verified using several radiometric methods.

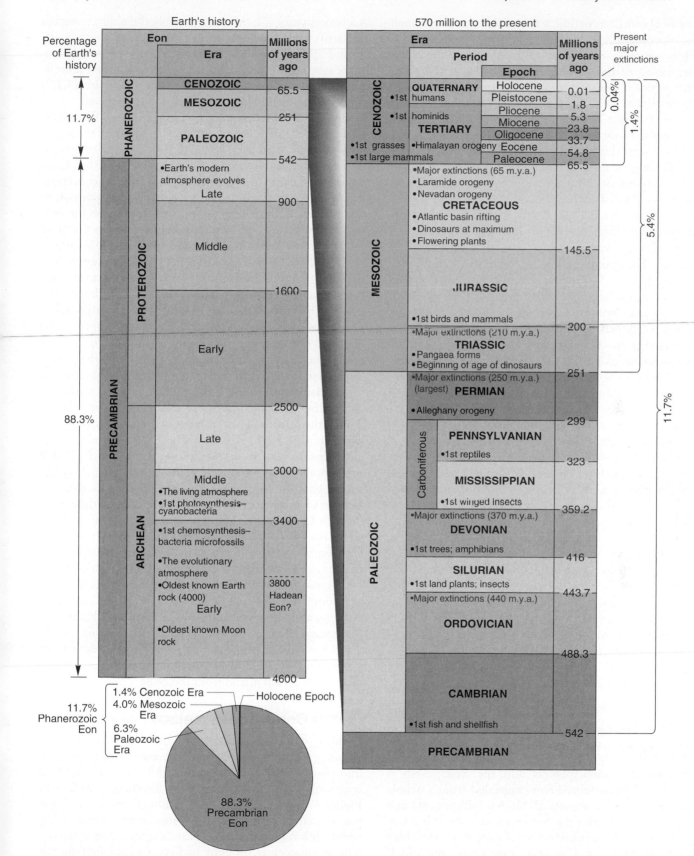

Percentage of Earth's history

Earth's history

570 million to the present

Present major extinctions

11.7% Phanerozoic Eon

88.3% Precambrian

1.4% Cenozoic Era
4.0% Mesozoic Era
6.3% Paleozoic Era
Holocene Epoch

88.3% Precambrian Eon

FIGURE 11.1 Geologic time scale.
Highlights of Earth's history are shown in the figure as bulleted items, including the six major extinctions or depletions of life forms, printed in red (the sixth extinction episode is underway at the present time). In the column at the left, note that 88% of geologic time occurred during the Precambrian Eon. Dates appear in m.y.a., million years ago. [Data and update from Geological Society of America and *Nature* 429 (May 13, 2004): 124–25.]

ANIMATION

Applying Relative Dating Principles

undcrway about designating a new name for the human epoch. A proposal is to call this time the *Anthropocene*.

However, when would such a time begin? Some think the beginning of the Industrial Revolution in the eighteenth century is the starting point, whereas others think an appropriate time is some 5000 years ago when Asians began flooding rice fields that released methane gas into the atmosphere. Or perhaps 8000 years ago is a better chronology with the spread of domesticated crops, widespread clearing of forests, and the human use of fire to modify environments. Human impacts on the atmosphere began at the roots of civilization.

A guiding principle of Earth science is **uniformitarianism**, which assumes that *the same physical processes active in the environment today have been operating throughout geologic time*. For example, if streams carve valleys now, they must have done so 500 m.y.a. The phrase "the present is the key to the past" describes this principle. Evidence from exploration and from the landscape record of volcanic eruptions, earthquakes, and exogenic processes support uniformitarianism. The concept was first proposed by James Hutton in his *Theory of the Earth* (1795) and was later amplified by Charles Lyell in *Principles of Geology* (1830). Today, the earlier Hutton version guides scientific thinking because of his view that rates and some processes do change, whereas essential underlying principles of biology, physics, geology, and chemistry remain uniform.

Within the principle of uniformitarianism, catastrophic events such as massive landslides, earthquakes, volcanic eruptions, and cyclic episodes of mountain building punctuate geologic time. Vast superfloods occurred as ice-dammed lakes tore through their frozen dams in the later days of the last ice age, such as glacial Lake Missoula, which left evidence across a broad region. These episodes occur as interruptions in the generally uniform processes that shape the slowly evolving landscape. Here, the *punctuated equilibrium* concept studied in the life sciences and paleontology might apply to aspects of Earth's long developmental history.

We start our journey deep within the planet. A knowledge of Earth's internal structure and energy is key to understanding the surface.

Earth's Structure and Internal Energy

Along with the other planets and the Sun, Earth is thought to have condensed and congealed from a nebula of dust, gas, and icy comets about 4.6 billion years ago (review this in Chapter 2). Scientists are observing this same formation process underway elsewhere in our Milky Way Galaxy and the Universe. Previously, the oldest surface rock discovered on Earth (known as the Acasta Gneiss) was found in northwestern Canada; it was radiometrically dated to an age of 3.96 billion years.

Recent research found detrital zircons (particles of preexisting zirconium silica oxides transported and deposited, forming rock) in Western Australia dating to about

4.3 billion years old; these are possibly the oldest materials in Earth's crust. These discoveries tell us something significant: Earth was forming continental crust at least 4 billion years ago, during the Archean Eon. Although not an official designation, note the Hadean Eon in Figure 11.1, a proposal to describe the period before the Archean.

As Earth solidified, gravity sorted materials by density. Heavier substances such as iron gravitated slowly to its center, and lighter elements such as silica slowly welled upward to the surface and became concentrated in the crust. Consequently, Earth's interior is sorted into roughly concentric layers, each one distinct in either chemical composition or temperature. Heat energy migrates outward from the center by conduction and by physical convection in the more fluid or plastic layers in the mantle and nearer the surface.

Our knowledge of Earth's *internal differentiation* into these layers is acquired entirely through indirect evidence because we are unable to drill more than a few kilometers into Earth's crust. There are several physical properties of Earth materials that enable us to know the nature of the interior. For example, when an earthquake or underground nuclear test sends shock waves through the planet, the cooler areas, which generally are more rigid, transmit these **seismic waves** at a higher velocity than do the hotter areas, where seismic waves are slowed to lower velocity. This is the science of *seismic tomography*, as if Earth is subjected to a kind of CAT scan with every earthquake.

Density also affects seismic-wave velocities. Plastic zones simply do not transmit some seismic waves; they absorb them. Some seismic waves are reflected as densities change, whereas others are refracted, or bent, as they travel through Earth. Thus, the distinctive ways in which seismic waves pass through Earth and the time they take to travel between two surface points help seismologists deduce the structure of Earth. Figure 11.2 is a model of Earth's interior.

Figure 11.3 illustrates the dimensions of Earth's interior compared with surface distances in North America to give you a sense of size and scale. An airplane flying from Anchorage, Alaska, to Fort Lauderdale, Florida, would travel the same distance as that from Earth's center to its surface. Note that, in this figure, Earth's thin crust extends only about 30 km inland from the coast.

Earth's Core and Magnetism

A third of Earth's entire mass, but only a sixth of its volume, lies in its dense core. The **core** is differentiated into two regions—*inner core* and *outer core*—divided by a transition zone several hundred kilometers wide (see Figure 11.2b). The inner core is thought to be solid iron that is well above the melting temperature of iron at the surface but remains solid because of tremendous pressure. The inner core is thought to have formed first shortly after Earth condensed. The iron in the core is impure, probably combined with silicon and possibly oxygen and sulfur. Recent research points to the conclusion that the inner core may be a single, enormous crystal of iron. The outer core is molten, metallic iron with a lighter density than the inner core.

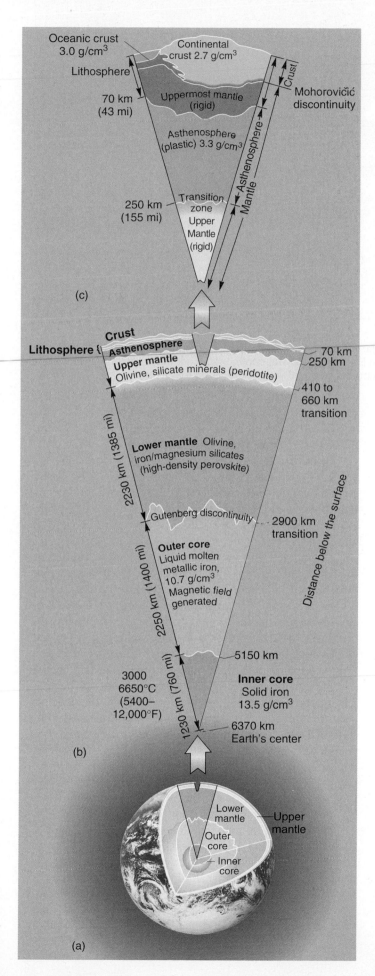

FIGURE 11.2 Earth in cross section.
(a) Cutaway showing Earth's interior. (b) Earth's interior in cross section, from the inner core to the crust. (c) Detail of the structure of the lithosphere and its relation to the asthenosphere. (For comparison to the densities noted, the density of water is 1.0 g/cm³, and mercury, a liquid metal, is 13.0 g/cm³.) A revised estimate of Earth's mass (or weight) made in 2000 set it at 5.972 sextillion metric tons (5972 followed by 18 zeros). [Photo from NASA.]

Earth's Magnetism   The fluid outer core generates at least 90% of Earth's magnetic field and the magnetosphere that surrounds and protects Earth from the solar wind and cosmic radiation. One hypothesis explains that circulation in the outer core converts thermal and gravitational energy into magnetic energy, producing Earth's magnetic field. The north magnetic pole (NMP) is near 83° N by 114° W (in 2005) in the Canadian Arctic, and the south magnetic pole is off the coast of Wilkes Land, Antarctica. These surface expressions of Earth's magnetic field migrate; for example, the NMP has moved 1100 km (685 mi) just this past century.

An intriguing feature of Earth's magnetic field is that its polarity sometimes fades to zero and then returns to full strength, with north and south magnetic poles reversed. In the process, the field does not blink on and off but instead diminishes slowly to low intensity, perhaps 25% strength, and then rapidly regains to full power. This **geomagnetic reversal** has taken place 9 times during the past 4 million years and hundreds of times over Earth's history. During the transition interval of low strength, Earth's surface receives higher levels of cosmic radiation and solar particles, but note that past reversals do not correlate with species extinctions. The evolution of life has weathered many of these transitions.

The average period of a magnetic reversal is about 500,000 years; several hundred documented occurrences varied from as short as 20 to 30 thousand years to 50 million years. Transition periods of low intensity last from 1000 to 10,000 years. The last reversal was 790,000 years ago, preceded by a transition ranging 2000 years in length along the equator to 10,000 years in the midlatitudes. Given present rates of magnetic field decay over the last 150 years, we are perhaps 1000 years away from entering the next phase of field changes, although there is no expected pattern to forecast the timing. An apparent trend in recent geologic time is toward more frequent reversals. The obvious question for us is, Why does this happen?

The reasons for these magnetic reversals are unknown. However, the spatial patterns they create at Earth's surface are a key tool in understanding the evolution of landmasses and the movements of the continents. When new iron-bearing rocks solidify from molten material (lava) at Earth's surface, the small magnetic particles in the rocks align according to the orientation of the magnetic poles at that time. As the rocks cool and solidify, this alignment locks in place. When Earth is without polarity in its magnetic field, a random pattern of magnetism in crustal rock results.

All across Earth, rocks of the same age bear an identical record of magnetic reversals in the form of measurable magnetic "stripes" in any magnetic material they contain, such as iron particles. These stripes illustrate global patterns of

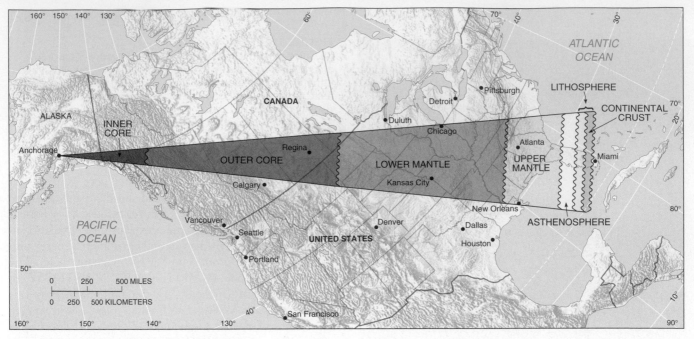

FIGURE 11.3 Distances from core to crust.
The distance from Anchorage, Alaska, to Fort Lauderdale, Florida, is the same distance as that from Earth's center to its outer crust. Earth's cross section overlays the surface map to illustrate Earth's radius. The continental crust thickness only extends an equivalent distance to the outskirts of Fort Lauderdale and its suburbs, some 30 km (18.6 mi) from the coast.

changing magnetism. Matching segments allow scientists to reassemble past continental arrangements. Later in this chapter we see the importance of these magnetic reversals.

Earth's Mantle

Figure 11.2b shows a transition zone several hundred kilometers wide, at an average depth of about 2900 km (1800 mi), dividing Earth's outer core from its mantle. Scientists at the California Institute of Technology analyzed more than 25,000 earthquakes and determined that this transition area is uneven, with ragged peak-and-valley-like formations. This is the *Gutenberg discontinuity*. A *discontinuity* is a place where physical differences occur between adjoining regions in Earth's interior, such as between the outer core and lower mantle. Scientists think that this zone of rough texture and low velocity creates some of the motion in the mantle.

The **mantle** (lower and upper together) represents about 80% of Earth's total volume. The mantle is rich in oxides of iron and magnesium and silicates (FeO, MgO, and SiO_2). They are dense and tightly packed at depth, grading to lesser densities toward the surface. A broad transition zone of several hundred kilometers, centered between 410 and 660 km (255–410 mi) below the surface, separates the upper mantle from the lower mantle. The entire mantle experiences a gradual temperature increase with depth and a stiffening due to increased pressures. The denser lower mantle is thought to contain a mixture of iron, magnesium, and silicates, with some calcium and aluminum. One hypothesis states that a tremendous amount of water is bound up with minerals in the mantle.

The upper mantle divides into three fairly distinct layers: upper mantle, asthenosphere, and *uppermost mantle*, shown in Figure 11.2c. Next to the crust is the uppermost mantle, a high-velocity zone just below the crust, where

seismic waves transmit through a rigid, cooler layer. This uppermost mantle, along with the crust, makes up the *lithosphere*, approximately 45–70 km (28–43 mi) thick.

Below the lithosphere, from about 70 km down to 250 km (43 mi to 155 mi), is the **asthenosphere**, or plastic layer (from the Greek *asthenos*, meaning "weak"). It contains pockets of increased heat from radioactive decay and is susceptible to slow convective currents in these hotter, and therefore less dense, materials.

Because of its dynamic condition, the asthenosphere is the least rigid region of the mantle, with densities averaging 3.3 g/cm³ (or 3300 kg/m³). About 10% of the asthenosphere is molten in asymmetrical patterns and hot spots. The resulting slow movement in this zone disturbs the overlying crust and creates tectonic activity—the folding, faulting, and general deformation of surface rocks. In return, movement of the crust apparently influences currents throughout the mantle.

The depth affected by convection currents is the subject of much scientific research. One body of evidence states that mixing occurs throughout the entire mantle, upwelling from great depths at the core–mantle boundary, sometimes in small blobs of material; other times, "megablobs" of mantle convect toward and away from the crust. Another view states that mixing in the mantle is layered, segregated above and below the 660-km boundary. Presently, evidence indicates some truth in both positions.

As an example, there are hot spots on Earth, such as those under Hawai'i, Easter Island, and Tahiti, that appear to be at the top of tall plumes of rising mantle rock that are anchored deep in the lower mantle, where the upward flow of warmer, less-dense material begins. The plume beneath Iceland appears to begin at the 660-km (410-mi) transition. Yet, there are other surface hot spots that sit atop smaller plumes going down some 200 km (124 mi). In other surface regions, slabs of crust descend and penetrate to the lower mantle.

Earth's Lithosphere and Crust

The lithosphere includes the **crust** and uppermost mantle to about 70 km (43 mi) in depth (Figure 11.2c). An important internal boundary between the crust and the high-velocity portion of the uppermost mantle is another discontinuity, called the **Mohorovičić discontinuity**, or **Moho** for short. It is named for the Yugoslavian seismologist who determined that seismic waves change at this depth owing to sharp contrasts of materials and densities.

Figure 11.2c illustrates the relation of the crust to the rest of the lithosphere and the asthenosphere below. Crustal areas beneath mountain masses extend deep, perhaps to 50–60 km (31–37 mi), whereas the crust beneath continental interiors averages about 30 km (19 mi) in thickness. Oceanic crust averages only 5 km (3 mi). The crust is only a fraction of Earth's overall mass. Drilling through the crust into the uppermost mantle remains an elusive scientific goal (see News Report 11.2).

News Report 11.2

Drilling the Crust to Record Depths

Scientists wanting to sample mantle material directly have unsuccessfully tried for decades to penetrate Earth's crust to the Moho discontinuity (the crust–mantle boundary). The longest-lasting deep-drilling attempt is on the northern Kola Peninsula near Zapolyarny, Russia, 250 km north of the Arctic Circle—the *Kola Borehole (KSDB)*. Twenty years of high-technology drilling (1970–1989) produced a hole 12.23 km deep (7.6 mi, or 40,128 ft), purely for exploration and science. Crystalline rock 1.4 billion years old at 180°C (356°F) was reached. The site has other active boreholes, and the fifth is underway. (See an analysis log at **http://www.icdp-online.de/**.) A recordholder for depth for a gas well is in Oklahoma; it

was stopped at 9750 m (32,000 ft) when the drill bit ran into molten sulfur.

Oceanic crust is thinner than continental crust and is the object of several drilling attempts. The Integrated Ocean Drilling Program (IODP), a cooperative effort between the governments of Japan and the United States, replaced the former ODP effort directed by Texas A & M University in 2003. However, the university program is still the Science Operator of the research ship *JOIDES Resolution* and lead coordinator.

Over the past 40 years, about 1700 bore holes were completed, yielding 160 km of sediment and rock cores and more than 35,000 samples for researchers. Much of this science

windfall helped put together the puzzle of geologic time and plate tectonics. See **http://www-odp.tamu.edu/** for background or **http://www.oceandrilling.org/** for the IODP home page information.

The first expeditions of the largest deep-ocean drilling ship, Japan's *Chikyu*, were in 2006 and 2007. The ship is the first capable of high-latitude drilling in the Arctic. The *Chikyu* is "riserless," in that it uses seawater as the primary drilling fluid (11.2.1b). The new IODP ship will be able to drill to 7 km, more than 3 times the *JOIDES Resolution* limits. The *JOIDES*, workhorse of the ODP, is being refurbished and converted to upgraded riserless status and reenters active duty in 2008 for the new IODP program.

(a)

(b)

FIGURE 11.2.1 Ocean drilling ship.
(a) A modern ocean-floor drilling ship, the *JOIDES Resolution*. The International Ocean Drilling Program operates the research ship. Since it began operations in 1984, *JOIDES* has spent more than 5000 days at sea through 2003 and recovered hundreds of kilometers of core for analysis.
(b) The new *Chikyu* research ship began drilling core samples from the ocean floor in 2006. [Photos by IODP and Texas A & M University.]

The composition and texture of continental and oceanic crusts are quite different, and this difference is a key to the concept of drifting continents. The oceanic crust is denser than continental crust. In collisions, the denser oceanic material plunges beneath the lighter, more buoyant continental crust.

- *Continental crust* is essentially **granite**; it is crystalline and high in silica, aluminum, potassium, calcium, and sodium. (Sometimes continental crust is called *sial*, shorthand for *si*lica and *al*uminum.) Continental crust is relatively low in density, averaging 2.7 g/cm³ (or 2700 kg/m³). Compare this with other densities given in Figure 11.2.

- *Oceanic crust* is **basalt**; it is granular and high in silica, magnesium, and iron. (Sometimes oceanic crust is called *sima*, shorthand for *si*lica and *ma*gnesium.) It is denser than continental crust, averaging 3.0 g/cm³ (or 3000 kg/m³).

Buoyancy is the principle that something less dense, such as wood, floats in something denser, such as water. The principles of buoyancy and balance were combined in the 1800s into the important principle of **isostasy**, which explains certain vertical movements of Earth's crust.

Think of Earth's crust as floating on the denser layers beneath, much as a boat floats on water. Where the load is greater, owing to glaciers, sediment, or mountains, the

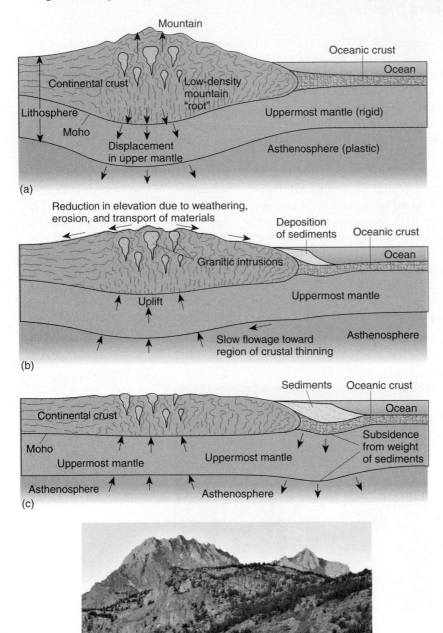

FIGURE 11.4 Isostatic adjustment of the crust.
Earth's entire crust is in a constant state of compensating adjustment, as suggested by these three sequential stages. In (a), the mountain mass slowly sinks, displacing mantle material. In (b), because of the loss of mass through erosion and transportation, the crust isostatically adjusts upward and sediments accumulate in the ocean. As the continental crust thins (c), the heavy sediment load offshore begins to deform the lithosphere beneath the ocean. (d) The melting of ice from the last ice age and losses of overlying sediments are thought to produce an ongoing isostatic uplift of portions of the Sierra Nevada batholith. [Photo by Bobbé Christopherson.]

crust tends to sink, or ride lower in the asthenosphere. Without that load (for example, when a glacier melts), the crust rides higher, in a recovery uplift known as *isostatic rebound*. Thus, the entire crust is in a constant state of compensating adjustment, or isostasy, slowly rising and sinking in response to its own burdens, as it is pushed, and dragged, and pulled over the asthenosphere (Figure 11.4).

An interesting development in Alaska relates this concept of isostatic rebound and climate change, as described in High Latitude Connection 11.1.

Earth's crust is the outermost shell: an irregular, brittle layer that resides restlessly on a dynamic and diverse interior. Let us examine the processes at work on this crust and the variety of rock types that compose the landscape.

High Latitude Connection 11.1

Isostatic Rebound in Alaska

The retreat of glacial ice following the last ice-age cycle unloaded much weight off the crust in Alaska, among those regions affected. Researchers from the Geophysical Institute at the University of Alaska–Fairbanks, using an array of global positioning system (GPS) receivers, measured isostatic rebound of the crust following glacial ice losses. They expected to find a slowed rate of crustal rebound in southeastern Alaska as compared to the more rapid response in the distant past when the ice first retreated.

Instead, they detected some of the most rapid vertical motion on Earth, averaging about 36 mm (1.42 in.) per year. Scientists attribute this isostatic rebound to the loss of glaciers in the region, especially the areas from Yakutat Bay and the Saint Elias Mountains in the north, through Glacier Bay, Juneau, and on south through the inland passage. Only about 1 mm of this uplift is from post–ice-age response. This rapid rebound is attributable to glacial melt and retreat over the past

150 years and correlates with accelerating ice losses and record warmth across Alaska (see News Report 17.1).

For instance, Glacier Bay experienced a retreat of ice of more than 97 km (60 mi) since 1794. The land that was beneath this ice has isostatically rebounded 5.5 m (18 ft). The vertical action appears to be affecting some of the fault systems in the region, although the extent of this and its consequences are unknown at this writing.

The Geologic Cycle

Earth's crust is in an ongoing state of change, being formed, deformed, moved, and broken down by physical, chemical, and biological processes. While the endogenic (internal) system is at work building landforms, the exogenic (external) system is busily wearing them down. This vast give-and-take at the Earth–atmosphere–ocean interface is the **geologic cycle**. It is fueled from two sources—Earth's internal heat and solar energy from space—influenced by the ever-present leveling force of Earth's gravity.

Figure 11.5 illustrates the geologic cycle, combining many of the elements presented in this text. The geologic cycle is composed of three subsystems:

- The *hydrologic cycle* is the vast system that circulates water, water vapor, ice, and energy throughout the Earth–atmosphere–ocean environment. This cycle rearranges Earth materials through erosion, transportation, and deposition, and it circulates water as the critical medium that sustains life. (Chapters 7, 8, and 9 discussed water's properties, weather, and the hydrologic cycle.)
- The *rock cycle*, through processes in the atmosphere, crust, and mantle, produces three basic rock types—igneous, sedimentary, and metamorphic. We examine this next.
- The *tectonic cycle* brings heat energy and new materials to the surface and recycles old materials to mantle

depths, creating movement and deformation of the crust. We discuss the tectonic cycle later in this chapter.

The Rock Cycle

To begin our look at the rock cycle, we see that only eight natural elements compose 99% of Earth's crust. Just two of these—oxygen and silicon—account for 74.3% of the crust (Table 11.1). Oxygen, the most reactive gas in the lower atmosphere, readily combines with other elements. For this reason, the percentage of oxygen is greater in the crust (about 47%) than in the atmosphere (about 21%). The internal differentiation process explains the relatively large percentages of lightweight elements such as silicon and aluminum in the crust. These less-dense elements migrate toward the surface, as discussed earlier.

Minerals and Rocks Earth's elements combine to form minerals. A **mineral** is an inorganic, or nonliving, natural compound having a specific chemical formula and usually possessing a crystalline structure. The combination of elements and the crystal structure give each mineral its characteristic hardness, color, density, and other properties. For example, the common mineral *quartz* is silicon dioxide, SiO_2, and has a distinctive six-sided crystal (see photo inset in Table 11.1).

Of the more than 4200 minerals, about 30 minerals comprise the *rock-forming minerals* most commonly encountered. *Mineralogy* is the study of the composition,

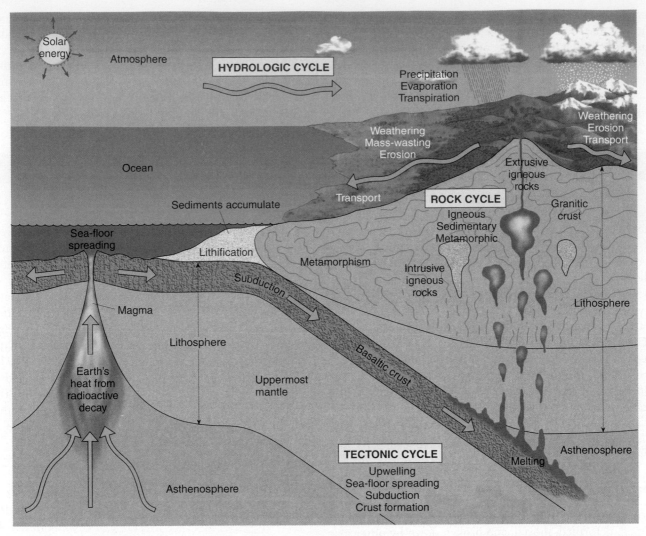

(a)

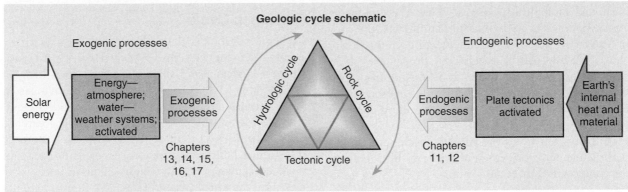

(b)

FIGURE 11.5 The geologic cycle.
(a) The geologic cycle is a model showing the interactive relation among the hydrologic cycle, rock cycle, and tectonic cycle. (b) Earth's surface is where two dynamic systems interact—the endogenic (internal) and exogenic (external).

Convection in a Lava Lamp
ANIMATION

properties, and classification of minerals. (Note that mercury, a liquid metal, is an exception to this definition, for it has no crystalline structure; also, that water is not a mineral, but ice does fit the definition of a mineral.) For more on minerals see **http://webmineral.com/**.

One of the most widespread mineral families on Earth is the *silicates* because silicon and oxygen are so common and because they readily combine with each other and with other elements. Roughly 95% of Earth's crust comprises silicates. This mineral family includes quartz, feldspar, clay minerals, and numerous gemstones. *Oxides* are another group of minerals in which oxygen combines with metallic elements, such as iron, to form hematite, Fe_2O_3. Also, there are the *sulfides* and *sulfates*

Table 11.1   Common Elements in Earth's Crust

Element	Percentage of Earth's Crust by Weight
Oxygen (O)	46.6
Silicon (Si)	27.7
Aluminum (Al)	8.1
Iron (Fe)	5.0
Calcium (Ca)	3.6
Sodium (Na)	2.8
Potassium (K)	2.6
Magnesium (Mg)	2.1
All others	1.5
Total	100.00

Note: A quartz crystal (SiO_2) consists of Earth's two most abundant elements, silicon (Si) and oxygen (O). Inset photo from *Laboratory Manual in Physical Geology*, 3rd ed., R. M. Busch, ed. © 1993 by Macmillan Publishing Co.

groups in which sulfur compounds combine with metallic elements to form pyrite (FeS_2) and anhydrite ($CaSO_4$), respectively.

Another important mineral family is the *carbonate* group, which features carbon in combination with oxygen and other elements such as calcium, magnesium, and potassium. An example is the mineral calcite ($CaCO_3$), a form of calcium carbonate.

A **rock** is an assemblage of minerals bound together (such as granite, a rock containing three minerals), or a mass of a single mineral (such as rock salt), or undifferentiated material (such as the noncrystalline glassy obsidian, or volcanic glass), or even solid organic material (such as coal). Thousands of different rocks have been identified. All can be sorted into one of three kinds, depending on the processes that formed them: *igneous* (melted), *sedimentary* (from settling out), and *metamorphic* (altered). Figure 11.6 illustrates these three processes and the interrelations among them that constitute the **rock cycle**. Let us examine each rock-forming process.

Igneous Processes

An **igneous rock** is one that solidifies and crystallizes from a molten state. Familiar examples are granite, basalt, and rhyolite. Igneous rocks form from **magma**, which is molten rock beneath the surface (hence the name *igneous*, which means "fire-formed" in Latin). Magma is fluid, highly gaseous, and under tremendous pressure. It either *intrudes* into crustal rocks, cools, and hardens, or it *extrudes* onto the surface as **lava**.

The cooling history of an igneous rock—how fast it cooled and how steadily its temperature dropped— determines its crystalline physical characteristics, or crystallization. Igneous rocks range from coarse-grained (slower cooling, with more time for larger crystals to form) to fine-grained or glassy (faster cooling).

Igneous rocks comprise approximately 90% of Earth's crust, although sedimentary rocks (sandstone, shale, limestone), soil, or oceans frequently cover them. Figure 11.7 illustrates the variety of occurrences of igneous rocks, both on and beneath Earth's surface.

Intrusive and Extrusive Igneous Rocks   Intrusive igneous rock that cools slowly in the crust forms a **pluton**, a general term for any intrusive igneous rock body, regardless of size or shape, that invaded layers of crustal rocks. The Roman god of the underworld, Pluto, is the namesake. The largest pluton form is a **batholith**, defined as an irregular-shaped mass with a surface greater than 100 km² (40 mi²) (Figure 11.8a). Batholiths form the mass of many large mountain ranges— for example, the Sierra Nevada batholith in California, the Idaho batholith, and the Coast Range batholith of British Columbia and Washington State.

Smaller plutons include the magma conduits of ancient volcanoes that have cooled and hardened. Those that form parallel to layers of sedimentary rock are *sills*; those that cross layers of the rock they invade are *dikes*. You see these two forms in Figure 11.7 at Petermann Island, Antarctica. Magma also can bulge between rock strata and produce a lens-shaped body called a *laccolith*, a type of sill. In addition, magma conduits themselves may solidify in roughly cylindrical forms that stand starkly above the landscape when finally exposed by weathering and erosion. Shiprock *volcanic neck* in New Mexico is such a feature, rising 518 m (1700 ft) above the surrounding plain, as suggested by the art in Figure 11.7 and shown in the inset photos; note the radiating dikes in the aerial photo. Weathering action of air, water, and ice can expose all of these intrusive forms.

Volcanic eruptions and flows produce extrusive igneous rock, such as lava that cools and forms basalt

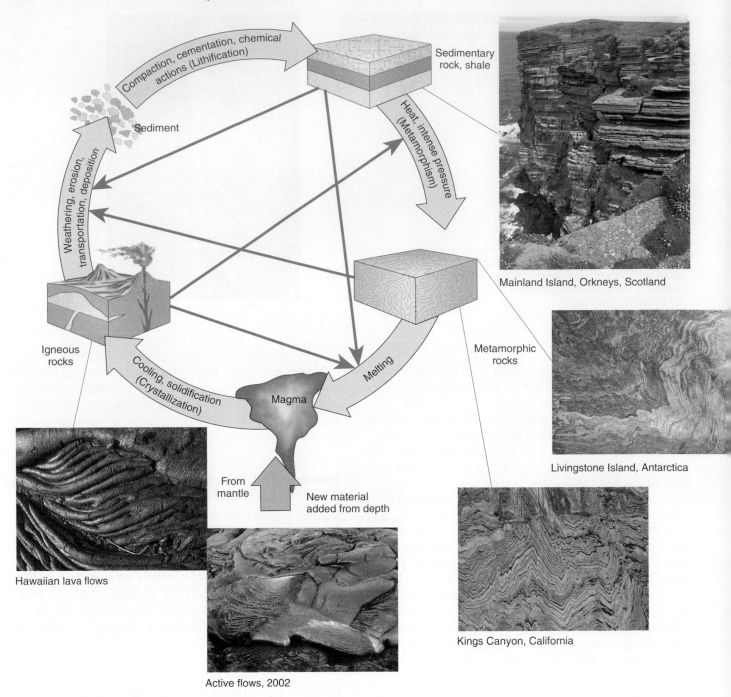

FIGURE 11.6 The rock cycle.
A rock-cycle schematic demonstrates the relation among igneous, sedimentary, and metamorphic processes. The arrows indicate that each rock type can enter the cycle at various points and be transformed into other rock types. [Art adapted by permission from R. M. Busch, ed., *Laboratory Manual in Physical Geology*, 3rd ed., © 1993 by Macmillan Publishing Co. Photos by Bobbé Christopherson.]

 The Rock Cyc

(Figure 11.8b). Chapter 12 presents volcanism and treats this in detail.

Classifying Igneous Rocks   Mineral composition and texture usually classify igneous rocks (Table 11.2). The two broad categories are:

1. *Felsic* igneous rocks—derived both in composition and name from *fel*dspar and *sili*ca. Felsic minerals are generally high in silica, aluminum, potassium, and sodium and have low melting points. Rocks formed from felsic minerals generally are lighter in color and are less dense than mafic mineral rocks.

2. *Mafic* igneous rocks—derived both in composition and name from *ma*gnesium and *fer*ric (Latin for iron). Mafic minerals are low in silica, high in magnesium and iron, and have high melting points.

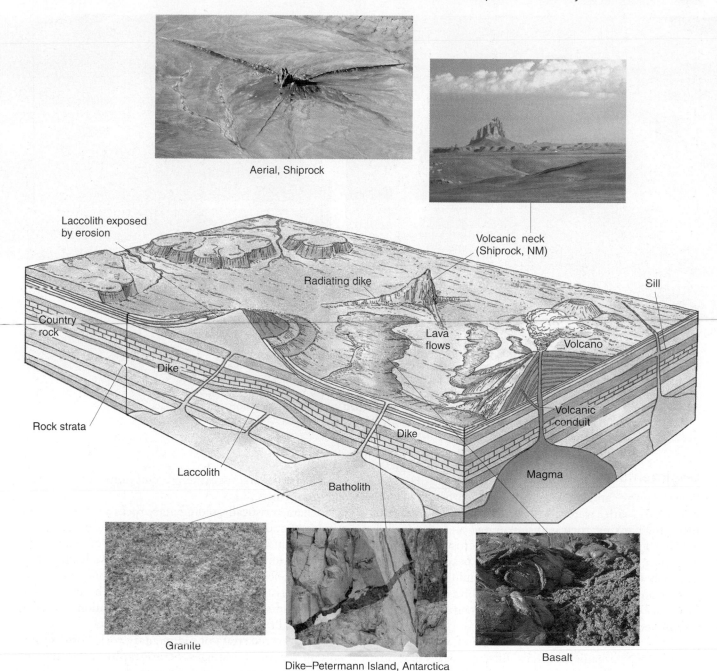

Aerial, Shiprock

Laccolith exposed by erosion

Volcanic neck (Shiprock, NM)

Radiating dike

Sill

Country rock

Lava flows

Volcano

Dike

Rock strata

Dike

Volcanic conduit

Laccolith

Batholith

Magma

Granite

Dike–Petermann Island, Antarctica

Basalt

FIGURE 11.7 Igneous rock types.
The variety of occurrences of igneous rocks, both intrusive (below the surface) and extrusive (on the surface). Inset photographs show samples of granite (intrusive) and basalt (extrusive) and other igneous forms. [Photos of rock samples and aerial by Bobbé Christopherson; surface photo of Shiprock volcanic neck by author.]

ANIMATION

Formation of Intrusive Igneous Features

Rocks formed from mafic minerals are darker in color and of greater density than felsic mineral rocks.

The same magma that produces coarse-grained granite (when it slowly cools beneath the surface) can form fine-grained *rhyolite* (when it cools above the surface—see inset photo in table). If it cools quickly, magma having silica content comparable to granite and rhyolite can form the dark, smoky, glassy-textured rock called *obsidian*, or volcanic glass (see inset photo). Another glassy rock,

pumice, forms when escaping gases bubble a frothy texture into the lava. Pumice is full of small holes, is light in weight, and is low enough in density to float in water (see inset photo).

On the mafic side, basalt is the most common fine-grained extrusive igneous rock. It makes up the bulk of the ocean floor, accounting for 71% of Earth's surface. It appears in lava flows such as those on the big island of Hawai'i (Figure 11.8b). An intrusive counterpart to basalt, formed by slow cooling of the parent magma, is *gabbro*.

(a)

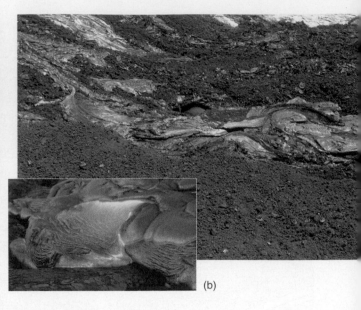

(b)

FIGURE 11.8 Intrusive and extrusive rocks.
(a) Exposed granites of the Sierra Nevada batholith. The boulders sitting on the granite are erratics left behind by glacial ice that melted and retreated thousands of years ago. (b) Basaltic lava flows on Hawai'i. The glowing opening is from a skylight into an active lava tube where molten lava is visible; the shiny surface is where lava recently flowed out of the skylight. (Do you see a suggestion of a dragon head and glowing red eye in the lava sculpture?). [Photos by Bobbé Christopherson.]

Sedimentary Processes

Solar energy and gravity drive the process of *sedimentation*, with water as the principal transporting medium. Existing rock is disintegrated and dissolved by weathering, picked up and moved by erosion and transportation, and deposited along river, beach, and ocean sites, where burial initiates the rock-forming process. The formation of **sedimentary rock** involves **lithification** processes of cementation, compaction, and a hardening of sediments.

Most sedimentary rocks derive from fragments of existing rock or organic materials. Bits and pieces of former rocks—principally quartz, feldspar, and clay minerals—erode and then are mechanically transported by water (lake, stream, ocean, overland flow), ice (glacial action), wind, and gravity. They are transported from "higher-energy" sites, where the carrying medium has the energy to pick up and move them, to "lower-energy" sites, where the material is dumped.

The common sedimentary rocks are *sandstone* (sand that became cemented together), *shale* (mud that became compacted into rock, as in the part-opening photo), *limestone* (bones and shells that became cemented or calcium carbonate that precipitated in ocean and lake waters), and *coal* (ancient plant remains that became compacted into rock). Some minerals, such as calcium carbonate, dissolve into solution and form sedimentary deposits by precipitating from those solutions to form rock. This is an important process in both the oceanic and karst (weathered limestone; Chapter 13) environments.

Characteristically, sedimentary rocks are laid down by wind, water, or ice in horizontally layered beds. Different environmental conditions produce a variety of sedimentary forms. Various cements, depending on availability, fuse rock particles together. Lime, or calcium carbonate ($CaCO_3$), is the most common, followed by iron oxides (Fe_2O_3) and silica (SiO_2). Drying (dehydration), heating, or chemical reactions can also unite particles.

The layered strata of sedimentary rocks form an important record of past ages. **Stratigraphy** is the study of the sequence (superposition), thickness, and spatial distribution of strata. These sequences yield clues to the age and origin of the rocks. Figure 11.9a shows a sandstone sedimentary rock in a desert landscape. Note the multiple layers in the formation and how differently they resist weathering processes. The climatic history of this rock's formation environment is disclosed in the stratigraphy—drier periods of former dunes near the top and wetter periods of horizontal deposits in the middle and lower portions. Figure 11.9c shows sedimentary rock strata on Mars that imply water deposition of sediments sometime in the distant Martian past.

The two primary sources of sedimentary rocks are *clastic sediments*, formed from the mechanically transported fragments of older rock, and *chemical sediments* from the dissolved minerals in solution, some having organic origins.

Table 11.2   Igneous Rock Minerals

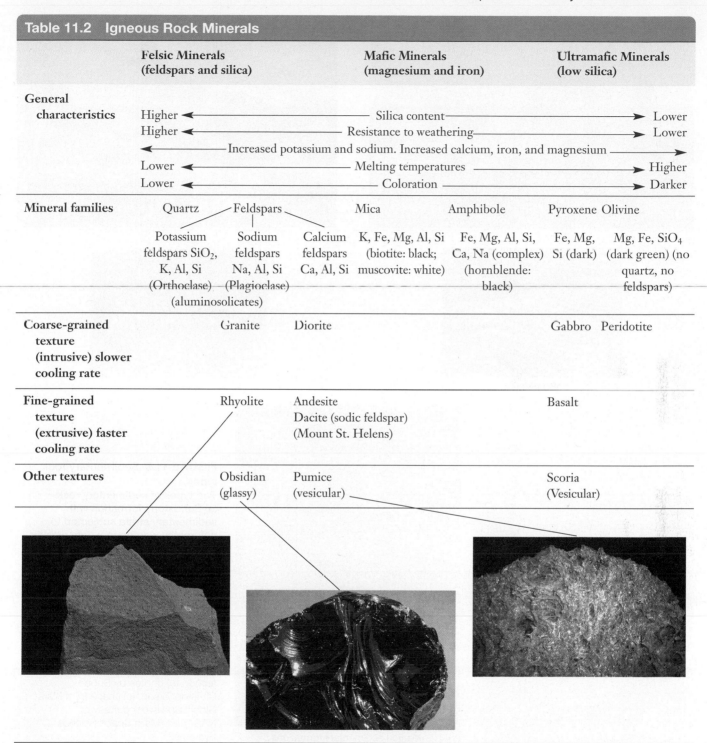

	Felsic Minerals (feldspars and silica)			Mafic Minerals (magnesium and iron)		Ultramafic Minerals (low silica)	
General characteristics	Higher ◄———————— Silica content ————————► Lower						
	Higher ◄———————— Resistance to weathering ————————► Lower						
	◄——— Increased potassium and sodium. Increased calcium, iron, and magnesium ———►						
	Lower ◄———————— Melting temperatures ————————► Higher						
	Lower ◄———————— Coloration ————————► Darker						
Mineral families	Quartz	Feldspars		Mica	Amphibole	Pyroxene	Olivine
	Potassium feldspars SiO₂, K, Al, Si (Orthoclase) (aluminosolicates)	Sodium feldspars Na, Al, Si (Plagioclase)	Calcium feldspars Ca, Al, Si	K, Fe, Mg, Al, Si (biotite: black; muscovite: white)	Fe, Mg, Al, Si, Ca, Na (complex) (hornblende: black)	Fe, Mg, Si (dark)	Mg, Fe, SiO₄ (dark green) (no quartz, no feldspars)
Coarse-grained texture (intrusive) slower cooling rate	Granite	Diorite				Gabbro	Peridotite
Fine-grained texture (extrusive) faster cooling rate	Rhyolite	Andesite Dacite (sodic feldspar) (Mount St. Helens)				Basalt	
Other textures	Obsidian (glassy)	Pumice (vesicular)				Scoria (Vesicular)	

Source: Photos from R. M. Busch, ed., *Laboratory Manual in Physical Geology*, 3rd ed., © 1993 by Macmillan Publishing Co.

Clastic Sedimentary Rocks   Weathered and fragmented rocks that are further worn in transport provide *clastic sediments.* Table 11.3 lists the range of *clast* sizes—everything from boulders to microscopic clay particles—and the form they take as *lithified rock.* Common sedimentary rocks from clastic sediments include siltstone or mudstone (silt-sized particles), shale (clay to silt-sized particles), and sandstone (sand-sized particles ranging from 0.06 to 2 mm in diameter).

Chemical Sedimentary Rocks   *Chemical sedimentary* rocks are not formed from physical pieces of broken rock but instead from dissolved minerals, transported in solution and chemically precipitated from solution (they are essentially nonclastic). The most common chemical sedimentary rock is **limestone**, which is lithified calcium carbonate, derived from inorganic and organic sources. A similar form is *dolomite*, which is lithified calcium–magnesium carbonate, $CaMg(CO_3)_2$. Limestone from

(a)

(b)

Sandstone

Limestone

(c) Martian sedimentary formations

FIGURE 11.9 Sedimentary rock types.
Two types of sedimentary rock.
(a) Sandstone formation with sedimentary strata subjected to differential weathering; note weaker underlying siltstone. (b) A limestone landscape, formed by chemical sedimentary processes, in south-central Indiana; inset samples have whole shells and some clasts (parts) cemented together. (c) Ancient sediment deposition through repeated cycles of erosion and deposition is evident in the **Martian western Arabia Terra (8° N and 7° W).** [(a) Photos by author; (b) photos by Bobbé Christopherson; (c) image from Mars Global Surveyor courtesy of NASA/JPL/Malin Space Science Systems.]

marine organic origins is most common; it is biochemical—derived from shell and bone produced by biological activity (see inset photo in Figure 11.9b). Once formed, these rocks are vulnerable to chemical weathering, which produces unique landforms, as discussed in the weathering section of Chapter 13.

Chemical sediments from inorganic sources are deposited when water evaporates and leaves behind a residue of salts. These *evaporites* may exist as common salts, such as gypsum or sodium chloride (table salt). They often appear as flat, layered deposits across a dry landscape. The pair of photographs in Figure 11.10 dramatically demonstrates this process; one was made in Death Valley National Park one day after a record 2.57-cm (1.01-in.) rainfall, and the other photo was made one month later at the exact same spot.

Chemical deposition also occurs in the water of natural hot springs from chemical reactions between minerals and oxygen. This is a sedimentary process related to *hydrothermal activity*. An example is the massive

Table 11.4	
Parent Rock	
Shale (clay mi	
Granite, slate	
Basalt, shale,	
Limestone, d	
Sandstone	

Plate Tectonic

Have you ever looked at
few of the continental
America and Africa—app
pieces of a jigsaw puzzl
Peninsula coastlines of th
is that the continental pi
gether! Continental land
present locations but co
6 cm (2.4 in.) per year.

We say that the cont
dense lithosphere is m
upper mantle, the asthen
tle that is active and in m
arrangement of continer
not permanent, but is in a

A continent, such as
lage of crustal pieces that
the landscape we know. T
let us trace the discoveri
theory—a major revolutio

A Brief History

As early mapping gained
the symmetry among
between South America
(1527–1598), a geographe
continental coastlines in h
In 1620, English philoso
gross similarities between
America (although he did
apart). Benjamin Frankli
must be a shell that can b
fluid below. Others wro

Table 11.3 Clastic Sediment Sizes and Related Rock Form		
Unconsolidated Sediment	**Grain Size**	**Rock Form**
Boulders, cobbles	>80 mm	Conglomerate (breccia, if pieces are angular)
Pebbles, gravel	>2 mm	
Coarse sand	0.5–2.0 mm	Sandstone
Medium-to-fine sand	0.062–0.5 mm	Sandstone
Silt	0.002–0.062 mm	Siltstone (mudstone)
Clay	<0.002 mm	Shale

deposit of travertine, a form of calcium carbonate, at Mammoth Hot Springs in Yellowstone National Park (Figure 11.11a).

Hydrothermal activity also occurs on the ocean floor along the rift valleys formed by sea-floor spreading. These "black smokers" belch dark clouds of hydrogen sulfides, minerals, and metals that hot water (in excess of 380°C; 716°F) leached from the basalt. In contact with seawater, metals and minerals precipitate into deposits (Figure 11.11b). Along the Mid-Atlantic Ridge, one region of hydrothermal activity has 30- to 60-m- (98- to 198-ft-) tall vents of calcium carbonate that are at least 30,000 years old and still active. Here, the interaction of the mineral-laden spring water reacts with seawater to produce rock. This area is nicknamed "The Lost City" because of the way the towering chimneys look from the observation submersibles.

Metamorphic Processes

Any rock, either igneous or sedimentary, may be transformed into a **metamorphic rock** by going through profound physical or chemical changes under pressure and increased temperature. (The name *metamorphic* comes from a Greek word meaning "to change form.") Metamorphic rocks generally are more compact than the original rock and therefore are harder and more resistant to weathering and erosion (Figure 11.12).

Several conditions can cause metamorphism. Most common is when subsurface rock is subjected to high temperatures and high compressional stresses occurring over millions of years. Igneous rocks become compressed during collisions between slabs of Earth's crust (see the discussion of plate tectonics later in this chapter). Sometimes rocks simply are crushed under great weight when a crustal area is thrust beneath other crust. In another setting, igneous rocks may be sheared and stressed along earthquake fault zones, causing metamorphism.

Metamorphic rocks comprise the ancient roots of mountains. Exposed at the bottom of the inner gorge of the Grand Canyon in Arizona, a Precambrian (Archean) metamorphic rock, the Vishnu Schist, is a remnant of such an ancient mountain root (Figure 11.12b). Despite the fact that the schist is harder than steel, the Colorado River has cut down into the uplifted Colorado Plateau, exposing and eroding these ancient rocks.

Another metamorphic condition occurs when sediments collect in broad depressions in Earth's crust and, because of their own weight, create enough pressure in the bottommost layers to transform the sediments into metamorphic rock, a process called *regional metamorphism*. Also, molten magma rising within the crust may "cook" adjacent rock, a process called *contact metamorphism*.

(a)

(b)

FIGURE 11.10 Death Valley, wet and dry.
A Death Valley landscape: (a) one day after a record rainfall when the valley was covered by several square kilometers of water only a few centimeters deep; (b) one month later the water had evaporated, and the same valley is coated with evaporites (borated salts) shown in close-up inset photo. [Photos by author.]

(a)

(b)

FIGURE 11.11 Hydrothe
(a) Mammoth Hot Springs
example of a hydrotherma
travertine (CaCO₃), depos
heated spring water as it
with black smokers rise a
[(a) Photo by author; (b) ima
of Oceanography.]

Metamorphic rocks
and chemically from th
some metamorphic rock
sultant textures. If the
particular alignment a
foliated, and some miner
(streaks or lines) in the
tographs demonstrate f
Which form do you se
metamorphic rock from
(center right), foliated o

Also in the table, no
comes metamorphic sla

of the way Earth's surface evolves. *Tectonic*, from the Greek *tektonikùs*, meaning "building" or "construction," refers to changes in the configuration of Earth's crust as a result of internal forces. **Plate tectonic** processes include upwelling of magma; lithospheric plate movements; sea-floor spreading and lithospheric subduction; earthquakes; volcanic activity; and lithospheric deformation such as warping, folding, and faulting.

Sea-Floor Spreading and Production of New Crust

The key to establishing the theory of continental drift was a better understanding of the seafloor. The seafloor has a remarkable feature: an interconnected worldwide mountain chain, forming a ridge some 64,000 km (40,000 mi) in extent and averaging more than 1000 km (600 mi) in width. A striking view of this great undersea mountain chain opens Chapter 12. How did this global mountain chain get there?

In the early 1960s, geophysicists Harry H. Hess and Robert S. Dietz proposed **sea-floor spreading** as the mechanism that builds this mountain chain and drives continental movement. Hess said that these submarine mountain ranges were **mid-ocean ridges** and the direct result of upwelling flows of magma from hot areas in the upper mantle and asthenosphere and perhaps from the deeper lower mantle. These are sites of intense hydrothermal activity, as mentioned earlier.

When mantle convection brings magma up to the crust, the crust fractures, and the magma extrudes onto the seafloor and cools to form new seafloor. This process builds the mid-ocean ridges and spreads the seafloor laterally, rifting as the two sides move apart. The concept is illustrated in Figure 11.13, which shows how the ocean floor is rifted (moves apart) and scarred along mid-ocean ridges. Figure 11.13b is a remote-sensing image of the mid-ocean ridge and Figure 11.13c is a map of the system in the Atlantic. Figure 11.13d shows a portion of the mid-ocean ridge system that surfaces at Thingvellir, Iceland, forming these rifts—the North American plate along one side and the European plate on the other. Iceland's first Althing (parliament) met here in A.D. 930; leaders used the acoustics of the rift walls to be heard by the gathering.

As new crust generates and the seafloor spreads, magnetic particles in the lava orient with the magnetic field in force at the time the lava cools and hardens. The particles become locked in this alignment as new seafloor forms, which creates a kind of magnetic tape recording in the seafloor. The continually forming oceanic crust records each magnetic reversal and reorientation of Earth's polarity.

Figure 11.14 illustrates just such a recording from the Mid-Atlantic Ridge south of Iceland. The colors denote the alternating magnetic polarization preserved in the minerals of the oceanic crust. Note the mirror images that develop on either side of the sea-floor rift as a result of the nearly symmetrical spreading of the seafloor; the

relative ages of the rocks increase with distance from the ridge. These periodic reversals of Earth's magnetic field are a valuable clue to understanding sea-floor spreading, helping scientists fit together pieces of Earth's crust.

These sea-floor recordings of Earth's magnetic-field reversals and other measurements allowed the age of the seafloor to be determined. The complex harmony of the two concepts of plate tectonics and sea-floor spreading thus became clearer. The youngest crust anywhere on Earth is at the spreading centers of the mid-ocean ridges, and with increasing distance from these centers, the crust gets steadily older (Figure 11.15). The oldest seafloor is in the western Pacific near Japan, dating to the Jurassic Period. Note on the map in the figure the distance between this basin and its spreading center in the South Pacific west of South America.

Overall, the seafloor is relatively young; nowhere does it exceed 208 m.y.a. (million years ago)—remarkable when you remember that Earth's age is 4.6 billion years. The reason is that oceanic crust is short-lived—the oldest sections, farthest from the mid-ocean ridges, are slowly plunging beneath continental crust along Earth's deep oceanic trenches. The discovery that the seafloor is young demolished earlier thinking that the oldest rocks would be found there.

Subduction of the Lithosphere

In contrast to the upwelling zones along the mid-ocean ridges are the areas of descending lithosphere elsewhere. On the left side of Figure 11.13a note how one plate of the crust is diving, or being dragged beneath another, into the mantle. Recall that the basaltic ocean crust has a density of 3.0 g/cm³, whereas continental crust averages a lighter 2.7 g/cm³. As a result, when continental crust and oceanic crust slowly collide, the denser ocean floor will grind beneath the lighter continental crust, thus forming a **subduction zone**, as shown in the figure. The subducting slab of crust exerts a gravitational pull on the rest of the plate—an important driving force in plate motion. This plate motion can trigger through shear traction along the base of the lithosphere a flow in the mantle material beneath. Thus, subduction and plate motion are a complex process.

The world's deep ocean trenches coincide with these subduction zones and are the lowest features on Earth's surface. Deepest is the Mariana Trench near Guam, which descends below sea level to −11,030 m (−36,198 ft); next in depth are the Puerto Rico Trench at −8605 m (−28,224 ft) and, in the Indian Ocean, the Java Trench at −7125 m (−23,376 ft).

The subducted portion of lithosphere travels down into the asthenosphere, where it remelts and eventually is recycled as magma, rising again toward the surface through deep fissures and cracks in crustal rock. Volcanic mountains such as the Andes in South America and the Cascade Range from northern California to the Canadian border form inland of these subduction zones as a result of rising plumes of magma, as suggested in Figure 11.13a.

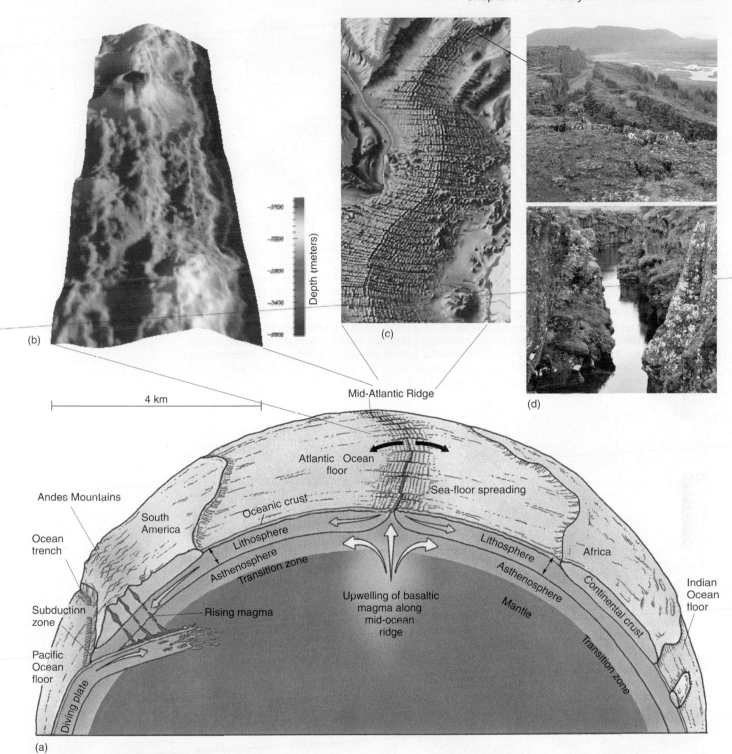

Depth (meters)

4 km

Mid-Atlantic Ridge

(b)

(c)

(d)

Atlantic Ocean floor

Sea-floor spreading

Andes Mountains

South America

Oceanic crust

Ocean trench

Lithosphere

Asthenosphere

Transition zone

Subduction zone

Rising magma

Pacific Ocean floor

Diving plate

Upwelling of basaltic magma along mid-ocean ridge

Lithosphere

Asthenosphere

Mantle

Continental crust

Africa

Indian Ocean floor

Transition zone

(a)

FIGURE 11.13 Crustal movements.
(a) Sea-floor spreading, upwelling currents, subduction, and plate movements, shown in cross section. Arrows indicate the direction of the spreading. (b) A 4-km-wide image of the Mid-Atlantic Ridge showing linear faults, a volcanic crater, a rift valley, and ridges. The image was taken by the TOBI (towed ocean-bottom instrument) at approximately 29° N latitude. (c) Detail from Tharp's ocean-floor map. (d) The mid-Atlantic rift surfaces through Iceland. This scene is at Thingvellir in southeastern Iceland, where the North American and European plates of Earth's crust meet. Only in Iceland does a portion of Earth's 64,000 km (40,000 mi) mid-ocean ridge system surface. [(a) After P. J. Wyllie, *The Way the Earth Works*, © 1976, by John Wiley & Sons, adapted by permission; (b) image courtesy of D. K. Smith, Woods Hole Oceanographic Institute, Woods Hole, Massachusetts. All rights reserved; (c) Bruce C. Heezen and Marie Tharp, courtesy Marie Tharp; (d) photos by Bobbé Christopherson.]

ANIMATION

Sea-floor Spreading, Subduction

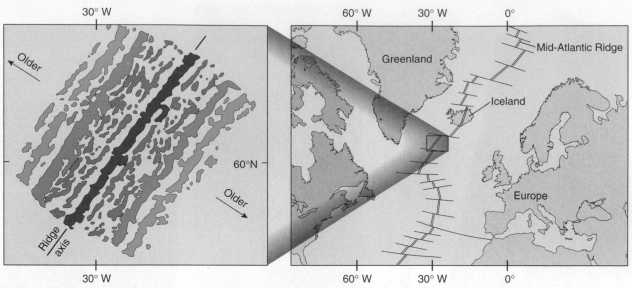

FIGURE 11.14 Magnetic reversals recorded in the seafloor.
Magnetic reversals recorded in the seafloor south of Iceland along the Mid-Atlantic Ridge. Note rocks with similar magnetic orientation are roughly the same distance from the spreading center. Colors indicate rocks of similar geologic age, with the rocks getting increasingly older with distance away from the spreading center. [Magnetic reversals reprinted from J. R. Heirtzler, S. Le Pichon, and J. G. Baron, *Deep-Sea Research* 13, © 1966, Pergamon Press, p. 247.]

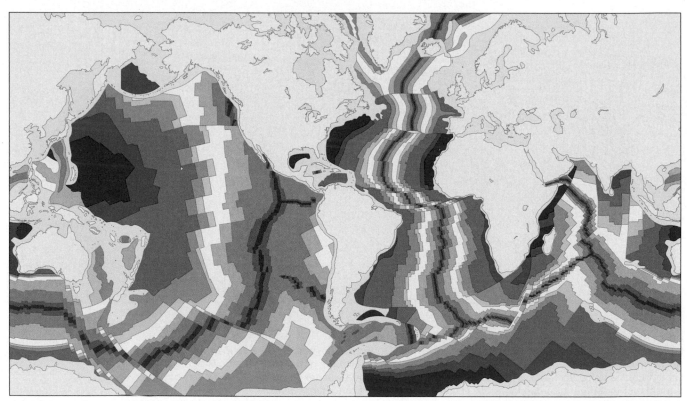

FIGURE 11.15 Relative age of the oceanic crust.
Compare the dark-red color near the East Pacific rise in the eastern Pacific Ocean with the same dark-red color along the Mid-Atlantic Ridge. What does the difference in width tell you about the rates of plate motion in the two locations? [Adapted with the permission of W. H. Freeman and Company from *The Bedrock Geology of the World* by R. L. Larson and others, © 1985.]

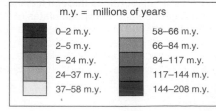

m.y. = millions of years

0–2 m.y.	58–66 m.y.
2–5 m.y.	66–84 m.y.
5–24 m.y.	84–117 m.y.
24–37 m.y.	117–144 m.y.
37–58 m.y.	144–208 m.y.

Sometimes the diving plate remains intact for hundreds of kilometers, whereas at other times it can break into large pieces, thought to be the case under the Cascade Range with its distribution of volcanoes.

The fact that spreading ridges and subduction zones are areas of earthquake and volcanic activity provides important proof of plate tectonics. Now, using current scientific findings, let us go back and reconstruct the past and Pangaea. See this in review at **http://geology.er.usgs.gov/eastern/tectonic.html**.

The Formation and Breakup of Pangaea

The supercontinent of Pangaea and its subsequent breakup into today's continents represent only the last 225 million years of Earth's 4.6 billion years, or only the most recent 1/23 of Earth's existence. During the other 22/23 of geologic time, other things were happening. The landmasses as we know them were unrecognizable.

Figure 11.16a begins with the pre-Pangaea arrangement of 465 m.y.a. (during the middle Ordovician Period). Researchers are discovering other, older arrangements of continents. One proposal is for an ancient supercontinent, tentatively called *Rodinia*, that formed approximately a billion years ago, with breakup at about 700 million.

Figure 11.16b illustrates an updated version of Wegener's Pangaea, 225–200 m.y.a. (Triassic–Jurassic Periods). The movement of plates that occurred by 135 m.y.a. (the beginning of the Cretaceous Period) is in Figure 11.16c. Figure 11.16d presents the arrangement 65 m.y.a. (shortly after the beginning of the Tertiary Period). Finally, the present arrangement in modern geologic time (the late Cenozoic Era) is in Figure 11.16e.

Earth's present crust is divided into at least 14 plates, of which about half are major and half are minor, in terms of area (Figure 11.17). Literally hundreds of smaller pieces and perhaps dozens of microplates that migrated together comprise these broad plates. The arrows in the figure indicate the direction in which each plate is presently moving, and the length of the arrows suggests the rate of movement during the past 20 million years. On the CD-ROM that accompanies this text you can manipulate this map to feature the speed arrows, or the plates, with or without the landmasses.

Compare Figure 11.17 with the image of the ocean floor in Figure 11.18. In the image, satellite radar-altimeter measurements determined the sea-surface height to a remarkable accuracy of 0.03 m, or 1 in. The sea-surface elevation is far from uniform; it is higher or lower in direct response to the mountains, plains, and trenches of the ocean floor beneath it. This sea-floor topography causes slight differences in Earth's gravity.

For example, a massive mountain on the ocean floor exerts a high gravitational field and attracts water to it, producing higher sea level above the mountain; over a trench there is less gravity, resulting in a drop in sea level. A mountain 2000 m (6500 ft) high that is 20 km (12 mi) across its base creates a 2-m rise in the sea surface. Until Scripps Institution of Oceanography and NOAA developed this use of satellite technology, these tiny changes in height were not detectable. Now we have a comprehensive map of the ocean *floor*, derived from remote sensing of the ocean *surface*!

The illustration of the seafloor that begins Chapter 12 also is helpful in identifying the plate boundaries in the following discussion. It is interesting to correlate the features illustrated in Figures 11.17 and 11.18 and Chapter 12's opening map. For more on the science of plate tectonics, see **http://www.ig.utexas.edu/research/**.

Plate Boundaries

The boundaries where plates meet clearly are dynamic places, although slow-moving within human time frames. The block diagram inserts in Figure 11.16e show the three general types of motion and interaction that occur along the boundary areas:

- *Divergent boundaries* (lower left in figure) are characteristic of sea-floor spreading centers, where upwelling material from the mantle forms new seafloor and lithospheric plates spread apart—a constructional process. The spreading makes these zones of tension. An example noted in the figure is the divergent boundary along the East Pacific rise, which gives birth to the Nazca plate (moving eastward) and the Pacific plate (moving northwestward). Whereas most divergent boundaries occur at mid-ocean ridges, there are a few within continents themselves. An example is the Great Rift Valley of East Africa, where crust is rifting apart.

- *Convergent boundaries* (upper left) are characteristic of collision zones, where areas of continental and oceanic lithosphere collide. These are zones of compression and crustal loss—a destructional process. Examples include the subduction zone off the west coast of South and Central America and the area along the Japan and Aleutian Trenches. Along the western edge of South America, the Nazca plate collides with and is subducted beneath the South American plate. This convergence creates the Andes Mountains chain and related volcanoes. The collision of India and Asia is another example of a convergent boundary, mentioned with Figure 11.17.

- *Transform boundaries* (lower right) occur where plates slide laterally past one another at right angles to a sea-floor spreading center, neither diverging nor converging, and usually with no volcanic eruptions. These are the right-angle fractures stretching across the mid-ocean ridge system worldwide.

Related to plate boundaries, another piece of the tectonic puzzle fell into place in 1965 when University of Toronto geophysicist Tuzo Wilson first described the nature of transform boundaries and their relation to *(text continued on page 349)*

FIGURE 11.16 Continents adrift, from 465 m.y.a. to the present. Observe the formation and breakup of Pangaea and the types of motions occurring at plate boundaries. [(a) From R. K. Bambach, "Before Pangaea: The geography of the Paleozoic world," *American Scientist* 68 (1980): 26–38, reprinted by permission; (b–e) from R. S. Dietz and J. C. Holden, *Journal of Geophysical Research* 75, no. 26 (September 10, 1970): 4939–4956, © The American Geophysical Union; (f) remote-sensing image courtesy of S. Tighe, University of Rhode Island, and R. Detrick, Woods Hole Oceanographic Institute, Woods Hole, Massachusetts.]

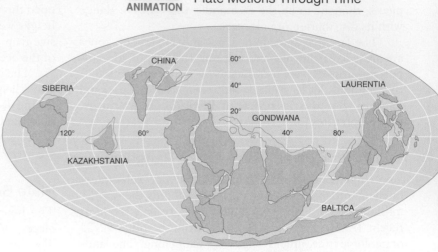

ANIMATION Plate Motions Through Time

(a) 465 million years ago

Pangaea—all Earth. Panthalassa ("all seas") became the Pacific Ocean, and the Tethys Sea (partly enclosed by the African and the Eurasian plates) became the Mediterranean Sea; trapped portions of former ocean became the present-day Caspian Sea. The Atlantic Ocean did not exist.

Africa shared a common connection with both North and South America. Today, the Appalachian Mountains in the eastern United States and the Lesser-Atlas Mountains of northwestern Africa reflect this common ancestry; they are, in fact, portions of the same mountain range, torn thousands of kilometers apart.

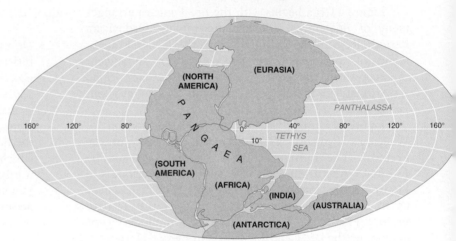

(b) 225 million years ago

New seafloor that formed since the last map is highlighted (gray tint). An active spreading center rifted North America away from landmasses to the east, shaping the coast of Labrador. India was farther along in its journey, with a spreading center to the south and a subduction zone to the north; the leading edge of the India plate was diving beneath Eurasia.

The outlines of South America, Africa, India, Australia, and southern Europe appear within the continent called *Gondwana* in the Southern Hemisphere. North America, Europe, and Asia comprise *Laurasia*, the northern portion.

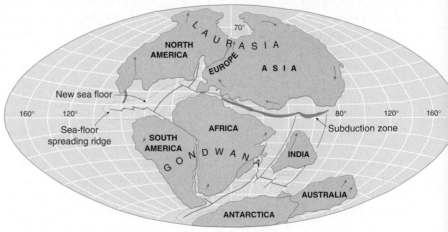

(c) 135 million years ago

ANIMATION Motions at Plate Boundaries

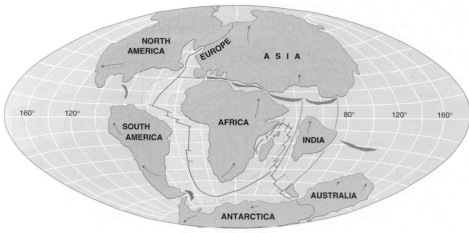

(d) 65 million years ago

Sea-floor spreading along the Mid-Atlantic Ridge grew some 3000 km (almost 1900 mi) in 70 million years. Africa moved northward about 10° in latitude, leaving Madagascar split from the mainland and opening up the Gulf of Aden. The rifting along what would be the Red Sea began. The India plate moved three-fourths of the way to Asia, as Asia continued to rotate clockwise. Of all the major plates, India traveled the farthest—almost 10,000 km.

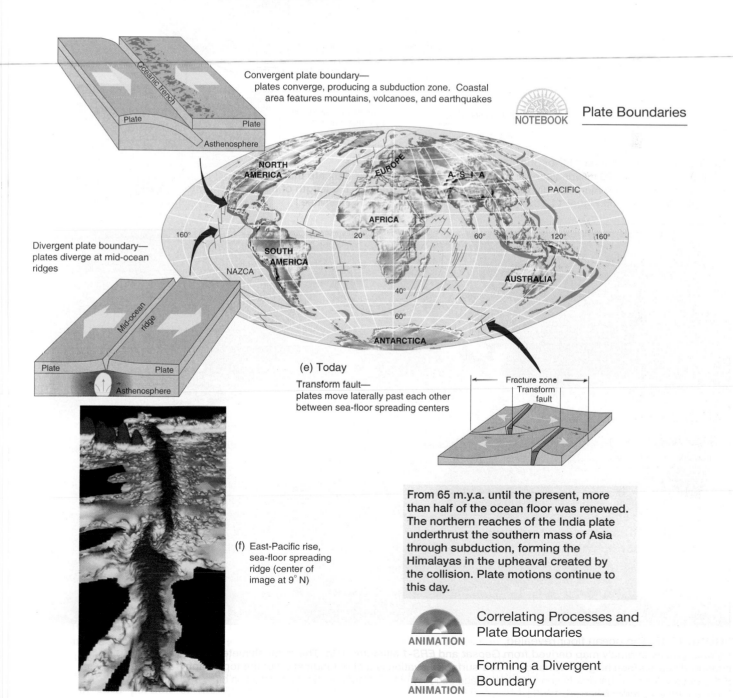

Convergent plate boundary—plates converge, producing a subduction zone. Coastal area features mountains, volcanoes, and earthquakes

Plate Boundaries

NOTEBOOK

Divergent plate boundary—plates diverge at mid-ocean ridges

(e) Today

Transform fault—plates move laterally past each other between sea-floor spreading centers

Fracture zone
Transform fault

(f) East-Pacific rise, sea-floor spreading ridge (center of image at 9° N)

From 65 m.y.a. until the present, more than half of the ocean floor was renewed. The northern reaches of the India plate underthrust the southern mass of Asia through subduction, forming the Himalayas in the upheaval created by the collision. Plate motions continue to this day.

ANIMATION   Correlating Processes and Plate Boundaries

ANIMATION   Forming a Divergent Boundary

Focus Study 11.1 (continued)

problems of decreased yields. The Geysers are operating below the capacity of the field potential at the time of this writing.

Some 2800 MWe of geothermal power capacity is operating in California, Hawai'i, Nevada, and Utah, with a new 185-MWe plant proposed in the desert near Brawley, California. Figure 11.1.4 maps low-, medium-, and high-temperature known geothermal resource areas (KGRA) in the United States. About 300 cities are within 8 km (5 mi) of a KGRA. The Department of Energy has identified about 9000 potential development sites. In Canada, a proposed project at Mount Meager, B.C., will produce electricity; geothermal direct heats buildings at Carleton University, Ottawa; and some direct operations are in development in Nova Scotia. For more information, see http://www1.eere.energy. gov/geothermal/, http://geothermal.marin. org/, http:// www.geothermal. org/, or http:// www.smu.edu/geothermal/.

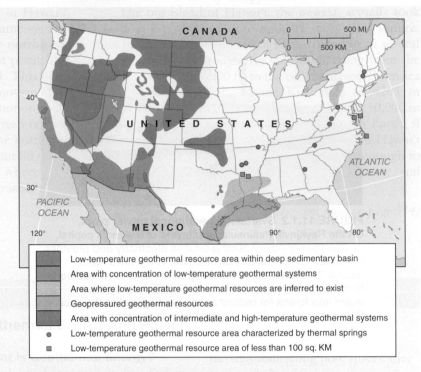

FIGURE 11.1.4 Geothermal resources.
Geothermal resources in the conterminous United States. [Map courtesy of USGS, from W. A. Duffield et al., *Tapping the Earth's Natural Heat*, USGS Circular 1125, 1994, p. 35.]

Legend:
- Low-temperature geothermal resource area within deep sedimentary basin
- Area with concentration of low-temperature geothermal systems
- Area where low-temperature geothermal resources are inferred to exist
- Geopressured geothermal resources
- Area with concentration of intermediate and high-temperature geothermal systems
- Low-temperature geothermal resource area characterized by thermal springs
- Low-temperature geothermal resource area of less than 100 sq. KM

The youngest island in the Hawaiian chain is still a seamount, a submarine mountain that does not reach the surface. It rises 3350 m (11,000 ft) from its base but is still 975 m (3200 ft) beneath the ocean surface. Even though this new island will not experience the tropical Sun for about 10,000 years, it is already named Lö'ihi (noted on the map).

Iceland is a result of an active hot spot sitting astride a mid-ocean ridge—visible on the different maps and images of the seafloor (see Figure 11.13d). It is an excellent example of a segment of mid-ocean ridge rising above sea level. This hot spot has generated enough material to form Iceland, and eruptions continue from deep in the mantle. As a result, Iceland is still growing in area and volume. This is further evidence that, indeed, Earth is a dynamic planet.

Summary and Review—The Dynamic Planet

■ *Distinguish* between the endogenic and exogenic systems, *determine* the driving force for each, and *explain* the pace at which these systems operate.

The Earth–atmosphere interface is where the **endogenic system** (internal), powered by heat energy from within the planet, interacts with the **exogenic system** (external), powered by insolation and influenced by gravity. These systems work together to produce Earth's diverse landscape. The **geologic time scale** is

an effective device for organizing the vast span of geologic time. It depicts the sequence of Earth's events (relative time) and the approximate actual dates (absolute time).

The most fundamental principle of Earth science is **uniformitarianism**. Uniformitarianism assumes that the *same physical processes active in the environment today have been operating throughout geologic time*, although dramatic episodes, such as massive landslides or volcanic eruptions, can interrupt the flow of events, leading to a state of *punctuated equilibrium*.

endogenic system (p. 321)
exogenic system (p. 321)
geologic time scale (p. 324)
uniformitarianism (p. 326)

1. To what extent is Earth's crust active at this time in its history?

2. Define the endogenic and the exogenic systems. Describe the driving forces that energize these systems.

3. How is the geologic time scale organized? What is the basis for the time scale in relative and absolute terms? What era, period, and epoch are we living in today?

4. Contrast uniformitarianism science and catastrophism beliefs as models for Earth's development.

■ *Diagram* Earth's interior in cross section, and *describe* each distinct layer.

We have learned about Earth's interior from indirect evidence—the way its various layers transmit **seismic waves**. The **core** is differentiated into an inner core and an outer core, divided by a transition zone. Earth's magnetic field is generated almost entirely within the outer core. Polarity reversals in Earth's magnetism are recorded in cooling magma that contains iron minerals. The patterns of **geomagnetic reversal** frozen in rock help scientists piece together the story of Earth's mobile crust.

Beyond Earth's core lies the **mantle**, differentiated into lower mantle and upper mantle. It experiences a gradual temperature increase with depth and a stiffening due to increased pressures. The upper mantle is divided into three fairly distinct layers. The uppermost mantle, along with the crust, makes up the *lithosphere*. Below the lithosphere is the **asthenosphere**, or *plastic layer*. It contains pockets of increased heat from radioactive decay and is susceptible to slow convective currents in these hotter materials. An important internal boundary between the **crust** and the high-velocity portion of the uppermost mantle is the **Mohorovičić discontinuity**, or **Moho**. *Continental crust* is basically **granite**; it is crystalline and high in silica, aluminum, potassium, calcium, and sodium. *Oceanic crust* is **basalt**; it is granular and high in silica, magnesium, and iron. The principles of buoyancy and balance were combined in the 1800s into the important principle of **isostasy**. Isostasy explains certain vertical movements of Earth's crust, such as isostatic rebound when the weight of ice is removed.

seismic waves (p. 326)
core (p. 326)
geomagnetic reversal (p. 327)
mantle (p. 328)
asthenosphere (p. 328)
crust (p. 329)
Mohorovičić discontinuity (Moho) (p. 328)
granite (p. 330)
basalt (p. 330)
isostasy (p. 330)

5. Make a simple sketch of Earth's interior, label each layer, and list the physical characteristics, temperature, composition, and range of size of each on your drawing.

6. How does Earth generate its magnetic field? Is the magnetic field constant, or does it change? Explain the implications of your answer.

7. Describe the asthenosphere. Why is it also known as the plastic layer? What are the consequences of its convection currents?

8. What is a discontinuity? Describe the principal discontinuities within Earth.

9. Define isostasy and isostatic rebound, and explain the crustal equilibrium concept.

10. Diagram the uppermost mantle and crust. Label the density of the layers in grams per cubic centimeter. What two types of crust were described in the text in terms of rock composition?

■ *Illustrate* the geologic cycle, and *relate* the rock cycle and rock types to endogenic and exogenic processes.

The **geologic cycle** is a model of the internal and external interactions that shape the crust. A **mineral** is an inorganic natural compound having a specific chemical formula and possessing a crystalline structure. A **rock** is an assemblage of minerals bound together (such as granite, a rock containing three minerals), or it may be a mass of a single mineral (such as rock salt). Thousands of different rocks have been identified.

The geologic cycle comprises three cycles: the *hydrologic cycle*, the *tectonic cycle*, and the **rock cycle**. The rock cycle describes the three principal rock-forming processes and the rocks they produce. **Igneous rocks** form from **magma**, which is molten rock beneath the surface. Magma is fluid, highly gaseous, and under tremendous pressure. It either *intrudes* into crustal rocks, cools, and hardens, or it *extrudes* onto the surface as **lava**. Intrusive igneous rock that cools slowly in the crust forms a **pluton**. The largest pluton form is a **batholith**.

The cementation, compaction, and hardening of sediments into **sedimentary rocks** is **lithification**. These layered strata form important records of past ages. **Stratigraphy** is the study of the sequence (superposition), thickness, and spatial distribution of strata that yield clues to the age and origin of the rocks.

Clastic sedimentary rocks are derived from the fragments of weathered rocks. Chemical sedimentary rocks are not formed from physical pieces of broken rock, but instead are dissolved minerals, transported in solution, and chemically precipitated out of solution (they are essentially nonclastic). The most common chemical sedimentary rock is **limestone**, which is lithified calcium carbonate, $CaCO_3$.

Any rock, either igneous or sedimentary, may be transformed into a **metamorphic rock** by going through profound physical or chemical changes under pressure and increased temperature. Examples of metamorphism include limestone becoming marble, or shale becoming slate.

geologic cycle (p. 331)
mineral (p. 331)
rock (p. 333)
rock cycle (p. 333)
igneous rock (p. 333)
magma (p. 333)
lava (p. 333)
pluton (p. 333)
batholith (p. 333)
sedimentary rock (p. 336)
lithification (p. 336)
stratigraphy (p. 336)
limestone (p. 337)
metamorphic rock (p. 339)

11. Illustrate the geologic cycle, and define each component: rock cycle, tectonic cycle, and hydrologic cycle.

12. What is a mineral? A mineral family? Name the most common minerals on Earth. What is a rock?

13. Describe igneous processes. What is the difference between intrusive and extrusive types of igneous rocks?

14. Characterize felsic and mafic minerals. Give examples of both coarse- and fine-grained textures.

15. Briefly describe sedimentary processes and lithification. Describe the sources and particle sizes of sedimentary rocks.

16. What is metamorphism, and how are metamorphic rocks produced? Name some original parent rocks and their metamorphic equivalents.

■ *Describe* Pangaea and its breakup, and *relate* several physical proofs that crustal drifting is continuing today.

The present configuration of the ocean basins and continents is the result of *tectonic processes* involving Earth's interior dynamics and crust. Alfred Wegener coined the phrase **continental drift** to describe his idea that the crust is moved by vast forces within the planet. **Pangaea** was the name he gave to a single assemblage of continental crust some 225 m.y.a. that subsequently broke apart. Earth's lithosphere is fractured into huge slabs or plates, each moving in response to their own motion of gravitational pull and to flowing currents in the mantle that create frictional drag on the plate. The all-encompassing theory of **plate tectonics** includes **sea-floor spreading** along **mid-ocean ridges** and denser oceanic crust diving beneath lighter continental crust along **subduction zones**.

> continental drift (p. 341)
> Pangaea (p. 341)
> plate tectonics (p. 342)
> sea-floor spreading (p. 342)
> mid-ocean ridge (p. 342)
> subduction zone (p. 342)

17. Briefly review the history of the theory of continental drift, sea-floor spreading, and the all-inclusive plate tectonics theory. What was Alfred Wegener's role?

18. Define upwelling, and describe related features on the ocean floor. Define subduction, and explain the process.

19. What was Pangaea? What happened to it during the past 225 million years?

20. Characterize the three types of plate boundaries and the actions associated with each type.

■ *Portray* the pattern of Earth's major plates, and *relate* this pattern to the occurrence of earthquakes, volcanic activity, and hot spots.

Occurrences of often-damaging earthquakes and volcanoes are correlated with plate boundaries. Three types of plate boundaries form: divergent, convergent, and transform. Along the offset portions of mid-ocean ridges, horizontal motions produce **transform faults**.

As many as 50 to 100 **hot spots** exist across Earth's surface, where plumes of magma, some anchored in the lower mantle, others originating from shallow sources in the upper mantle, generate a flow upward. **Geothermal energy** literally refers to heat from Earth's interior, whereas *geothermal power* relates to specific applied strategies of geothermal electric or geothermal direct applications.

> transform faults (p. 349)
> hot spots (p. 350)
> geothermal energy (p. 353)

21. What is the relation between plate boundaries and volcanic and earthquake activity?

22. What is the nature of motion along a transform fault? Name a famous example of such a fault.

NetWork

The *Geosystems* Student Learning Center provides on-line resources for this chapter on the World Wide Web. To begin: Once at the Center, click on the cover of this textbook, scroll the Table of Contents menu, and select this chapter. You will find self-tests that are graded, review exercises, specific

updates for items in the chapter, and in "Destinations" many links to interesting related pathways on the Internet. *Geosystems* Student Learning Center is found at **http://www.prenhall.com/ christopherson/**.

Critical Thinking

A. Using the maps in this chapter, determine your present location relative to Earth's crustal plates. Now, using Figure 11.16b, approximately identify where your present location was 225 m.y.a.; express it in a rough estimate using the equator and the longitudes noted on the map.

B. Relative to the motion of the Pacific plate shown in Figure 11.21 (note the map's graphic scale in the lower-left corner), the island of Midway formed 27.7 m.y.a. over the hot spot that is active under the southeast coast of the big island of Hawai'i today. Given the scale of the map, roughly determine the average annual speed of the Pacific plate in centimeters per year for Midway to have traveled this distance.

C. Under the heading "The Pace of Change," the text discusses a possible "Anthropocene" Epoch. The section states, in part,

> Holocene is the name given to the youngest epoch in the geologic time scale, characteristic of postglacial conditions since the retreat of the continental glaciers, approximately the last 11,500 years. Because of the impact of human society on planetary systems, discussion is underway about designating a new name for the human epoch. A proposal is to call this time the *Anthropocene*.

Take a moment to explore this concept of naming our human epoch in the geologic time scale. Determine some arguments for such a change, and describe some that argue against such a move to rename the late Holocene. What do you think?

Floor of the Oceans, 1975, by Bruce C. Heezen and Marie Tharp.

The scarred ocean floor is clearly visible: sea-floor spreading centers marked by oceanic ridges that stretch over 64,000 km (40,000 mi), subduction zones indicated by deep oceanic trenches, and transform faults slicing across oceanic ridges.

Follow the East Pacific rise (an ocean ridge and spreading center) northward as it trends beneath the west coast of the North American plate, disappearing under earthquake-prone California. The continents, offshore-submerged continental shelves, and the expanse of the sediment-covered abyssal plain are all identifiable in this illustration.

On the floor of the Indian Ocean, you see the wide track along which the India plate traveled northward to its collision with the Eurasian plate. Vast deposits of sediment cover the Indian Ocean floor, south of the Ganges River to the east of India. Sediments derived from the Himalayan Range blanket the floor of the Bay of Bengal (south of Bangladesh) to a depth of 20 km (12.4 mi). These sediments result from centuries of soil erosion in the land of the monsoons.

Isostasy can be examined in action in central and west-central Greenland, where the weight of the ice sheet has depressed portions of the land far below sea level—now visible with the ice artificially removed by the cartographer (see the upper-left corner of the map). In contrast, the region around Canada's Hudson Bay became ice-free about 8000 years ago and has isostatically rebounded 300 m (1100 ft).

In the area of the Hawaiian Islands, the hot-spot track is marked by a chain of islands and seamounts that you follow along the Pacific plate from Hawai'i to the Aleutians. Subduction zones south and east of Alaska and Japan, and along the western coast of South and Central America, are visible as dark trenches. Along the left margin of the map, you find Iceland's position on the Mid-Atlantic Ridge. [© 1980 by Marie Tharp. Reproduced by permission of Marie Tharp.]

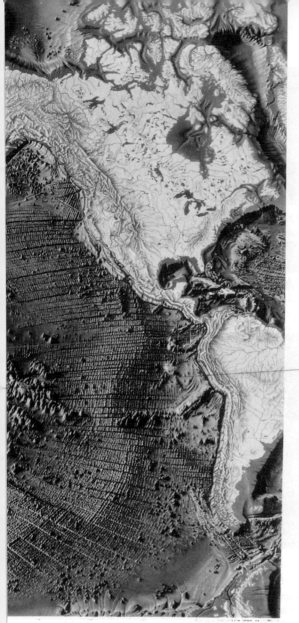

12

Tectonics, Earthquakes, and Volcanism

■ Key Learning Concepts

After reading the chapter, you should be able to:

- ■ *Describe* first, second, and third orders of relief, and *relate* examples of each from Earth's major topographic regions.

- ■ *Describe* the several origins of continental crust, and *define* displaced terranes.

- ■ *Explain* compressional processes and folding, and *describe* four principal types of faults and their characteristic landforms.

- ■ *Relate* the three types of plate collisions associated with orogenesis, and *identify* specific examples of each.

- ■ *Explain* the nature of earthquakes, their measurement, and the nature of faulting.

- ■ *Distinguish* between an effusive and an explosive volcanic eruption, and *describe* related landforms, using specific examples.

arth's endogenic systems produce flows of heat and material toward the surface to form crust. Ongoing processes produce continental landscapes and oceanic sea-floor crust, sometimes in dramatic episodes. Earth's physical systems move to the front pages whenever an earthquake strikes or after one of the nearly 50 volcanic eruptions a year that threaten a city.

In 1999, more than 30,000 people perished in quakes in Turkey, Greece, Taiwan, and Mexico; in 2003, in one earthquake alone in Iran 30,000 were killed (Figure 12.1a); and more than 80,000 were killed by

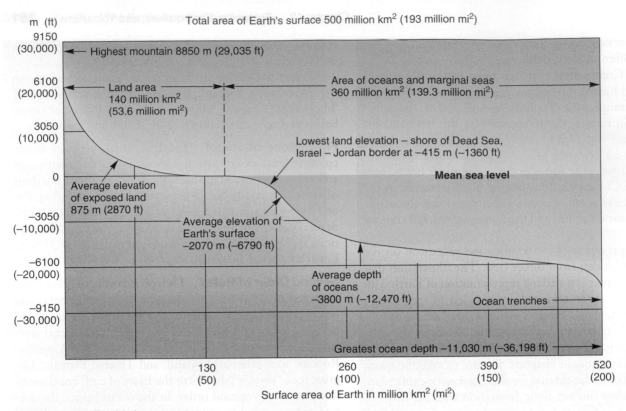

FIGURE 12.2 Earth's hypsometry.
Hypsographic curve of Earth's surface, charting area and elevation as related to mean sea level. From the highest point above sea level (Mount Everest) to the deepest oceanic trench (Mariana Trench), Earth's overall relief is almost 20 km (12.5 mi).

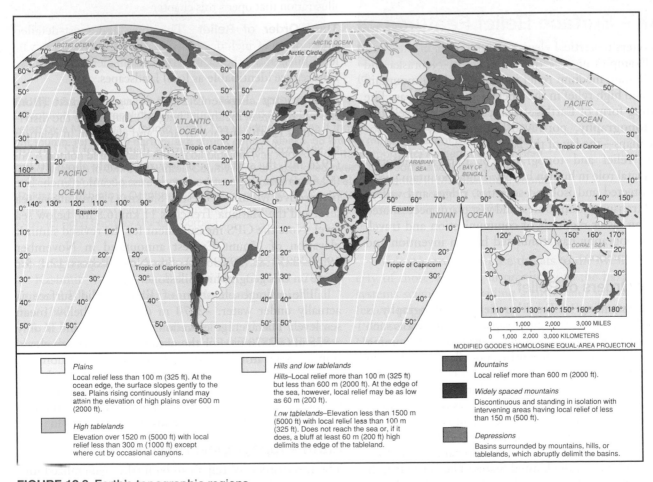

FIGURE 12.3 Earth's topographic regions.
Earth's topography is characterized as plains, high and low tablelands, hills, mountains, and depressions [After R. E. Murphy, "Landforms of the world," *Annals of the Association of American Geographers* 58, no. 1 (March 1968). Adapted by permission.]

Mount Everest at New Heights

The announcement came November 11, 1999, in Washington, DC, at the opening reception of the 87th annual meeting of the American Alpine Club. Mount Everest had a new revised elevation, determined by direct Global Positioning System (GPS) placement on the mountain's icy summit!

In May 1999, mountaineers Pete Athans and Bill Crouse reached the summit with five Sherpas. The climbers operated Global Positioning System satellite equipment on the top of Mount Everest and determined the precise height of the world's tallest mountain. In Chapter 1, the photograph in News Report 1.2 shows a GPS unit placed by climber Wally Berg 18 m below the summit in 1998. This site is still Earth's highest benchmark, with a permanent metal marker installed in rock. Other GPS instruments in the region permit differential calibration for exact measurement (Figure 12.1.1). This is a continuation of a measurement effort begun in 1995 by the late Bradford Washburn, renowned mountain photographer/explorer, who was honorary director of Boston's Museum of Science.

Astronaut Thomas D. Jones made the photo of Mount Everest in Figure 12.1.2 during Space Shuttle flight STS-80 in 1996 aboard Space Shuttle

FIGURE 12.1.1 GPS installation measures Mount Everest.
A global positioning system (GPS) installation near Namche Bazar in the Khumbu (Everest) region of Nepal. The station is at latitude 27.8° N, longitude 86.7° E, at 3523 m (11,558 ft). The summit of Mount Everest is 30 km (18.6 mi) distant—visible above the right side of the weather dome. A network of such GPS installations correlated with the unit the climbers took to Everest's summit.
[Photo by Charles Corfield, science manager to the 1998 and 1999 expeditions.]

Columbia. This startling view caught the mountain's triangular East Face bathed in morning light (east is to the lower left, north is to the lower right, south at the upper left; see locator map direction arrow). The North Face is in shadow. Several glaciers are visible flowing outward from the mountains, darkened by rock and debris.

Washburn announced the new measurement of 8850 m (29,035 ft). Everest's new elevation is close to the previous official measure of 8848 m (29,028 ft) set in 1954 by the Survey of India. Washburn stated an interesting note, "... from these GPS readings it appears that the horizontal position of Everest seems to be moving steadily and slightly northeastward; between 3 and 6 mm a year [up to 0.25 in. a year]." The mountain range is being driven farther into Asia by plate tectonics—the continuing collision of Indian and Asian landmasses.

FIGURE 12.1.2 Mount Everest in morning light from orbit aboard Space Shuttle *Columbia*. [Photo courtesy of NASA, made by Dr. Thomas Jones, astronaut, 1996.]

descriptive limit that is in common use defines each type of topography (see the map legend).

Four of the continents possess extensive *plains*, areas with local relief of less than 100 m (325 ft) and slope angles of 5° or less. Some plains have high elevations of more than 600 m (2000 ft); in the United States, the high plains attain elevations above 1220 m (4000 ft). The Colorado Plateau, Greenland, and Antarctica are notable *high tablelands* (the latter two of ice), with elevations exceeding 1520 m (5000 ft). *Hills* and *low tablelands* dominate Africa.

Mountain ranges, characterized by local relief exceeding 600 m (2000 ft), occur on each continent. Earth's relief and topography are undergoing constant change as a result of processes that form crust.

Crustal Formation Processes

How did Earth's continental crust form? What gave rise to the three orders of relief just discussed? Ultimately, the answers to both questions are the combined effects of tectonic activity, which is driven by our planet's internal energy, and the exogenic processes of weathering and erosion, powered by the Sun through the actions of air, water, waves, and ice.

Tectonic activity generally is slow, requiring millions of years. Endogenic (internal) processes result in gradual uplift and new landforms, with major mountain building occurring along plate boundaries. These uplifted crustal regions are quite varied, but we think of them in three general categories, all discussed in this chapter:

* Residual mountains and stable continental cratons, formed from inactive remnants of ancient tectonic activity;
* Tectonic mountains and landforms, produced by active folding, faulting, and crustal movements;
* Volcanic features, formed by the surface accumulation of molten rock from eruptions of subsurface materials.

Thus, several distinct processes operate in concert to produce the continental crust we see around us.

Continental Shields

All continents have a nucleus of ancient crystalline rock on which the continent "grows" with the addition of crustal fragments and sediments. This nucleus is the *craton*, or heartland region, of the continental crust. Cratons generally have been eroded to a low elevation and relief. Most date to the Precambrian and can exceed 2 billion years of age. The lack of basaltic components in these cratons offers a clue to their stability. In cratonic regions, which consist of crust and the lithospheric uppermost mantle, the lithosphere is thicker than beneath younger portions of continents and oceanic crust.

A **continental shield** is a region where a craton is exposed at the surface. Figure 12.4 shows the principal areas of exposed shields and a photo of the Canadian shield. Layers of younger sedimentary rock surround these shields and appear quite stable over time. Examples of

such a stable *platform* are the region that stretches from east of the Rockies to the Appalachians and northward into central and eastern Canada, a large portion of China, eastern Europe to the Ural Mountains, and across portions of Siberia.

Building Continental Crust and Terranes

The formation of continental crust is complex and takes hundreds of millions of years. It involves the entire sequence of sea-floor spreading and formation of oceanic crust, its later subduction and remelting, and its subsequent rise as new magma, all summarized in Figure 12.5.

To understand this process, study Figure 12.5 (and Figure 11.13). Begin with the magma that originates in the asthenosphere and wells up along the mid-ocean ridges. Basaltic magma is formed from minerals in the upper mantle that are rich in iron and magnesium. Such magma has less than 50% silica and has a low-viscosity (thin) texture—it tends to flow. This mafic material rises to erupt at spreading centers and cools to form new basaltic seafloor, which spreads outward to collide with continental crust along its far edges. This denser oceanic crust plunges beneath the lighter continental crust, into the mantle, where it remelts. The new magma then rises and cools, forming more continental crust, in the form of intrusive granitic igneous rock.

As the subducting oceanic plate works its way under a continental plate, it takes with it trapped seawater and sediment from eroded continental crust. The remelting incorporates the seawater, sediments, and surrounding crust into the mixture. As a result, the magma, generally called a *melt*, that migrates upward from a subducted plate contains 50%–75% silica and aluminum (called andesitic or silicic, depending on silica content). The melt has a high-viscosity (thick) texture—it tends to block and plug conduits to the surface.

Bodies of such silica-rich magma may reach the surface in explosive volcanic eruptions, or they may stop short and become subsurface intrusive bodies in the crust, cooling slowly to form granitic crystalline plutons such as batholiths (see Figures 11.7 and 11.8a). Note that this composition is quite different from the magma that rises directly from the asthenosphere at sea-floor spreading centers. In these processes of crustal formation, you can literally follow the cycling of materials through the tectonic cycle.

Each of Earth's major lithospheric plates actually is a collage of many crustal pieces acquired from a variety of sources. Crustal fragments of ocean floor, curving chains (or arcs) of volcanic islands, and other pieces of continental crust all have been forced against the edges of continental shields and platforms. These slowly migrating crustal pieces, which have become attached or accreted to the plates, are called **terranes** (not to be confused with "terrain," which refers to the topography of a tract of land). These displaced terranes, sometimes called *microplate* or *exotic terranes*, have histories different from

FIGURE 12.4 Continental shields.
(a) Portions of major continental shields that have been exposed by erosion. Adjacent portions of these shields remain covered by younger sedimentary layers.
(b) Canadian shield landscape in central Labrador interior, stable for hundreds of millions of years, stripped by past glaciations and marked by intrusive igneous dikes (magmatic intrusions). [(a) After R. E. Murphy, "Landforms of the world," *Annals of the Association of American Geographers* 58, no. 1 (March 1968). Adapted by permission. (b) Photo by John Eastcott/Yva Momatiuk/The Image Works.]

(b)

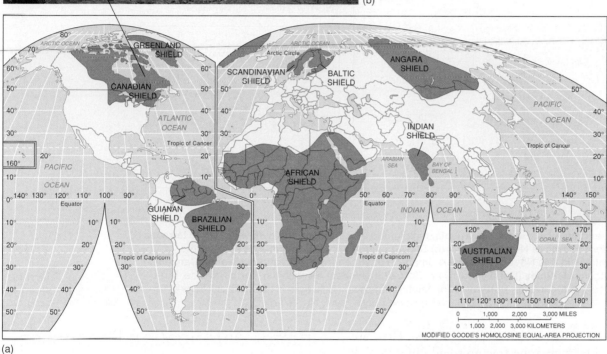

(a)

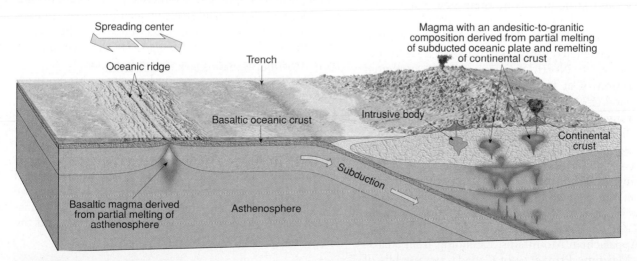

FIGURE 12.5 Crustal formation.
Material from the asthenosphere upwells along sea-floor spreading centers. Basaltic ocean floor is subducted beneath lighter continental crust, where it melts, along with its cargo of sediments, water, and minerals. This melting generates magma, which makes its way up through the crust to form igneous intrusions and extrusive eruptions. [After E. J. Tarbuck and F. K. Lutgens, *Earth, An Introduction to Physical Geology*, 5th ed., © Prentice Hall, 1996, Figure 20.20, p. 502.]

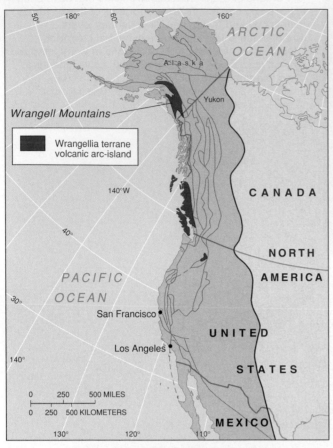

(a)

(b) Wrangell Mountains, Alaska

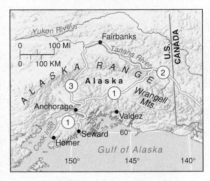

FIGURE 12.6 North American terranes.
(a) Wrangellia terranes occur in four segments, highlighted in red among the outlines of other terranes along the western margin of North America. (b) Snow-covered Wrangell Mountains north of the Chugach Mountains in this November 7, 2001, image in east-central Alaska and across the Canadian border. Note Mount McKinley (Denali) at 6194 m (20,322 ft), highest elevation in North America, in the upper-left part of the image, north of Cook Inlet. [(a) Based on data from U.S. Geological Survey. (b) *Terra* MODIS image, courtesy of MODIS Land Rapid Response Team, NASA/GSFC.]

 Terrane Formation

those of the continents that capture them. They are usually framed by fault-zone fractures and differ in rock composition and structure from their new continental homes.

In the region surrounding the Pacific, accreted terranes are particularly prevalent. At least 25% of the growth of western North America can be attributed to the accretion of at least 50 terranes since the early Jurassic Period (190 million years ago). A good example is the Wrangell Mountains, which lie just east of Prince William Sound and the city of Valdez, Alaska. The *Wrangellia terranes*—a former volcanic island arc and associated marine sediments from near the equator—migrated approximately 10,000 km (6200 mi) to form the Wrangell Mountains and three other distinct formations along the western margin of the continent (Figure 12.6).

The Appalachian Mountains, extending from Alabama to the Maritime Provinces of Canada, possess bits of land once attached to ancient Europe, Africa, South America, Antarctica, and various oceanic islands. The discovery of terranes, made only in the 1980s, demonstrates one of the ways continents are assembled.

Crustal Deformation Processes

Rocks, whether igneous, sedimentary, or metamorphic, are subjected to powerful stress by tectonic forces, gravity, and the weight of overlying rocks. There are three types of stress: *tension* (stretching), *compression* (shortening), and *shear* (twisting or tearing), as shown in Figure 12.7.

Strain is how rocks respond to stress. Strain is expressed in rocks by *folding* (bending) or *faulting* (breaking). Think of stress as a force and the resulting strain as the deformation in the rock. Whether a rock bends or breaks depends on several factors, including composition and how much pressure is on the rock. An important quality is whether the rock is *brittle* or *ductile*. The patterns created by these processes are clearly visible in the landforms we see today, especially in mountain areas. Figure 12.7 illustrates each type of stress and its resulting strain and the surface expressions that develop.

Folding and Broad Warping

When rock strata that are layered and flat are subjected to compressional forces, they become deformed (Figure 12.8). Convergent plate boundaries intensely compress rocks,

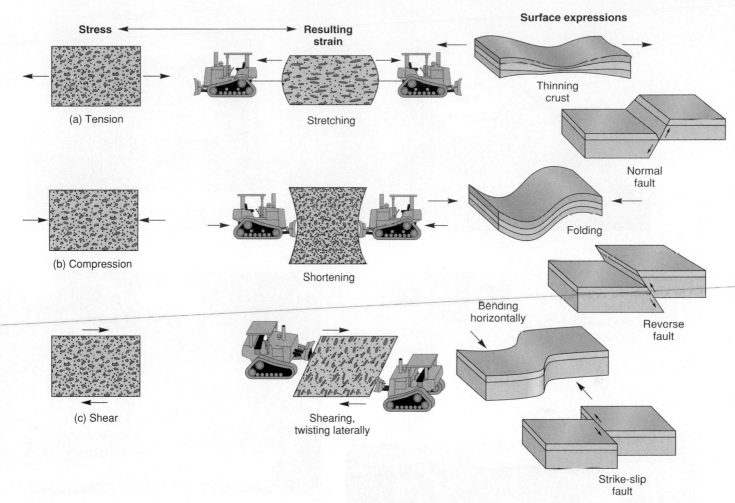

Stress ← → **Resulting strain**

Surface expressions

(a) Tension
Stretching
Thinning crust
Normal fault

(b) Compression
Shortening
Folding
Reverse fault

Bending horizontally

(c) Shear
Shearing, twisting laterally
Strike-slip fault

FIGURE 12.7 Three kinds of stress and strain, and resulting surface expressions. The bulldozers represent the stress, or force, on the rock strata. (a) Tension stress produces a stretching and thinning of the crust and a *normal fault.* (b) Compression produces shortening and folding and *reverse faulting* of the crust. In a back-and-forth horizontal motion, (c) shear stress produces a bending of the crust and, on breaking, a *strike-slip fault.*

deforming them in a process known as **folding**. As an analogy, if we take sections of thick fabric, stack them flat on a table, and then slowly push on opposite ends of the stack, the cloth layers will bend and rumple into folds similar to those shown in Figure 12.8a.

If we then draw a line down the center axis of a resulting ridge, and a line down the center of a resulting trough, we see how the names of the folds are assigned. Along the *ridge* of a fold, layers *slope downward away from the axis*, which is called an **anticline**. In the *trough* of a fold, however, layers *slope downward toward the axis*, called a **syncline**.

If the axis of either type of fold is not "level" (horizontal, or parallel to Earth's surface), the fold axis then *plunges*, inclined (dipped down) at an angle. Knowledge of how folds are angled to Earth's surface and where they are located is important for the petroleum industry. For example, petroleum geologists know that oil and natural gas collect in the upper portions of anticlinal folds in permeable rock layers such as sandstone. This is the "anticlinal theory of petroleum accumulation"

attributed to I. C. White, founder of the West Virginia Geological Survey.

Figure 12.8a further illustrates various folds that have been weathered and reduced by exogenic processes:

- A residual "synclinal ridge" may form within a syncline because different rock strata offer greater resistance to weathering processes. An actual synclinal ridge is exposed dramatically in an interstate highway roadcut in Figure 12.8b.
- Compressional forces often push folds far enough that they actually overturn upon their own strata ("overturned anticline").
- Further stress eventually fractures the rock strata along distinct lines, and some overturned folds are thrust upward, causing a considerable shortening of the original strata ("thrust fault"). Areas of intense stress, compressional folding, and faulting are visible in a roadcut where the San Andreas fault in Southern California passes beneath the Antelope Valley Freeway (Figure 12.8c).

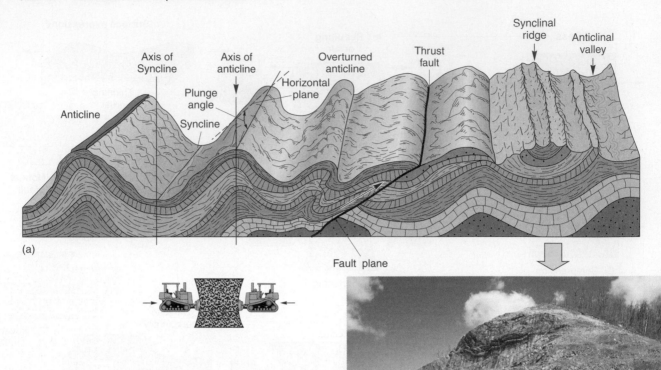

(a)

(c)

(b)

FIGURE 12.8 Folded landscapes.
(a) Folded landscape and the basic types of fold structures. (b) A roadcut exposes a synclinal ridge in western Maryland. This syncline is a natural outdoor classroom, and the state of Maryland built an interpretive center with a walkway above the highway. (c) Folding, squeezing, and uplift from strong compressional and shearing forces show well in a roadcut along the San Andreas fault zone, along the Antelope Valley Freeway. [Photos by (b) Mike Boroff/Photri-Microstock; (c) Bobbé Christopherson.]

ANIMATION Folds, Anticline and Synclines

The Canadian Rocky Mountains, the Appalachian Mountains, and areas of the Middle East illustrate the complexity of folded landscapes. Satellites allow us to view many of these structures from an orbital perspective, as in Figure 12.9, north of the Persian Gulf and the Zagros Mountains of Iran. This area was a dispersed terrane that separated from the Eurasian plate. However, the collision produced by the northward push of the Arabian block is now shoving this terrane back into Eurasia and forming an active margin known as the Zagros crush zone, a zone of continuing collision more than 400 km (250 mi) wide. In the satellite image, anticlines form the parallel ridges; active weathering and erosion processes are exposing the underlying strata.

In addition to the rumpling of rock strata just discussed, broad warping actions also affect Earth's continental crust. These actions produce similar up-and-down bending of strata, but the bends are far greater in extent than those produced by folding. Warping forces include mantle convection, isostatic adjustment such as that caused by the weight of previous ice loads across northern Canada, or crustal swelling above an underlying hot spot. Warping features can be small, individual, foldlike structures called *basins* and *domes* (Figure 12.10a, b). They can also range up to regional features the size of the Ozark Mountain complex in Arkansas and Missouri, the Colorado Plateau in the West, the Richat dome in Mauritania, or the Black Hills of South Dakota (Figure 12.10c, d).

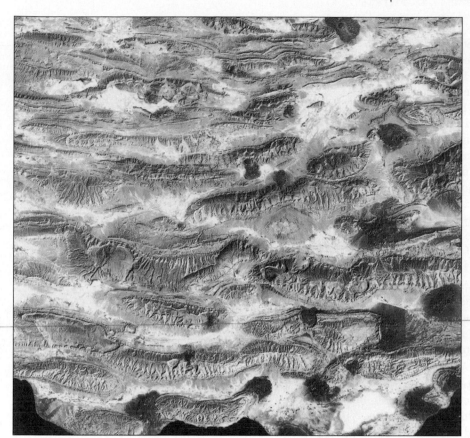

FIGURE 12.9 Folding in the Zagros crush zone, Iran.
The Zagros Mountains are a product of the Zagros crush zone between the Arabian and Eurasian plates, where the northward push of the Arabian block is causing the folded mountains shown.
[NASA image.]

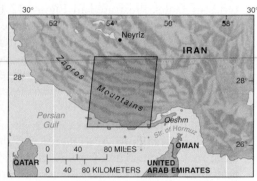

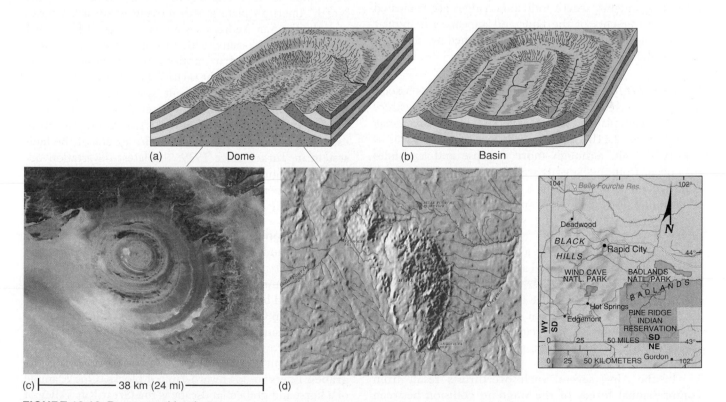

(a)   Dome   (b)   Basin

(c) ├────── 38 km (24 mi) ──────┤   (d)

FIGURE 12.10 Domes and basins.
(a) An upwarped dome. (b) A structural basin. (c) The Richat dome structure of Mauritania. (d) The Black Hills of South Dakota is a dome structure, shown here on a digitized relief map. [(c) Image from *Terra*, October 7, 2000, courtesy of NASA/GSFC/MITI and U.S./Japan ASTER team; (d) from USGS digital terrain map I-2206, 1992.]

Faulting

A freshly poured concrete sidewalk is smooth and strong. But stress the sidewalk by driving heavy equipment over it, and the resulting strain might cause a fracture. Pieces on either side of the fracture may move up, down, or horizontally, depending on the direction of stress. Similarly, when rock strata are stressed beyond their ability to remain a solid unit, they express the strain as a fracture. Rocks on either side of the fracture displace relative to the other side in a process known as **faulting**. Thus, *fault zones* are areas where fractures in the rock demonstrate crustal movement. At the moment of fracture, a sharp release of energy occurs, called an **earthquake** or *quake*.

The fracture surface along which the two sides of a fault move is the *fault plane*. The names of the three basic types of faults, illustrated in Figure 12.11, are based on the tilt and orientation of the fault plane. A normal fault forms when rocks are pulled apart by tensional stress. A thrust or reverse fault results when rocks are forced together by compressional stress. A strike-slip fault forms when rocks are torn by lateral-shearing stress.

Normal Fault   When forces pull rocks apart, the tension causes a **normal fault**, or *tension fault*. When the break occurs, rock on one side moves vertically along an inclined fault plane (Figure 12.11a). The downward-shifting side is the *hanging wall*; it drops relative to the *footwall block*. The exposed fault plane sometimes is visible along the base of faulted mountains, where individual ridges are truncated by the movements of the fault and appear as triangular facets at the ends of the ridges. A cliff formed by faulting is commonly called a *fault scarp*, or *escarpment*.

Reverse (Thrust) Fault   Compressional forces associated with converging plates force rocks to move *upward* along the fault plane. This is a **reverse fault**, or *compression fault* (Figure 12.11b). On the surface, it appears similar to a normal fault, although more collapse and landslides may occur from the hanging wall component. In England, when miners worked along a reverse fault, they would stand on the lower side (footwall) and hang their lanterns on the upper side (hanging wall), giving rise to these terms.

If the fault plane forms a low angle relative to the horizontal, the fault is termed a **thrust fault**, or *overthrust fault*, indicating that the overlying block has shifted far over the underlying block (see Figure 12.8, "thrust fault"). Place your hands palms-down on your desk, with fingertips together, and slide one hand up over the other—this is the motion of a low-angle thrust fault, with one side pushing over the other.

In the Alps, several such overthrusts result from compressional forces of the ongoing collision between the African and Eurasian plates. Beneath the Los Angeles Basin, overthrust faults produce a high risk of earthquakes and caused many quakes in the twentieth century, including the $30-billion 1994 Northridge earthquake. These blind (unknown until they rupture) thrust faults beneath the Los Angeles region remain a major earthquake threat.

Strike-Slip Fault   If movement along a fault plane is horizontal, such as produced along a transform fault, it forms a **strike-slip fault** (Figure 12.11c; refer to Figure 11.19 for review). The movement is *right-lateral* or *left-lateral*, depending on the motion perceived when you observe movement on one side of the fault relative to the other side.

Although strike-slip faults do not produce cliffs (scarps), as do the other types of faults, they can create linear rift valleys. This is the case with the San Andreas fault system of California. The rift valley is clearly visible in Figure 12.11c, where the edges of the North American and Pacific plates are grinding past one another as a result of transform-fault movement. The evolution of this fault system is shown in Figure 12.12.

Note in the figure how the East Pacific rise developed as a spreading center with associated transform faults (1), while the North American plate was progressing westward after the breakup of Pangaea. Forces then shifted the transform faults toward a northwest–southeast alignment along a weaving axis (2). Finally, the western margin of North America overrode those shifting transform faults (3). Note the convergence rate is a rapid 4 cm (1.6 in.) per year.

In *relative* terms, the motion along this series of transform faults is right-lateral, whereas in *absolute* terms the North American plate is still moving westward. Consequently, the San Andreas system is a series of faults that are *transform* (associated with a former spreading center), *strike-slip* (horizontal in motion), and *right-lateral* (one side is moving to the right relative to the other side).

The North Anatolian fault system in Turkey is a strike-slip fault and has a right-lateral motion similar to the San Andreas system (Figure 12.13). A progression of earthquakes have hit along the entire extent of the fault system in Turkey since 1939; the latest devastation occurred in August and November 1999 and again in February 2002. There is more on the nature of such strike-slip faulting in Figure 12.21.

Faults in Concert   Combinations of faults can produce distinctive landscapes. In the U.S. interior west, the *Basin and Range Province* (featuring a roughly parallel series of faulted mountains and valleys) experienced tensional forces caused by uplifting and thinning of the crust (illustrated later in Chapter 15, Figure 15.24). This movement cracked the surface to form aligned pairs of normal faults and a distinctive landscape (Figure 12.14a).

The term **horst** applies to upward-faulted blocks; **graben** refers to downward-faulted blocks. One example of a horst and graben landscape is the Great Rift Valley of East Africa (associated with crustal spreading); it extends northward to the Red Sea, which fills the rift formed by parallel normal faults (Figure 12.14b). Another example is the Rhine graben, through which the Rhine River flows in Europe.

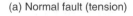
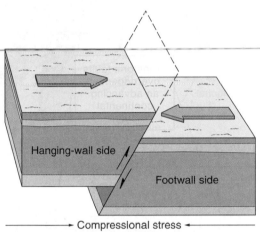

(a) Normal fault (tension)

(b) Thrust or reverse fault (compression)

Right-lateral*   Left-lateral**

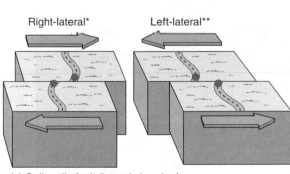

(c) Strike-slip fault (lateral shearing)

* Viewed from either dot on each road,
movement to opposite side is *to the right*.
** Viewed from either dot on each road,
movement to opposite side is *to the left*.

(a)

(b)

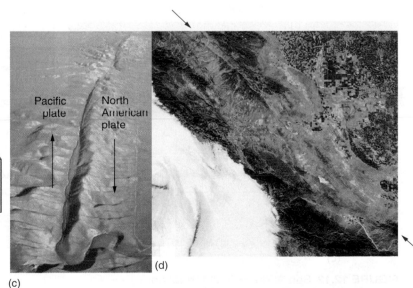

(c)

(d)

FIGURE 12.11 Types of faults.
(a) A normal fault produced by tension in the crust, visible along the edge in Tashkent, Uzbekistan. (b) A thrust, or reverse fault, produced by compression in the crust, visible in these offset strata in coal seams and volcanic ash in British Columbia. (c) A strike-slip fault produced by lateral shearing, clearly seen looking north along the San Andreas fault rift zone on the eastern edge of the Coast Ranges in California—Pacific plate to the left, North American plate to the right. (d) Can you find the San Andreas fault on the satellite image? Look for a linear rift stretching from southeast to northwest, at the edge of the coastal mountains and the San Joaquin Valley. Use the arrows as guides. [(a) Photo by Fred McConnaughey/Photo Researchers, Inc.; (b) photo by Fletcher and Baylis/Photo Researchers, Inc.; (c) Kevin Schafer/Peter Arnold, Inc.; (d) *Terra* MISR image courtesy of MISR Team, NASA/GSFC/JPL.]

ANIMATION   Fault Types, Transform Faults, Plate Margins

(a)

(b)

FIGURE 12.22 1995 earthquake in Kobe, Japan.
(a) Catastrophic failure of an elevated freeway; note the failure of the supporting pillars. (b) Ground failure, mainly liquefaction, tilts this building to near collapse; note the space between the building and the ground on the right side. [Photos by Haruyoshi Yamaguchi/Sygma.]

felling buildings and initiating a firestorm fed by broken gas pipes. Movement along a fault was evident over 435 km (270 mi), prompting intensive research to discover the nature of such faulting. The elastic-rebound theory developed as a result of research done on this rupture. Realize that this quake occurred 6 years before Wegener proposed his continental drift hypothesis.

The fault system that devastated San Francisco in 1906, and coastal and inland California throughout recorded history, is the San Andreas. The San Andreas fault system provides a good example of a spreading center overridden by an advancing continental plate (refer to Figure 12.12 for its evolution).

The tectonic history of the San Andreas was brought home to hundreds of millions of television viewers as they watched the 1989 baseball World Series played in Candlestick Park near San Francisco. Less than half an hour before the call to "play ball," a powerful earthquake rocked the region, turning sportscasters into newscasters and sports fans into disaster witnesses. This magnitude 7.0 earthquake involved a portion of the San Andreas fault, near Loma Prieta, approximately 16 km (9.9 mi) east of Santa Cruz and 95 km (59 mi) south of San Francisco (see Figure 12.20). A fault had ruptured at a focus unusually deep for the San Andreas system, more than 18 km (11.5 mi) below the surface.

Unlike previous earthquakes—such as the one in 1906, when the plates shifted a maximum of 6.4 m (21 ft) relative to each other—there was no evidence of a fault plane or rifting at the surface in the Loma Prieta area. Instead, the fault plane suggested in Figure 12.20 shows the two plates moving horizontally approximately 2 m (6 ft) past each other deep below the surface, with the Pacific plate thrusting 1.3 m (4.3 ft) upward. This vertical motion is unusual for the San Andreas fault and indicates that this portion of the San Andreas system is more complex than previously thought. Damage totaled $8 billion, 14,000 people were displaced from their homes, 4000 were injured, and 67 killed.

In the San Francisco Bay Area the present forecast places a 62% chance of a M 6.7, or higher, striking in the next 30 years (2000–2030) along the San Andreas system of faults. Added to this is a 14% chance of a significant earthquake occurring along some yet unknown fault.

Los Angeles Region

Since the mid-1980s, an area east of Los Angeles has experienced seven earthquakes greater than M 6.0. After a M 6.1 quake in 1992, a M 7.4 tremor rocked the lightly populated area near Landers, California (see the enlargement map of southern California in Figure 12.12). Although it was the single largest quake in California in 30 years, the remote location kept injuries and damage slight. Involved in this series were four different faults and some unknown segments. The main faulting caused a displacement of 6.1 m (20 ft).

For yet unexplained reasons, related earthquakes over the next few weeks struck in Mammoth Lakes, about 645 km (400 mi) to the north; at Mount Shasta in northern California; in southern Nevada and Utah; and 1810 km (1125 mi) distant in Yellowstone National Park, Wyoming. Evidence is growing that in complex fault systems, such as in the Los Angeles Region, active faults can interact, transferring stress to adjoining systems, with one fault perhaps triggering activity along another fault. This complicates analysis and forecasting.

The 1971 San Fernando, 1987 Whittier, 1988 Pasadena, 1991 Sierra Madre, and 1994 Northridge (Reseda) earthquakes are a few of the many quakes associated with deeply buried thrust faults. Although not directly aligned with the San Andreas, scientists think that southern California will be affected by more of these thrust-fault actions as strain continues to build along the nearby San Andreas system of faults. The fault line that marks the contact between the Pacific and the North American plates is indeed restless.

Earthquake Forecasting and Planning

The map in Figure 12.23 plots the epicenters of earthquakes in the United States and southern Canada between 1899 and 1990. These occurrences give some indication of relative risk by region. The challenge is to discover how to predict the *specific time and place* for a quake in the short term. See News Report 12.2 for more on forecasting.

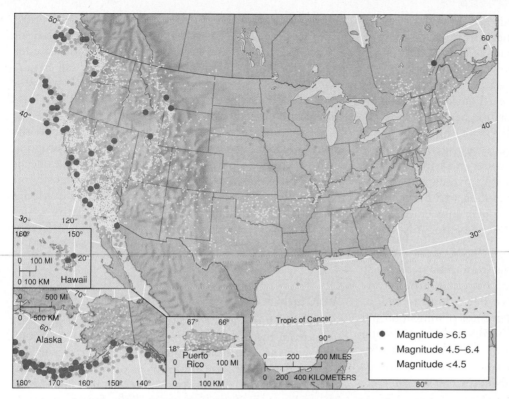

FIGURE 12.23 Seismicity of the United States and southern Canada, 1899–1990. Earthquake occurrences of magnitude 4.5, or greater, indicate areas of greatest seismic risk. Note the contrast between stable and unstable regions, related to tectonic activity. Active seismic regions include the West Coast, the Wasatch Front of Utah northward into Canada, the central Mississippi Valley, southern Appalachians, and portions of South Carolina, upstate New York, and Ontario. [From the USGS, National Earthquake Information Center; see http://eqhazmaps.usgs.gov/.]

News Report 12.2

Seismic Gaps, Nervous Animals, Dilitancy, and Radon Gas

How do you forecast an earthquake? One approach is to examine the history of each plate boundary and determine the frequency of past earthquakes, a study called *paleoseismology*. Paleoseismologists construct maps that provide an estimate of expected earthquake activity based on past performance. An area that is quiet and overdue for an earthquake is a *seismic gap*; such an area forms a gap in the earthquake occurrence record and is therefore a place that possesses accumulated strain. The area along the Aleutian Trench subduction zone had three such gaps until the great 1964 Alaskan earthquake filled one of them.

The areas around San Francisco and northeast of Los Angeles represent other such gaps where the fault system appears to be locked by friction and stress is producing accumulating strain. The U.S. Geological Survey in 1988 made a prediction that there was a 30% chance of an earthquake occurring with a magnitude 6.5 within 30 years in the Loma Prieta portion of the fault system. The actual 1989 quake dramatically filled a portion of the seismic gap in that region. The Nojima fault in the area of Kobe occurred in another gap.

One potentially positive discovery from the Loma Prieta disaster was that Stanford University scientists found unusually great changes in Earth's magnetic field—about 30 times more than normal—3 hours before the main shock. These measurements were made 7 km (4.3 mi) from the epicenter, raising hopes that short-term warnings might be possible in the future. A valid research question is whether animals have the ability to detect these minute changes in magnetic fields, an ability that was perhaps lost in humans. If so, strange prequake animal behavior, often reported, might provide forecasting clues.

Dilitancy refers to the slight increase in volume of rock produced by small cracks that form under stress and accumulated strain. The affected region may tilt and swell in response to strain. Tiltmeters measure these changes in suspect areas. Another indicator of dilitancy is an increase in radon (a naturally occurring, slightly radioactive gas) dissolved in groundwater. At present, earthquake hazard zones have thousands of radon monitors taking samples in test wells.

A seismic network in operation in Mexico City is providing 70-second warnings of arriving seismic-wave energy from distant epicenters. Following the Loma Prieta earthquake, freeway-repair workers were linked to a broadcast system that gave them 20-second alerts of aftershocks from distant epicenters. In southern California, the USGS is coordinating a Southern California Seismographic Network (SCSN at http://www.scsn.org/) to correlate 350 instruments for immediate earthquake location analysis and disaster coordination. Also, for the southern California region, see http://earthquake.usgs.gov/regional/sca/.

Someday, accurate earthquake forecasting may be a reality, but several questions remain. How will humans respond to a forecast? Can a major metropolitan region be evacuated for short periods of time? Can cities relocate after a disaster to areas of lesser risk?

Actual implementation of an action plan to reduce death, injury, and property damage from earthquakes is difficult. The political environment adds complexity since studies show that an accurate earthquake prediction may threaten a region's economy. Research examining the potential socioeconomic impact of earthquake prediction on an urban community shows surprising negative economic impacts in the period before the quake hits. Imagine a chamber of commerce, bank, real estate agent, tax assessor, or politician who would welcome an earthquake prediction and such negative publicity for their city. These factors work against effective planning and preparation.

Long-range planning is a complex subject. After the Loma Prieta earthquake in 1989, the National Research Council concluded in a report:

> One of the most jarring lessons from these quakes may be that earthquake professionals have long known many of the things that could have been done to reduce devastation.... The cost-effectiveness of mitigation and the importance of closing the knowledge gap among researchers, building professionals, government officials, and the public are only two of the many lessons from the Loma Prieta that need immediate action.

A valid and applicable generalization is that *humans and their institutions are unable or unwilling to perceive hazards in a familiar environment*. In other words, we tend to feel secure in our homes and communities, even if they are sitting on a quiet fault zone. Such an axiom of human behavior certainly helps explain why large populations continue to live and work in earthquake-prone settings. Similar questions also can be raised about populations in areas vulnerable to floods, droughts, hurricanes and coastal storm surge, and settlements on barrier islands. (See the Natural Hazards Observer at **http://www. colorado.edu/hazards/o/** as part of the Natural Hazards Center at the University of Colorado at **http://www. colorado.edu/hazards/index.html**.)

Volcanism

Volcanic eruptions across the globe remind us of Earth's internal energy. Ongoing eruption activity matches plate tectonic activity, as shown on the map in Figure 11.20. A sample of recent eruptions includes: Kīlauea in Hawai'i, Soufriere Hills on Montserrat (Lesser Antilles), Mount Etna (Sicily, Italy), several on the Kamchatka Peninsula (Russia), Rabaul (Papua, New Guinea), Grímsvötn (Iceland), Axial Summit and the Jackson Segment (sea-floor spreading center off the coast of Oregon), Shishaldin (Unimak Island, Alaska; see Figure 12.36), Lascar (Chile), White Island (New Zealand; see Figure 12.31), Nyiragongo (Congo, Africa), and Mayon (Philippines), among many.

Over 1300 identifiable volcanic cones and mountains exist on Earth, although fewer than 600 are active (have had at least one eruption in recorded history). In an average year, about 50 volcanoes erupt worldwide, varying from modest activity to major explosions. (An index to the world's volcanoes is at **http://vulcan.wr.usgs.gov/Volcanoes/**

framework.html, and the National Museum of Natural History Global Volcanism Program lists information for more than 8500 eruptions at **http://www.nmnh.si.edu/ gvp/**. For more volcanic eruption listings, see **http:// volcano.und.nodak.edu/**, or check **http://www.geo. mtu.edu/volcanoes/links/observatories.html** for links to volcano observatories across the globe.)

Eruptions in remote locations and at depths on the seafloor go largely unnoticed, but the occasional eruption of great magnitude near a population center makes headlines. North America has about 70 volcanoes (mostly inactive) along the western margin of the continent. Mount St. Helens in Washington State is a famous active example, and over 1 million visitors a year travel to Mount St. Helens Volcanic National Monument to see volcanism for themselves.

Volcanic Features

A **volcano** forms at the end of a central vent or pipe that rises from the asthenosphere and upper mantle through the crust into a volcanic mountain. A **crater**, or circular surface depression, usually forms at or near the summit. Magma rises and collects in a magma chamber deep below the volcano until conditions are right for an eruption. This subsurface magma emits tremendous heat; in some areas it boils groundwater, producing **geothermal energy**, as seen in the thermal springs and geysers of Yellowstone National Park and elsewhere in the world where it is harnessed for geothermal power production.

Lava (molten rock), gases, and **pyroclastics**, or *tephra* (pulverized rock and clastic materials of various sizes ejected violently during an eruption), pass through the vent to the surface and build the volcanic landform. There are two principal forms of flowing basaltic lava, both named in Hawaiian terms (Figure 12.24). A basaltic lava, **aa** is rough and jagged with sharp edges. This texture happens because the lava loses trapped gases, flows slowly, and develops a thick skin that cracks into the jagged surface. Another principal lava texture that is more fluid than aa is **pahoehoe**. As this lava flows it forms a thin crust that develops folds and appears "ropy," like coiled, twisted rope. Both forms can come from the same eruption, and sometimes pahoehoe will become aa as the flow progresses. Other types of basaltic magma are described later in this section.

The fact that lava can occur in many different textures and forms accounts for the varied behavior of volcanoes and the different landforms they build. In this section, we look at five volcanic landforms and their origins: cinder cones, calderas, shield volcanoes, plateau basalts, and composite volcanoes.

A **cinder cone** is a small, cone-shaped hill usually less than 450 m (1500 ft) high, with a truncated top formed from cinders that accumulate during moderately explosive eruptions. Cinder cones are made of pyroclastic material and *scoria* (cindery rock, full of air bubbles).

A **caldera** (Spanish for "kettle") is a large, basin-shaped depression. It forms when summit material on a volcanic mountain collapses inward after an eruption or other loss of

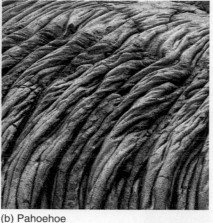

(a) Aa   (b) Pahoehoe

FIGURE 12.24 Two types of basaltic lava—Hawaiian examples.
(a) A rough, sharp-edged lava, aa is said to get its name from the sounds people make when they take off their shoes and attempt to walk on it. The gold-colored strands are a clump of natural spun glass formed by the lava as it blew out into the air—called *Pele's hair*. (b) Pahoehoe forms ropy chords in twisted folds. [Photos by Bobbé Christopherson.]

 ANIMATION   Tectonic Settings and Volcanic Activity

magma. A caldera may fill with rainwater, as it did in beautiful Crater Lake in southern Oregon. News Report 12.3 discusses the Long Valley Caldera in California.

In the Cape Verde Islands (15° N 24.5° W), Fogo Island has such a caldera in the midst of a volcanic cone, opening to the sea on the eastern side. The best farmland is in the caldera, so the people choose to live and work there. Unfortunately, eruptions began in 1995, destroying farmland and forcing the evacuation of 5000 people. Previously, eruptions occurred between A.D. 1500 and 1750, followed by six more in the nineteenth century, each destroying houses and crops.

 ANIMATION   Formation of Crater Lake

Location and Types of Volcanic Activity

The location of volcanic mountains on Earth is a function of plate tectonics and hot-spot activity. Volcanic activity occurs in three settings (with representative examples):

1. Along subduction boundaries at continental plate–oceanic plate convergence (such as Mount St. Helens and Kliuchevskoi, Siberia) or oceanic plate–oceanic plate convergence (Philippines and Japan).
2. Along sea-floor spreading centers on the ocean floor (Iceland on the Mid-Atlantic Ridge or off the coast of Oregon and Washington) or areas of rifting on continental plates (the rift zone in east Africa).
3. At hot spots, where individual plumes of magma rise to the crust (such as Hawai'i and Yellowstone National Park).

Figure 12.25 illustrates these three types of volcanic activity, which you can compare with the active volcano sites and plate boundaries shown in Figure 11.20. Figures 12.26 to 12.31 illustrate the various aspects of volcanism presented in Figure 12.25 and in the following discussion.

The variety of forms among volcanoes makes them difficult to classify; most fall in a middle ground between one type and another. Even during a single eruption, a volcano may behave in several different ways. The primary factors in determining an eruption type are: (1) the magma's chemistry, which is related to its source; and (2) the magma's viscosity. Viscosity is the magma's resistance to flow ("thickness"), ranging from low viscosity (very fluid) to high viscosity (thick and flowing slowly). We consider two types of eruptions—effusive and explosive—and the characteristic landforms they build.

Effusive Eruptions

Effusive eruptions are the relatively gentle ones that produce enormous volumes of lava annually on the seafloor and in places such as Hawai'i and Iceland. These direct eruptions from the asthenosphere and upper mantle produce a low-viscosity magma that is fluid and cools to form a dark, basaltic rock low in silica (less than 50% silica and rich in iron and magnesium). Gases readily escape from this magma because of its low viscosity, adding to a gentle **effusive eruption** that pours out on the surface, with relatively small explosions and few pyroclastics. However, dramatic fountains of basaltic lava sometimes shoot upward, powered by jets of rapidly expanding gases.

An effusive eruption may come from a single vent or from the flank of a volcano, through a side vent. If such vents form a linear opening, they are called *fissures*; these sometimes erupt in a dramatic *curtain of fire* (sheets of molten rock spraying into the air). Rift zones capable of erupting tend to converge on the central crater, or vent, as they do in Hawai'i. The interior of such a crater, often a sunken caldera, may fill with low-viscosity magma during an eruption, forming a molten lake, which then may overflow lava downslope in dramatic rivers and falls of molten rock.

On the island of Hawai'i, the continuing Kīlauea eruption is the longest in recorded history, active since January 3, 1983—the 55th episode of this series began in February 1997. More than 60 separate eruptions occurred at Kīlauea's summit between 1778 (the arrival of westerners) and 1982. To date, Kīlauea produced 3.1 km³ (0.7 mi³) of lava, covering 117 km² (45 mi²). Fourteen km (7 mi) of the Chain-of-Craters road are covered up to 35-m (115-ft) deep in lava flows. Needless to say, the NPS Rangers keep their end-of-road checkpoint for tourists movable.

FIGURE 12.25 Tectonic settings of volcanic activity.
Magma rises and lava erupts from rifts, through crust above subduction zones, and where thermal plumes at hot spots break through the crust. [After U.S. Geological Survey, *The Dynamic Planet* (Washington, DC: Government Printing Office, 1989).]

Forming Types of Volcanoes

ANIMATION

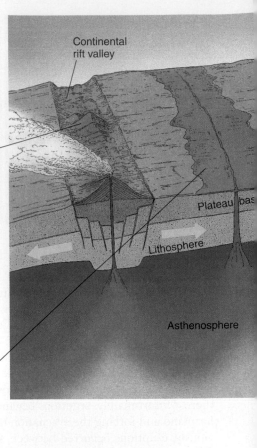

FIGURE 12.26 Rift valley.
Along the rift in Iceland, at the dramatic Gullfoss (waterfalls). Iceland is the only place where the Mid-Atlantic Ridge system is seen above sea level. Iceland is growing as new material increases the size of this island country. [Photo by Bobbé Christopherson.]

FIGURE 12.27 Two expressions of volcanic activity.
In the foreground are plateau (flood) basalts characteristic of the Columbia Plateau in Oregon. Mount Hood, in the background to the west of the plateau, is a composite volcano, part of the Cascade Range of active and dormant volcanoes that result from subduction of the Juan de Fuca plate beneath the North American plate. [Photo by author.]

FIGURE 12.28 Volcanic fountaining, boiling surf, and lava falls in Hawai'i
From Earth's most active volcano: (a) Kīlauea erupts a fountain of lava from East Rift spatter cone; (b) lava flows reach the sea, adding new land to the b island of Hawai'i; (c) continuing flows form spectacular lava falls over the pa (cliff) in 2002. [(a) Photo by Kepa Maly/U.S. Geological Survey; (b) and (c) photos Bobbé Christopherson.]

FIGURE 12.32 Dramatic Hawaiian active lava flows form
The glowing lava is 1200°C (2200°F) as it sluggishly makes i
of this aerial photo, the radiated heat from the lava flow is in
blown out as the ocean quenches the lava. The present epis
Christopherson.]

(a)

(b)

1

200

FIGURE 12.30 Extrusive igneous
rocks compared.
A dacite volcanic bomb (lighter rock
on left, higher in silica) from Mount St.
Helens and a basalt lava rock (darker
rock on right, lower in silica) from
Hawai'i. What does this coloration tell
you about the rocks' history and
composition? Find these rock types in
Table 11.2. [Photos by Bobbé
Christopherson.]

Ontong Java Plateau, which covers an area of the seafloor
in the Pacific (Figure 12.35). These regions are sometimes
referred to as *plateau basalt provinces.* Of these extensive
igneous provinces, no presently active sites come close in
size to the largest of the extinct igneous provinces, some of
which formed more than 200 million years ago.

GURE 12.29 Kīlauea landscape.
rial photo of the newest land on the planet produced by the massive
ws of basaltic lavas from the Kīlauea volcano, Hawai'i Volcano
tional Park. In the distance, ocean water quenches the intense
00°C (2200°F) lava on contact, producing steam and hydrochloric
id mist. [Photo by Bobbé Christopherson.]

FIGURE 12.31 Volcanic island off the coast of
New Zealand.
White Island, New Zealand (37.5° S, 177° E), actively erupted
through 2001 without warning, sometimes emitting gas
and steam, other times bursts of pyroclastics and ash.
[Photo by Harvey Lloyd/Stock Market.]

News Report 12.3

Is the Long Valley Caldera Ne...

Trees are dying in the forests on Mammoth Mountain, in a portion of the Long Valley Caldera, near the California–Nevada border. The oval caldera is about 15 by 30 km (9 by 19 mi) long in a north–south direction and sits at 2000 m (6500 ft) elevation, rising to 2600 m (8500 ft) along the western side.

About 1200 tons of carbon dioxide is coming up through the soil in the old caldera each day. Carbon dioxide levels reach 30% to 96% of gases in some soil samples. The source: active and moving magma at some 3-km depth. Elsewhere in the world, this and other gas emissions assist forecasts of potential volcanic activity. At Mammoth Mountain, these

gases signify ... be a portent ... 12.3.1). The ... dead trees in ...

A powe... 760,000 year... Valley Calder... exceeded the ... 1980 Mount ... than 2000 tim... lesser extent ... over the past... sand years, ... A.D. 1720 and ... area north of ...

In the la... 1996, contin... swarms of ear...

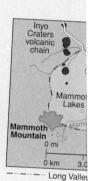

During 1989–1990, lava flows from Kīlauea actually consumed several visitor buildings in the Hawai'i Volcanoes National Park and two housing subdivisions in Kalapana, Kapa'ahu, and Kaimū. Kīlauea has produced more lava than any other vent on Earth in recorded history (Figures 12.28 and 12.29). Beginning on Mothers Day 2002, an aggressive eruption sequence intensified (Figure 12.28c), flowing over Chain of Craters Road and continuing for several years to send streams of lava to the ocean (Figure 12.32). Pu'u O'o is the active crater on Kīlauea and is presently continuing to erupt. In 2007, a portion of Pu'u O'o crater's west side collapsed, filled with lava and collapsed again (Figure 12.33)—see **http://hvo.wr.usgs.gov/**.

A typical mountain landform built from effusive eruptions is gently sloped, gradually rising from the surrounding landscape to a summit crater. The shape is similar in outline to a shield of armor lying face up on the ground and therefore is called a **shield volcano**. The shield shape and size of Mauna Loa in Hawai'i is distinctive when compared with Mount Rainier in Washington,

---

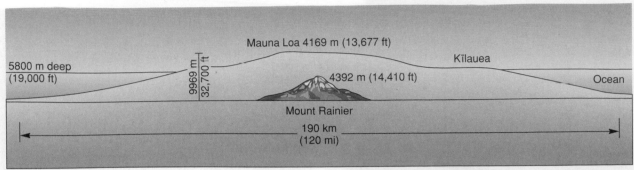

FIGURE 12.34 Shield and composite volcanoes compared.
Comparison of Mauna Loa in Hawai'i, a shield volcano, and Mount Rainier in Washington State, a composite volcano. Their strikingly different profiles signify their different tectonic origins. Note the gently sloped shield shape in the inset photo. [After U.S. Geological Survey, *Eruption of Hawaiian Volcanoes* (Washington, DC: Government Printing Office, 1986). Inset photo by Bobbé Christopherson.]

it tends to block the magma conduit inside the volcano. The blockage traps and compresses gases, causing pressure and conditions to build for a possible **explosive eruption**.

From this magma, a lighter dacitic rock forms at the surface, as illustrated in the comparison in Figure 12.30 (see *dacite* in Table 11.2). Unlike the volcanoes in Hawai'i Volcanoes National Park, where tourists gather at observation platforms or hike out across the cooling lava flows to watch the relatively calm effusive eruptions, these explosive eruptions do not invite close inspection and can explode with little warning.

The term **composite volcano** describes these explosively formed mountains. (They are sometimes called *stratovolcanoes* because they are built up in alternating layers of ash, rock, and lava, but shield volcanoes also can exhibit a stratified structure, so *composite* is the preferred term.) Composite volcanoes tend to have steep sides, are more conical in shape than shield volcanoes, and therefore are also known as *composite cones*. If a single summit vent erupts repeatedly, a remarkable symmetry may develop as the mountain grows in size, as demonstrated by Mount Orizaba in Mexico, Mount Shishaldin in Alaska (Figure 12.36), Mount Fuji in

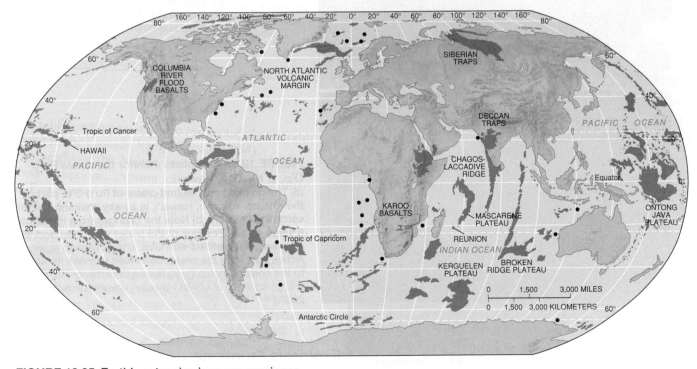

FIGURE 12.35 Earth's extensive igneous provinces.
Plateau basalts form large igneous provinces across the globe. Many are ancient and inactive. The largest known is the Ontong Java Plateau, which covers nearly 2 million km² (0.8 million mi²) on the Pacific Ocean floor. Black dots are areas of ancient volcanism along continental margins. [After M. F. Coffin and O. Eldholm, "Large igneous provinces," *Scientific American* (October 1993): 42–43. © Scientific American, Inc.]

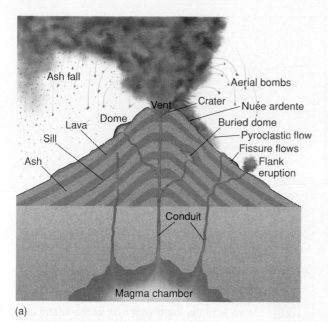

(a)

(b)

FIGURE 12.36 A composite volcano.
(a) A typical composite volcano with its cone-shaped form. (b) Mount Shishaldin, Unimak Island in the eastern Aleutian Islands, Alaska, is a 2857-m (9372-ft) composite volcano. Here shown during a 1995 eruption; it was active through 2001. [(b) Photo courtesy of AeroMap U.S., Inc., Anchorage, Alaska.]

Japan, the pre-1980-eruption shape of Mount St. Helens in Washington, and Mount Mayon in the Philippines.

Owing to its chemical makeup and high viscosity, rising magma in a composite volcano forms a plug near the surface. The blockage causes tremendous pressure to build, keeping the trapped gases compressed and liquefied. When the blockage can no longer hold back this pressurized inferno, explosions equivalent to megatons of TNT blast the tops and sides off these mountains. This type of eruption produces much less lava than effusive eruptions but larger amounts of *pyroclastics*, which include volcanic *ash* (<2 mm, or 0.08 in., in diameter), dust, cinders, *lapilli* (up to 32 mm, or 1.26 in., in diameter), *scoria* (volcanic slag), pumice, and *aerial bombs* (explosively ejected blobs of incandescent lava). A *nuée ardente*, French for "glowing cloud," is an incandescent, hot, turbulent gas, ash, and pyroclastic cloud that can jet across the landscape in an eruption.

Europe's largest volcano is of this type—Mount Etna on the island of Sicily in the Mediterranean Sea, with its 3340-m (10,958-ft) summit. A volcano has been active at this site for roughly half a million years, with the first reported eruption in 1500 B.C. The current cycle began with huge amounts of magma moving upward and inflating the summit in 1994. Dramatic Strombolian-type eruptions of spraying and jetting fountains of lava were visible from a 30-km (18.6-mi) distance. Summit and flank eruptions have marked the ensuing years, such as the latest in the sequence during October 2002 (Figure 12.37). The summit crater emitted ash and some lava flows in episodes from 2004 to 2007; magma was again on the move in the volcano's upper conduit and local villages on alert. Rather than being perceived as an immediate danger, such activity is stimulating tourism.

Table 12.3 presents a sampling of some notable composite volcano eruptions. Focus Study 12.1 details the highly publicized 1980 eruption of Mount St. Helens.

Mount Pinatubo Eruption   In June 1991, after 600 years of dormancy, Mount Pinatubo in the Philippines erupted. The summit of the 1460-m (4795-ft) volcano exploded, devastating many surrounding villages and permanently closing Clark Air Force Base operated by the United States. Fortunately, scientists from the U.S. Geological Survey and local scientists accurately predicted the eruption. A timely evacuation of the surrounding countryside saved thousands of lives, but 800 people were killed.

Although volcanoes are regional events, their spatial implications can be worldwide. The single volcanic eruption of Mount Pinatubo was significant to the global environment, as discussed in Chapters 1, 3, 4, 5, and 10 (see Figure 1.6). Here is a summary of effects:

- 15–20 million tons of ash and sulfuric acid mist were blasted into the atmosphere, concentrating at 16–25 km (10–15.5 mi) altitude.
- 12 km^3 (3.0 mi^3) of material was ejected and extruded by the eruption (12 times the volume from Mount St. Helens).
- 60 days after the eruption, about 42% of the globe was affected (from 20° S to 30° N) by the thin, spreading aerosol cloud in the atmosphere.
- Colorful twilight and dawn skies were observed worldwide.
- An increase in atmospheric albedo of 1.5% (4.3 W/m^2) occurred.
- An increase in the atmospheric absorption of insolation followed (2.5 W/m^2).

FIGURE 12.37 Ageless Mount Etna, on the Isle of Sicily.
Mount Etna erupts in view of the International Space Station (ISS). Visible is the ash and steam from a vigorous eruption. Gas emissions are along the north slope (lower left) through a series of vents. Smoke on the lower slope is from wildfires set by lava flows. View is to the southeast across the island of Sicily. Ashfall from this episode reached North Africa. [October 30, 2002, ISS photo courtesy of Earth Science and Image Analysis Laboratory, JSC, NASA.]

Table 12.3 Sample of Notable Composite Volcano Eruptions

Date	Location	Number of Deaths	Amount Extruded (mostly pyroclastics) in km³ (mi³)
Prehistoric	Yellowstone, Wyoming	Unknown	2400 (576)
4600 B.C.	Mount Mazama (Crater Lake, Oregon)	Unknown	50–70 (12–17)
1900 B.C.	Mount St. Helens	Unknown	4 (0.95)
A.D. 79	Mount Vesuvius, Italy	20,000	3 (0.7)
1500	Mount St. Helens	Unknown	1 (0.24)
1815	Tambora, Indonesia	66,000	80–100 (19–24)
1883	Krakatau, Indonesia	36,000	18 (4.3)
1902	Mont Pelée, Martinique	29,000	Unknown
1912	Mount Katmai, Alaska	Unknown	12 (2.9)
1943–1952	Paricutín, Mexico	0	1.3 (0.30)
1980	Mount St. Helens	54	4 (0.95)
1985	Nevado del Ruiz, Colombia	23,000	1 (0.24)
1991	Mount Unzen, Japan	10	2 (0.5)
1991	Mount Pinatubo, Philippines	800	12 (3.0)
1992	Mount Spurr/Mount Shishaldin, Alaska	0	1 (0.24)
1993	Galeras Volcano, Colombia	5	1 (0.24)
1993, 2003, 2004	Mount Mayon, Philippines	0	1 (0.24)
2002	Nyiragongo, Congo	70	1 (0.24)
1994–2007	Kliuchevskoi, Russia Soufrière Hills Montserrat, West Indies	0	Ash, steam, pyroclastic flows
2007	Arenal, Costa Rica Karymsky, Kamchatka, Russia	0	Ash, pyroclastics

- A decrease in net radiation at the surface and a lowering of Northern Hemisphere average temperatures of 0.5 C° (0.9 F°) were measured.
- Atmospheric scientists and volcanologists were able to study the eruption aftermath using satellite-borne orbiting sensors and general-circulation-model computer simulations.

Volcano Forecasting and Planning

The USGS and the Office of Foreign Disaster Assistance operate the Volcano Disaster Assistance Program (VDAP). For a listing of more than two dozen projects, see **http://vulcan.wr.usgs.gov/Vdap/framework.html**. The need for such a program is evident in that over the past 30 years, volcanic activity has killed 29,000 people, forced more than 800,000 to evacuate their homes, and caused more than $3 billion in damage. The program was established after 23,000 people died in the eruption of Nevado del Ruiz, Colombia, in 1985. The U.S. VDAP is in place to help local scientists with eruption forecasts by

setting up mobile volcano-monitoring systems at the most-threatened sites.

An effort such as this led to the life-saving evacuation of 60,000 people hours before Mount Pinatubo exploded. In addition, satellite remote sensing is helping VDAP to monitor eruption cloud dynamics, atmospheric emissions and climatic effects, and lava and thermal measurements; to make topographic measurements; to estimate volcanic hazard potential; and to enhance geologic mapping—all in an effort to better understand our dynamic planet. Integrated seismographic networks and monitoring are making possible early warning systems.

In this era of the Internet you can access "volcano cams" positioned around the world to give you 24-hour surveillance of many volcanoes. For an exciting visual adventure, go to the following URL and add it to your bookmarks (note whether it is day or night for the location you are checking): **http://vulcan.wr.usgs.gov/Photo/volcano_cams.html**.

Focus Study 12.1

The 1980 Eruption of Mount St. Helens

Probably the most studied and photographed composite (explosive) volcano on Earth is Mount St. Helens, located 70 km (45 mi) northeast of Portland, Oregon, and 130 km (80 mi) south of the Tacoma–Seattle area of Washington. Mount St. Helens is the youngest and most active of the Cascade Range of volcanoes, which form a line from Mount Lassen

in California to Mount Meager in British Columbia (Figure 12.1.1). The Cascade Range is the product of the Juan de Fuca sea-floor spreading center off the coast of northern California, Oregon, Washington, and British Columbia and the plate subduction that occurs offshore, as identified in the upper right of Figure 12.12.

The mountain had been quiet since 1857. New activity began in March 1980, with a sharp earthquake registering a magnitude 4.1. The first eruptive outburst occurred one week later, beginning with a magnitude 4.5 quake and continuing with a thick black plume of ash and the development of a small summit crater. Ten days later, the first volcanic earth-

Debris Avalanche and Eruption of Mt. St. Helens
ANIMATION

(a)   (b)   (c)

FIGURE 12.1.1 Mount St. Helens before and after the eruption—days and years later.
(a) Mount St. Helens prior to the 1980 eruption. (b) The devastated and scorched land shortly after the eruption in 1980. The scorched earth and tree blow-down area covered some 38,950 hectares (95,000 acres). (c) A landscape in recovery as life moves back in and takes hold to establish new ecosystems in 1999 along the Toutle River and debris flow. [(a) Photo by Pat and Tom Lesson/Photo Researchers; (b) photo by Krafft-Explorer/Photo Researchers, Inc.; (c) photo by Bobbé Christopherson.]

(continued)

Focus Study 12.1 (continued)

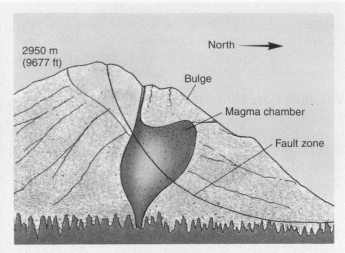

North →

2950 m
(9677 ft)

Bulge

Magma chamber

Fault zone

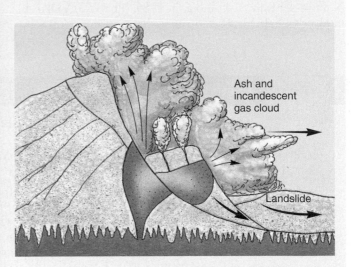

Ash and
incandescent
gas cloud

Landslide

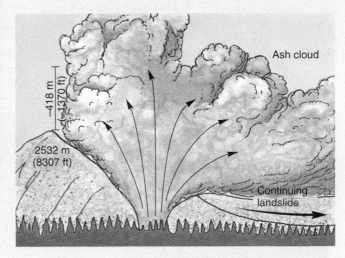

Ash cloud

418 m
(1370 ft)

2532 m
(8307 ft)

Continuing
landslide

FIGURE 12.1.2 The Mount St. Helens eruption sequence and corresponding schematics. [Photo sequence by Keith Ronnholm. All rights reserved.]

quake, called a *harmonic tremor*, registered on the many instruments that had been hurriedly placed around the volcano. Harmonic tremors are slow, steady vibrations, unlike the sharp releases of energy associated with tectonic earthquakes and faulting. Harmonic tremors told scientists that magma was on the move within the mountain.

Also developing was a massive bulge on the north side of the mountain. This bulge indicated the direction of the magma flow within the volcano. A bulge represents the greatest risk from a composite volcano, for it could signal a potential lateral burst through the bulge and across the landscape.

Early on Sunday, May 18, the area north of the mountain was rocked by a magnitude 5.0 quake, the strongest to date. The mountain, with its distended 245-m (800-ft) bulge, was shaken, but nothing happened. Then a second quake (magnitude 5.1) hit a few minutes later, loosening the bulge and launching the eruption. David Johnston, a volcanologist with the U.S. Geological Survey, was only 8 km (5 mi) from the mountain, servicing instruments, when he saw the eruption begin. He radioed headquarters in Vancouver, Washington, saying, "Vancouver, Vancouver, this is it!" He perished in the eruption. For a continuously updated live picture of Mount St Helens from Johnston's approximate observation point and other specifics, go to http://www.fs.fed.us/gpnf/mshnvm/volcanocam/. The camera is mounted below the roofline at the Johnston Ridge Observatory. The observatory is open to the public.

As the contents of the mountain exploded, a surge of hot gas (about 300°C, or 570°F), steam-filled ash, pyroclastics, and a *nuée ardente* (rapidly moving, very hot, explosive ash and incandescent gases) moved northward, hugging the ground and traveling at speeds up to 400 kmph (250 mph) for a distance of 28 km (17 mi).

The slumping north face of the mountain produced the greatest landslide witnessed in recorded history; about 2.75 km³ (0.67 mi³) of rock, ice, and trapped air, all fluidized with steam, surged at speeds approaching 250 kmph (155 mph). Landslide materials traveled for 21 km (13 mi) into the valley, blanketing the forest, covering a lake, and filling the rivers below. A series of photographs, taken at 10-second intervals from the east looking west, records this sequence (Figure 12.1.2). The eruption continued with intensity for 9 hours, first clearing out old rock from the throat of the volcano and then blasting new material.

As destructive as such eruptions are, they also are constructive, for this is the way in which a volcano eventually builds its height. Before the eruption, Mount St. Helens was 2950 m (9677 ft) tall; the eruption blew away 418 m (1370 ft). Today, Mount St. Helens is building a lava dome within its crater. The thick lava rapidly and repeatedly plugs and breaks in a series of lesser dome eruptions that may continue for several decades. The dome already is more than 300 m (1000 ft) high, so a new mountain is being born from the eruption of the old.

The lava dome is covered with a dozen survey benchmarks and several tiltmeters to monitor the status of the volcano. Swarms of minor earthquakes along the north flank of the mountain began in November 2001, marking this mountain as still unstable. Dome eruptions of varying intensities occurred through 2007. Intensive scientific research and monitoring has paid off, as every eruption since 1980 was successfully forecasted from days to as long as 3 weeks in advance, with the exception of one small eruption in 1984.

An ever-resilient ecology is recovering as plants and animals reclaim the devastated landscape. Please refer to the dramatic comparison photos from 1983 and 1999 in Figure 19.28. More than two decades have passed, yet strong interest in the area continues; scientists and more than 1 million tourists visit the Mount St. Helens Volcanic National Monument each year. For more information, see the Cascades Volcano Observatory at http://vulcan.wr.usgs.gov/ and specifically Mount St. Helens at http://vulcan.wr.usgs.gov/Volcanoes/MSH/framework.html.

Summary and Review—Tectonics, Earthquakes, and Volcanism

■ *Describe* first, second, and third orders of relief, and *relate* examples of each from Earth's major topographic regions.

Tectonic forces generated within the planet dramatically shape Earth's surface. **Relief** is the vertical elevation difference in a local landscape. The undulating physical surface of Earth, including relief, is **topography**. Convenient descriptive categories are termed *orders of relief*. The coarsest level of landforms includes the **continental landmasses** and **ocean basins**; the finest comprises local hills and valleys.

relief (p. 361)
topography (p. 361)
continental landmasses (p. 361)
ocean basins (p. 361)

1. How does the map of the ocean floor (chapter-opening illustration) exhibit the principles of plate tectonics? Briefly analyze.

2. What is meant by an "order of relief"? Give an example from each order.
3. Explain the difference between relief and topography.

■ *Describe* the several origins of continental crust, and *define* displaced terranes.

The continents are formed as a result of several processes, including upwelling material from below the crust and migrating portions of crust. The source of material generated by plate tectonic processes determines the behavior and the composition of the crust that results.

A continent has a nucleus of ancient crystalline rock called a *craton*. A region where a craton is exposed is a **continental shield**. As continental crust forms, it is enlarged through accretion of dispersed **terranes**. An example is the *Wrangellia terrane* of the Pacific Northwest and Alaska.

continental shield (p. 364)
terranes (p. 364)

4. What is a craton? Relate this structure to continental shields and platforms, and describe these regions in North America.

5. What is a migrating terrane, and how does it add to the formation of continental masses?

6. Briefly describe the journey and destination of the Wrangellia Terrane.

■ *Explain* compressional processes and folding, and *describe* four principal types of faults and their characteristic landforms.

Folding, broad warping, and faulting deform the crust and produce characteristic landforms. Compression causes rocks to deform in a process known as **folding**, during which rock strata bend and may overturn. Along the *ridge* of a fold, layers *slope downward away from the axis*, which is an **anticline**. In the *trough* of a fold, however, layers *slope downward toward the axis*; this is a **syncline**.

When rock strata are stressed beyond their ability to remain a solid unit, they express the strain as a fracture. Rocks on either side of the fracture are displaced relative to the other side in a process known as **faulting**. Thus, *fault zones* are areas where fractures in the rock demonstrate crustal movement. At the moment of fracture, a sharp release of energy, an **earthquake** or *quake*, occurs.

When forces pull rocks apart, the tension causes a **normal fault**, sometimes visible on the landscape as a scarp, or escarpment. Compressional forces associated with converging plates force rocks to move upward, producing a **reverse fault**. A low-angle fault plane is referred to as a **thrust fault**. Horizontal movement along a fault plane that produces a linear rift valley is a **strike-slip fault**. In the U.S. interior west, the Basin and Range Province is an example of aligned pairs of normal faults and a distinctive horst-and-graben landscape. The term **horst** is applied to upward-faulted blocks; **graben** refers to downward-faulted blocks.

> folding (p. 367)
> anticline (p. 367)
> syncline (p. 367)
> faulting (p. 370)
> earthquake (p. 370)
> normal fault (p. 370)
> reverse fault (p. 370)
> thrust fault (p. 370)
> strike-slip fault (p. 370)
> horst (p. 370)
> graben (p. 370)

7. Diagram a simple folded landscape in cross section, and identify the features created by the folded strata.

8. Define the four basic types of faults. How are faults related to earthquakes and seismic activity?

9. How did the Basin and Range Province evolve in the western United States? What other examples exist of this type of landscape?

■ *Relate* the three types of plate collisions associated with orogenesis, and *identify* specific examples of each.

Orogenesis is the birth of mountains. An orogeny is a mountain-building episode, occurring over millions of years, that thickens continental crust. It can occur through large-scale deformation and uplift of the crust. It also may include the capture of migrating terranes and cementation of them to the continental margins and the intrusion of granitic magmas to form plutons.

Oceanic plate–continental plate collision orogenesis is now occurring along the Pacific coast of the Americas and has formed the Andes, the Sierra of Central America, the Rockies, and other western mountains. *Oceanic plate–oceanic plate* collision orogenesis produces either simple volcanic island arcs or more complex arcs such as Japan, the Philippines, the Kurils, and portions of the Aleutians. The region around the Pacific contains expressions of each type of collision in the **circum-Pacific belt**, or the **ring of fire**.

Continental plate–continental plate collision orogenesis is quite mechanical; large masses of continental crust, such as the Himalayan Range, are subjected to intense folding, overthrusting, faulting, and uplifting.

> orogenesis (p. 374)
> circum-Pacific belt (p. 376)
> ring of fire (p. 376)

10. Define *orogenesis*. What is meant by the birth of mountain chains?

11. Name some significant orogenies.

12. Identify on a map Earth's two large mountain chains. What processes contributed to their development?

13. How are plate boundaries related to episodes of mountain building? Explain how different types of plate boundaries produce differing orogenic episodes and different landscapes.

14. Relate tectonic processes to the formation of the Appalachians and the Alleghany orogeny.

■ *Explain* the nature of earthquakes, their measurement, and the nature of faulting.

Earthquakes generally occur along plate boundaries; major ones can be disastrous. Earthquakes result from faults, which are under continuing study to learn the nature of faulting, stress and the buildup of strain, irregularities along fault-plane surfaces, the way faults rupture, and the relationship among active faults. Earthquake prediction and improved planning are active concerns of *seismology*, the study of earthquake waves and Earth's interior. Seismic motions are measured with a **seismograph**.

Charles Richter developed the **Richter scale**, a measure of earthquake magnitude. A more precise and quantitative scale that assesses the *seismic moment* is now used, especially for larger quakes; this is the **moment magnitude scale**. The specific mechanics of how a fault breaks are under study, but the **elastic-rebound theory** describes the basic process. In general, two sides along a fault appear to be locked by friction, resisting any movement. This stress continues to build strain along the fault surfaces, storing elastic energy like a wound-up spring. When energy is released abruptly as the rock breaks, both sides of the fault return to a condition of less strain.

> seismograph (p. 380)
> Richter scale (p. 381)
> moment magnitude scale (p. 381)
> elastic-rebound theory (p. 381)

15. Describe the differences in human response between the two earthquakes that occurred in China in 1975 and 1976.

16. Differentiate between the Mercalli and moment magnitude and amplitude scales. How are these used to describe an earthquake? Why has the Richter scale been updated and modified?

17. What is the relationship between an epicenter and the focus of an earthquake? Give examples from the Loma Prieta, California, and Kobe, Japan, earthquakes.

18. What local soil and surface conditions in San Francisco severely magnify the energy felt in earthquakes?

19. How do the elastic-rebound theory and asperities help explain the nature of faulting? In your explanation, relate the concepts of stress (force) and strain (deformation) along a fault. How does this lead to rupture and earthquake?

20. Describe the San Andreas fault and its relationship to ancient sea-floor spreading movements along transform faults.

21. How is the seismic gap concept related to expected earthquake occurrences as described in News Report 12.3? Are any gaps correlated with earthquake events in the recent past? Explain.

22. What do you see as the biggest barrier to effective earthquake prediction?

■ *Distinguish* between an effusive and an explosive volcanic eruption, and *describe* related landforms, using specific examples.

Volcanoes offer direct evidence of the makeup of the asthenosphere and uppermost mantle. A **volcano** forms at the end of a central vent or pipe that rises from the asthenosphere through the crust into a volcanic mountain. A **crater**, or circular surface depression, usually forms at the summit. Areas where magma is near the surface may heat groundwater, producing **geothermal energy**.

Eruptions produce **lava** (molten rock), gases, and **pyroclastics** (pulverized rock and clastic materials ejected violently during an eruption) that pass through the vent to openings and fissures at the surface and build volcanic landforms. Basaltic lava flows occur in two principal textures: **aa**, rough and sharp-edged lava, and **pahoehoe**, smooth, ropy folds of lava. Volcanic activity produces landforms such as a **cinder cone**, a small hill, or a large basin-shaped depression sometimes caused by the collapse of a volcano's summit, forming a **caldera**.

Volcanoes are of two general types, based on the chemistry and the viscosity of the magma involved. **Effusive eruption** produces a **shield volcano** (such as Kīlauea in Hawai'i) and extensive deposits of **plateau basalts**, or flood basalts. Magma of higher viscosity leads to an **explosive eruption** (such as Mount Pinatubo in the Philippines), producing a **composite volcano**. Volcanic activity has produced some destructive moments in history but constantly creates new seafloor, land, and soils.

volcano (p. 386)
crater (p. 386)
geothermal energy (p. 386)
lava (p. 386)
pyroclastics (p. 386)
aa (p. 386)
pahoehoe (p. 386)
cinder cone (p. 386)
caldera (p. 386)
effusive eruption (p. 387)
shield volcano (p. 390)
plateau basalts (p. 390)
explosive eruption (p. 392)
composite volcano (p. 392)

23. What is a volcano? In general terms, describe some related features.

24. Where do you expect to find volcanic activity in the world? Why?

25. Compare effusive and explosive eruptions. Why are they different? What distinct landforms are produced by each type? Give examples of each.

26. Describe several recent volcanic eruptions.

NetWork

The *Geosystems* Student Learning Center provides on-line resources for this chapter on the World Wide Web. To begin: Once at the Center, click on the cover of this textbook, scroll the Table of Contents menu, and select this chapter. You will find self-tests that are graded, review exercises, specific updates for items in the chapter, and in "Destinations" many links to interesting related pathways on the Internet. *Geosystems* Student Learning Center is found at **http://www.prenhall.com/christopherson/**.

Critical Thinking

A. Using the topographic regions map in Figure 12.3, assess the region within 100 km (62 mi) and 1000 km (620 mi) of your campus. Describe the topographic character and the variety of relief within these two regional scales. You may want to consult local maps and atlases in your analysis. Do you perceive that this type of topographic region influences lifestyles? Economic activities? Transportation? History of the region?

B. Determine from the library, Internet, and local agencies the seismic potential of the region in which your campus is sited. If a hazard exists, does it influence regional planning? Availability of property insurance? If appropriate, are there existing disaster or emergency plans for your campus? For your community, state, or province? Explain any other significant issues you found in working on this critical thinking challenge.

As water freezes it exerts tremendous force. Here marble, a metamorphic rock, is shattered in the freeze–thaw physical weathering process of frost action. The harsh climate in a high-latitude landscape takes its toll on the rock formation. [Photo by Bobbé Christopherson.]

Weathering, Karst Landscapes, and Mass Movement

■ Key Learning Concepts

After reading the chapter, you should be able to:

- ■ *Define* the science of geomorphology.

- ■ *Illustrate* the forces at work on materials residing on a slope.

- ■ *Define* weathering, and *explain* the importance of the parent rock and joints and fractures in rock.

- ■ *Describe* frost action, salt-crystal growth, pressure-release jointing, and the role of freezing water as physical weathering processes.

- ■ *Describe* the susceptibility of different minerals to the chemical weathering processes called hydration, hydrolysis, oxidation, carbonation, and solution.

- ■ *Review* the processes and features associated with karst topography.

- ■ *Portray* the various types of mass movements, and *identify* examples of each in relation to moisture content and speed of movement.

A benefit you receive from a physical geography course is a new appreciation of the scenery. Whether you go by foot, bicycle, car, train, or plane, travel is an opportunity to experience Earth's varied landscapes and to witness the active processes that produce them.

In the last two chapters, we discussed the endogenic (internal) processes of our planet and how they produce landforms. However, as the landscape is formed, several exogenic (external) processes simultaneously

FIGURE 13.1 Weathered landform. Delicate Arch—a dramatic example of differential weathering in Arches National Park, Utah. Resistant rock strata at the top of the structure have helped preserve the arch beneath as surrounding rock was eroded away. Note the person standing at the base in the inset photo for a sense of scale. In the distance are the snow-covered Manti LaSal Mountains, extinct volcanoes. [Photos by author.]

wear and waste it. This ongoing struggle between tectonics and climate shapes the landscape upon which living systems play their drama.

Here we begin a five-chapter examination of exogenic processes at work on the landscape. This chapter examines weathering and mass movement of the lithosphere. The next four chapters look at specific exogenic agents and their handiwork—river systems, wind-influenced landscapes, deserts in water-deficit regions, coastal processes and landforms, and regions worked by ice and glaciers. Whether you enjoy time along a river, love the desert or the waves along a coastline, or live in a place where glaciers once carved the land, you will find something of interest in these chapters.

In this chapter: We look at physical (mechanical) and chemical weathering processes that break up, dissolve, and generally reduce the landscape. Such weathering releases essential minerals from bedrock for soil formation or enrichment. Perhaps you live in a region that has caves and caverns. Water has dissolved enormous underground worlds of mystery and darkness, yet many caves remain undiscovered. In addition, we examine mass-movement processes that continually operate in and upon the landscape. News broadcasts may describe an avalanche in Colombia; mudflows in Mexico, China, or Peru; tragic landslides in China, Turkey, Indonesia, or Colombia; massive rockfall in Yosemite; or prediction of an imminent mudflow down the slopes of a volcano in New Zealand or in a southern California suburb. We seek to understand the processes leading to these events.

Landmass Denudation

Geomorphology is the science of landforms—their origin, evolution, form, and spatial distribution. This science is an important aspect of physical geography. **Denudation** is any process that wears away or rearranges landforms. The principal denudation processes affecting surface materials include *weathering*, *mass movement*, *erosion*, *transportation*, and *deposition*, as produced by the agents of moving water, air, waves, and ice—all influenced by the pull of gravity.

Interactions between the structural elements of the land and denudation processes are complex. They represent a continuing struggle between Earth's internal and external processes, between the resistance of materials and weathering and erosional processes. The 15-story-tall Delicate Arch in Utah is dramatic evidence of this struggle (Figure 13.1). Differing resistances of the rocks, coupled with variations in the processes at work on the rock, are carving this delicate sculpture—an example of **differential weathering**, where a more resistant cap rock protects supporting strata below.

Ideally, endogenic processes build *initial landscapes*, whereas exogenic processes develop *sequential landscapes* of low relief, gradual change, and stability. These countering processes must be viewed as going on simultaneously. Several hypotheses have been proposed to model denudation processes and to account for the appearance of the landscape.

Dynamic Equilibrium Approach to Landforms

A landscape is an open system, with highly variable inputs of energy and materials: Uplift creates the *potential energy of position* above sea level and therefore disequilibrium, an imbalance, between relief and energy. The Sun provides radiant energy that converts into *heat energy*. The hydrologic cycle imparts *kinetic energy* through mechanical motion. *Chemical energy* is made available from the atmosphere and various reactions within the crust. In response to this input of energy and materials, landforms constantly adjust toward *equilibrium*.

Most geomorphologists accept the **dynamic equilibrium model**, which emphasizes a balance among force, form, and process. The dynamic equilibrium model summarizes this balancing act between tectonic uplift and reduction by weathering and erosion, between the resistance of rocks and the ceaseless attack of weathering processes. A dynamic equilibrium demonstrates a trend over time. According to current thinking, landscapes in a dynamic equilibrium feature ongoing adaptations to the ever-changing conditions of rock structure, climate, local relief, and elevation. In a sense, every landscape is unique unto itself.

Endogenic events (such as earthquakes and volcanic eruptions) or exogenic events (such as heavy rainfall or forest fire) may provide new sets of relationships for the landscape. Following such destabilizing events, a landform system arrives at a **geomorphic threshold**—the point at which there is enough energy to overcome resistance against movement. At this threshold, the system breaks through to a new equilibrium as the landform adjusts. The pattern over time follows a sequence: (1) equilibrium stability (fluctuating around some average), (2) a destabilizing event, (3) a period of adjustment, and (4) development of a new and different condition of equilibrium stability.

The disturbed hillslope in Figure 13.2 is in the midst of compensating adjustment. The failure of saturated slopes caused a landslide into the river and set a disequilibrium condition. As a consequence, the new dam of material threw the stream into disequilibrium between its flow and sediment load.

Slow, continuous-change events, such as soil development and erosion, tend to maintain an approximate equilibrium condition. Dramatic events such as a major landslide or dam collapse require longer recovery times before equilibrium is reestablished. (Figure 1.5 graphically illustrates both the distinction between a steady-state equilibrium and a dynamic equilibrium and the occurrence of a geomorphic threshold.)

Slopes   Material loosened by weathering is susceptible to erosion and transportation. However, if gravity is to move loosened material downslope, the agents of erosion must overcome the forces of friction, inertia (the tendency of objects at rest to remain at rest), and the cohesion of particles to each other (Figure 13.3a). If the slope angle is steep enough for gravity to overcome frictional forces, or if material is dislodged by the impact of raindrops, hail, falling branches, moving animals, wind, trail bikes, off-road vehicles, logging, or construction equipment, then particle erosion and transport downslope occur.

Slopes or *hillslopes* are curved, inclined surfaces that form the boundaries of landforms. Figure 13.3b illustrates basic slope components that vary among slopes with conditions of rock structure and climate. Slopes generally feature an upper *waxing slope* near the top (*waxing* means "increasing"). This convex surface curves downward and grades into the *free face* below. The presence of a free face indicates an outcrop of resistant rock that forms a steep scarp or cliff.

Downslope from the free face is a *debris slope*, which receives rock fragments and materials from above. The condition of a debris slope reflects the local climate. In humid climates, continually moving water carries material away, lowering the debris slope. But in arid climates, debris slopes accumulate. A debris slope transitions into a *waning slope*, a concave surface along the base of the slope. This waning surface of erosional, coarser materials gently slopes at a continuously decreasing angle to the valley floor, where finer materials are deposited.

A slope is an open system seeking an *angle of equilibrium*. Conflicting forces work simultaneously on slopes to establish an optimum compromise incline that balances these forces. You can identify these slope components and conditions on the actual hillslope shown in Figure 13.3c. When any condition in the balance is altered, all forces on the slope compensate by adjusting to a new dynamic equilibrium.

The relation between rates of weathering and breakup of slope materials, coupled with the rates of mass movement and material erosion, forms the shape of the slope. A slope is *stable* if its strength exceeds these denudation processes and *unstable* if materials are weaker than these processes. Why are hillslopes shaped in certain ways? How do slope elements evolve? How do hillslopes behave during rapid, moderate, or slow uplift? These are topics of active scientific study and research.

Now, with the concepts of landmass denudation, dynamic equilibrium, and slope development in mind, let us examine specific processes that operate to wear away landforms.

FIGURE 13.2 A slope in disequilibrium.
Unstable, saturated soils gave way, leaving a debris dam partially blocking the river. The hillslope, river, and forest ecosystem are in disequilibrium as adjustments to new conditions proceed. [Photo by author.]

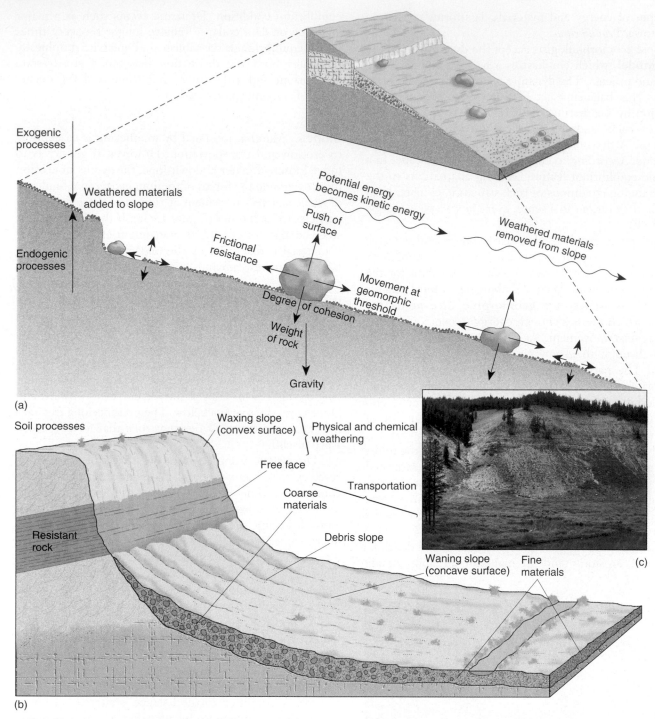

FIGURE 13.3 Slope mechanics and form.
(a) Directional forces (noted by arrows) act on materials along an inclined slope. (b) The principal elements of a slope. (c) A hillslope example; note the role of the rock outcrop as a slope interruption and the rock fragments loosened from the outcrop by frost action. [Photo by author.]

Weathering Processes

Weathering processes attack rocks at Earth's surface and to some depth below the surface. **Weathering** processes either disintegrate rock into mineral particles or dissolve them in water. Weathering processes are both physical (mechanical) and chemical and the interplay of the two is complex. There is a synergy, a combined action, between physical and chemical weathering as the suite of processes work the rock.

Weathering does not transport the materials; it simply generates them for erosion and transport by the agents of water, wind, waves, and ice—all influenced by gravity. In most areas, the upper surface of bedrock undergoes continual weathering, creating broken-up rock called **regolith**. Loose surface material comes from further weathering of regolith and from transported and deposited regolith (Figure 13.4a). In some areas, regolith may be missing or undeveloped, thus exposing an outcrop

FIGURE 13.4 Regolith, soil, and parent materials.
(a) A cross section of a typical hillside. (b) A cliff exposes hillside components. (c) These reddish-colored dunes in the Navajo Tribal Park, near the Utah–Arizona border, derive their color from the red-sandstone parent materials in the background. (d) This image, from a larger panoramic scene, was made by the Mars Exploration Rover *Spirit* on March 12, 2004. Looking toward the Columbia Hills, you can see weathered rocks and windblown sand in approximate true colors. [Photos (b) and (c) by author; Mars image (d) courtesy of NASA/JPL and Cornell University.]

of unweathered bedrock. The thickness of the soil cover depends on a balance between soil production rates and erosion (removal of soil particles)—an equilibrium between competing processes.

Bedrock is the *parent rock* from which weathered regolith and soils develop. Wherever a soil is relatively youthful, its parent rock is traceable through similarities in composition. For example, the sand in Figure 13.4c derives its color and character from the parent rock in the background, just as the sediments on Mars derive their

characteristics from the weathered parent material present (Figure 13.4d). Such sandy, unconsolidated fragmental material, known as **sediment**, combines with weathered rock to form the **parent material** from which soil evolves.

Factors Influencing Weathering Processes

Weathering is greatly influenced by the character of the bedrock: hard or soft, soluble or insoluble, broken or unbroken. *Jointing* in rock is important for weathering processes. **Joints** are fractures or separations in rock that occur without displacement of the sides (as would be the case in faulting). The presence of these usually flat spaces increases the surface area of rock exposed to both physical and chemical weathering.

Important controls on weathering rates are climatic elements—precipitation, temperature, and freeze–thaw cycles. There is a relation among climate (annual precipitation and temperature), physical weathering, and chemical weathering processes. In general, physical weathering dominates in drier, cooler climates, whereas chemical weathering dominates in wetter, warmer climates. Extreme dryness reduces weathering rates, as is experienced in desert climates (*low- and midlatitude hot desert*). In the hot, wet, tropical and equatorial rain forest climates (*tropical rain forest*), most rocks weather rapidly, and the weathering extends deep below the surface. Also related to climate and significant weathering is the position of the water table and water movement on and within rock structures and the subsurface environment.

Another control over weathering rates is the *geographic orientation* of a slope—whether it faces north, south, east, or west. Orientation controls the slope's exposure to Sun, wind, and precipitation. Slopes facing away from the Sun's rays tend to be cooler, moister, and more vegetated than slopes in direct sunlight. This effect of orientation is especially noticeable in middle and higher latitudes.

Vegetation is also a factor in weathering. Although vegetative cover can protect rock by shielding it from raindrop impact and providing roots to stabilize soil, it also produces organic acids from the partial decay of organic matter; these acids contribute to chemical weathering. Plant roots can enter crevices and break up a rock, exerting enough pressure to drive rock segments apart, thereby exposing greater surface area to other weathering processes (Figure 13.5a). You may have observed how tree roots can heave the sections of a sidewalk or driveway sufficiently to raise and crack the concrete.

The scale at which we analyze weathering processes is important. Research at *microscale* levels reveals greater complexity in the relation of climate and weathering, which disproves earlier assumptions about weathering rates in dry climates. At the small scale of actual reaction sites on the rock surface, both physical and chemical weathering processes can occur across varied climate types. Hygroscopic water (a molecule-thin water layer on soil particles) and capillary water (soil water) activate chemical weathering processes, even in the driest landscape. (Review sections in Chapter 9, Figure 9.8, for these water types.)

Imagine all the factors that influence weathering rates as operating in concert: rock composition and structure (jointing), climate influence (precipitation and temperature), groundwater and water movement, slope orientation, vegetation, and microscopic boundary-layer conditions at reaction sites. We separate weathering processes here for convenience of study, but they operate in a complex synergy. Of course in all of this, *time* is the crucial factor, for these processes require long periods of time to operate.

ANIMATION
Physical Weathering

(a)

(b)

FIGURE 13.5 Physical weathering examples.
(a) Roots exert a force on the sides of this joint in the rock.
(b) Frost action shattered this granite rock; ice expansion forced the rock segments apart. [Photos by (a) author; (b) Bobbé Christopherson.]

Physical Weathering Processes

When rock is broken and disintegrated without any chemical alteration, the process is **physical weathering** or *mechanical weathering*. By breaking up rock, physical weathering produces more surface area on which chemical weathering may operate. A single rock that becomes broken into eight pieces has doubled its surface area susceptible to weathering processes. We look briefly at three physical weathering processes: frost action, crystallization, and pressure-release jointing.

Frost Action When water freezes, its volume expands as much as 9% (see Chapter 7). Such expansion creates a powerful mechanical force called **frost action**, or *freeze–thaw action*, which can exceed the tensional strength of rock (Figure 13.5b). Repeated freezing (expanding) and thawing (contracting) of water breaks rocks apart, as shown in this chapter's opening photograph and the Part III opening photograph. Freezing actions are

(a)

(b)

FIGURE 13.6 Physical weathering as joint-block.
(a) Physical weathering along joints in rock produces discrete blocks in the back country of Canyonlands National Park, Utah. (b) Joint-block separation in slate at Alkehornet, Isfjord, on Spitsbergen Island in the Arctic Ocean, where freezing is intense. [Photos by (a) author; (b) Bobbé Christopherson.]

FIGURE 13.7 Rockfall.
Shattered rock debris from a large rockfall in Yosemite National Park. A larger rockfall involving 162,000 tons of granite shocked Yosemite in July 1996, and another in 1999. Freshly exposed, light-colored rock shows where the rockfall originated. [Photo by author.]

important in the humid microthermal climates (*humid continental* and *subarctic*) and polar climates, and they occur at higher elevations in mountains worldwide in the highland climates. In arctic and subarctic climates, frost action dominates soil conditions (discussed in further detail in Chapter 17).

The work of ice begins in small openings, gradually expanding until rocks are cleaved, or split. Figure 13.6 shows blocks of rock on which this *joint-block separation* occurs along existing joints and fractures. This weathering action, called *frost-wedging*, pushes portions of the rock apart. Cracking and breaking create varied shapes in the rocks, depending on the rock structure. In Figure 13.6a, the softer supporting rock underneath the slabs is weathering at a faster pace in *differential weathering*.

Frost action was important to various cultures as a force to quarry rock. Pioneers in the early American West drilled holes in rock, poured water in the holes, and then plugged them. During the cold winter months, expanding ice broke off large blocks along lines determined by the drill-hole patterns. In spring, they hauled the blocks to town for

construction material. Frost action also produces unwanted fractures that damage road pavement and burst water pipes. Perhaps you have noticed rough and broken highways in areas that experience freezing temperatures. The pavement breaks into chunks each winter and develops potholes.

Spring can be a risky time to venture into mountainous terrain. As rising temperatures melt the winter's ice, newly fractured rock pieces fall without warning and may even start rock slides. The falling rock pieces may physically shatter on impact—another form of physical weathering (Figure 13.7). One such rockfall in Yosemite National Park in 1996 involved a 670-m (2200-ft) crashing drop of a 162,000-ton granite slab at 260 kmph (160 mph). The impact shattered the rock, covering 50 acres in powdered rock and felling 500 trees.

Salt-Crystal Growth (Salt Weathering)   Especially in arid climates, dry weather draws moisture to the surface of rocks. As the water evaporates, dissolved minerals in

(a)

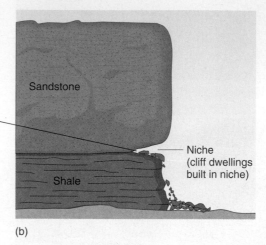

(b)

the water grow crystals—crystallization. Over time, as the crystals grow and enlarge, they exert a force great enough to spread apart individual mineral grains and begin breaking up the rock. Such *salt-crystal growth*, or *crystallization*, is a form of physical weathering.

In the Colorado Plateau of the Southwest, salty water slowly flows from rock strata. As this salty water evaporates, crystallization loosens the sand grains. Subsequent erosion and transportation by water and wind complete the sculpturing process. Deep indentations develop in sandstone cliffs, especially where the sandstone lies above an impervious layer such as shale. More than 1000 years ago, Native Americans built entire villages in these weathered niches at several locations, including Mesa Verde in Colorado and Arizona's Canyon de Chelly (pronounced "canyon duh shay," Figure 13.8).

Pressure-Release Jointing Recall from Chapter 11 how rising magma that is deeply buried and subjected to high pressure forms intrusive igneous rocks called plutons. These plutons cool slowly and produce coarse-grained, crystalline, granitic rocks. As the landscape is subjected to uplift, the regolith overburden is weathered, eroded, and transported away, eventually exposing the

pluton as a mountainous batholith (a plutonic batholith, illustrated in Figure 11.7).

As the tremendous weight of overburden is removed from the granite, the pressure of deep burial is relieved. Over millions of years, the granite slowly responds with an enormous physical heave. In a process known as *pressure-release jointing*, layer after layer of rock peels off in curved slabs or plates, thinner at the top of the rock structure and thicker at the sides. As these slabs weather, they slip off in the process of **sheeting**. This *exfoliation process* creates arch-shaped and dome-shaped features on the exposed landscape, sometimes forming an **exfoliation dome** (Figure 13.9). Such domes are probably the largest weathering features on Earth (in areal extent).

Chemical Weathering Processes

Chemical weathering refers to actual decomposition and decay of the constituent minerals in rock due to chemical alteration of those minerals, always in the presence of water. The chemical breakdown becomes more intense as both temperature and precipitation increase. Although individual minerals vary in susceptibility, no rock-forming minerals are completely unresponsive to

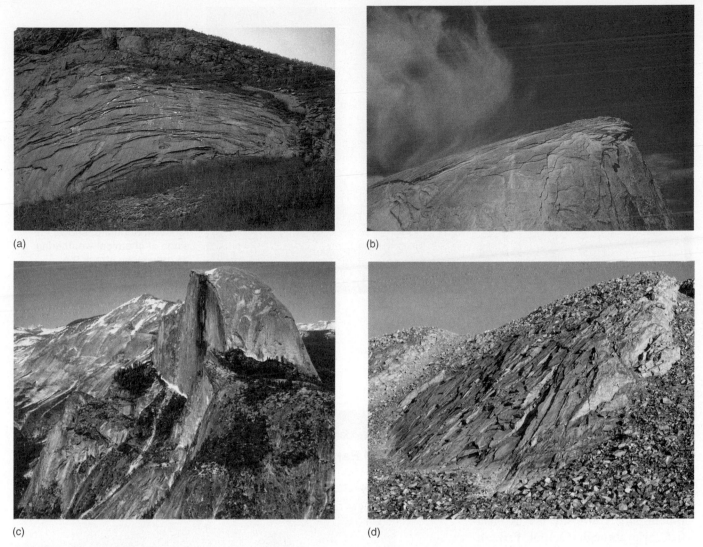

(a)

(b)

(c)

(d)

FIGURE 13.9 Exfoliation in granite.
Exfoliation processes loosen slabs of granite, freeing them for further weathering and downslope movement. (a) Great arches form in the White Mountains of New Hampshire. (b) Exfoliated layers of rock are visible in characteristic dome formations in granites. The loosened slabs of rock are susceptible to further weathering and downslope movement. This view is from the east side of Half Dome in Yosemite National Park, California. (c) Half Dome perspective from the west; relief is approximately 1500 m (5000 ft) from the top of the dome to the glaciated valley below. (d) Rock sheeting exposed along Beverly Sund, Nordaustlandet Island in the Arctic Ocean. [Photos by (a) and (d) Bobbé Christopherson, (b) and (c) by author.]

chemical weathering. A familiar example of chemical weathering is the eating away of cathedral façades and the etching of tombstones by acid precipitation.

In Europe, where increasingly acidic rains resulted from the burning of coal, chemical weathering processes are visible on many buildings. As an example, Saint Magnus Cathedral, in Kirkwell, Orkney Islands, north of Scotland, was built of red and yellow sandstone. Construction began in A.D. 1137—this is 872 years before the copyright date of this book. Almost nine centuries of chemical weathering dissolved cementing materials in the sandstone, breaking down the rock and making many of the sculptured design elements unrecognizable, as if they "melted" or became "out-of-focus" (Figure 13.10). News Report 13.1 discusses the bridges of Central Park in New York City with similar weathering.

As an example of the way chemical weathering attacks rock, consider **spheroidal weathering**. The sharp edges and corners of rocks are rounded as the alteration of minerals progresses through the rock. Joints in the rock offer more surfaces of opportunity for more weathering. Water penetrates joints and fractures and dissolves the rock's weaker minerals or cementing materials. A boulder can be attacked from all sides, shedding spherical shells of decayed rock, like the layers of an onion. The resulting rounded edges are the basis for the name *spheroidal*. Spheroidal weathering of rock resembles exfoliation, but it does not result from pressure-release jointing.

We now look at chemical weathering processes: hydration and hydrolysis, oxidation, and dissolution of carbonates.

FIGURE 13.10 Chemical weathering attacks sandstone cathedral.
(a) The west-front entrance of Saint Magnus Cathedral, in Kirkwell, Scotland, shows the signs of chemical weathering as processes attack the cementing materials in the rock. Construction of this edifice began in A.D. 1137, so the weathering forces have been at work for almost nine centuries. [Photos by Bobbé Christopherson.]

News Report 13.1

Weathering on Bridges in Central Park, NYC

Geographers Greg Pope of Montclair State University and Patricia Beyer of Bloomsburg University led us on a field trip through Central Park in New York City—called "Cultural Stones of Central Park: Outcrops, Arches, and Statues." They prepared a detailed field guide for our urban expedition.

The popular park has gone through many stages of design and construction over the last 150 years, including a variety of arches and bridges (the terms seem interchangeable here) to accommodate different forms of traffic. Thirty-six bridges remain of the original thirty-nine, built from a great variety of rock. Dr. Pope said, "The stones used by the designers and builders in the art and architecture of the park are truly 'cultural stones.' Sources for all this building stone cover a wide geography, from many locales across the Northeast and into Canada."

All this imported rock material presented a diverse array of surfaces on which physical and chemical

FIGURE 13.1.1 Weathered sandstone in Central Park, New York City.
Here on the sides of one of the many arches, constructed in 1859, we see the ravages of both physical and chemical weathering. The patterns etched into this Alberta sandstone, found in New Brunswick, are completely gone in several squares. Encrusted salts, leading to salt-crystal growth (crystallization) processes, are visible. [Photo by Bobbé Christopherson.]

weathering processes could work (Figure 13.1.1). With fossil-fuel consumption and resulting air pollution, increased acidity in rain and snow hastened rates of weathering. In several bridges weathering processes were worsened by auto traffic and the use of salt on the roads in winter.

Some sandstone blocks so poorly resisted weathering that they were actually replaced by cast concrete. Researchers, along with geography students, study the rates of building-stone weathering around Central Park structures and New York City and then map the spatial results.

(a)

(b)

FIGURE 13.11 Chemical weathering and spheroidal weathering. (a) Chemical weathering processes act on the joints in granite to dissolve weaker minerals, leading to a rounding of the edges of the cracks. (b) Rounded granite outcrop demonstrates spheroidal weathering and the disintegration of rock. The surface is actually crumbly. [Photos by Robbé Christopherson.]

Hydration and Hydrolysis Although these two processes are different, we group them together because they both involve water and both work to decompose rock—one a simple combination of water with a mineral in chemical actions and the other a chemical reaction of water with a mineral.

Hydration, meaning "combination with water," involves little chemical change. Water becomes part of the chemical composition of the mineral (such a hydrate is gypsum, which is hydrous calcium sulfate: $CaSO_4 \cdot 2H_2O$). When some minerals hydrate, they expand, creating a strong mechanical effect, a wedging pressure, that stresses the rock, forcing grains apart.

A cycle of hydration and dehydration can lead to granular disintegration and further susceptibility of the rock to chemical weathering. Hydration works together with carbonation and oxidation to convert feldspar, a common mineral in many rocks, to clay minerals and silica. The feldspar gains water in hydration. The hydration process is also at work on the sandstone niches shown in the cliff-dwelling photo; the water and minerals are consumed in the process, becoming a new compound.

When minerals chemically react with water, the process is **hydrolysis**. Hydrolysis is a decomposition process that breaks down silicate minerals in rocks. Compared with hydration, in which water combines with minerals in the rock, the hydrolysis process involves water and elements in chemical reactions to produce different compounds.

For example, the weathering of feldspar minerals in granite can be caused by a reaction to the normal mild acids dissolved in precipitation:

feldspar (K, Al, Si, O) + carbonic acid and water →
residual clays + dissolved minerals + silica

So the by-products of chemical weathering of feldspar in granite include clay (such as kaolinite) and silica. As clay forms from some minerals in the granite, quartz (SiO_2) particles are left behind. The resistant quartz may wash downstream, eventually becoming sand on some distant beach. Clay minerals become a major component in soil and in shale, a common sedimentary rock.

When weaker minerals in rock are changed by hydrolysis, the interlocking crystal network breaks down, so the rock fails and *granular disintegration* takes place. Such disintegration in granite may make the rock appear etched, corroded, and softened, even crumbly (Figure 13.11b).

In Table 11.2, the second line shows resistance to chemical weathering in igneous rocks. On the ultramafic side of the table (right side), the low-silica minerals olivine and peridotite are most susceptible to chemical weathering. Stability gradually increases toward the high-silica minerals such as feldspar. On the far left side of the table, quartz is resistant to chemical weathering. You can see the nature of the constituent minerals, where basalt weathers faster chemically than does granite.

Oxidation Another example of chemical weathering occurs when certain metallic elements combine with oxygen to form oxides. This is a chemical weathering process known as **oxidation**. Perhaps the most familiar oxidation form is the "rusting" of iron in rocks or soil that produces a reddish-brown stain of iron oxide (Fe_2O_3). We have all left a tool or nails outside only to find them weeks later, coated with iron oxide. The rusty color is visible on the surfaces of rock and in heavily oxidized soils such as those in the southeastern United States, southwestern deserts, or the tropics (Figure 13.12). Here is a simple oxidation reaction in iron:

iron (Fe) + oxygen (O_2) → iron oxide (hematite; Fe_2O_3)

As iron is removed from the minerals in a rock, the disruption of the crystal structures in the rock's minerals makes the rock more susceptible to further chemical weathering and disintegration.

Dissolution of Carbonates A third form of chemical weathering occurs when a mineral dissolves into *solution*—for example, when sodium chloride (common table salt)

(a)

(b)

FIGURE 13.12 Oxidation processes in rock and soil.
(a) Oxidation of iron minerals produces these brilliant red colors in the sandstone formations of Red Rock Canyon, Nevada. (b) Ultisols, soils produced by warm, moist conditions, in Sumter County, Georgia, bear the color of iron and aluminum oxides. The crop is peanuts. [Photos by Bobbé Christopherson.]

dissolves in water. Water is the universal solvent because it is capable of dissolving at least 57 of the natural elements and many of their compounds.

Water vapor readily dissolves carbon dioxide, thereby yielding precipitation containing carbonic acid (H_2CO_3). This acid is strong enough to dissolve many minerals, especially limestone, in the process **carbonation**.

FIGURE 13.13 Dissolution of Limestone.
A marble tombstone is chemically weathered beyond recognition in a Scottish churchyard. Marble is a metamorphic form of limestone. Based upon readable dates on surrounding tombstones, this one is about 225 years old. [Photo by Bobbé Christopherson.]

Carbonation simply means reactions whereby carbon combines with minerals, dissolving them.

Such chemical weathering transforms minerals that contain calcium, magnesium, potassium, and sodium. When rainwater attacks formations of limestone (which is calcium carbonate, $CaCO_3$), the constituent minerals dissolve and wash away with the mildly acidic rainwater:

$$\text{calcium carbonate} + \text{carbonic acid and water} \rightarrow$$
$$\text{calcium bicarbonate } (Ca_2^{2+}CO_2H_2O)$$

Walk through an old cemetery and you can observe the dissolution of marble, a metamorphic form of limestone (Figure 13.13). Weathered limestone and marble, in tombstones or in rock formations, appear pitted and worn wherever adequate water is available for dissolution. In this era of human-induced increases of acid precipitation, carbonation processes are greatly enhanced (see Focus Study 3.2, "Acid Deposition: Damaging to Ecosystems").

Chemical weathering involving dissolution of carbonates dominates entire landscapes composed of limestone. These are the regions of karst topography, which we examine next.

Karst Topography and Landscapes

Limestone is so abundant on Earth that many landscapes are composed of it (Figure 13.14). These areas are quite susceptible to chemical weathering. Such weathering creates a specific landscape of pitted, bumpy surface topography, poor surface drainage, and well-developed solution channels (dissolved openings and conduits) underground. Remarkable mazes of underworld caverns also may develop, owing to weathering and erosion caused by groundwater.

These are the hallmarks of **karst topography**, named for the Krš Plateau in Slovenia (formerly Yugoslavia), where karst processes were first studied. Approximately

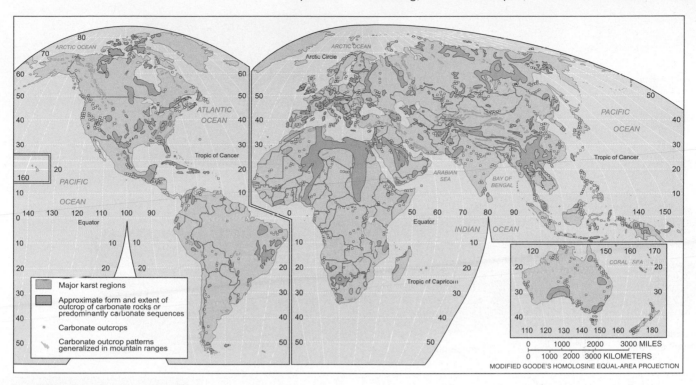

FIGURE 13.14 Karst landscapes and limestone regions.
Major karst regions exist on every continent. The outcrops of carbonate rocks or predominantly
carbonate sequences are limestone and dolomite (calcium, magnesium carbonate) but may contain
other carbonate rocks. [Map adapted by Pam Schaus, after R. E. Snead, *Atlas of the World Physical Features*,
p. 76. © 1972 by John Wiley & Sons; and D. C. Ford and P. Williams, *Karst Geomorphology and Hydrology*,
p. 601. © 1989 by Kluwer Academic Publishers. Adapted by permission.]

15% of Earth's land area has some karst features, with out-standing examples found in southern China, Japan, Puerto Rico, Cuba, the Yucatán of Mexico, Kentucky, Indiana, New Mexico, and Florida. As an example, approximately 38% of Kentucky has sinkholes and related karst features noted on topographic maps.

Formation of Karst

For a limestone landscape to develop into karst topography, there are several necessary conditions:

- The limestone formation must contain 80% or more calcium carbonate for dissolution processes to proceed effectively.
- Complex patterns of joints in the otherwise imper-meable limestone are needed for water to form routes to subsurface drainage channels.
- There must be an aerated (containing air) zone between the ground surface and the water table.
- Vegetation cover is required to supply varying amounts of organic acids that enhance the dissolution process.

The role of climate in providing optimum conditions for karst processes remains under debate, although the amount and distribution of rainfall appear important. Karst occurs in arid regions, but it is primarily due to for-mer climatic conditions of greater humidity. Karst is rare in the Arctic and Antarctic regions because the water, although present, is generally frozen.

As with all weathering processes, time is a factor. Early in the last century, karst landscapes were thought to progress through evolutionary stages of development, as if they were aging. Today, these landscapes are thought to be locally unique, a result of specific conditions, and there is little evidence that different regions evolve sequentially along similar lines. Nonetheless, mature karst landscapes do display certain characteristic forms.

Lands Covered with Sinkholes

The weathering of limestone landscapes creates many **sinkholes**, which form in circular depressions. (Traditional studies may call a sinkhole a *doline*.) A *collapse sinkhole* forms if a solution sinkhole collapses through the roof of an underground cavern. A gently rolling limestone plain might be pockmarked by slow subsidence of surface mate-rials in *solution sinkholes* with depths of 2–100 m (7–330 ft) and diameters of 10–1000 m (33–3300 ft), as shown in Figure 13.15a. Through continuing solution and collapse, sinkholes may coalesce to form a *karst valley*—an elongated depression up to several kilometers long.

The area southwest of Orleans, Indiana, has an average of 1022 sinkholes per 2.6 km² (1 mi²). In this area, the Lost River, a "disappearing stream," flows more than 13 km (8 mi) underground before it resurfaces at its Lost River rise near the Orangeville rise shown in Figure 13.15e. The Lost River flow diverts from the surface through sinkholes and solution channels. Its dry bed can be seen on the lower left of the topographic map in Figure 13.15b.

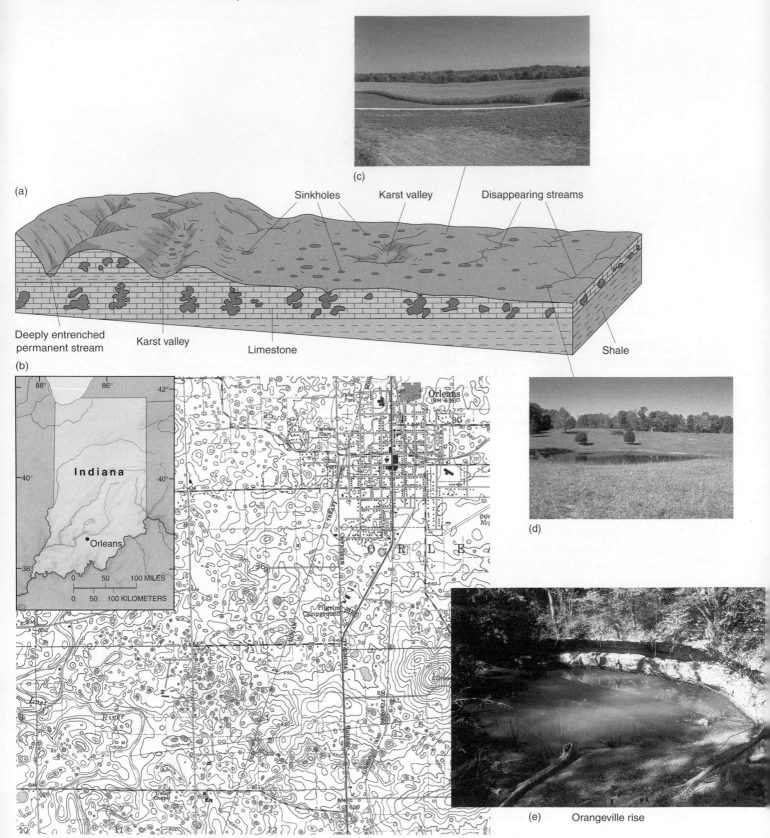

(a)

(c)

Sinkholes   Karst valley   Disappearing streams

Deeply entrenched permanent stream   Karst valley   Limestone   Shale

(b)

(d)

(e)   Orangeville rise

FIGURE 13.15 Features of karst topography in Indiana.
(a) Idealized features of karst topography in southern Indiana. (b) Karst topography southwest of the town of Orleans, Indiana. On the map, note the contour lines: Depressions are indicated with small hachures (tick marks) on the downslope side of contour lines. (c) Gently rolling karst landscape and cornfields near Orleans, Indiana. (d) This pond is in a sinkhole depression near Palmyra, Indiana. (e) The Orangeville rise, near Orangeville, Indiana, is just north of the Lost River rise. During periods of high rainfall, this rise is almost filled with water. [(a) Adapted from W. D. Thornbury, *Principles of Geomorphology,* illustration by W. J. Wayne, p. 326; © 1954 by John Wiley & Sons; (b) Mitchell, Indiana quadrangle, USGS; photos (c), (d), and (e) by Bobbé Christopherson.]

 NOTEBOOK   Karst Farm Park, Indiana

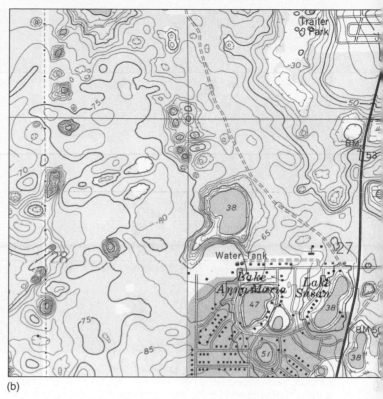

(a)

FIGURE 13.16 Sinkholes.
(a) Florida sinkhole, formed in 1981 in Winter Park, a suburb of Orlando.
(b) Karst area 25 km (15.5 mi) north of Winter Park depicted on a topographic map. Note that the depressions are marked by small hachures (tick marks). [(a) Photo by Jim Tuten/Black Star; (b) Orange City quadrangle, USGS.]

(b)

In Florida, several sinkholes have made news because lowered water tables (lowered by pumping from municipal wells) caused their collapse into underground solution caves, taking with them homes, businesses, and even new cars from an auto dealership. One such sinkhole collapsed in a suburban area in 1981, and others in 1993 and 1998 (Figure 13.16).

In Stockertown, Northampton County, Pennsylvania, more than two dozen sinkholes formed since October 2000 along Bushkill Creek. The timing correlated with dewatering pumping at the nearby Hercules cement quarry that lowered water tables. This coincidence led residents with related property damage to assume a link between the mine pumping and the sinkholes in the limestone landscape. The collapse of a portion of the bridge for State Route 33 and another bridge made problems worse (Figure 13.17). A new bridge was completed in 2006 for the main highway. However, understanding the problem was hindered by political delays in the drilling of test wells for proper geophysical analysis. A group of citizens is pressing the case for mitigating action. The point to be made in Florida and Pennsylvania, and elsewhere, is that limestone landscapes are complex and require integrated spatial analysis involving hydrology and geology before any mining and related pumping, municipal water wells, or other disruptions in groundwater are allowed.

A complex landscape in which sinkholes intersect is a *cockpit karst*. The sinkholes can be symmetrically shaped in certain circumstances; one at Arecibo, Puerto

FIGURE 13.17 Sinkhole swallows a bridge.
Downstream from the Hercules quarry several dozen sinkholes have collapsed since 2000, taking down one of the bridges from State Route 33 and part of this small bridge in Stockertown, Pennsylvania. [Photo by Bobbé Christopherson.]

Rico, is shaped perfectly for a radio telescope installation (Figure 13.18).

Another type of karst topography forms in the wet tropics, where deeply jointed, thick limestone beds are weathered into gorges, leaving isolated resistant blocks standing. These resistant cones and towers are most remarkable in several areas of China, where *tower karst* up to 200 m (660 ft) high interrupts an otherwise lower-level plain (Figure 13.19).

FIGURE 13.18 Deep-space research from a sinkhole. Cockpit karst topography near Arecibo, Puerto Rico, provides a natural depression for the dish antenna of a giant radio telescope. The Arecibo Observatory is part of the National Astronomy and Ionosphere Center, which is operated by Cornell University under contract with the National Science Foundation (http://www.naic.edu/). [Photo courtesy of Cornell University.]

FIGURE 13.19 Tower karst of the Guangxi (Kwangsi) Province, China.
Resistant strata protect each tower as weathering removes the surrounding limestone. [Photo by Wolfgang Kaehler/Wolfgang Kaehler Photography.]

Caves and Caverns

Caves form in limestone because it is so easily dissolved by carbonation. The largest limestone caverns in the United States are Mammoth Cave in Kentucky (also the longest surveyed cave in the world at 560 km, 350 mi), Carlsbad Caverns in New Mexico, and Lehman Cave in Nevada.

Carlsbad Caverns are in 200-million-year-old limestone formations deposited when shallow seas covered the region. Regional uplifts associated with the building of the Rockies (the Laramide orogeny, 40–80 million years ago) elevated the region above sea level, subsequently leading to active cave formation (Figure 13.20).

Caves generally form just beneath the water table, where later lowering of the water level exposes them to further development. *Dripstones* form as water containing dissolved minerals slowly drips from the cave ceiling. Calcium carbonate precipitates out of the evaporating solution, literally one molecular layer at a time, and accumulates at a point below on the cave floor. Forming depositional features, *stalactites* grow from the ceiling and *stalagmites* build from the floor; sometimes the two grow until they connect and form a continuous *column* (Figure 13.20b). A dramatic subterranean world is thus created.

Cave science is an aspect of geomorphology where amateur cavers make important discoveries about these unique habitats and new life forms—*biospeleology*. More than 90% of known caves are not biologically surveyed, and some 90%

of possible caves worldwide still lie undiscovered, making this a major research frontier (see News Report 13.2). (For more on caves and related formations, see **http://www.goodearthgraphics.com/virtcave/virtcave.html**.)

Mass-Movement Processes

Nevado del Ruiz, northernmost of two dozen dormant (not extinct, sometimes active) volcanic peaks in the Cordilleran Central of Colombia, had erupted six times during the past 3000 years, killing 1000 people during its last eruption in 1845. On November 13, 1985, at 11 P.M., after a year of earthquakes and harmonic tremors, a growing bulge on its northeast flank, and months of small summit eruptions, Nevado del Ruiz violently erupted in a lateral explosion. The mountain was back in action.

On this night, the familiar pyroclastics, lava, and blast were not the worst problem. The hot eruption quickly melted ice on the mountain's snowy peak, liquefying mud and volcanic ash, sending a hot mudflow downslope. Such a flow is a *lahar*, an Indonesian word referring to mudflows of volcanic origin. This lahar moved rapidly down the Lagunilla River toward the villages below. The wall of mud was at least 40 m (130 ft) high as it approached Armero, a regional center with a population of 25,000. The city slept as the lahar buried its homes: 23,000 people were killed; thousands were injured; 60,000 were left homeless across the region. The volcanic debris flow is now a permanent grave for its victims. Not all mass movements are this destructive, but such processes play an important role in the denudation of the landscape.

For more on mass-movement hazards, including landslides, see the web site of the Natural Hazards Center at the University of Colorado, Boulder, at **http://www.Colorado.edu/hazards/** or the USGS Geologic Hazards page at **http://landslides.usgs.gov/**.

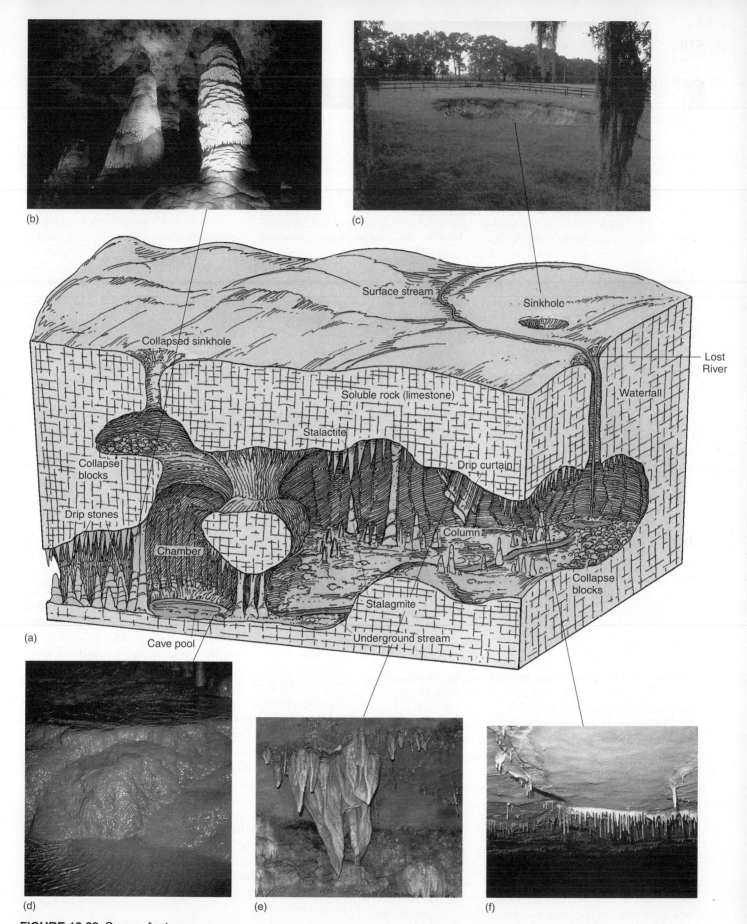

FIGURE 13.20 Cavern features.
(a) An underground cavern and related forms in limestone. (b) A column in Carlsbad Caverns, New Mexico, where a series of underground caverns includes rooms more than 1200 m (4000 ft) long and 190 m (625 ft) wide. Although a national park since 1930, unexplored portions remain. (c) A sinkhole in a Florida pasture. From Marengo Caves, Marengo, Indiana: (d) a flowstone and pool of water; (e) dripstone drapery formations; (f) soda straws hanging from ceiling cracks, forming one molecular layer at a time. [(b) Photo by author; (c) photo by Thomas M. Scott, Florida Geological Survey; photos (d), (e), (f) by Bobbé Christopherson.]

Amateurs Make Cave Discoveries

The exploration and scientific study of caves is *speleology*. Although professional physical and biological scientists carry on investigations, amateur cavers, or "spelunkers," have made many important discoveries. As an example, in the early 1940s George Colglazier, a farmer southwest of Bedford, Indiana, awoke to find his farm pond at the bottom of a deep collapsed sinkhole. This sinkhole is now the entrance to an extensive cave system that includes a subterranean navigable stream.

Cave habitats are unique. They are nearly closed, self-contained ecosystems with simple food chains and great stability. In total darkness, bacteria synthesize inorganic elements and produce organic compounds that sustain many types of cave life, including algae, small invertebrates, amphibians, and fish.

In a cave discovered in 1986, near Movile in southeastern Romania, cave-adapted invertebrates were discovered after millions of years of sunless isolation. Thirty-one of these organisms were previously unknown. Without sunlight, the ecosystem in Movile is sustained on sulfur-metabolizing bacteria that synthesize organic matter using energy from oxidation processes. These chemosynthetic bacteria feed other bacteria and fungi that in turn support cave animals. The sulfur bacteria produce sulfuric acid compounds that may prove to be important in the chemical weathering of some caves.

The mystery, intrigue, and excitement of cave exploration lie in the variety of dark passageways, enormous chambers that narrow to tiny crawl spaces, strange formations, and underwater worlds that can be accessed only by cave diving. Private-property owners and amateur adventurers discovered many of the major caves, a fact that keeps this popular science/sport very much alive. (For nearly a thousand worldwide links and information, see the **http://www.cbel.com/speleology** web site, and the Russian "SpeleoInfo-Centre" at **http://fadr.msu.ru/~sigalov/ldlists.html**.)

Mass-Movement Mechanics

Physical and chemical weathering processes create an overall weakening of surface rock, which makes it more susceptible to the pull of gravity. The term **mass movement** applies to any unit movement of a body of material, propelled and controlled by gravity, such as the lahar just described. Mass movements can be surface processes or they can be submarine landslides beneath the ocean. Mass-movement content can range from dry to wet, slow to fast, or small to large, and from free-falling to gradual or intermittent (see Figure 13.22).

The term *mass movement* is sometimes used interchangeably with **mass wasting**, which is the general process involved in mass movements and erosion of the landscape. To combine the concepts, we can say that the mass movement of material works to waste slopes and provide raw material for erosion, transportation, and deposition.

The Role of Slopes All mass movements occur on slopes under the influence of gravitational stress. If we pile dry sand on a beach, the grains will flow downslope until equilibrium is achieved. The steepness of the resulting slope depends on the size and texture of the grains; this steepness is the **angle of repose**. This angle represents a balance of the driving force (gravity) and resisting force (friction and shearing). The angle of repose for various materials commonly ranges between 33° and 37° (from horizontal) and between 30° and 50° for snow avalanche slopes.

The *driving force* in mass movement is gravity. It works in conjunction with the weight, size, and shape of the surface material; the degree to which the slope is oversteepened (how far it exceeds the angle of repose); and the amount and form of moisture available (frozen or fluid). The greater the slope angle, the more susceptible the surface material is to mass wasting processes.

The *resisting force* is the shearing strength of slope material, that is, its cohesiveness and internal friction, which work against gravity and mass wasting. To reduce shearing strength is to increase shearing stress, which eventually reaches the point at which gravity overcomes friction, initiating slope failure.

Clays, shales, and mudstones are highly susceptible to hydration (physical swelling in response to the presence of water). If such materials underlie rock strata in a slope, the strata will move with less driving-force energy. When clay surfaces are wet, they deform slowly in the direction of movement, and when saturated they form a viscous fluid with little shearing strength (resistance to movement) to hold back the slope. However, if the rock strata are such that material is held back from slipping, then more driving-force energy may be required, such as that generated by an earthquake.

Madison River Canyon Landslide In the Madison River Canyon near West Yellowstone, Montana, a blockade of dolomite (a magnesium-rich carbonate rock) held back a deeply weathered and *over-steepened slope* (40° to 60° slope

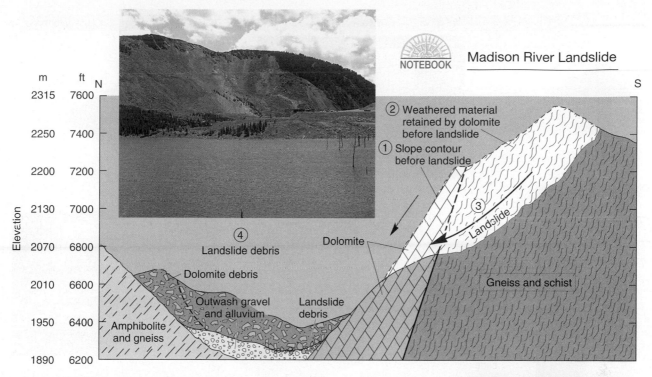

FIGURE 13.21 Madison River landslide.
Cross section showing geologic structure of the Madison River Canyon in Montana, where an earthquake triggered a landslide in 1959: (1) prequake slope contour; (2) weathered rock that failed; (3) direction of landslide; and (4) landslide debris blocking the canyon and damming the Madison River. [After J. B. Hadley, *Landslides and Related Phenomena Accompanying the Hebgen Lake Earthquake of 17 August 1959*, USGS Professional Paper 435-K, p. 115. Inset photo by Bobbé Christopherson.]

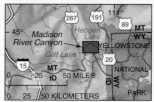

angle) for untold centuries (white area in Figure 13.21). Then, shortly after midnight on August 17, 1959, a magnitude 7.5 earthquake broke the dolomite structure along the foot of the slope. The break released 32 million m³ (1.13 billion ft³) of mountainside, which moved downslope at 95 kmph (60 mph), causing gale-force winds through the canyon. Momentum carried the material more than 120 m (about 400 ft) up the opposite canyon slope, trapping several hundred campers with about 80 m (260 ft) of rock, killing 28 people.

The mass of material also effectively dammed the Madison River and thus created a new lake, dubbed Quake Lake. The landslide debris dam established a new temporary equilibrium for the river and canyon. A channel was quickly excavated by the U.S. Army Corps of Engineers to prevent a disaster below the dam, for if Quake Lake overflowed the landslide dam, the water would quickly erode a channel and thereby release the entire contents of the new lake onto farmland downstream. This event conveys a dramatic example of the role of slopes and tectonic forces in creating massive land movements.

Classes of Mass Movements

In any mass movement, gravity pulls on a mass until the critical shear-failure point is reached—*a geomorphic threshold*. The material then can *fall*, *slide*, *flow*, or *creep*—the four classes of mass movement. Figure 13.22 summarizes these classes. Note the temperature and moisture gradients in the margins of the illustration, which show the relation between water content and movement rate. The Madison River Canyon event was a type of slide, whereas the Nevado del Ruiz lahar mentioned earlier was a flow. We now look at specific mass-movement classes.

For a comprehensive look at an ongoing landslide situation in southern California, see **http://geology. wr.usgs.gov/wgmt/elnino/scampen/examples.html**. This site includes general background on mass movements, including many diagrams, and before-and-after photos.

Falls and Avalanches   This class of mass movement includes rockfalls and debris avalanches. A **rockfall** is simply a volume of rock that falls through the air and hits a surface. During a rockfall, individual pieces fall independently and characteristically form a cone-shaped pile of irregular broken rocks in a **talus slope** at the base of a steep incline, where several *talus cones* coalesce (Figure 13.23).

A **debris avalanche** is a mass of falling and tumbling rock, debris, and soil. It is differentiated from a slower debris slide or landslide by the velocity of onrushing material. This speed often results from ice and water

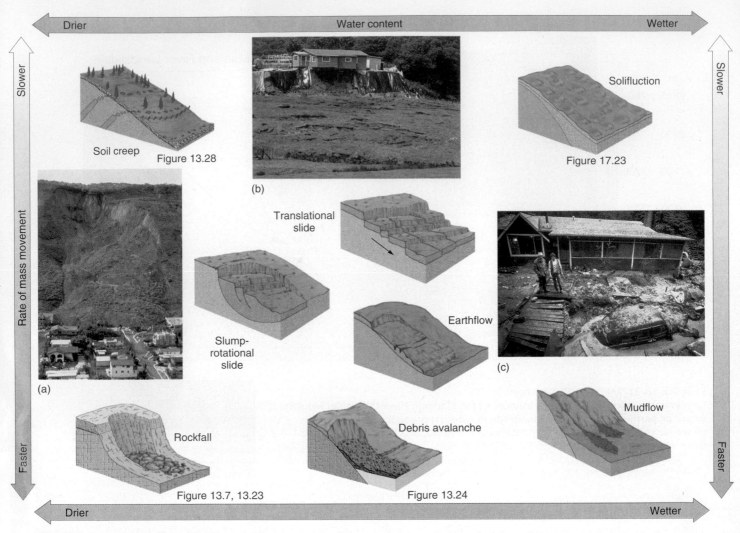

Drier ← Water content → Wetter

Slower ← Rate of mass movement → Faster

Soil creep Figure 13.28

(b)

Translational slide

Slump-rotational slide

Earthflow

(a)

(c)

Solifluction Figure 17.23

Rockfall

Debris avalanche

Mudflow

Figure 13.7, 13.23

Figure 13.24

Drier ← → Wetter

 ANIMATION Mass Movements

FIGURE 13.22 Mass-movement classes.
Principal types of mass-movement and mass-wasting events. Variations in water content and rates of movement produce a variety of forms. (a) A 1995 slide in La Conchita, California, also site of a 2005 mudslide event; (b) saturated hillsides fail; (c) mudflows 2 m deep, both in Santa Cruz County, California. [Photos by (a) Robert L. Schuster/USGS; (b) Alexander Lowry/Photo Researchers, Inc.; (c) James A. Sugar.]

FIGURE 13.23 Talus slope.
Rockfall and talus deposits at the base of a steep slope along Duve Fjord, Nordaustlandet Island. Can you see the lighter rock strata above that are the source for the three talus cones? [Photo by Bobbé Christopherson.]

that fluidize the debris. The extreme danger of a debris avalanche results from its tremendous speed and consequent lack of warning.

In 1962 and again in 1970, debris avalanches roared down the west face of Nevado Huascarán, the highest peak in the Peruvian Andes. The 1962 debris avalanche contained an estimated 13 million m³ (460 million ft³) of material, burying the city of Ranrahirca and eight other towns, killing 4000 people.

An earthquake initiated the 1970 event. Upward of 100 million m³ (3.53 billion ft³) of debris buried the city of Yungay, where 18,000 people perished (Figure 13.24). This avalanche attained velocities of 300 kmph (185 mph), which is especially incredible when you consider the quantity of material involved and the fact that some boulders weighed thousands of tons. The avalanche covered a vertical drop of 4144 m (13,600 ft) and a horizontal distance of 16 km (10 mi) in just a few minutes.

FIGURE 13.24 Debris avalanche, Peru.
A 1970 debris avalanche falls more than 4100 m (2.5 mi)
down the west face of Nevado Huascarán, burying the city
of Yungay, Peru. The same area was devastated by a similar
avalanche in 1962 and by others in pre-Columbian times.
A great danger remains for the towns in the valley from
possible future mass movements. [Photo by George Plafker.]

ANIMATION Debris Avalanche and Eruption of Mt. St. Helens

Another dramatic example of a debris avalanche, but
one that did not kill, occurred west of the Saint Elias
Range, north of Yakutat Bay in Alaska. A magnitude
7.1 earthquake triggered several dozen rockfalls, large
debris avalanches, and snow avalanches. The largest single avalanche covered approximately 4.7 km² (1.8 mi²) of
the Cascade Glacier (Figure 13.25). The photo reveals
characteristic grooves, lobes, and large rocks associated
with these fluid avalanches.

Landslides A sudden rapid movement of a cohesive
mass of regolith or bedrock that is not saturated with
moisture is a **landslide**—a large amount of material failing simultaneously. Surprise creates the danger, for the
downward pull of gravity wins the struggle for equilibrium
in an instant. Focus Study 13.1 describes one such surprise event that struck near Longarone, Italy, in 1963.

To eliminate the surprise element, scientists are using
the global positioning system (GPS) to monitor landslide

FIGURE 13.25 Debris avalanche, Alaska.
A debris avalanche covers portions of the Cascade Glacier,
west of the Saint Elias Range in Alaska. The one pictured was
the largest of several dozen triggered by a 1979 earthquake.
[Photo by George Plafker.]

movement. With GPS, scientists measure slight land
shifts in suspect areas for clues to possible mass wasting.
GPS was applied in two cases in Japan and effectively
identified prelandslide movements of 2–5 cm per year,
providing information to expand the area of hazard concern and warning.

Slides occur in one of two basic forms: translational or
rotational (see Figure 13.22 for an idealized view of each).
Translational slides involve movement along a planar (flat)
surface roughly parallel to the angle of the slope, with no
rotation. The Madison Canyon landslide described earlier
was a translational slide. Flow and creep patterns also are
considered translational in nature.

Rotational slides occur when surface material moves
along a concave surface. Frequently, underlying clay presents an impervious surface to percolating water. As a
result, water flows along the clay surface, undermining the
overlying block. The surface may rotate as a single unit, or
it may present a stepped appearance. Continuing rotational
mudslides plague La Conchita, California (Figure 13.22a).
For thousands of years this area has been unstable; then a
slump landslide hit in 1995, burying some homes, and most
recently in January 2005 when debris buried 30 homes and
took 10 lives in a mudslide episode that followed a period of
heavy rains.

In October 2007, mass-movement realities struck up-
scale La Jolla, California, damaging more than 100 homes
along Soledad Mountain Road (Figure 13.26). The rotational slump along a hillside pulled material from beneath
the highway, causing a sinkhole-like collapse. The area is
inherently unstable, experiencing three other collapse
events since 1961, with early warning signs of this occurrence beginning in July. In an area such as this it is difficult to imagine that standard cut-and-fill housing pads
were allowed in the planning stage prior to construction,

Focus Study 13.1

Vaiont Reservoir Landslide Disaster

Place: Northeastern Italy, Vaiont Canyon in the Italian Alps, rugged scenery, 680 m (2200 ft) above sea level.

Location: 46.3° N, 12.3° E, near the border of the Veneto and Friuli-Venezia Giulia regions.

Situation: Centrally located in the region, an ideal situation for hydro-electric power production.

Site: Steep-sided, narrow, glaciated canyon, opening to populated lowlands to the west; ideal site for a narrow-crested, high dam and deep reservoir.

The Plan: Build the second-highest dam in the world, using a new thin-arch design, 262 m (860 ft) high, 190 m (623 ft) crest length, impounding a reservoir capacity of 150 million m³

(5.3 billion ft³) of water (third-largest reservoir in the world), and generate hydroelectric power for distribution.

Geological Analysis

1. Steep canyon walls composed of interbedded limestone and shale; badly cracked and deformed structures; open fractures in the shale inclined toward the future reservoir body. The steepness of the canyon walls enhances the strong driving forces (gravitational) at work on rock structures.

2. High potential for bank storage (water absorption by canyon walls) into the groundwater system that will increase water pressure on all rocks in contact with the reservoir.

3. The nature of the shale beds is such that cohesion will be reduced as their clay minerals become saturated.

4. Evidence of ancient rockfalls and landslides on the north side of the canyon.

5. Evidence of creep activity along the south side of the canyon.

Political/Engineering Decision

Begin design and construction of a thin-arch dam at this site.

Events During Construction and Filling

Large volumes of concrete were injected into the bedrock as "dental work" in an attempt to strengthen the fractured rock. During reservoir filling in 1960, 700,000 m³ (2.5 million ft³) of

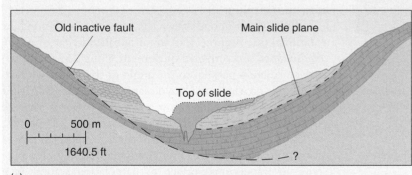

(a)

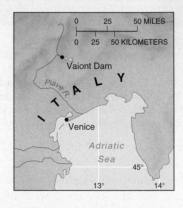

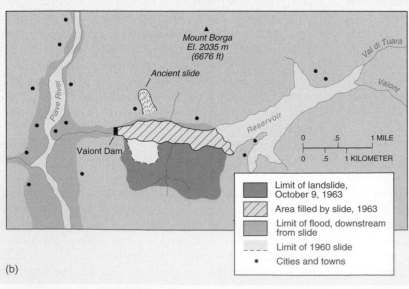

(b)

FIGURE 13.1.1 Vaiont Reservoir disaster. (a) Cross section of the Vaiont River Valley. (b) Map of the disaster site noted with evidence of previous activity. [After G. A. Kiersch, "The Vaiont Reservoir Disaster," California Division of Mines and Geology, Mineral Information Service, vol. 18, no. 7, 1964, pp. 129–138.]

rock and soil slid into the reservoir from the south side. A slow creep of slope materials along the entire south side began shortly thereafter, increasing to about 1 cm per week by January 1963. Creep rate increased as the level of the reservoir rose. By mid-September 1963, the creep rate exceeded 40 cm (16 in.) a day.

This sequence of events set the stage for one of the worst dam disasters in history. Heavy rains began on September 28, 1963. Alarmed at the runoff from the rain into the reservoir and the increase in creep rate along the south wall, engineers opened the outlet tunnels on October 8 in an attempt to bring down the reservoir level—but it was too late.

The next evening, it took only 30 seconds for a landslide of 240 million m³ (8.5 billion ft³) to crash into the reservoir, shaking seismographs across Europe. A 150-m-thick (500-ft) slab of mountainside gave way (2 km by 1.6 km in area; 1.2 mi by 1 mi), sending shock waves of wind and water up the canyon 2 km (1.2 mi) and splashing a 100-m (330-ft) wave over the dam. The former reservoir was effectively filled with bedrock, regolith, and soil that almost entirely displaced its water content (Figure 13.1.1). Amazingly, the experimental dam design held.

Downstream, in the unsuspecting town of Longarone, near the mouth of Vaiont Canyon on the Piave River, people heard the distant rumble and were quickly drowned by the 69-m (226-ft) wave of water that came out of the canyon. That night 3000 people perished. As for a lesson or recommendation, we need only to reread the preconstruction geologic analysis that began this Focus Study. The Italian courts eventually prosecuted the persons responsible.

FIGURE 13.26 Landslide under major highway and homes.
More than 100 homes and a portion of Soledad Mountain Road were damaged by this October 2007 rotational slide and collapse. The area experienced such failures in the past. [AP Photo/Chris Park.]

sandstone formations rested on weak shale and siltstone, which became moistened and soft, offering little resistance to the overlying strata. The slide is still visible after 75 years, as you can see in Figure 13.27.

Because of melted snow and rain, the water content of the Gros Ventre landslide was great enough to classify it as an *earthflow*. About 37 million m³ (1.3 billion ft³) of wet soil and rock moved down one side of the canyon and surged 30 m (100 ft) up the other side. The earthflow dammed the river and formed a lake, as did the landslide across the Madison River in the Hebgen Lake area in 1959. However, in 1925 equipment was not available to excavate a channel, so the new lake filled. Two years later, the lake water broke through the temporary earthflow dam, transporting a tremendous quantity of debris over the region downstream.

Creep A persistent, gradual mass movement of surface soil is **soil creep**. In creep, individual soil particles are lifted and disturbed by the expansion of soil moisture as it freezes, by cycles of moistness and dryness, by diurnal temperature variations, or by grazing livestock or digging animals.

In the freeze–thaw cycle, particles are lifted at right angles to the slope by freezing soil moisture, as shown in Figure 13.28. When the ice melts, however, the particles fall straight downward in response to gravity. As the process repeats, the surface soil gradually creeps its way downslope.

The overall wasting of a slope may cover a wide area and may cause fence posts, utility poles, and even trees to lean downslope. Various strategies are used to arrest the mass movement of slope material—grading the terrain, building terraces and retaining walls, planting ground cover—but the persistence of creep nearly always wins.

or that homeowners were allowed to irrigate lawns and gardens with no restrictions, feeding drainage water into subsurface strata.

Flows Flows include *earthflows* and more fluid **mudflows**. When the moisture content of moving material is high, the suffix *-flow* is used (see Figure 13.22). Heavy rains can saturate barren mountain slopes and set them moving, as was the case east of Jackson Hole, Wyoming, in the spring of 1925. Material above the Gros Ventre River (pronounced "grow vaunt") broke loose and slid downslope as a unit. The slide occurred because

FIGURE 13.27 The Gros Ventre earthflow near Jackson, Wyoming.
Evidence of the 1925 earthflow is still visible after more than 80 years. [Photo by Steven K. Huhtala.]

This forested mass is the main portion of the earthflow.

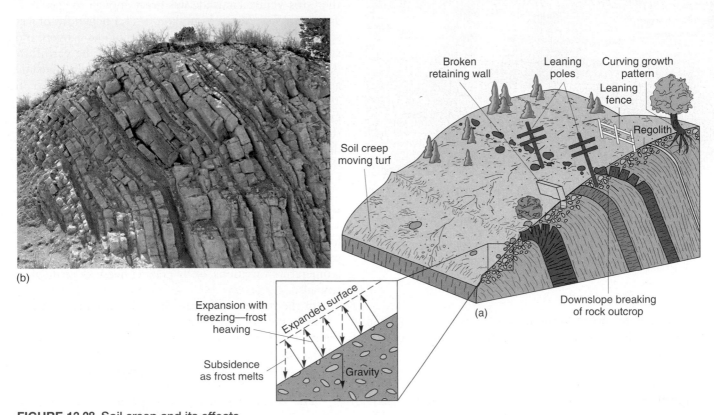

FIGURE 13.28 Soil creep and its effects.
(a) Typical soil-creep features. (b) Note the subsurface effects on rock strata caused by surface creep.
[(b) Photo by Bobbé Christopherson.]

Chapter 13

Human-Induced Mass Movements (Scarification)

Every human disturbance of a slope—highway roadcut, surface mining, or building of a shopping mall, housing development, or home—can hasten mass wasting. The newly destabilized and oversteepened surfaces are thrust into a search for a new equilibrium. Imagine the disequilibrium in slope relations created by the highway roadcut pictured in Figure 12.8b.

Large open-pit surface mines—such as the Bingham Copper Mine west of Salt Lake City, the abandoned Berkeley Pit in Butte, Montana, and numerous large coal surface mines in the U.S. East and West, such as Black Mesa (Figure 13.29a)—are examples of human-induced mass movements, generally known as **scarification**. At the Berkeley Pit, toxic drainage water is threatening the regional aquifers, the Clark Fork River, and the local freshwater supply. Residues of copper, zinc, lead, and

(a)

(b)

(c)

(d)

(e)

FIGURE 13.29 Scarification.
(a) Black Mesa, Arizona, strip-mining for coal. (b) Bingham Canyon, Utah, west of Salt Lake City, strip-mining for copper and other minerals. (c) Mountaintop removal coal strip-mining on Kayford Mountain, West Virginia. (d) Spoil banks of tailings in a West Virginia coal-mining area. (e) The 55-m-deep Dally slate quarry in Pen Argyl, Pennsylvania. [Photos (a) and (d) by author; (b) and (e) by Bobbé Christopherson; and (c) by Dr. Chris Mayda, Eastern Michigan University, all rights reserved.]

arsenic lace one creek, now devoid of life, and are in the soil on Butte Hill, where children play. In other areas, wind dispersal is a particular problem with radioactive uranium tailings in the West.

At the Bingham Copper Mine, a mountain literally was removed since mining began in 1906, forming a pit 4-km wide and 1-km deep, 2.5 by 0.6-mi (Figure 13.29b). This is easily the largest human-made excavation on Earth, with ores containing copper, gold, silver, and molybdenum. The disposal of tailings (mined ore of little value) and waste material is a significant problem at any mine. Such large excavations produce tailing piles that are unstable and susceptible to further weathering, mass wasting, or wind dispersal. The leaching of toxic materials from tailings and waste piles poses an ever-increasing problem to streams, aquifers, and public health across the country.

Where underground mining is common, particularly for coal in the Appalachians, land subsidence and collapse produce further mass movements. Homes, highways, streams, wells, and property values are severely affected. Another controversial form of mining called *mountaintop removal* is done by removing ridges and summits and dumping the debris into stream valleys, thereby exposing the coal seams for mining and burying the stream channels. A change in the Clean Water Act in 2002 removed a 20-year-old restriction that barred dumping within 100 feet of a stream to protect flows. Figure 13.29c is an aerial view of mountaintop removal on Kayford Mountain, West Virginia. The 20-story-tall dragline machine is visible center-right. In the background, forest cover and litter are removed using fire to prepare the site for minimal reclamation, although this is difficult since the landscape is essentially lost. Some 2000 km (1245 mi) of streams are now filled with tailings (Figure 13.29d) and mountaintop scenery is gone.

In eastern Pennsylvania commercial slate deposits formed over millions of years. The Dally slate quarry in Pen Argyl is now working a second pit that is 55 m (180 ft) deep (Figure 13.29e). The grain of the metamorphic rock guides the slate removal so as to produce the flattest pieces. Slate is mined for shingles and countertops. The local county turned the first quarry into a landfill—a practical use for the pit.

Scientists can informally quantify the scale of human-induced scarification for comparison with natural denudation processes. Geologist R. L. Hooke used estimates of U.S. excavations for new housing, mineral production (including the three largest—stone, sand and gravel, and coal), and highway construction. He then prorated these quantities of moved earth for all countries, using their gross domestic product (GDP), energy consumption, and agriculture's effect on river sediment loads. From these he calculated a global estimate for human earth moving. Hooke estimated that humans, as a geomorphic agent, annually move 40–45 billion tons (40–45 Gt) of the planet's surface. Compare this quantity with natural river sediment transfer (14 Gt/yr), movement through stream meandering (39 Gt/yr), haulage by glaciers (4.3 Gt/yr), movement due to wave action and erosion (1.25 Gt/yr), wind transport (1 Gt/yr), sediment movement by continental and oceanic mountain building (34 Gt/yr), or deep-ocean sedimentation (7 Gt/yr). As Hooke concluded about us,

Homo sapiens has become an impressive geomorphic agent. Coupling our earth-moving prowess with our inadvertent adding of sediment load to rivers and the visual impact of our activities on the landscape, one is compelled to acknowledge that, for better or for worse, this biogeomorphic agent may be the premier geomorphic agent of our time.*

*R. L. Hooke, "On the efficacy of humans as geomorphic agents," *GSA Today*, The Geological Society of America, 4, no. 9 (September 1994): 217–226.

Summary and Review—Weathering, Karst Landscapes, and Mass Movement

■ *Define* the science of geomorphology.

Geomorphology is the science that analyzes and describes the origin, evolution, form, and spatial distribution of landforms. The exogenic system, powered by solar energy and gravity, tears down the landscape through processes of landmass **denudation** involving weathering, mass movement, erosion, transportation, and deposition. Different rocks offer differing resistance to these weathering processes and produce a pattern on the landscape of **differential weathering**.

Agents of change include moving air, water, waves, and ice. Since the 1960s, research and understanding of the processes of denudation have moved toward the **dynamic equilibrium model**, which considers slope and landform stability to be consequences of the resistance of rock materials to the attack of denudation processes.

geomorphology (p. 402)
denudation (p. 402)
differential weathering (p. 402)
dynamic equilibrium model (p. 403)

1. Define geomorphology, and describe its relationship to physical geography.
2. Define landmass denudation. What processes are included in the concept?
3. What is the interplay between the resistance of rock structures and weathering variabilities?
4. Describe what is at work to produce the landform in Figure 13.1.
5. What are the principal considerations in the dynamic equilibrium model?

■ *Illustrate* the forces at work on materials residing on a slope.

Slopes are shaped by the relation between rate of weathering and breakup of slope materials and the rate of mass movement and erosion of those materials. A slope is considered stable if it is stronger than these denudation processes; it is unstable if it is weaker. In this struggle against gravity, a slope may reach a **geomorphic threshold**—the point at which there is enough energy to overcome resistance against movement. **Slopes** that form the boundaries of landforms have several general components: *waxing slope*, *free face*, *debris slope*, and *waning slope*. Slopes seek an angle of equilibrium among the operating forces.

> geomorphic threshold (p. 403)
> slopes (p. 403)

6. Describe conditions on a hillslope that is right at the geomorphic threshold. What factors might push the slope beyond this point?

7. Given all the interacting variables, do you think a landscape ever reaches a stable, old-age condition? Explain.

8. What are the general components of an ideal slope?

9. Relative to slopes, what is meant by an "angle of equilibrium"? Can you apply this concept to the photograph in Figure 13.2?

■ *Define* weathering, and *explain* the importance of parent rock and joints and fractures in rock.

Weathering processes disintegrate both surface and subsurface rock into mineral particles or dissolve them in water. The upper layers of surface material undergo continual weathering and create broken-up rock called **regolith**. Weathered **bedrock** is the *parent rock* from which regolith forms. The unconsolidated, fragmented material that develops after weathering is **sediment**, which along with weathered rock forms the **parent material** from which soil evolves.

Important in weathering processes are **joints**, the fractures and separations in the rock. Jointing opens up rock surfaces on which weathering processes operate. Factors that influence weathering include character of the bedrock (hard or soft, soluble or insoluble, broken or unbroken), climatic elements (temperature, precipitation, freeze–thaw cycles), position of the water table, slope orientation, surface vegetation and its subsurface roots, and time.

> weathering (p. 404)
> regolith (p. 404)
> bedrock (p. 405)
> sediment (p. 405)
> parent material (p. 405)
> joints (p. 406)

10. Describe weathering processes operating on an open expanse of bedrock. How does regolith develop? How is sediment derived?

11. Describe the relationship between mesoscale climatic conditions and rates of weathering activities.

12. What is the relation among parent rock, parent material, regolith, and soil?

13. What role do joints play in the weathering process? Give an example from one of the illustrations in this chapter.

■ *Describe* frost action, salt-crystal growth, pressure-release jointing, and the role of freezing water as physical weathering processes.

Physical weathering refers to the breakup of rock into smaller pieces with no alteration of mineral identity. The physical action of water when it freezes (expands) and thaws (contracts) is a powerful agent in shaping the landscape. This **frost action** may break apart any rock. Working in joints, expanded ice can produce *joint-block separation* through the process of *frost-wedging*. Another process of physical weathering is *salt-crystal growth* (*salt weathering*); as crystals in rock grow and enlarge over time, they force apart mineral grains and break up rock.

As overburden is removed from a granitic batholith, the pressure of deep burial is relieved. The granite slowly responds with *pressure-release jointing*, with layer after layer of rock peeling off in curved slabs or plates. As these slabs weather, they slip off in the process of **sheeting**. This *exfoliation process* creates an arch-shaped or dome-shaped feature on the exposed landscape, forming an **exfoliation dome**.

> physical weathering (p. 406)
> frost action (p. 406)
> sheeting (p. 408)
> exfoliation dome (p. 408)

14. What is physical weathering? Give an example.

15. Why is freezing water such an effective physical weathering agent?

16. What weathering processes produce a granite dome? Describe the sequence of events.

■ *Describe* the susceptibility of different minerals to the chemical weathering processes called hydration, hydrolysis, oxidation, carbonation, and solution.

Chemical weathering is the chemical decomposition of minerals in rock. It can cause **spheroidal weathering**, in which chemical weathering occurs in cracks in the rock. As cementing and binding materials are removed, the rock begins to disintegrate, and sharp edges and corners become rounded.

Hydration occurs when a mineral absorbs water and expands, thus creating a strong mechanical force that stresses rocks. **Hydrolysis** breaks down silicate minerals in rock, as in the chemical weathering of feldspar into clays and silica. Water actively participates in chemical reactions. **Oxidation** is the reaction of oxygen with certain metallic elements, the most familiar example being the rusting of iron, producing iron oxide. The dissolution of materials into *solution* is considered chemical weathering. For instance, a mild acid such as carbonic acid in rainwater will cause **carbonation**, wherein carbon combines with certain minerals, such as calcium, magnesium, potassium, and sodium.

> chemical weathering (p. 408)
> spheroidal weathering (p. 409)
> hydration (p. 411)
> hydrolysis (p. 411)
> oxidation (p. 411)
> carbonation (p. 412)

17. What is chemical weathering? Contrast this set of processes to physical weathering.

18. What is meant by the term *spheroidal weathering*? How does spheroidal weathering occur?

19. What is hydration? What is hydrolysis? Differentiate between these processes. How do they affect rocks?

20. Iron minerals in rock are susceptible to which form of chemical weathering? What characteristic color is associated with this type of weathering?

21. With what kind of minerals do carbon compounds react, and under what circumstances does the dissolution of carbonate minerals occur? What is this weathering process called?

■ *Review* the processes and features associated with karst topography.

Karst topography refers to distinctively pitted and weathered limestone landscapes. Surface circular **sinkholes** form and may extend to form a *karst valley*. A sinkhole may collapse through the roof of an underground cavern, forming a *collapse sinkhole*. The formation of caverns is part of karst processes and groundwater erosion. Limestone caves feature many unique erosional and depositional features, producing a dramatic subterranean world.

> karst topography (p. 412)
> sinkholes (p. 413)

22. Describe the development of limestone topography. What is the name applied to such landscapes? From what area was this name derived?

23. Differentiate among sinkholes, karst valleys, and cockpit karst. Within which form is the radio telescope at Arecibo, Puerto Rico?

24. In general, how would you characterize the region southwest of Orleans, Indiana?

25. What are some of the unique erosional and depositional features you find in a limestone cavern?

■ *Portray* the various types of mass movements, and *identify* examples of each in relation to moisture content and speed of movement.

Any movement of a body of material, propelled and controlled by gravity, is **mass movement**, also called **mass wasting**. The **angle of repose** of loose sediment grains represents a balance of driving and resisting forces on a slope. Mass movement of Earth's surface produces some dramatic incidents, including:

rockfalls (a volume of rock that falls) that can form a **talus slope** of loose rock along the base of the cliff; **debris avalanches** (a mass of tumbling, falling rock, debris, and soil moving at high speed); **landslides** (a large amount of material failing simultaneously); **mudflows** (material in motion with a high moisture content); and **soil creep** (a persistent movement of individual soil particles that are lifted by the expansion of soil moisture as it freezes, by cycles of wetness and dryness, by temperature variations, or by the impact of grazing animals). In addition, human mining and construction activities have created massive **scarification** of landscapes.

> mass movement (p. 418)
> mass wasting (p. 418)
> angle of repose (p. 418)
> rockfall (p. 419)
> talus slope (p. 419)
> debris avalanche (p. 419)
> landslide (p. 421)
> mudflows (p. 423)
> soil creep (p. 423)
> scarification (p. 425)

26. Define the role of slopes in mass movements, using the terms *angle of repose*, *driving force*, *resisting force*, and *geomorphic threshold*.

27. What events occurred in the Madison River Canyon in 1959?

28. What are the classes of mass movement? Describe each briefly and differentiate among these classes.

29. Name and describe the type of mudflow associated with a volcanic eruption.

30. Describe the difference between a landslide and what happened on the slopes of Nevado Huascarán.

31. What is scarification, and why is it considered a type of mass movement? Give several examples of scarification. Why are humans a significant geomorphic agent?

NetWork

The *Geosystems* Student Learning Center provides on-line resources for this chapter on the World Wide Web. To begin: Once at the Center, click on the cover of this textbook, scroll the Table of Contents menu, and select this chapter. You will find self-tests that are graded, review exercises, specific updates for items in the chapter, and in "Destinations" many links to interesting related pathways on the Internet. *Geosystems* Student Learning Center is found at **http://www.prenhall.com/ christopherson/**.

Critical Thinking

A. Locate a slope, possibly near campus, near your home, or a local roadcut. Using Figure 13.3a, b, and c, can you identify the forces and forms of a hillslope at your site? How would you go about assessing the stability of the slope? Is there any evidence of the mass wasting of materials, soil creep, or other processes discussed in this chapter?

B. The USGS has completed a landslide hazards potential map for the conterminous United States. You can check out this resource at **http://www.usgs.gov/hazards/landslides/**. Note the zoom-in capability to look at specific areas in greater detail. The map legend rates landslide incidence and susceptibility. How does the map portray landslide incidences? In what way does the map rate landslide susceptibility to an area? Are you able to find a location that you have visited and determine its vulnerability?

C. Examine the photo of a Russian coal-fired power plant (see below). The plant lacks scrubbers to reduce stack emissions, as do many coal plants in other parts of the world including in the United States. We discussed acid deposition in Focus Study 3.2 and, in this chapter, chemical weathering of cultural stones such as tombstones or bridges in Central Park. Describe the interrelation of such power plant emissions and these concepts. What are the links? The chemical reactions involved?

To keep the Mississippi River flowing past New Orleans to the Gulf of Mexico, the government built several dams (see Figure 14.28 for a map). The Atchafalaya River is a more direct route to the coast and the Mississippi kept favoring it. In 1963, artificial barriers, the Low Sill and Overbank structures, part of the Old River Control Project, went into operation. The effort is to block the Atchafalaya from full access to the Mississippi discharge; about 30% of the Mississippi flow does pass through this alternative route. Here is the Old River Control Auxiliary Structure (intake side) with its gates and abutments, which went into service in 1986. [Photo by Bobbé Christopherson.]

River Systems and Landforms

■ Key Learning Concepts

After reading the chapter, you should be able to:

■ **Define** the term *fluvial*, and **explain** processes of fluvial erosion, transportation, and deposition.

■ **Construct** a basic drainage basin model, and **identify** different types of drainage patterns and internal drainage, with examples.

■ **Describe** the relation among velocity, depth, width, and discharge, and **explain** the various ways that a stream erodes and transports its load.

■ **Develop** a model of a meandering stream, including point bar, undercut bank, and cutoff, and **explain** the role of stream gradient in these flow characteristics.

■ **Define** a floodplain, and **analyze** the behavior of a stream channel during a flood.

■ **Differentiate** the several types of river deltas, and **describe** each.

■ **Explain** flood probability estimates, and **review** strategies for mitigating flood hazards.

E arth's rivers and waterways form vast arterial networks that drain the continents. They also shape the landscape by removing the products of weathering, mass movement, and erosion and transporting them downstream. To call rivers "Earth's lifeblood" is no exaggeration, inasmuch as rivers redistribute mineral nutrients important for soil formation and plant growth and serve society in many ways.

Rivers not only provide essential water supplies, they also process waste (diluting and transporting it), provide critical cooling water for manufacturing and power generation, and form essential transportation networks. Rivers have been important in the geography of human history, influencing where settlements were built, where livelihoods were made, and where borders were drawn. This chapter discusses the dynamics of river systems and their landforms.

Hydrology is the science of water, its global circulation, distribution, and properties, specifically water at and below Earth's surface. For hydrology links on the web, see **http://www.worldwater.org/links.htm** or the Global Hydrology and Climate Center at **http://weather.msfc. nasa.gov/surface_hydrology/**; also see the Amazon River at **http://boto.ocean.washington.edu/eos/index. html**.

In this chapter: We begin with a look at the largest rivers on Earth. Essential fluvial concepts of base level, drainage basin, and drainage density and patterns follow. With this foundation, we define streamflow discharge, "Q," the essential element in fluvial processes. We discuss factors that affect streamflow characteristics and the work performed by flowing water, including erosion and transport. Stream depositional features such as deltaic forms are illustrated with a detailed look at the Mississippi River delta. We look at the effects of urbanization on hydrology. Human response to floods and floodplain management are important aspects of river management, with details of the New Orleans levee and floodwall disaster and tragedy concluding the chapter.

Fluvial Processes and Landscapes

At any moment, approximately 1250 km³ (300 mi³) of water is flowing through Earth's waterways. Even though this volume is only 0.003% of all freshwater, the work performed by this energetic flow makes it a dominant agent of landmass denudation. Of the world's rivers, those with the greatest *discharge* (streamflow volume past a point in a given unit of time) are the Amazon of South America (Figure 14.1), the Congo of Africa, the Chàng Jiang (Yangtze) of Asia, and the Orinoco of South America (Table 14.1). In North America, the greatest discharges are from the Missouri–Ohio–Mississippi, Saint Lawrence, and Mackenzie River systems.

Stream-related processes are **fluvial** (from the Latin *fluvius*, meaning "river"). Geographers analyze stream patterns and the fluvial processes that created them. Fluvial systems, like all natural systems, have characteristic processes and produce recognizable landforms. Yet a stream system can behave with randomness and seeming disorder. The term *river* is applied to a trunk, or main stream, or an entire river system. *Stream* is a more general term not necessarily related to size. There is some overlap in usage between the terms *river* and *stream*.

Insolation and gravity power the hydrologic cycle and are the driving forces of fluvial systems. Individual streams vary greatly from one another, depending on the climate in which they operate, the composition of the surface, topography over which they flow, the nature of vegetation and plant cover, and the length of time they have been operating in a specific setting.

Water dislodges, dissolves, or removes surface material in the **erosion** process. Streams produce *fluvial erosion*, in which weathered sediment is picked up for transport to new locations. Thus, a stream is a mixture of water and solids; the solids are carried in suspension, by mechanical **transport**, and in dissolved solution. Materials are laid down by another process, **deposition**. **Alluvium** is the general term for the clay, silt, sand, gravel, and mineral fragments deposited by running water as sorted or semi-sorted sediment on a floodplain, delta, or streambed.

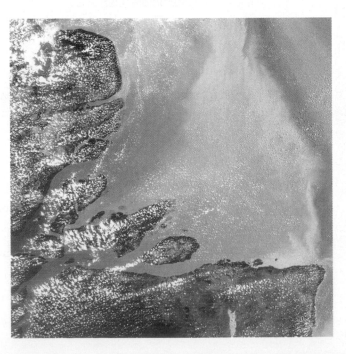

FIGURE 14.1 Mouth of the Amazon River.
The mouth of the Amazon River discharges a fifth of all the freshwater that enters the world's oceans. The mouth of the Amazon is 160 km (100 mi) wide. Millions of tons of sediments are derived from the Amazon's drainage basin, which is as large as the Australian continent. Large islands of sediment are left where the river's discharge leaves the mouth and flows into the Atlantic Ocean. [*Terra* image courtesy of NASA/GSFC/JPL and the MISR Team.]

Table 14.1   Largest Rivers on Earth Ranked by Discharge

Rank by Volume	Average Discharge at Mouth in Thousands of m³/s (cfs)	River (with Tributaries)	Outflow/Location	Length km (mi)	Rank by Length
1	180 (6350)	Amazon (Ucayali, Tambo, Ene, Apurimac)	Atlantic Ocean/ Amapá-Pará, Brazil	6570 (4080)	2
2	41 (1460)	Congo (Lualaba)	Atlantic Ocean/ Angola, Congo	4630 (2880)	10
3	34 (1201)	Yangtze (Chàng Jiang)	East China Sea/Kiangsu, China	6300 (3915)	3
4	30 (1060)	Orinoco	Atlantic Ocean/Venezuela	2737 (1700)	27
5	21.8 (779)	La Plata estuary (Paraná)	Atlantic Ocean/Argentina	3945 (2450)	16
6	19.6 (699)	Ganges (Brahmaputra)	Bay of Bengal/India	2510 (1560)	23
7	19.4 (692)	Yenisey (Angara, Selenga or Selenge, Ider)	Gulf of Kara Sea/Siberia	5870 (3650)	5
8	18.2 (650)	Mississippi (Missouri, Ohio, Tennessee, Jefferson, Beaverhead, Red Rock)	Gulf of Mexico/Louisiana	6020 (3740)	4
9	16.0 (568)	Lena	Laptev Sea/Siberia	4400 (2730)	11
17	9.7 (348)	St. Lawrence	Gulf of St. Lawrence/Canada and United States	3060 (1900)	21
36	2.83 (100)	Nile (Kagera, Ruvuvu, Luvironza)	Mediterranean Sea/Egypt	6690 (4160)	1

Base Level of Streams

Base level is a level below which a stream cannot erode its valley. In general, the *ultimate base level* is sea level, the average level between high and low tides. Imagine base level as a surface extending inland from sea level, inclined gently upward under the continents. Ideally, this is the lowest practical level for all denudation processes (Figure 14.2a).

American geologist and ethnologist John Wesley Powell (1834–1902) put forward the idea in 1875. He was a director of the U.S. Geological Survey, first director of the U.S. Bureau of Ethnology, explorer of the Colorado River, and a pioneer in understanding the western landscape.

Of course, Powell recognized that not every landscape has degraded all the way to sea level; clearly, other intermediate base levels are in operation. A *local base level*, or temporary one, may control the lower limit of local streams for a region. The local base level may be a river, a lake, hard and resistant rock, or a human-made dam (Figure 14.2b). The local base level set by Hoover Dam

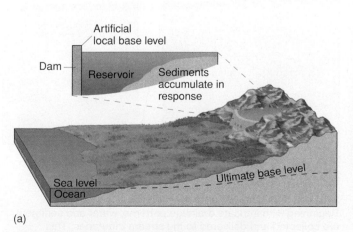

(a)

(b)

FIGURE 14.2 Ultimate and local base levels.
(a) The concepts of ultimate base level (sea level) and local base level (natural, such as a lake, or artificial, such as a dam). Note how base level curves gently upward from the sea as it is traced inland; this is the theoretical limit for stream erosion. (b) A local base level was introduced to the landscape above Glen Canyon Dam when it closed off the canyon. [Photo by Bobbé Christopherson.]

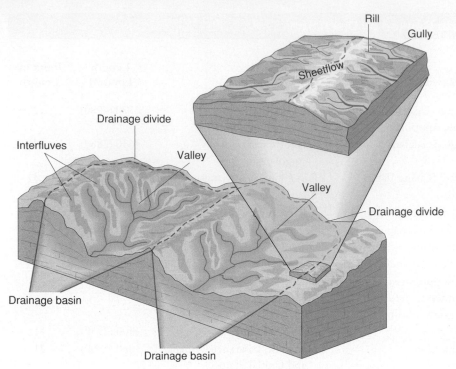

FIGURE 14.3 A drainage basin. A drainage divide separates the drainage basin and its watershed from other basins.

since 1936 on the Arizona–Nevada border backed up the Colorado River into Lake Mead, filling sinuous canyons. The multiyear western drought has brought the reservoir down to record low levels, as shown in the Chapter 9 chapter-opening photo comparison. In this arid landscape, Hoover Dam is the local base level for the impounded Colorado River above the dam.

Over time, the work of streams modifies the landscape dramatically. Landforms are produced by two basic processes: (1) erosive action of flowing water and (2) deposition of stream-transported materials. Before we introduce streamflow characteristics, let us begin by examining a basic fluvial unit—the drainage basin.

Drainage Basins

Figure 14.3 illustrates drainage basin concepts. Every stream has a **drainage basin**, ranging in size from tiny to vast. Ridges form *drainage divides* that define every drainage basin. That is, the ridges are the dividing lines that control into which basin runoff water from precipitation drains. Drainage divides define a drainage basin, the catchment (water-receiving) area of the drainage basin that delivers water. In any drainage basin, water initially moves downslope in a thin film called **sheetflow**, or *overland flow*. High ground that separates one valley from another and directs sheetflow is an *interfluve*. Surface runoff concentrates in *rills* that are small-scale downhill grooves, which may develop into deeper *gullies* and then into a stream in the valley.

The drainage basin acts as a collection system of water comprised of many subsystems, from erosion in the headwaters, transportation throughout, to deposition in the lower portions. Figure 14.4 illustrates the drainage

basin system operation from the smallest rill and gully, to the main tributaries and stream trunk, to the dispersing lower extremities and mouth, possibly forming a delta or depositing sediments in a water body.

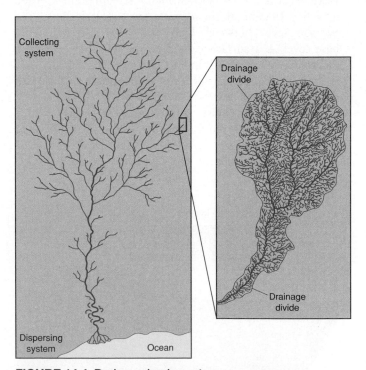

FIGURE 14.4 Drainage basin system. Beginning with intricate drainage patterns, water and sediment are collected and delivered to the stream tributaries. This fluidized mass transports to the main trunk and the dispersing system at the end of the stream, where it enters another water body. [Adapted from W.K. Hamblin and E.H. Christiansen, *Earth's Dynamic System*, 10th ed., Upper Saddle River, NJ: Pearson Prentice Hall, 2004; Figure 12.3, p. 300.]

Drainage Divides and Basins   Several high drainage divides, called **continental divides**, are situated in the United States and Canada. These are extensive mountain and highland regions separating drainage basins, sending flows to the Pacific, the Gulf of Mexico, the Atlantic, Hudson Bay, or the Arctic Ocean. The principal drainage divides and drainage basins in the United States and Canada are mapped in Figure 14.5. These divides form water-resource regions and provide a spatial framework for water-management planning.

A major drainage basin system is made up of many smaller drainage basins. Each drainage basin gathers and delivers its precipitation and sediment to a larger basin, concentrating the volume into the main stream. A good example from Figure 14.5 is the great Mississippi–Missouri–Ohio River system, draining some 3.1 million km² (1.2 million mi²), or 41% of the continental United States.

Consider the travels of rainfall in north-central Pennsylvania. This water feeds hundreds of small streams that flow into the Allegheny River. At the same time, rainfall in

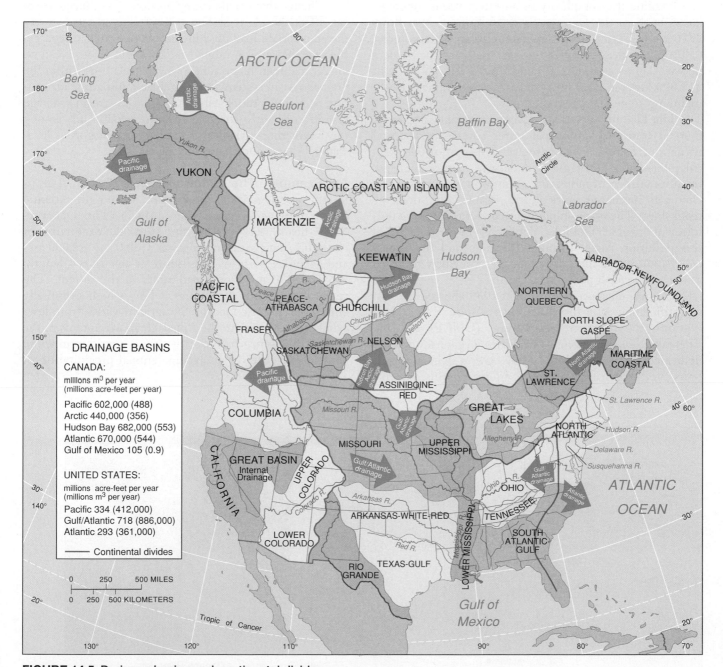

FIGURE 14.5 Drainage basins and continental divides.
Continental divides (blue lines) separate the major drainage basins that empty into the Pacific, Atlantic, Gulf of Mexico, and to the north through Canada into Hudson Bay and the Arctic Ocean. Subdividing these large-scale basins are major river basins. [After U.S. Geological Survey; *The National Atlas of Canada*, 1985, "Energy, Mines, and Resources Canada"; and Environment Canada, *Currents of Change—Inquiry on Federal Water Policy—Final Report 1986*.]

southern Pennsylvania feeds hundreds of streams that flow into the Monongahela River. The two rivers then join at Pittsburgh to form the Ohio River. The Ohio flows southwestward and at Cairo, Illinois, connects with the Mississippi River, which eventually flows on past New Orleans and disperses into the Gulf of Mexico. Each contributing tributary, large or small, adds its discharge, pollution, and sediment load to the larger river. In our example, sediment weathered and eroded in north-central and southern Pennsylvania is transported thousands of kilometers and accumulates on the floor of the Gulf of Mexico, where it forms the Mississippi River delta.

Imagine the complexity of an international drainage basin, such as the Danube River in Europe, that flows 2850 km (1770 mi) from western Germany's Black Forest to the Black Sea. The river crosses or forms the borders between nine countries (Figure 14.6). A total area of 817,000 km² (315,000 mi²) fall within the drainage basin, including some 300 tributaries.

The Danube serves many economic functions: commercial transport, municipal water source, agricultural irrigation, fishing, and hydroelectric-power production. An international struggle is underway to save the river from its burden of industrial and mining wastes, sewage, chemical discharge, agricultural runoff, and drainage from ships. The many shipping canals actually spread pollution and worsen biological conditions in the river. All of this pollution passes through Romania and the deltaic ecosystems in the Black Sea. The river is widely regarded as one of the most polluted on Earth.

Political changes in Europe in 1989 allowed the first scientific analysis of the entire river system. The United Nations Environment Programme (UNEP) and the European Union, along with other organizations, are dedicated to clearing the Danube, saving deltaic ecosystems, and restoring the environment of this valuable resource; see **http://www.icpdr.org/**. The river delta is one of the International Biosphere Reserves; see **http://www.unesco.org/mab/** and **http://www.blacksea-environment.org/**.

Drainage Basins as Open Systems Drainage basins are open systems. Inputs include precipitation and the minerals and rocks of the regional geology. Energy and materials are redistributed as the stream constantly adjusts to its landscape. System outputs of water and sediment disperse through the mouth of the river, into a lake, another river, or the ocean, as shown in Figure 14.4.

Change that occurs in any portion of a drainage basin can affect the entire system. The stream adjusts to carry the appropriate load of sediment relative to its discharge. If a river system is brought to a threshold where it can no longer maintain its present form, the relations within the drainage basin system are destabilized, initiating a transition period to a more stable condition. A stream drainage system constantly struggles toward equilibrium among the interacting variables of discharge, transported load, channel shape, and channel steepness.

An Example: The Delaware River Basin Let us look at the Delaware River basin, within the Atlantic Ocean drainage region (Figure 14.7). The Delaware River headwaters are in the Catskill Mountains of New York. This basin encompasses 33,060 km² (12,890 mi²) and includes parts of five states in the river's length, 595 km (370 mi) from headwaters to the mouth. The river system ends at Delaware Bay, which eventually enters the Atlantic Ocean. Topography varies from low-relief coastal plains to the Appalachian Mountains in the north. The entire basin lies within a humid, temperate climate and receives an average annual precipitation of 120 cm (47.2 in.).

Danube River delta

FIGURE 14.6 An international drainage basin—the Danube River.
The Danube crosses or forms the border of nine countries as it flows across Europe to the Black Sea.
The river spews polluted discharge into the Black Sea through its arcuate-form delta. [*Terra* image, June 15, 2002, courtesy of MODIS Land Rapid Response Team, NASA/GSFC.]

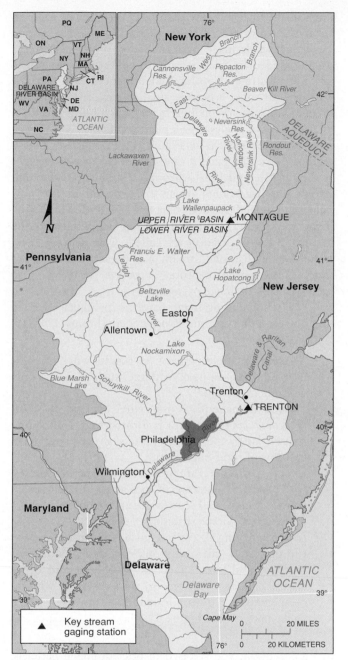

FIGURE 14.7 The Delaware River drainage basin.
This basin is studied by the USGS to assess the potential
impact of global warming on a river system. [Map adapted
from USGS, 1986, "Hydrologic events and surface water
resources," *National Water Summary 1985*, Water Supply Paper
2300 (Washington, DC: Government Printing Office), p. 30.]

The river provides water for an estimated 20 million
people, not only within the basin but to cities outside the
basin as well. Several major conduits export water from
the Delaware River. Note the Delaware Aqueduct to
New York City (in the north) and the Delaware & Raritan
Canal (near Trenton). Several reservoirs in the drainage
basin allow some control over streamflow and storage for
dry periods. The sustainability of this water resource is
critical to the entire region. Clearly, effective planning for
a drainage basin requires regional cooperation and careful
spatial analysis of all variables.

Internal Drainage Most streams find their way to
progressively larger rivers and eventually into the ocean.
In some regions, however, stream drainage does not
reach the ocean. Instead, the water leaves the drainage
basin by means of evaporation or subsurface gravitational
flow. Such streams terminate in areas of **internal
drainage**.

Portions of Asia, Africa, Australia, Mexico, and the
western United States have regions with such internal
drainage patterns. Internal drainage defines the region of
the Great Basin of the western United States, as shown
on maps in Figures 14.5 and 15.24a. For example, the
Humboldt River flows across Nevada to the west, where
evaporation and seepage losses to groundwater make it
disappear into the Humboldt "sink." Many streams and
creeks flow into the Great Salt Lake, but its only outlet
is evaporation, so it exemplifies a region of internal
drainage. In the Middle East, the Dead Sea region, and in
Asia, the areas around the Aral Sea and Caspian Sea, have
no outlet to the ocean.

Drainage Density and Patterns

A primary feature of any drainage basin is its drainage
density. **Drainage density** is determined by dividing the
total length of all stream channels in the basin by the area
of the basin. The number and length of channels in a
given area reflect the landscape's regional topography and
surface appearance. For example, Figure 14.8 reveals a
very high drainage density in a humid climate. In contrast,
the typical desert has a very low drainage density.

The *Landsat* image and topographic map in Figure
14.8 are of the Ohio River drainage near the junction of
West Virginia, Ohio, and Kentucky. The high-density
drainage pattern and intricate dissection of the land occur
because the region has generally level, easily eroded sandstone, siltstone, and shale strata and a humid mesothermal
climate. Fluvial action and other denudation processes are
responsible for this dissected topography.

The **drainage pattern** is the arrangement of channels in an area. Patterns are quite distinctive, for they are
determined by the combination of regional steepness,
variable rock resistance, variable climate, variable hydrology, relief of the land, and structural controls imposed by
the underlying rocks. Consequently, the drainage pattern
of any land area on Earth is a remarkable visual summary
of every characteristic—geologic and climatic—of that
region.

Common Drainage Patterns The seven most common drainage patterns are shown in Figure 14.9. A most
familiar pattern is *dendritic drainage* (Figure 14.9a). This
treelike pattern (Greek *dendron*, or "tree") is similar to
that of many natural systems, such as capillaries in the
human circulatory system, the vein patterns in leaves, and
tree roots. Energy expended by this drainage system is
efficient because the overall length of the branches is minimized. Figure 14.8 shows dendritic drainage; on the

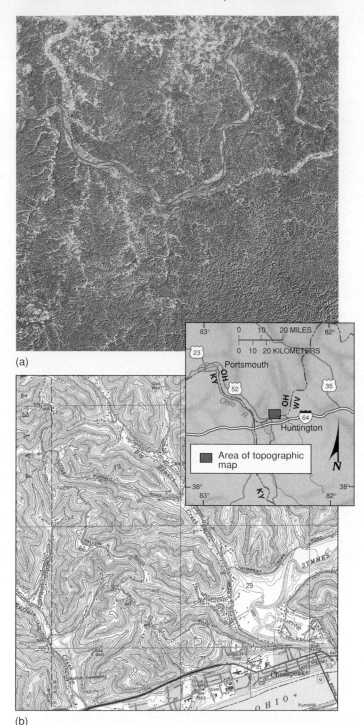

(a)

(b)

FIGURE 14.8 A stream-dissected landscape.
(a) A highly dissected topography (featuring a dendritic drainage pattern shown in Figure 14.9) around the junction of West Virginia, Ohio, and Kentucky. All of these borders are formed by rivers. (b) A portion of the USGS topographic map covering the area north of Huntington, West Virginia, reveals the intricate complexity of dissected landscapes. (The image and map are not at the same scale.) [(a) *Landsat* image from NASA; (b) Huntington Quadrangle, USGS.]

satellite image and topographic map, you can trace the branching pattern of streams.

The *trellis drainage* pattern (Figure 14.9b) is characteristic of dipping or folded topography. Such drainage exists in the nearly parallel mountain folds of the Ridge

and Valley Province in the eastern United States. Refer to Figure 12.18, where a satellite image of this region shows its distinctive drainage pattern. Here, drainage patterns are influenced by rock structures of variable resistance and folded strata. Parallel folded structures direct the principal streams, whereas smaller dendritic tributary streams are at work on nearby slopes, joining the main streams at right angles, like a plant trellis.

The inset sketch in Figure 14.9b suggests that a headward-eroding part of one stream (to the lower right of the inset) could break through a drainage divide and *capture* the headwaters of another stream in the next valley, and indeed this does happen. The dotted line is the abandoned former channel. The sharp bends in two of the streams in the illustration are called *elbows of capture* and are evidence that one stream has breached a drainage divide. This type of capture, or *stream piracy*, can also occur in other drainage patterns.

The remaining drainage patterns in Figure 14.9 are responses to other specific structural conditions:

- A *radial* drainage pattern (c) results when streams flow off a central peak or dome, such as occurs on a volcanic mountain.
- *Parallel* drainage (d) is associated with steep slopes.
- A *rectangular* pattern (e) is formed by a faulted and jointed landscape, which directs stream courses in patterns of right-angle turns.
- *Annular* patterns (f) are produced by structural domes, with concentric patterns of rock strata guiding stream courses. Figure 12.10c provides an example of annular drainage on a dome structure.
- In areas having disrupted surface patterns, such as the glaciated shield regions of Canada, northern Europe, and some parts of Michigan and other states, a *deranged* pattern (g) is in evidence, with no clear geometry in the drainage and no true stream valley pattern (see Figure 17.14a).

In Figure 14.9h, you see two distinct drainage patterns. Of the seven types illustrated, which two patterns are most like those in the aerial photo, made in central Montana? Looking back at the example in Figure 4.4, which drainage pattern do you assign it?

Occasionally, drainage patterns occur that are discordant with the landscape through which they flow. For example, a drainage system may flow in apparent conflict with older, buried structures that have been uncovered by erosion, so that the streams appear to be *superimposed*. Where an existing stream flows as rocks are uplifted, the stream keeps its original course, cutting into the rock in a pattern contrary to its structure. Such a stream is a *superposed stream* (the stream cuts across weak and resistant rocks alike). A few examples include Wills Creek, cutting a water gap through Haystack Mountain at Cumberland, Maryland; the Columbia River through the Cascade Mountains of Washington; and the River Arun that cuts across the Himalayas.

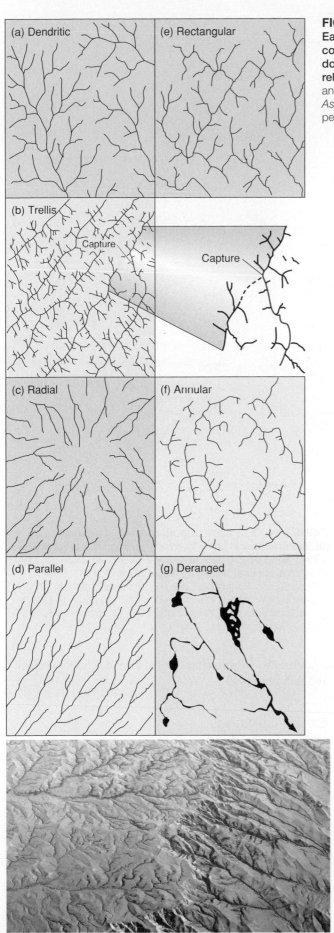

(a) Dendritic

(e) Rectangular

(b) Trellis

Capture

Capture

(c) Radial

(f) Annular

(d) Parallel

(g) Deranged

(h)

FIGURE 14.9 The seven most common drainage patterns. Each pattern is a visual summary of all the geologic and climatic conditions of its region. (h) Two of these drainage patterns dominate this scene from central Montana, in response to local relief and rock structure. [(a) through (g) after A. D. Howard, "Drainage analysis in geological interpretation: A summation," *Bulletin of American Association of Petroleum Geologists* 51 (1967): 2248, adapted by permission; (h) photo by Bobbé Christopherson.]

Streamflow Characteristics

A mass of water positioned above base level in a stream has potential energy. As the water flows downslope, or downstream, under the influence of gravity, this energy becomes kinetic energy. The rate of this conversion from potential to kinetic energy depends on the steepness of the stream channel and volume of water involved.

Streams vary in *width* and *depth*. The streams that flow in them vary in *velocity* and in the *sediment load* they carry. All of these factors increase with increasing **discharge**, or the stream's volume of flow per unit of time. Discharge is calculated by multiplying the velocity of the stream by its width and depth for a specific cross section of the channel, as stated in the simple expression

$$Q = wdv$$

where Q = discharge, w = channel width, d = channel depth, and v = stream velocity. As Q increases, some combination of channel width, depth, and stream velocity increases. Discharge is expressed either in cubic meters per second (m^3/s) or cubic feet per second (cfs).

Figure 14.10 illustrates the relation of discharge to width, depth, and velocity. The graphs show that mean velocity increases with greater discharge, despite the common misperception that downstream flow becomes more sluggish. The increased velocity downstream often is masked by the apparent smooth, quiet flow of the water.

Given the interplay of channel width and depth and stream velocity with discharge, the cross section of a stream varies over time, especially during heavy floods. Figure 14.11 shows changes in the San Juan River channel in Utah that occurred during a flood. Greater discharge increases the velocity and therefore the capacity of the river to transport sediment as the flood progresses. As a result, the river's ability to scour materials from its bed is enhanced. Such scouring represents a powerful clearing action, especially in the excavation of alluvium. Scouring might provide new recreational beach areas and wildlife habitats, as explored by the Grand Canyon experiment in News Report 14.1.

You can see in Figure 14.11 that the San Juan River's channel was deepest on October 14, when floodwaters were highest (blue line). Then, as the discharge returned to normal, the kinetic energy of the river was reduced, and the bed again filled as sediments were redeposited. You can see this process in the progression of the red, green, and purple lines. The flood and scouring process

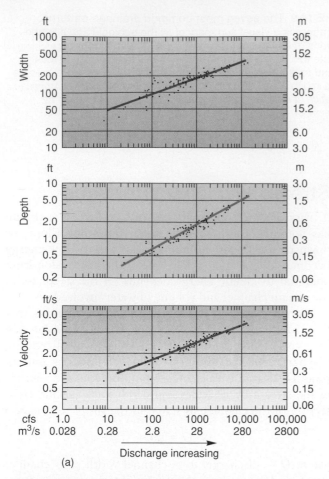

FIGURE 14.10 Effects of stream discharge.
(a) Relation of stream width, depth, and velocity to stream discharge of the Powder River at Locate, Montana. Discharge is shown in cubic meters per second (m³/s) and cubic feet per second (cfs). (b) The Powder River area near the Wyoming border in southeastern Montana. [(a) After L. Leopold and T. Maddock, Jr., *The Hydraulic Geometry of Stream Channels and Some Physiographic Implications*, USGS Professional Paper 252 (Washington, DC: Government Printing Office, 1953), p. 7; (b) photo by Joyce Wilson.]

graphed in Figure 14.11 moved a depth of about 3 m (10 ft) of sediment from this cross section of the stream channel. Such adjustments in a stream channel occur as the stream system continuously works toward

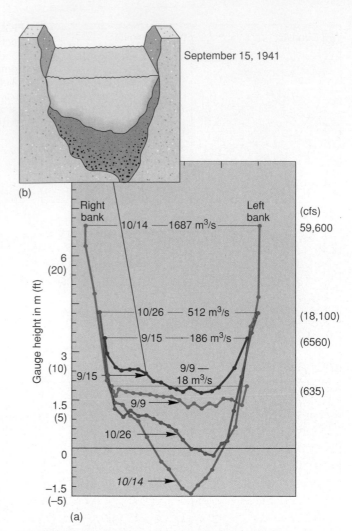

FIGURE 14.11 A flood affects a stream channel.
(a) Stream-channel cross sections showing the progress of a 1941 flood on the San Juan River near Bluff, Utah. (b) Detail of stream-channel profile on September 15, 1941. [After L. Leopold and T. Maddock, Jr., *The Hydraulic Geometry of Stream Channels and Some Physiographic Implications*, USGS Professional Paper 252 (Washington, DC: Government Printing Office, 1953), p. 32.]

equilibrium, in an effort to balance discharge, velocity, and sediment load.

Exotic Streams

Most streamflows increase discharge downstream because the area being drained increases. The Mississippi River is typical. It starts as many small brooks and grows to a mighty river pouring into the Gulf of Mexico. In contrast, a stream can originate in a humid region and subsequently flow through an arid region. In that case, the *discharge usually decreases with distance*, because of high potential evapotranspiration rates in the arid area. Such a stream is an **exotic stream** (*exotic* means "of foreign origin").

The Nile River exemplifies exotic streams. This great river, Earth's longest, drains much of northeastern Africa. But as it courses through the deserts of Sudan and Egypt, it loses water instead of gaining it, because of evaporation and withdrawal for agriculture. By the time it empties into

News Report 14.1

Scouring the Grand Canyon for New Beaches and Habitats

Glen Canyon Dam sealed the Colorado River gorge north of the Grand Canyon, near Lees Ferry and the Utah–Arizona border (see the map in Focus Study 15.1, Figure 15.1.1). Impoundment of water by the new Lake Powell, and the inundation of many canyons upstream from the dam, began in 1963. The lake began collecting the tremendous sediment load of the Colorado River. Dam operations reduced the volume of water in the river downstream, eliminating the natural seasonal fluctuations in river discharge. The production of hydroelectricity at Glen Canyon further affected the Grand Canyon with highly variable water releases keyed to electrical turbine operations. Water in the canyon rose and fell as much as 4.3 m (14 ft) in a 24-hour period as the distant lights and air-conditioning of Las Vegas and Phoenix went on and off.

All of these changes affected the canyon. Over the years, the beaches were starved for sand, channels filled with sediment, fisheries were disrupted, and backwater channels were depleted of nutrients. In 1996, an unprecedented experiment on the Colorado River took place. The Grand Canyon was artificially flooded with a 1260-m³/s (45,000-cfs) release from Glen Canyon (Figure 14.1.1). The flow lasted for 7 days and then was reduced to 224 m³/s (8000 cfs). Flows were decreased gradually to allow the newly formed beaches to drain, in contrast to the rapid opening of the pipes that initiated the flood. Lake Powell dropped 1 m (3.3 ft), and Lake Mead, downstream behind Hoover Dam, rose 0.8 m (2.6 ft).

The results were initially thought to be positive: 35% more beach area

FIGURE 14.1.1 Glen Canyon Dam floods the Grand Canyon, on purpose. Bypass tubes dramatically discharge water from Lake Powell in a 1996 experiment to correct the damaging effects of the dam on the Grand Canyon. Note for scale the workers on the walkway, who are experiencing a deafening roar and vibration. Another modified experiment took place in 2004. [Photo by Tom Smart, *Deseret News*.]

was created by the scouring of sand and sediments from the channel. Approximately 80% of the aggradation (building up of beaches) took place in the first 40 hours and was completed by the 100-hour mark. Numerous backwater channels were created, flush with fresh nutrients for the humpback chub and other endangered fish species. However, benefits turned out to be limited, and negatives, in the form of existing ecosystem disruption and immediate losses of some sediment deposits, marred the experiment. Present thinking is that the lack of sediment to supply the floodwaters for downstream deposition and beach building was the problem. Some of the initial beach deposits quickly disappeared. For more on this experiment, see http:// water.usgs.gov/pubs/FS/FS-060-99/.

In November 2004, hydrologists conducted a second test. Scientists timed the artificial flood with natural peak runoff that supplies fresh sediment to the system. Tributaries deliver more than 3 million metric tons of sediment to the Colorado River. Also, instead of 7 days, the artificial flood releases from Glen Canyon lasted 60 hours (2.5 days). The release coincided with local storms that fed tributaries, such as the Paria River, with sediment to accomplish the task of rebuilding beaches downstream. Results of the experiment in sediment redistribution are being monitored. (See Grand Canyon Monitoring and Research Center for progress reports and analysis at http:// www.gcmrc.gov/.) Record low river discharge and low reservoir levels are testing the entire Colorado River basin in this period of drought.

the Mediterranean Sea, the Nile's flow has dwindled so much that it ranks only 36th in discharge.

The United States has exotic streams too, notably the Colorado River. Its *Q* (discharge) decreases with distance from its source; in fact, the river no longer produces

enough natural discharge to reach its mouth in the Gulf of California—only some agricultural runoff remains at its delta. The exotic Colorado River is depleted not only by passage across dry, desert lands but also by upstream removal of water for agriculture and municipal uses; see

(a)

(b)

FIGURE 14.12 Stream velocity and discharge increase together. (a) A low-discharge, low-velocity (but turbulent) mountain stream in the White Mountains, New Hampshire. (b) The high-discharge, high-velocity (or smooth water flow) portion of the Ocmulgee River near Jacksonville, Georgia. [Photos by Bobbé Christopherson.]

Focus Study 15.1 in Chapter 15 for a detailed discussion of this river and a satellite image of the river's former mouth.

Stream Erosion

A stream's erosional turbulence and abrasion carve and shape the landscape through which it flows. **Hydraulic action** is the work of flowing water alone. Running water causes hydraulic squeeze-and-release action that loosens and lifts rocks. As this debris moves along, it mechanically erodes the streambed further, through the process of **abrasion**, with rock particles grinding and carving the streambed like liquid sandpaper.

The upstream tributaries in a drainage basin usually have small and irregular discharges, and most of the stream's energy is expended in turbulent eddies. As a result, hydraulic action in these upstream sections is at a maximum, whereas the coarse-textured load of such a stream is small. The downstream portions of a river, however, move much larger volumes of water past a given point and carry larger suspended loads of sediment (Figure 14.12).

Stream Transport

You may have watched a river or creek after a rainfall, the water colored brown by the high sediment load being transported. The amount of material available to a stream depends on topographic relief, the nature of rock and soil through which the stream flows, climate, vegetation, and human activity in a drainage basin. *Competence*, which is a stream's ability to move particles of a specific size, is a function of stream velocity and the energy available to suspend materials. *Capacity* is the total possible load that a stream can transport. Four processes transport eroded materials: solution, suspension, saltation, and traction; each is shown in action in Figure 14.13.

Solution refers to the **dissolved load** of a stream, especially the chemical solution derived from minerals such as limestone or dolomite or from soluble salts. The main contributor of material in solution is chemical weathering. Sometimes the undesirable salt content that hinders human use of some rivers comes from dissolved rock formations and from springs in the stream channel; as an example, the San Juan and Little Colorado rivers that flow into the Colorado River near the Utah–Arizona border add dissolved salts to the system.

The **suspended load** consists of fine-grained, clastic particles (bits and pieces of rock). They are held aloft in the stream, with the finest particles not deposited until the stream velocity slows nearly to zero. Turbulence in the water, with random upward motion, is an important mechanical factor in holding a load of sediment in suspension.

Bed load refers to coarser materials that are dragged, rolled, or pushed along the streambed by **traction** or by **saltation**, a term referring to the way particles bounce along in short hops and jumps (from the Latin *saltim*, which means "by leaps or jumps"). Particles transported by saltation are too large to remain in suspension but are not limited to the sliding and rolling motion of traction. These processes relate directly to a stream's velocity and

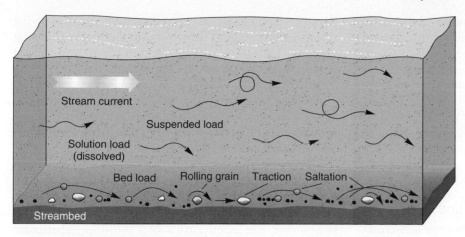

FIGURE 14.13 Fluvial transport. Fluvial transportation of eroded materials through suspension, solution, saltation, and traction.

its ability to retain particles in suspension. With increased kinetic energy, parts of the bed load are rafted upward and become suspended load.

The early explorers who visited the Grand Canyon reported in their journals that they were kept awake at night by the thundering sound of the river and rapids. Imagine the tremendous quantity of material being moved along by the natural Colorado River before any

dams were built. Dams and reservoirs now trap sediments that formerly contributed to the river's bed load.

If the load (bed and suspended) exceeds a stream's capacity, sediments accumulate as **aggradation** (the opposite of degradation) and the stream channel builds up through deposition. With excess sediment, a stream becomes a maze of interconnected channels that form a **braided stream** pattern (Figure 14.14). Braiding often

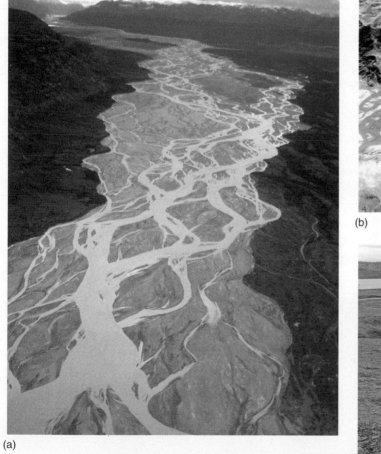

FIGURE 14.14 Braided streams.
(a) Braided stream pattern in Chitina River, Wrangell–Saint Elias National Park, Alaska. (b) A 15-km (9.3-mi) stretch of the braided Brahmaputra River channel some 35 km (22 mi) south of Lhasa, Tibet, in a narrow valley south of the Tibetan Plateau. (c) A braided stream of meltwaters on Spitsbergen Island in the Arctic Ocean. [Photos by (a) Tom Bean; (b) International Space Station astronaut, October 13, 2001, courtesy of Earth Science and Image Analysis Lab, JSC/NASA; and (c) Bobbé Christopherson.]

occurs when reduced discharge lowers a stream's transporting ability, such as after flooding, or when a landslide occurs upstream, or from increased load where weak banks of sand or gravel exist. Locally, braiding also may result from a new sediment load from glacial meltwaters, as in three examples in Figure 14.14: Alaska's Chitina River, a river in Tibet, and a stream on Spitsbergen Island in the Arctic Ocean. Glacial action and erosion produced the materials that exceed stream capacity in all three locations.

Flow and Channel Characteristics

Flow characteristics of a stream are best seen in a cross-sectional view. The greatest velocities in a stream are near the surface at center channel (Figure 14.15), corresponding to the deepest part of the stream channel. Velocities decrease closer to the sides and bottom of the channel because of the frictional drag on the water flow. The portion of the stream flowing at maximum velocity moves diagonally across the stream from bend to bend in a meandering stream.

Flow becomes *turbulent* in shallow streams, or where the channel is rough, as in a section of rapids. Small eddies are caused by friction between streamflows and the channel sides and bed. Complex turbulent flows propel sand, pebbles, and even boulders, increasing the suspension, traction, and saltation.

Three types of stream channels are notable: *braided*, *straight*, and *meandering*. Where slope is gradual, stream channels develop a sinuous (snakelike) form, weaving across the landscape. This action produces a **meandering stream**. The term *meander* comes from the ancient Greek Maiandros River in Asia Minor (the present-day Menderes River in Turkey), which had what came to be known as a meandering channel pattern. The tendency to meander is evidence of a river system's struggle to operate with least effort, between self-organizing order (equilibrium) and chaotic disorder in nature.

The outer portion of each meandering curve is subject to the fastest water velocity and therefore the greatest scouring erosive action; it can be the site of a steep bank, the **undercut bank**, or *cutbank* (Figures 14.15 and 14.16). In contrast, the inner portion of a meander experiences the slowest water velocity and thus receives sediment fill, forming a **point bar** deposit. As meanders develop, these scour-and-fill features gradually work at stream banks. As a result, the landscape near a meandering river bears meander scars of residual deposits from previous river channels (see Figure 14.22). The photograph of the Itkillik River in Alaska, Figure 14.16a, shows both meanders and meander scars.

Meandering streams create a remarkable looping pattern on the landscape, as shown in the four-part sequence in Figure 14.16b. In (1), the stream erodes its outside bank as the curve migrates downstream, forming a neck. In (2), the narrowing neck of land created by the looping meander eventually erodes through and forms a *cutoff* in (3). A cutoff marks an abrupt change in the stream's lateral movements—the stream becomes straighter.

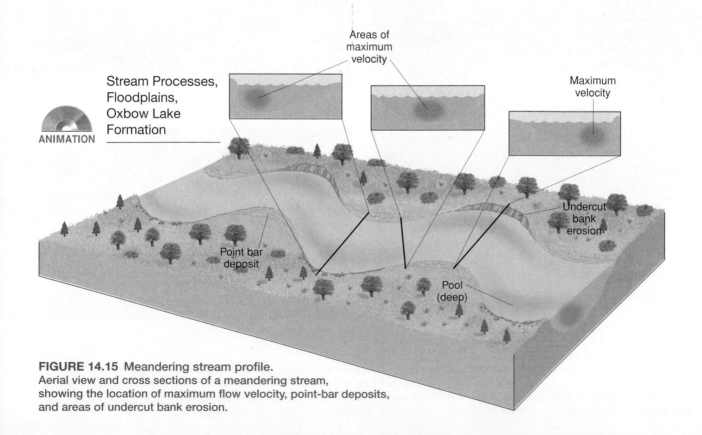

FIGURE 14.15 Meandering stream profile.
Aerial view and cross sections of a meandering stream, showing the location of maximum flow velocity, point-bar deposits, and areas of undercut bank erosion.

(a)

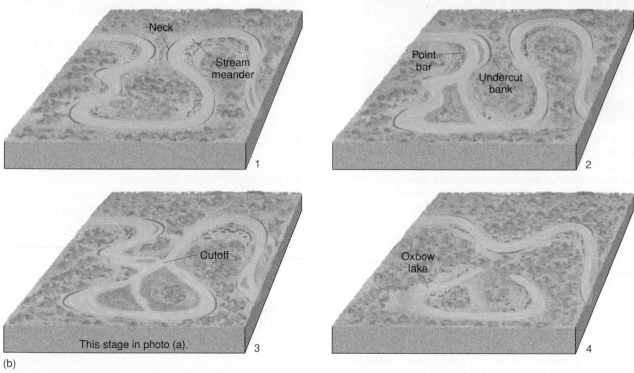

(b)

FIGURE 14.16 Meandering stream development.
(a) Itkillik River in Alaska. (b) Development of a river meander and oxbow lake simplified in four stages.
[(a) U.S. Geological Survey photo.]

When the former meander becomes isolated from the rest of the river, the resulting **oxbow lake** (4) may gradually fill with organic debris and silt or may again become part of the river when it floods. The Mississippi River is many miles shorter today than it was in the 1830s because of artificial cutoffs that were dredged across meander necks to improve navigation and safety. How many of these stream features—meanders, oxbow lakes, cutoffs—can you spot in the aerial photo in Figure 14.17, looking east across northern Argentina? News Report 14.2 looks at the difficulty of using a meandering stream as a political boundary.

Stream Gradient

Every stream runs downhill under the pull of gravity. In the course of this process, each stream develops its own **gradient**, which is the rate of elevation decline from its headwaters to its mouth. This decline is far from linear; characteristically, the *longitudinal profile* (side view) of a stream features a steeper slope upstream and a gentler slope downstream, forming an uneven concave shape (Figure 14.18). This gradient is concave for complex reasons related to the stream's having just enough energy to

FIGURE 14.17 Meandering stream features.
Meandering streams catch the morning light over northern Argentina. Identify where former meanders were cut off, oxbow lakes, and former stream channels. Several inversion layers are visible in the far distance, as the burning of tropical rain forest to the north fills the air with smoke. [Photo by Bobbé Christopherson.]

transport the sediment load it receives. The longitudinal profile of streams can be expressed mathematically, enabling scientists to categorize and predict stream behavior.

An important fluvial concept is that of the graded stream, nicely defined by J. H. Mackin, a geomorphologist:

A graded stream is one in which, over a period of years, slope is delicately adjusted to provide, with available discharge and with prevailing channel characteristics,

just the velocity required for transportation of the load supplied from the drainage basin.*

In other words, a **graded stream** is one that attains a graded condition; the term does not mean that the stream is at its lowest gradient. Rather, graded represents a present balance, a dynamic equilibrium among erosion, transportation, and deposition over time along a portion of the stream. Both high-gradient and low-gradient streams can achieve a graded condition. The longitudinal profile of a graded stream is its *profile of equilibrium*—a parabolic curve, gently flattening toward the mouth. Stream dynamics can then be compared against this ideal balance.

One problem with applying the graded-stream concept is that an individual stream can have both graded and ungraded portions and may have graded sections without having an overall graded slope. A profile of equilibrium may not be smooth throughout its course and cannot exist for long, for it represents a theoretical perfect balance. With streams, as in all of nature, change is the only constant.

Stream gradient may be affected by tectonic uplift of the landscape, which changes the base level. If tectonic forces slowly lift the landscape, the stream gradient will increase, stimulating renewed erosional activity. A meandering stream flowing through the uplifted landscape becomes *rejuvenated*; that is, the river actively returns to downcutting and can eventually form *entrenched meanders* in the landscape. Figure 14.19 depicts actual rejuvenated landscapes.

Nickpoints When the longitudinal profile of a stream shows an abrupt change in gradient, such as at a waterfall or an area of rapids, the point of interruption is a **nickpoint** (also spelled *knickpoint*). At a nickpoint, the conversion of potential energy in the water at the lip of the falls to concentrated kinetic energy at the base works to eliminate the nickpoint interruption and smooth out the gradient. Figure 14.20a shows a stream with two such interruptions. The photos show nickpoints and falls in New Hampshire and South Dakota.

*J. H. Mackin, "Concept of the graded river," *Geological Society of America Bulletin* 59 (1948): 463.

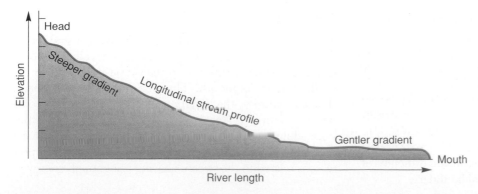

FIGURE 14.18 An ideal longitudinal profile.
Idealized longitudinal profile of a stream, showing its gradient. Upstream segments have a steeper gradient; downstream, the gradient is gentler. The middle and lower portions in the illustration appear graded, or in dynamic equilibrium.

Rivers Make Poor Political Boundaries

Streams often serve the role of natural political boundaries. Commonly, where a stream forms a boundary, the boundary line is drawn down the middle of the stream. It is easy to see how boundary disputes might arise when they are based on river channels that shift their positions quite rapidly during times of flood. The boundaries based on the original stream channels will not match new channel positions.

Carter Lake, Iowa, provides a fascinating example (Figure 14.2.1). The Nebraska–Iowa border originally was placed midchannel in the Missouri River. But in 1877, the meander loop that curved around the town of Carter Lake in Iowa was cut off by the river, leaving the town "captured" by Nebraska! The old boundary marked along the former meander bend still is used as the state line. The oxbow lake created by the cutoff was named Carter Lake.

This event illustrates why boundaries should be fixed by surveys independent of river locations. Such surveys have been completed along the Rio Grande near El Paso, Texas, and along the Colorado River between Arizona and California, permanently establishing political boundaries separate from shifting river channels. In the latter example, a midpoint between the bluffs on either side of the floodplain is used as the state line. Meanwhile, Carter Lake is still in Iowa.

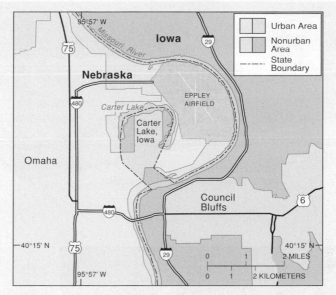

FIGURE 14.2.1 A stranded town.
Carter Lake, Iowa, sits within the curve of a former meander that was cut off by the Missouri River. The city and oxbow lake remain part of Iowa even though they are stranded within Nebraska.

(a)

(b)

(c)

FIGURE 14.19 Entrenched meanders.
(a) and (b), The San Juan River near Mexican Hat, Utah, cuts down into the uplifted Colorado Plateau landscape, producing entrenched meanders called the Goosenecks of the San Juan. (c) Entrenched meanders merge in central Montana. [Photos (a) and (c) by Bobbé Christopherson; (b) by Randall M. Christopherson.]

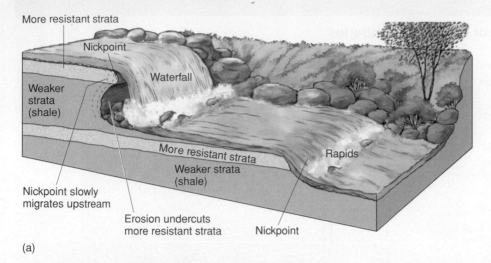

(a)

(b)

(c)

FIGURE 14.20 Nickpoints interrupt a stream profile.
(a) Longitudinal stream profile showing nickpoints produced by resistant rock strata. Potential energy is converted into kinetic energy and concentrated at the nickpoint, accelerating erosion, which will eventually eliminate the feature. (b) A nickpoint and granite pothole interrupt the stream gradient on the Pemigewassey River, Franconia Notch Park, New Hampshire. (c) Falls and rapids break the flow of the Big Sioux River in Sioux Falls, South Dakota. [Photos by Bobbé Christopherson.]

Nickpoints can result when a stream flows across a zone of hard, resistant rock or from various tectonic uplift episodes, such as might occur along a fault line. Temporary blockage in a channel, caused by a landslide or a logjam, also could be considered a nickpoint; when the logjam breaks, the stream quickly readjusts its channel to its former grade.

Waterfalls are interesting and beautiful gradient breaks. At the edge of a fall, a stream is free-falling, moving at high velocity under the acceleration of gravity, causing increased abrasion and hydraulic action in the channel below. The increased action generally undercuts the waterfall. Eventually, the excavation will cause the rock ledge at the lip of the fall to collapse, and the waterfall will shift a bit farther upstream. The height of the waterfall is gradually reduced as debris accumulates at its

base (see Figure 14.21b). Thus, a nickpoint migrates upstream, sometimes for kilometers, until it becomes a series of rapids and is eventually eliminated.

At Niagara Falls on the Ontario–New York border, glaciers advanced over the region and then receded some 13,000 years ago. In doing so, they exposed resistant rock strata that are underlain by less-resistant shales. This tilted formation is a *cuesta*, which is a ridge with a steep slope on one side and beds gently sloping away on the other side (Figure 14.21a). This Niagara escarpment actually stretches across more than 700 km (435 mi) from east of the falls, northward through Ontario, Canada, the Upper Peninsula of Michigan, curving south through Wisconsin along the western shore of Lake Michigan and the Door Peninsula. As this less-resistant material continues to weather away, the overlying rock

FIGURE 14.21 Retreat of Niagara Falls.
(a) Headward retreat of Niagara Falls from the Niagara escarpment. It has taken the falls about 12,000 years to reach this position at a pace of about 1.3 m (4.3 ft) per year. (b) The escarpment seen from below American Falls. (c) Niagara Falls, with the American Falls portion almost completely shut off by upstream controls for engineering inspection. Horseshoe Falls in the background is still flowing over its 57-m (188-ft) plunge. (d) In contrast to (c), American Falls in full discharge. [(a) After W. K. Hamblin, *Earth's Dynamic Systems*, 6th ed. (Upper Saddle River, NJ: Pearson Prentice Hall, Inc. © 1992), Figure 12.15, p. 246. Photos: (c) courtesy of the New York Power Authority; (b) and (d) Bobbé Christopherson.]

Niagara Falls

strata collapse, allowing Niagara Falls to erode farther upstream toward Lake Erie.

Niagara Falls is a place where natural processes labor to eliminate a nickpoint and reduce this portion of the river to a series of mere rapids. In fact, the falls have retreated more than 11 km (6.8 mi) from the steep face of the Niagara escarpment during the last 12,000 years. A nickpoint is a relatively temporary and mobile feature on the landscape. In the past, engineers have used control facilities upstream to reduce flows over the American Falls at Niagara for inspection of the cliff to assess the progress of natural processes that are working to eliminate the Niagara Falls nickpoint (Figure 14.21c); compare to normal discharge in (d).

Stream Gradient and Landscape Forms   William Morris Davis, a geomorphologist who founded the Association of American Geographers in 1904, introduced evolutionary concepts of erosion that included fluvial processes. Davis incorporated the graded-stream concept into his model, identifying erosion stages in a cyclic model he named *youth*, *maturity*, and *old age*—old age is approached as the floodplain broadens and a low stream gradient produces a wide, meandering flow pattern.

Today, geomorphologists support a *functional model of dynamic equilibrium*. The dynamic equilibrium model emphasizes the effects of individual processes interacting on streams and hillslope systems. Stream form and behavior result from complex interactions of slope, discharge, and sediment load, all of which are variable within different climates and with different rock types. Landscapes simply do not provide enough clear evidence to support a cyclic model of evolution. Regardless, Davis's work was a breakthrough in understanding landscapes, and many of his terms are still in use.

As suggested by S. A. Schumm and R. W. Lichty, two geomorphologists, the validity of cyclic or functional landscape models may depend on the *time frame*. Let us consider three time frames: geologic time, graded time, and steady time. Over the long span of *geologic time*, cyclic models of evolutionary development might explain the disappearance of entire mountain ranges through, for example, denudation. At the other extreme of time, *steady time* applies to short-term adjustments ongoing in a drainage basin. *Graded time* is between the two time frames, and within it lies the realm of dynamic equilibrium conditions.

Stream Deposition and Landforms

After weathering, mass movement, erosion, and transportation, deposition is the next logical event in a sequence. In *deposition*, a stream deposits alluvium, or unconsolidated sediments, thereby creating depositional landforms, such as floodplains, terraces, or deltas.

As discussed earlier, stream meanders tend to migrate downstream through the landscape. Over time, the landscape near a meandering river comes to bear meander scars of residual deposits from former, abandoned channels. Former point-bar deposits leave low-lying ridges, creating a *bar-and-swale relief* (a swale is a gentle low area), forming a *scroll topography*. The *Landsat* image in Figure 14.22 exhibits characteristic meandering scars: meander bends, oxbow lakes, natural levees, point bars, and undercut banks.

Floodplains The flat, low-lying area flanking many stream channels that is subjected to recurrent flooding is a **floodplain**. It is formed when the river overflows its channel during times of high flow. Thus, when floods occur, the floodplain is inundated. When the water recedes, it leaves behind alluvial deposits that generally mask the underlying rock with their accumulating thickness. The present river channel is embedded in

these alluvial deposits. Figure 14.23 illustrates a characteristic floodplain, related features in photos, and a representative topographic map of an area near Philipp, Mississippi.

On either bank of some streams, **natural levees** develop as by-products of flooding. When floodwaters rise, the river overflows its banks, loses stream competence and capacity as it spreads out, and drops a portion of its sediment load to form the levees. Larger, sand-sized particles drop out first, forming the principal component of the levees, with finer silts and clays deposited farther from the river. Successive floods increase the height of the levees (*levée* is French for "raising"). The levees may grow in height until the river channel becomes elevated, or *perched*, above the surrounding floodplain.

On the topographic map (Figure 14.23d), you can see the natural levees represented by several contour lines that run immediately adjacent to the Tallahatchie River. These contour lines (5-ft interval) denote a height of 10–15 ft (3–4.5 m) above the river and the adjoining floodplain. Next time you have an opportunity to see a river and its floodplain, look for levees. They may be low and subtle.

Notice in Figure 14.23a an area labeled backswamp and a stream called a yazoo tributary. The natural levees

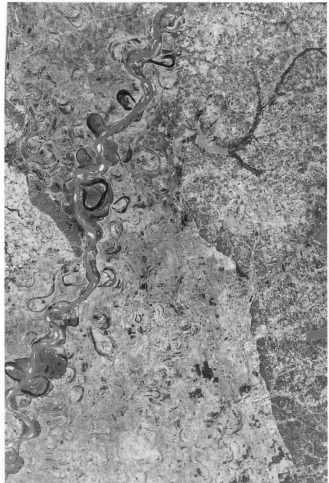

(a)

FIGURE 14.22 Meander scars.
The Mississippi River forms a portion of the Mississippi–Arkansas border near Senatobia, Mississippi. Characteristic meander patterns and scars of former channels are visible in the *Landsat* image. [Image by GEOPIC, Earth Satellite Corporation. Used by permission.]

(b)

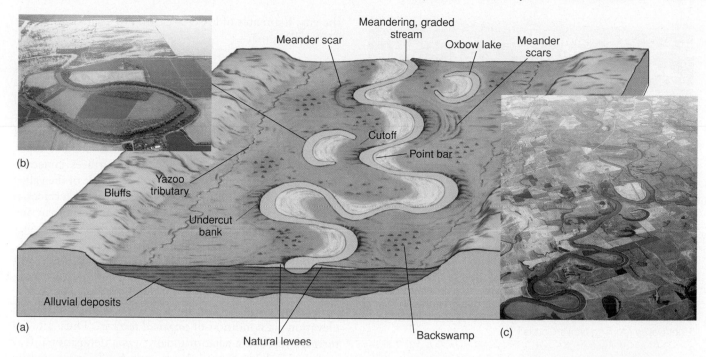

(a)

(b)

(c)

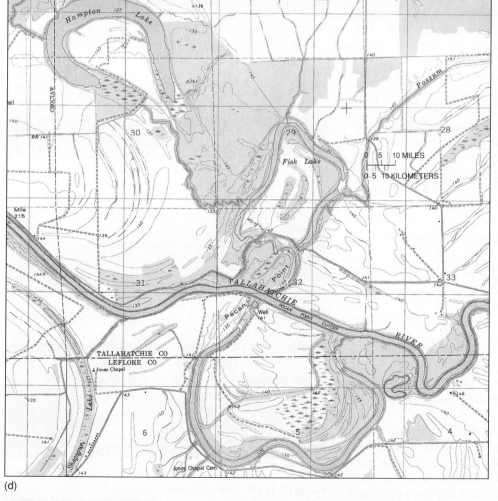

(d)

MISS.

QUADRANGLE LOCATION

Stream Processes, Floodplains

Oxbow Lake Formation

FIGURE 14.23 A floodplain.
(a) Typical floodplain landscape and related landscape features. (b) An oxbow lake.
(c) Levees, oxbow lakes, and farmland on a floodplain. (d) A portion of the Philipp,
Mississippi, topographic map quadrangle. [(b) and (c) photos by Bobbé Christopherson;
(d) topographic map from USGS.]

FIGURE 14.24 Levee break and devastation in New Orleans.
August 30, 2005, view across the Inner Harbor Navigation Canal, as broken levees allow floodwaters to drown the city. A dramatic failure of engineering, levee design and construction, and proper planning. [Photo by © Smiley N. Pool/*Dallas Morning News*/Corbis.]

and elevated channel of the river prevent this **yazoo tributary** from joining the main channel, so it flows parallel to the river and through the **backswamp** area. The name comes from the Yazoo River in the southern part of the Mississippi floodplain.

People build cities on floodplains despite the threat of flooding because floodplains are nearly level and they are next to water. There may be historic momentum for settlement. People often are encouraged by government assurances of artificial protection from floods and disaster assistance if floods occur. Government assistance may provide for the building of artificial levees on top of natural levees. Artificial levees do increase the capacity in the channel, but they also lead to even greater floods when they are overtopped by floodwaters or when they fail, as happened in New Orleans (Figure 14.24).

Perhaps the best use of some floodplains is for crop agriculture, because inundation generally delivers nutrients to the land with each new alluvial deposit. A significant example is the Nile River in Egypt, where annual flooding enriches the soil. However, floodplains that are covered with coarse sediment—sand and gravel—are less suitable for agriculture. If there are river or creek floodplains where you live, what is your impression of present land-use patterns, local planning and zoning, and people's hazard perception overall?

Nationally, in 2001, some two-thirds of disaster losses were attributable to floods. Tropical storm Allison left $6 billion in damage in its wandering visit to Texas in June 2001, producing the most flood damage along occupied floodplains. The 2002 floods in Europe produced estimated losses exceeding US $50 billion. Of course, the 2005 levee and floodwall breaks resulted in catastrophic loss of life and property in New Orleans and will forever be in our memories, as floodwaters engulfed 80% of the city. Estimates of losses are climbing to more than the $100 billion mark.

Stream Terraces As explained earlier, several factors may rejuvenate stream energy and stream–landscape relations so that a stream can scour downward with renewed vigor and increased erosion. The resulting entrenchment of the river deeper into its own floodplain produces **alluvial terraces** on either side of the valley, which look like topographic steps above the river. Alluvial terraces generally appear paired at similar elevations on each side of the valley (Figure 14.25). If more than one set of paired terraces is present, the valley probably has undergone more than one episode of rejuvenation. The flat terrace areas along a river have always been a location for settlement.

If the terraces on either side of the valley do not match in elevation, then entrenchment actions must have been continuous as the river meandered from side to side, with each meander cutting a terrace slightly lower in elevation—a condition of *unpaired terraces*. Thus, alluvial terraces represent what originally was a depositional feature (a floodplain) that subsequently has been eroded by the rejuvenated stream.

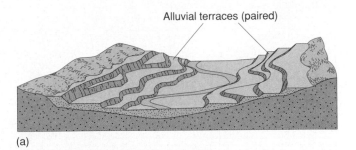

Alluvial terraces (paired)

(a)

(b)

FIGURE 14.25 Alluvial stream terraces.
(a) Alluvial terraces are formed as a stream cuts into a valley.
(b) Alluvial terraces along the Rakaia River in New Zealand.
[(a) After W. M. Davis, *Geographical Essays* (New York: Dover, 1964 [1909]), p. 515; (b) photo by Bill Bachman/ Photo Researchers, Inc.]

ANIMATION   Stream Terrace Formation

River Deltas   The mouth of a river is where it reaches a base level. The river's forward velocity rapidly decelerates as it enters a larger, standing body of water. The reduced stream competence and capacity causes the sediment load to quickly exceed the river's ability to transport it. Coarse sediments such as sand and gravel drop out first and are deposited closest to the river's mouth. Finer clays are carried farther and form the extreme end of the deposit. The level or nearly level depositional plain that forms at the mouth of a river is a **delta** for its characteristic triangular shape, after the Greek letter delta (Δ).

Each flood stage deposits a new layer of alluvium over portions of the delta, some of which is deposited below water or *subaqueous*, growing the delta outward. At the same time, river channels divide into smaller courses known as *distributaries*, which appear as a reverse of the dendritic drainage pattern of tributary streams discussed earlier. Here are a few examples:

• The Ganges River delta features an extensive lower delta plain formed in relation to high tidal ranges in an *arcuate* (arc-shaped) pattern. It is covered by an intricate maze of distributaries in *braided* forms. Bountiful alluvium carried from deforested slopes

upstream provides excess sediment that is deposited to form many deltaic islands (Figure 14.26). The combined Ganges–Brahmaputra River delta complex is the largest in the world at some 60,000 km² (23,165 mi²).

• The Nile River delta is an *arcuate* delta (Figure 14.27). Also arcuate are the Danube River delta in Romania, where it enters the Black Sea, and the Indus River delta. (See News Report 14.3 for an update on the condition of the disappearing Nile delta.)

• The Tiber River in Italy has an *estuarine delta*, one that is in the process of filling an **estuary**, which is the seaward mouth of a river where the river's freshwater encounters seawater.

Mississippi River Delta   The Mississippi River delta has an interesting history. Over the past 120 million years, the Mississippi has collected sediments throughout its vast basin and deposited them into the Gulf of Mexico. During the past 5000 years, the river has formed a succession of seven distinct deltaic complexes along the Louisiana coast (Figure 14.28a).

Each generalized lobe in the illustration reflects distinct course changes in the Mississippi River, probably

FIGURE 14.26 The Ganges River enters the Bay of Bengal.
The complex distributary pattern in the "many mouths" of the Ganges River delta in Bangladesh and extreme eastern India from the *Terra* satellite. Imagine the devastation to this barely-above-sea-level delta when Tropical Storm SIDR, a category 4, plowed through in November 2007, killing thousands of people on the low-lying islands. [*Terra* MODIS sensor image courtesy of MODIS Land Team, NASA.]

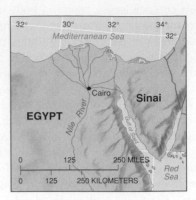

FIGURE 14.27 The Nile River delta. The arcuate Nile River delta. Intensive agricultural activity and small settlements are visible on the delta and along the Nile River floodplain in this true-color image. Cairo is at the apex of the delta. You can see the two main distributaries: Damietta to the east and Rosetta to the west. [January 30, 2001, *Terra* image courtesy of MISR Team, NASA/GSFC/JPL.]

where the river broke through its natural levees during episodes of severe flooding, thus changing the configuration of the delta. The seventh and current delta has been building for at least 500 years and is a classic example of a *bird's-foot delta*—a long channel with many distributaries and sediments carried beyond the tip of the delta into the Gulf of Mexico.

The Mississippi River delta clearly is dynamic over time as sediments accumulate on the floor of the Gulf of Mexico and the distributaries shift (Figure 14.28b). The main channel persists because of much effort and expense directed at maintaining the artificial levee system. The 3.25-million-km² (1.25-million-mi²) Mississippi drainage basin produces enough sediment to extend the Louisiana coast 90 m (295 ft) a year—550 million metric tons a year—however, several factors are actually causing losses to the delta each year.

Compaction and the tremendous weight of the sediments in the Mississippi River create isostatic adjustments in Earth's crust. These adjustments are causing the entire region of the delta to subside, thereby placing ever-increasing stress on natural and artificial levees and other structures along the lower Mississippi. The many canals and waterways excavated through the delta by the oil and gas industry and shipping interests have worked to deprive the delta of new alluvial building materials. Also, the pumping of tremendous quantities of oil and gas from

thousands of onshore and offshore wells is thought to be a cause of regional land subsidence and lowering.

An additional problem for the lower Mississippi Valley is the possibility, in a worst-case flood, that the river could break from its existing channel and seek a new route to the Gulf of Mexico. If you examine the map in Figure 14.28c and look at the sediment plume to the west

FIGURE 14.28 The Mississippi River delta.
(a) Evolution of the present delta, from 5000 years ago (1) to present (7). (b) The bird's-foot delta of the Mississippi River receives a continuous supply of sediments, focused by controlling levees, although subsidence of the delta and rising sea level have diminished the overall surface area. (c) Location map of the Old Control Structures and potential capture point (arrow) where the Atchafalaya River may one day divert the present channel. (d) Where the Atchafalaya River enters the Gulf. (e) Old River Control Auxiliary Structure, one of the dams to keep the Mississippi in its channel. (f) The end of the bird's-foot delta stretches far out in the Gulf. (g) Mississippi delta waterscape; note the house with the raised first floor on stilts. [(a) Adapted from C. R. Kolb and J. R. Van Lopik, "Depositional environments of the Mississippi River deltaic plain," in *Deltas in Their Geologic Framework* (Houston: Houston Geological Society, 1966); (b) *Terra* MODIS sensor image, March 5, 2001, courtesy of Liam Gumley, Space Science and Engineering Center, University of Wisconsin, and the MODIS Science Team, NASA; photos (f) by astronaut aboard the International Space Station, NASA, and (d), (e) and (g) by Bobbé Christopherson.]

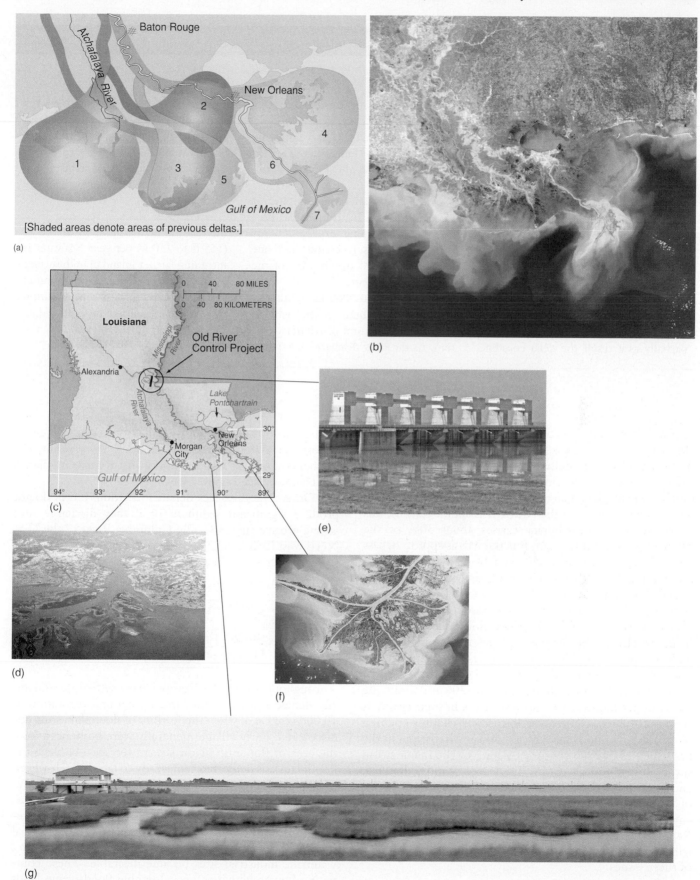

(a)

[Shaded areas denote areas of previous deltas.]

(b)

(c)

(d)

(e)

(f)

(g)

FIGURE 14.28

News Report 14.3

The Nile Delta Is Disappearing

People along the Nile River have depended on its regular flow and annual floods for millennia. Herodotus noted in the fifth century B.C. how the people farmed the fields in the floodplain and delta regions. After harvesting their crops, they retreated from the area to their homes. They would await the annual floods that brought fresh silt and nutrients for next year's planting. This cycle of fertility continued until the completion of the Aswân High Dam in 1964. This structure caused a decrease in the supply of sediment to the delta, and partially as a result the delta coastline

continues to actively recede. Herodotus stated in *The History*, Book Two, "In the part called the Delta, it seems to me that if the Nile no longer floods . . . for all time to come, the Egyptians will suffer."

J. Stanley, an oceanographer at the Smithsonian Institution, has proposed an intriguing explanation for the Nile Delta's recession, and one that goes beyond the impacts of the Aswân High Dam. Over the centuries, more than 9000 km (5500 mi) of canals were built in the delta to augment the natural distributary system. As the river discharge enters the

network of canals, flow velocity is reduced, stream competence and capacity are decreased, and sediment load is deposited far short of where the delta touches the Mediterranean Sea. River flows no longer effectively reach the sea.

The Nile delta is receding from the coast at an alarming 50 to 100 m (165 ft to 330 ft) per year. Seawater is intruding farther inland in both surface water and groundwater. Human action and reaction to this evolving situation will no doubt determine the delta's future. If Herodotus could only see the delta he described as it looks today!

---

of the main delta in the satellite image, an obvious alternative to the Mississippi's present channel is the Atchafalaya River (its delta is shown in Figure 14.28d). The Atchafalaya would provide a much shorter route to the Gulf of Mexico, less than one-half the present distance, and it has a steeper gradient than the Mississippi. Presently, this alternative-route distributary carries about 30% of the Mississippi's total discharge. For the Mississippi to bypass New Orleans entirely would be a blessing, for it would remove the flood threat. However, this shift would be a financial disaster, as a major U.S. port would silt in and seawater would intrude into freshwater resources.

At present, artificial barriers block the Atchafalaya from reaching the Mississippi at the point shown; without the floodgates the two rivers do connect. The Old River Control Project (1963) maintains three structures and a lock about 320 km (200 mi) from the Mississippi's mouth to keep these rivers in their channels (Figure 14.28e; a larger version is our chapter-opening photo). Many floods have hit the lower Mississippi in the past, such as in 2003 (Figure 14.28g), or the major storm surge and flooding from hurricanes Cindy, Katrina, and Rita in 2005. Another major flood is only a matter of time, one that might cause the river channel to change back into the Atchafalaya. A case study in News Report 14.4 about Bayou Lafourche, a former major distributary of the Mississippi until 1904, highlights many of these issues.

Rivers Without Deltas The Amazon River, Earth's highest-discharge stream, exceeds 175,000 m³/s (6.2 million cfs) in discharge and carries sediments far into the deep Atlantic offshore. Yet the Amazon lacks a true delta.

Its mouth, 160 km (100 mi) wide, has formed a subaqueous deposit on a sloping continental shelf. As a result, the Amazon's mouth is braided into a broad maze of islands and channels (see Figure 14.1).

Other rivers also lack deltaic formations if they do not produce significant sediment or if they discharge into strong erosive currents. The Columbia River of the U.S. Northwest lacks a delta because offshore currents remove sediment before it can accumulate into a delta.

Floods and River Management

 Stream Processes,
Floodplains

Throughout history, civilizations have settled floodplains and deltas, especially since the agricultural revolution of 10,000 years ago, when the fertility of floodplain soils was discovered. Early villages generally were built away from the area of flooding, or on stream terraces, because the floodplain was dedicated exclusively to farming. However, as commerce grew, competition for sites near rivers grew, because these locations were important for transportation. Port and dock facilities were built, as were river bridges. Because water is a basic industrial raw material used for cooling and for diluting and removing wastes, waterside industrial sites became desirable. These competing human activities on vulnerable flood-prone lands place lives and property at risk during floods.

The abuse and misuse of river floodplains brought catastrophe to North Carolina in 1999. In short succession during September and October, hurricanes Dennis,

News Report 14.4

What Once Was Bayou Lafourche—Analysis of a Photo

Bayou Lafourche was the main distributary of the Mississippi River between the Civil War and 1904, when a dam was built that stopped the flow, creating water losses in the bayou and stagnation of the former marshlands. The bayou was closed to navigation by 1930. In 1955, attempts began to reverse the losses to this bayou by restoring some water flow with pumps. Meanwhile, the overall losses to the delta continue.

Bayou is a general term for several water features in the Lower Mississippi

River system, including creeks and secondary waterways. These sometimes stagnant, winding watercourses pass through coastal marshlands and swamps and allow tidal waters access to deltaic lowlands.

The human impact on Bayou Lafourche's 175-km (110-mi) stretch of the delta has been severe and is mirrored by similar impacts across southern Louisiana. In fact, most of coastal Louisiana, which represents about 40% of the coastal marshes in the United States, is profoundly

altered and disrupted by dam and levee construction, flow alterations, oil and gas exploration, wells, pumping, pipelines, and dredging for navigation, logging, and industry needs. Approximately 65 km² (25 mi²) of these Louisiana coastal lowlands are submerged by the Gulf of Mexico in an average year. These coastal marshes are important ecologically and they help protect the coast from storm surge and tropical storms.

Figure 14.4.1 illustrates a portion of Bayou Lafourche in 2003. The

(a)

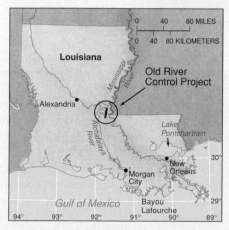

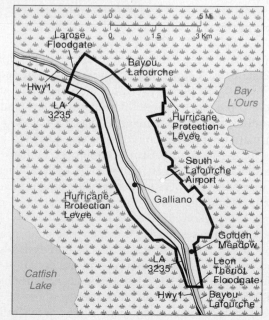

FIGURE 14.4.1 A hurricane-protection levee compound in the Mississippi delta.
(a) Development of settlements, farms, and industry within a 65-km (40-mi) hurricane-protection levee system. Bayou Lafourche is the waterway that runs along the highway, with locks at the north and south ends of the levee system to seal off the interior in times of storm surges. (b) A map of this general region. [Photo by Bobbé Christopherson.]

(b)

(continued)

News Report 14.4 (continued)

towns of Galliano and Golden Meadow and a total population of more than 10,000 people are protected from rising water, tidal fluctuations, and hurricane storm surge within the compound surrounded by levees. The 65-km (40-mi) hurricane protection levee system, some 4 m (13 ft) in height, 28 m (92 ft) wide at the base, was begun in the late 1970s. You can see the long-lot land survey system dividing lands at right angles to Bayou Lafourche and Highway 1, some plots planted in sugar cane. Floodgates at either end of this enclave are closed when the flood threat is imminent. Water tables in the protected area have decreased since the levee enclosure was completed and surface water flows have been reduced.

Hurricane Katrina carried shrimp boats and other marine craft far inland and set them down in the debris of broken structures blocking roads and fields. Almost a month later, Hurricane Rita surrounded the levee compound with 2.4 m (8 ft) of water, but during both events the levee system held. However, the wind caused damage, which is evidenced by the familiar blue tarps covering torn roofs. Businesses remained on electrical generators months after the event. A multiunit trailer park provides temporary housing.

FIGURE 14.4.2 Port Fourchon. Port Fourchon is the service center for the offshore oil industry, with about 6000 workers using the port as a depot for commuting to offshore drilling rigs. [Photo by Bobbé Christopherson.]

With delta subsidence and rising sea levels, increasing levee height is a continuing process. Ongoing construction, known as a "lift" to the levee, is adding another meter of height in an attempt to maintain 100-year storm-surge protection—a project about 90% completed in 2005. The real irony is that the former coastal wetlands were a buffer between the Gulf storm surges and interior lands, yet today because of all the human disruptions, there is much open water between Golden Meadow and the former shoreline along the margins of the delta.

Only a few marsh mitigation and restoration projects are underway. One is at Port Fourchon at the extreme southern end of the former coast (south of the Figure 14.4.1 photo). This is a major oil and gas industry transshipment port and helicopter base

where workers connect for their jobs on drilling rigs in the Gulf of Mexico (Figure 14.4.2).

Hurricane Katrina picked up huge oil barges, equipment, and debris and dumped them inland. Grand Isle, to the east of the port on the Gulf Coast, experienced major destruction. Some 80% of homes, businesses, and fishing camps were destroyed. Many of the homes that are built on stilts were simply gone after Katrina. And weeks later, Hurricane Rita passed to the west and further damaged many of these same areas.

Meanwhile, Galliano is presently 122 cm (48 in.) above sea level. The Mississippi delta lands are slowly subsiding and sea level is continuing to rise. The story continues to unfold.

Floyd, and Irene delivered several feet of precipitation to the state, each storm falling on already saturated ground. About 50,000 people were left homeless and at least 50 died, while more than 4000 homes were lost and an equal amount were badly damaged (Figure 14.29a). The dollar estimate for the ongoing disaster exceeded $10 billion.

Hogs, in factory farms, outnumber humans in North Carolina. More than 10 million hogs, each producing 2 tons of waste per year, were located in about 3000 agricultural factories. These generally unregulated operations collect almost 20 million tons of manure into nearly 500 open lagoons, many set on river floodplains. The hurricane downpour flushed out these waste lagoons into wetlands, streams, and eventually Pamlico Sound and the ocean—a spreading "dead zone" of an environmental catastrophe (Figure 14.29b). There has been little progress toward resolving these waste/floodplain issues since these events in 1999.

Catastrophic floods continue to be a threat, especially in poor nations. Bangladesh is perhaps the most persistent example. Bangladesh is one of the most densely populated countries on Earth, and more than *three-fourths* of its land area is a floodplain. The country's vast alluvial plain sprawls over an area the size of Alabama (130,000 km², or 50,000 mi²).

The flooding severity in Bangladesh is a consequence of human economic activities, along with heavy precipitation episodes. Excessive forest harvesting in the upstream portions of the Ganges–Brahmaputra River watersheds increased runoff. Over time, the increased sediment load carried by the river was deposited in the Bay of Bengal, creating new islands (see Figure 14.26). These islands, barely above sea level, became sites for new farming villages. As a result, about 150,000 people perished in the 1988 and 1991 floods and storm surge. (For information

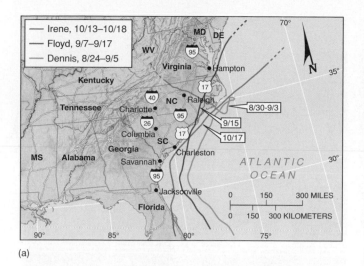

(a)

(b)

FIGURE 14.29 North Carolina 1999 floodplain disaster. (a) Three hurricanes deluged North Carolina with several feet of rain during September and October 1999, Hurricane Floyd being the worst. (b) Hundreds of thousands of livestock were killed in the floods, which also washed out hundreds of animal sewage lagoons into wetlands, streams, and the ocean. Many of these factory farms and lagoons were sited on floodplains. [(b) Photo by Mel Nathanson, *Raleigh News & Observer*.]

on floods worldwide, see http://www.dartmouth.edu/~floods/Resources.html/ or a daily flood summary at http://www.nws.noaa.gov/oh/hic/current/fln/fln_sum.shtml.)

Engineering Failures in New Orleans, 2005

If all is well, the city of New Orleans is almost entirely below river level but dry, with sections of the city below sea level in elevation. Severe flooding is a certainty for existing and planned settlements unless further intervention or urban relocation occurs. The building of multiple flood-control structures and extensive reclamation efforts by the U.S. Army Corps of Engineers apparently have only delayed the peril, as demonstrated by recent flooding. The events of 2005, following the passage of Hurricane Katrina, will remain for many years as one of the greatest engineering failures by the government. Four levee breaks and at least four dozen levee breaches (topping) permitted the inundation of a major city, which confirms this conclusion. The polluted water remained for days.

Figure 14.30a and b from *Landsat*–7 are matching images of New Orleans along the shore of Lake Pontchartrain on April 24 and August 30, 2005. The locator map helps you position the levee/floodwalls, areas that are below sea level, and the four major breaches. The extensive

floods are the darker areas of the city, where some neighborhoods were submerged up to 6.1 m (20 ft). At the time of image (b), New Orleans was approximetly 80% under water.

Six investigations by civil engineers, scientists, and political bodies agreed that in concept, design, construction, and maintenance, the "protection" system was flawed. Many of the structures failed before they reached design-failure limits. After all, Hurricane Katrina was diminished in power as it moved onshore. Several of the levee floodwalls and pilings were not anchored to the design depth, such as along the 17th Street Canal (Figure 14.31). This inadequate anchoring depth made parts of the system inherently weak. Unexplained are findings in 2007 by civil engineers inspecting levee and floodwall repair and placement that the composition of fill materials is too low in clays and too high in sand. As of 2007, work in New Orleans was not completed, and elements of the completed sections appear inadequate and positioned for repeat disasters.

As we contemplate all this, we must consider the ongoing rise in sea level coupled with subsidence along this portion of the Gulf Coast. In Chapter 16, a new map portrays a 1-m rise in sea level forecasted for this area. Physical geography has much to study, analyze, and conclude about this enigmatic situation that we will read about throughout our lives.

(a) April 24, 2005

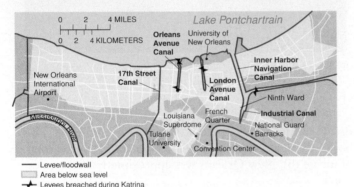

(b) August 30, 2005

FIGURE 14.30 New Orleans before and after disaster.
Landsat–7 images made (a) April 24, 2005, and (b) August 30, 2005, clearly show canal locations and flooded portions of the city; compare features to the locator map. Numerous levee breaks and breaches put 80% of the city under water for more than a week. [*Landsat–7*, Bands 7, 5, and 3, courtesy of USGS *Landsat* Image Gallery.]

Rating Floodplain Risk

A **flood** is a high water level that overflows the natural riverbank along any portion of a stream. Both floods and the floodplains they might occupy are rated statistically for the expected time intervals between floods, based upon historic data. Thus, you hear about "10-year floods," "50-year floods," and so on. A *10-year flood* is the greatest level of flooding that is likely to occur once every 10 years. This also means that such flooding has only a 10% likelihood of occurring in any one year and is likely to occur about 10 times each century. For any given floodplain, such a frequency indicates a moderate threat.

A 50-year or 100-year flood is of greater and perhaps catastrophic consequence, but it is also less likely to occur in a given year. These probability ratings of flood levels are mapped for an area, and the defined floodplains

17th St. Canal

FIGURE 14.31 Levee breach and break.
The Lakeview neighborhood is under water (to the right) from the 17th Street Canal levee failure, whereas the Metairie neighborhood is essentially dry (to the left). You see debris caught up north of the bridge where steel beams were placed in an attempt to curb flows. Flotsam, or floating debris, is visible in the photo. [Courtesy of NOAA aerial photography program.]

that result are then labeled as a "50-year floodplain" or a "100-year floodplain."

These statistical estimates are probabilities that events will occur randomly during any single year of the specified period. Of course, two decades might pass without a 50-year flood, or a 50-year level of flooding could occur 3 years in a row. The record-breaking Mississippi River Valley floods in 1993 easily exceeded a 1000-year flood probability, as did the 2005 catastrophe. See Focus Study 14.1 for more about floodplain hazards and management strategies.

Streamflow Measurement

Flood patterns in a drainage basin are as complex as the weather, for floods and weather are equally variable and both include a level of unpredictability. Measuring and analyzing the behavior of each large watershed and stream enables engineers and concerned parties to develop the best possible flood-management strategy. Unfortunately, reliable data often are not available for small basins or for the changing landscapes of urban areas.

The key to flood avoidance or management is to possess extensive measurements of *streamflow*, a stream's discharge, and how it performs during a precipitation event

(text continued on page 464)

Focus Study 14.1

Floodplain Strategies

Detailed measurements of stream-flows and flood records have been kept rigorously in the United States for only about 100 years, in particular since the 1940s. At any selected location along any given stream, the *probable maximum flood* (PMF) is a hypothetical flood of such a magnitude that there is virtually no possibility it will be exceeded. Because the collection and concentration of rainfall produce floods, hydrologists speak of a corollary, the *probable maximum precipitation* (PMP) for a given drainage basin, which is an amount of rainfall so great that it will never be exceeded.

Hydrologic engineers use these parameters to establish a design flood against which to take protective measures. For urban areas near creeks, planning maps often include survey lines for a 50-year or a 100-year floodplain; such maps have been completed for most U.S. urban areas. The design flood usually is used to enforce planning restrictions and special insurance requirements. Restrictive zoning using these floodplain designations is an effective way of avoiding potential damage.

Such political action is not always enforced, and the scenario sometimes goes like this: (1) Minimal zoning precautions are not carefully supervised, (2) a flooding disaster occurs, (3) the public is outraged at being caught off guard, (4) businesses and homeowners are surprisingly resistant to stricter laws and enforcement, (5) eventually another flood refreshes the memory and promotes more planning meetings and questions. As strange as it seems, there is little indication that our risk perception improves as the risk increases. (The floodplain managers' organization at **http://www.floods. org/** is a source of information.)

Bypass Channels as a Planning Strategy

A planning strategy used in some large river systems is to develop artificial floodplains; this is done by constructing

(a)

(b)

(c)

ANIMATION

Stream Processes, Floodplains

FIGURE 14.1.1 Weir and bypass channels to divert floodwaters.
(a) The Bonnet Carré spillway (a weir), on the lower Mississippi, New Orleans District, is opened in flood events to let water flow to Lake Pontchartrain and bypass New Orleans to the south. (b) The Sacramento weir awaits floodwaters from the Sacramento River, which it will direct through a channel into the Yolo Bypass visible in the distance in (c). When not flooded, the bypass is used for farmland crops. [Photos by Bobbé Christopherson.]

(continued)

Focus Study 14.1 (continued)

bypass channels to accept seasonal or occasional floods. When not flooded, the bypass channel can serve as farmland, often benefiting from the occasional soil-replenishing inundation. When the river reaches flood stage, large gates called *weirs* are opened, allowing the water to enter the bypass channel. This alternate route relieves the main channel of the burden of carrying the entire discharge.

Along the Lower Mississippi north of New Orleans, the Bonnet Carré spillway is such a weir, in operation in the photo (Figure 14.1.1a). In high-water flood events, water is let through a floodway into a bypass between a pair

FIGURE 14.1.2 A delta island.
Levees surround this delta island and its farms. The central portion of the island is below the river level. Note the former water channel and an oxbow on the land. [Photo by Bobbé Christopherson.]

of guide levees to relieve the flood discharge flow. The bypass diverts the flood 9.7 km (6 mi) into Lake Pontchartrain, as it has done seven times since 1937, to protect New Orleans. For more details, see **http://www.mvn.usace.army.mil/pao/bcarre/bcarre.htm**.

In the Sacramento Valley, California, such weirs lead to bypass channels for taking floodwaters past the metropolitan areas. In Figure 14.1.1b we see the Sacramento weir, and in Figure 14.1.1c the bypass is filled with floodwaters. Farther downstream in the Sacramento–San Joaquin delta, artificial levees and pumps attempt to keep farmlands dry that are actually below river level (Figure 14.1.2). To make matters worse, the area in the aerial photo is subsiding due to oxidation of peat in the soils, withdrawal of groundwater, and compaction. One of these delta islands flooded in a levee break in 2004.

Reservoir Considerations

Dams and reservoirs (an impoundment of water behind a dam) are common streamflow-control methods within a watershed. For conservation purposes, a dam holds back seasonal peak flows for distribution during low-water periods. In this way, streamflows are regulated to assure

(a)

(b)

FIGURE 14.1.3 Reservoir extremes.
Comparative photographs of the New Hogan reservoir, central California, during (a) dry and (b) wet weather conditions. Reservoirs help regulate runoff variability. [Photos by author.]

year-round water supplies. Dams also are constructed for flood control, to hold back excess flows for later release at more moderate discharge levels. Adding hydroelectric-power production to these functions of conservation and flood control can define a modern multipurpose reclamation project. Of all types there are approximately 45,000 large dams in the world and innumerable smaller ones. The World Commission on Dams is at **http://www.dams.org/**.

The function of reservoir impoundment is to provide flexible storage capacity within a watershed to regulate river flows, especially in a region with variable precipitation. Figure 14.1.3 shows one reservoir during drought conditions and during a time of wetter weather 6 years later.

Unfortunately, the multipurpose benefits of reservoir construction are countered by some negative consequences. The area upstream from a dam becomes permanently drowned. In mountainous regions, this may mean loss of white-water rapids and recreational sections of a river. In agricultural areas, the ironic end result may be that a hectare of farmland is inundated upstream to preserve a hectare of farmland downstream. Furthermore, dams built in warm and arid climates lose substantial water to evaporation, compared with the free-flowing streams they replace. Reservoirs in the southwestern United States can lose 3–4 m (10–13 ft) of water a year; the larger the surface area, the greater the loss. Also, sedimentation can reduce the effective capacity of a reservoir and can shorten a dam's life span.

A Final Thought About Floods

The benefit of any levee, bypass, or other project intended to prevent flood destruction is measured in avoided damage and is used to justify the cost of the protection facility. Thus, ever-increasing damage leads to the justification of ever-increasing flood-control structures. All such strategies are subjected to cost–benefit analysis, but bias is a serious drawback because such an analysis usually is prepared by an agency or bureau with a vested interest in building more flood-control projects.

As suggested in an article titled "Settlement Control Beats Flood Control,"* published over 50 years ago, there are other ways to protect populations than with enormous, expensive, sometimes environmentally disruptive projects. Strictly zoning the floodplain is one approach. However, the flat, easily developed floodplains near pleasant rivers are desirable for housing and thus weaken political resolve (Figure 14.1.4). A reasoned zoning strategy would set aside the floodplain for farming or passive recreation, such as a riverine park, golf course, or plant and wildlife sanctuary, or for other uses that are not hurt by natural floods. This study concludes that "urban and industrial losses would be largely obviated [avoided] by set-back levees and zoning and thus cancel the biggest share of the assessed benefits which justify big dams."

(a)

(b)

FIGURE 14.1.4 Living and building in a floodplain. (a) Near Harpers Ferry, Iowa, homes are located on a forested sandbar in the Mississippi River. (b) New fill is being dumped into the river to extend the land for further development. The river is at a high-water level in the photo. [Photos by Bobbé Christopherson.]

*Walter Kollmorgen, *Economic Geography* 29, no. 3 (July 1953): 215.

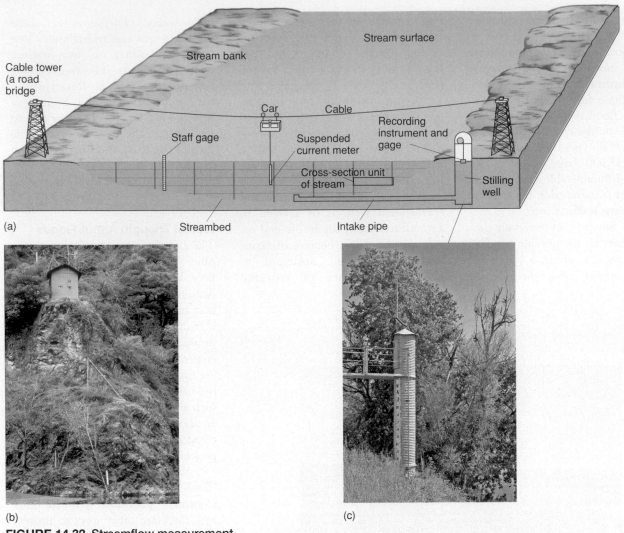

FIGURE 14.32 Streamflow measurement.
(a) A typical streamflow measurement installation may use a variety of devices: staff gage, stilling well with recording instrument, and suspended current meter. (b) An automated hydrographic station and (c) stilling well send telemetry to a satellite for collection by the USGS. [Photos (b) courtesy of California Department of Water Resources; (c) by Bobbé Christopherson.]

(Figure 14.32a). Once the cross section of a stream is fully measured, only the stream level is needed to estimate discharge (using the calculation: discharge = width × depth × velocity). A *staff gage* (a pole marked with water levels) is placed in a stream, as shown in the figure, to measure stream level. Another method involves a *stilling well* on the stream bank with a gage mounted in it to measure stream level. A movable current meter can be used to sample stream velocity at various locations.

Approximately 11,000 stream gaging stations are in use in the United States (an average of more than 200 per state). Of these, 7000 are operated by the U.S. Geological Survey and have continuous recorders for stage (level) and discharge (see **http://water.usgs.gov/pubs/circ/circ1123/**). Many of these stations automatically send telemetry data to satellites, from which information is retransmitted to regional centers (Figure 14.32b and c). Environment Canada's Water Survey of Canada maintains more than 3000 gaging stations (see **http://www.wsc.ec.gc.ca/**). Scientists hope that adequate funding to maintain such hydrologic monitoring by stream gaging networks will continue, for the data are essential to hazard assessment and proper planning, yet we find river monitoring systems fighting for their budgets.

Hydrographs A graph of stream discharge over time for a specific place is a **hydrograph**. The hydrograph in Figure 14.33a shows the relation between precipitation input (the bar graph) and stream discharge (the curves). During dry periods, at low-water stages, the flow is described as base flow and is largely maintained by input from local groundwater (dark blue line).

When rainfall occurs in some portion of the watershed, the runoff collects and is concentrated in streams and tributaries. The amount, location, and duration of the rainfall episode determine the *peak flow*. Also important is the nature of the surface in a watershed, whether permeable or impermeable. A hydrograph for a specific portion of a stream changes after a forest fire or following urbanization of the watershed.

Human activities have enormous impact on water flow in a basin. The effects of urbanization are quite dramatic,

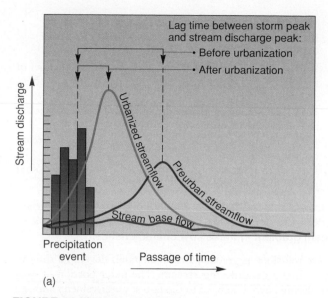

(b)

FIGURE 14.33 Urban flooding.
(a) Effect of urbanization on a typical stream hydrograph. Normal base flow is indicated with a dark blue line. The purple line indicates discharge after a storm, before urbanization. Following urbanization, stream discharge dramatically increases, as shown by the light blue line. (b) Severe flooding of an urban area in Linda, California, after a levee break on the Sacramento River in 1986. It flooded again in 1997. [(b) Photo from California Department of Water Resources.]

both increasing and hastening peak flow, as you can see by comparing preurban streamflow (purple curve) and urbanized streamflow (light blue) in Figure 14.33a. In fact, urban areas produce runoff patterns quite similar to those of deserts. The sealed surfaces of the city drastically reduce infiltration and soil-moisture recharge; their effect is similar to that of the usually hard surfaces in the desert. A significant part of urban flooding occurs because of the alteration in surfaces, which produces shortened concentration times; peak flows may strike with little warning. These issues will intensify as urbanization of vulnerable areas continues. The flooded region in Figure 14.33b was hit in 1986 and again in 1997.

Summary and Review—River Systems and Landforms

■ *Define* the term *fluvial*, and *explain* processes of fluvial erosion, transportation, and deposition.

River systems, fluvial processes and landscapes, floodplains, and river control strategies are important to human populations as demands for limited water resources increase. **Hydrology** is the science of water—its global circulation, distribution, and properties, specifically water at and below Earth's surface. Stream-related processes are **fluvial**. Water dislodges, dissolves, or removes surface material in the **erosion** process. Streams produce *fluvial erosion*, in which weathered sediment is picked up for **transport**, movement to new locations. Sediments are laid down by another process, **deposition**. **Alluvium** is the general term for the clay, silt, sand, gravel, or other unconsolidated rock and mineral fragments deposited by running water as sorted or semisorted sediment on a floodplain, delta, or streambed.

　Base level is the lowest elevation limit of stream erosion in a region. A *local base level* occurs when something interrupts the stream's ability to achieve base level, such as is created by a dam or a landslide that blocks a stream channel.

　　hydrology (p. 432)
　　fluvial (p. 432)
　　erosion (p. 432)
　　transport (p. 432)
　　deposition (p. 432)
　　alluvium (p. 432)
　　base level (p. 433)

1. What role is played by rivers in the hydrologic cycle?
2. What are the five largest rivers on Earth in terms of discharge? Relate these to the weather patterns in each area and to regional potential evapotranspiration (POTET) and precipitation (PRECIP)—concepts discussed in Chapter 9.
3. Define the term *fluvial*. What is a fluvial process?
4. What is the sequence of events that takes place as a stream dislodges material?
5. Explain the base-level concept. What happens to a local base level when a reservoir is constructed?

■ *Construct* a basic drainage basin model, and *identify* different types of drainage patterns and internal drainage, with examples.

The basic fluvial system is a **drainage basin**, which is an open system. *Drainage divides* define the drainage basin catchment (water-receiving) area of the drainage basin. In any drainage basin, water initially moves downslope in a thin film called **sheetflow**, or *overland flow*. This surface runoff concentrates in *rills*, or small-scale downhill grooves, which may develop into deeper *gullies* and a stream course in a valley. High ground that separates one valley from another and directs sheetflow is an *interfluve*. Extensive mountain and highland regions act as **continental divides** that separate major drainage basins. Some regions, such as the Great Salt Lake Basin, have **internal drainage** that does not reach the ocean, the only outlets being evaporation and subsurface gravitational flow.

Drainage density is determined by the number and length of channels in a given area and is an expression of a landscape's topographic surface appearance. **Drainage pattern** refers to the arrangement of channels in an area as determined by the steepness, variable rock resistance, variable climate, hydrology, relief of the land, and structural controls imposed by the landscape. Seven basic drainage patterns are generally found in nature: dendritic, trellis, radial, parallel, rectangular, annular, and deranged.

drainage basin (p. 434)
sheetflow (p. 434)
continental divides (p. 435)
internal drainage (p. 437)
drainage density (p. 437)
drainage pattern (p. 437)

6. What is the spatial geomorphic unit of an individual river system? How is it determined on the landscape? Define the several relevant key terms used.

7. In Figure 14.5, follow the Allegheny–Ohio–Mississippi River systems to the Gulf of Mexico. Analyze the pattern of tributaries and describe the channel. What role do continental divides play in this drainage?

8. Describe drainage patterns. Define the various patterns that commonly appear in nature. What drainage patterns exist in your hometown? Where you attend school?

■ *Describe* the relation among velocity, depth, width, and discharge, and *explain* the various ways that a stream erodes and transports its load.

Stream channels vary in *width* and *depth*. The streams that flow in them vary in velocity and in the *sediment load* they carry. All of these factors may increase with increasing discharge. **Discharge**, a stream's volume of flow per unit of time, is calculated by multiplying the velocity of the stream by its width and depth for a specific cross section of the channel. Most streams increase discharge downstream. But some streams originate in a humid region and flow through an arid region, such that discharge decreases with distance. Such an **exotic stream** is exemplified by the Nile River or the Colorado River.

Hydraulic action is the work of *turbulence* in the water. Running water causes hydraulic squeeze-and-release action to loosen and lift rocks and sediment. As this debris moves along, it mechanically erodes the streambed further through a process of **abrasion**.

Solution refers to the **dissolved load** of a stream, especially the chemical solution derived from minerals such as limestone or dolomite or from soluble salts. The **suspended load** consists of fine-grained, clastic particles held aloft in the stream, with the finest particles not deposited until the stream velocity slows nearly to zero. **Bed load** refers to coarser materials that are dragged and pushed and rolled along the streambed by **traction** or that bounce and hop along by **saltation**. If the load in a stream exceeds its capacity, sediments accumulate as **aggradation** as the stream channel builds through deposition. With excess sediment, a stream becomes a maze of interconnected channels that form a **braided stream** pattern.

discharge (p. 439)
exotic stream (p. 440)
hydraulic action (p. 442)
abrasion (p. 442)
dissolved load (p. 442)
suspended load (p. 442)
bed load (p. 442)
traction (p. 442)

saltation (p. 442)
aggradation (p. 443)
braided stream (p. 443)

9. What was the impact of flood discharge on the channel of the San Juan River near Bluff, Utah? Why did these changes take place?

10. How does stream discharge do its erosive work? What are the processes at work in the channel?

11. Differentiate between stream competence and stream capacity.

12. How does a stream transport its sediment load? What processes are at work?

■ *Develop* a model of a meandering stream, including point bar, undercut bank, and cutoff, and *explain* the role of stream gradient in these flow characteristics.

Where the slope is gradual, stream channels develop a sinuous form called a **meandering stream**. The outer portion of each meandering curve is subject to the fastest water velocity and can be the site of a steep **undercut bank**. On the other hand, the inner portion of a meander experiences the slowest water velocity and forms a **point-bar** deposit. When a meander neck is cut off as two undercut banks merge, the meander becomes isolated and forms an **oxbow lake**.

Every stream develops its own **gradient** and establishes a longitudinal profile. A portion of the stream is designated a **graded stream** when that stream portion is balanced among available discharge, channel characteristics, its velocity, and the load supplied from the drainage basin. An interruption in a stream's longitudinal profile is called a **nickpoint**. A nickpoint can occur as the stream flows across hard, resistant rock or after tectonic uplift episodes.

meandering stream (p. 444)
undercut bank (p. 444)
point bar (p. 444)
oxbow lake (p. 445)
gradient (p. 445)
graded stream (p. 446)
nickpoint (p. 446)

13. Describe the flow characteristics of a meandering stream. What is the pattern of flow in the channel? What are the erosional and depositional features and the typical landforms created?

14. Explain these statements: (a) All streams have a gradient, but not all streams are graded. (b) Graded streams may have ungraded segments.

15. Why is Niagara Falls an example of a nickpoint? Without human intervention, what do you think will eventually take place at Niagara Falls?

16. What is meant by "the validity of cyclic or equilibrium models depends on which of three time frames is being considered"? Explain and discuss.

■ *Define* a floodplain, and *analyze* the behavior of a stream channel during a flood.

Floodplains have been an important site of human activity throughout history. Rich soils, bathed in fresh nutrients by floodwaters, attract agricultural activity and urbanization. Despite our historical knowledge of devastation by floods, floodplains are settled, raising issues of human perception of hazard. The flat low-lying area along a stream channel that is subjected to recurrent flooding is a **floodplain**. It is formed when the river overflows its channel during times of high flow. On either bank of some streams, **natural levees** develop as by-products of flooding. On the floodplain, backswamps and yazoo tributaries may develop. The natural levees and elevated channel of the river prevent a **yazoo tributary** from joining the main channel, so it flows parallel to the river and through the **backswamp** area. Entrenchment of a river into its own floodplain forms **alluvial terraces**.

> floodplain (p. 450)
> natural levees (p. 450)
> yazoo tributary (p. 452)
> backswamp (p. 452)
> alluvial terraces (p. 452)

17. Describe the formation of a floodplain. How are natural levees, oxbow lakes, backswamps, and yazoo tributaries produced?

18. Identify any of the features listed in Question 17 on the Philipp, Mississippi, topographic quadrangle in Figure 14.23d.

19. Describe any floodplains near where you live or where you go to college. Have you seen any of the floodplain features discussed in this chapter? If so, which ones?

■ *Differentiate* the several types of river deltas, and *detail* each.

A depositional plain formed at the mouth of a river is called a **delta**. When the mouth of a river enters the sea and is inundated by seawater in a mix with freshwater, it is called an **estuary**.

> delta (p. 453)
> estuary (p. 453)

20. What is a river delta? What are the various deltaic forms? Give some examples.

21. Based on materials in this chapter and in Chapter 8, assess what occurred in New Orleans. How might New Orleans change later in this century? Explain. In your opinion did

Hurricane Katrina cause the disaster or was it a result of engineering failures and hazard-planning shortfalls?

22. Describe the Ganges River delta. What factors upstream explain its form and pattern? Assess the consequences of settlement on this delta.

23. What is meant by the statement "the Nile River delta is disappearing"?

■ *Explain* flood probability estimates, and *review* strategies for mitigating flood hazards.

A **flood** occurs when high water overflows the natural riverbank along any portion of a stream. Both floods and the floodplains they occupy are rated statistically for the expected time interval between floods. A 10-year flood is the greatest level of flooding that is likely once every 10 years. A graph of stream discharge over time for a specific place is called a **hydrograph**.

Collective efforts by government agencies undertake to reduce flood probability. Such management attempts include the construction of artificial levees, bypasses, straightened channels, diversions, dams, and reservoirs. Society is still learning how to live in a sustainable way with Earth's dynamic river systems.

> flood (p. 460)
> hydrograph (p. 464)

24. Specifically, what is a flood? How are such flows measured and tracked?

25. Differentiate between a hydrograph from a natural terrain and one from an urbanized area.

26. What do you see as the major consideration regarding floodplain management? How would you describe the general attitude of society toward natural hazards and disasters?

27. What do you think the author of the article "Settlement Control Beats Flood Control" meant by the title? Explain your answer, using information presented in the chapter.

NetWork

The *Geosystems* Student Learning Center provides on-line resources for this chapter on the World Wide Web. To begin: Once at the Center, click on the cover of this textbook, scroll the Table of Contents menu, and select this chapter. You will find self-tests that are graded, review exercises, specific

updates for items in the chapter, and in "Destinations" many links to interesting related pathways on the Internet. *Geosystems* Student Learning Center is found at **http://www.prenhall.com/christopherson/**.

Critical Thinking

A. Determine the name of the drainage basin within which your campus is located. Where are its headwaters? Where is the river's mouth? If you are in the United States or Canada, use Figure 14.5 to locate the larger drainage basins and divides for your region. Is there any regulatory organization that oversees planning and coordination for the drainage basin you identified? Does the library or your geography department have topographic maps on file? After examining your region's map, can you discern a prominent drainage pattern for the area (Figure 14.9)?

B. Using your search engine or browser on the Internet, type in "New Orleans flooding, 2005," or "Gulf Coast damage, Katrina," or "New Orleans levee failures." Allow some time

to sample through at least 10 of the links. Describe what you found. Did you find photo galleries, engineering reports, or news coverage? What is your opinion about this region of the lower Mississippi River, New Orleans, and the Gulf Coast cities and the direction they should take at this point in history: rebuild things as they were; rethink the hazards and adopt a new strategy of replacing urban with low-density recreation zones; establish possible set-back hazard zoning; walk away except for necessary port functions; continue pumping oil and gas and order corporations to repressurize the fields to raise ground levels; or abandon the region? Are there other options you think are important? Remember, as you think this through, sea level is still rising at an unprecedented rate.

Ancient sand dunes in the late Jurassic Period (150 m.y.a.) lithified, forming this colorful sandstone. Weathering and wind processes continue to sculpt these rock surfaces in myriad shapes, textures, and colors. Such is the beauty to be found in Earth's arid lands, here in Valley of Fire State Park, southern Nevada. [Photos by Bobbé Christopherson.]

Eolian Processes and Arid Landscapes

■ Key Learning Concepts_____

After reading the chapter, you should be able to:

- ■ *Characterize* the unique work accomplished by wind and eolian processes.

- ■ *Describe* eolian erosion, including deflation, abrasion, and the resultant landforms.

- ■ *Describe* eolian transportation, and *explain* saltation and surface creep.

- ■ *Identify* the major classes of sand dunes, and *present* examples within each class.

- ■ *Define* loess deposits and their origins, locations, and landforms.

- ■ *Portray* desert landscapes, and *locate* these regions on a world map.

W ind is an agent of geomorphic change. Like moving water, moving air causes erosion, transportation, and deposition of materials; moving air is a fluid and it behaves similarly, although it has a lower viscosity (it is less dense) than water. Although lacking the lifting ability of water, wind processes can modify and move quantities of materials in deserts, along coastlines, and elsewhere.

Wind processes modify rocks and move sediment in deserts and along coastlines in a variety of climates. Wind may contribute to soil formation in places far distant from the point of origin by bringing fine material from regions where glaciers deposited it. Elsewhere, fallow fields give up their soil resource to wind erosion. Scientists are only now getting an accurate picture of the amount of windblown dust that fills the atmosphere and crosses the oceans between continents. Chemical fingerprints and satellites are used to trace windblown dust from African soils to North and South America and from Asian landscapes to Europe. Winds even spread living organisms. One study found related mosses, liverworts, and lichens distributed over thousands of kilometers between islands in the Southern Ocean.

Earth's dry lands stand out in stark contrast on the water planet. Arid landscapes display unique landforms and life forms: "Instead of finding chaos and disorder the observer never fails to be amazed at a simplicity of form, an exactitude of repetition and a geometric order in the desert."* Although perpetually cold, the polar regions are deserts as well, receiving little precipitation, so there should be no surprise that Yuma, Arizona, and weather stations in Antarctica receive the same amount of annual precipitation.

In this chapter: We examine the work of wind, associated erosion, transport, and depositional processes, and resulting landforms. Windblown fine particles form vast loess deposits, the basis for rich agricultural soils. We include the discussion of arid lands in this chapter. Most deserts are rocky and covered with desert pavement, whereas other dry landscapes are covered in sand dunes. In desert environments, landscapes lacking moisture and stabilizing vegetation are set off in sharp, sun-baked relief. Water is the major erosional agent in the desert, yet water is the limiting resource for human development. We examine the causes and distribution of Earth's desert landscapes. The troubled Colorado River and the western drought are the subject of an important Focus Study.

The Work of Wind

The work of the wind—erosion, transportation, and deposition—is called **eolian** (also spelled *aeolian*). The word comes from Aeolus, ruler of the winds in Greek mythology. British Army major Ralph Bagnold, an engineering officer, while stationed in Egypt in 1925 accomplished much eolian research. In the deserts west of the Nile, Bagnold measured, sketched, and developed hypotheses about the wind and desert forms. Imagine the scene in the 1920s: A Model-T Ford chugs across the desert west of Cairo taking Bagnold to his research area. He laid rolls of chicken wire over treacherous stretches of sand to prevent getting stuck. Bagnold's often-cited work *The Physics of Blown Sand and Desert Dunes* was published in 1941.

The ability of wind to move materials is actually small compared with that of other transporting agents such as water and ice, because air is so much less dense than those other media. Yet, over time, wind accomplishes enormous

*R. A. Bagnold, *The Physics of Blown Sand and Desert Dunes* (London: Methuen, 1941).

FIGURE 15.1 Sand movement and wind velocity.
Sand movement relative to wind velocity, as measured over a meter cross section of ground surface. [After R. A. Bagnold, *The Physics of Blown Sand and Desert Dunes* (London: Methuen, 1941). Adapted by permission.]

work. Bagnold studied the ability of wind to transport sand over the surface of a dune. Figure 15.1 shows that a steady wind of 50 kmph (30 mph) can move approximately one-half ton of sand per day over a meter cross section of dune. The graph also demonstrates how rapidly the amount of transported sand increases with wind speed.

Grain size is important in wind erosion. Intermediate-sized grains move most easily—they bounce along. It is the largest and the smallest sand particles that require the strongest winds to move. The large particles are heavier and thus require stronger winds. Small particles are difficult to move because they exhibit a mutual cohesiveness and because they usually present a smooth surface (aerodynamic) to the wind. Yet, the finest dust, once aloft, is carried from continent to continent (Figure 15.2). In addition, wind can prune and shape vegetation, especially where winds are strong in a consistent direction.

Eolian Erosion

Two principal wind-erosion processes are **deflation**, the removal and lifting of individual loose particles, and **abrasion**, the grinding of rock surfaces by the "sandblasting" action of particles captured in the air. Deflation and abrasion produce a variety of distinctive landforms and landscapes.

Deflation Deflation literally blows away loose or noncohesive sediment and works with rainwater to form a surface resembling a cobblestone street: a **desert pavement**

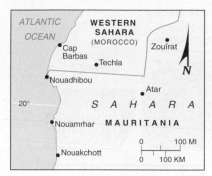

FIGURE 15.2 The work of the wind.
Streaks of windblown sands cover bedrock south of Atar, in central Mauritania. Note the circular Richat dome structure, just east of the Western Sahara political border corner (refer to Figure 12.10c). [*Terra* image courtesy of NASA Earth Observatory.]

that protects underlying sediment from further deflation and water erosion. Traditionally, deflation was regarded as a key formative process, eroding fine dust, clay, and sand away, leaving behind a concentration of pebbles and gravel as desert pavement (Figure 15.3a).

Another hypothesis that better explains some desert pavement surfaces states that deposition of windblown sediments is the formative agent, not removal. Windblown particles settle between and below coarse rocks and pebbles that are gradually displaced upwards. Rainwater is involved as wetting and drying episodes swell and shrink clay-sized particles. The gravel fragments are gradually lifted to surface positions to form the pavement (Figure 15.3b).

Desert pavements are so common that many provincial names are used for them—for example, *gibber plain* in Australia; *gobi* in China; and in Africa, *lag gravels* or *serir*, or *reg* desert if some fine particles remain. Large depressions in the Sahara Desert are at least partially formed by deflation. The enormous Munkhafad el Qaṭṭâra (Qaṭṭâra Depression), which covers 18,000 km² (6950 mi²) just inland from the Mediterranean Sea in the Western Desert of Egypt, is now about 130 m (427 ft) below sea level at its lowest point.

Heavy recreational activity damages fragile desert landscapes, especially in the arid lands of the United States, where more than 15 million off-road vehicles (ORVs) are now in use. Such vehicles crush plants and animals; disrupt desert pavement, leading to greater deflation; and create ruts that easily concentrate sheetwash to form gullies.

Military activities can also threaten desert landscapes. A serious environmental impact of both the 1991 Persian Gulf War and ongoing Iraq War is the disruption of desert pavement. Thousands of square kilometers of stable desert pavement have been shattered by the bombardment of hundreds of thousands of tons of explosives and disrupted by the movement of heavy vehicles. The resulting silt made available for deflation has plagued cities and farms ever since with increased dust and sand accumulations.

Wherever wind encounters loose sediment, deflation may remove enough material to form basins called **blowout depressions**. These depressions range from small indentations less than a meter wide up to areas hundreds of meters wide and many meters deep. Chemical weathering, although slow in the desert owing to the lack of water, is important in the formation of a blowout, for it removes the cementing materials that give particles their cohesiveness. In arid climates chemical weathering is active on the surfaces of particles at the microscopic level, utilizing hygroscopic and capillary water in reactions.

Abrasion You may have seen work crews sandblasting surfaces on buildings, bridges, or streets to clean them or to remove unwanted markings. Sandblasting uses a stream of compressed air filled with sand grains to quickly abrade a surface. Abrasion by windblown particles is nature's slower version of sandblasting, and it is especially effective at polishing exposed rocks when the abrading particles are hard and angular. Variables that affect the rate of abrasion include the hardness of surface rocks, wind velocity, and wind constancy. Abrasive action is restricted to the area immediately above the ground, usually no more than a meter or two in height, because sand grains are lifted only a short distance.

Rocks exposed to eolian abrasion appear pitted, fluted (grooved), or polished. They usually are aerodynamically shaped in a specific direction, according to the consistent flow of airborne particles carried by prevailing winds. Rocks that have such evidence of eolian erosion are **ventifacts** (literally, "artifacts of the wind"). On a larger scale, deflation and abrasion are capable of streamlining rock structures that are aligned parallel to the most effective wind direction, leaving behind distinctive, elongated ridges or formations called **yardangs**. These wind-sculpted features can range from meters to kilometers in length and up to many meters in height.

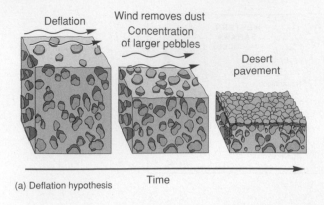

Deflation

Wind removes dust
Concentration
of larger pebbles

Desert
pavement

Time

(a) Deflation hypothesis

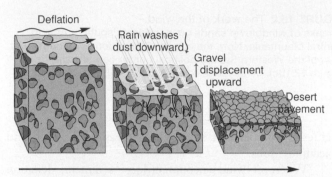

Deflation

Rain washes
dust downward

Gravel
displacement
upward

Desert
pavement

(b) Sediment-accumulation hypothesis

(c)

FIGURE 15.3 Desert pavement.
(a) Desert pavement may be formed when larger rocks and fragments are left after deflation and sheetwash remove finer dust, silt, and clays. (b) Other desert pavements seem to be formed by wind delivering fine particles that settle and wash downward; through cycles of swelling and shrinking, the gravels gradually migrate upward to form pavement. (c) A typical desert pavement. [Photo by Bobbé Christopherson.]

On Earth, some yardangs are large enough to be detected on satellite imagery. The Ica Valley of southern Peru contains yardangs reaching 100 m (330 ft) in height and several kilometers in length, and yardangs in the Lûut Desert of Iran attain 150 m (490 ft) in height. Abrasion is concentrated on the windward end of each yardang, with deflation operating on the leeward portions (Figure 15.4).

FIGURE 15.4 A yardang.
A small wind-sculpted rock formation located in Snow Canyon outside St. George, Utah. [Photo by Bobbé Christopherson.]

The Sphinx in Egypt was perhaps partially formed as a yardang, suggesting a head and body. Some scientists think this shape led the ancients to complete the bulk of the sculpture artificially with masonry.

Eolian Transportation

Atmospheric circulation can transport fine material, such as volcanic debris, fire soot and smoke, and dust, worldwide within days. Wind exerts a drag, or frictional pull, on surface particles until they become airborne, just as water in a stream picks up sediment (again, think of air as a fluid). The distance that wind is capable of transporting particles in suspension varies greatly with particle size. Only the finest dust particles travel significant distances, so the finer material suspended in a dust storm is lifted much higher than the coarser particles of a sandstorm.

People living in areas of frequent dust storms are faced with infiltration of very fine particles into their homes and businesses through even the smallest cracks. Figure 3.9 illustrates such dust storms in the Nevada desert, a dust plume from orbit, and alkali dust near Mono Lake, California. People living in desert regions and along sandy beaches, where frequent sandstorms occur, contend with the sandblasting of painted surfaces and etched window glass.

Both human and natural sand erosion and transport from a beach are slowed by conservation measures such as the introduction of stabilizing native plants, the use of fences, and the restriction of pedestrian traffic to walkways (Figure 15.5). As human settlement encroaches on coastal dunes, sand transport becomes problematic. As an example, at Nags Head, North Carolina, the Jocky's

FIGURE 15.5 Preventing sand transport.
(a) Popham Beach State Park, Maine. Further erosion and transport of coastal dunes can be controlled by stabilizing strategies, planting native plants, and by confining pedestrian traffic to walkways. (b) Tufts of grass are planted to stabilize this Atlantic City, New Jersey, beach with a fence installed to prevent human foot traffic. [Photos by Bobbé Christopherson.]

(a)

(b)

Ridge dune, some 43 m (140 ft) in height, is actively migrating over roads and yards.

The term *saltation* was used in Chapter 14 to describe movement of particles by water. The term also describes the wind transport of grains, usually larger than 0.2 mm (0.008 in.), along the ground. About 80% of wind transport of particles is accomplished by this skipping and bouncing action (Figure 15.6). Compared with fluvial transport, in which saltation is accomplished by hydraulic lift, eolian saltation is executed by aerodynamic lift, elastic bounce, and impact (compare Figure 15.6a with Figure 14.13).

On impact, grains hit other grains and knock them into the air. Saltating particles crash into other particles, knocking them both loose and forward. This type of movement is **surface creep**, which slides and rolls particles too large for saltation and affects about 20% of the material being transported. Once in motion, particles continue to be transported by lower wind velocities. In a desert or along a beach, sometimes you can hear a slight hissing sound, almost like steam escaping, produced by the myriad saltating grains of sand as they bounce along and collide with surface particles.

Through processes of weathering, erosion, and transportation, mineral grains are removed from parent rock and redistributed elsewhere. In Chapter 13, Figure 13.4c, you saw the relation between the composition and color

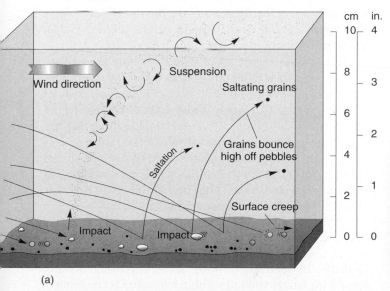

(a)

(b)

FIGURE 15.6 How the wind moves sand.
(a) Eolian suspension, saltation, and surface creep are mechanisms of sediment transportation. Compare with the saltation and traction that occur in another fluid, water, in Figure 14.13. (b) Sand grains saltating along the surface in the Stovepipe Wells dune field, Death Valley. [Photo by author.]

 ANIMATION

How Wind
Moves
Sediment

FIGURE 15.7 Sand ripples.
Sand ripple patterns later may become lithified into fixed patterns in rock. The area in the photo looks vast but is only about 1 m (3.3 ft) wide. [Photo by author.]

of the sandstone in the background and the derived sandy surface in the foreground. Wind action is not that significant in the weathering process that frees individual grains of sand from the parent rock, but it is active in relocating the weathered grains.

Eolian Depositional Landforms

The smallest features shaped by individual saltating grains are ripples (Figure 15.7). Ripples form in crests and troughs, positioned transversely (at a right angle) to the direction of the wind. Their formation with each wind-blown bounce is influenced by the length of time saltating particles are airborne. Eolian ripples are similar to fluvial ripples, although the impact of saltating grains is slight in water.

Many people have never been in a desert; they know this arid landscape only from the movies, which leave the impression that most deserts are covered by sand. Instead, desert pavements predominate across most subtropical arid landscapes; only about 10% of desert areas are covered with sand. Sand grains generally are deposited as transient ridges or hills called dunes.

A **dune** is a wind-sculpted accumulation of sand. An extensive area of dunes, such as that found in North Africa, is characteristic of an **erg desert**, or **sand sea**. The Grand Erg Oriental in the central Sahara exceeds 1200 m (4000 ft) in depth and covers 192,000 km² (75,000 mi²), comparable to the area of Nebraska. This sand sea has been active for more than 1.3 million years and has average dune heights of 120 m (400 ft). The Sahara Marzūq sand sea shown in Figure 15.8a is wider than 300 km (185 mi). On the Martian surface, eolian processes are at work forming similar parallel ridges along the floor of Melas Chasma in Figure 15.8b. Similar sand seas, such as the Grand Ar Rub'al Khālī Erg, are active in Saudi Arabia.

Many types of dunes in arid regions or along coastlines tend to migrate in the direction of strong prevailing winds. Strong seasonal winds or winds from a passing storm may sometimes prove more effective than average prevailing winds. When saltating sand grains encounter small patches of sand, their kinetic energy (motion) is dissipated and they start to accumulate; a dune is born. As height increases above 30 cm (12 in.), a steeply sloping slipface on the lee side of the dune and characteristic dune features form.

By studying the dune model in Figure 15.9 you can see that winds characteristically create a gently sloping *windward side* (stoss side), with a more steeply sloped **slipface** on the *leeward side*. A dune usually is asymmetrical in one or more directions. The angle of a slipface is the steepest angle at which loose material is stable—its *angle of repose*. Thus, the constant flow of new material makes a slipface a type of *avalanche slope*. Sand builds up as it moves over the crest of the dune to the brink; then it avalanches, falling and cascading as the slipface continually adjusts, seeking its angle of repose (usually 30° to 34°). In this way, a dune migrates downwind, as suggested by the successive dune profiles in Figure 15.9.

Dunes have many wind-shaped styles that make classification difficult. We simplify dune forms into three classes—*crescentic* (crescent, curved shape), *linear* (straight forms), and massive *star dunes*. Each is summarized in Figure 15.10. The ever-changing form of these eolian deposits is part of their beauty, eloquently described by one author: "I see hills and hollows of sand like rising and falling waves. Now at midmorning, they appear paper white. At dawn they were fog gray. This evening they will be eggshell brown."[*]

Star dunes are the mountainous giants of the sandy desert. They form in response to complicated, changing wind patterns and have multiple slipfaces. They are pinwheel-shaped, with several radiating arms rising and joining to form a common central peak. The best examples of star dunes are in the Sahara and the Namib Desert, where they approach 200 m (650 ft) in height (Figure 15.11). Some in the Sonoran Desert approach this size.

The map in Figure 15.12 shows the correlation of active sand regions with deserts (tropical, continental interior, and coastal). Note the limited extent of desert area covered by active sand dunes—only about 10% of all continental land between 30° N and 30° S. Also noted on the map are dune fields in humid climates outside of this range, such as along coastal Oregon, the south shore of Lake Michigan (Figure 15.12b), along the Gulf and Atlantic coastlines, in Europe, and elsewhere. The remarkable coastal desert sands of Namibia are pictured in Figure 15.12c.

These same dune-forming principles and terms (for example, *dune* and *slipface*) apply to snow-covered landscapes. *Snow dunes* are formed as wind deposits snow in drifts. In semiarid farming areas, capturing drifting snow with fences and tall stubble left in fields contributes significantly to soil moisture when the snow melts.

[*]J. E. Bowers, *Seasons of the Wind* (Flagstaff, AZ: Northland Press, 1985), p. 1.

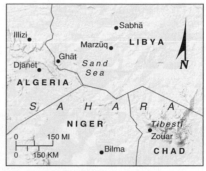

(a)

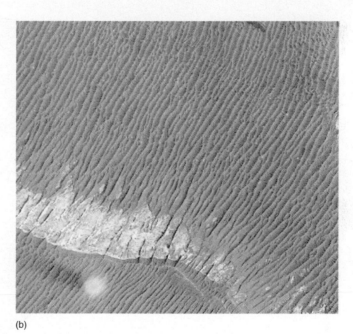

(b)

FIGURE 15.8 A sand sea.
(a) The Sahara Marzūq, an erg desert that dominates southwestern Libya. Effective northwesterly winds shape the pattern and direction of the transverse and barchanoid (series of connected barchans) dunes. This sand sea exceeds 300 km across. (b) A similar pattern of dunes appears on the Martian surface in the southern area of Melas Chasma in Valles Marineris. Note the dust devil in the lower left. The area covered is about 2 km wide. [(a) *Terra* image courtesy of MODIS Land Rapid Response Team, NASA/GSFC, November 9, 2001. (b) Mars Global Surveyor, Mars Orbiter Camera image courtesy of NASA/JPL/Malin Space Science Systems, July 11, 1999.]

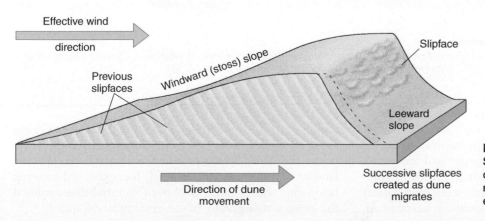

Effective wind direction

Previous slipfaces

Windward (stoss) slope

Slipface

Leeward slope

Direction of dune movement

Successive slipfaces created as dune migrates

ANIMATION

Formation of Cross Bedding

FIGURE 15.9 Dune cross section. Successive slipfaces exhibit a distinctive pattern as the dune migrates in the direction of the effective wind.

Class	Type	Description
Crescentic	Barchan	Crescent-shaped dune with horns pointed downwind. Winds are constant with little directional variability. Limited sand available. Only one slipface. Can be scattered over bare rock or desert pavement or commonly in dune fields.
	Transverse	Asymmetrical ridge, transverse to wind direction (right angle). Only one slipface. Results from relatively ineffective wind and abundant sand supply.
	Parabolic	Role of anchoring vegetation important. Open end faces upwind with U-shaped "blowout" and arms anchored by vegetation. Multiple slipfaces, partially stabilized.
	Barchanoid ridge	A wavy, asymmetrical dune ridge aligned transverse to effective winds. Formed from coalesced barchans; looks like connected crescents in rows with open areas between them.

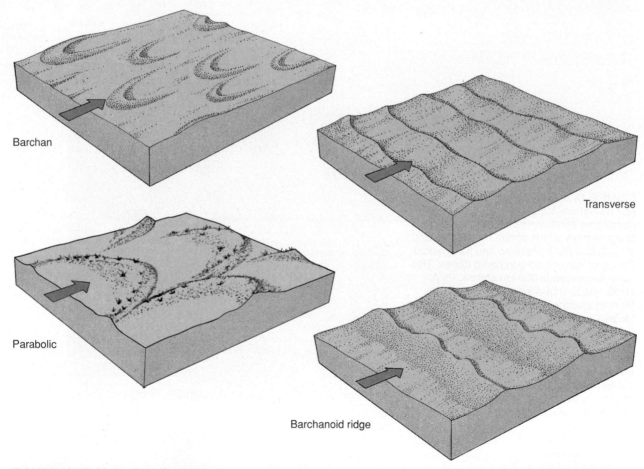

Barchan

Transverse

Parabolic

Barchanoid ridge

FIGURE 15.10 Major dune forms.
Please see table for dune descriptions. Arrows show wind direction. [Adapted from E. D. McKee, *A Study of Global Sand Seas*, USGS Professional Paper 1052 (Washington, DC: U.S. Government Printing Office, 1979).]

Loess Deposits

Approximately 15,000 years ago, in several episodes, Pleistocene glaciers retreated in many parts of the world, leaving behind large glacial outwash deposits of fine-grained clays and silts. These materials were blown great distances by the wind and redeposited in unstratified, homogeneous (evenly mixed) deposits. Peasants working along the Rhine River Valley in Germany gave these deposits the name **loess** (pronounced "luss"). No specific landforms were created; instead, loess covered existing landforms with a thick blanket of material that assumed the general topography of the existing landscape.

Class	Type	Description
Linear	Longitudinal	Long, slightly sinuous, ridge-shaped dune, aligned parallel with the wind direction; two slipfaces. Average 100 m high and 100 km long, and the "draa" at the extreme in size can reach to 400 m high. Results from strong effective winds varying in one direction.
	Seif	After Arabic word for "sword"; a more sinuous crest and shorter than longitudinal dunes. Rounded toward upwind direction and pointed downwind. (Not illustrated.)
Star dune		The giant of dunes. Pyramidal or star-shaped with three or more sinuous, radiating arms extending outward from a central peak. Slipfaces in multiple directions. Results from effective winds shifting in all directions. Tend to form isolated mounds in high effective winds and connected sinuous arms in low effective winds.
Other	Dome	Circular or elliptical mound with no slipface. Can be modified into barchanoid forms; vegetation can play role in coastal locales.
	Reversing	Asymmetrical ridge form intermediate between star dune and transverse dune. Wind variability can alter shape between forms.

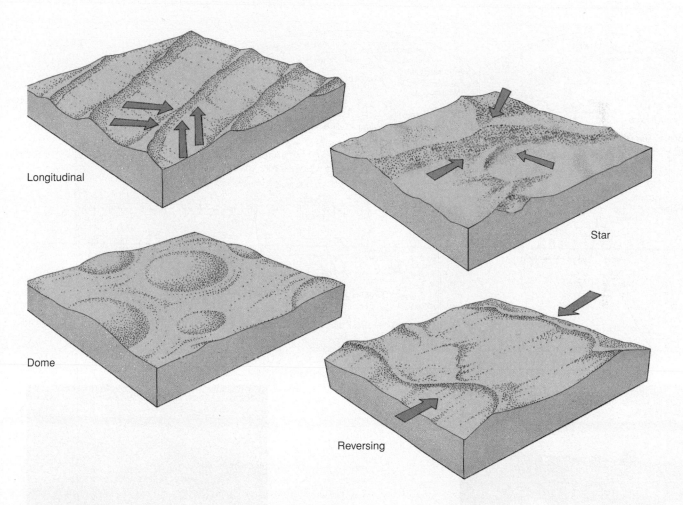

Because of its own binding strength and its internal coherence, loess weathers and erodes into steep bluffs, or vertical faces. At Xi'an, Shaanxi Province, China, a loess wall was excavated for dwelling space (Figure 15.13a). When a bank is cut into a loess deposit, it generally will stand vertically, although it can fail if saturated (Figure 15.13b). There are historical accounts of Civil War soldiers in the Vicksburg, Mississippi, area excavating places to live in loess banks.

Figure 15.14a (p. 480) shows the worldwide distribution of loess deposits. Significant accumulations throughout the Mississippi and Missouri river valleys form continuous deposits 15–30 m (50–100 ft) thick. Loess deposits also occur in eastern Washington State and Idaho. The hills and gullies in the Loess Hills of Iowa in Figure 15.14b demonstrate the erosion potential of these soils. They reach heights of about 61 m (200 ft) above the nearby

FIGURE 15.11 Sand mountains of the desert.
Star dune in the Namib Desert in southwestern Africa. [Photo by Comstock.]

prairie farmlands and Missouri River and run north and south more than 322 km (200 mi). Only China has deposits that exceed these dimensions. This silt explains the fertility of the soils in these regions, for loess deposits are well drained and deep and have excellent moisture retention. Loess deposits also cover much of Ukraine, central Europe, China, the Pampas–Patagonia regions of Argentina, and lowland New Zealand. The soils derived from loess help form some of Earth's "breadbasket" farming regions. Transport of loess soils occurred in the catastrophic 1930s Dust Bowl in the United States; see News Report 15.1.

In Europe and North America, loess is thought to be derived mainly from glacial and periglacial sources. The vast deposits of loess in China, covering more than 300,000 km² (116,000 mi²), are derived from windblown desert sediment rather than glacial sources. Accumulations in the Loess Plateau of China exceed 300 m (1000 ft) in thickness, forming complex weathered badlands and

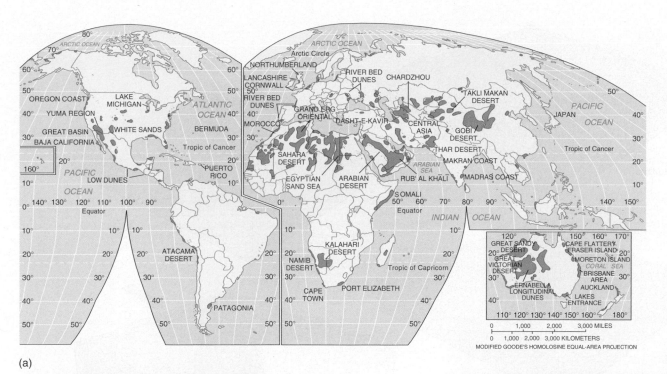

(a)

(b)

(c)

FIGURE 15.12 Sandy regions of the world.
(a) Worldwide distribution of active and stable sand regions. (b) Sand dunes along the shore of Lake Michigan in Indiana Dunes State Park, Indiana. (c) Sandy area in the Namib-Naukluft Park, Namib Desert, Namibia, Africa. [(a) After R. E. Snead, *Atlas of World Physical Features*, p. 134, © 1972 by John Wiley & Sons. Adapted by permission. (b) photo by Bobbé Christopherson; (c) photo by Nigel J. Dennis/Photo Researchers, Inc.]

(a)

(b)

FIGURE 15.13 Example of loess deposits.
(a) Loess formation in Xi'an, Shaanxi Province, China, has sufficient structural strength to permit excavation for dwelling rooms. (b) A loess bluff in western Iowa. [(a) Photo by Betty Crowell; (b) photo by Bobbé Christopherson.]

some good agricultural land. These windblown deposits are interwoven with much of Chinese history and society.

Overview of Desert Landscapes

Dry climates occupy about 26% of Earth's land surface. If all semiarid climates are considered, perhaps as much as 35% of all land area tends toward dryness, constituting the largest single climatic region on Earth (see Figures 10.4 and 10.5 for the location of these *arid deserts* and *semiarid steppe* climate regions and Figure 20.4 for the distribution of these desert environments).

Deserts occur worldwide as topographic plains, such as the Great Sandy and Simpson deserts of Australia, the Arabian and Kalahari deserts of Africa, and portions of the extensive Taklimakan Desert, which covers some 270,000 km² (105,000 mi²) in the central Tarim Basin of China. Deserts also are found in mountainous regions:

News Report 15.1

The Dust Bowl

Deflation and wind transport of soils including loess produced a catastrophe in the American Great Plains in the 1930s—the Dust Bowl. More than a century of overgrazing and intensive agriculture left soil susceptible to drought and eolian deflation processes. Reduced precipitation amounts and above-normal temperatures triggered the multiyear disaster. The deflation of many centimeters of soil occurred in southern Nebraska, Kansas, Oklahoma, Texas, eastern Colorado, and even in southern Canada and northern Mexico.

Fine sediments were lifted by winds to form severe dust storms. The transported dust darkened the skies of Midwestern cities and drifted over farmland to depths that covered failing crops. Streetlights were left on throughout the day in Kansas City, St. Louis, and other Midwestern cities and towns. Such episodes can devastate economies, cause tremendous loss of topsoil, and even bury farmsteads. Southeastern Australia experienced severe dust storms in 1993 that included consequences similar to those of the American Dust Bowl. For more on the Dust Bowl, see **http://www.pbs.org/wgbh/amex/dustbowl/**.

Scientists are seeking answers as to why this took place when it did.

Tree-ring analysis shows that a drought of that intensity happened about twice a century over the past 400 years. One such drought in what is now called Nebraska in the thirteenth century appears to have been as intense as that in the 1930s; however, it lasted nearly 40 years. The Dust Bowl weather conditions seem to correlate with cool Pacific Ocean sea-surface temperatures (SST) and warm Atlantic Ocean SST. Chapter 6 discussed the Pacific Decadal Oscillation (PDO) and its relation to drought occurrences in the United States.

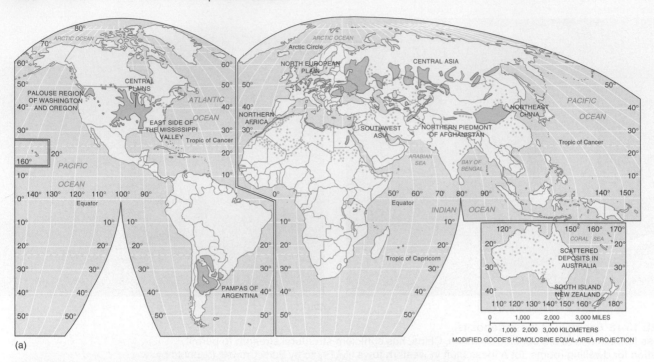

(a)

(b)

FIGURE 15.14 Loess deposits worldwide and in Iowa.
(a) Areas of major accumulations of loess. Dots represent small, scattered loess formations. (b) The Loess Hills of western Iowa give evidence of the last ice age, as the finely ground sediments left by the glaciers were windblown to this setting. [(a) After R. E. Snead, *Atlas of World Physical Features*, p. 138, © 1972 by John Wiley & Sons. Adapted by permission; (b) photos by Bobbé Christopherson.]

interior Asia, from Iran to Pakistan, and in China and Mongolia. In South America, the rugged Atacama Desert lies between the ocean and the Andes. And Earth's deserts are expanding, as we will discuss shortly. Now, let us look at the link between climate and Earth's deserts.

Desert Climates

The spatial distribution of dry lands is related to three climatological settings:

- subtropical high-pressure cells between 15° and 35°, both N and S latitudes (see Figures 6.10 and 6.12);
- the rain shadow on the lee side of mountain ranges (see Figure 8.9); and,
- areas at great distance from moisture-bearing air masses, such as central Asia.

Figure 15.15 portrays this distribution according to the climate classification used in this text and includes photographs of four major desert regions.

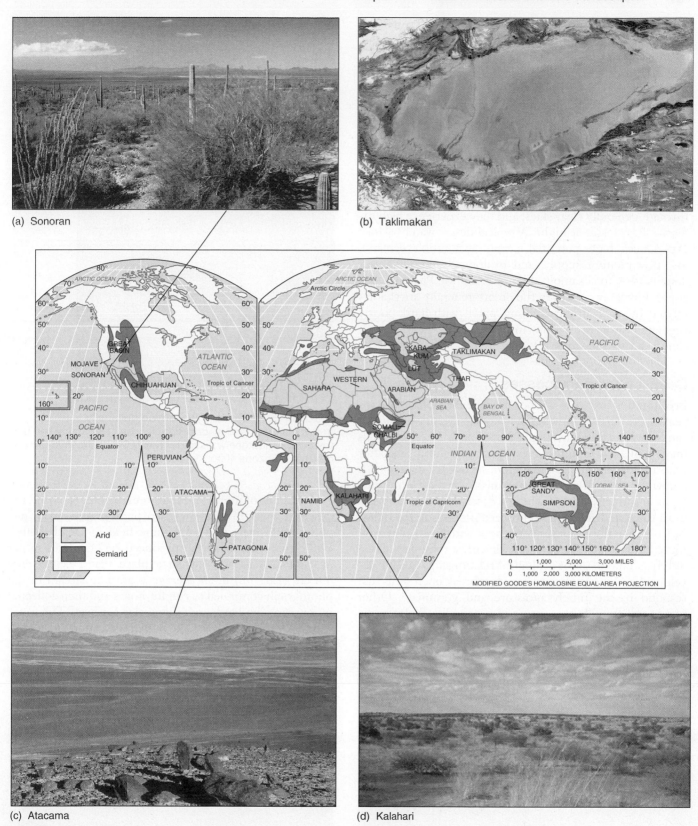

(a) Sonoran

(b) Taklimakan

(c) Atacama

(d) Kalahari

FIGURE 15.15 The world's dry regions.
Worldwide distribution of arid lands (*arid desert* climates) and semiarid lands (*semiarid steppe* climates).
(a) Sonoran Desert in the American Southwest. (b) Taklimakan Desert in central Asia. (c) Atacama Desert
in subtropical Chile near Baquedano. (d) Kalahari Desert in south-central Africa. [(a) Photo by author;
(b) *Terra* image, October 27, 2001, courtesy of MODIS Land Rapid Response Team, NASA/GSFC; (c) photo
by Jacques Jangoux/Photo Researchers, Inc.; (d) photo by Nigel J. Dennis/Photo Researchers, Inc.]

Desert areas possess unique landscapes created by the interaction of intermittent precipitation events, weathering processes, and wind. Rugged, hard-edged desert landscapes of cliffs and scarps contrast sharply with the vegetation-covered, rounded, and smoothed slopes characteristic of humid regions. Vast arid plains with limitless horizons stand in contrast to dense forests and their enclosed environments.

The daily surface energy balance for El Mirage, California, presented in Figure 4.20a and b, highlights the high sensible heat conditions and intense ground heating in the desert. Such areas receive a high input of insolation through generally clear skies, and they experience high radiative heat losses at night. A typical desert water balance experiences high potential evapotranspiration demand, low precipitation supply, and prolonged summer water deficits (see, for example, Figure 9.12e for Phoenix, Arizona). Fluvial processes in the desert generally are characterized by intermittent running water, with hard, poorly vegetated surfaces of thin, immature soils yielding high runoff during rainstorms (Figure 15.16).

Desert Fluvial Processes

Precipitation events in a desert may be rare indeed, even a year or two apart, but when they do occur, a dry streambed can fill with a torrent called a **flash flood**. These channels may fill in a few minutes and surge briefly during and after a storm. Depending on the region, such a dry streambed is known as a **wash**, an *arroyo* (Spanish), or a *wadi* (Arabic). Signs are usually posted where a desert highway crosses a wash to warn drivers not to proceed if rain is in the vicinity, for a flash flood can suddenly arise and sweep away anything in its path. When washes fill with surging flash flood waters, a unique set of ecological relationships quickly develops. Crashing rocks and boulders break open seeds that respond to the timely moisture and germinate. Other plants and animals also spring into brief life cycles as the water irrigates their limited habitats.

FIGURE 15.16 The land in Canyon de Chelly.
In this marginal land, undependable water flows between the towering sandstone walls surrounding Chinle Wash in Canyon de Chelly, Arizona. [Photo by author.]

At times of intense rainfall, remarkable scenes fill the desert. Figure 15.17 shows two photographs taken just one month apart in a sand dune field in Death Valley, California. A rainfall event produced 2.57 cm (1.01 in.) of precipitation in one day, in a place that receives only 4.6 cm (1.83 in.) in an average year. The stream in the photograph continued to run for hours and then collected in low spots on hard, underlying clay surfaces. The water was quickly consumed by the high evaporation demand so

(a)   (b)

FIGURE 15.17 An improbable river in Death Valley.
The Stovepipe Wells dune field of Death Valley, California, shown (a) the day after a 2.57-cm (1.01-in.) rainfall and (b) one month later (identical location). This 1985 rainfall event was exceeded by one in 2004. [Photos by author.]

that, in just a month, these short-lived watercourses were dry and covered with accumulations of alluvium.

The 1985 precipitation event just described was exceeded by one on August 15, 2004, when several thunderstorm cells stalled over the mountains along the east side of Death Valley. Rainfall totals reached 6.35 cm (2.5 in.) in little more than an hour. Cars were swept away from the Furnace Creek Inn parking lot, about 4.8 km (3 mi) of Highways 190 and 178 washed out, and there were two fatalities in the Zabriskie Point area. One annex building had a 3-m- (10-ft-) high water mark on it. Again, water is sparse in the deserts, yet it is the major erosional force—sometimes dramatically so. For recent information and maps of Death Valley, see http://www.nps.gov/deva/.

As runoff water evaporates, salt crusts may be left behind on the desert floor in a **playa**. This intermittently wet and dry lowest area of a closed drainage basin is the site of an *ephemeral lake* when water is present. Accompanying our earlier discussion of evaporites, Figure 11.10 in Chapter 11 shows such a playa in Death Valley, covered with salt-crust precipitate over fine clays just one month after the 1985 record rainfall event.

Permanent lakes and continuously flowing rivers are uncommon features in the desert, although the Nile River and the Colorado River are notable exceptions. Both these rivers are *exotic streams*, having their headwaters in a wetter region and the bulk of their course through arid regions. Focus Study 15.1 discusses the Colorado River, the ongoing drought, and the problem of increasing water demands in an arid land.

In arid climates, a prominent landform is an **alluvial fan**, which occurs at the mouth of a canyon where it exits into a valley. The fan is produced by flowing water that abruptly loses velocity as it leaves the constricted channel of the canyon and therefore drops layer upon layer of sediment along the base of the mountain block. Water then flows over the surface of the fan and produces a braided drainage pattern, sometimes shifting from channel to channel (Figure 15.18). A continuous apron, or **bajada** (Spanish for "slope"), may form if individual alluvial fans coalesce into one sloping surface (see Figure 15.24d). Fan formation of any sort is reduced in humid climates because perennial streams constantly carry away sediment, preventing its deposition.

(text continued on page 488)

FIGURE 15.18 An alluvial fan.
The photo shows an alluvial fan in a desert landscape. The topographic map shows the Cedar Creek alluvial fan. The topographic map is the Ennis Quadrangle, 15-minute series, scale 1:62,500, contour interval = 40 ft; latitude/longitude coordinates for the mouth of the canyon are 45°2' N 111°35' W. [Photo by Bobbé Christopherson; USGS map.]

The Colorado River: A System Out of Balance

The headwaters of an exotic stream rise in a region of water surpluses, but the stream flows mostly through arid lands for the rest of its journey to the sea. Exotic streams have few incoming tributaries. Consequently, an exotic stream has a discharge pattern that is different than that of a typical humid-region stream: Instead of discharge increasing downstream, it decreases (see Chapter 14). The Nile and Colorado rivers are prominent examples. These natural conditions are further affected by human impacts of water withdrawals and reservoir construction, along with high evaporation rates.

In the case of the Nile, the East African mountains and plateaus to the south provide a humid source area. The Nile first rises as the remote headwaters of the Kagera River in the eastern portion of the Lake Plateau country of East Africa. On its way to Lake Victoria, it forms the partial boundary of Tanzania, Rwanda, and Uganda. The Nile itself then flows from the lake and continues on its 6650-km (4132-mi) course, with decreasing discharge all the way to its mouth on the Mediterranean Sea near Cairo.

The Colorado River Basin

This river rises in Colorado's Rocky Mountain National Park (Figure 15.1.1a) and flows almost 2317 km (1440 mi) to where a trickle of water disappears in the sand, kilometers short of its former mouth in the Gulf of California.

Orographic precipitation totaling 102 cm (40 in.) per year falls mostly as snow in the Rockies, feeding the Colorado headwaters. But at Yuma, Arizona, near the river's end, annual precipitation is a scant 8.9 cm (3.5 in.), an extremely small amount when compared with the high annual potential evapotranspiration demand in the Yuma region of 140 cm (55 in.).

From its source region, the Colorado River quickly leaves the humid Rockies and spills out into the arid desert of western Colorado and eastern Utah. At Grand Junction, Colorado, near the Utah border, annual precipitation is only 20 cm (8 in.) (upstream from Figure 15.1.1b). After carving its way through the intricate labyrinth of canyonlands in Utah, the river enters Lake Powell, behind Glen Canyon Dam (Figure 15.1.1c). The Colorado then flows through the Grand Canyon, formed by the river's own erosive power.

West of the Grand Canyon, the river turns southward, marking its final 644 km (400 mi) as the Arizona–California border. Along this stretch sits Hoover Dam, just east of Las Vegas (Figure 15.1.1d); Davis Dam, built to control the releases from Hoover (Figure 15.1.1e); Parker Dam, for the water needs of Los Angeles; three more dams for irrigation water (Palo Verde, Imperial, and Laguna, Figure 15.1.1f); and finally, Morelos Dam at the Mexican border (Figure 15.1.1g). Mexico owns the end of the river and whatever water is left. Hydrologically exhausted, the river no longer reaches its

mouth in the Gulf of California (Figure 15.1.1i).

Figure 15.1.1j shows the annual water discharge and suspended sediment load for the Colorado River at Yuma, Arizona, from 1905 to 1964. The completion of Hoover Dam in the 1930s dramatically reduced suspended sediment. The addition of Glen Canyon Dam upstream from Hoover Dam in 1963 further reduced discharge.

Overall, the drainage basin encompasses 641,025 km^2 (247,500 mi^2) of mountain, basin-and-range, plateau, canyon, and desert landscapes in parts of seven states and two countries. A discussion of the Colorado River is included in this chapter because of its crucial part in the history of the Southwest and its role in the future of this drought-plagued arid and semi-arid region.

Dividing Up the Colorado's Dammed Water

John Wesley Powell (1834–1902) was the first Euro-American of record to successfully navigate the Colorado River through the Grand Canyon. Lake Powell is named after him. Powell perceived that the challenge of the West was too great for individual efforts and believed that solutions to problems such as water availability could be met only through private cooperative efforts. His 1878 study (reprinted 1962), *Report of the Lands of the Arid Region of the United States*, is a conservation landmark.

FIGURE 15.1.1 The Colorado River drainage basin.
The Colorado River basin, showing division of the upper and lower basins near Lees Ferry in northern Arizona. (a) Headwaters of the Colorado River near Mount Richthofen in the Colorado Rockies. (b) The river near Moab, Utah. (c) Glen Canyon Dam, a regulatory, administrative facility near Lees Ferry, Arizona. (d) Hoover Dam spillways in rare operation during 1983 floods. (e) Davis Dam in full release during flood. (f) Irrigated fresh-cut flowers in the Imperial Valley, refrigerated trucks in the field, and a cargo jet at a local airport await shipments to distant markets. (g) Morelos Dam at the Mexican border is the final stop as the river dwindles to a mere canal. (h) Central Arizona Project aqueduct west of Phoenix. (i) The Colorado River (far upper left) stops short of its former delta. The bluish-purple water in the former channel is actually an inlet for Gulf of California water; the gray deposits are sediments in mud flats along this inlet. (j) Dam construction affects river discharge and sediment yields. [Photos by (a and f) Bobbé Christopherson; (b–e, g) photos by author; (h) photo by Tom Bean/DRK Photo; (i) *Terra* image courtesy of NASA/GSFC/MITI/ERSDAC/JAROS and the U.S./Japan ASTER Science Team, September 8, 2000; and (j) data from USGS, 1985, *National Water Summary 1984*, Water Supply Paper 2275 (Washington, DC: Government Printing Office, p. 55).]

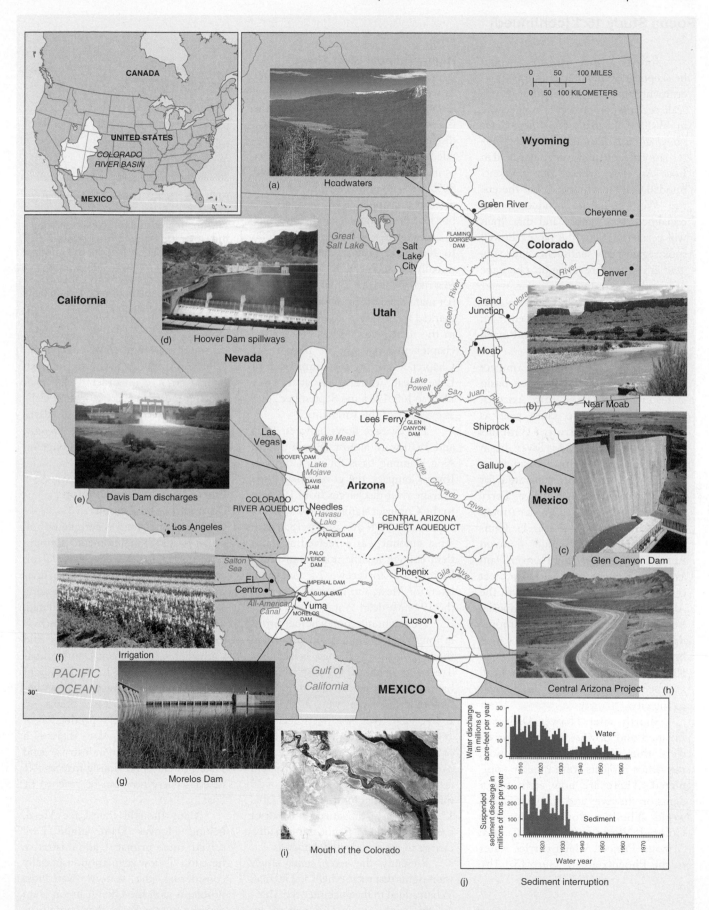

(a) Headwaters

CANADA

UNITED STATES

COLORADO
RIVER BASIN

MEXICO

(d) Hoover Dam spillways

(e) Davis Dam discharges

(f) Irrigation

(g) Morelos Dam

(i) Mouth of the Colorado

(b) Near Moab

(c) Glen Canyon Dam

(h) Central Arizona Project

(j) Sediment interruption

Wyoming

Colorado

Utah

Nevada

California

Arizona

New
Mexico

MEXICO

PACIFIC
OCEAN

Green River

Cheyenne

Denver

Grand
Junction

Moab

Salt
Lake
City

Great
Salt Lake

Lake
Powell

Lees Ferry

Shiprock

Gallup

Las
Vegas

Lake Mead

Needles

Los Angeles

El
Centro

Yuma

Phoenix

Tucson

Salton
Sea

Gulf of
California

FLAMING
GORGE
DAM

GLEN
CANYON
DAM

HOOVER DAM

DAVIS
DAM

PARKER DAM

PALO
VERDE
DAM

IMPERIAL DAM

LAGUNA DAM

MORELOS
DAM

COLORADO
RIVER AQUEDUCT

CENTRAL ARIZONA
PROJECT AQUEDUCT

All-American
Canal

Lake
Mojave

Havasu
Lake

Green River

Colorado River

San Juan River

Little Colorado River

Gila River

30°

0   50   100 MILES
0   50   100 KILOMETERS

Water discharge
in millions of
acre-feet per year

Water

Suspended
sediment discharge in
millions of tons per year

Sediment

Water year

(continued)

Focus Study 15.1 (continued)

Today, Powell probably would be skeptical of the intervention by government agencies in building large-scale reclamation projects. An anecdote in Wallace Stegner's book *Beyond the Hundredth Meridian* tells of an 1893 international irrigation conference held in Los Angeles, where development-minded delegates bragged that the entire West could be conquered and reclaimed from nature and that "rain will certainly follow the plow." Powell spoke against that sentiment: "I tell you, gentlemen, you are piling up a heritage of conflict and litigation over water rights, for there is not sufficient water to supply the land."* He was booed from the hall, but history has shown Powell to be correct.

The Colorado River Compact was signed by six of the seven basin states in 1923. (The seventh, Arizona, signed in 1944, the same year as the Mexican Water Treaty.) With this compact, the Colorado River basin was divided into an upper basin and a lower basin, arbitrarily separated for administrative purposes at Lees Ferry near the Utah–Arizona border (noted on the map in Figure 15.1.1). Congress adopted the Boulder Canyon Act in 1928, authorizing Hoover Dam as the first major reclamation project on the river. Also authorized was the All-American Canal into the Imperial Valley, which required an additional dam (Figure 15.1.2). Los Angeles then began its project to bring Colorado River water 390 km (240 mi) from still another dam and reservoir on the river to the city.

Shortly after Hoover Dam was finished and downstream enterprises were thus offered flood protection, the other projects were quickly completed. There are now eight major dams on the river and many irrigation works. The last effort to redistribute Colorado River water is the Central Arizona Project, which carries water to the Phoenix area (Figure 15.1.1h).

*W. Stegner, *Beyond the Hundredth Meridian* (Boston: Houghton Mifflin, 1954), p. 343.

Highly Variable River Flows

The flaw in all this planning and water distribution is that exotic streamflow discharges are highly variable, and the Colorado is no exception. In 1917, the discharge measured at Lees Ferry totaled 24 million acre-feet (maf), whereas in 1934 it was only 5.03 maf, about 80% lower than in 1917. In 1977, the discharge dropped again to 5.02 maf, but in 1984 it rose to an all-time high of 24.5 maf. Yet, in contrast, from 2000 to 2006 river discharge fell to exceptionally low levels, with 2002 dropping to a record low of 3.1 maf, and 2003 to 6.4 maf. The 2002 to 2006 water years marked the lowest 5-year average flow in the total record. Please revisit the chapter-opening pair of photos for Chapter 9 showing the comparison of Lake Mead water levels at Hoover Dam in July 1983 and September 2007 as evidence of this variability.

The average flows between 1906 and 1930 were almost 18 maf a year. As a planning basis for the Colorado River Compact, the government used average river discharges from 1914 up to the Compact signing in 1923, an exceptionally high average of 18.8 maf. That amount was perceived as more than enough for the upper and lower basins, each to receive 7.5 maf, and, later, for Mexico to receive 1.5 maf in the 1944 Mexican Water Treaty.

We might question whether proper long-range planning should rely on the providence of high variability. Tree-ring analyses of past climates have disclosed that the only other time Colorado discharges were at the high 1914–1923 level was between A.D. 1606 and 1625. The dependable flows of the river have been consistently overestimated. This shortfall problem is shown in an estimated budget of 20 million acre-feet for the river (Table 15.1.1): Clearly, the situation is out of balance, for there is not enough discharge to meet budgeted demands.

Presently, the seven states *ideally* want rights that total as high as 25.0 maf. When added to the guarantee for Mexico, this comes up to 26.5 maf of demands (more than in Table 15.1.1). And

FIGURE 15.1.2 Colorado water in a canal.
All-American canal brings water from the Colorado River to the irrigated lands of the Salton Trough: the Imperial and Coachella valleys. Only a small portion of this canal is lined, losing water to seepage in transit. [Photo by Bobbé Christopherson.]

six states share one common opinion: California's right to the water must be limited to a court-ordered 4.4 maf, which it exceeded every year until California's access to surplus Colorado River water was stopped in 2002. At present, there is no surplus during the ongoing western drought.

Water Loss at Glen Canyon Dam and Lake Powell

Glen Canyon Dam was completed and began water impoundment, forming Lake Powell, in 1963, 27.4 km (17 mi) north of the upper and lower basin division point at Lees Ferry. The advancing water slowly flooded the deep, fluted, inner gorges of many canyons, including Glen, Navajo, Labyrinth, and Cathedral. Glen Canyon Dam's primary purpose, according to the Bureau of Reclamation, was to regulate flows between the upper and lower basins. An additional benefit is the production of hydroelectric power. Also, there is a recreation and tourism industry around Lake Powell, as previously inaccessible desert scenery now can be reached by boat.

Many thought that Lake Mead, behind Hoover Dam, could have served the primary administrative function of flow regulation, especially considering the serious water-loss problems with the Glen Canyon Dam and the porous rocks that contain its reservoir. Porous Navajo sandstone

Table 15.1.1   Estimated Colorado River Budget through 2006

Water Demand	Quantity (maf)[a]
Upper Basin (7.5) Lower Basin (7.5)[b]	15.0
Central Arizona Project (rising to 2.8 maf)	1.0
Mexican allotment (1944 Treaty)	1.5
Evaporation from reservoirs	1.5
Bank storage at Lake Powell	0.5
Phreatophytic losses (water-demanding plants)	0.5
Budgeted total demand	**20.0 maf**

Average Flows at Lees Ferry	
1906–1930	17.70 maf
1930–2003 average flow of the river	14.10 maf
1990–2006	11.34 maf
2000	11.06 maf
2001	10.75 maf
2002	3.1 maf
2003	6.4 maf
2004	5.6 maf
2005	8.3 maf
2006	8.4 maf

A water year runs from October 1 to September 30.

[a]1 million acre-feet = 325,872 gallons = 1.24 million liters = 43,560 ft^3.

[b]In-basin consumptive uses 75% agricultural.

Source: Bureau of Reclamation and states of Arizona, California, Nevada.

underlies most of the Lake Powell reservoir. This sandstone absorbs an estimated 0.5 maf of the river's overall annual discharge as bank storage—the higher the lake level, the greater the loss into the sandstone. Second, Lake Powell is an open body of water in an arid desert, where hot, dry winds accelerate evaporative losses. Another 0.5 maf of the Colorado's overall discharge is lost annually from the reservoir in this manner, equaling about a third of all Colorado River reservoir evaporation losses. Third, now-permanent sand bars and banks have stabilized along the regulated river, allowing water-demanding plants called phreatophytes to establish and extract an additional 0.5 maf of the river flow. Therefore, if Lake Powell ever returns to capacity, some 1.5 maf, or 13.2%, of the overall river discharge (1990–2006 average) will be lost from the system budget.

Colorado River Floods—1983

Intense precipitation and heavy snowpack in the Rockies, attributable to the 1982–1983 El Niño (Focus Study 10.1), led to record-high discharge rates on the Colorado, testing the controllability of one of the most regulated rivers in the world. Federal reservoir managers were not prepared for the high discharge, since they had set aside Lake Mead's primary purpose (flood control) in favor of competing water and power interests. What followed was the 1983 human-caused flood on the world's most engineered river.

The only time the spillways at Hoover Dam ever operated was more than 40 years earlier, when the reservoir capacity was artificially raised for a test; now they were opened to release actual floodwaters (Figure 15.1.1d; Chapter 9 opener). Davis Dam, which regulates releases from Hoover Dam, was within 30 cm (1 ft) of overflow, a real problem for a structure made partially of earth fill (Figure 15.1.1e). In addition, Glen Canyon Dam was over capacity and at risk and was damaged by the volume of discharge tearing through its spillways. The decision to increase releases from Hoover to protect the dam doomed towns and homeowners living on floodplains along the river to the south, especially inundating subdivisions near Needles, California.

The Western Drought and Beyond

The severity of the western drought that emerged in 1999 surpasses anything in the historical record and is approaching the driest in the record according to tree-ring analyses. Satellites can spot the symptoms of this western drought: reduced snowpack in the Rockies, reduced soil moisture, the drought-stressed conditions of vegetation, the prevalence of wildfires, and lowered reservoir levels. A seemingly permanent shift of the subtropical high-pressure system is being observed.

Since A.D. 1226, nine droughts lasting 10 to 20 years and four lasting more than 20 years occurred. Whether the current drought will be of this duration or be more like the three droughts of the twentieth century that ran 4 to 11 years is unknown. However, this is the first drought to occur in the Colorado River system in the presence of a human demand for water that is increasing at such record levels.

Water inflow to Lake Powell is running less than one-third of the 30-year average (1961–1990)—with 2002 at just 25%. System reservoirs are at record lows. The production of hydroelectric power has fallen to a third of capacity from these dams. The system is in an official "drought condition." There is speculation that Glen Canyon Dam might cease operations before the end of the decade

(continued)

Focus Study 15.1 (continued)

because of these low-water conditions, leading some citizen groups to suggest decommissioning, possible removal, and restoration of the canyons to predevelopment status.

The worst appears yet to come. As of 2006, the total system stands at less than 50% of capacity and engineers estimate that it will take 13 normal winters of snowpack in the Rockies and basin-wide precipitation to "reset" the system. This recovery is much slower than previous drought recoveries because of the rapid increase in water demand across the Southwest. Arizona's population doubled between 1990 and 2007, reaching 6.2 million. Las Vegas went from 368,000 in 1985 to 1.8 million by 2007 (490% increase). Present probabilities for the period after 2016 put the chances of surplus discharge

occurring in any given year at only 20%. Imagine what the population in the region will be after 2016 if no steps are taken to control growth.

Government stopgap measures include, among others: buying water from farmers to meet contractual obligations for Colorado River water; reopening the desalination plant in Yuma, Arizona, to treat agricultural runoff water for its return to the river to meet international agreements with Mexico; and increasing storage, when possible, in upstream reservoirs in the system where evaporation rates are lower than at Lake Powell.

Conservation (using less water) and efficiency (using water more effectively) are beginning to emerge as ways to reduce the tremendous demand for water. Southern Nevada has launched a

campaign to replace lawns with drought-tolerant xeroscaping—desert landscaping. However, limiting or halting further metropolitan construction and population growth in the Colorado River Basin are not part of present strategies to lower demand. Despite the conditions along the river, we still seem to be "supply" focused in our approach and not focusing on "demand." Remember, it is the tremendous increase in the demand for this water that stands in the way of resetting the Colorado River system in the foreseeable future.

We might wonder what John Wesley Powell would think if he were alive today to witness such attempts to control the mighty and variable Colorado. He foretold such a "heritage of conflict and litigation."

A logical aspect of an alluvial fan is the natural sorting of materials by size. Near the mouth of the canyon at the apex of the fan, coarse materials are deposited, grading slowly to pebbles and finer gravels with distance out from the mouth. Then sands and silts are deposited, with the finest clays and dissolved salts carried in suspension and solution all the way to the valley floor. Dissolved minerals accumulate as evaporite deposits on the valley floor left after evaporation of the water from the playa and from salt-laden upward-moving groundwater. Figure 15.19 shows the edge of a playa and active dune field; there is a natural sorting of materials by grain size, with sands dropped in the dunes, silts and clays moving farther toward the playa, and the finest and dissolved materials in the playa.

Well-developed alluvial fans also can be a major source of groundwater. Some cities—San Bernardino, California, for example—are built on alluvial fans and extract their municipal water supplies from them. In other parts of the world, such water-bearing alluvial fans are known as *qanat* (Iran), *karex* (Pakistan), or *foggara* (western Sahara).

Desert Landscapes

Deserts are not wastelands, for they abound in specially adapted plants and animals. Moreover, the limited vegetation, intermittent rainfall, intense insolation, and distant vistas produce starkly beautiful landscapes (see Death Valley photographs in Focus Study 10.1, Figure 10.1.1). And all deserts are not the same: For example, North American deserts have more vegetation cover than do the

generally more barren Asian desert expanses, as we saw in Figure 15.15.

The shimmering heat waves and related mirage effects in the desert are products of light refraction through layers of air that have developed a temperature gradient near the hot ground. The desert's enchantment is captured in the book *Desert Solitaire*:

FIGURE 15.19 Sand and playa—natural sorting of materials.
Playa of dissolved and finer materials (top of photo), sand dunes (middle), and coarser particles on slopes along the mountains (lower area) demonstrate the natural sorting that occurs in desert landscapes. [Photo by Bobbé Christopherson.]

The upper layers of sandstone along the top of an arch or butte are more resistant to weathering and protect the sandstone rock beneath.

The removal of surrounding rock through differential weathering leaves enormous buttes as residuals on the landscape. If you imagine a line intersecting the tops of the Mitten Buttes shown in Figure 15.21, you can gain some idea of the quantity of material that has been removed. These buttes exceed 300 m (1000 ft) in height, similar to the Chrysler Building in New York City or First Canadian Place in Toronto.

Desert landscapes are places where stark erosional remnants can stand above the surrounding terrain as knobs or hills. Such a bare, exposed rock is an *inselberg* (island mountain), as exemplified by Uluru (Ayers) Rock in Australia (Figure 15.22). The formation is 348 m (1145 ft) high and 2.5 km long by 1.6 km wide. Uluru Rock is sacred to Aboriginal peoples and is protected as part of Uluru National Park, established in 1950.

In a desert area, weak surface material may weather to a complex, rugged topography, usually of relatively low and varied relief. Such a landscape is called a *badland*, probably so named because in the American West it offered little economic value and was difficult to traverse in nineteenth-century wagons. The Badlands region of the Dakotas and north-central Arizona (the Painted Desert) are so named.

Sand dunes that existed in some ancient deserts have lithified, forming sandstone structures that bear the imprint of *cross-stratification*. When such a dune was accumulating, sand cascaded down its slipface and distinct bedding planes (layers) were established that remained after the dune lithified (Figure 15.23). Ripple marks, animal tracks, and fossils also are found preserved in these sandstones, which originally were eolian-deposited sand dunes.

Basin and Range Province

A physiographic province is a large region that is identified by several geologic or physiographic traits. We find alternating basins (valleys) and mountain ranges that lie in the rain shadow of mountains to the west (Figure 15.24) characterizing the **Basin and Range Province** of the western United States. The physiography and geography combine to give the province a dry climate, few permanent streams, and *internal drainage patterns*—drainage basins that lack any outlet to the ocean (see Figure 14.5, "Great Basin" and its internal drainage).

The vast Basin and Range Province—almost 800,000 km² (300,000 mi²)—was a major barrier to early settlers in their migration westward. The combination of desert climate and north–south-trending mountain ranges presented harsh challenges. Today, when you follow U.S. Highway 50 across Nevada, you cross five passes of more than 1950 m (6400 ft) and traverse numerous basins. Throughout the drive you are reminded where you are by prideful signs that plainly state: "The Loneliest Road in America." It is difficult to imagine crossing this topography with wagons and oxen (Figure 15.24e).

FIGURE 15.20   A balanced rock—differential weathering. Balanced Rock in Arches National Park, Utah, where writer-naturalist Edward Abbey (quoted in text) worked as a ranger years before it became a park. The overall feature is 39 m (128 ft) tall and composed of Entrada sandstone. The balanced-rock portion is 17 m (55 ft) tall and weighs 3255 metric (3577 short) tons. [Photo by author.]

Around noon the heat waves begin flowing upward from the expanses of sand and bare rock. They shimmer like transparent, filmy veils between my sanctuary in the shade and all the sun-dazzled world beyond. Objects and forms viewed through this tremulous flow appear somewhat displaced or distorted. . . . The great Balanced Rock floats a few inches above its pedestal, supported by a layer of superheated air. The buttes, pinnacles, and fins in the windows area bend and undulate beyond the middle ground like a painted backdrop stirred by a draft of air.[*]

The buttes, pinnacles, and mesas of arid landscapes are resistant horizontal rock strata that have eroded differentially. Removal of the less-resistant sandstone strata produces unusual desert sculptures—arches, windows, pedestals, and delicately balanced rocks (Figure 15.20).

[*]E. Abbey, *Desert Solitaire* (New York: McGraw-Hill, 1968), p. 154. Copyright © 1968 by Edward Abbey.

(a)

FIGURE 15.21 Monument Valley landscape.
(a) Mitten Buttes, Merrick Butte, and rainbow in Monument Valley, Navajo Tribal Park, along the Utah–Arizona border, which have appeared as a backdrop in many movies. (b) A schematic of the tremendous removal of material by weathering, erosion, and transport. (c) Aerial view of Monument Valley. [Photos by (a) author; and (c) Bobbé Christopherson.]

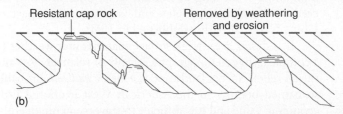
Resistant cap rock   Removed by weathering and erosion
(b)

(c)

FIGURE 15.22 Australian landmark.
Uluru (Ayers) Rock in Northern Territory, Australia, is an isolated mass of weathered rock. [Photo by Porterfield/Chickering.]

FIGURE 15.23 Cross-bedding in sedimentary rocks.
The bedding pattern, called cross-stratification, in these sandstone rocks tells us about patterns that were established in the dunes before lithification (hardening into rock). [Photo by Bobbé Christopherson.]

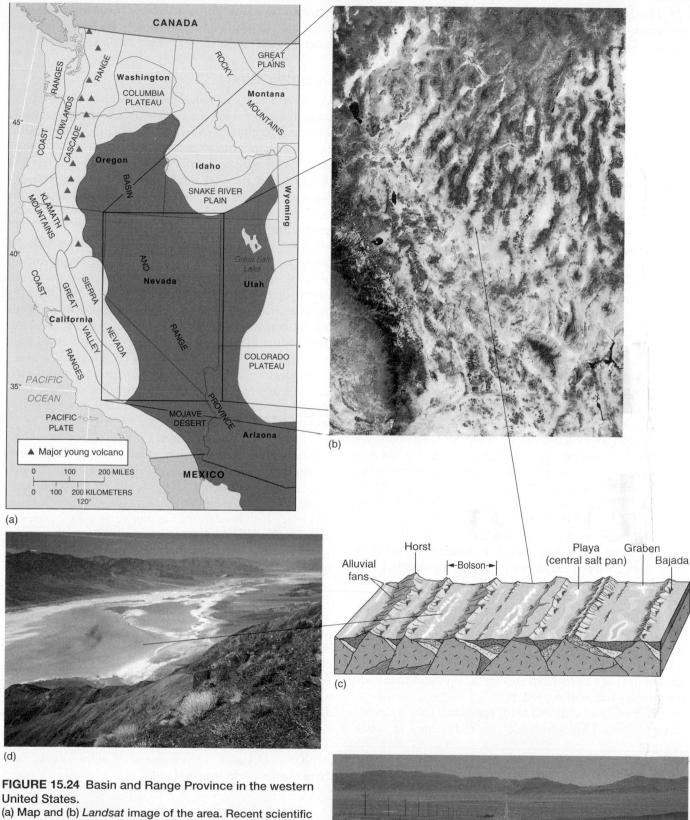

FIGURE 15.24 Basin and Range Province in the western United States.
(a) Map and (b) *Landsat* image of the area. Recent scientific discoveries demonstrate that this province extends south through northern and central Mexico. (c) A bolson in the mountainous desert landscape of the Basin and Range Province. Parallel faults produce a series of ranges and basins. (d) Death Valley features a central playa, parallel mountain ranges, alluvial fans, and *bajada* along the base of the ranges. (e) "The Loneliest Road in America," Highway U.S. 50, slashes westward through the province. [(b) Image from NASA; (d) photo by author; (e) photo by Bobbé Christopherson.]

The Basin and Range formed as a result of tectonic processes. As the North American plate lumbered westward, it overrode former oceanic crust and hot spots at such a rapid pace that slabs of subducted material literally were run over. The crust was uplifted and stretched, creating a tensional landscape fractured by many faults. The present landscape consists of nearly parallel sequences; some are tilted-fault blocks, others have pairs of faults on either side of *horsts* of upward-faulted blocks, which are the "ranges," and *grabens* of downward fault-bounded blocks, which are the "basins." Figure 15.24c shows this pattern of normal faults and tensional stretching. John McPhee captured the feel of this desert province in his book *Basin and Range*:

> Supreme over all is silence. Discounting the cry of the occasional bird, the wailing of a pack of coyotes, silence—a great spatial silence—is pure in the Basin and Range. It is a soundless immensity with mountains in it. You stand . . . and look up at a high mountain front, and turn your head and look fifty miles down the valley, and there is utter silence.*

Basin-and-range relief is abrupt, and rock structures are angular and rugged. As the ranges erode, transported materials accumulate to great depths in the basins, gradually producing extensive desert plains. The basin's elevation averages roughly 1200–1500 m (4000–5000 ft) above sea level, with mountain crests rising higher by another 900–1500 m (3000–5000 ft). Death Valley, California, is the lowest of these basins, with an elevation of −86 m (−282 ft). However, to the west of the valley, the Panamint Range rises to 3368 m (11,050 ft) at Telescope Peak—almost 3.5 vertical kilometers (2.2 mi) of desert mountain relief.

In Figure 15.24c and d, note the **bolson**, a slope-and-basin area between the crests of two adjacent ridges in a dry region of internal drainage. Death Valley provides a dramatic example of these arid-land features. Figure 15.24c also identifies a *playa* (central pan), a *bajada* (coalesced alluvial fans), and a mountain front in retreat from weathering and erosional attack. A *pediment* is an area of bedrock that is layered with a thin veneer, or coating, of alluvium. It is an erosional surface, as opposed to the depositional surface of the *bajada*.

Vast arid and semiarid lands remain an enigma on the water planet. They challenge our technology to access

*J. McPhee, *Basin and Range* (New York: Farrar, Straus, Giroux, 1981), p. 46.

water, our courage, and our personal need for water. Yet, these lands hold a fascination, perhaps because they are so lacking in the moisture that infuses our lives.

Desertification

We are witnessing an unwanted expansion of Earth's desert lands in a process known as **desertification**, a worldwide phenomenon along the margins of semiarid and arid lands. Desertification is due principally to poor agricultural practices (overgrazing and agricultural activities that abuse soil structure and fertility), improper soil-moisture management, erosion and salinization, deforestation, and the ongoing global climatic change, which is shifting temperature and precipitation patterns.

The southward expansion of Saharan conditions through portions of the *Sahel region* has left many African peoples on land that no longer experiences the rainfall of just two decades ago (outlined on Figure 15.25). Other regions at risk of desertification stretch from Asia and central Australia to portions of North and South America. The United Nations estimates that degraded lands have covered some 800 million hectares (2 billion acres) since 1930; many millions of additional hectares are added each year. An immediate need is to improve the database for a more accurate accounting of the problem and a better understanding of what is occurring.

Figure 15.25 is drawn from a map prepared for a U.N. Conference on Desertification; although dated, the patterns persist to this day. Desertification areas are ranked: A moderate hazard area has an average 10%–25% drop in agricultural productivity; a high hazard area has a 25%–50% drop; and a very high hazard area has more than a 50% decrease. Because human activities and economies, especially unwise grazing practices, appear to be the major cause of desertification, solutions to slow the process are readily available but progress is slow. The severity of this problem is magnified by the poverty in many of the affected regions, for the poor lack the capital to change practices and implement conservation strategies.

One of the successes coming from the 1992 Earth Summit in Rio de Janeiro was an initiative for empowering a Convention to Combat Desertification, which began operations in 1994 and is still active. International conferences are convened every other year and many programs are underway. For more on these global arid lands, see the U.N. site at **http://www.unccd.int/main.php**; and for the UNDPs Dryland Development Centre work, see **http://www.undp.org/drylands/**.

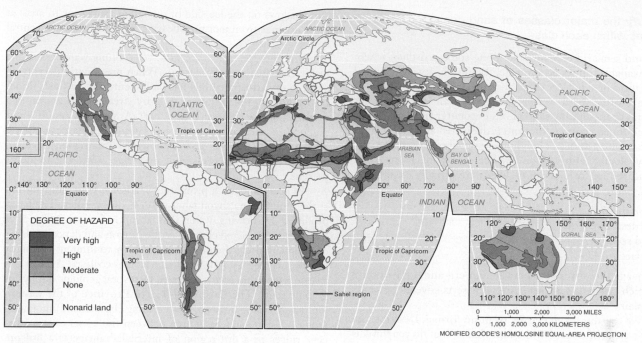

FIGURE 15.25 The desertification hazard.
Worldwide desertification estimates by the United Nations. The climate-stressed Sahel region is
outlined on the map. [Data from U.N. Food and Agricultural Organization (FAO), World Meteorological
Organization (WMO), and United Nations Educational, Scientific, and Cultural Organization (UNESCO), Nairobi,
Kenya.]

Summary and Review—Eolian Processes and Arid Landscapes

■ *Characterize* the unique work accomplished by wind and eolian processes.

The movement of the atmosphere in response to pressure differences produces wind. Wind is a geomorphic agent of erosion, transportation, and deposition. **Eolian** processes modify and move sand accumulations along coastal beaches and deserts. Wind's ability to move materials is small compared with that of water and ice.

eolian (p. 470)

1. Who was Ralph Bagnold? What was his contribution to eolian studies?
2. Explain the term *eolian* and its application in this chapter. How would you characterize the ability of the wind to move material?

■ *Describe* eolian erosion, including deflation, abrasion, and the resultant landforms.

Two principal wind-erosion processes are **deflation**, the removal and lifting of individual loose particles, and **abrasion**, the "sandblasting" of rock surfaces with particles captured in the air. Deflation literally blows away loose or noncohesive sediment and works with rainwater to form a surface resembling a cobblestone street: a **desert pavement** that protects underlying sediment from further deflation and water erosion. Wherever wind encounters loose sediment, deflation may remove enough material to form basins. Called **blowout depressions**, they range from small indentations less than a meter wide up to areas hundreds of meters wide and many meters deep. Rocks that bear evidence of eolian erosion are called **ventifacts**. On a larger scale, deflation and abrasion are

capable of streamlining rock structures, leaving behind distinctive rock formations or elongated ridges called **yardangs**.

deflation (p. 470)
abrasion (p. 470)
desert pavement (p. 470)
blowout depressions (p. 471)
ventifacts (p. 471)
yardangs (p. 471)

3. Describe the erosional processes associated with moving air.
4. Explain deflation and overview each hypothesis on the processes that produce desert pavement.
5. How are ventifacts and yardangs formed by the wind?

■ *Describe* eolian transportation, and *explain* saltation and surface creep.

Wind exerts a drag or frictional pull on surface particles until they become airborne. Only the finest dust particles travel significant distances, so the finer material suspended in a dust storm is lifted much higher than the coarser particles of a sandstorm. Saltating particles crash into other particles, knocking them both loose and forward. The motion called **surface creep** slides and rolls particles too large for saltation.

surface creep (p. 473)

6. Differentiate between a dust storm and a sandstorm.
7. What is the difference between eolian saltation and fluvial saltation?
8. Explain the concept of surface creep.

■ *Identify* the major classes of sand dunes, and *present* examples within each class.

In arid and semiarid climates and along some coastlines where sand is available, dunes accumulate. A **dune** is a wind-sculpted accumulation of sand. An extensive area of dunes, such as that found in North Africa, is characteristic of an **erg desert**, or **sand sea**. When saltating sand grains encounter small patches of sand, their kinetic energy (motion) is dissipated and they start to accumulate into a dune. As height increases above 30 cm (12 in.), a steeply sloping **slipface** on the lee side and characteristic dune features are formed. Dune forms are broadly classified as *crescentic*, *linear*, and *star*.

> dune (p. 474)
> erg desert (p. 474)
> sand sea (p. 474)
> slipface (p. 474)

9. What is the difference between an erg and a reg desert? Which type is a sand sea? Are all deserts covered by sand? Explain.
10. What are the three classes of dune forms? Describe the basic types of dunes within each class. What do you think is the major shaping force for sand dunes?
11. Which form of dune is the mountain giant of the desert? What are the characteristic wind patterns that produce such dunes?

■ *Define* loess deposits and their origins, locations, and landforms.

Eolian-transported materials contribute to soil formation in distant places. Windblown **loess** deposits occur worldwide and can develop into good agricultural soils. These fine-grained clays and silts are moved by the wind many kilometers and are redeposited as an unstratified, homogeneous blanket of material. The binding strength of loess causes it to weather and erode in steep bluffs, or vertical faces.

Significant accumulations throughout the Mississippi and Missouri river valleys form continuous deposits 15–30 m (50–100 ft) thick. Loess deposits also occur in eastern Washington State, Idaho, much of Ukraine, central Europe, China, the Pampas–Patagonia regions of Argentina, and lowland New Zealand.

> loess (p. 476)

12. How are loess materials generated? What form do they assume when deposited?
13. Name a few examples of significant loess deposits on Earth.

■ *Portray* desert landscapes, and *locate* these regions on a world map.

Dry and semiarid climates occupy about 35% of Earth's land surface. The spatial distribution of these dry lands is related to subtropical high-pressure cells between 15° and 35° N and S, to rain shadows on the lee side of mountain ranges, or to areas at great distance from moisture-bearing air masses, such as central Asia.

Precipitation events are rare, yet running water is still the major erosional agent in deserts. When precipitation events do occur, a dry streambed fills with a torrent called a **flash flood**. Depending on the region, such a dry streambed is known as a **wash**, an *arroyo* (Spanish), or a *wadi* (Arabic). As runoff water evaporates, salt crusts may be left behind on the desert floor. This intermittently wet and dry low area in a region of closed drainage is called a **playa**, site of an *ephemeral lake* when water is present.

In arid climates, a prominent landform is the **alluvial fan** at the mouth of a canyon where it exits into a valley. The fan is produced by flowing water that abruptly loses velocity as it leaves the constricted channel of the canyon and deposits a layer of sediment along the mountain block. A continuous apron, or **bajada**, may form if individual alluvial fans coalesce. A *province* is a large region that is characterized by several geologic or physiographic traits. The **Basin and Range Province** of the western United States consists of alternating basins and mountain ranges. A slope-and-basin area between the crests of two adjacent ridges in a dry region of internal drainage is a **bolson**. **Desertification** is the ongoing process that leads to an unwanted expansion of the Earth's desert lands.

> flash flood (p. 482)
> wash (p. 482)
> playa (p. 483)
> alluvial fan (p. 483)
> bajada (p. 483)
> Basin and Range Province (p. 489)
> bolson (p. 492)
> desertification (p. 492)

14. Characterize desert energy and water-balance regimes. What are the significant patterns of occurrence for arid landscapes?
15. How would you describe the water budget of the Colorado River? What was the basis for agreements regarding distribution of the river's water? Why has thinking about the river's discharge been so optimistic? Overview current conditions of the Colorado River system in light of the ongoing drought and increasing water demands.
16. Describe a desert bolson from crest to crest. Draw a simple sketch with the components of the landscape labeled.
17. Where is the Basin and Range Province? Briefly describe its appearance and character.
18. What is meant by desertification? Using the maps in Figures 15.15 and 15.24, and the text description, locate several of the regions affected by desert expansion.

NetWork

The *Geosystems* Student Learning Center provides on-line resources for this chapter on the World Wide Web. To begin: Once at the Center, click on the cover of this textbook, scroll the Table of Contents menu, and select this chapter. You will find self-tests that are graded, review exercises, specific updates for items in the chapter, and in "Destinations" many links to interesting related pathways on the Internet. *Geosystems* Student Learning Center is found at **http://www.prenhall.com/christopherson/**.

Critical Thinking

A. "Water is the major erosion and transport medium in the desert." Respond to this quotation. How is it possible for this to be true when deserts are so dry? What factors have you learned from this chapter that prove this statement true?

B. Where are eolian features (coastal, lakeshore, or desert dunes, or loess deposits) nearest to your present location? Which causative factors discussed in this chapter explain the features you identified? Are you able to visit the site?

C. Relative to the Colorado River system, define, compare, and contrast what you think is meant by a "supply strategy" or a "demand strategy" in dealing with the present Colorado River water budget (Table 15.1.1, Focus Study 15.1). What political and economic forces do you think are at play in planning for the river? Who are the stakeholders? With your analysis, briefly speculate on which strategies you think would be most effective in dealing with this situation.

A beach along Engelskbukta (Engelsk Bay), west coast of Spitsbergen. In the background, stormy clouds hug the mountains and active glaciers extend to the bay. Such Arctic coastlines are increasingly vulnerable to erosion as the sea ice, formerly protective of the coast by its presence, retreats. A beautiful, moody coastal scene, yet a watchful eye must be maintained for any polar bears. [Photo by Bobbé Christopherson.]

16

The Oceans, Coastal Processes, and Landforms

■ Key Learning Concepts_____

After reading the chapter, you should be able to:

■ *Describe* the chemical composition of seawater and the physical structure of the ocean.

■ *Identify* the components of the coastal environment, and *list* the physical inputs to the coastal system, including tides and mean sea level.

■ *Describe* wave motion at sea and near shore, and *explain* coastal straightening as a product of wave refraction.

■ *Identify* characteristic coastal erosional and depositional landforms.

■ *Describe* barrier islands and their hazards as they relate to human settlement.

■ *Assess* living coastal environments: corals, wetlands, salt marshes, and mangroves.

■ *Construct* an environmentally sensitive model for settlement and land use along the coast.

Walk along a shoreline and you witness the dramatic interaction of Earth's vast oceanic, atmospheric, and lithospheric systems. At times, the ocean attacks the coast in a stormy rage of erosive power; at other times, the moist sea breeze, salty mist, and

FIGURE 16.1 Where the land confronts the sea.
The Nubble Light on Cape Neddick, Maine, was built in 1879, near the town of York. Its 27-m (88-ft) tower has withstood storms and winds for almost 130 years. At such a place we feel the dramatic interaction of Earth's oceanic, atmospheric, and lithospheric systems. A photo of Nubble Light was included on the recorded disk of planetary scenes and sounds carried aboard the *Voyager* spacecraft, launched into deep space in 1977. [Photo by Bobbé Christopherson.]

repetitive motion of the water are gentle and calming (Figure 16.1). The coastlines are areas of dynamic change and beauty. Few have captured this confrontation between land and sea as well as biologist Rachel Carson:

> The edge of the sea is a strange and beautiful place. All through the long history of Earth it has been an area of unrest where waves have broken heavily against the land, where the tides have pressed forward over the continents, receded, and then returned. For no two successive days is the shoreline precisely the same. Not only do the tides advance and retreat in their eternal rhythms, but the level of the sea itself is never at rest. It rises or falls as the glaciers melt or grow, as the floors of the deep ocean basins shift under its increasing load of sediments, or as the earth's crust along the continental margins warps up or down in adjustment to strain and tension. Today a little more land may belong to the sea, tomorrow a little less. Always the edge of the sea remains an elusive and indefinable boundary.*

Despite such variability between sea and land, many people live and work near the ocean because of commerce, shipping, fishing, and tourism. A 1995 scientific assessment estimated that about 40% of Earth's population lives within 100 km (62 mi), and 49% within 200 km (145 mi), of coastlines. In the United States, about 50% of the people live in areas designated as *coastal* (this includes the Great Lakes).

Therefore, an understanding of coastal processes and landforms is important to nearly half the world's population. And because these processes along coastlines often produce dramatic change, they are essential to consider in planning and development. A World Resources Institute study found as much as 50% of the world's coastlines at some risk of loss through things like erosion and rising sea level or disruption from pollution. The National Ocean Service coordinates many scientific activities related to the ocean. You find information about these activities at **http://www.nos.noaa.gov/**.

The ocean is a vast ecosystem, intricately linked to life on the planet and to life-sustaining systems in the atmosphere, the hydrosphere, and the lithosphere. In the *Atlas of the Oceans—The Deep Frontier*, Jean-Michel Cousteau on "The Future of the Ocean" states:

> Today, we are coming to better appreciate the extent to which our actions affect an ecosystem—and the people who depend on it—thousands of miles away. The reef fisherman in Fiji is not undone by the local poacher, but by global warming intensified by the driving of a car in downtown Toronto, Canada. Yet these connections are not all bad news. The web of interdependence is built with strands of responsibility and hope. Our ever-expanding ability to communicate across borders and oceans is helping to drive a truly global dialogue about the planet's most pressing environmental challenges.[†]

In this chapter: We begin the chapter with a brief look at our global oceans and seas—1998 was celebrated as The International Year of the Ocean by all United Nations countries (the web site is active at **http://www.yoto98.noaa.gov/**). The physical and chemical properties of the sea distinguish it from the waters of the continent. Our coverage includes discussions about tides, waves, coastal erosional and depositional landforms, beaches, and barrier islands. Important organic processes produce corals, wetlands, salt marshes, and mangroves. A systems framework of specific inputs (components and driving forces), actions (movements and processes), and outputs (results and consequences) organizes our discussion of coastal processes. We conclude with a look at the considerable human interaction with and impact on coastal environments.

Global Oceans and Seas

The ocean is one of Earth's last great scientific frontiers and is of great interest to geographers. Remote sensing from orbit, aircraft, surface vessels, and submersibles is providing a wealth of data and a new capability to understand the oceanic system. The pattern of sea-surface temperatures is presented in Figure 5.11 and ocean currents, both surface and deep, in Figures 6.21 and 6.22. The

---

*"The Marginal World," in *The Edge of the Sea* by Rachel Carson. © 1955 by Rachel Carson, © renewed 1983 by R. Christie (Boston: Houghton Mifflin), p. 11.

[†]*Atlas of the Oceans—The Deep Frontier* by Sylvia Earle. © 2001 by National Geographic Society, text © 2001 by Sylvia Earle (Washington: National Geographic Society), p. 171.

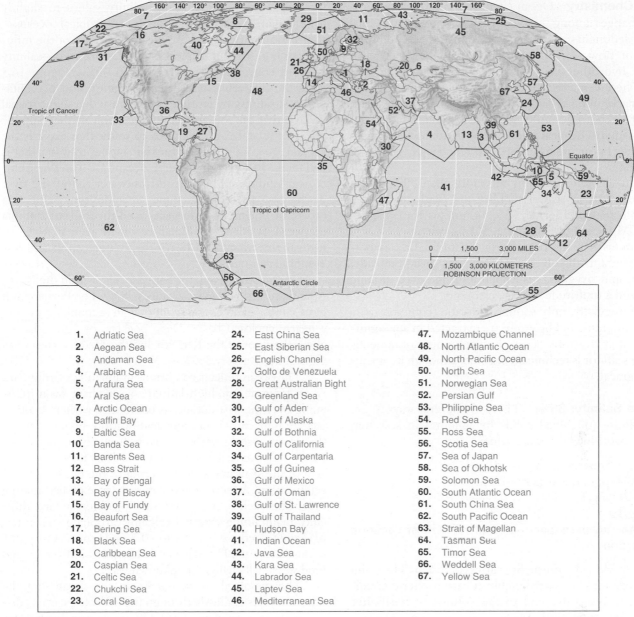

FIGURE 16.2 Principal oceans and seas of the world.
A sea is generally smaller than an ocean and is near a landmass; sometimes the term refers to a large, inland, salty body of water. Match the alphabetized name and number with its location on the map.

1.	Adriatic Sea	24.	East China Sea	47.	Mozambique Channel
2.	Aegean Sea	25.	East Siberian Sea	48.	North Atlantic Ocean
3.	Andaman Sea	26.	English Channel	49.	North Pacific Ocean
4.	Arabian Sea	27.	Golfo de Venezuela	50.	North Sea
5.	Arafura Sea	28.	Great Australian Bight	51.	Norwegian Sea
6.	Aral Sea	29.	Greenland Sea	52.	Persian Gulf
7.	Arctic Ocean	30.	Gulf of Aden	53.	Philippine Sea
8.	Baffin Bay	31.	Gulf of Alaska	54.	Red Sea
9.	Baltic Sea	32.	Gulf of Bothnia	55.	Ross Sea
10.	Banda Sea	33.	Gulf of California	56.	Scotia Sea
11.	Barents Sea	34.	Gulf of Carpentaria	57.	Sea of Japan
12.	Bass Strait	35.	Gulf of Guinea	58.	Sea of Okhotsk
13.	Bay of Bengal	36.	Gulf of Mexico	59.	Solomon Sea
14.	Bay of Biscay	37.	Gulf of Oman	60.	South Atlantic Ocean
15.	Bay of Fundy	38.	Gulf of St. Lawrence	61.	South China Sea
16.	Beaufort Sea	39.	Gulf of Thailand	62.	South Pacific Ocean
17.	Bering Sea	40.	Hudson Bay	63.	Strait of Magellan
18.	Black Sea	41.	Indian Ocean	64.	Tasman Sea
19.	Caribbean Sea	42.	Java Sea	65.	Timor Sea
20.	Caspian Sea	43.	Kara Sea	66.	Weddell Sea
21.	Celtic Sea	44.	Labrador Sea	67.	Yellow Sea
22.	Chukchi Sea	45.	Laptev Sea		
23.	Coral Sea	46.	Mediterranean Sea		

world's oceans and their area, volume, and depth are listed in a table in Figure 7.3. The locations of oceans and major seas are shown and listed alphabetically in Figure 16.2.

Chemical Composition of Seawater

Water is the "universal solvent," dissolving at least 57 of the 92 elements found in nature. In fact, most natural elements and the compounds they form are found in the seas as dissolved solids, or *solutes*. Thus, seawater is a solution, and the concentration of dissolved solids is **salinity**.

The ocean remains a remarkably homogeneous mixture. The ratio of individual salts does not change, despite minor fluctuations in overall salinity. In 1874, the British HMS *Challenger* sailed around the world taking surface and depth measurements and collecting samples of seawater. Analyses of those samples first demonstrated the uniform composition of seawater.

Although ocean chemistry was thought to be fairly constant during the Phanerozoic Eon (the past 542 million years), recent evidence suggests that seawater chemistry has varied over time within a narrow range. The variations are consistent with changes in sea-floor spreading rates, volcanism, and sea level. Evidence is gathered from fluid inclusions in marine formations such as in limestone and evaporite deposits, which contain ancient seawater. The ocean reflects conditions in Earth's environment. As an example, the high-latitude oceans have been freshening over the past decade in response to the large volume of freshwater coming from glacial ice melt, increased precipitation in response to warmer temperatures, and increasing discharge from the Russian rivers that drain into the Arctic Ocean.

Ocean Chemistry   Ocean chemistry is a result of complex exchanges among seawater, the atmosphere, minerals, bottom sediments, and living organisms. In addition, significant flows of mineral-rich water enter the ocean through hydrothermal (hot water) vents in the ocean floor. (These vents are the "black smokers" noted for the dense, black, mineral-laden water that spews from them, Figure 11.11b.) The uniformity of seawater results from complementary chemical reactions and continuous mixing—after all, the ocean basins interconnect, and water circulates among them.

Seven elements account for more than 99% of the dissolved solids in seawater. They are (with their ionic form) chlorine (as chloride, Cl⁻), sodium (as Na⁺), magnesium (as Mg^{2+}), sulfur (as sulfate, SO_4^{2-}), calcium (as Ca^{2+}), potassium (as K⁺), and bromine (as bromide, Br⁻). Seawater also contains dissolved gases (such as carbon dioxide, nitrogen, and oxygen), suspended and dissolved organic matter, and a multitude of trace elements.

Commercially, only sodium chloride (common table salt), magnesium, and bromine are extracted in any significant amount from the ocean. Future mining of minerals from the seafloor is technically feasible, although it remains uneconomical.

Average Salinity: 35‰

There are several ways to express salinity (dissolved solids by volume) in seawater, using the worldwide average value:

- 3.5% (% parts per hundred)
- 35,000 ppm (parts per million)
- 35,000 mg/l
- 35 g/kg
- 35‰ (‰ parts per thousand), the most common notation

Salinity worldwide normally varies between 34‰ and 37‰; variations are attributable to atmospheric conditions above the water and to the volume of freshwater inflows. In equatorial water, precipitation is great throughout the year, diluting salinity values to slightly lower than average (34.5‰). In subtropical oceans—where evaporation rates are greatest because of the influence of hot, dry, subtropical high-pressure cells—salinity is more concentrated, increasing to 36‰. Figure 16.3 plots the difference between evaporation and precipitation and salinity by latitude to illustrate this slight spatial variability. Can you answer the question in the caption?

The term **brine** is applied to water that exceeds the average of 35‰ salinity. **Brackish** applies to water that is less than 35‰ salts. In general, oceans are lower in salinity near landmasses because of freshwater runoff and river discharges. Extreme examples include the Baltic Sea (north of Poland and Germany) and the Gulf of Bothnia (between Sweden and Finland), which average 10‰ or less salinity because of heavy freshwater runoff and low evaporation rates.

On the other hand, the Sargasso Sea, within the North Atlantic subtropical gyre, averages 38‰. The Persian Gulf has a salinity of 40‰ as a result of high evaporation rates in a nearly enclosed basin. Deep pockets, or "brine lakes," along the floor of the Red Sea and the Mediterranean Sea register up to a salty 225‰.

The ocean reflects conditions in Earth's environment. As an example, the high-latitude oceans have been freshening over the past decade, as mentioned earlier. Another change underway is oceanic acidification, taking place as the ocean absorbs carbon dioxide from the atmosphere, where it is increasing rapidly. Hydrolysis forms carbonic acid in the seawater and a lowering of the ocean pH—an acidification. A more acid ocean will cause certain marine organisms such as corals and some plankton to have difficulty maintaining external calcium carbonate structures. pH could decrease by –0.4 to –0.5 units this century; the oceans' average pH today is 8.2. Oceanic biodiversity and food webs will change in unknown ways.

The dissolved solids remain in the ocean, but the water recycles endlessly through the hydrologic cycle, driven by energy from the Sun. The water you drink today

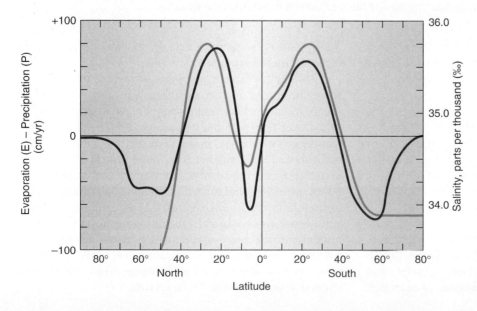

FIGURE 16.3 Variation in ocean salinity by latitude.
Salinity (green line) is principally a function of climatic conditions. Specifically important is the moisture relation expressed by the difference between evaporation and precipitation (E – P) (purple line). Why is salinity higher in the subtropics and lower along the equator? [G. Wüst, 1936.]

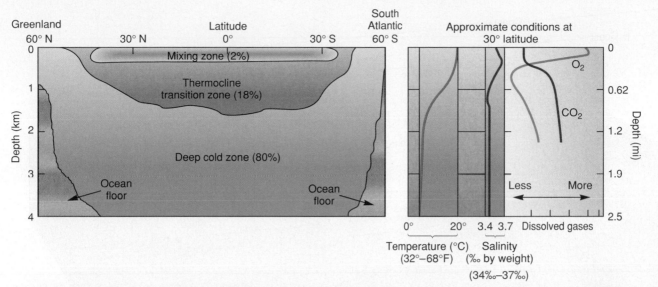

Midlatitude
Productivity

FIGURE 16.4 The ocean's physical structure.
Schematic of average physical structure observed throughout the ocean's vertical profile as sampled along a line from Greenland to the South Atlantic. Temperature, salinity, and dissolved gases are shown plotted by depth.

may have water molecules in it that not long ago were in the Pacific Ocean, in the Yangtze River, in groundwater in Sweden, or airborne in the clouds over Peru.

Physical Structure of the Ocean

The basic physical structure of the ocean is layered, as shown in Figure 16.4. The figure also graphs four key aspects of the ocean, each of which varies with increasing depth: average temperature, salinity, dissolved carbon dioxide, and dissolved oxygen.

The ocean's surface layer is warmed by the Sun and is wind-driven. Variations in water temperature and solutes are blended rapidly in a *mixing zone* that represents only 2% of the oceanic mass. Below the mixing zone is the *thermocline transition zone*, a more than 1-km-deep region of decreasing temperature gradient that lacks the motion of the surface. Friction at these depths dampens the effect of surface currents. In addition, colder water temperatures at the lower margin tend to inhibit any convective movements.

From a depth of 1–1.5 km (0.6–0.9 mi) to the ocean floor, temperature and salinity values are quite uniform. Temperatures in this *deep cold zone* are near 0°C (32°F). Water in the deep cold zone does not freeze, however, because of its salinity and intense pressures at those depths; seawater freezes at about –2°C (28.4°F) at the surface. Generally, the coldest water is along the bottom. Now let us shift to the edge of the sea and examine Earth's coastlines.

Coastal System Components

We know that the continents were formed over many millions of years. However, most of Earth's coastlines are relatively new, existing in their present state as the setting for

continuous change. A dynamic equilibrium exists among the energy of waves, tides, wind, and currents; the supply of materials; the slope of the coastal terrain; and the fluctuation of relative sea level. These interactions produce an infinite variety of erosional and depositional features and coastlines of diverse beauty.

Inputs to the Coastal System

Inputs to the coastal environment include many elements we have already discussed:

- *Solar energy* input drives the atmosphere and the hydrosphere. The conversion of insolation to kinetic energy produces prevailing winds, weather systems, and climate.
- *Atmospheric winds*, in turn, generate ocean currents and waves, key inputs to the coastal environment.
- *Climatic regimes*, which result from insolation and moisture, strongly influence coastal geomorphic processes.
- The nature of *coastal geomorphology* (rock type, landforms, tectonic activity) is important in determining rates of erosion and sediment production.
- *Human activities* are an increasingly significant input to coastal change.

All of these inputs occur within the ever-present influence of gravity's pull, not only from Earth but also from the Moon and Sun. Gravity provides the potential energy of position for materials in motion and generates the tides.

The Coastal Environment and Sea Level

The coastal environment is the **littoral zone**, from the Latin word for "shore." Figure 16.5 illustrates the littoral zone and includes specific components discussed later in

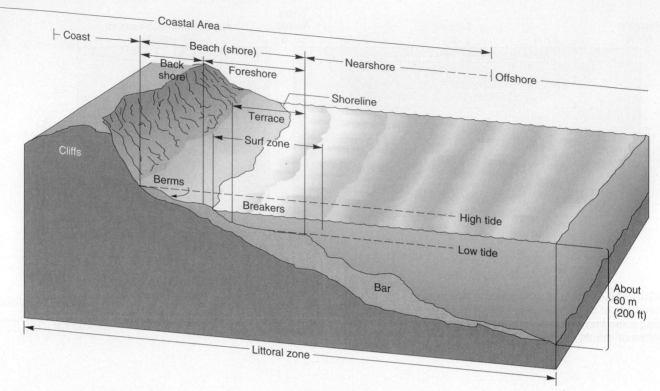

FIGURE 16.5 The littoral zone.
The littoral zone includes the coast, beach, and nearshore environments.

the chapter. The littoral zone spans some land as well as water. Landward, it extends to the highest waterline that occurs on shore during a storm. Seaward, it extends to where water is too deep for storm waves to move sediments on the seafloor—usually around 60 m, or 200 ft, in depth. The specific contact line between the sea and the land is the *shoreline*, although this line shifts with tides, storms, and sea-level adjustments. The *coast* continues inland from high tide to the first major landform change and may include areas considered to be part of the coast in local usage.

Because the level of the ocean varies, the littoral zone naturally shifts position from time to time. A rise in sea level causes submergence of land, whereas a drop in sea level exposes new coastal areas. In addition, uplift and subsidence of the land itself initiates changes in the littoral zone.

Sea level is an important concept. Every elevation you see in an atlas or on a map is referenced to mean sea level. Yet this average sea level changes daily with the tides and over the long term with changes in climate, tectonic plate movements, and glaciation. Thus, *sea level* is a relative term. At present, there exists no international system to determine exact sea level over time. The Global Sea Level Observing System (GLOSS) is an international group actively working on sea-level issues and is part of the larger Permanent Service for Mean Sea Level, which you can find at **http://www.pol.ac.uk/psmsl/ programmes/**.

Mean sea level (MSL) is a value based on average tidal levels recorded hourly at a given site over many years. MSL varies spatially because of ocean currents and waves, tidal variations, air temperature and pressure differences, ocean temperature variations, slight variations in Earth's gravity, and changes in oceanic volume. At present, sea level is rising at a historically high rate related to global climate change. In Chapter 10, you read excerpts from scientists who say sea level is responding more quickly to global warming than any of the forecast models predict. For more information, see News Report 16.1 and the dramatic maps of coastal impacts from a rising sea.

At present, the overall U.S. MSL is calculated at approximately 40 locations along the coastal margins of the continent. These sites are being upgraded with new equipment in the Next Generation Water Level Measurement System, using next-generation tide gauges, specifically along the U.S. and Canadian Atlantic coasts, Bermuda, and the Hawaiian Islands. The *NAVSTAR* satellites that make up the Global Positioning System (GPS) make possible the correlation of data within a network of ground- and ocean-based measurements.

Remote-sensing technology augments these measurements, including the *TOPEX/Poseidon* satellite launched in 1992 and still in service (see **http://topex-www.jpl.nasa.gov/**). This satellite has two radar altimeters that measure changes in mean sea level at any one location every 10 days between 66° N and 66° S latitudes. These measurements are made to an astonishing precision

News Report 16.1

Sea-Level Variations and the Present MSL Increase

Sea level varies along the full extent of North American shorelines. The mean sea level (MSL) of the U.S. Gulf Coast is about 25 cm (10 in.) higher than that of Florida's east coast, which is the lowest in North America. MSL rises northward along the eastern coast, to 38 cm (15 in.) higher in Maine than in Florida. Along the U.S. western coast, MSL is higher than Florida's by about 58 cm (23 in.) in San Diego and by about 86 cm (34 in.) in Oregon.

Overall, North America's Pacific coast MSL averages about 66 cm (26 in.) higher than the Atlantic coast MSL. MSL is affected by differences in ocean currents, air pressure and wind patterns, water density, and water temperature.

Over the long term, sea-level fluctuations expose a range of coastal landforms to tidal and wave processes. As average global temperatures cycle through cold or warm climatic spells, the quantity of ice locked up in the ice sheets of Antarctica and Greenland and in hundreds of mountain glaciers can increase or decrease and result in sea-level changes accordingly. At the peak of the most recent Pleistocene glaciation about 18,000 B.P. (years before the present), sea level was about 130 m (430 ft) lower than it is today. On the other hand, if Antarctica and Greenland ever became ice-free (ice sheets fully melted), sea level would rise at least 65 m (215 ft) worldwide.

Just 100 years ago sea level was 38 cm (15 in.) lower along the coast of southern Florida. Venice, Italy, has experienced a rise of 25 cm (10 in.) since 1890. During the last century, sea level rose 10–20 cm (4–8 in.), a rate 10 times higher than the average rate during the last 3000 years. The present sea-level rise is spatially uneven; for instance, the rate along the coast of Argentina is nearly 10 times the rate along the coast of France. For the United States, see http:// tidesandcurrents.noaa.gov/sltrends/ sltrends.shtml.

Given these trends and the predicted climatic change, sea level will continue to rise and be potentially devastating for many coastal locations. A rise of only 0.3 m (1 ft) would cause shorelines worldwide to move inland an average of 30 m (100 ft). This elevated sea level would inundate some valuable real estate along coastlines all over the world. Some 20,000 km² (7800 mi²) of land along North American shores alone would be drowned, at a staggering loss of more than a trillion dollars. A 95-cm (3.1-ft) sea-level rise could inundate 15% of Egypt's arable land, 17% of Bangladesh, and many island nations and communities. However, uncertainty exists in these forecasts, and the pace of the rise should be slow.

The 2007 IPCC forecast for global mean sea-level rise this century,

given regional variations, is a range from 0.18 to 0.59 m (7.1 to 23.2 in.). Unfortunately, the new 2006–2007 measurements of Greenland's ice loss were not included in the forecasts, and data appear to suggest an acceleration of ice melt and rising sea level. A fair working estimate for this century now stands at 1.0 to 1.2 m (3.28 to 3.94 ft).

The University of Arizona's Environmental Studies Laboratory prepared a dramatic map series. Using USGS DEM data for elevations, the scientists plotted 1-m to 6-m increases in sea level and the amount of coastal inundation (Figure 16.1.1). As you examine the three maps, consider what adjustments will be required. What is your assessment of southern Louisiana? Certainly any calculations on the costs of slowing increasing greenhouse gas emissions should be balanced against damage estimates from such coastal inundation.

Review the "Global Climate Change" section in Chapter 10 for specific forecasts. Despite any uncertainty, planning should start now along coastlines worldwide because preventive strategies are cheaper than recovery costs from possible destruction. Insurance underwriters have begun the process by refusing coverage for shoreline properties vulnerable to rising sea level.

FIGURE 16.1.1 Coastal inundation by a 1-m (3.28-ft) sea-level rise.
A sampling of three portions of U.S. coastline from a worldwide survey completed by scientists at the University of Arizona. Using USGS digital elevation data, they produced these maps to show a 1-m sea-level rise and its impact on coastal environs. [Maps prepared by Jeremy Weiss and Jonathan Overpeck, Environmental Studies Laboratory, Department of Geosciences, University of Arizona. Used by permission.]

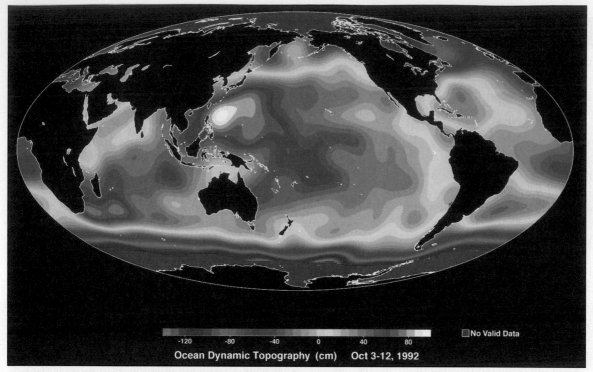

-120   -80   -40   0   40   80   ☐ No Valid Data

Ocean Dynamic Topography (cm)   Oct 3-12, 1992

FIGURE 16.6 Ocean topography as revealed by satellite.
Sea-level data recorded by the radar altimeter aboard the *TOPEX/Poseidon* satellite, October 3–12,
1992. The color scale is given in centimeters above or below Earth's geoid. The overall relief portrayed
in the image is about 2 m (6.6 ft). The maximum sea level is in the western Pacific Ocean (white). The
minimum is around Antarctica (blue and purple). [Image from JPL and NASA/GSFC; *TOPEX/Poseidon* is a
joint U.S.–French mission.]

of 4.2 cm, or 1.7 in. (Figure 16.6). Spectacular *TOPEX/Poseidon* portraits of El Niño and La Niña in the Pacific are in Focus Study 10.1. Another ocean surface topography satellite launched in December 2001, named *Jason-1*, is extending the science of determining mean sea level, ocean topography, atmosphere–ocean interactions, and ocean circulation. *Jason-2* is set for a mid-2008 launch.

Coastal System Actions

The coastal system is the scene of complex tidal fluctuation, winds, waves, ocean currents, and the occasional impact of storms. These forces shape landforms ranging from gentle beaches to steep cliffs, and they sustain delicate ecosystems.

Tides

Tides are complex daily oscillations in sea level, ranging worldwide from barely noticeable to several meters. They are experienced to varying degrees along every ocean shore around the world. Tidal action is a relentless energy agent for geomorphic change. As tides flood (rise) and ebb (fall), the daily migration of the shoreline landward and seaward causes significant changes that affect sediment erosion and transportation.

Tides are important in human activities, including navigation, fishing, and recreation. Tides are especially important to ships because the entrance to many ports is limited by shallow water, and thus high tide is required for passage. Tall-masted ships may need a low tide to clear overhead bridges. Tides also exist in large lakes, but because the tidal range is small, tides are difficult to distinguish from changes caused by wind. Lake Superior, for instance, has a tidal variation of only about 5 cm (2 in.).

Causes of Tides Tides are produced by the gravitational pull of both the Sun and the Moon. Chapter 2 discusses Earth's relation to the Sun and the Moon and the reasons for the seasons. The Sun's influence is only about half that of the Moon's because of the Sun's greater distance from Earth, although it is a significant force. Figure 16.7 illustrates the relation among the Moon, the Sun, and Earth and the generation of variable tidal bulges on opposite sides of the planet.

The gravitational pull of the Moon tugs on Earth's atmosphere, oceans, and lithosphere. The same is true for the Sun, to a lesser extent. Earth's solid and fluid surfaces all experience some stretching as a result of this gravitational pull. The stretching raises large *tidal bulges* in the atmosphere (which we can't see), smaller tidal bulges in the ocean, and very slight bulges in Earth's rigid crust. Our concern here is the tidal bulges in the ocean.

Gravity and inertia are essential elements in understanding tides. *Gravity* is the force of attraction between two bodies. *Inertia* is the tendency of objects to stay still if

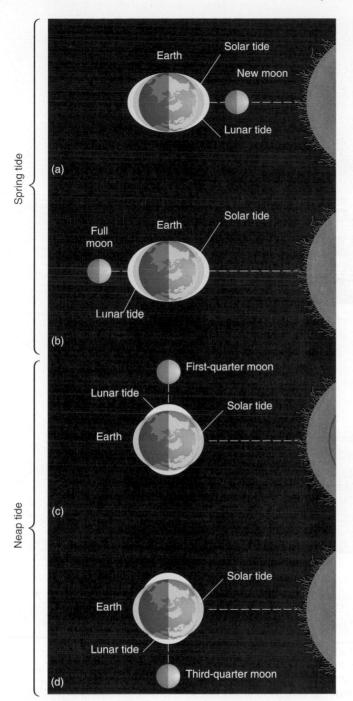

FIGURE 16.7 The cause of tides.
Gravitational relations of Sun, Moon, and Earth combine
to produce spring tides (a, b) and neap tides (c, d).
(Tides are greatly exaggerated for
illustration.)

ANIMATION   Monthly Tidal
Cycles

motionless or to keep moving in the same direction if in
motion. The gravitational effect on the side of Earth facing the Moon or Sun is greater than that experienced by
the far side, where inertial forces are slightly greater. This
difference exists because gravitational influences decrease
with distance. It is this difference in the net force of gravitational attraction and inertia that generates the tides.

Figure 16.7a shows the Moon and the Sun in
conjunction (lined up with Earth), a position in which the
sum of their gravitational forces produces a large tidal
bulge. The corresponding tidal bulge on Earth's opposite
side is primarily the result of the farside water's remaining
in position (being left behind) because its inertia exceeds
the gravitational pull of the Moon and Sun. In effect,
from this inertial point of view, as the nearside water and
Earth are drawn toward the Moon and Sun, the farside
water is left behind because of the slightly weaker gravitational pull. This arrangement produces the two opposing
tidal bulges on opposite sides of Earth.

Tides appear to move in and out along the shoreline,
but they do not actually do so. Instead, Earth's surface
rotates into and out of the relatively "fixed" tidal bulges as
Earth changes its position in relation to the Moon and
Sun. Every 24 hours and 50 minutes, any given point on
Earth rotates through two bulges as a direct result of this
rotational positioning. Thus, every day, most coastal locations experience two high (rising) tides, known as **flood
tides**, and two low (falling) tides, known as **ebb tides**.
The difference between consecutive high and low tides is
considered the *tidal range*.

Spring and Neap Tides   The combined gravitational
effect of the Sun and Moon is strongest in the conjunction
alignment and results in the greatest tidal range between
high and low tides, known as **spring tides**. (*Spring* means
to "spring forth"; it has no relation to the season of
the year.) Figure 16.7b shows the other alignment that
gives rise to spring tides, when the Moon and Sun are
at *opposition*. In this arrangement, the Moon and Sun cause
separate tidal bulges, affecting the water nearest to each of
them. In addition, the left-behind water resulting from the
pull of the body on the opposite side augments each bulge.

When the Moon and the Sun are neither in conjunction nor in opposition but are more or less in the positions
shown in Figure 16.7c and d, their gravitational influences
are offset and counteract each other, producing a lesser
tidal range known as **neap tide**. (*Neap* means "without the
power of advancing.")

Tides also are influenced by other factors, including
ocean basin characteristics (size, depth, and topography),
latitude, and shoreline shape. These factors cause a great
variety of tidal ranges. For example, some locations may
experience almost no difference between high and low
tides. The highest tides occur when open water is forced
into partially enclosed gulfs or bays. The Bay of Fundy in
Nova Scotia records the greatest tidal range on Earth, a difference of 16 m (52.5 ft) (Figure 16.8a, b). The comparison
in 16.8c shows a smaller tidal range along England's east
coast. (For more on tides and tide prediction, and a complete listing of links, see **http://ocean.peterbrueggeman.
com/tidepredict.html**.)

Tidal Power   The fact that sea level changes daily with
the tides suggests an opportunity: Could these predictable
flows be harnessed to generate electricity? The answer is
yes, given the right conditions. Bays and estuaries tend to
focus tidal energy, concentrating it in a smaller area than

FIGURE 16.8 Tidal range and tidal power.
Tidal range is great in some bays and estuaries, such as Halls Harbor near the Bay of Fundy at flood
tide (a) and ebb tide (b). (c) Tidal variation of about 3 m (9.8 ft) in the harbor at Bamburgh, Northumbrian,
England, over several hours. (d) The flow of water between high and low tide is ideal for turning turbines
and generating electricity, as is done at the Annapolis Tidal Generating Station, near the Bay of Fundy in
Nova Scotia, in operation since 1984. [Photos by (a), (b), (d) Jeff Newbery; (c) Bobbé Christopherson.]

in the open ocean. This circumstance provides an opportunity to construct a dam with water gates, locks to let ships through, and turbines to generate power.

Only 30 locations in the world are suited for tidal power generation. At present, only 3 are producing electricity. Two are outside North America—a 4-megawatt-capacity station in Russia in operation since 1968 (at Kislaya-Guba Bay on the White Sea) and a facility in France operating since 1967 (on the Rance River estuary on the Brittany coast). The tides in the Rance estuary fluctuate up to 13 m (43 ft), and power production has been almost continuous there, providing an electrical-generating capacity of a moderate 240 megawatts (about 20% of the capacity of Hoover Dam).

The third area is the Bay of Fundy in Nova Scotia, Canada. At one of several favorable sites on the bay, the Annapolis Tidal Generating Station was built in 1984. Nova Scotia Power Incorporated operates this 20-megawatt plant (Figure 16.8d). According to the Canadian government, tidal power generation at ideal sites is economically competitive with fossil-fuel plants.

Another kind of tidal power taking form in Norway involves sea-floor devices that harness the motion of coastal currents. Windmill-like turbines are set in motion by the tides and currents, as are paddle-driven turbines. Electrical production using this method began in 2003.

Waves

Friction between moving air (wind) and the ocean surface generates undulations of water called **waves**. Waves travel in groups of *wave trains*. Waves vary widely in scale: On a small scale, a moving boat creates a wake of small waves; at a larger scale, storms generate large groups of wave trains. At the extreme is the wind wake produced by the presence of the Hawaiian Islands, traceable westward across the Pacific Ocean surface for 3000 km (1865 mi). The islands disrupt the steady trade winds and produce related surface temperature and wind changes.

A stormy area at sea is a *generating region* for large wave trains, which radiate outward in all directions. The ocean is crisscrossed with intricate patterns of these multi-directional waves. The waves seen along a coast may be the product of a storm center thousands of kilometers away.

Regular patterns of smooth, rounded waves, the mature undulations of the open ocean, are **swells**. As waves leave the generating region, wave energy continues to run in these swells, which can range from small ripples to very large flat-crested waves. A wave leaving a deep-water generating region tends to extend its wavelength horizontally for many meters. Tremendous energy occasionally accumulates to form unusually large waves. One moonlit night in 1933, the U.S. Navy tanker *Ramapo* reported a wave in the Pacific higher than its mainmast, at about 34 m (112 ft)!

As you watch waves in open water, it appears that water is migrating in the direction of wave travel, but only a slight amount of water is actually advancing. It is the *wave energy* that is moving through the flexible medium of water. Water within a wave in the open ocean is simply transferring energy from molecule to molecule in simple cyclic undulations; these are *waves of transition* (Figure 16.9). Individual water particles move forward only slightly, forming a vertically circular pattern.

The diameter of the paths formed by the orbiting water particles decreases with depth. As a deep-ocean wave approaches the shoreline and enters shallower water (10–20 m, or 30–65 ft), the orbiting water particles are vertically restricted. This restriction causes more-elliptical, flattened orbits to form near the bottom. This change from circular to elliptical orbits slows the entire wave, although more waves continue arriving. The resultant effects are closer-spaced waves, growing in height and steepness, with sharper wave crests. As the crest of each wave rises, a point is reached when its height exceeds its vertical stability and the wave falls into a characteristic **breaker**, crashing onto the beach (Figure 16.9b).

In a breaker, the orbital motion of transition gives way to elliptical *waves of translation* in which both energy and water move toward shore. The slope of the shore determines wave style. Plunging breakers indicate a steep bottom profile, whereas spilling breakers indicate a gentle, shallow bottom profile. In some areas, unexpected high waves can arise suddenly. It is a good idea to learn to recognize severe wave conditions before you venture along the shore in these areas.

When the backwash of water flows to the ocean from the beach in a concentrated column, usually at a right angle to the line of breakers, it is a *rip current*. A person caught in one of these can be swept offshore, but usually only a short distance. These brief, short torrents of water can be dangerous (Figure 16.9c).

As various wave trains move along in the open sea, they interact by *interference*. These interfering waves sometimes align so that the wave crests and troughs from one wave train are in phase with those of another. When this in-phase condition occurs, the height of the waves is increased, sometimes dramatically. The resulting "killer waves" or "sleeper waves" can sweep in unannounced and overtake unsuspecting victims. Signs along portions of the California, Oregon, Washington, and British Columbia coastline warn beachcombers to watch for "killer waves." On the other hand, out-of-phase wave trains will dampen wave energy at the shore. When you observe the breakers along a beach, the changing beat of the surf actually is produced by the patterns of *wave interference* that occurred in far-distant areas of the ocean.

Wave Refraction In general, wave action tends to straighten a coastline. Where waves approach an irregular coast, they bend around headlands, which are protruding landforms generally composed of resistant rocks (Figure 16.10). The submarine topography refracts, or bends, approaching waves. The refracted energy is focused around headlands and dissipates energy in coves, bays, and the submerged coastal valleys between headlands. Thus, headlands receive the brunt of wave attack along a coastline. This **wave refraction** redistributes wave energy so that different sections of the coastline vary in erosion potential, with the long-term effect of straightening the coast.

Figure 16.11 shows waves approaching the coast at an angle, for they usually arrive at some angle other than parallel. As the waves enter shallow water, they are refracted as the shoreline end of the wave slows. The wave portion in deeper water moves faster in comparison, thus producing a current parallel to the coast, zigzagging in the prevalent direction of the incoming waves. This **longshore current**, or *littoral current*, depends on wind direction and resultant wave direction. A longshore current is generated only in the surf zone and works in combination with wave action to transport large amounts of sand, gravel, sediment, and debris along the shore as *longshore drift*, or **littoral drift**, the more comprehensive term.

Particles on the beach also are moved along as **beach drift**, shifting back and forth between water and land with each *swash* and *backwash* of surf. Individual sediment

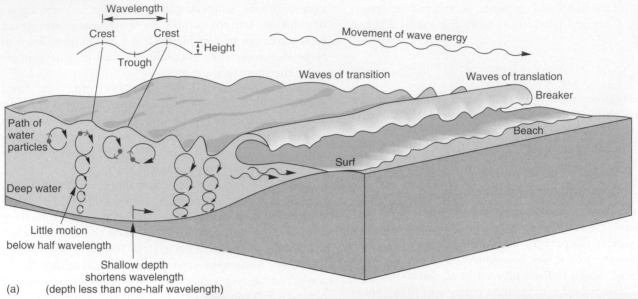

Wave Motion/Wave Refraction

FIGURE 16.9 Wave formation and breakers. (a) The orbiting tracks of water particles change from circular motions and swells in deep water (waves of transition) to more elliptical orbits near the bottom in shallow water (waves of translation). (b) Cascades of waves attack the shore along Baja California, Mexico. (c) A dangerous rip current interrupts approaching breakers. Note the churned-up water where the rip current enters the surf. [Photos by Bobbé Christopherson.]

grains trace arched paths along the beach. You have perhaps stood on a beach and heard the sound of myriad sand grains and seawater in the backwash of surf. These dislodged materials are available for transport and eventual deposition in coves and inlets and can represent a significant volume.

Tsunami, or Seismic Sea Wave An occasional wave that momentarily but powerfully influences coastlines is the tsunami. **Tsunami** is Japanese for "harbor wave," named for its devastating effect when its energy is focused in harbors. Often, tsunami are reported incorrectly as "tidal waves," but they have no relation to the tides.

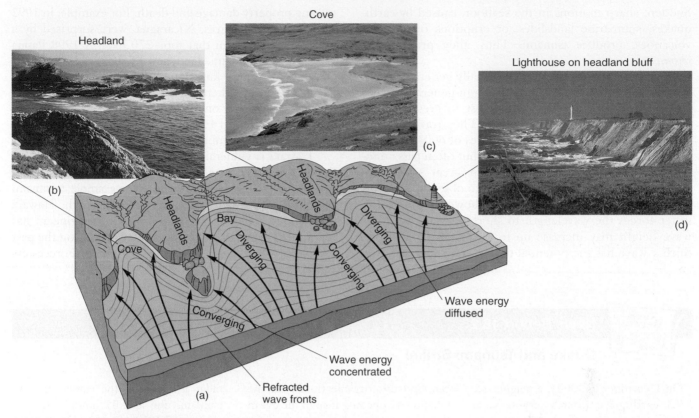

FIGURE 16.10 Coastal straightening.
(a) The process of coastal straightening is brought about by wave refraction. Wave energy is concentrated as it converges on headlands (b) and is diffused as it diverges in coves and bays (c). Headlands are frequent sites for lighthouses such as the light at Point Arena, California (d). [Photos by Bobbé Christopherson.]

ANIMATION   Beach Drift, Coastal Erosion

FIGURE 16.11 Longshore current and beach drift.
(a) Longshore currents are produced as waves approach the surf zone and shallower water. Longshore and beach drift results as substantial volumes of material are moved along the shore. (b) Processes at work along Point Reyes Beach, Point Reyes National Seashore, California. [(b) Photo by author.]

Sudden, sharp motions in the seafloor, caused by earthquakes, submarine landslides, or eruptions of undersea volcanoes, produce tsunami. Thus, they properly are *seismic sea waves*.

A large undersea disturbance usually generates a solitary wave of great wavelength. Tsunami generally exceed 100 km (60 mi) in wavelength (crest to crest) but are only a meter (3 ft) or so in height. They travel at great speeds in deep-ocean water—velocities of 600–800 kmph (375–500 mph) are not uncommon—but often pass unnoticed on the open sea because their great wavelength makes the rise and fall of water hard to observe.

As a tsunami approaches a coast, however, the shallow water forces the wavelength to shorten. As a result, the wave height may increase up to 15 m (50 ft) or more. Such a wave has the potential to devastate a coastal area,

causing property damage and death. For example, in 1992 the citizens of Casares, Nicaragua, were surprised by a 12-m (39-ft) tsunami that took 270 lives. A 1998 Papua New Guinea tsunami, launched by a massive undersea landslide of some 4 km³ (1 mi³), killed 2000. During the twentieth century, there were 141 damaging tsunami and perhaps 900 smaller ones, with a total death toll of about 70,000. News Report 16.2 offers some details of the deadly Indian Ocean tsunami that hit in December 2004.

Hawai'i is vulnerable to tsunami because of its position in the open central Pacific, surrounded by the ring of fire that outlines the Pacific Basin. As an example, a tsunami produced near the Philippines would reach Hawai'i in 10 hours. The U.S. Army Corps of Engineers has reported 41 damaging tsunami in Hawai'i during the past 142 years—statistically, 1 every 3.5 years. Forecasters

News Report 16.2

Quake and Tsunami Strike!

On December 26, 2004, a magnitude 9.3 earthquake (fourth largest in a century) struck off the west coast of northern Sumatra, triggering a massive tsunami across the Indian Ocean—the Sumatra-Andaman earthquake and tsunami. The earthquake trigger was the continuing underthrust of the Indo-Australian plate beneath the Burma plate along the Sunda Trench subduction zone. Imagine the island of Sumatra springing up about 13.7 m (45 ft) in elevation! You

can find this oceanic trench on the Chapter 12 opening map of the ocean floor, stretching along the coast of Indonesia in the eastern Indian Ocean Basin.

Energy from this wave actually traveled several times around the global ocean basins before settling down. With Earth's mid-ocean mountain chain acting as a guide, related large waves arrived at the shores of Nova Scotia, Antarctica, and Peru. Figure 16.2.1 shows two satellite

radar images and the extent of the tsunami, one at 2:05 hours and the other at 7:10 hours after the event. Geophysicists even picked up slight changes in Earth's rotation speed (slowed 3 microseconds) and a shift in the North Pole location by 2.5 cm (1 in.) in response to this event. Total deaths from the quake and tsunami exceeded 150,000, and a final count may never be known. Recovery will take years.

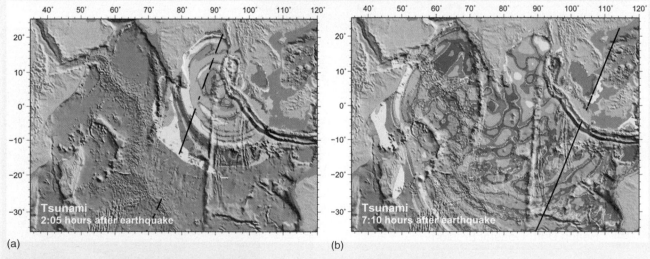

(a) (b)

FIGURE 16.2.1 Satellites track tsunami waves across the Indian Ocean, 2004.
(a) A *Jason-1* image 2 hours after the magnitude 9.3 earthquake showing wave heights of 60 cm (24 in.) radiating outward from the epicenter off Sumatra. (b) The *GFO-satellite* radar captured the wave patterns 7 hours and 10 minutes after the earthquake. [Images courtesy of NOAA, Laboratory for Satellite Altimetry.]

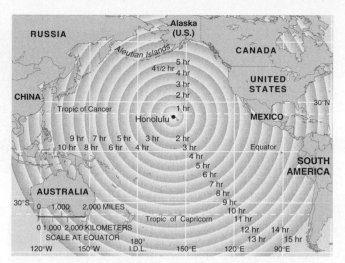

FIGURE 16.12 Tsunami travel times to Honolulu, Hawai'i. [After NOAA.]

watch the unstable Kīlauea region of the southeast coast of Hawai'i, where the possible collapse of a 40-km by 20-km by 2000-m-thick (25-mi × 12-mi × 6560-ft) portion of relatively new and unstable crustal basalt could affect the Pacific Basin. In the Atlantic Basin, there is continuing threat from volcanic island collapses, such as the closely watched Cumbre Vieja volcano in the Canary Islands.

Because tsunami travel such great distances so quickly and are undetectable in the open ocean, accurate forecasts are difficult and occurrences are often unexpected. A warning system now is in operation for nations surrounding the Pacific, where the majority of tsunami occur. A warning is issued whenever seismic stations detect a significant quake or landslide under water, where it might generate a tsunami. A new tsunami hazard-mitigation network began operating in 2004 with the deployment of six pressure sensors on the ocean floor with accompanying surface buoys—this is the Deep ocean Assessment and Reporting of Tsunamis (DART). Three instruments are along the Aleutian Islands, south of Alaska, two along the U.S. West Coast, and one near the equator off South America. The challenge to the world community is to apply this technology so that early warnings are possible in all vulnerable oceans, such as the Indian.

No warning system was in place in the Indian Ocean as it is in the Pacific to detect the December 2004 tsunami disaster. NOAA provided images from four Earth-orbiting radar satellites at the time, including *Jason-1*, *GFO* (Geosat Follow-On), *Envisat*, and *TOPEX/Poseidon*, that clearly showed the tsunami wave energy traveling across the Indian Ocean (shown in News Report 16.2).

Warnings always should be heeded, despite the many false alarms, for the causes lie hidden beneath the ocean and are difficult to monitor in any consistent manner, although remote-sensing capabilities are improving accuracy. (For the tsunami research program, see **http://nctr.pmel.noaa.gov/**. The tsunami home page is at **http://www.ess.washington.edu/tsunami/index.html**.)

Coastal System Outputs

As you can see, coastlines are active places, with energy and sediment being continuously delivered to a narrow environment. The action of tides, currents, wind, waves, and changing sea level produces a variety of erosional and depositional landforms. We look first at erosional coastlines such as the U.S. West Coast, then at depositional coastlines as found along the East and Gulf Coasts. In this era of rising sea level, coastlines are becoming more dynamic.

Erosional Coastal Processes and Landforms

The active margin of the Pacific Ocean along North and South America is a typical erosional coastline. *Erosional coastlines* tend to be rugged, of high relief, and tectonically active, as expected from their association with the leading edge of drifting lithospheric plates (review the plate tectonics discussion in Chapter 11). Figure 16.13 presents features commonly observed along an erosional coast.

Sea cliffs are formed by the undercutting action of the sea. As indentations slowly grow at water level, a sea cliff becomes notched and eventually will collapse and retreat. Other erosional forms evolve along cliff-dominated coastlines, including *sea caves*, *sea arches*, and *sea stacks*. As erosion continues, arches may collapse, leaving isolated stacks in the water (Figure 16.13b, c). The coasts of southern England and Oregon are prime examples of such erosional landscapes. As an example, *Stac an Armin* is in the St. Kilda Archipelago, 165 km (103 mi) west of northern Scotland in the Atlantic Ocean (Figure 16.14).

Wave action can cut a horizontal bench in the tidal zone, extending from a sea cliff out into the sea. Such a structure is a **wave-cut platform**, or *wave-cut terrace*. If the relation between the land and sea level has changed over time, multiple platforms or terraces may rise like stair steps back from the coast. These marine terraces are remarkable indicators of a changing relation between the land and sea, with some terraces more than 370 m (1200 ft) above sea level. A tectonically active region, such as the California coast, has many examples of multiple wave-cut platforms, which at times can be unstable and vulnerable to failure (Figure 16.13d, e).

Depositional Coastal Processes and Landforms

Depositional coasts generally are along land of gentle relief, where sediments from many sources are available. Such is the case with the Atlantic and Gulf coastal plains of the United States, which lie along the relatively passive, trailing edge of the North American lithospheric plate. Erosional processes and inundation influence depositional coasts, particularly during storm activity.

Figure 16.15 illustrates characteristic landforms deposited by waves and currents. One notable deposition landform is the **barrier spit**, which consists of material deposited in a long ridge extending out from a coast.

FIGURE 16.13 Erosional coastal features.
(a) Characteristic coastal erosional landforms: (b) headlands and an arch, Boraray, Scotland; (c) stacks, debris, and headlands; (d) collapsing cliffs and failing houses; and (e) wave-cut platforms near Bixby Bridge, Cabrillo Highway, California. [Photos by (b) and (c) author; (d) Lowell Georgia/Photo Researchers, Inc.; (e) Bobbé Christopherson.]

It partially crosses and blocks the mouth of a bay. Classic examples of a barrier spit include Sandy Hook, New Jersey (south of New York City), and Cape Cod, Massachusetts. The Little Sur River in California (16.15a) and Prion Bay in Southwestern National Park in Tasmania (16.15b) show barrier spits forming partway across the mouths of their respective rivers.

If a spit grows to completely cut off the bay from the ocean and form an inland lagoon, it becomes a **bay barrier**, or *baymouth bar*. Spits and barriers are made up of materials that have been eroded and transported by *littoral drift* (considered beach and longshore drift combined). For much sediment to accumulate, offshore currents must be weak; strong currents carry material away before it can be

deposited. Tidal flats and salt marshes are characteristic low-relief features wherever tidal influence is greater than wave action. If these deposits completely cut off the bay from the ocean, an inland **lagoon** is formed. A **tombolo** occurs when sediment deposits connect the shoreline with an offshore island or sea stack by accumulating on an underwater wave-built terrace (Figure 16.15c).

Coastal erosion and depositional features operate through a complex history as the relation between the land and the ocean changes. High Latitude Connection 16.1 examines isostatic rebound of the crust, causing such shoreline features to rise in relation to sea level.

Not all the beaches of the world are composed of sand, for they can be made up of shingles (beach gravel)

FIGURE 16.14 Sea stack.
Stac an Armin is the highest sea stack in the British Isles at 191 m (627 ft), 57.7° N 8.5° W. Note the thousands of nesting Northern Gannets, the nutrients from their wastes fertilizing the green moss growth. [Photo by Bobbé Christopherson.]

and shells, among other materials (Figure 16.15d). Let's examine these transient coastal deposits in more detail—let's go to the beach.

Beaches Of all the features associated with a depositional coastline, beaches probably are the most familiar. Beaches vary in type and permanence, especially along coastlines dominated by wave action. Technically, a **beach** is that place along a coast where sediment is in motion, deposited by waves and currents. Material from the land temporarily resides on the beach while it is in active transit along the shore. You probably have experienced a beach at some time, along a seacoast, a lakeshore, or even along a stream. Perhaps you have even built your own

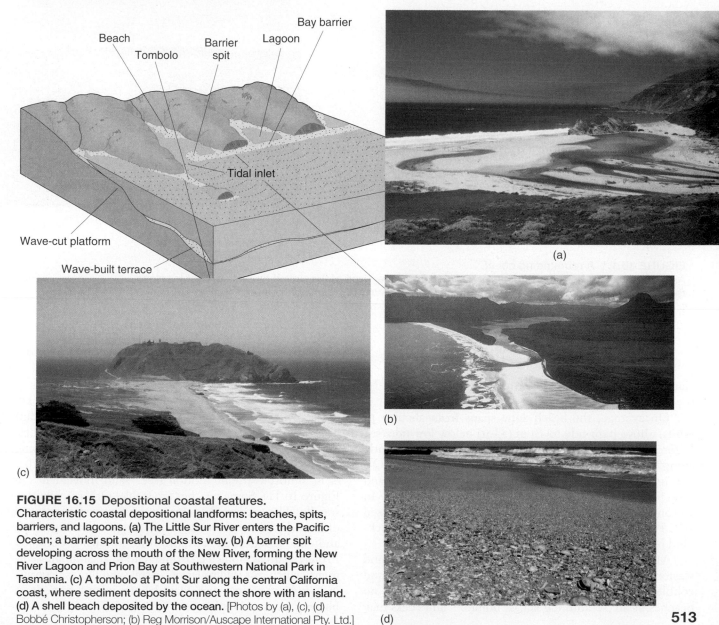

FIGURE 16.15 Depositional coastal features.
Characteristic coastal depositional landforms: beaches, spits, barriers, and lagoons. (a) The Little Sur River enters the Pacific Ocean; a barrier spit nearly blocks its way. (b) A barrier spit developing across the mouth of the New River, forming the New River Lagoon and Prion Bay at Southwestern National Park in Tasmania. (c) A tombolo at Point Sur along the central California coast, where sediment deposits connect the shore with an island. (d) A shell beach deposited by the ocean. [Photos by (a), (c), (d) Bobbé Christopherson; (b) Reg Morrison/Auscape International Pty. Ltd.]

513

High Latitude Connection 16.1

A Rebounding Shoreline and Coastal Features

We learned from Chapter 11, Figure 11.4, that the lighter crust rests on denser materials beneath—the principal of buoyancy applies. Where the load is greater, owing to glaciers, sediment, or mountains, the crust tends to sink, or ride lower. When a glacier melts, for instance, the crust rides higher with the decreased load in a recovery uplift known as *isostatic rebound*. Thus, the entire crust is in a constant state of compensating adjustment.

As Earth systems came out of the last ice age and the massive glaciers and continental ice retreated, the unburdened crust began its rebound, or *postglacial uplift*. The last continental ice retreated about 6000 years ago. The floor of Hudson Bay, Canada, for example, is rising at about 15 mm a year. Recently, as we see in Chapter 17, southeastern Alaska is experiencing an accelerated rise beyond the postglacial uplift associated with the last ice age—a rebound rate of 36 mm a year. This new uplift is thought to be a result of the substantial losses of glacial ice related to increasing temperatures across Alaska.

On Nordaustlandet Island in the Arctic Ocean north of Norway, along Duve Fjord, the land is rising at approximately 2.5 m (8.3 ft) per century. Figure 16.1.1 shows landscape features caused by this phenomenon. The former shoreline appears in successive terraces. Where a depositional tombolo had formed, the neck of land connecting to the offshore island is now stranded above sea level (Figure 16.1.1b).

(a)

(b)

FIGURE 16.1.1 A rebounding coast.
Along Duve Fjord, Nordaustlandet Island, Arctic Ocean: (a) Each of the curved terraces of cobbles along this coast is a former shoreline, formed as the island isostatically rebounded following the retreat of the glaciers. (b) A raised tombolo on the uplifting coast. [Photos by Bobbé Christopherson.]

"landforms" in the sand, only to see them washed away by the waves: a lesson in erosion.

On average, the beach zone spans from about 5 m (16 ft) above high tide to 10 m (33 ft) below low tide (see Figure 16.5). However, the specific definition varies greatly along individual shorelines. Worldwide, quartz (SiO_2) dominates beach sands because it resists weathering and therefore remains after other minerals are removed. In volcanic areas, beaches are derived from wave-processed lava. Hawai'i and Iceland, for example, feature some black-sand beaches.

Many beaches, such as those in southern France and western Italy, lack sand and are composed of pebbles and cobbles—a type of *shingle beach*. Some shores have no beaches at all; scrambling across boulders and rocks may be the only way to move along the coast. The coasts of Maine and portions of Canada's Atlantic provinces are classic examples. These coasts, composed of resistant granite rock, are scenically rugged and have few beaches.

A beach acts to stabilize a shoreline by absorbing wave energy, as is evident by the amount of material that is in almost constant motion (see "sand movement" in Figure 16.11). Some beaches are stable. Others cycle seasonally: They accumulate during the summer, are moved offshore by winter storm waves, forming a submerged bar, and are redeposited onshore the following summer. Protected areas along a coastline tend to accumulate sediment, which can lead to large coastal sand dunes. Prevailing winds often drag such coastal dunes inland, sometimes burying trees and highways.

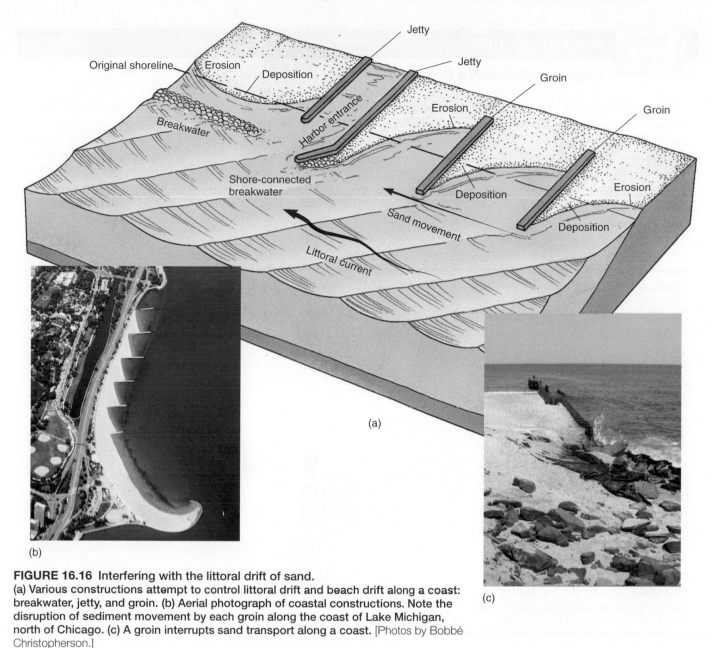

FIGURE 16.16 Interfering with the littoral drift of sand.
(a) Various constructions attempt to control littoral drift and beach drift along a coast: breakwater, jetty, and groin. (b) Aerial photograph of coastal constructions. Note the disruption of sediment movement by each groin along the coast of Lake Michigan, north of Chicago. (c) A groin interrupts sand transport along a coast. [Photos by Bobbé Christopherson.]

ANIMATION   Coastal Stabilization Structures

Maintaining Beaches   Changes in coastal sediment transport can disrupt human activities—beaches are lost, harbors are closed, and coastal highways and beach houses can be inundated with sediment. Thus, people use various strategies to interrupt littoral drift and beach drift. The goal is either to halt sand accumulation or to force accumulation in a desired way through construction of engineered structures, called "hard" shoreline protection.

Figure 16.16 illustrates common approaches: a *jetty* to block material from harbor entrances, a *groin* to slow drift action along the coast, and a *breakwater* to create a zone of still water near the coastline. However, interrupting the littoral drift that is the natural replenishment for beaches may lead to unwanted changes in sediment distribution downcurrent. Careful planning and impact assessment should be part of any strategy for preserving or altering a beach.

Beach nourishment refers to the artificial replacement of sand along a beach (see News Report 16.3). Through such efforts, a beach that normally experiences a net loss of sediment will be "nourished" with new sand. Years of human effort and expense to build beaches can be erased by a single storm, as occurred when hurricane Camille completely eliminated offshore islands along the Gulf Coast in 1969, or in 2004 when hurricanes Charley, Frances, Ivan, and Jeanne devastated coastal Florida beaches and barrier islands.

News Report 16.3

Engineers Nourish a Beach

The city of Miami, Florida, and surrounding Dade County have spent almost $70 million since the 1970s in a continuing effort to rebuild their beaches. Sand is transported to the replenishment area from a source area. To maintain a 200-m- (660-ft-) wide beach, planners determine net sand loss per year and set a schedule for replenishment. In Miami Beach, an 8-year replenishment cycle is maintained. During Hurricane Andrew in 1992, the replenished Miami Beach is thought to have prevented millions of dollars in shoreline structural damage. The cost of beach replenishment following the 2004 and 2005 hurricane onslaught and sand loss is still undetermined. Following Hurricane Wilma in 2005, massive replenishment of sands along Cancún, Mexico, tourist beaches took place. This work is already eroding away from active tropical storms and surge.

Unforeseen environmental impact may accompany the addition of sand to a beach, especially if the sand is from an unmatched source, in terms of its ecological traits. If the new sands do not physically and chemically match the existing varieties, disruption of coastal marine life is possible. The U.S. Army Corps of Engineers, which operates the Miami replenishment program, is running

out of "borrowing areas" for sand that matches the natural sand of the beach—carbonate sand is used to replace quartz sand. A proposal to haul a different type of sand from the Bahamas is being studied as to possible environmental consequences.

In contrast to the term *hard structures*, this hauling of sand to replenish a beach is considered "soft" shoreline protection (Figure 16.3.1). Enormous energy and material must be committed to counteract the relentless energy that nature expends along the coast.

The four hurricanes that hit Florida in 2004 removed enormous quantities of sand from beaches and barrier islands. Each loss makes the coast that much more vulnerable. Nearly $2 billion has been spent from New York to Florida through 2005 to replenish beaches, with costs over the next 10 years forecasted to rise to $5.5 billion more. For more on beach nourishment see **http://www.csc. noaa.gov/beachnourishment/** and **http://www.brynmawr.edu/geology/ geomorph/beachnourishmentinfo. html.**

FIGURE 16.3.1 Beach sand replenishment. Efforts to create protection for the beach and former primary dune must be continuous. Note the sign to help keep the sand in place. One bad storm can wipe out these efforts. [Photo by Bobbé Christopherson.]

Barrier Formation   Barrier chains are long, narrow, depositional features, generally of sand, that form offshore roughly parallel to the coast. Common forms are **barrier beaches** or the broader, more extensive landform **barrier islands**. Tidal variation in the area usually is moderate to low, with adequate sediment supplies coming from nearby coastal plains. Figure 16.17 illustrates the many features of barrier chains, using North Carolina's famed Outer Banks as an example, including Cape Hatteras, across Pamlico Sound from the mainland. The area presently is designated as one of 10 national seashore reserves supervised by the National Park Service.

On the landward side of a barrier formation are tidal flats, marshes, swamps, lagoons, coastal dunes, and beaches, visible in Figure 16.17. Barrier beaches appear to

adjust to sea level and may naturally shift position from time to time in response to wave action and longshore currents. A break in a barrier forms an inlet that connects a bay with the ocean. The name *barrier* is appropriate, for these formations take the brunt of storm energy and actually shield the mainland. Various hypotheses have been proposed to explain the formation of barrier islands. They may begin as offshore bars or low ridges of submerged sediment near shore and then gradually migrate toward shore as sea level rises.

Barrier beaches and islands are quite common worldwide, lying offshore of nearly 10% of Earth's coastlines. Examples are found off the shores of Africa, India's eastern coast, Sri Lanka, Australia, Alaska's northern slope, and the shores of the Baltic and Mediterranean seas.

(a)

(b)

(c)

(d)

(e)

FIGURE 16.17 Barrier-island chain.
(a) *Landsat* image of barrier-island chain along the North Carolina coast. You can see key depositional forms: spit, island, beach, lagoon, and inlet. A sound is a large inlet of the ocean; Pamlico Sound is an example. (b) View of Cape Hatteras lighthouse at its old location illustrates the narrow strand of sand that stands between the ocean and the mainland. (c) The path along which the Cape Hatteras lighthouse was moved inland in 1999 to safer ground. Answer to Chapter 1, Figure 1.16 photo question: The lighthouse moved South 12″, West 48″. (d) Construction continues along Hatteras and the Outer Banks despite the great risk. (e) Salt spray from Hurricane Dennis "burned" these trees south of the lighthouse. [(a) *Terra* image courtesy of NASA/GSFC; photos by (b) Eric Horan/Liaison Agency, Inc.; (c) (d) (e) Bobbé Christopherson.]

(b)

(a)

(c)

FIGURE 16.18 Coastal barrier islands along the Texas Gulf Coast. (a) A barrier island stands between Texas and the Gulf of Mexico. Note the Intercoastal Waterway for shipping along Bolivar Island, near Galveston. (b) In the portion of Padre Island not set aside in the National Seashore, houses line up behind the primary dunes and people drive vehicles on the beach. (c) Typical construction places houses on stilts to deal with storm surges. [Photos by Bobbé Christopherson.]

Earth's most extensive chain of barrier islands is along the U.S. Atlantic and Gulf Coasts, extending some 5000 km (3100 mi) from Long Island to Texas and Mexico.

Barrier Island Vulnerabilities and Hazards Because many barrier islands seem to be migrating landward, they are an unwise choice for home sites or commercial building. Nonetheless, they are a common choice, even though they take the brunt of storm energy. The hazard represented by the settlement of barrier islands was made graphically clear when Hurricane Hugo assaulted South Carolina in 1989. In the Charleston area, Hugo swept away beachfront houses, barrier-island developments, and millions of tons of sand; the hurricane destroyed up to 95% of the single-family homes in one community. The southern portion of one island was torn away. With increased development and continuing real estate appreciation, each future storm can be expected to cause ever-increasing capital losses.

Because of the continuing loss to the barrier islands of North Carolina's Outer Banks, the famous Cape Hatteras lighthouse was moved inland in 1999 to safer ground (Figure 16.17c). Its new position is about 488 m (1600 ft) from the ocean, or approximately the distance it was in 1870 when it was built—that much sand has been lost back to the sea. The change in latitude and longitude coordinates and a photo of this lighthouse are in Figure 1.16 in Chapter 1. (See http://www.ncsu.edu/coast/chl/ for details of the extraordinary effort to save this landmark.) Vulnerable settlements (Figure 16.17d) fill the Outer Banks and Cape Hatteras and are repeatedly slammed by tropical storms, causing damage and beach erosion. There have been more than 30 tropical storms since 1851; more recently: Emily, 1993; Dennis, Floyd, and Irene, 1999; Isabel in 2003; Alex and Bonnie in 2004; and Ophelia in 2005.

And previous damage is visible, such as the salt-spray-burned vegetation from Hurricane Dennis (Figure 16.17e).

Figure 16.18a is an aerial photo of development along the Texas Gulf Coast, near Galveston. Portions of the barrier island chain along the Texas coast are experiencing beach erosion rates averaging 2 m (6.5 ft) per year as sea level continues to rise. Just south of the developed area in Figure 16.18b near Corpus Christi is the Padre Island National Seashore, which is the longest stretch of protected barrier island in the world (177 km, 110 mi), encompassing more than 130,000 acres (52,610 hectares).

The barrier islands off the Louisiana shore are disappearing. They are affected by subsidence through compaction of Mississippi delta sediments, a changing sea level that is rising at 1 cm (0.4 in.) per year in the region, and an increase in tropical storm intensity. In 1998, Hurricane Georges destroyed large tracts of the Chandeleur Islands, located 30 to 40 km (19 to 25 mi) from the Louisiana and Mississippi Gulf Coast, leaving the mainland with reduced protection (Figure 16.19). Louisiana's increasingly exposed wetlands are disappearing at rates of 65 km^2 (25 mi^2) per year. Hurricane Katrina (2005) alone removed this amount of wetlands and swept away most of the remaining Chandeleur Islands (Figure 16.19c, d). A USGS coastal researcher, Abby Sallenger, said when viewing the scene after Katrina, "I've never seen it this bad, the sand is just gone." The regional office of the USGS predicts that in a few decades the barrier islands may be gone.

The same questions about hazard perception and risk arise with the barrier islands along the Florida coast. Figure 16.20a is an aerial photo of the chain along the Gulf Coast near Clearwater. North along the Florida coast, near Cedar Key, a residence sits on the barrier island at sea level (Figure 16.20b). One of the problems of

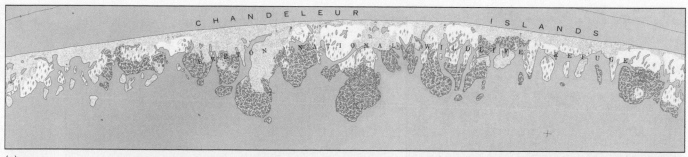

(a)

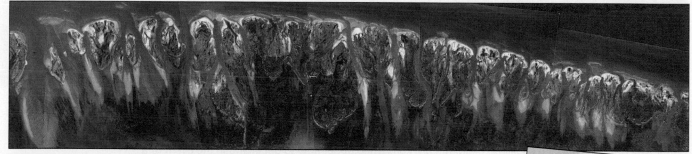

(b) 1988

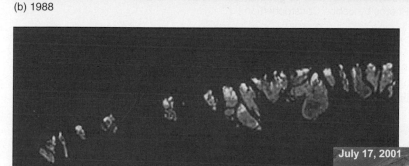

(c) September 16, 2005

July 17, 2001

August 31, 2005

SATELLITE   Hurricane Isabel 9/6–9/19/04

SATELLITE   Hurricane Georges

FIGURE 16.19 Hurricane Georges (1998) and Katrina (2005) take their toll.
Hurricane Georges eroded huge amounts of sand from the Chandeleur Islands, just off the Louisiana and Mississippi Gulf Coast. Compare the *before* topographic map (a) with the *after* aerial photo composite (b). (c) Little of the barrier-island chain is visible after Katrina's passage removed some 80% of the sand. (d) Aerial photo comparison between 2001 and two days after Katrina shows the devastation in the Northern Chandeleur Islands. The yellow arrow points to the same location in each photo. [(a) Topographic map provided by USGS; (b) photo by Aerial Data Service, Earth Imaging; (c) and (d) USGS Hurricane Impact Studies aerial photos—see *Sound Waves Newsletter*, September 2005.]

(d)

(a)

(b)

(c)

FIGURE 16.20 Florida Gulf Coast barrier islands.
(a) Marinas and development along the barrier islands near Clearwater, Florida. (b) A house on the Cedar Key barrier island. (c) All island residents are concerned about how to get inland in the event of a hurricane—evacuation signs mark certain highways. [Photos by Bobbé Christopherson.]

barrier-island living is being able to evacuate when a hurricane is on the way: Access is restricted, traffic snarls are common, and roadways and bridges may be washed out (Figure 16.20c).

Biological Processes: Coral Formations

Not all coastlines form by purely physical processes. Some form as the result of biological processes, such as coral growth. A **coral** is a simple marine animal with a small, cylindrical, saclike body called a *polyp*; it is related to other marine invertebrates, such as anemones and jellyfish. Corals secrete calcium carbonate ($CaCO_3$) from the lower half of their bodies, forming a hard, calcified external skeleton.

Corals live in a *symbiotic* relationship with algae: They live together in a mutually helpful arrangement, each dependent on the other for survival. Corals cannot photosynthesize, but they do obtain some of their own nourishment. Algae perform photosynthesis and convert solar energy to chemical energy in the system, providing the coral with about 60% of its nutrition and assisting the coral with the calcification process. In return, corals provide the algae with nutrients. Coral reefs are the most diverse marine ecosystems. Preliminary estimates of coral species place the number at a million worldwide, yet, as in most ecosystems in water or on land, biodiversity is declining in these communities.

Figure 16.21 shows the distribution of living coral formations. Corals mostly thrive in warm tropical oceans, so the difference in ocean temperature between the western coasts and eastern coasts of continents is critical to their distribution. Western coastal waters tend to be cooler, thereby discouraging coral activity, whereas eastern coastal currents are warmer and thus enhance coral growth.

Living colonial corals range in distribution from about 30° N to 30° S. Corals occupy a very specific ecological zone: 10–55 m (30–180 ft) depth, 27‰–40‰ (parts per thousand) salinity, and 18° to 29°C (64°–85°F) water temperature—30°C (86°F) is the upper threshold; above that temperature the corals begin to bleach and die. However, an interesting exception to this are unique species of cool corals that exist in cold, dark water and

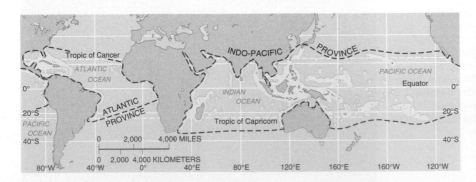

FIGURE 16.21 Worldwide distribution of living coral formations.
Yellow areas include prolific reef growth and atoll formation. The red dotted line marks the geographical limits of coral activity. Colonial corals range in distribution from about 30° N to 30° S.
[After J. L. Davies, *Geographical Variation in Coastal Development*, Essex, England: Longman House, 1973. Adapted by permission.]

deep ocean, at temperatures as low as 4°C (39°F) and in depths to 2000 m (6562 ft), beyond the expected range. Scientists study these remarkable oddities, which do not rely on algae but harvest nutrients from plankton and particulate matter. Corals require clear, sediment-free water and consequently do not locate near the mouths of sediment-charged freshwater streams. For example, note the lack of these structures along the U.S. Gulf Coast.

Coral Reefs There are both solitary and colonial corals. It is the colonial corals that produce enormous structures. Their skeletons accumulate, forming coral rock. Through many generations, live corals near the ocean's surface build on the foundation of older coral skeletons, which in turn may rest upon a volcanic seamount or some other submarine feature built up from the ocean floor. *Coral reefs* form by this process. Thus, a coral reef is a biologically derived sedimentary rock. It can assume one of several distinctive shapes.

In 1842, Charles Darwin hypothesized an evolution of reef formation. He suggested that as reefs developed around a volcanic island and the island itself gradually subsided, an equilibrium between the subsidence of the island and the upward growth of the corals is maintained. This idea, generally accepted today, is portrayed in Figure 16.22. Note the specific examples of each reef stage: *fringing reefs* (platforms of surrounding coral rock), *barrier reefs* (forming enclosed lagoons), and *atolls* (circular, ring-shaped).

Earth's most extensive fringing reef is the Bahamian platform in the western Atlantic (Figure 16.22d), covering some 96,000 km² (37,000 mi²). The largest barrier reef, the Great Barrier Reef along the shore of the state of Queensland, Australia, exceeds 2025 km (1260 mi) in length, is 16–145 km (10–90 mi) wide, and includes at least 700 coral-formed islands and keys (coral islets or barrier islands).

Coral Bleaching A troubling phenomenon is occurring among corals around the world as normally colorful corals turn stark white by expelling their own nutrient-supplying, colorful algae (red-brown to green), a phenomenon known as *bleaching*. Exactly why the corals eject their symbiotic partner is unknown, for without algae the corals die. Scientists are tracking this unprecedented bleaching and dying worldwide. Locations in the Caribbean Sea and the Indian Ocean, as well as off the shores of Australia, Indonesia, Japan, Kenya, Florida, Texas, and Hawai'i, are experiencing this phenomenon.

Possible causes include local pollution, disease, sedimentation, and changes in salinity. One acknowledged cause is the 1 to 2 C° (1.8 to 3.6 F°) warming of sea-surface temperatures, as stimulated by greenhouse warming of the atmosphere. In a report, the *Status of Coral Reefs of the World: 2000*, from the Global Coral Reef Monitoring Network, warmer water was found to be a greater threat to corals than local pollution or other environmental problems (see http://www.coris.noaa.gov/).

Coral bleaching is continuing all over the world as average ocean temperatures climb higher, thus linking the issue of climate change to the health of all living coral formations. By the end of 2000, approximately 30% of reefs were lost, especially following the record El Niño event of 1998. As sea-surface temperatures continue to rise, forecasts state that coral losses will continue. The 2007 IPCC *Fourth Assessment Report*, Working Group II, states, "Increases in sea surface temperatures of about 1 to 3 C° are projected to result in more frequent coral bleaching events and widespread mortality." For more information and Internet links, see http://www.usgs.gov/coralreef.html and the Coral Reef Monitoring Network at http://www.coral.noaa.gov/gcrmn/. The *World Atlas of Coral Reefs* in 2001 summarized:

Humans are thus bringing new pressures to bear on the world's coral reefs and driving more profound changes, more rapidly, than any natural impact has ever done. Over fishing has become so widespread that there are few, if any, reefs in the world which are not threatened.... From onshore a much greater suite of damaging activities is taking place. Often remote from reefs, deforestation, urban development, and intensive agriculture are now producing vast quantities of sediments and pollutants which are pouring into the sea and rapidly degrading coral reefs.... A further specter overshadowing the world of coral reefs is that of global climate change.*

Wetlands, Salt Marshes, and Mangrove Swamps

Some coastal areas have great *biological productivity* (plant growth; spawning grounds for fish, shellfish, and other organisms) stemming from trapped organic matter and sediments. Such a rich coastal marsh environment can greatly outproduce a wheat field in raw vegetation per acre. Thus, coastal marshes can support rich wildlife habitats. Unfortunately, these wetland ecosystems are quite fragile and are threatened by human development.

Wetlands are saturated with water enough of the time to support *hydrophytic vegetation* (plants that grow in water or wet soil). Wetlands usually occur on poorly drained soils. Geographically, they occur not only along coastlines but also as northern bogs (peatlands with high water tables), as potholes in prairie lands, as cypress swamps (with standing or gently flowing water), as river bottomlands and floodplains, and as arctic and subarctic environments that experience permafrost during the year. We studied the loss to coastal wetlands in the Mississippi Delta in Chapter 14.

*M. D. Spalding, C. Ravilious, E. P. Green, and UNEP/WCMC, *World Atlas of Coral Reefs* (Berkeley: University of California Press, 2001), p. 11.

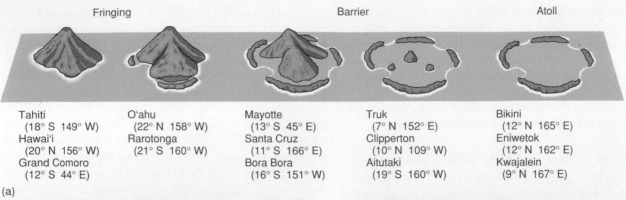

Fringing   Barrier   Atoll

Tahiti
(18° S 149° W)
Hawai'i
(20° N 156° W)
Grand Comoro
(12° S 44° E)

O'ahu
(22° N 158° W)
Rarotonga
(21° S 160° W)

Mayotte
(13° S 45° E)
Santa Cruz
(11° S 166° E)
Bora Bora
(16° S 151° W)

Truk
(7° N 152° E)
Clipperton
(10° N 109° W)
Aitutaki
(19° S 160° W)

Bikini
(12° N 165° E)
Eniwetok
(12° N 162° E)
Kwajalein
(9° N 167° E)

(a)

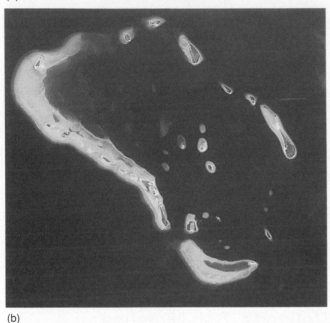

(b)

(c)

(d)

FIGURE 16.22 Coral forms.
(a) Common coral formations in a sequence of reef growth formed around a subsiding volcanic island: fringing reefs, barrier reefs, and an atoll. (b) Satellite image of a portion of the Maldive Islands, in the Indian Ocean (5° N 75° E). (c) Aerial photograph of the atolls in Bora Bora, Society Islands (16° S 152° W). (d) The Bahamas; the lighter areas feature ocean depths to 10 m (33 ft); the darker blues indicate depths dropping off to 4000 m (13,100 ft). [(a) After D. R. Stoddart, *The Geographical Magazine* 63 (1971): 610; (b) *Landsat-7* image courtesy of NASA; (c) photo by Harvey Lloyd/Stock Market; (d) *Terra* image courtesy of NASA/GSFC, MODIS Land Response Team, April, 18, 2000.]

Coastal Wetlands

Coastal wetlands are of two general types—salt marshes and mangrove swamps. In the Northern Hemisphere, **salt marshes** tend to form north of the 30th parallel, whereas **mangrove swamps** form equatorward of that line. This distribution is dictated by the occurrence of freezing conditions, which control the survival of mangrove seedlings. Roughly the same latitudinal limits apply in the Southern Hemisphere.

Salt marshes usually form in estuaries and behind barrier beaches and spits. An accumulation of mud produces a site for the growth of *halophytic* (salt-tolerant) plants. This vegetation then traps additional alluvial sediments and adds to the salt marsh area. Because salt marshes are in the intertidal zone (between the farthest reaches of high and low tides), sinuous, branching channels are produced as tidal waters flood into and ebb from the marsh (Figure 16.23).

FIGURE 16.23 Coastal salt marsh.
Salt marshes are productive ecosystems commonly occurring poleward of 30° latitude in both hemispheres. This is Gearheart Marsh, part of the Arcata Marsh system on the Pacific Coast in northern California. [Photo by Bobbé Christopherson.]

(a)

FIGURE 16.24 Mangroves.
Mangroves tend to grow equatorward of 30° latitude.
(a) Mangroves along the East Alligator River (12° S latitude) in Kakadu National Park, Northern Territory, Australia. (b) Mangroves retain sediments and can form anchors for island formations, seen here at sunset in "Ding" Darling National Wildlife Refuge, Sanibel Island, Florida. (c) Mangroves in the Florida Keys. [Photos by (a) Belinda Wright/DRK Photo; (b) Bobbé Christopherson; and (c) author.]

(b)

(c)

Sediment accumulation on tropical coastlines provides the site for mangrove trees, shrubs, and other small trees. The prop roots of the mangrove are constantly finding new anchorages. The roots are visible above the waterline but reach below the water surface, providing a habitat for a multitude of specialized life forms. Mangrove swamps often secure and fix enough material to form islands (Figure 16.24).

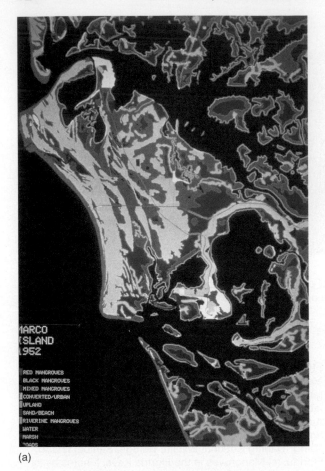

(a)

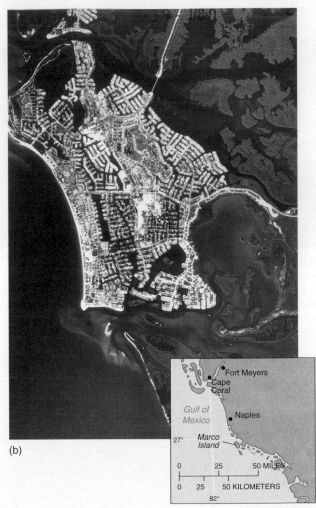

(b)

FIGURE 16.25 Mangrove loss at Marco Island.
Digitized inventory of Marco Island, Florida: (a) 1952 survey map and (b) 1984 color-infrared photograph of the island. This type of analysis is similar to GIS (Geographic Information System) studies. For more on GIS, see Figure 1.29. [From S. Patterson, *Mangrove Community Boundary Interpretation and Detection of Areal Changes on Marco Island, Florida,* Biological Report 86 (10), for National Wetlands Research Center, U.S. Fish and Wildlife Service, August 1986, pp. 23, 49.]

Marco Island The development of Florida's Marco Island is an example of mangrove loss due to urbanization. Marco Island is on the southwestern Florida coast about 24 km (15 mi) south of Naples. Figure 16.25a shows Marco Island as it was in 1952: approximately 5300 acres of subtropical mangrove habitat and barrier-island terrain, formed from river sediments and old shell and reef fragments.

Development began in 1962, despite tropical storms that occasionally assault the area, as Hurricane Donna did in 1960. It included artificial landfill for housing sites and general urbanization of the entire island. Scientists regarded the island and estuarine habitat as highly productive and unique in its mixture of mangrove species, bird species, and a productive aquatic environment.

Consequently, numerous challenges and court actions were initiated in attempts to halt development. But construction continued, and by 1984 Marco Island was completely developed (Figure 16.25b). Compare the 1952 digitized inventory and 1984 color-infrared aerial photograph to see the extent of urban development on this island. Note that the only remaining natural community is restricted to a very limited portion along the extreme perimeter. Today, geographic information system models greatly facilitate such comparative analyses.

The World Resources Institute and the U.N. Environment Programme estimate that from preagricultural times to today, mangrove losses are running between 40% (examples, Cameroon and Indonesia) and nearly 80% (examples, Bangladesh and Philippines). Deliberate removal was a common practice by many governments in the early days of settlement because of a falsely conceived fear of disease or pestilence in these swamplands.

Human Impact on Coastal Environments

Development of modern societies is closely linked to estuaries, wetlands, barrier beaches, and coastlines. Estuaries are important sites of human settlement because they provide natural harbors, a food source, and convenient sewage and waste disposal. Society depends on daily tidal flushing of estuaries to dilute the pollution created by waste disposal.

FIGURE 16.26 Barrier-island subdivisions.
The town of Barnegat Light abruptly stops where Long Beach Island is protected in Barnegat Lighthouse State Park, New Jersey. The protected area is almost entirely composed of replenished sand. [Photo by Bobbé Christopherson.]

Thus, the estuarine and coastal environment is vulnerable to abuse and destruction if development is not carefully planned. We are approaching a time when most barrier islands will be developed and occupied (Figure 16.26).

Barrier islands and coastal beaches do migrate over time. Houses that were more than a mile from the sea in the Hamptons along the southeastern shore of Long Island, New York, are now within only 30 m (100 ft) of it. The time remaining for these homes can be quickly shortened by a single hurricane or by a sequence of storms (Figure 16.27). Despite our understanding of beach and barrier-island migration, the effect of storms, and warnings from scientists and government agencies, coastal development proceeds. Society behaves as though beaches and barrier islands are stable, fixed features, or as though they can be engineered to be permanent. Experience has shown that severe erosion generally cannot be prevented. Shoreline planning is the topic of Focus Study 16.1.

A reasonable conclusion seems to be that the key to protective environmental planning and zoning is to *allocate responsibility and cost in the event of a disaster*. An ideal system places a hazard tax on land, based on assessed risk, and restricts the government's responsibility to fund reconstruction or an individual's right to reconstruct on frequently damaged sites. Comprehensive mapping of erosion-hazard areas would help avoid the ever-increasing costs from recurring disasters.

(a)

(b)

FIGURE 16.27 Beach erosion dooms houses in Massachusetts and New York.
(a) Two homes on stilts sit atop a primary berm made of cobbles. The ocean is just beyond the houses along this Massachusetts shore. (b) Placing houses on stilts along the shore here on Long Island offers little protection from an attacking storm surge.
[Photos by (a) Bobbé Christopherson; (b) Mark Wexler/Woodfin Camp & Associates.]

Focus Study 16.1

An Environmental Approach to Shoreline Planning

Coastlines are places of wonderful opportunity. They also are zones of specific constraints. Poor understanding of this resource and a lack of environ- mental analysis often go hand in hand. The late ecologist and landscape architect Ian McHarg, in *Design with Nature*, discusses the New Jersey shore. He shows how proper understanding of a coastal environment could have avoided problems from major storms and coastal development.

(continued)

Focus Study 16.1 (continued)

Much of what is presented here applies to coastal areas elsewhere. Figure 16.1.1 illustrates the New Jersey shore from ocean to back bay. Let us walk across this landscape and discover how it should be treated under ideal conditions.

Beaches and Dunes—Where to Build?

We begin our walk at the water's edge and proceed across the beach. Sand beaches are the primary natural defense against the ocean; they act as bulwarks against the pounding of a stormy sea. The shoreline tolerates recreation, but not construction, because of its shifting, changing nature during storms, daily tidal fluctuations, and the potential effects of rising sea level. Beaches are susceptible to pollution and require environmental

Ocean	Beach	Primary dune	Trough	Secondary dune	Backdune	Bayshore	Bay
Tolerant	Tolerant	Intolerant	Relatively tolerant	Intolerant	Tolerant	Intolerant	Tolerant
Intensive recreation	Intensive recreation	No passage, breaching, or building	Limited recreation	No passage, breaching, or building	Most suitable for development	No filling	Intensive recreation
Subject to pollution controls	Intolerant of construction		Limited structures				

FIGURE 16.1.1 Planning along a coast.
(a) Coastal environment—a planning perspective from ocean to bay along the New Jersey shore. Note the placement of each letter on the illustration to identify photo location. (b) Light trucks and SUVs drive along the beach in the protected Island Beach State Park, whereas they are banned from the stretches of beach that are developed with houses and business. (c) Houses once distant from the surf are hit by waves. Sand is brought in to strengthen defenses. (d) The primary dune in a somewhat natural state. (e) An undeveloped trough behind the primary dune. (f) Development along the primary dunes—an amusement park—challenges the natural setting. (g) The former trough, paved and developed, barely above high-tide levels. (h) Development and commerce in the bayshore and bay. [(a) After *Design with Nature* by Ian McHarg. Copyright © 1969 by Ian L. McHarg. Adapted by permission. All photos by Bobbé Christopherson.]

protection to control nearshore dumping of dangerous materials.

Walking inland, we encounter the primary dune. It is even more sensitive than the beach: It is fragile, easily disturbed, and vulnerable to erosion, and it cannot tolerate the passage of people trekking to the beach. Some vegetation is present, but it is easily disturbed and has only a small effect on sand stabilization. Primary dunes are like human-made dikes in the Netherlands: They are the primary defense against the sea, so development or heavy traffic should not disturb them. Carefully controlled access points to the beach should be designated, and restricted access should be enforced, for even foot traffic can cause destruction. (See two examples of beach stabilization in Figure 15.5.)

The trough behind the primary dune is relatively tolerant of limited recreation and building. The plants that fix themselves to the surface send roots down to fresh groundwater reserves and anchor the sand in the process. Thus, if construction should inhibit the surface recharge of that water supply, the natural protective ground cover could die and destabilize the environment. Or subsequent saltwater intrusion might contaminate well water. Clearly, groundwater resources and the location of recharge aquifers must be considered in planning. The

continuing use of septic sewage systems in the trough is not sustainable in these environments.

Behind the trough is the secondary dune, a second line of defense against the sea. It, too, is tolerant of some use yet is vulnerable to destruction. Next is the backdune, more suitable for development than any zone between it and the sea. Further inland are the bayshore and the bay, where no dredging or filling and only limited dumping of treated wastes and toxics should be permitted. This zone is tolerant to intensive recreation. In reality, of course, the opposite of such careful assessment and planning prevails.

A Scientific View and a Political Reality

Before an intensive government study of the New Jersey shore was completed in 1962, no analysis of coastal hazards had been done outside of academic circles. What is common knowledge to geographers, botanists, biologists, and ecologists in the classroom and laboratory still has not filtered through to the general planning and political processes. As a result, on the New Jersey shore and along much of the Atlantic and Gulf Coasts, improper development of the fragile coastal zone (primary dunes, trough, and secondary dunes) led to extensive destruction during storms in

1962, 1992, and numerous other times.

Scientists estimate that the coastlines will continue their retreat, in some cases tens of meters in a few decades. Such estimates include the entire East Coast around to the Gulf Coast. Society should reconcile ecology and economics if these coastal environments are to be sustained.

South Carolina enacted the Beach Management Act of 1988 (modified 1990) to apply some of McHarg's principles. In its first two years, more than 70 lawsuits protested the act as an invalid seizure of private property without compensation. Destructive or threatening hurricane activity is helping to drive the importance of such cooperative planning and proactive laws. Similar measures in other states have faced the same difficult path. Implementation of any planning process is problematic, political pressure is intense, and results are mixed. Ian McHarg voiced an optimistic hope: "May it be that these simple ecological lessons will become known and incorporated into ordinance [law] so that people can continue to enjoy the special delights of life by the sea."*

*From *Design with Nature* by Ian McHarg, p. 17. © 1969 by Ian L. McHarg. Published by Bantam Doubleday Dell Publishing Group, Inc.

Summary and Review—The Oceans, Coastal Processes, and Landforms

■ *Describe* the chemical composition of seawater and the physical structure of the ocean.

Water is the "universal solvent," dissolving at least 57 of the 92 elements found in nature. Most natural elements and the compounds they form are found in the seas as dissolved solids. Seawater is a solution, and the concentration of dissolved solids is **salinity**. **Brine** exceeds the average 35‰ (parts per thousand) salinity; **brackish** applies to water that is less than 35‰. The ocean is divided by depth into a narrow mixing zone at the surface, a thermocline transition zone, and the deep cold zone.

salinity (p. 499)
brine (p. 500)
brackish (p. 500)

1. Describe the salinity of seawater: its composition, amount, and distribution.
2. Analyze the latitudinal distribution of salinity as shown in Figure 16.3. Why is salinity less along the equator and greater in the subtropics?
3. What are the three general zones relative to physical structure within the ocean? Characterize each by temperature, salinity, dissolved oxygen, and dissolved carbon dioxide.

■ *Identify* the components of the coastal environment, and *list* the physical inputs to the coastal system, including tides and mean sea level.

The coastal environment is the **littoral zone** and exists where the tide-driven, wave-driven sea confronts the land. Inputs to the coastal environment include solar energy, wind and weather, climatic variation, the nature of coastal geomorphology, and human activities.

Mean sea level (MSL) is based on average tidal levels recorded hourly at a given site over many years. MSL varies spatially because of ocean currents and waves, tidal variations, air temperature and pressure differences, ocean temperature variations, slight variations in Earth's gravity, and changes in oceanic volume. MSL worldwide is rising in response to global warming of the atmosphere and oceans.

Tides are complex daily oscillations in sea level, ranging worldwide from barely noticeable to many meters. Tides are produced by the gravitational pull of both the Moon and the Sun. Most coastal locations experience two high (rising) **flood tides** and two low (falling) **ebb tides** every day. The difference between consecutive high and low tides is the tidal range. **Spring tides** exhibit the greatest tidal range, when the Moon and Sun are either in conjunction or opposition. **Neap tides** produce a lesser tidal range.

> littoral zone (p. 501)
> mean sea level (MSL) (p. 502)
> tide (p. 504)
> flood tide (p. 505)
> ebb tide (p. 505)
> spring tide (p. 505)
> neap tide (p. 505)

4. What are the key terms used to describe the coastal environment?
5. Define mean sea level. How is this value determined? Is it constant or variable around the world? Explain.
6. What interacting forces generate the pattern of tides?
7. What characteristic tides are expected during a new Moon or a full Moon? During the first-quarter and third-quarter phases of the Moon? What is meant by a flood tide? An ebb tide?
8. Is tidal power being used anywhere to generate electricity? Explain briefly how such a plant would utilize the tides to produce electricity. Are there any sites in North America? Where are they?

■ *Describe* wave motion at sea and near shore, and *explain* coastal straightening as a product of wave refraction.

Friction between moving air (wind) and the ocean surface generates undulations of water that we call **waves**. Wave energy in the open sea travels through water, but the water itself stays in place. Regular patterns of smooth, rounded waves, the mature undulations of the open ocean, are **swells**. Near shore, the restricted depth of water slows the wave, forming *waves of translation*, in which both energy and water actually move forward toward shore. As the crest of each wave rises, the wave falls into a characteristic **breaker**.

Wave refraction redistributes wave energy so that different sections of the coastline vary in erosion potential. Headlands are eroded, whereas coves and bays receive materials, with the long-term effect of straightening the coast. As waves approach a shore at an angle, refraction produces a **longshore current** of water moving parallel to the shore. This current produces the *longshore drift* of sand, sediment, and gravel and assorted materials. **Littoral drift** is transport of materials along the shore and a more comprehensive term. Particles move along the beach as **beach drift**, back and forth between water and land. A **tsunami**

is a seismic sea wave triggered by an undersea landslide or earthquake. It travels at great speeds in the open sea and gains height as it comes ashore, posing a coastal hazard.

> wave (p. 507)
> swell (p. 507)
> breaker (p. 507)
> wave refraction (p. 507)
> longshore current (p. 507)
> littoral drift (p. 507)
> beach drift (p. 507)
> tsunami (p. 508)

9. What is a wave? How are waves generated, and how do they travel across the ocean? Does the water travel with the wave? Discuss the process of wave formation and transmission.
10. Describe the refraction process that occurs when waves reach an irregular coastline. Why is the coastline straightened?
11. Define the components of beach drift and the longshore current and longshore drift.
12. Explain how a seismic sea wave attains such tremendous velocities. Why is it given a Japanese name?

■ *Identify* characteristic coastal erosional and depositional landforms.

An *erosional coast* features wave action that cuts a horizontal bench in the tidal zone, extending from a sea cliff out into the sea. Such a structure is a **wave-cut platform**, or *wave-cut terrace*. In contrast, *depositional coasts* generally are located along land of gentle relief, where depositional sediments are available from many sources. Characteristic landforms deposited by waves and currents are a **barrier spit** (material deposited in a long ridge extending out from a coast); a **bay barrier**, or *baymouth bar* (a spit that cuts off the bay from the ocean and forms an inland **lagoon**); a **tombolo** (where sediment deposits connect the shoreline with an offshore island or sea stack); and a **beach** (land along the shore where sediment is in motion, deposited by waves and currents). A beach helps to stabilize the shoreline, although it may be unstable seasonally.

> wave-cut platform (p. 511)
> barrier spit (p. 511)
> bay barrier (p. 512)
> lagoon (p. 512)
> tombolo (p. 512)
> beach (p. 513)

13. What is meant by an erosional coast? What are the expected features of such a coast?
14. What is meant by a depositional coast? What are the expected features of such a coast?
15. How do people attempt to modify littoral drift? What strategies are used? What are the positive and negative impacts of these actions?
16. Describe a beach—its form, composition, function, and evolution.
17. What success has Miami had with beach replenishment? Is it a practical strategy?

■ *Describe* barrier islands and their hazards as they relate to human settlement.

Barrier chains are long, narrow, depositional features, generally of sand, that form offshore roughly parallel to the coast.

Common forms are **barrier beaches** and the broader, more extensive **barrier islands**. Barrier formations are transient coastal features, constantly on the move, and they are a poor, but common, choice for development.

> barrier beach (p. 516)
> barrier island (p. 516)

18. On the basis of the information in the text and any other sources at your disposal, do you think barrier islands and beaches should be used for development? If so, under what conditions? If not, why not?
19. After Hurricane Hazel destroyed the Grand Strand off South Carolina in 1954, settlements were rebuilt, yet were hit again by Hurricane Hugo 35 years later, in 1989. Why does this type of recurring damage happen to human populations?

■ *Assess* living coastal environments: corals, wetlands, salt marshes, and mangroves.

A **coral** is a simple marine invertebrate that forms a hard, calcified, external skeleton. Over generations, corals accumulate in large reef structures. Corals live in a *symbiotic* (mutually helpful) relationship with algae; each is dependent on the other for survival.

Wetlands are lands saturated with water that support specific plants adapted to wet conditions. They occur along coastlands and inland in bogs, swamps, and river bottomlands. Coastal wetlands form as **salt marshes** poleward of the 30th parallel in each hemisphere and as **mangrove swamps** equatorward of these parallels.

> coral (p. 520)
> wetlands (p. 521)
> salt marsh (p. 522)
> mangrove swamp (p. 522)

20. How are corals able to construct reefs and islands?
21. Describe a trend in corals that is troubling scientists, and discuss some possible causes.
22. Why are the coastal wetlands poleward of 30° N and S latitude different from those that are equatorward? Describe the differences.

■ *Construct* an environmentally sensitive model for settlement and land use along the coast.

Coastlines are zones of specific constraints. Poor understanding of this resource and a lack of environmental analysis often go hand in hand, producing frequent disasters to coastal ecosystems and real property losses. Society must reconcile ecology and economics if these coastal environments are to be sustained.

23. Describe the condition of Marco Island, Florida. Was a rational model used to assess the environment prior to development? What economic and political forces were involved?
24. What type of environmental analysis is needed for rational development and growth in a region like the New Jersey shore? Evaluate South Carolina's approach to coastal hazards and protection.

NetWork

The *Geosystems* Student Learning Center provides on-line resources for this chapter on the World Wide Web. To begin: Once at the Center, click on the cover of this textbook, scroll the Table of Contents menu, and select this chapter. You will find self-tests that are graded, review exercises, specific updates for items in the chapter, and in "Destinations" many links to interesting related pathways on the Internet. *Geosystems* Student Learning Center is found at **http://www.prenhall.com/ christopherson/**.

Critical Thinking

A. This chapter includes the following statement:

> The key to protective environmental planning and zoning is to *allocate responsibility and cost in the event of a disaster.* An ideal system places a hazard tax on land, based on assessed risk, and restricts the government's responsibility to fund reconstruction or an individual's right to reconstruct at frequently damaged sites.

What do you think about this as a policy statement? How would you approach implementing such a strategy? In what way could you use geographic information systems (GIS), as described in Chapter 1, to survey, assess, list owners, and follow taxation status for a vulnerable stretch of coastline?

B. Under "Destinations" in Chapter 16 of the *Geosystems* Home Page there is a link called "Coral Reefs." Sample some of the links on this page. Do you find any information about the damage to and bleaching of coral reefs reported in 1998 or at present? In which places in the world? Are there some suspect causes presented in "Bleaching Hot Spots" or any of the "Coral Reef Alliance" references?

A dramatic moment at the 14th July Glacier (Fjortende Julibreen), west coast of Spitsbergen Island near 79° N. An active glacial front calves blocks of glacial ice as it moves forward into a body of water. Five seconds after the main photo, an enormous calving happened, shown in the inset. Wave energy quickly spread across the bay, producing breakers. The glacial front is approximately 35 m (115 ft) high, yet when the ice plunges into the water, heaving up and down to establish equilibrium, the ice block reaches a point where it is 1/7th exposed and 6/7th submerged—an iceberg. [Photos by Bobbé Christopherson.]

Glacial and Periglacial Processes and Landforms

■ Key Learning Concepts

After reading the chapter, you should be able to:

- ■ **Differentiate** between alpine and continental glaciers, and **describe** their principal features.

- ■ **Describe** the process of glacial ice formation, and **portray** the mechanics of glacial movement.

- ■ **Describe** characteristic erosional and depositional landforms created by alpine glaciation and continental glaciation.

- ■ **Analyze** the spatial distribution of periglacial processes, and **describe** several unique landforms and topographic features related to permafrost and frozen ground phenomena.

- ■ **Explain** the Pleistocene ice-age epoch and related glacials and interglacials, and **describe** some of the methods used to study paleoclimatology.

About 77% of Earth's freshwater is frozen, with the bulk of that ice sitting restlessly in just two places—Greenland and Antarctica. The remaining ice covers various mountains and fills some alpine valleys. A volume of more than 32.7 million km³ (7.8 million mi³) of water is tied up as ice in Greenland (2.38 million km³), Antarctica (30.1 million km³), and ice caps and mountain glaciers worldwide (180,000 km³). For comparison, the Vatnajökull ice cap in Figure 17.5 has a volume of 3100 km³.

These deposits provide an extensive frozen record of Earth's climatic history over the past several million years and perhaps some clues to its climatic future. This is Earth's *cryosphere*, the portion of the hydrosphere and groundwater that is perennially frozen, generally at high latitudes and elevations.

Worldwide, glacial ice is in retreat, melting at rates exceeding anything in the ice record. In the European Alps alone some 75% of the glaciers have receded in the past 50 years, losing more than 50% of their ice mass since 1850. At this rate, the European Alps will have only 20% of their preindustrial glacial ice left by 2050.

In this chapter: We focus on Earth's extensive ice deposits—their formation, movement, and the ways in which they produce various erosional and depositional landforms. Glaciers, transient landforms themselves, leave in their wake a variety of landscape features. The fate of glaciers is intricately tied to change in global temperature, which ultimately concerns us all. We discuss the methods used to decipher past climates—the science of *paleoclimatology*—and the clues for understanding future climate patterns. Exciting discoveries are arising from ice cores taken from Earth's two ice sheets, one dating back 800,000 years from the present.

We examine the cold world of permafrost and periglacial geomorphic processes. These periglacial environments may be separated from actual glacial ice across time (glaciers from past ages) and space (physical distance in the present). Approximately 25% of Earth's land area is subject to freezing conditions and frost action characteristic of periglacial regions, including areas that have specific related features of relict, or past, permafrost.

Rivers of Ice

A **glacier** is a large mass of ice resting on land or floating as an ice shelf in the sea adjacent to land. Glaciers are not frozen lakes or groundwater ice. Instead, they form by the continual accumulation of snow that recrystallizes under its own weight into an ice mass. Glaciers are not stationary; they move slowly under the pressure of their own great weight and the pull of gravity. In fact, they move slowly in streamlike patterns, merging as tributaries into large rivers of ice, as you can see in the satellite image and aerial photos in Figure 17.1. In Greenland and Antarctica

vast sheets of ice dominate, slowly flowing outwards towards the ocean.

Today, these slowly flowing rivers and sheets of ice dominate about 11% of Earth's land area. As much as 30% of continental land was covered by glacial ice during colder episodes in the past. Through these "ice ages," below-freezing temperatures prevailed at lower latitudes more than they do today, allowing snow to accumulate year after year.

(a)

(b)

(c)

FIGURE 17.1 Rivers and sheets of ice.
(a) Alpine glaciers merge from adjoining glacial valleys in the northeast region of Ellesmere Island in the Canadian Arctic. (b) Numerous glaciers carve the coast of Greenland. (c) These peaks, known as *nunataks*, rise above the Greenland Ice Sheet. There are no clouds in the photo; this is all ice at approximately 2500 m (8200 ft) elevation—an accumulation perhaps 100,000 years in the making. [(a) *Terra* image courtesy of University of Alberta, NASA/GSFC/ERSDAC/JAROS and the U.S./Japan ASTER Science Team, July 31, 2000; (b) and (c) photos by Bobbé Christopherson.]

Glaciers form in areas of permanent snow, both at high latitudes and at high elevations at any latitude. A **snowline** is the lowest elevation where snow can survive year-round; specifically, it is the lowest line where winter snow accumulation persists throughout the summer. Glaciers form on some high mountains along the equator, such as in the Andes Mountains of South America and on Mount Kilimanjaro in Tanzania, Africa. On equatorial mountains, the snowline is around 5000 m (16,400 ft); on midlatitude mountains, such as the European Alps, snowlines average 2700 m (8850 ft); and in southern Greenland, snowlines are as low as 600 m (1970 ft). (For Internet links to Global Land Ice Measurements from Space, go to **http://www.glims.org/**, which includes an inventory of world glaciers, or the National Snow and Ice Data Center at **http://nsidc.org/**. *Landsat*-7 images are listed at **http://www.emporia.edu/earthsci/gage/glacier7.htm**.)

Glaciers are as varied as the landscape itself. They fall within two general groups, based on their form, size, and flow characteristics: alpine glaciers and continental glaciers.

Alpine Glaciers

With few exceptions, a glacier in a mountain range is an **alpine glacier**, or *mountain glacier*. The name comes from the Alps of central Europe, where such glaciers abound. Alpine glaciers form in several subtypes. One prominent type is a *valley glacier*, literally a river of ice confined within a valley that originally was formed by stream action. Such glaciers range in length from only 100 m (325 ft) to more than 100 km (60 mi). In Figure 17.2, at least a half dozen valley glaciers are identifiable in the high-altitude photograph of the Alaska Range. Several are named on the map, specifically the Eldridge and Ruth Glaciers, which fill valleys as they flow from source areas near Mount McKinley.

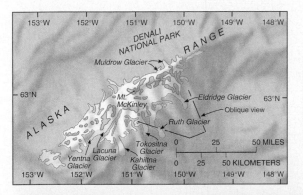

FIGURE 17.2 Glaciers in south-central Alaska.
Oblique infrared (false-color) image of Eldridge and Ruth Glaciers, with Mount McKinley at upper left, in the Alaska Range of Denali National Park. Photo made at 18,300 m (60,000 ft). [Alaska High Altitude Aerial Photography from EROS Data Center, USGS.]

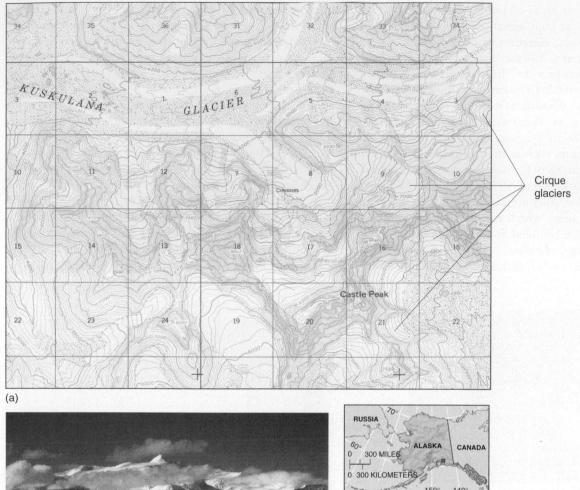

(a)

Cirque
glaciers

(b)

FIGURE 17.3 Topographic map from southeastern Alaska. (a) Numerous cirque glaciers depicted on a topographic map of southeastern Alaska (McCarthy C-7 Quadrangle, 1:63,360 scale, 100-ft contour interval). The blue areas represent active glaciers, and the brown-striped areas on the ice are moraines. (b) Aerial photo of the active Kuskulana glacier in the Wrangell Range. [(a) Courtesy of U.S. Geological Survey; (b) photo by Steve McCutcheon/Visuals Unlimited.]

As a valley glacier flows slowly downhill, the mountains, canyons, and river valleys beneath its mass are profoundly altered by its erosive passage. Some of the debris created by the glacier's excavation is transported on the ice, visible as dark streaks and bands being transported for deposition elsewhere; other portions of its debris load are carried within or along its base.

Most alpine glaciers originate in a mountain *snowfield* that is confined in a bowl-shaped recess. This scooped-out erosional landform at the head of a valley is a **cirque**. A glacier that forms in a cirque is a *cirque glacier.* Several cirque glaciers may jointly feed a valley glacier, as shown on the topographic map and photograph in Figure 17.3. The Kuskulana Glacier, which flows off the map to the west, is supplied by numerous cirque glaciers feeding several tributary valley glaciers. In the aerial photos in Figure 17.1, how many valley glaciers join the main glacier in (a) and (b)?

Wherever several valley glaciers pour out of their confining valleys and coalesce at the base of a mountain range, a *piedmont glacier* is formed and spreads freely over the lowlands, such as the Malaspina Glacier that flows into Yakutat Bay, Alaska. A *tidewater glacier*, or *tidal glacier*, ends in the sea, calving (breaking off) to form floating ice as **icebergs** (Figure 17.4a and b). Icebergs usually form wherever glaciers meet an ocean, or bay, or fjord. Icebergs are inherently unstable, as their center of gravity shifts with melting and further breaking. Figure 17.4c shows such adjustments; note the former waterline, melt channels, and delicately sculpted underwater surfaces, far above water level. In Figure 17.4d, a magnificent ice tower brings to mind the ratio of 1/7 (14%) exposed and 6/7 (86%) submerged—these proportions varying depending on the age of the ice and air content.

(a)

(b)

(c)

(d)

FIGURE 17.4 Glacial ice ready to calve into the sea and icebergs.
(a) When glaciers reach the sea, large pieces break off, forming icebergs, such as these massive 30-m-(98-ft-) high blocks ready to fall at Neko Harbor, Antarctica. (b) A tidewater glacier calving numerous icebergs into the ocean in southern Greenland. (c) A weathered iceberg off Antarctica; see the former waterline marking where the berg floated lower in the water when it was greater in mass. (d) This iceberg tower in East Greenland is about 60 m (200 ft) tall; imagine its keel extending some 360 m (1180 ft) below the surface. [All photos by Bobbé Christopherson.]

Continental Glaciers

On a much larger scale than individual alpine glaciers, a continuous mass of ice is a **continental glacier**. In its most extensive form, it is an **ice sheet**. Most of Earth's glacial ice exists in the ice sheets that blanket 81% of Greenland—1,756,000 km² (678,000 mi²) of ice—and 90% of Antarctica—14.2 million km² (5.48 million mi²) of ice. Antarctica alone has 92% of all the glacial ice on the planet. Refer to Figure 1.26 in Chapter 1 for an image of Antarctica.

The Antarctic and Greenland ice sheets have such enormous mass that large portions of each landmass beneath the ice are isostatically depressed (pressed down by weight) below sea level. Each ice sheet reaches depths of more than 3000 m (10,000 ft), with average depths around 2000 m, burying all but the highest peaks.

Two additional types of continuous ice cover associated with mountain locations are *ice caps* and *ice fields*. An **ice cap** is roughly circular and, by definition, covers an area of less than 50,000 km² (19,300 mi²). An ice cap completely buries the underlying landscape. The volcanic island of Iceland features several ice caps, such as the Vatnajökull Ice Cap in the *Landsat* image in Figure 17.5a. Volcanoes lie beneath these icy surfaces. Iceland's Grímsvötn Volcano erupted in 1996 and again in 2004, producing large quantities of melted glacial water and floods, a flow Icelanders call a *jökulhlaup*.

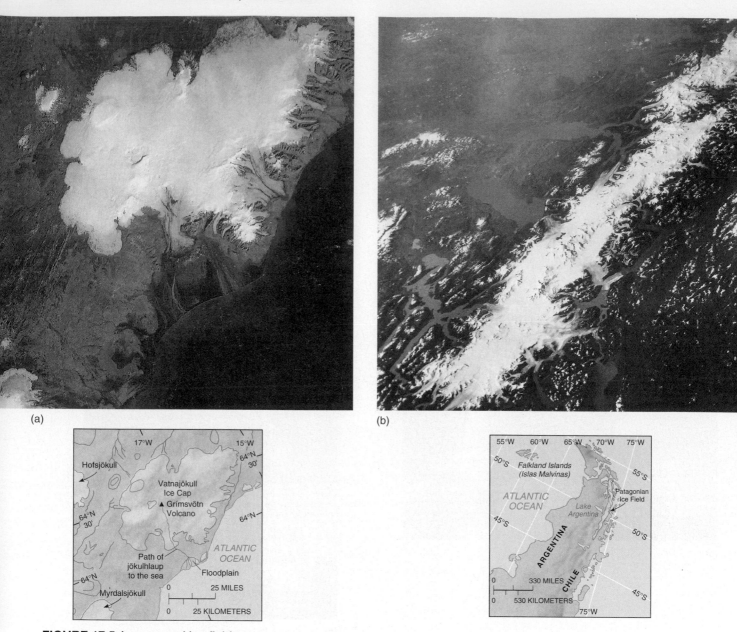

FIGURE 17.5 Ice cap and ice field.
(a) The Vatnajökull ice cap in southeastern Iceland (*jökull* means "ice cap" in Danish). Note the location of the Grímsvötn Volcano on the map and the path of the floods to the sea. (b) The southern Patagonian ice field of Argentina. [(a) *Landsat* image from NASA; (b) photo by Cosmonauts G. M. Greshko and Yu V. Romanenko, *Salyut 6.*]

An **ice field** is not extensive enough to form the characteristic dome of an ice cap; instead, it extends in a characteristic elongated pattern in a mountainous region. A fine example is the Patagonian ice field of Argentina and Chile, one of Earth's largest. It attains only 90 km (56 mi) width but stretches 360 km (224 mi), from 46° to 51° S latitude (equivalent in latitude from South Dakota to central Manitoba in the Northern Hemisphere). In an ice field, ridges and peaks are visible above the buried terrain; the term *nunatak* refers to these peaks, visible in Figures 17.5b and 17.1c.

Continuous ice sheets or ice caps are drained by rapidly moving, solid *ice streams* that form around their periphery, moving to the sea or to lowlands. Such frozen ice streams flow from the edges of Greenland and Antarctica through stationary and slower-moving ice. An *outlet glacier* flows out from an ice sheet or ice cap but is constrained by a mountain valley or pass.

Glacial Processes

A glacier is a dynamic body, moving relentlessly downslope at rates that vary within its mass, excavating the landscape through which it flows. The mass is dense ice that is formed from snow and water through a process of compaction, recrystallization, and growth. A glacier's *mass budget* consists of net gains or losses of this glacial ice, which determine whether the glacier expands or retreats. Let us now look at glacial ice formation, mass balance, movement, and erosion before we discuss the fascinating landforms produced by these processes.

Formation of Glacial Ice

Consider for a moment the nature of ice. You may be surprised to learn that ice is both a mineral (an inorganic natural compound of specific chemical makeup and crystalline structure) and a rock (a mass of one or more minerals). Ice is a frozen fluid, a trait that it shares with igneous rocks. The accumulation of snow in layered deposits is similar to sedimentary rock formations. To give birth to a glacier, snow and ice are transformed under pressure, recrystallizing into a type of metamorphic rock. Glacial ice is a remarkable material!

The essential input to a glacier is snow that accumulates in a snowfield, a glacier's accumulation zone (Figure 17.6a and c). Snowfields typically are at the highest elevation of

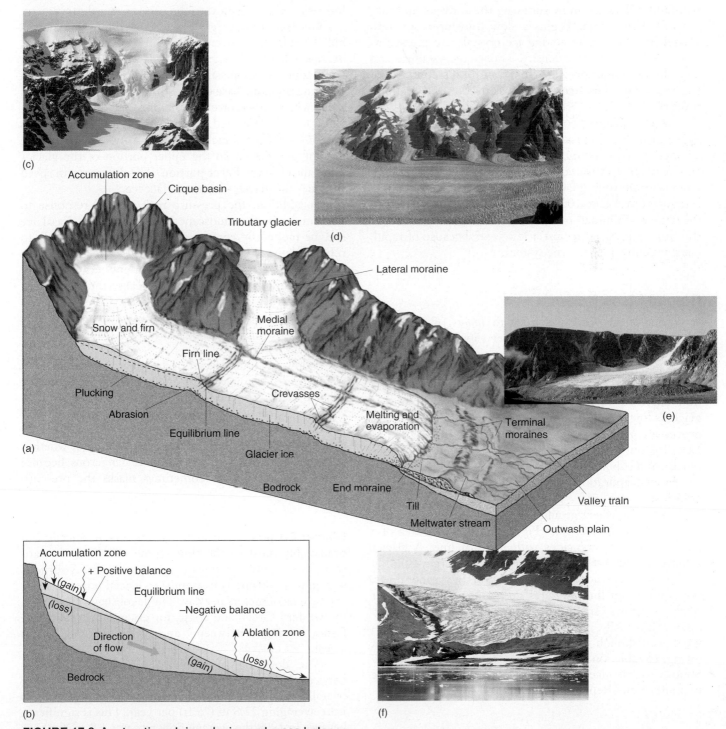

FIGURE 17.6 A retreating alpine glacier and mass balance.
(a) Cross section of a typical retreating alpine glacier. (b) Annual mass balance of a glacial system, showing how the relation between accumulation and ablation controls the location of the equilibrium line. (c) The accumulation zone for a glacier in Antarctica. (d) Four tributary glaciers flow into a compound valley glacier in Greenland. (e) A terminal moraine marks the farthest advance of this glacier on Nordaustlandet Island, Arctic Ocean. (f) Moraine rock and debris form a terminal moraine and lateral moraines left by a retreating glacier.
[All photos by Bobbé Christopherson.]

ANIMATION

Budget of a Glacier, Mass Balance, Flow of Ice Within a Glacier

an ice sheet, ice cap, or head of a valley glacier, usually in a cirque. Avalanches from surrounding mountain slopes can add to the snowfield.

As the snow accumulation deepens in sedimentary-like layers, the increasing thickness results in increased weight and pressure on underlying ice. Rain and summer snowmelt then contribute water, which stimulates further melting, and that meltwater seeps down into the snowfield and refreezes. Snow surviving the summer and into the following winter begins a slow transformation into glacial ice. Air spaces among ice crystals are pressed as snow packs to a greater density. The ice recrystallizes and consolidates under pressure. In a transition step to glacial ice, snow becomes **firn**, which has a compact, granular texture.

As this process continues, many years pass before dense glacial ice is produced. Formation of **glacial ice** is analogous to metamorphic processes: Sediments (snow and firn) are pressured and recrystallized into a dense metamorphic rock (glacial ice). In Antarctica, glacial ice formation may take 1000 years because of the dryness of the climate (minimal snow input), whereas in wet climates the time is reduced to just a few years because of rapid, constant snow input to the system.

Glacial Mass Balance

A glacier is an open system, with *inputs* of snow and *outputs* of ice, meltwater, and water vapor. Snowfall and other moisture in the *accumulation zone* feed the glacier's upper reaches (Figure 17.6b). This area ends at the **firn line**, indicating where the winter snow and ice accumulation survived the summer melting season. Toward a glacier's lower end, it is wasted (reduced) through several processes: melting on the surface, internally, and at its base; ice removal by deflation (wind); the calving of ice blocks; and sublimation (recall from Chapter 7 that this is the direct evaporation of ice). Collectively, these losses are **ablation**.

The zone where accumulation gain balances ablation loss is the *equilibrium line* (Figure 17.6a). This area of a glacier generally coincides with the firn line. A glacier achieves *positive net balance* of mass—grows larger—during cold periods with adequate precipitation. In warmer times, the equilibrium line migrates up-glacier, and the glacier retreats—grows smaller—because of its *negative net balance*. Internally, gravity continues to move a glacier forward even though its lower terminus might be in retreat owing to ablation. The mass-balance losses from the South Cascade glacier and Alaskan glaciers provide a case in point (News Report 17.1).

Glacial Movement

Like all minerals, ice has specific properties of hardness, color, melting point (quite low in the case of ice), and brittleness. We know the properties of ice best from those brittle little cubes in the freezer. But glacial ice has different properties, depending on its location in a glacier. In a glacier's depths, glacial ice behaves in a plastic manner, distorting and flowing in response to weight and pressure from above and the degree of slope below. In contrast, the glacier's upper portion is more like the everyday ice we know, quite brittle. A glacier's rate of flow ranges from almost nothing to a kilometer or two per year on a steep slope. The rate of snow accumulation in the formation area is critical to the pace of glacial movement.

Glaciers are not rigid blocks that simply slide downhill. The greatest movement within a valley glacier occurs *internally*, below the rigid surface layer, where the underlying zone moves plastically forward (Figure 17.7a). At the same time, the base creeps and slides along, varying its speed with temperature and the presence of any lubricating water or saturated sediment beneath the ice. This *basal slip* usually is much slower than the internal plastic flow of the glacier, so the upper portion of the glacier flows ahead of the lower portion. The difference in speed stretches the glacier's brittle surface ice.

In addition, the pressure may vary in response to unevenness in the landscape beneath the ice. Basal ice may be melted by compression at one moment, only to refreeze later. This process is *ice regelation*, meaning to refreeze or regel. Regelation is important because it facilitates downslope movement and because the process incorporates rock debris into the glacier. Consequently, a glacier's basal ice layer, which can extend tens of meters above its base, has a much greater debris content than the ice above.

A flowing alpine glacier or ice stream in a continental glacier can develop vertical cracks known as **crevasses** (Figure 17.7b, c, d). Crevasses result from friction with valley walls, or tension from stretching as the glacier passes over convex slopes, or compression as the glacier passes over concave slopes. Traversing a glacier, whether an alpine glacier or an ice sheet, is dangerous because a thin veneer of snow sometimes masks the presence of a crevasse.

Glacier Surges Although glaciers flow plastically and predictably most of the time, some will lurch forward with little or no warning in a **glacier surge**. A surge is not quite as abrupt as it sounds; in glacial terms, a surge can be tens of meters per day. The Jakobshavn Glacier on the western Greenland coast, for example, is one of the fastest moving at between 7 and 12 km (4.3 and 7.5 mi) a year.

The doubling of ice-mass loss from Greenland between 1996 and 2005 means that glaciers are surging overall. In southern Greenland in 2007, outlet glaciers were averaging 22.8 m (75 ft) per year. This is significant because in 1999 the average flow rate was 1.8 m (6 ft) per year. In Greenland, scientists determined that 2006–2007 experienced more days of melting snow at higher elevations than average. The fear is that the meltwater works its way to the basal layer, lubricating underlying *soft beds* of clay.

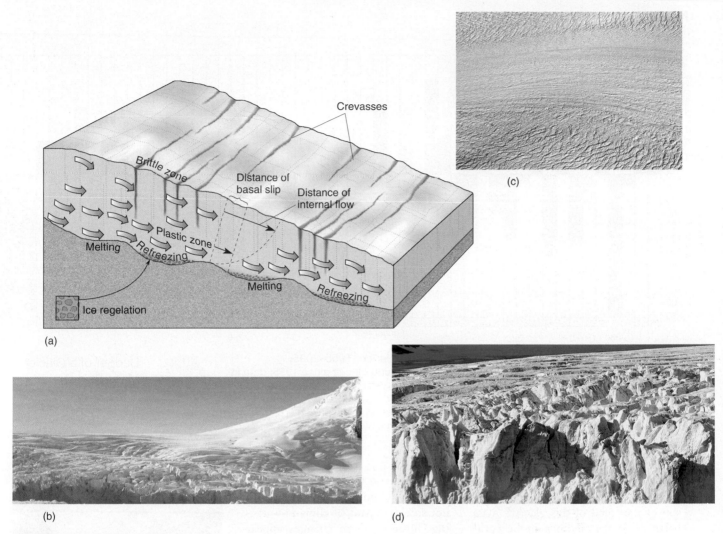

FIGURE 17.7 Glacial movement.
(a) Cross section of a glacier, showing its forward motion and brittle cracking at the surface and flow along its basal layer. (b) Surface crevasses and cracks are evidence of a glacier's forward motion on Yalour Island, Antarctica. (c) Crevasses frame this ice stream flowing off the Greenland Ice Sheet, evidence of great stress forces at work far inland in the ice sheet. (d) Crevasses mark the jumbled surface of this glacier in Hornsund, Spitsbergen, Arctic Ocean. [All photos by Bobbé Christopherson.]

News Report 17.1

South Cascade and Alaskan Glaciers Lose Mass

As part of a global trend, glacial mass budgets are running in the negative, on average, with significant ice losses in Alaska, the Andes, the European Alps, and the Himalayas. The net mass balance of the South Cascade Glacier in Washington State demonstrated significant losses between 1955 and 2005. In just 1 year (September 1991 to October 1992), the terminus of the glacier retreated 38 m (125 ft), resulting in major changes to the surface and sides of the glacier. More than a 2% loss of the glacier's mass occurred in just that 1 year. The net accumulation and net wastage are illustrated in Figure 17.1.1.

Even though the graph shows a slight positive mass balance for 2002, the glacier's terminus retreated 4 m (13.1 ft) during that year. The years 2003–2005 showed significant losses as regional temperatures increased. The glacier has retreated every year in the record except 1972. Figure 17.1.2 is a photo comparison between 1979 and 2003.

The reasons for these losses are not fully understood. Increasing average air temperature and decreasing precipitation are the leading possible causes. The heavy snowfall in the Northwest associated with the La Niña

(continued)

News Report 17.1 (continued)

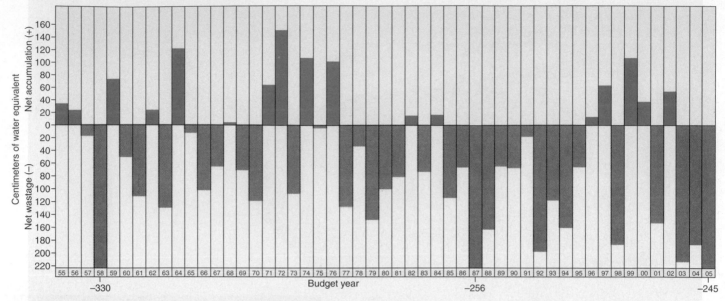

FIGURE 17.1.1 South Cascade Glacier net mass balance, 1955–2005.
A negative net mass balance has dominated this shrinking glacier since 1955; data in
centimeters (2.54 cm per in.). [Data from R. M. Krimmel, *Water, Ice, Meteorological, and
Speed Measurements at South Cascade Glacier, Washington, 2002 Balance Year.*
USGS Water Resources Report, Tacoma, Washington; personal communication and
http://ak.water.usgs.gov/glaciology/all_bmg/3glacier_balance.htm]

Budget of a Glacier,
Mass Balance

episode is reflected in the positive mass
balance for 1999 and 2000. A comparison of the trend of this glacier's mass
balance with that of others in the world
shows that temperature changes apparently are causing widespread reductions in middle- and lower-elevation
glacial ice. The present wastage (ice
loss) from alpine glaciers worldwide is
thought to contribute over 25% to the
rise in sea level.

Research covering almost 50 years
of 67 glaciers in Alaska was published
in 2002. The study estimated volume
changes in these glaciers, which were
grouped into seven regions, shown in
Figure 17.1.3a. The rates of glacierwide thickness loss for 67 glaciers between the mid-1950s and 1995 and for
28 glaciers between 1995 and 2001 are
in Figure 17.1.3b.

A total volume change was estimated to be −52 ± 15 km³ per year
(−12.5 ± 3.6 mi³). These losses alone
account for a rise in global sea level.
About 9% of global sea-level rise is
coming from the Alaskan meltdown.

The reasons for these losses are
complex. Thinning and wastage are
not solely a result of climatic warming; however these changes appear
to be initiated by negative mass
balances due to the warming. The
individual dynamics of each glacier
must be considered. Relative to the
contribution of the Alaskan meltdown to sea-level rise, the scientists
summarized:

Compared with estimated inputs
from the Greenland ice sheet and
other sources, Alaskan glaciers
have, over the past 50 years,
made the largest single glaciolog-
ical contribution to rising sea
level yet measured... the different rates of thinning observed in
the various Alaskan regions may
be important in characterizing
patterns of climate change.*

*A. A. Arendt et al., "Rapid Wastage of
Alaska Glaciers and Their Contribution to
Rising Sea Level," *Science* 297 (July 19,
2002): 382–86.

FIGURE 17.1.2 Photo comparison
of 24 years at South Cascade.
Photos visually remind us of the
tremendous quantities of ice that have
ablated from the South Cascade
Glacier. [Photos courtesy of USGS Water
Resources for Alaska, Glacier and Snow
Program.]

FIGURE 17.1.3 Negative mass balances for Alaska's glaciers.
(a) Sixty-seven Alaskan glaciers divided into seven regions: 55 are in Alaska, 11 cross the Alaskan–Canadian border, and one is in the Yukon. (b) Rate of change (meters per year) in glacierwide thickness for 67 glaciers in the earlier period (1950s to 1995, solid brown bars) and 28 glaciers in the recent period (1995 to 2001, red bars). [Adapted by permission of AAAS, from Anthony A. Arendt, et al., Figures 1 and 3, "Rapid wastage of Alaska glaciers and their contribution to rising sea level," *Science* 297 (July 19, 2002): 382–86.]

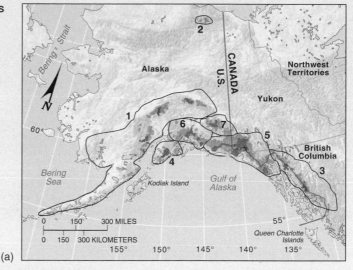

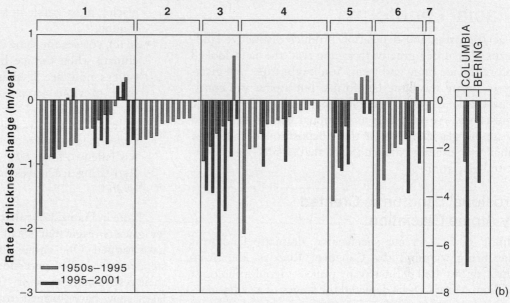

The exact cause of such a glacier surge is being studied. Some surge events result from a buildup of water pressure under the glacier, sometimes enough to actually float the glacier slightly, detaching it from its bed during the surge. As a surge begins, ice quakes are detectable and ice faults are visible. Surges can occur in dry conditions as well, as the glacier plucks (picks up) rock from its bed and moves forward. Another cause of glacier surges is the presence of a water-saturated layer of sediment *soft bed* beneath the glacier. This is a deformable layer that cannot resist the tremendous sheer stress produced by the moving ice of the glacier. Scientists examining cores taken from several ice streams now accelerating through the West Antarctic Ice Sheet think they have identified this cause of glacial surges—although water pressure is still important.

Glacial Erosion The way in which a glacier erodes the land is similar to a large excavation project, with the glacier hauling debris from one site to another for deposition. The passing glacier mechanically plucks rock material and carries it away. Debris is carried on its surface and is also transported internally, or *englacially*, embedded within the glacier itself. There is evidence that rock pieces actually freeze to the basal layers of the glacier in a *glacial plucking*, or a picking up process, and once embedded, enable the glacier to scour and sandpaper the landscape as it moves—a process of **abrasion**. This abrasion and gouging produce a smooth surface on exposed rock, which shines with *glacial polish* when the glacier retreats. Larger rocks in the glacier act much like chisels, gouging the underlying surface and producing glacial striations parallel to the flow direction (Figure 17.8).

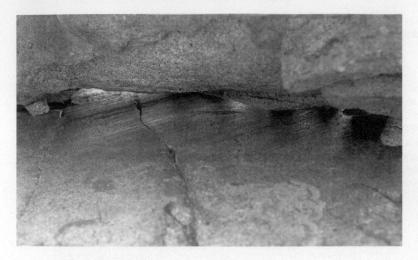

FIGURE 17.8 Glacial sandpapering polishes rock.
Glacial polish and striations are examples of glacial abrasion and erosion. The polished, marked surface is seen beneath a glacial erratic—a rock left behind by a retreating glacier. [Photo by Bobbé Christopherson.]

Glacial Landforms

Glacial erosion and deposition produce distinctive landforms that differ greatly from the way the land looked before the ice came and went. You might expect all glaciers to create the same landforms, but alpine and continental glaciers each generate their own characteristic landscapes. We look first at erosional landforms created by alpine glaciers, then at their depositional landforms. Finally, we examine the landscape that results from continental glaciation.

Erosional Landforms Created by Alpine Glaciation

Alpine glaciers create spectacular, dramatic landforms that bring to mind the Canadian Rockies, the Swiss Alps, or vaulted Himalayan peaks. Geomorphologist William Morris Davis depicted the stages of a valley glacier in drawings published in 1906 and redrawn here in Figure 17.9. Study of these figures reveals the handiwork of ice as sculptor:

- In (a), you see typical stream-cut valleys as they exist before glaciation. Note the prominent **V**-shaped valley.
- In (b), you see the same landscape during subsequent glaciation. Glacial erosion and transport actively remove much of the regolith (weathered bedrock) and the soils that covered the stream-valley landscape. As the *cirque* walls erode away, sharp ridges form, dividing adjacent cirque basins. These **arêtes** ("knife-edge" in French) become the *sawtooth*, serrated ridges in glaciated mountains. Two eroding cirques may reduce an arête to a saddlelike depression or pass, called a **col**. A **horn** (pyramidal peak) results when several cirque glaciers gouge an individual mountain summit from all sides. Most famous is the Matterhorn in the Swiss Alps, but many others occur worldwide. A **bergschrund** forms when a crevasse or wide crack opens along the headwall of a glacier, most visible in summer when covering snow is gone.

- In (c), you see the same landscape at a time of warmer climate when the ice has retreated. The glaciated valleys now are **U**-shaped, greatly changed from their previous stream-cut **V** form. You can see the steep sides and the straightened course of the valleys. Physical weathering from the freeze-thaw cycle has loosened rock along the steep cliffs, where it has fallen to form *talus slopes* along the valley sides. Retreating ice leaves behind transported rocks as *erratics*.

Note in Figure 17.9c that in the cirques where the valley glaciers originated, small mountain lakes called **tarns** have formed. One cirque contains small, circular, stair-stepped lakes called **paternoster** ("our father") **lakes** for their resemblance to rosary (religious) beads. Paternoster lakes may have formed from the differing resistance of rock to glacial processes or from damming by glacial deposits.

The valleys carved by tributary glaciers are left stranded high above the valley floor, because the primary glacier eroded the valley floor so deeply. These *hanging valleys* are the sites of spectacular waterfalls.

FIGURE 17.9 The geomorphic handiwork of alpine glaciers.
(a) A preglacial landscape with V-shaped stream-cut valleys. (b) The same landscape filled with valley glaciers. Note inset photos of a *horn*, a *cirque basin*, *sawtooth ridge*, and *bergschrund*. (c) When the glaciers retreat, the new landscape is unveiled. Note inset photos of *glacial erratics*, a *hanging valley* and waterfall, a *tarn* (lake), and a characteristic U-shaped glacial valley, or trough in surface and aerial views. [After W. M. Davis, in Tarbuck, Lutgens, and Pinzke, *Applications and Investigations in Earth Science* (New York: Macmillan, an imprint of Prentice Hall, Inc., 1994), p. 85. Photos: waterfall by author; all other photos by Bobbé Christopherson.]

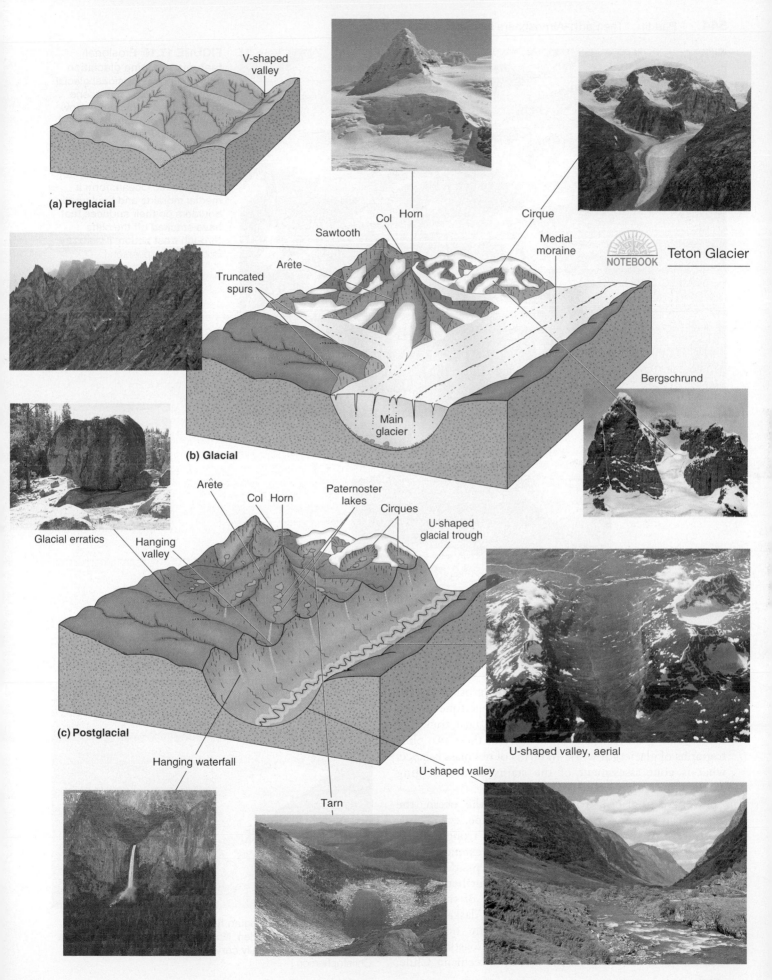

V-shaped valley

(a) Preglacial

Col   Horn

Sawtooth

Arête

Truncated spurs

Cirque

Medial moraine

NOTEBOOK   Teton Glacier

Bergschrund

Main glacier

(b) Glacial

Arête

Col   Horn

Paternoster lakes

Cirques

U-shaped glacial trough

Hanging valley

Glacial erratics

(c) Postglacial

Hanging waterfall

Tarn

U-shaped valley, aerial

U-shaped valley

543

(a)

(b)

FIGURE 17.10 Erosional features of Alpine glaciation. (a) Many of the erosional glacial features appear in this scene from southern Greenland. How many of the erosional features illustrated in Figure 17.9 can you find in the photo? (b) These merging glaciers along Beverlysundet, Nordaustlandet Island, Arctic Ocean, form a medial moraine and haul erratic boulders on their surfaces that have cracked off the cliffs through frost action. [Photos by Bobbé Christopherson.]

In Figure 17.10 see how many of the erosional forms from Figure 17.9 (arêtes, col, horn, cirque, cirque glacier, sawtooth ridge, U-shaped valleys, erratics, tarn, truncated spurs, among others) you can identify in these two photographs of glaciers at work. In terms of net mass balance, what is your assessment of the status of the merging glaciers in Figure 17.10b?

Where a glacial trough intersects the ocean, the glacier can continue to erode the landscape, even below sea level. As the glacier retreats, the trough floods and forms a deep **fjord** in which the sea extends inland, filling the lower reaches of the steep-sided valley (Figure 17.11). The fjord may be flooded further by rising sea level or by changes in the elevation of the coastal region. All along the glaciated coast of Alaska, glaciers are now in retreat, thus opening many new fjords that previously were blocked by ice. Coastlines with notable fjords include those of Norway, Greenland, Chile,

FIGURE 17.11 Norwegian fjord.
The beautiful Fjoerfjorden, Norway, where an inlet of the sea fills a U-shaped, glacially carved valley. [Photo by Bobbé Christopherson.]

the South Island of New Zealand, Alaska, and British Columbia.

Depositional Landforms Created by Alpine Glaciation

You have just seen how glaciers excavate tremendous amounts of material and create fascinating landforms in the process. Glaciers produce a different set of distinctive landforms when they melt and deposit their debris cargo at the glacier's *terminus*. Figure 17.6a, e, and f show that after the glacier melts, debris accumulates in moraines to mark the former margins of the glacier, both its end and sides.

Glacial Drift   The general term for all glacial deposits, both unsorted and sorted, is **glacial drift**. Sediments deposited by glacial meltwater are sorted by size and are **stratified drift**. **Till** describes direct ice deposits left as unstratified and unsorted debris.

As a glacier flows to a lower elevation, a wide assortment of rock fragments becomes *entrained* (carried along) on its surface or embedded within its mass or in its base. As the glacier melts, this unsorted cargo is deposited on the ground surface. Such till is poorly sorted and is difficult to cultivate for farming, but the clays and finer particles can provide a basis for soil development.

Retreating glaciers leave behind large rocks (sometimes house-sized), boulders, and cobbles that are "foreign" in composition and origin from the ground on which they were deposited. These *glacial erratics*, lying in strange locations with no obvious means of transport, were an early clue that blankets of ice once had covered the land (pictured in Figures 17.9c and 17.10b).

Moraines   The deposition of glacial sediments produces a specific landform called a **moraine**. Several types of moraines are exhibited in Figure 17.12. A **lateral moraine** forms along each side of a glacier. If two glaciers with lateral moraines join, a **medial moraine** may form (see Figures 17.1 and 17.6f). Both medial and lateral moraine forms are clearly visible in Figure 17.12. A deposition of till that is generally spread across a surface is called a *ground moraine*, or *till plain*, and may hide the former landscape. Such plains are found in portions of the U.S. Midwest.

Eroded debris that is dropped at the glacier's farthest extent is a **terminal moraine**. However, there also may be *end moraines*, formed at other points where a glacier paused after reaching a new equilibrium between growth and ablation. Lakes may form behind terminal and end moraines after a glacier's retreat, with the moraine acting as a dam. The large terminal moraine deposit in

(a)

(b)

(c)

(d)

FIGURE 17.12 Depositional features of alpine glaciation.
Medial, lateral, and ground moraine deposits are evident in eastern and southern Greenland. (a) As these tributary glaciers merge, their lateral moraines combine to form a medial moraine. (b) Retreating ice leaves behind these lateral moraines. (c) Note how each lateral moraine remains identifiable as they combine in this valley glacier. How many moraines can you count? Note the medial moraine flowing from the upper-left corner of the photo. (d) A terminal moraine forms this island separated from its ice-cap glacier by more than a kilometer of ocean, at Isispynten Island, Arctic Ocean. [All photos by Bobbé Christopherson.]

Figure 17.12d forms an island, when just 9 years earlier it was a peninsula at the edge of an ice cap. The ice cap rapidly retreated more than a kilometer in this time frame.

Tributary valley glaciers merge to form a *compound valley glacier*. The flowing movement of a compound valley glacier is different from that of a river with tributaries. Tributary glaciers flow into a compound glacier and merge alongside one another by extending and thinning rather than by blending, as do rivers. Each tributary maintains its own patterns of transported debris (dark streaks, visible in the photo in Figure 17.12c).

All of these till types are unsorted and unstratified. In contrast, streams of glacial meltwater can carry and deposit sorted and stratified glacial drift beyond a terminal moraine. Meltwater-deposited material downvalley from a glacier is a *valley train deposit*. Peyto Glacier in Alberta, Canada, produces such a valley train that continues into Peyto Lake (Figure 17.13). Distributary stream channels appear braided across its surface. The picture also shows the milky meltwater associated with glaciers, laden with finely ground "rock flour." Meltwater comes from glaciers at all times, not just when they are retreating. However, the Peyto Glacier (far left in photo) has experienced massive ice losses since 1966 and is in retreat owing to increased ablation and decreased accumulation.

Erosional and Depositional Landforms Created by Continental Glaciation

The extent of the most recent continental glaciation in North America and Europe, 18,000 years ago, is portrayed several pages ahead in Figure 17.27. When these huge sheets of ice advanced and retreated, they produced some of the erosional and depositional features characteristic of alpine glaciation. However, because continental glaciers form under different circumstances—not in mountains, but across broad, open landscapes—the intricately carved alpine features, lateral moraines, and medial moraines all are lacking in continental glaciation. Table 17.1 compares the erosional and depositional features of alpine (valley) and continental glaciers.

Figure 17.14 illustrates some of the most common erosional and depositional features associated with the retreat of a continental glacier. A **till plain** forms behind an end moraine; it features *unstratified* coarse till, has low and rolling relief, and has a deranged drainage pattern (Figure 17.14a, and see Figure 14.9). Beyond the morainal deposits lies the **outwash plain** of *stratified drift* featuring stream channels that are meltwater-fed, braided, and overloaded with sorted and deposited materials.

Figure 17.14b shows a sinuously curving, narrow ridge of coarse sand and gravel called an **esker**. It forms along the channel of a meltwater stream that flows beneath a glacier, in an ice tunnel, or between ice walls. As a glacier retreats, the steep-sided esker is left behind in a

FIGURE 17.13 A valley train deposit.
Peyto Glacier in Alberta, Canada, is in extensive retreat. Note valley train, braided stream, and milky-colored glacial meltwater. [Photo by author.]

Table 17.1   Features of Valley Glaciation and Continental Glaciation Compared

Features and Landforms	Alpine (Valley) Glacier	Continental Glaciation
Erosional		
Striations, polish, etc.	Common	Common
Cirques	Common	Absent
Horns, arêtes, cols	Common	Absent
U-shaped valleys, truncated spurs, hanging valleys	Common	Rare
Fjords	Common	Absent
Roche moutonnée	Common	Common
Depositional		
Till	Common	Common
Terminal moraines	Common	Common
Recessional moraines	Common	Common
Ground moraines	Common	Common
Lateral moraines	Common	Absent
Medial moraines	Common, easily destroyed	Absent
Drumlins	Rare or absent	Locally common
Erratics	Common	Common
Stratified drift	Common	Common
Kettles	Common	Common
Eskers, crevasse fillings	Rare	Common
Kames	Common	Common
Kame terraces	Common	Present in hilly country

Source: Adapted from L. D. Leet, S. Judson, and M. Kauffman, *Physical Geology*, 5th ed., © 1978, p. 317. Upper Saddle River, NJ.: Prentice Hall, Inc.

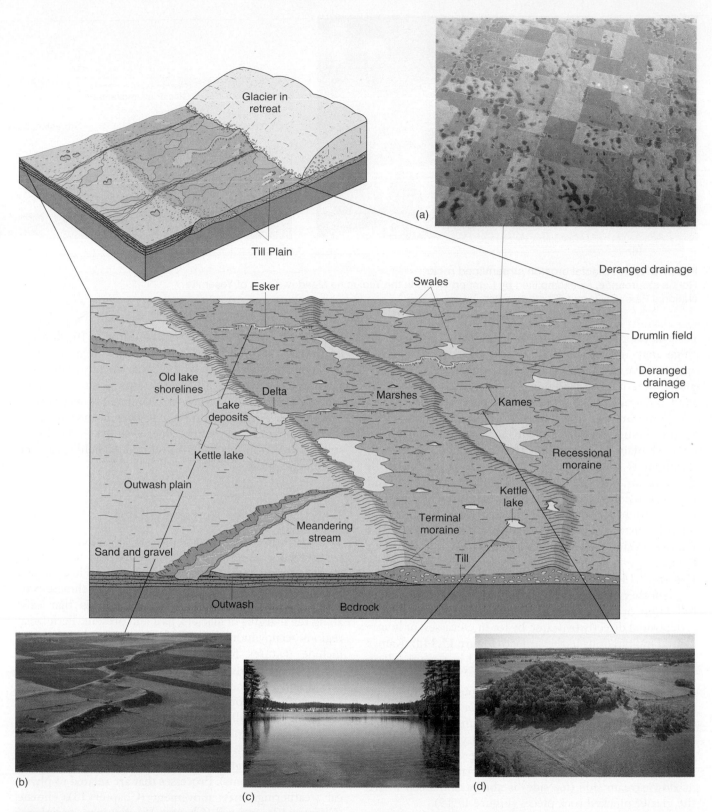

FIGURE 17.14 Continental glacier depositional features.
Common depositional landforms produced by glaciers. (a) Deranged drainage tells us continental
glaciers advanced and retreated across the prairies of central Saskatchewan, Canada. (b) An esker
through farmland near Dahlen, North Dakota. (c) Walden Pond is a kettle surrounded by mixed forest
near Boston, Massachusetts. (d) A kame covered by a woodlot near Campbellsport, Wisconsin.
[Illustration from R. M. Busch, ed., *Laboratory Manual in Physical Geology*, 3rd ed. (New York: Macmillan,
an imprint of Prentice Hall, Inc., 1993), p. 188. Photos by (a) and (c) Bobbé Christopherson; (b) and (d) Tom
Bean/DRK.]

(a)

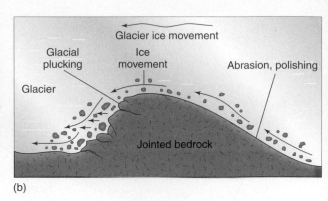

(b)

FIGURE 17.15 Glacial erosion streamlined rock.
Roche moutonnée, as exemplified by Lembert Dome in the Tuolumne Meadows area of Yosemite
National Park, California. [Photo by author.]

pattern roughly parallel to the path of the glacier. The ridge may not be continuous and in places may even appear to be branched, following the path set by the subglacial watercourse. Commercially valuable deposits of sand and gravel are quarried from some eskers.

Sometimes an isolated block of ice, perhaps more than a kilometer across, remains in a ground moraine, an outwash plain, or valley floor after a glacier has retreated. As much as 20 to 30 years is required for it to melt. In the interim, material continues to accumulate around the melting ice block. When the block finally melts, it leaves behind a steep-sided hole. Such a feature then frequently fills with water. This feature is a **kettle**. Thoreau's famous Walden Pond, mentioned in the quotation in Chapter 7, is such a glacial kettle and is pictured in Figure 17.14c.

Another feature of outwash plains is a **kame**, a small hill, knob, or mound of poorly sorted sand and gravel that is deposited directly by water, by ice in crevasses, or in ice-caused indentations in the surface (Figure 17.14d). Kames also can be found in deltaic forms and in terraces along valley walls.

Glacial action also forms two types of streamlined hills. One is erosional, called a *roche moutonnée*, and the other is depositional, called a *drumlin*. A **roche moutonnée** ("sheep rock" in French) is an asymmetrical hill of exposed bedrock. Its gently sloping upstream side (stoss side) has been polished smooth by glacial action, whereas its downstream side (lee side) is abrupt and steep where the glacier plucked rock pieces (Figure 17.15).

A **drumlin** is deposited till that has been streamlined in the direction of continental ice movement, blunt end upstream and tapered end downstream (the opposite of a roche moutonnée). "Swarms" of drumlins occur across the landscape in portions of New York and Wisconsin, among other areas. Sometimes their shape is that of an elongated teaspoon bowl, lying face down. They attain lengths of 100 to 5000 m (300 ft to more than 3 mi) and

heights up to 200 m (650 ft). Figure 17.16 shows a portion of a topographic map for the area south of Williamson, New York, which experienced continental glaciation during the last ice age. In studying the map, can you identify the numerous drumlins? In what direction do you think the continental glaciers moved across this region? Look for gentle, tapered slopes in the downstream direction of each drumlin, left behind as the glacier retreated.

Periglacial Landscapes

In 1909, Polish geologist W. Lozinski coined the term **periglacial** to describe frost weathering and freeze-thaw rock shattering in the Carpathian Mountains. He expanded the concept in 1910 to include several cold-climate processes, landforms, and topographic features that exist along the margins of glaciers, past and present. Periglacial regions occupy more than 20% of Earth's land surface and include features formed under nonglacial conditions. Under these conditions, a unique combination of frost action and periglacial processes operate, including permafrost, frost action, and ground ice.

Climatologically, these regions are in *subarctic* and *polar* climates, especially *tundra* climate. Such climates occur either at high latitude (tundra and boreal forest environments) or at high elevation in lower-latitude mountains (alpine environments). Processes that are related to physical weathering, mass movement (Chapter 13), climate (Chapter 10), and soil (Chapter 18) dominate periglacial regions.

Geography of Permafrost

When soil or rock temperatures remain below 0°C (32°F) for at least 2 years, **permafrost** ("permanent frost") develops. An area of permafrost that is not covered by glaciers

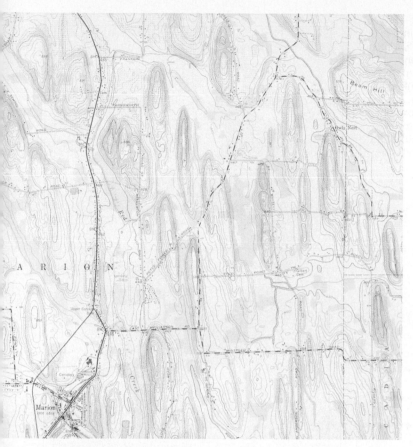

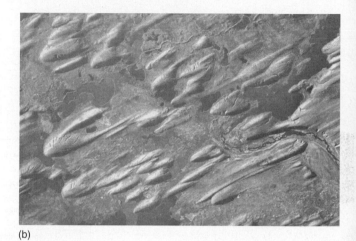

(b)

FIGURE 17.16 Glacially deposited streamlined features.
(a) Topographic map south of Williamson, New York, featuring numerous drumlins. (7.5-minute
series quadrangle map, originally produced at a 1:24,000 scale, 10-ft contour interval.)
(b) A drumlin field, Snare Lake, Canada. [(a) USGS map; (b) photo by National Air Photo
Library, NRC, Ottawa, Canada.]

is considered periglacial, with the largest extent of such lands in Russia. Approximately 80% of Alaska has permafrost beneath its surface. Canada, China, Scandinavia, Greenland, and Antarctica, in addition to alpine mountain regions of the world, also are affected. Note that this criterion is based solely on *temperature* and has nothing to do with how much or how little water is present. Two other factors also contribute to permafrost: the presence of fossil permafrost from previous ice-age conditions and the insulating effect of snow cover or vegetation that inhibits heat loss.

Continuous and Discontinuous Zones   Permafrost regions are divided into two general categories: continuous and discontinuous, which merge along a general transition zone. *Continuous permafrost* is the region of severest cold and is perennial, roughly poleward of the −7°C (19°F) mean annual temperature isotherm (purple area in Figure 17.17). Continuous permafrost affects all surfaces except those beneath deep lakes or rivers. The depth of continuous permafrost may exceed 1000 m (3300 ft), averaging approximately 400 m (1300 ft).

Unconnected patches of *discontinuous permafrost* gradually coalesce poleward toward the continuous zone. Permafrost becomes scattered or sporadic until it gradually disappears equatorward of the −1°C (30.2°F) mean annual temperature isotherm (dark blue area on map). In the discontinuous zone, permafrost is absent on sun-exposed south-facing slopes, areas of warm soil, or areas insulated by snow. In the Southern Hemisphere, north-facing slopes experience increased warmth.

Areas of discontinuous permafrost feature a mixture of *cryotic* (frozen) and *noncryotic* ground. In addition to these two types of ground, zones of high-altitude *alpine permafrost* extend to lower latitudes, as shown on the map. Microclimatic factors such as slope orientation and snow cover are important in the alpine environment.

In the Mackenzie Mountains of Canada (62°N), continuous permafrost extends down to an elevation of 1200 m (4000 ft), and discontinuous permafrost occurs throughout the range. The Colorado Rockies (40°N) experience continuous permafrost down to an elevation of 3400 m (11,150 ft) and discontinuous permafrost to 1700 m (5600 ft).

FIGURE 17.17 Permafrost distribution.
Distribution of permafrost in the Northern Hemisphere. Alpine permafrost is noted except for small occurrences in Hawai'i, Mexico, Europe, and Japan. Sub-sea permafrost occurs in the ground beneath the Arctic Ocean along the margins of the continents, as shown. Note the towns of Resolute and Coppermine (Kugluktuk) in Nunavut (formerly part of NWT) and Hotchkiss in Alberta. A cross section of the permafrost beneath these towns is shown in Figure 17.18. [Adapted from T. L. Péwé, "Alpine permafrost in the contiguous United States: A review," *Arctic and Alpine Research* 15, no. 2 (May 1983): 146. © University of Colorado. Used by permission.]

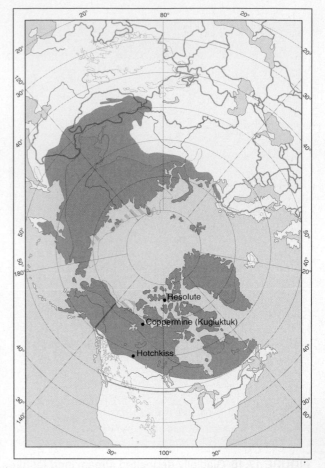

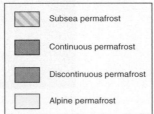

Subsea permafrost

Continuous permafrost

Discontinuous permafrost

Alpine permafrost

Behavior of Permafrost We looked at the spatial distribution of permafrost; now let us examine how permafrost behaves. Figure 17.18 is a stylized cross section from approximately 75° N to 55° N, using the three sites located on the map in Figure 17.17. The **active layer** is the zone of seasonally frozen ground that exists between the subsurface permafrost layer and the ground surface.

The active layer is subjected to consistent daily and seasonal freeze-thaw cycles. This cyclic melting of the active layer affects as little as 10 cm (4 in.) of depth in the north (Ellesmere Island, 78° N), up to 2 m (6.6 ft) in the southern margins (55° N) of the periglacial region, and 15m (50 ft) in the alpine permafrost of the Colorado Rockies (40° N).

The depth and thickness of the active layer and permafrost zone change slowly in response to climatic change. Higher temperatures degrade (reduce) permafrost and increase the thickness of the active layer; lower temperatures gradually aggrade (increase) permafrost depth and reduce active-layer thickness. Although somewhat sluggish in response, the active layer is a dynamic open system driven by energy gains and losses in the subsurface environment. As you might expect, most permafrost exists in disequilibrium with environmental conditions and therefore actively adjusts to inconstant climatic conditions.

With the incredibly warm temperatures recorded in the Canadian and Siberian Arctic since 1990, more disruption of surfaces is occurring—leading to highway, railway, and building damage. This warming is causing a large release of carbon from peat-rich thawed ground that in turn influences the global greenhouse—a real-time positive feedback mechanism. In Siberia, many lakes have disappeared in the discontinuous permafrost region as subsurface drainage opens, yet new lakes have formed in the continuous region as thawed soils become waterlogged. In Canada, hundreds of lakes have disappeared simply from excessive evaporation into the warm air. These trends are measurable from satellite imagery. The thaw in the active layer in the Alaskan tundra has shortened the number of days that oil exploration equipment can venture out from 200 to only 100 days a year. The weight of the trucks requires a frozen, more solid surface to operate.

A *talik* is unfrozen ground that may occur above, below, or within a body of discontinuous permafrost or beneath a water body in the continuous region. Taliks occur beneath deep lakes and may extend to bedrock and noncryotic soil beneath large, deep lakes (see Figure 17.18). Taliks form connections between the active layer and groundwater, whereas in continuous permafrost groundwater is essentially cut off from surface water. In this way, permafrost disrupts aquifers and taliks, leading to water-supply problems.

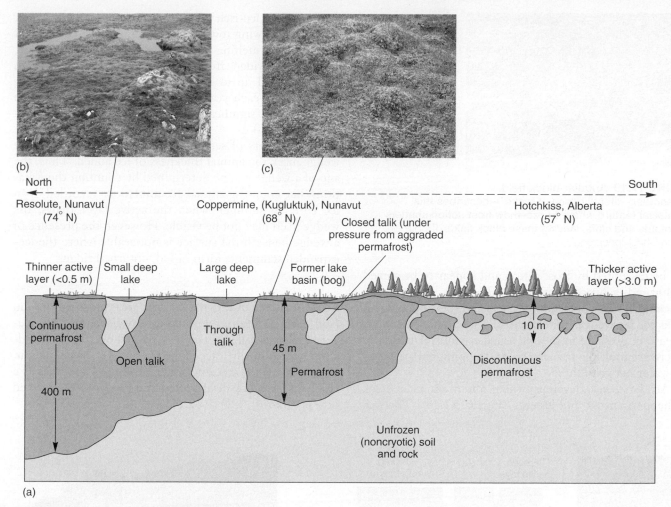

FIGURE 17.18 Periglacial environments.
(a) Cross section of a periglacial region in northern Canada, showing typical forms of permafrost, active layer, talik, and ground ice. The three sites noted are shown on the map in Figure 17.17. (b) Poor drainage, with some standing water and hummocks, or (c) irregular ice-forced bumps of tundra turf, indicate permafrost. [Photos by Bobbé Christopherson.]

Ground Ice and Frozen Ground Phenomena

In regions of permafrost, frozen subsurface water is termed *ground ice*. The moisture content of areas with ground ice varies from nearly none in drier regions to almost 100% in saturated soils. From the area of maximum energy loss, freezing progresses through the ground along a *freezing front*, or boundary between frozen and unfrozen soil. The presence of frozen water in the soil initiates geomorphic processes associated with *frost action* and the expansion of water as it freezes. Ground ice may occur as:

- Common *pore ice* (subsurface water frozen in the soil's pore spaces)
- *Lenses* (horizontal bodies) and *veins* (channels extending in any direction)
- *Segregated ice* (layers of buried ice that increase in mass by accreting water as the ground freezes, producing layers of relatively pure ice)
- *Intrusive ice* (the freezing of water injected under pressure, as in a pingo, discussed shortly)
- *Wedge ice* (surface water entering a crack and freezing)

Various aspects of frost action as a geomorphic agent are discussed in Chapter 13 ("Physical Weathering Processes" and the section "Classes of Mass Movements," which describes soil creep). Some forms of ground ice may occur at the surface during episodes of *icing*, in which a river or spring forms freezing layers of surface ice. Such icings can occur on slopes as well as level ground. Spring flooding along rivers can result from the presence of icings; it is a particular problem in rivers draining into the Arctic Ocean.

Frost Action Processes   The 9% expansion of water as it freezes produces strong mechanical forces. Such frost action shatters rock, producing angular pieces that form a *block field*, or *felsenmeer*. The felsenmeer accumulates as part of the arctic and alpine periglacial landscape, particularly on mountain summits and slopes (Figure 17.19).

If sufficient water freezes, the saturated soil and rocks are subjected to *frost-heaving* (vertical movement) and *frost-thrusting* (horizontal movement). Boulders and rock slabs may be thrust to the surface. Soil horizons (layers)

FIGURE 17.19 Angular block field.
Felsenmeer—literally, "sea of rocks"—describes this periglacial feature where freeze–thaw frost action shatters mountains and cliffs, leaving these block fields. [Photo by Bobbé Christopherson.]

may be disrupted by frost action and appear to be stirred or churned, a process termed *cryoturbation*. Frost action also can produce contractions in soil and rock, opening up cracks for ice wedges to form. Also, there is a tremendous increase in pressure in the soil as ice expands, particularly if there are multiple freezing fronts trapping unfrozen soil and water between them.

An *ice wedge* develops when water enters a crack in the permafrost and freezes (Figure 17.20). Thermal contraction in ice-rich soil forms a tapered crack—wider at the top, narrowing toward the bottom. Repeated seasonal freezing and melting progressively enlarge the wedge, which may widen from a few millimeters to 5–6 m (16–20 ft) and up to 30 m (100 ft) in depth. Widening may be small each year, but after many years the wedge can become significant, like the sample shown in Figure 17.20b.

Thin layers of sediment that form foliations in the wedge mark the annual thickness of ice added. Thus, the age of a wedge can be determined by counting the foliations, like counting annual accumulations recorded in an ice core. In summer, when the active layer thaws, the wedge itself may not be visible. However, the presence of a wedge beneath the surface is noticeable where the ice-expanded sediments form raised, upturned ridges.

Frost-Action Landforms   Large areas of frozen ground (soil-covered ice) can develop a heaved-up, circular, ice-cored mound called a *pingo*. It rises above the flat landscape, occasionally exceeding 60 m (200 ft) in height. Pingos rise when freezing water expands, sometimes as a result of pressure developed by artesian water injected into permafrost (Figure 17.21).

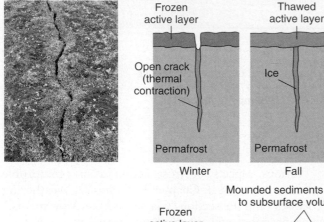

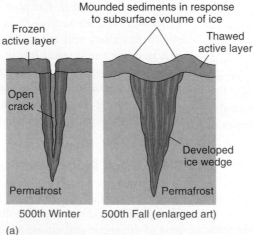

FIGURE 17.20 Evolution of an ice wedge.
(a) Sequential illustration of ice-wedge formation beginning with a crack in the tundra shown in the photo. (b) An ice wedge and ground ice in northern Canada. [(a) Illustration adapted from A. H. Lachenbruch, "Mechanics of thermal contraction and ice-wedge polygons in permafrost," *Geological Society of America Bulletin Special Paper 70* (1962). Photos by (a) Bobbé Christopherson; (b) H. M. French.]

FIGURE 17.21 A pingo.
An ice-cored pingo resulting from hydraulic pressures that pushed the mound upward above the landscape. Coastal erosion exposed the ice core. This pingo is near Tuktoyaktuk, Mackenzie Delta, NWT, Canada. [Photo by H. M. French.]

Frost-action expansion and contraction results in the transport of stones and boulders. As the water-ice volume changes and an ice wedge deepens, coarser particles are moved toward the surface. An area with a system of ground ice and frost action develops sorted and unsorted accumulations of rock at the surface that take the shape of polygons called **patterned ground**.

Patterned ground may include polygons of sorted rocks that coalesce into stone polygon nets (Figure 17.22a and b). Various terms are in use to describe such ice-wedge and stone polygon forms: nets, circles, vegetation-covered hummocks, and stripes (elongated polygons formed on a hillslope, Figure 17.22c). Ice wedges sometimes are found beneath the perimeter of each cell; however, questions still exist as to how such patterned ground actually develops and the degree to which such ground forms are the vestiges of past eras. Such features may take centuries to form.

(a)

(b)

(c)

(d)

FIGURE 17.22 Patterned ground phenomena.
(a) Patterned ground in Beacon Valley of the McMurdo Dry Valleys of East Antarctica. (b) Rock nets form polygons and circles (about a meter across) near Duvefjord, Nordaustlandet Island, Arctic Ocean. (c) Patterned ground in stripes, near Sundbukta, Barents Island, Arctic Ocean. (d) Polygons in the Martian northern plains; each cell averages a little more than 100 m across (300 ft). [Photos by (a) writer Joan Myers; (b) and (c) Bobbé Christopherson; and (d) image from the *Mars Global Surveyor,* Mars Orbiter Camera, courtesy of NASA/JPL/Malin Space Science Systems, May 1999.]

Such polygon nets in patterned ground provide vivid evidence of frozen subsurface water on Mars (Figure 17.22d). In this 1999 image, the Mars Global Surveyor captured a region of these features on the Martian northern plains. To date, some 600 different sites on Mars demonstrate such patterned ground.

Hillslope Processes: Gelifluction and Solifluction

Soil drainage is poor in areas of permafrost and ground ice. The active layer of soil and regolith is saturated with soil moisture during the thaw cycle (summer), and the whole layer commences to flow from higher to lower elevation if the landscape is even slightly inclined. This flow of soil is generally called *solifluction* and may occur in all climate conditions. In the presence of ground ice or permafrost, the more specific term *gelifluction* is applied. In this ice-bound type of soil flow, movement up to 5 cm (2 in.) per year can occur on slopes as gentle as a degree or two.

The cumulative effect of this landflow can be an overall flattening of a rolling landscape, with identifiable sagging surfaces and scalloped and lobed patterns in the downslope soil movements (Figure 17.23). Other types of periglacial mass movement include failure in the active layer, producing translational and rotational slides and rapid flows associated with melting ground ice. Periglacial mass movement processes are related to slope dynamics and processes discussed in Chapter 13.

Humans and Periglacial Landscapes

In areas of permafrost and frozen-ground phenomena, people face several related problems. Because thawed ground above the permafrost zone frequently shifts,

FIGURE 17.23 Soil flowage in periglacial environments. Gelifluction lobes of soil underlain by frozen ground in permafrost conditions, on a hillside on Barents Island, Arctic Ocean. [Photo by Bobbé Christopherson.]

highways and rail lines become warped, twisted, and fail and utility lines are disrupted. In addition, any building placed directly on frozen ground will "melt" into the defrosting soil, creating subsidence in structures (Figure 17.24).

Construction in periglacial regions dictates placing structures above the ground to allow air circulation beneath. This airflow allows the ground to cycle through its normal annual temperature pattern. Utilities such as water and sewer lines must be built aboveground in "utilidors" to protect them from freezing and thawing ground (Figure 17.25). Likewise, the Trans-Alaska oil pipeline

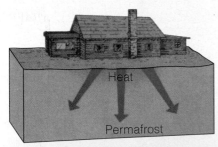

FIGURE 17.24 Permafrost melting and structure collapse.
Building failure due to improper construction and the melting of permafrost, south of Fairbanks, Alaska. [Adapted from U.S. Geological Survey. Photo by Steve McCutcheon; illustration based on U.S. Geological Survey pamphlet "Permafrost" by L. L. Ray.]

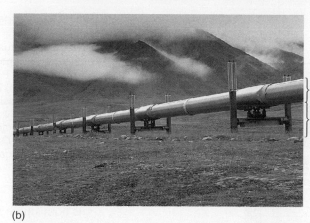

Pipeline
1.2 m diameter

Average height
1.5 to 3.0 m

(b)

FIGURE 17.25 Special structures for permafrost.
(a) Proper construction in periglacial environments requires raising of buildings aboveground and running water and sewage lines in elevated "utilidors," here in Barentsburg, a Russian settlement on Spitsbergen Island. (b) Supporting the Trans-Alaska oil pipeline on racks protects the permafrost from heat. [Photos by (a) Bobbé Christopherson; (b) Galen Rowell/Mountain Light Photography, Inc.]

(a)

was constructed aboveground on racks for 675 of its 1285-km (420 of its 800 mi) length to avoid melting the frozen ground, causing shifting that could rupture the line. The pipeline that is underground uses a cooling system to keep the permafrost around the pipeline stable. Earthquakes remain a threat, as happened in November 2002—see Figure 12.1b for a photo of this pipeline with broken support brackets.

The effects of global warming are already showing up in Siberia, where the active layer is now more than twice the depth it was in the past. Buildings constructed with shallow pilings for support will certainly fail if this trend continues. In an era of global change, the regions of permafrost provide further climatic indicators.

The Pleistocene Ice-Age Epoch

Imagine almost a third of Earth's land surface buried beneath ice sheets and glaciers—most of Canada, the northern Midwest, England, and northern Europe, and many mountain ranges beneath thousands of meters of ice. This is how it was at the height of the Pleistocene Epoch of the late Cenozoic Era. In addition, periglacial regions along the margins of the ice during the last ice age covered about twice their present areal extent.

The Pleistocene is thought to have begun about 1.65 million years ago and is one of the more prolonged cold periods in Earth's history. It featured not just one glacial advance and retreat, but at least 18 expansions of ice over Europe and North America, each obliterating and confusing the evidence from the one before. Apparently, glaciation can take about 90,000 years, whereas deglaciation is rapid, requiring less than about 10,000 years to melt away the accumulation.

The term *ice age*, or *glacial age*, is applied to any extended period of cold (not a single brief cold spell), which may last several million years. An **ice age** is a time

of generally cold climate that includes one or more *glacials*, interrupted by brief warm spells known as *interglacials*. Each glacial and interglacial is given a name that is usually based on the location where evidence of the episode is prominent, for example, "Wisconsinan glacial."

Modern research techniques to understand past climates include the examination of ancient ratios of oxygen isotopes, depths of coral growth in the tropics, analysis of ocean and lake sediments worldwide, and analysis of the latest ice cores from Greenland and Antarctica, including the Dome C ice core that plunges the record back 800,000 years. These techniques have opened the way for a new chronology and understanding of past climates and for gaining perspective on the present warming trends. The record is clear and correlates across all proxy methods: Present levels of (CO_2) and methane (CH_4) are the highest in 800,000 years, as shown in Chapter 10. Earth systems have no experience in dealing with levels this high during this time span, and these concentrations are increasing.

Glaciologists currently recognize the Illinoian glacial and Wisconsinan glacial interval, with the Sangamon interglacial between them. These events span the 300,000-year interval prior to our present Holocene Epoch (Figure 17.26). The chart shows that the Illinoian glacial actually consisted of two glacials occurring during Marine (oxygen) Isotope Stages (MIS) 6 and 8, as did the Wisconsinan (MIS 2 and 4), which are dated at 10,000 to 35,000 years ago. The oxygen isotope glacial/interglacial stages on the chart are numbered back to MIS 23 at approximately 900,000 years ago. To overcome local bias in core records, investigators correlate oxygen isotope data with other indicators worldwide. Note the label on the chart in the figure at MIS 20.2 for the Dome C extent of 800,000 years. (For web links on glaciers and the Pleistocene, see **http://userpages.umbc.edu/~miller/geog111/glacierlinks. htm**.)

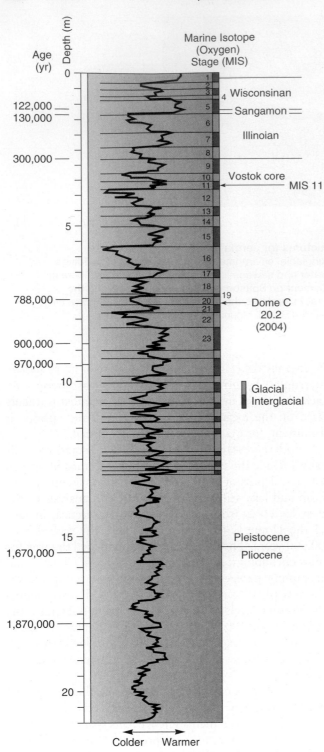

FIGURE 17.26 Temperature record of the past 2 million years.
Oxygen isotope fluctuations in fossil planktonic foraminifera (tiny marine organisms having a calcareous shell) from deep-sea cores determine Pleistocene temperatures. Twenty-three Marine (oxygen) Isotope Stages (MIS) cover 900,000 years, with names assigned for the past 300,000 years. [After N. J. Shackleton and N. D. Opdyke, *Oxygen-Isotope and Paleomagnetic Stratigraphy of Pacific Core V28-239, Late Pliocene to Latest Pleistocene.* Geological Society of America Memoir 145. ©1976 by the GSA. Adapted by permission.]

As both alpine and continental glaciers retreated, they exposed a drastically altered landscape: the rocky soils of New England, the polished and scarred surfaces of Canada's Atlantic Provinces, the sharp crests of the Sawtooth Range and Tetons of Idaho and Wyoming, the scenery of the Canadian Rockies and the Sierra Nevada, the Great Lakes of the United States and Canada, the Matterhorn of Switzerland, and much more. In the Southern Hemisphere, there is evidence of this ice age in the form of fjords and sculpted mountains in New Zealand and Chile.

The continental glaciers came and went several times over the region we know as the Great Lakes (Figure 17.28). The ice enlarged and deepened stream valleys to form the basins of the future lakes. This complex history produced five lakes that today cover 244,000 km² (94,000 mi²) and hold some 18% of all the lake water on Earth. Figure 17.28 shows the final formation of the Great Lakes, which involved two advancing and two retreating stages—between 13,200 and 10,000 years before the present. During the final retreat, tremendous quantities of glacial meltwater flowed into the isostatically (weight of the ice) depressed basins. Drainage at first was to the Mississippi River via the Illinois River, to the St. Lawrence River via the Ottawa River, and to the Hudson River in the east. In recent times, drainage has shifted through the St. Lawrence system.

Lowered Sea Levels and Lower Temperatures

Sea levels 18,000 years ago were approximately 100 m (330 ft) lower than they are today because so much of Earth's water was frozen and tied up in the glaciers instead of being in the ocean. Imagine the coastline of New York being 100 km farther east, Alaska and Russia connected by land across the Bering Straits, and England and France joined by a land bridge. In fact, sea ice extended southward into the North Atlantic and Pacific and northward in the Southern Hemisphere about 50% farther than it does today.

Sea-surface temperatures 18,000 years ago averaged between 1.4 C° and 1.7 C° (2.5 F° and 3.1 F°) lower than today. During the coldest portion of the Pleistocene ice age, air temperatures were as much as 12 C° (22 F°) colder than today's average, although milder periods ranged to within 5 C° (9 F°) of present air temperatures.

Changes in the Landscape

The continental ice sheets covered portions of Canada, the United States, Europe, and Asia about 18,000 years ago, as illustrated on the polar map projection in Figure 17.27. Ice sheets ranged in thickness to more than 2 km (1.2 mi). In North America, the Ohio and Missouri River systems mark the southern terminus of continuous ice at its greatest extent during the Pleistocene. The ice sheet disappeared by 7000 years ago.

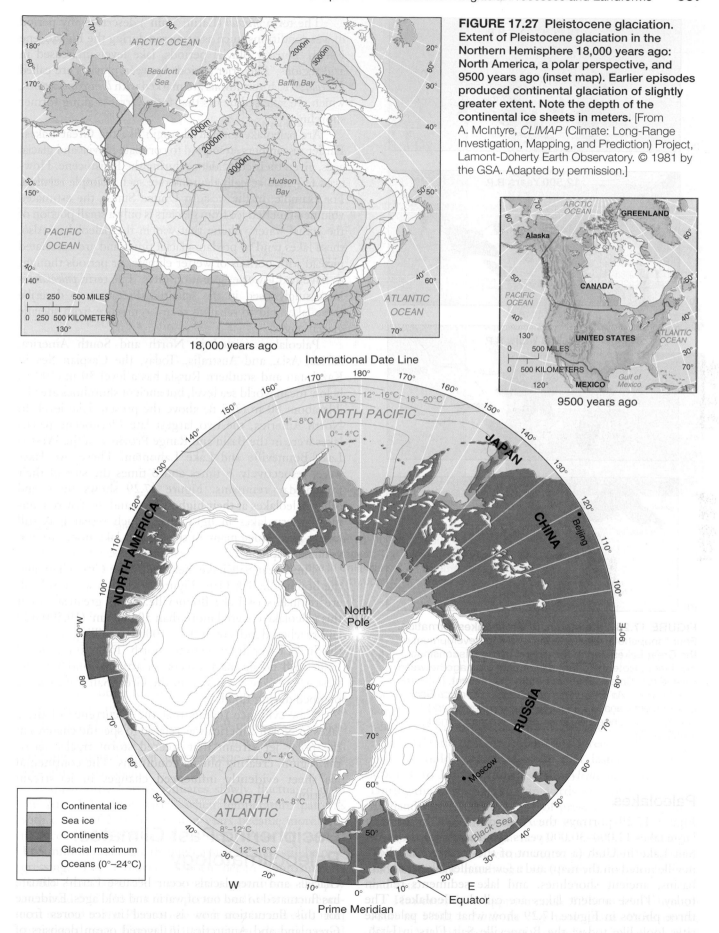

FIGURE 17.27 Pleistocene glaciation. Extent of Pleistocene glaciation in the Northern Hemisphere 18,000 years ago: North America, a polar perspective, and 9500 years ago (inset map). Earlier episodes produced continental glaciation of slightly greater extent. Note the depth of the continental ice sheets in meters. [From A. McIntyre, *CLIMAP* (Climate: Long-Range Investigation, Mapping, and Prediction) Project, Lamont-Doherty Earth Observatory. © 1981 by the GSA. Adapted by permission.]

18,000 years ago

9500 years ago

International Date Line

Prime Meridian

Equator

Continental ice
Sea ice
Continents
Glacial maximum
Oceans (0°–24°C)

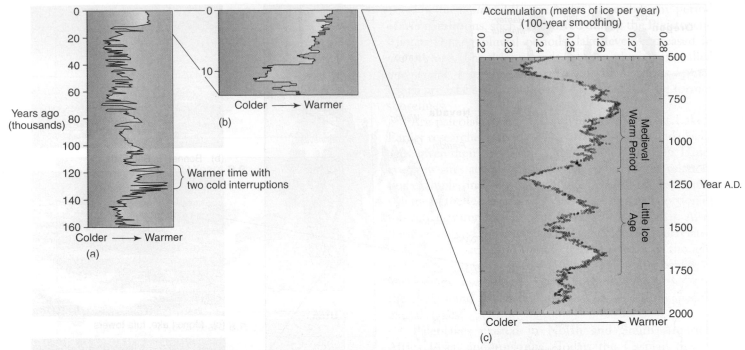

FIGURE 17.30 Recent climates determined from an ice core.
(a) Temperature patterns during the past 160,000 years, back to MIS 6. Note the two cold spells that interrupted the earlier interglacial (between 115,000 and 135,000 years ago). (b) Higher resolution of the last 12,000 years. The cold period known as the "Younger Dryas" intensified at the beginning of this record. Warming was underway by 11,700 years ago and abruptly began to increase by the Holocene.
(c) Note the Medieval Warm Period and consistent temperatures as compared with the chaotic record of the Little Ice Age. [(a) and (b) courtesy of the Greenland Ice Core Project (GRIP); (c) from D. A. Meese et al., "The accumulation record from the GISP-2 core as an indicator of climate change throughout the Holocene," *Science* 266 (December 9, 1994): 1681, © AAAS.]

Europe lowered about 200 m (650 ft) in the coldest years. The Greenland colonies were deserted. Cropping patterns changed, and northern forests declined, along with human population in those regions. In the winter of 1779–1780, New York's Hudson and East Rivers and the entire Upper Bay froze over. People walked and hauled heavy loads across the ice between Staten and Manhattan islands!

But the Little Ice Age was not consistently cold throughout its 700-year reign. Reading the record of the Greenland ice cores, scientists have found many mild years among the harsh. More accurately, this was a time of rapid, short-term climate fluctuations that lasted only decades.

Ice cores drilled in Greenland have revealed a record of annual snow and ice accumulation that, when correlated with other aspects of the core sample, was indicative of air temperature. Figure 17.30c presents this record since A.D. 500. The warmth of the Medieval Warm Period is evident, whereas the Little Ice Age appears mixed with colder conditions around 1200, 1500, and after about 1800.

Mechanisms of Climate Fluctuation

What mechanisms cause short-term fluctuations? And why is Earth pulsing through long-term climatic changes that span several hundred million years? The ice-age concept is being researched and debated with an unprecedented intensity for three principal reasons: (1) Continuous ice cores from Greenland and Antarctica are providing a new, detailed record of weather and climate patterns, volcanic eruptions, and trends in the biosphere (discussed in Focus Study 17.1); (2) to understand present and future climate change and to refine general circulation models, we must understand the natural variability of the atmosphere and climate; (3) anthropogenic global warming, and its relation to ice ages, is a major concern.

Because past occurrences of low temperature appear to have followed a pattern, researchers have looked for causes that also are cyclic in nature. They have identified a complicated mix of interacting variables that appear to influence long-term climatic trends. Let us take a look at several of them.

Climate and Celestial Relations As our Solar System revolves around the distant center of the Milky Way, it crosses the plane of the galaxy approximately every 32 million years. At that time, Earth's plane of the ecliptic aligns parallel to the galaxy's plane, and we pass through regions in space of increased interstellar dust and gas, which may have some climatic effect.

Milutin Milankovitch (1879–1954), a Yugoslavian astronomer who studied Earth–Sun orbital relations, proposed other possible astronomical factors. Milankovitch

Focus Study 17.1

GRIP, GISP-2, and Dome-C: Boring Ice for Exciting History

The Greenland Ice Core Project (GRIP) was launched in 1989. A site was selected near the summit of the Greenland Ice Sheet at 3200 m (10,500 ft) so that the maximum thickness of ice history would be accessed (Figure 17.1.1). After 3 years, the drills hit bedrock 3030 m (9940 ft) below the site—or, in terms of time, 250,000 years into the past. The core is 10 cm (4 in.) in diameter.

In 1990, about 32 km (20 mi) west of the summit, the Greenland Ice Sheet Project (GISP-2) began to bore back through time (see Figure 17.1.1). GISP-2 reached bedrock in 1993. This core is slightly larger in diameter, at 13.2 cm (5.2 in.), and collects about twice the data as GRIP. The existence of a second core helped scientists compensate for any folds or disturbed sections they encountered below 2700 m, or about 115,000 years, in the first core. Also clarifying the record are correlations with antarctic ice cores, core samples of marine sediments from the oceans and Lake Baykal in Siberia, and coral studies that indicate past sea levels.

What is being discovered from a 3030-m ice core? Locked into the core is a record of past precipitation and air bubbles of past atmospheres, which indicate ancient gas concentrations. Of special interest are the greenhouse gases, carbon dioxide and methane. Chemical and physical properties of the atmosphere and the

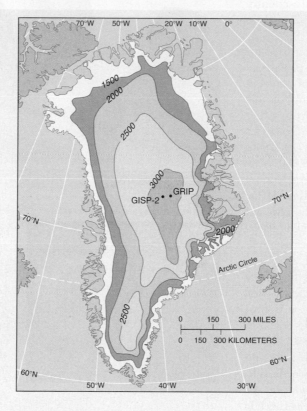

FIGURE 17.1.1 Greenland ice-core locations. GRIP is at 72.5° N, 37.5° W. GISP-2 is at 72.6° N, 38.5° W. [Map courtesy of GISP-2 Science Management Office.]

snow that accumulated each year are frozen in place.

Pollutants are locked into the ice-core record. For example, during cold periods, high concentrations of dust were present, brought by winds from distant dry lands. An invaluable record of past volcanic eruptions is included in the layers, as if on a calendar. Even the exact beginning of the Bronze Age is recorded in the ice core—about 3000 B.C. When the Greeks and later the Romans began

smelting copper, they produced ash and smoke that the winds carried to this distant place. The presence of ammonia indicates ancient forest fires at lower latitudes. And, as an important analog of past temperatures on the ice-sheet surface when each snowfall occurred, the ratio between stable forms of oxygen is measured.

Dome-C Ice Core

On the East Antarctic Plateau, 75.1° S, 123.3° E, approximately 900 km (560 mi) from the coast and 1750 km (1087 mi) from the South Pole, sits Dome C. This is the point on the Antarctic continent where the ice sheet is the thickest (see Figure 17.32c for location, and Figure 17.1.2). Here the ice is 3309-m (10,856-ft) deep. The mean annual surface temperature is −54.5°C (−66.1°F); however, the science crews experience temperatures that range from −50°C when they arrive to −25°C by midsummer. Dome C is 560 km (348 mi) from the Vostok base,

FIGURE 17.1.2 Dome-C station.
An overview of the Dome-C scientific station, with its gathering of tents and semipermanent structures.
[Photo courtesy of British Antarctic Survey.]

(continued)

Focus Study 17.1 (continued)

where the previous coring record of 400,000 years was set. The Dome-C 8-year project is part of a 10-country European Project for Ice Coring in Antarctica (EPICA).

The mechanical drill bit used for coring is 10 cm wide and provides a high-resolution record of ancient atmospheric gases, ash from volcanic eruptions, and materials from other atmospheric events. By December 2004, coring to a 3270.20-m (10,729-ft) depth has brought up 800,000 years of Earth's past climate history (Figure 17.1.3). In Figure 17.26, this is Marine Isotope Stage (MIS) 20.2. This remarkable record contains eight glacial and interglacial cycles. As far as scientists have been able to analyze results, the Dome-C record affirms the Vostok findings and correlates perfectly with the deep-sea core of oxygen isotope fluctuations in foraminifera shells (microfossils) from the Atlantic Ocean. Confirmed is the finding that the present concentration of carbon dioxide, methane, and nitrous oxide in the atmosphere is the highest it has been in the past 800,000 years and that these changing levels have marched in step with higher and lower temperatures throughout this time span.

Go, for example, to Figure 17.26 and find MIS 11. This warm spell at MIS 11, from 425,000 to 395,000 years ago, is perfectly recorded in the

(a)

(b)

FIGURE 17.1.3 Dome-C ice-core analysis.
(a) Two members of the British Antarctic Survey carefully inspect one of the ice-core segments. A quarter-section of each ice core is kept at Dome C in case there is an accident in transport back to labs in Europe. (b) Light shining through a thin section from the ice core; Earth's ancient atmosphere is trapped in the ice awaiting analysis. [Both photos courtesy of British Antarctic Survey, http://www.antarctica.ac.uk.]

Dome-C core and is of interest because CO_2 in the atmosphere during MIS 11 was similar to our preindustrial level of 280 ppm. Earth's orbital alignment was similar to that of today as well. The MIS 11 analogy means that a return to the next glacial period is perhaps 16,000 years in the future. Understanding the past helps decode future trends. The challenge for physical geographers is to put this data through spatial analysis and find the linkages to other Earth systems and to human economies and societies.

---

wondered whether the development of an ice age relates to seasonal astronomical factors—Earth's revolution around the Sun, rotation, and tilt—extended over a longer time span (Figure 17.31). In summary:

- Earth's elliptical orbit about the Sun is not constant. The shape of the ellipse varies by more than 17.7 million kilometers (11 million miles) during a 100,000-year cycle, from nearly circular to an extreme ellipse (Figure 17.31a).
- Earth's axis "wobbles" through a 26,000-year cycle, in a movement much like that of a spinning top winding down. Earth's wobble is called *precession*. As you can see in Figure 17.31b, precession changes the orientation of hemispheres and landmasses to the Sun.
- Earth's present axial tilt of 23.5° varies from 22° to 24° during a 41,000-year period (Figure 17.31c).

Milankovitch calculated, without the aid of today's computers, that the interaction of these Earth–Sun relations creates a 96,000-year climatic cycle. His glaciation model assumes that changes in astronomical relations affect the amounts of insolation received.

Milankovitch died in 1954, his ideas still not accepted by a skeptical scientific community. Now, in the era of computers, remote-sensing satellites, and worldwide efforts to decipher past climates, Milankovitch's valuable work has stimulated much research to explain climatic cycles and has experienced some confirmation. A roughly 100,000-year climatic cycle is confirmed in such diverse places as ice cores in Greenland and the accumulation of sediment in Lake Baykal, Siberia.

Climate and Solar Variability   If the Sun significantly varies its output over the years, as some other stars do, that variation would seem a convenient and plausible

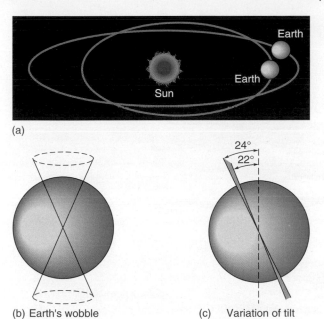

(a)

(b) Earth's wobble   (c)   Variation of tilt

FIGURE 17.31 Astronomical factors that may affect broad climatic cycles.
(a) Earth's elliptical orbit varies widely during a 100,000-year cycle, stretching out to an extreme ellipse. (b) Earth's 26,000-year axial wobble. (c) Variation in Earth's axial tilt every 41,000 years.

cause of ice-age timing. However, lack of evidence that the Sun's radiation output varies significantly over long cycles argues against this hypothesis. Nonetheless, inquiry about the Sun's variability continues. As we saw in Chapter 10, in the IPCC *Fourth Assessment Report* in Figure 10.31, solar variability is not correlated to temperature increases occurring in the atmosphere and oceans.

Climate and Tectonics   Major glaciations also can be associated with plate tectonics because some landmasses have migrated to higher, cooler latitudes. Chapters 11 and 12 explained that the shape and orientation of landmasses and ocean basins have changed greatly during Earth's history. Continental plates have drifted from equatorial locations to polar regions, and vice versa, thus exposing the land to a gradual change in climate. Gondwana (the southern half of Pangaea) experienced extensive glaciation that left its mark on the rocks of parts of present-day Africa, South America, India, Antarctica, and Australia. Landforms in the Sahara, for example, bear the markings of even earlier glacial activity. These markings are partly explained by the fact that portions of Africa were centered near the South Pole during the Ordovician Period, 465 million years ago (see Figure 11.16a).

Episodes of mountain building over the past billion years have forced mountain summits above the snowline, where snow remains after the summer melt. Mountain chains influence downwind weather patterns and jet-stream circulation, which in turn guides weather systems. More dust was present during glacial periods, suggesting drier weather and more extensive deserts beyond the frozen regions.

Climate and Atmospheric Factors   Some events alter the atmosphere and produce climate change. A volcanic eruption might produce lower temperatures for a year or two. The lower temperatures could initiate a buildup of long-term snow cover at high latitudes. These high-albedo snow surfaces then would reflect more insolation away from Earth to further enhance cooling in a positive feedback system.

The fluctuation of atmospheric greenhouse gases could trigger higher or lower temperatures. Ice cores taken at Vostok, the Russian research station near the geographic center of Antarctica, Greenland, and Dome-C contained trapped air samples that show carbon dioxide levels varying from more than 290 ppm to a low of nearly 180 ppm. Higher levels of carbon dioxide generally correlate with each interglacial, or warmer, period. In these first few years of a new century, atmospheric carbon dioxide reached 387 ppm in May 2007, higher than at any time in the past 800,000 years, principally owing to anthropogenic (human-caused) forcing—refer to the carbon dioxide records in Figures 10.32 (10,000-year record) and 10.29 (1000-year record).

Climate and Oceanic Circulation   Finally, oceanic circulation patterns have changed. For example, the Isthmus of Panama formed about 3 million years ago and effectively separated the circulation of the Atlantic and Pacific Oceans. Changes in ocean basin configuration, surface temperatures, and salinity and in upwelling and downwelling rates affect air mass formation and air temperature.

Our understanding of Earth's climate—past, present, and future—is unfolding. We are learning that climate is a multicyclic system controlled by an interacting set of cooling and warming processes, all founded on celestial relations, tectonic factors, changes in oceanic circulation, and atmospheric variables and human impacts.

Arctic and Antarctic Regions

Climatologists use environmental criteria to define the Arctic and the Antarctic regions. The *Arctic region* is defined in Figure 17.32 by the green line. This is the 10°C (50°F) isotherm for July, the Northern Hemisphere summer. This line coincides with the visible tree line—the boundary between the northern forests and tundra. The Arctic Ocean is covered by two kinds of ice: *floating sea ice* (frozen seawater) and *glacier ice* (frozen freshwater), a permanent multiyear ice and seasonal ice. This pack ice thins in the summer months and sometimes breaks up (Figure 17.32b).

As mentioned in Chapter 10, almost half of the Arctic ice pack by volume has disappeared since 1970 due to regional-scale warming. The year 2007 broke the record for lowest sea-ice extent, falling below the previous record low in 2005 by 24% (see Figure 5.1.1 in Chapter 5, High Latitude Connection 5.1). The fabled Northwest Passage across the Arctic from the Atlantic to the Pacific was ice-free in September 2007, as the Arctic ice continues

(a)

(b)

(c)

(d)

FIGURE 17.32 The Arctic and Antarctic regions.
In (a), note the 10°C (50°F) isotherm in midsummer, which designates the Arctic region, dominated by pack ice. (b) Do you see any wildlife in the photo, about 965 km (600 mi) from the North Pole? In (c), the Antarctic convergence designates the Antarctic region. Arrows on the ice sheet show the general direction of ice movement on Antarctica. Note the location of the Dome C and Vostok bases. [Photos (b) and (d) by Bobbé Christopherson.]

to melt. The Northeast passage, north of Russia, has been ice-free for the past several years.

The *Antarctic region* is defined by the Antarctic convergence, a narrow zone that extends around the continent as a boundary between colder Antarctic water and warmer water at lower latitudes. This boundary follows roughly the 10°C (50°F) isotherm for February, the Southern Hemisphere summer, and is located near 60° S latitude (green line in the figure). The Antarctic region that is covered just with sea ice represents an area greater than North America, Greenland, and Western Europe

combined. (For more information on polar-region ice, see the National Ice Center at **http://www.natice.noaa.gov/** and the Canadian Ice Service at **http://ice-glaces.ec.gc. ca/app/WsvPageDsp.cfm**.)

Antarctica is a continent-sized landmass surrounded by ocean and is much colder overall than the Arctic, which is an ocean surrounded by land. In simplest terms, Antarctica can be thought of as a continent covered by a single enormous glacier, although it contains distinct regions such as the East Antarctic and West Antarctic ice sheets, which respond differently to slight climatic variations. These ice sheets are in constant motion, as indicated in Figure 17.32c.

Changes Are Underway in the Polar Regions

Review the "Global Climate Change" section in Chapter 10, where a portrait of high-latitude temperatures and ongoing physical changes to pack ice, ice shelves, and glaciers is presented. News Report 17.2 looks at meltpond occurrence as an indicator of changing surface energy budgets. Throughout this text, several High Latitude

Connection features described additional dynamics in these polar regions. High Latitude Connection 17.1 covers the Ward Hunt Ice Shelf breakup.

Surrounding the margins of Antarctica, and constituting about 11% of its surface area, are numerous ice shelves, especially where sheltering inlets or bays exist. Covering many thousands of square kilometers, these ice shelves extend over the sea while still attached to continental ice. The loss of these ice shelves does not significantly raise sea level, for they already displace seawater. The concern is for the possible surge of grounded continental ice that the ice shelves hold back from the sea.

News Report 17.2

Increase in Meltponds Indicates Changing Surface Energy Budgets

An increase in meltponds across the Arctic and Antarctic regions indicates changing surface energy conditions. In Chapter 1, High Latitude Connection 1.1 describes these meltponds as positive feedback mechanisms. The *Landsat*-7 satellite in tandem with

aircraft equipped with video cameras spotted this increase in *meltpond* occurrence on glaciers, icebergs, ice shelves, and the Greenland ice sheet.

Figure 17.2.1 compares scenes from west Greenland in June 2001 and June 2003, using the *Terra* satellite.

Examine the bare rock along the eastern margin, indicating higher snow levels. The images show rapidly increasing numbers of meltponds and areas of water-saturated ice in the *melt zone* during this 2-year interval. A positive feedback system, the meltponds are darker in color than surrounding glacial ice and snow, absorbing more sunlight, heating up, melting more ice, forming more meltponds, and so on. The meltponds look like small blue dots across the ice.

A chance to view these phenomena in September 2004 provided this author firsthand evidence of these trends. The action of such melting is evident in the potholes mottling a glacier in southern Greenland (Figure 17.2.2a) and the actual meltponds on the ice sheet (Figure 17.2.2b, c). If you look at the Larsen-B ice shelf breakup in Figure 17.33b, you can identify the presence of meltponds as the ice decayed.

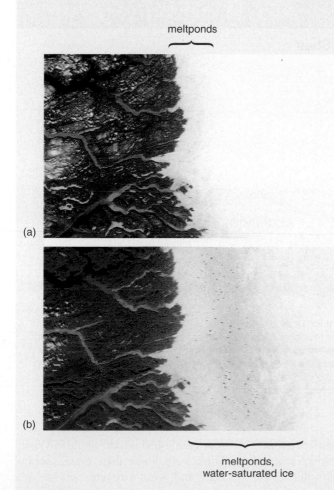

FIGURE 17.2.1 Meltponds on the increase in Greenland.
Meltponds (blue specks) are on the increase in the Arctic on icebergs, ice shelves, and the Greenland Ice Sheet. (a) Images of west Greenland, June 14, 2001 and (b) June 17, 2003, demonstrate these changes. [Images by *Terra* courtesy of MODIS Rapid Response Team, NASA/GSFC.]

(continued)

News Report 17.2 (continued)

(b)

(c)

(a)

FIGURE 17.2.2 Meltponds close up.
(a) Glacial ice loss is visible in the pockmarked appearance of this Greenland glacier. (b) An aerial photo of meltponds on the ice sheet near the area of the satellite images in Figure 17.2.1. (c) A meltpond close up. [All photos by Bobbe Christopherson.]

High Latitude Connection 17.1

Climate Change Impacts an Arctic Ice Shelf

By late fall 2003, news spread through the climate science community that the Ward Hunt Ice Shelf—largest in the Arctic, on the north coast of Ellesmere Island in Nunavut (noted on the map in Figure 17.32a)—was breaking up. This was shocking news, for the shelf had been stable for at least 4500 years. What scientists observed was that after three decades of mass losses, the ice shelf reached a systems threshold and rapidly broke up between 2000 and 2003. (Figure 17.1.1).

Backed up behind the Ward Hunt Ice Shelf was an *epishelf lake*, an extensive freshwater accumulation some 43 m deep, sitting on about 360 m of salty seawater in the Disraeli Fjord. As the ice-shelf dam failed, the freshwater epishelf lake drained. This was the largest such epishelf lake in the Northern Hemisphere. A unique biological community of freshwater and brackish-water plankton was lost and delicate ecosystems disrupted.

Scientists noted that regional temperatures increased between 1967 and

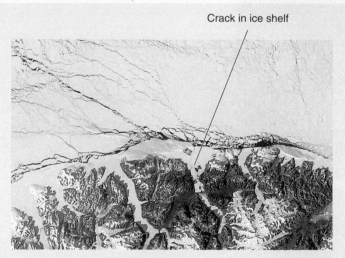

Crack in ice shelf

FIGURE 17.1.1 Ward Hunt Ice Shelf breakup.
The Ward Hunt Ice Shelf is near center on the north coast, with a large crack down its center and losses along its seaward front. The northwest coast of Greenland is to the east and Nansen Sound and Greely Fjord are to the west. The center of the image is at 80° N latitude—noted on the map in Figure 17.32a. [*Terra* image courtesy of MODIS Land Rapid Response Team, NASA/GSFC, September 10, 2003.]

2002 to an average July surface temperature of 1.3°C (34.3°F), which is above the threshold for ice-shelf failure established for Antarctica. Some other factors that pushed this threshold in Ward Hunt were freeze-thaw cycles, ocean tides and wind strength, sea-surface temperatures, and variations in salinity.

Although ice shelves constantly break up to produce icebergs, some large sections have recently broken free. In 1998, an iceberg the size of Delaware broke off the Ronne Ice Shelf, southeast of the Antarctic Peninsula. In March 2000, an iceberg tagged B-15 broke off the Ross Ice Shelf (some 90° longitude west of the Antarctic Peninsula), measuring *twice* the area of Delaware, 300 km by 40 km, or 190 mi by 25 mi. Remotely reporting weather instruments are in place on B-15 as the tabular iceberg breaks into pieces (Figure 17.33a).

Since 1993, six ice shelves have disintegrated in Antarctica. More than 8000 km² (3090 mi²) of ice shelf is gone, changing maps, freeing up islands to circumnavigation, and creating thousands of icebergs. The Larsen Ice Shelf, along the east coast of the Antarctic Peninsula, has been retreating slowly for years. Larsen-A suddenly disintegrated in 1995. In only 35 days in early 2002, Larsen-B collapsed into icebergs (Figure 10.33b). Larsen-B was at least 11,000 years old, meaning this collapse was farther south than any previously during the Holocene. Larsen-C, the next segment to the south, is losing mass on its underside since the water temperature is warmer by 0.65 C° (1.17 F°) than the melting point for ice at 300-m depth—warmth is melting the shelf from both the ocean and atmosphere sides. This ice loss is likely a result of the 2.5 C° (4.5 F°) temperature increase in the peninsula region

(a)

FIGURE 17.33 Disintegrating ice shelves along the Antarctic coast.
(a) 2003 image of a broken B-15 iceberg that calved off the Ross Ice Shelf, near Roosevelt Island, Antarctica, in 2000; originally twice the size of Delaware. (b) Disintegration and retreat of Larsen-B ice shelf between January 31 and March 7, 2002. Note the meltponds in the January image. [(a) and (b) *Terra* image courtesy Rapid Response Team, NASA.]

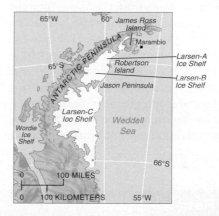

(b) January 31, 2002   March 7, 2002

FIGURE 17.34 Scientific base at the South Pole.
Aerial view of the Amundsen-Scott South Pole Station in 2001. The geodesic dome, completed in 1975, is 50 m wide and 16 m high (165 ft by 52 ft) and shelters buildings from the frigid cold and winds. On the left side you see the new modular buildings, which stand 3 m (10 ft) above the surface to accommodate snow accumulation and protect the ice under the building from melting. Above this module is the ceremonial South Pole; slightly to the right is the geographic South Pole. This station is at an elevation of 2835 m (9301 ft). [Photo courtesy of Kristan Hutchison, U.S. Antarctic Program.]

during the last 50 years. In response to the increasing warmth, the Antarctic Peninsula is supporting previously unseen vegetation growth, reduced sea ice, and disruption of penguin feeding, nesting, and fledging activities.

The Pine Island Glacier flows from the West Antarctic Ice Sheet (WAIS) as an outlet glacier to the Amundsen Sea. This glacier accelerated 3.5% in flow rate between April 2001 and April 2003—since the 1970s this brings the increase to more than 25% over previous averages for glacial movement. The Pine Island Glacier is accelerating toward the ocean and calving at a greater rate than previously recorded along its 40-km-wide ice shelf. Scientists determined that the acceleration is sensed through the ice

in the glacier; far back into the WAIS, flows of grounded ice upstream in the glacier that are not yet displacing ocean water are moving toward the sea.

The fact that Antarctica is a place so remote from civilization makes it an excellent laboratory for sampling past and present human and natural variables that are transported by atmospheric and oceanic circulation. High altitude, winter cold and darkness, and distance from pollution sources make the polar region an ideal location for certain astronomical and atmospheric observations (Figure 17.34). Twenty-eight scientists and support people work through the winter (February to October) and 130 personnel, or more, research and work there in the brief summer.

Summary and Review—Glacial and Periglacial Processes and Landforms

■ *Differentiate* between alpine and continental glaciers, and *describe* their principal features.

More than 77% of Earth's freshwater is frozen. Ice covers about 11% of Earth's surface, and periglacial features occupy another 20% of ice-free but cold-dominated landscapes. A **glacier** is a mass of ice sitting on land or floating as an ice shelf in the ocean next to land. Glaciers form in areas of permanent snow. A **snowline** is the lowest elevation where snow occurs year-round and its altitude varies by latitude—higher near the equator, lower poleward.

A glacier in a mountain range is an **alpine glacier**. If confined within a valley, it is termed a *valley glacier*. The area of origin is a snowfield, usually in a bowl-shaped erosional landform called a **cirque**. Where alpine glaciers flow down to the sea, they calve and form **icebergs**. A **continental glacier** is a continuous mass of ice on land. Its most extensive form is an **ice sheet**; a smaller, roughly circular form is an **ice cap**; and the least extensive form, usually in mountains, is an **ice field**.

glacier (p. 532)
snowline (p. 533)
alpine glacier (p. 533)
cirque (p. 534)

iceberg (p. 534)
continental glacier (p. 535)
ice sheet (p. 535)
ice cap (p. 535)
ice field (p. 536)

1. Describe the location of most freshwater on Earth today.
2. What is a glacier? What is implied about existing climate patterns in a glacial region?
3. Differentiate between an alpine glacier and the three types of continental glaciers. Which occurs in mountains? Which covers Antarctica?
4. How are icebergs generated? Describe their buoyancy characteristics from this discussion and the section on ice in Chapter 7. Why do you think an iceberg overturns from time to time as it melts?

■ *Describe* the process of glacial ice formation, and *portray* the mechanics of glacial movement.

Snow becomes glacial ice through accumulation, increasing thickness, pressure on underlying layers, and recrystallization. Snow progresses through transitional steps from **firn** (compact, granular) to a denser **glacial ice** after many years.

A glacier is an open system with inputs and outputs that can be analyzed through observation of the growth and wasting of the glacier itself. A **firn line** is the lower extent of a fresh snow-covered area. A glacier is fed by snowfall and is wasted by **ablation** (losses from its upper and lower surfaces and along its margins). Accumulation and ablation achieve a mass balance in each glacier.

As a glacier moves downhill, vertical **crevasses** may develop. Sometimes a glacier will move rapidly in a **glacier surge**. The presence of water along the basal layer appears to be important in glacial movements. As a glacier moves, it plucks rock pieces and debris, incorporating them into the ice, and this debris scours and sandpapers underlying rock through **abrasion**.

firn (p. 538)
glacial ice (p. 538)
firn line (p. 538)
ablation (p. 538)
crevasse (p. 538)
glacier surge (p. 538)
abrasion (p. 541)

5. Trace the evolution of glacial ice from fresh fallen snow.
6. What is meant by glacial mass balance? What are the basic inputs and outputs underlying that balance?
7. What is meant by a glacier surge? What do scientists think produces surging episodes?

■ *Describe* characteristic erosional and depositional landforms created by alpine glaciation and continental glaciation.

Extensive valley glaciers have profoundly reshaped mountains worldwide, carving **V**-shaped stream valleys into **U**-shaped glaciated valleys, producing many distinctive erosional and depositional landforms. As cirque walls erode away, sharp **arêtes** (sawtooth, serrated ridges) form, dividing adjacent cirque basins. Two eroding cirques may reduce an arête to a saddlelike **col**. A **horn** results when several cirque glaciers gouge an individual mountain summit from all sides, forming a pyramidal peak. A **bergschrund** forms when a crevasse or wide crack opens along the headwall of a glacier, most visible in summer when covering snow is gone. An ice-carved rock basin left as a glacier retreats may fill with water to form a **tarn**; tarns in a string separated by moraines are called **paternoster lakes**. Where a glacial valley trough joins the ocean, and the glacier retreats, the sea extends inland to form a **fjord**.

All glacial deposits, whether ice-borne or meltwater-borne, constitute **glacial drift**. Glacial meltwater deposits are sorted and are called **stratified drift**. Direct deposits from ice, called **till**, are unstratified and unsorted. Specific landforms produced by the deposition of drift are **moraines**. A **lateral moraine** forms along each side of a glacier; merging glaciers with lateral moraines form a **medial moraine**; and eroded debris dropped at the glacier's terminus is a **terminal moraine**.

Continental glaciation leaves different features than does alpine glaciation. A **till plain** forms behind end moraines, featuring unstratified coarse till, low and rolling relief, and deranged drainage. Beyond the moraidal deposits, **outwash plains** of stratified drift feature stream channels that are meltwater-fed, braided, and overloaded with debris that is sorted and deposited across the landscape. An **esker** is a sinuously curving, narrow ridge of coarse sand and gravel that forms along the channel of a meltwater stream beneath a glacier. An isolated block of ice left by a retreating glacier becomes surrounded by debris; when the block finally melts, it leaves a steep-sided **kettle**. A **kame** is a small hill, knob, or mound of poorly sorted sand and gravel that is deposited directly by water or by ice in crevasses.

Glacial action forms two types of streamlined hills: the erosional **roche moutonnée** is an asymmetrical hill of exposed bedrock, gently sloping upstream and abruptly sloping downstream; the depositional **drumlin** is deposited till, streamlined in the direction of continental ice movement (blunt end upstream and tapered end downstream).

arête (p. 542)
col (p. 542)
horn (p. 542)
bergschrund (p. 542)
tarn (p. 542)
paternoster lakes (p. 542)
fjord (p. 544)
glacial drift (p. 545)
stratified drift (p. 545)
till (p. 545)
moraine (p. 545)
lateral moraine (p. 545)
medial moraine (p. 545)
terminal moraine (p. 545)
till plain (p. 546)
outwash plain (p. 546)
esker (p. 546)
kettle (p. 548)
kame (p. 548)
roche moutonnée (p. 548)
drumlin (p. 548)

8. How does a glacier accomplish erosion?
9. Describe the evolution of a **V**-shaped stream valley to a **U**-shaped glaciated valley. What features are visible after the glacier retreats?
10. How is an arête formed? A col? A horn? Briefly differentiate among them.
11. Differentiate between two forms of glacial drift—till and outwash.
12. What is a morainal deposit? What specific moraines are created by alpine and continental glaciers?
13. What are some common depositional features encountered in a till plain?
14. Contrast a roche moutonnée and a drumlin regarding appearance, orientation, and the way each forms.

■ *Analyze* the spatial distribution of periglacial processes, and *describe* several unique landforms and topographic features related to permafrost and frozen ground phenomena.

The term **periglacial** describes cold-climate processes, landforms, and topographic features that exist along the margins of glaciers, past and present. When soil or rock temperatures remain below 0°C (32°F) for at least 2 years, **permafrost** ("permanent frost") develops. Note that this criterion is based solely on temperature and has nothing to do with how much or how little water is present. The **active layer** is the zone of seasonally frozen ground that exists between the subsurface permafrost layer and the ground surface. **Patterned ground** forms in the periglacial environment where freezing and thawing of the ground create polygonal forms of circles, polygons, stripes, nets, and steps.

periglacial (p. 548)
permafrost (p. 548)
active layer (p. 550)
patterned ground (p. 553)

15. In terms of climatic types, describe the areas on Earth where periglacial landscapes occur. Include both higher-latitude and higher-altitude climate types.

16. Define two types of permafrost, and differentiate their occurrence on Earth. What are the characteristics of each?

17. Describe the active zone in permafrost regions, and relate the degree of development to specific latitudes.

18. What is a talik? Where might you expect to find taliks, and to what depth do they occur?

19. What is the difference between permafrost and ground ice?

20. Describe the role of frost action in the formation of various landform types in the periglacial region, such as a pingo and patterned ground.

21. Relate some of the specific problems humans encounter in developing periglacial landscapes.

■ *Explain* the Pleistocene ice-age epoch and related glacials and interglacials, and *describe* some of the methods used to study paleoclimatology.

An **ice age** is any extended period of cold. The late Cenozoic Era featured pronounced ice-age conditions in an epoch called the Pleistocene. During this time, alpine and continental glaciers covered about 30% of Earth's land area in at least 18 glacials, punctuated by interglacials of milder weather. Beyond the ice, **paleolakes** formed because of wetter conditions. Evidence of ice-age conditions is gathered from ice cores drilled in Greenland and Antarctica, from ocean sediments, from coral growth in relation to past sea levels, and from rock. The study of past climates is *paleoclimatology*.

The apparent pattern followed by these low-temperature episodes indicates cyclic causes. A complicated mix of interacting variables appears to influence long-term climatic trends: celestial relations, solar variability, tectonic factors, atmospheric variables, and oceanic circulation. The present anthropogenic forcing of temperatures is not part of a natural cycle.

ice age (p. 555)
paleolake (p. 558)

22. What is paleoclimatology? Describe Earth's past climatic patterns. Are we experiencing a normal climate pattern in this era, or have scientists noticed any significant trends?

23. Define an ice age. When was the most recent? Explain "glacial" and "interglacial" in your answer.

24. Summarize what science has learned about the causes of ice ages by listing and explaining at least four possible factors in climate change.

25. Describe the role of ice cores in deciphering past climates. What record do they preserve? Where were they drilled?

26. Explain the relationship between the criteria defining the Arctic and Antarctic regions. Is there any coincidence between the Arctic criteria and the distribution of Northern Hemisphere forests on the continents?

27. Based on information in this chapter and in Chapter 10's "Global Climate Change" section, summarize a few of the changing conditions underway in each polar region. Are there any conditions that seem of greatest concern to you after studying them?

NetWork

The *Geosystems* Student Learning Center provides on-line resources for this chapter on the World Wide Web. To begin: Once at the Center, click on the cover of this textbook, scroll the Table of Contents menu, and select this chapter. You will find self-tests that are graded, review exercises, specific updates for items in the chapter, and in "Destinations" many links to interesting related pathways on the Internet. *Geosystems* Student Learning Center is found at **http://www.prenhall.com/christopherson/**.

Critical Thinking

A. After reading my summary of the *New South Polar Times*, 1993–1998, at the Geosystems Student Learning Center, imagine duty there for yourself. Of the 28 people who winter over at the station, some serve as scientists, technicians, and support staff. The station commander and reporter of the 1998 season was Katy McNitt-Jensen, in her third tour at the pole. Remember, the last airplane leaves mid-February and the first airplane lands mid-October—such is the isolation. What do you see as the positives and negatives of such service? How would you combat the elements? The isolation? The cold and dark conditions?

B. Hypothetically, speculate on the relationship between the Alaskan meltdown discussed in this chapter (pp. 540–41) and the occurrence of earthquakes, such as the magnitude 7.9 quake near Denali National Park, Alaska, the largest on Earth during 2002. The rapid melting of glacial ice is causing an isostatic rebound of the landscape (see Figure 11.4). Vertical motion in Alaska is averaging 36 mm (1.42 in.) per year of uplift—the fastest rate anywhere on Earth. Scientists see a tie between the rebounding crust and loss of glacial ice weight over the past 150 years and especially over the past 30 years. How might isostatic rebound, in response to the unloading of glacial ice, produce strain along faults?

Soils, Ecosystems, and Biomes

PART IV

Natural systems function in beauty at the Sacramento National Wildlife Refuge in the Sacramento Valley. Here tens of thousands of birds stop over on their long migrations. Snow Geese predominate in the scene, but we also see Cinnamon Teal and Pintail ducks and White-fronted Geese. The entire refuge complex of six tracts comprises 35,000 acres and some 30,000 acres of conservation easements. Unfortunately, the spread of West Nile Virus is having an impact on bird populations, and several thousand birds have died. Scientists are monitoring events closely. [Photo by Bobbé Christopherson.]

Earth is the home of the only known biosphere in the Solar System— a unique, complex, and interactive system of abiotic (nonliving) and biotic (living) components working together to sustain a tremendous diversity of life. Energy enters the biosphere through conversion of solar energy by photosynthesis in the leaves of plants. Soil is the essential link among the lithosphere, plants, and the rest of Earth's physical systems. Thus, soil helps sustain life and is a bridge between Part III and Part IV of this text.

Life is organized into a feeding hierarchy from producers to consumers, ending with decomposers. Taken together, the soils, plants, animals, and all abiotic components produce aquatic and terrestrial ecosystems, generally grouped together in various biomes. Today we face crucial issues, principally the preservation of the diversity of life in the biosphere and the survival of the biosphere itself. Patterns of land and ocean temperatures, precipitation, weather phenomena, and stratospheric ozone, among many elements, are changing as global climate systems shift. The resilience of the biosphere as we know it is being tested in a real-time, one-time experiment. These important issues of biogeography are considered in Part IV.

573

Peat, a Histosol organic soil, is cut and dried in blocks for fuel on Mainland Island, in the Orkneys, north of Scotland (59° N). More than 160 varieties of sphagnum moss grow in moist, cool climates (lower-right inset photo). As layer after layer of fibrous material compresses and chemically breaks down, peat is formed. Poor drainage and high water content is important to the process. Traditionally a family and community activity, the peat is hand cut and set out to dry. Once dried, the peat blocks burn hot and smoky. Peat is the first stage in the formation of lignite, an intermediate step toward coal. [Photos by Bobbé Christopherson.]

The Geography of Soils

■ Key Learning Concepts

After reading the chapter, you should be able to:

■ *Define* soil and soil science, and *describe* a pedon, polypedon, and typical soil profile.

■ *Describe* soil properties including color, texture, structure, consistence, porosity, and soil moisture.

■ *Explain* basic soil chemistry, including cation-exchange capacity, and *relate* these concepts to soil fertility.

■ *Evaluate* the principal soil formation factors, including the human element.

■ *Describe* the 12 soil orders of the Soil Taxonomy classification system, and *explain* their general occurrence.

arth's landscape generally is covered with soil. **Soil** is a dynamic natural material composed of fine particles in which plants grow, and it contains both mineral fragments and organic matter. The soil system includes human interactions and supports all human, other animal, and plant life. If you have ever planted a garden, tended a houseplant, or been concerned about famine and soil loss, this chapter will interest you. A knowledge of soil is at the heart of agriculture and food production.

You kneel down and scoop a handful of prairie soil, compressing it and breaking it apart with your fingers. You are holding a historical object, one that bears the legacy of the last 15,000 years or more. This lump of soil contains information about the last ice age and intervening warm intervals, about distinct and distant source materials, about several physical processes. We are using and abusing this legacy at rates much faster than it was formed. Soils do not reproduce, nor can they be re-created.

Soil science is interdisciplinary, involving physics, chemistry, biology, mineralogy, hydrology, taxonomy, climatology, and cartography. Physical geographers are interested in the spatial patterns formed by soil types and the environmental factors that interact to produce them. As an integrative science, physical geography is well suited to the study of soils.

Pedology concerns the origin, classification, distribution, and description of soil (*ped* from the Greek *pedon*, meaning "soil" or "earth"). Pedology is at the center of learning about soil as a natural body, but it does not dwell on its practical uses. *Edaphology* (from the Greek *edaphos*, meaning "soil" or "ground") focuses on soil as a medium for sustaining higher plants. Edaphology emphasizes plant growth, fertility, and the differences in productivity among soils. Pedology gives us a general understanding of soils and their classification, whereas edaphology reflects society's concern for food and fiber production and the management of soils to increase fertility and reduce soil losses.

In many locales, an *agricultural extension service* can provide specific information and perform a detailed analysis of local soils. Soil surveys and local soil maps are available for most counties in the United States and for the Canadian provinces. Your local phone book may list the U.S. Department of Agriculture, Natural Resources Conservation Service (**http://www.nrcs.usda.gov/**), or Agriculture Canada's Soil Information System (**http://sis.agr.gc.ca/cansis/intro.html**, with more information at **http://www.metla.fi/info/vlib/soils/old.htm**).

In this chapter: The geography of soils deals spatially with a complex substance, the characteristics of which vary from kilometer to kilometer, and even centimeter to centimeter. We begin with soil characteristics and the basic soil sampling and soil mapping units. The soil profile is a dynamic structure, mixing and exchanging materials and moisture across its horizons. Properties of soil include texture, structure, porosity, moisture, and chemistry—all integrating to form soil types. The chapter discusses both natural and human factors that affect soil formation. A global concern exists over the loss of soils to erosion, mistreatment, and conversion to other uses. The chapter concludes with a brief examination of the Soil Taxonomy, the 12 principal soil orders, and their spatial distribution.

Soil Characteristics

Classifying soils is similar to classifying climates because both involve interacting variables. Before we look at soil classification, let us examine the physical properties that distinguish soils as they develop through time in response to climate, relief, and topography.

Soil Profiles

As a book cannot be judged by its cover, so soils cannot be evaluated at the surface only. Instead, a soil profile should be studied from the surface to the deepest extent of plant roots, or to where regolith or bedrock is encountered. Such a profile is a **pedon**, a hexagonal column measuring 1 to 10 m² in top surface area (Figure 18.1). At the sides of the pedon, the various layers of the soil profile are visible

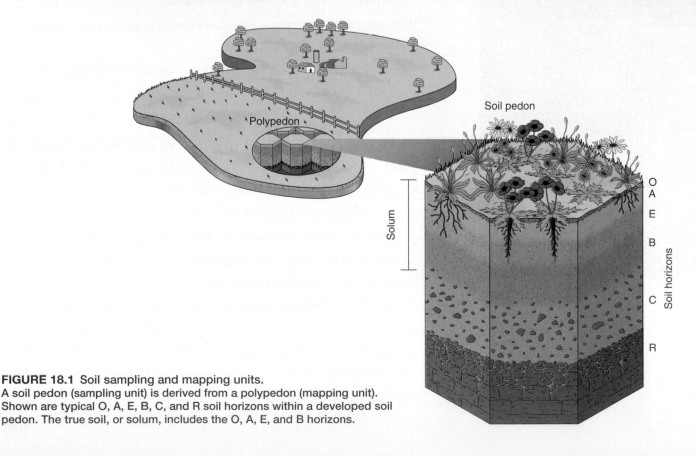

FIGURE 18.1 Soil sampling and mapping units.
A soil pedon (sampling unit) is derived from a polypedon (mapping unit). Shown are typical O, A, E, B, C, and R soil horizons within a developed soil pedon. The true soil, or solum, includes the O, A, E, and B horizons.

in cross section and are labeled with letters. *A pedon is the basic sampling unit used in soil surveys.*

Many pedons together in one area make up a **polypedon**, which has distinctive characteristics differentiating it from surrounding polypedons. A polypedon is an essential soil individual, comprising an identifiable series of soils in an area. It can have a minimum dimension of about 1 m² and no specified maximum size. *The polypedon is the basic mapping unit used in preparing local soil maps.*

Soil Horizons

Each distinct layer exposed in a pedon is a **soil horizon**. A horizon is roughly parallel to the pedon's surface and has characteristics distinctly different from horizons directly above or below. The boundary between horizons usually is distinguishable when viewed in profile, such as along a road cut, using the properties of color, texture, structure, consistence (meaning soil consistency or cohesiveness), porosity, the presence or absence of certain minerals, moisture, and chemical processes (Figure 18.2). Soil horizons are the building blocks of soil classification.

FIGURE 18.2 A typical soil profile.
This is a Mollisol pedon in southeastern South Dakota. The parent material is till, and the soil is well drained. The dark O and A horizons above the #1 transition into an E horizon. Distinct carbonate nodules are visible in the lower B and upper C horizons. [Photo from the Marbut Collection, Soil Science Society of America.]

At the top of the soil profile is the *O (organic) horizon*, named for its organic composition, derived from plant and animal litter that was deposited on the surface and transformed into humus. **Humus** is not just a single material; it is a mixture of decomposed and synthesized organic materials, usually dark in color. Microorganisms work busily on this organic debris, performing a portion of the *humification* (humus-making) process. The O horizon is 20%–30% or more organic matter, which is important because of its ability to retain water and nutrients and for the way it acts in a complementary manner to clay minerals.

At the bottom of the soil profile is the *R (rock) horizon*, consisting of either unconsolidated (loose) material or consolidated bedrock. When bedrock physically and chemically weathers into regolith, it may or may not contribute to overlying soil horizons. The A, E, B, and C horizons mark differing mineral strata between O and R. These middle layers are composed of sand, silt, clay, and other weathered by-products. (Table 11.3 in Chapter 11 presents a description of grain sizes for these weathered soil particles.)

In the *A horizon*, humus and clay particles are particularly important, for they provide essential chemical links between soil nutrients and plants. This horizon usually is richer in organic content, and hence darker, than lower horizons. Here is where human disruption through plowing, pasturing, and other uses takes place. The A horizon grades into the *E horizon*, made up of coarse sand, silt, and resistant minerals; generally, the E horizon is lighter in color.

From the lighter-colored E horizon, silicate clays and oxides of aluminum and iron are leached (removed by water) and carried to lower horizons with the water as it percolates through the soil. This process of removal of fine particles and minerals by water, leaving behind sand and silt, is **eluviation**—thus the E designation for this horizon. As precipitation increases, so does the rate of eluviation.

In contrast to the A and E horizons, *B horizons* accumulate clays, aluminum, and iron. B horizons are dominated by **illuviation**, a depositional process. (Eluviation is an erosional removal process; illuviation is depositional.) B horizons may exhibit reddish or yellowish hues because of the illuviated presence of minerals (silicate clays, iron and aluminum, carbonates, gypsum) and organic oxides. Some materials occurring in the B horizon may have formed in place from weathering processes rather than arriving there by *translocation*, or migration. In the humid tropics, these layers often develop to some depth. Likewise, clay losses in an A horizon may be caused by destructive processes and not eluviation. Research to better understand erosion and deposition of clays between soil horizons is one of the challenges in modern soil science.

The combination of the A and E horizons and the B horizon is designated the **solum**, considered the true definable soil of the pedon. The A, E, and B horizons experience active soil processes (labeled in Figure 18.1).

Below the solum is the *C horizon* of weathered bedrock or weathered parent material. This zone is identified as *regolith* (although the term sometimes is used to include the solum as well). The C horizon is not much affected by soil operations in the solum and lies outside the biological influences experienced in the shallower horizons. Plant roots and soil microorganisms are rare in the C horizon. It lacks clay concentrations and generally is made up of carbonates, gypsum, or soluble salts, or of iron and silica, which form cemented soil structures. In dry climates, calcium carbonate commonly forms the cementing material of these hardened layers.

Soil scientists using the U.S. classification system employ letter suffixes to further designate special conditions within each soil horizon. A few examples include Ap (A horizon has been plowed), Bt (B horizon is of illuviated clay), Bf (permafrost or frozen soil), and Bh (illuviated humus as a dark coating on sand and silt particles).

In 1995, an international organization—the International Committee on Anthropogenic Soils (ICOMANTH, **http://clic.cses.vt.edu/icomanth/**)—formed to better deal with classification of anthropogenic soils, those with major properties created by human activities. The designation "M" is used in the NRCS *Keys to Soil Taxonomy*, 10th edition (2006), to indicate the presence of horizontal artifacts such as concrete and asphalt pavement and plastic, rubber, and textile liners placed by humans for some practical purpose. An *M* tells us that a subsoil layer is present that limits root growth. This is an innovation for soil science to assist in assessing construction or septic system monitoring, among other considerations.

Soil Properties

Soils are complex and varied, as this section reveals. Observing a real soil profile will help you identify color, texture, structure, and other soil properties. A good opportunity to observe soil profiles is at a construction site or excavation, perhaps on your campus, or at a road cut along a highway. The USDA Natural Resources Conservation Service *Soil Survey Manual* (U.S. Department of Agriculture Handbook No. 18, October 1993) presents information on all soil properties. The NRCS publication *Soil Survey Laboratory Methods Manual* (U.S. Department of Agriculture Soil Survey Investigations Report No. 42, v. 4.0, November 2004) details specific methods and practices in conducting soil analysis.

Soil Color

Color is important, for it sometimes suggests composition and chemical makeup. If you look at exposed soil, color may be the most obvious trait. Among the many possible hues are the reds and yellows found in soils of the southeastern United States (high in iron oxides), the blacks of prairie soils in portions of the U.S. grain-growing regions and Ukraine (richly organic), and white-to-pale hues found in soils containing silicates and aluminum oxides. However, color can be deceptive: Soils of high humus content are often dark, yet

FIGURE 18.3 A Munsell Soil Color Chart page. A soil sample is viewed through the hole to match it with a color on the chart. Hue, value, and chroma are the characteristics of color assessed by this system. [Photo courtesy of Gretag Macbeth, Munsell Color.]

clays of warm–temperate and tropical regions with less than 3% organic content are some of the world's blackest soils.

To standardize color descriptions, soil scientists describe a soil's color by comparing it with a *Munsell Color Chart* (developed by artist and teacher Albert Munsell in 1913). These charts display 175 colors arranged by *hue* (the dominant spectral color, such as red), *value* (degree of darkness or lightness), and *chroma* (purity and saturation of the color, which increase with decreasing grayness). A Munsell notation identifies each color by a name, so soil scientists can make worldwide comparisons of soil color. Soil color is checked against the chart at various depths within a pedon (Figure 18.3).

Soil Texture

Soil texture, perhaps a soil's most permanent attribute, refers to the mixture of sizes of its particles and the proportion of different sizes. Individual mineral particles are *soil separates*. All particles smaller in diameter than 2 mm (0.08 in.), such as very coarse sand, are considered part of the soil. Larger particles, such as pebbles, gravel, or cobbles, are not part of the soil. (Sands are graded from coarse, to medium, to fine, down to 0.05 mm; silt to 0.002 mm; and clay at less than 0.002 mm.) Figure 18.4 is a *soil texture triangle* showing the relation of sand, silt, and clay concentrations in soil. Each corner of the triangle represents a soil consisting solely of the particle size noted (although rarely are true soils composed of a single separate). Every soil on Earth is defined somewhere in this triangle.

Figure 18.4 includes the common designation **loam**, which is a balanced mixture of sand, silt, and clay that is beneficial to plant growth. Farmers consider ideal a sandy loam with clay content below 30% (lower left) because of its water-holding characteristics and ease of cultivation. Soil texture is important in determining water-retention and water-transmission traits.

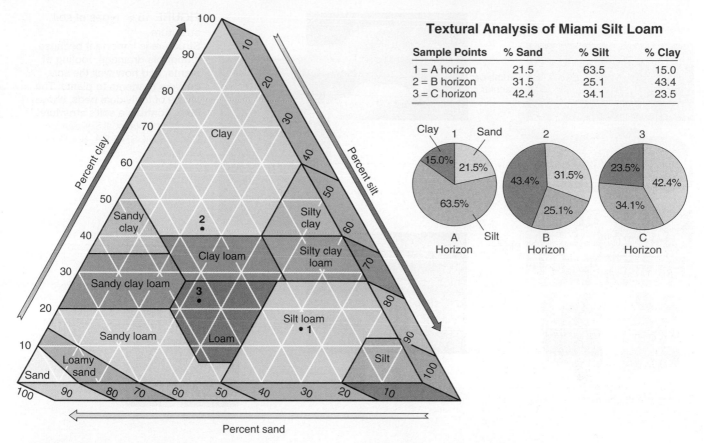

Textural Analysis of Miami Silt Loam

Sample Points	% Sand	% Silt	% Clay
1 = A horizon	21.5	63.5	15.0
2 = B horizon	31.5	25.1	43.4
3 = C horizon	42.4	34.1	23.5

FIGURE 18.4 Soil texture triangle.
Measures of the ratio of clay, silt, and sand determine soil texture. As an example, points 1 (horizon A), 2 (horizon B), and 3 (horizon C) designate samples taken at three different horizons in the Miami silt loam in Indiana. Note the ratio of sand to silt to clay shown in the three pie diagrams and table. [After USDA–NRCS, *Soil Survey Manual,* Agricultural Handbook No. 18, p. 138 (Washington, DC: U.S. Government Printing Office, 1993).]

To see how the soil texture triangle works, consider a *Miami silt loam* soil type from Indiana. Samples from this soil type are plotted on the soil texture triangle as points 1, 2, and 3. A sample taken near the surface in the A horizon is recorded at point 1, in the B horizon at point 2, and in the C horizon at point 3. Textural analyses of these samples are summarized in the table and in the three pie diagrams to the right of the triangle. Note that silt dominates the surface, clay the B horizon, and sand the C horizon. The *Soil Survey Manual* presents guidelines for estimating soil texture by feel, a relatively accurate method when used by an experienced person. However, laboratory methods using graduated sieves and separation by mechanical analysis in water allow more precise measurements.

Soil Structure

Soil *texture* describes the size of soil particles, but soil *structure* refers to their *arrangement*. Structure can partially modify the effects of soil texture. The smallest natural lump or cluster of particles is a *ped*. The shape of soil peds determines which of the structural types the soil exhibits: crumb or granular, platy, blocky, prismatic, or columnar (Figure 18.5).

Peds separate from each other along zones of weakness, creating voids, or pores, that are important for moisture storage and drainage. Rounded peds have more pore space between them and greater permeability than other shapes. They are therefore better for plant growth than are blocky, prismatic, or platy peds, despite comparable fertility. Terms used to describe soil structure include *fine, medium,* or *coarse.* Adhesion among peds ranges from weak to strong.

Soil Consistence

In soil science, the term *consistence* is used to describe the consistency of a soil or cohesion of its particles. Consistence is a product of texture (particle size) and structure (ped shape). Consistence reflects a soil's resistance to breaking and manipulation under varying moisture conditions:

• A *wet soil* is sticky between the thumb and forefinger, ranging from a little adherence to either finger, to sticking to both fingers, to stretching when the fingers are moved apart. *Plasticity,* the quality of being moldable, is roughly measured by rolling a

FIGURE 18.5 Types of soil structure.
Structure is important because it controls drainage, rooting of plants, and how well the soil delivers nutrients to plants. The shape of individual peds, shown here, controls a soil's structure. [Photos by National Soil Survey Center, NRCS, USDA, Soil Survey Staff.]

Crumb or granular

Platy

Blocky

Prismatic or columnar

piece of soil between your fingers and thumb to see whether it rolls into a thin strand.

- A *moist soil* is filled to about half of field capacity (the usable water capacity of soil), and its consistence grades from loose (noncoherent), to *friable* (easily pulverized), to firm (not crushable between thumb and forefinger).
- A *dry soil* is typically brittle and rigid, with consistence ranging from loose, to soft, to hard, to extremely hard.

Soil particles are sometimes cemented together to some degree. Soils are described as weakly cemented, strongly cemented, or *indurated* (hardened). Calcium carbonate, silica, and oxides or salts of iron and aluminum all can serve as cementing agents. The cementation of soil particles that occurs in various horizons is a function of consistence and may be continuous or discontinuous.

Soil Porosity

Soil porosity, permeability, and moisture storage were discussed in Chapter 9. Pores in the soil horizon control the movement of water—its intake, flow, and drainage—and air ventilation. Important porosity factors are pore *size*, pore *continuity* (whether they are interconnected), pore *shape* (whether they are spherical, irregular, or tubular), pore *orientation* (whether pore spaces are vertical, horizontal, or random), and pore *location* (whether they are within or between soil peds).

Porosity is improved by the biotic actions of plant roots, animal activity such as the tunneling of gophers

or worms, and human intervention through soil manipulation (plowing, adding humus or sand, or planting soil-building crops). Much of a farmer's soil preparation work before planting, and for the home gardener as well, is done to improve soil porosity.

Soil Moisture

Reviewing Figures 9.8 and 9.9 in Chapter 9 (soil moisture types and availability) will help you understand this section. Plants operate most efficiently when the soil is at *field capacity*, which is the maximum water availability for plant use after large pore spaces have drained of gravitational water. Soil type determines field capacity. The depth to which a plant sends its roots determines the amount of soil moisture to which the plant has access. If soil moisture is removed below field capacity, plants must exert increased energy to obtain available water. This moisture removal inefficiency worsens until the plant reaches its wilting point. Beyond this point, plants are unable to extract the water they need, and they die.

Soil moisture regimes and their associated climate types shape the biotic and abiotic properties of the soil more than any other factor (Table 18.1). The NRCS recognizes five soil moisture regimes based on Thornthwaite's water-balance principles (see "The Soil-Water-Budget Concept" in Chapter 9).

Soil Chemistry

Recall that soil pores may be filled with air, water, or a mixture of the two. Consequently, soil chemistry involves both air and water. The atmosphere within soil pores is

Table 18.1 Principal Soil-Moisture Regimes	
Regime	**Description**
Aquic (L. *aqua*, "water")	*The groundwater table lies at or near the surface*, so the soil is almost constantly wet, as in bogs, marshes, and swamps, a reducing environment with virtually no dissolved oxygen present. Commonly, groundwater levels fluctuate seasonally; a small borehole will produce standing, stagnant water.
Aridic (torric) (L. *aridis*, "dry," and L. *torridus*, "hot and dry")	*Soils in this regime are dry more than half the time*, with soil temperatures at a depth of 50 cm (19.7 in.) above 5°C (41°F). In some or all areas, soils are never moist for as long as 90 consecutive days. This regime occurs mainly in arid climates, although where surface structure inhibits infiltration and recharge or where soils are thin and therefore dry, this regime occurs in semiarid regions as well.
Udic (L. *udus*, "humid")	*Soils have little or no moisture deficiency throughout the year*, specifically during the growing season. The water balance exhibits soil moisture surpluses that flush through the soils during one season of the year. If the water balance exhibits a moisture surplus in all months of the year, the regime is called *perudic*, with adequate soil moisture always available to plants.
Ustic (L. *ustus*, "burnt," implying dryness)	*This regime is intermediate between aridic and udic regimes*; it includes the semiarid and tropical wet–dry climates. Moisture is available but is limited, with a prolonged deficit period following the period of soil moisture utilization during the early portion of the growing season. Temperature is important in determining the moisture efficiency of available water in this borderline moisture regime.
Xeric (L. *xeros*, "dry")	*This regime applies to those few areas that experience a Mediterranean climate*: dry and warm summer, rainy and cool winter. There are at least 45 consecutive dry days during the 4 months following the summer solstice. Winter rains effectively leach the soils.

Source: U.S. Department of Agriculture, *Soil Taxonomy*, Agricultural Handbook No. 436 (Washington, DC: U.S. Government Printing Office, 1975).

mostly nitrogen, oxygen, and carbon dioxide. Nitrogen concentrations are about the same as in the atmosphere, but oxygen is less and carbon dioxide is greater because of ongoing respiration processes.

Water present in soil pores is the *soil solution*. It is the medium for chemical reactions in soil. This solution is critical to plants as their source of nutrients, and it is the foundation of *soil fertility*. Carbon dioxide combines with the water to produce carbonic acid, and various organic materials combine with the water to produce organic acids. These acids are then active participants in soil processes, as are dissolved alkalies and salts.

To understand how the soil solution behaves, let us go through a quick chemistry review. An *ion* is an atom, or group of atoms, that carries an electrical charge (examples: Na^+, Cl^-, HCO_3^-). An ion has either a positive charge or a negative charge. For example, when NaCl (sodium chloride) dissolves in solution, it separates into two ions: Na^+, a *cation* (positively charged ion), and Cl^-, an *anion* (negatively charged ion). Some ions in soil carry single charges, whereas others carry double or even triple charges (e.g., sulfate, SO_4^{2-}; and aluminum, Al^{3+}).

Ions in soil are retained by **soil colloids**. These tiny particles of clay and organic material (humus) carry a negative electrical charge and consequently attract any positively charged ions in the soil (Figure 18.6). The positive ions, many metallic, are critical to plant growth. If it were

not for the negatively charged soil colloids, the positive ions would be leached away in the soil solution and thus would be unavailable to plant roots.

Individual clay colloids are thin and platelike, with parallel surfaces that are negatively charged. They are

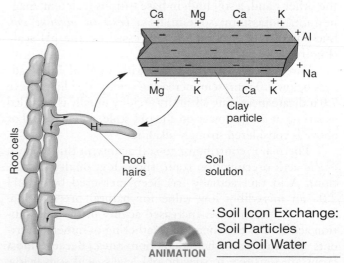

Soil Icon Exchange:
Soil Particles
and Soil Water

FIGURE 18.6 Soil colloids and cation-exchange capacity (CEC).
This typical soil colloid retains mineral ions by adsorption to its surface (opposite charges attract). This process holds the ions until they are absorbed by root hairs.

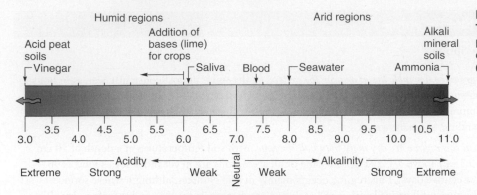

FIGURE 18.7 pH scale. The pH scale measures acidity (lower pH) and alkalinity (higher pH). (The complete pH scale ranges between 0 and 14.)

more chemically active than silt and sand particles but less active than organic colloids. Metallic cations attach to the surfaces of the colloids by *adsorption* (not *ab*sorption, which means "to enter"). Colloids can exchange cations between their surfaces and the soil solution, an ability called **cation-exchange capacity (CEC)**, which is the measure of soil fertility. A high CEC means that the soil colloids can store or exchange more cations from the soil solution, an indication of good soil fertility (unless there is a complicating factor, such as a soil that is too acid).

Therefore, **soil fertility** is the ability of soil to sustain plants. Soil is fertile when it contains organic substances and clay minerals that absorb water and adsorb certain elements needed by plants. Billions of dollars are expended to create fertile soil conditions, yet the future of Earth's most fertile soils is threatened because soil erosion is on the increase worldwide.

Soil Acidity and Alkalinity

A soil solution may contain significant hydrogen ions (H^+), the cations that stimulate acid formation. The result is a soil rich in hydrogen ions, or an *acid soil*. On the other hand, a soil high in base cations (calcium, magnesium, potassium, sodium) is a *basic* or *alkaline soil*. Such acidity or alkalinity is expressed on the pH scale (Figure 18.7).

Pure water is nearly neutral, with a pH of 7.0. Readings below 7.0 represent increasing acidity. Readings above 7.0 indicate increasing alkalinity. Acidity usually is regarded as strong at 5.0 or lower on the pH scale, whereas 10.0 or above is considered strongly alkaline.

The major contributor to soil acidity in this modern era is acid precipitation (rain, snow, fog, or dry deposition). Acid rain actually has been measured below pH 2.0—an incredibly low value for natural precipitation, as acid as lemon juice. Increased acidity in the soil solution accelerates the chemical weathering of mineral nutrients and increases their depletion rates. Because most crops are sensitive to specific pH levels, acid soils below pH 6.0 require treatment to raise the pH. This soil treatment is accomplished by the addition of bases in the form of minerals that are rich in base cations, usually lime (calcium carbonate, $CaCO_3$).

Soil Formation Factors and Management

Soil is an open system involving physical inputs and outputs. Soil-forming factors are both *passive* (parent material, topography and relief, and time) and *dynamic* (climate, biology, and human activities). These factors work together as a system to form soils. The roles of these factors are considered here and in the soil-order discussions that follow.

Natural Factors

Physical and chemical weathering of rocks in the upper lithosphere provides the raw mineral ingredients for soil formation. These rocks supply the parent materials, and their composition, texture, and chemical nature help determine the type of soil that forms. Clay minerals are the principal weathered by-products in soil.

Climate types correlate closely with soil types worldwide. The moisture, evaporation, and temperature regimes of climates determine the chemical reactions, organic activity, and eluviation rates of soils. Not only is the present climate important but many soils exhibit the imprint of past climates, sometimes over thousands of years. Most notable is the effect of glaciations. Among other contributions, glaciation produced the loess soil materials that have been windblown thousands of kilometers to their present locations (discussed in Chapter 15).

Vegetation, animal, and bacterial activity determine the organic content of soil, along with all that is living in soil—algae, fungi, worms, and insects. The chemical makeup of the vegetation contributes to the acidity or alkalinity of the soil solution. For example, broadleaf trees tend to increase alkalinity, whereas needleleaf trees tend to produce higher acidity. Thus, when civilization moves into new areas and alters the natural vegetation by logging or plowing, the affected soils are likewise altered, often permanently.

Topography also affects soil formation. Slopes that are too steep cannot have full soil development because gravity and erosional processes remove materials. Lands that are nearly level inhibit soil drainage and can become waterlogged. The compass orientation of slopes is important because it controls exposure to sunlight. In the Northern Hemisphere, a south-facing slope is warmer overall through the year because it receives direct sunlight. Water-balance relations are affected because north-facing slopes are colder, causing slower snowmelt and lower evaporation rate, providing more moisture for plants than is available on south-facing slopes, which tend to dry faster.

All of the identified natural factors in soil development (parent material, climate, biological activity, landforms and topography) require *time* to operate. Over geologic time plate tectonics has redistributed landscapes and thus subjected soil-forming processes to diverse conditions.

The Human Factor

Human intervention has a major impact on soils. Millennia ago, farmers in most cultures learned to plant slopes "on the contour"—to make rows or mounds around a slope at the same elevation, not vertically up and down the slope. Planting on the contour prevents water from flowing straight down the slope and thus reduces soil erosion. It was common to plant and harvest a floodplain but to live on higher ground nearby. Floods were celebrated as blessings that brought water, nutrients, and more soil to the land. Society is drifting away from these commonsense strategies.

A few centimeters' thickness of prime farmland soil may require *500 years* to mature. Yet, this same thickness

is being lost annually through soil erosion when the soil-holding vegetation is removed and the land is plowed regardless of topography. Flood control structures block sediments and nutrients from replenishing floodplain soils, leading to additional soil losses. Over the same period, exposed soils may be completely leached of needed cations, thereby losing their fertility. Unlike living species, soils do not reproduce, nor can they be re-created. Some 35% of farmlands are losing soil faster than it can form—a loss exceeding 23 billion metric tonnes (25 billion tons) per year. Soil depletion and loss are at record levels from Iowa to China, Peru to Ethiopia, the Middle East to the Americas. The impact on society is potentially disastrous as population and food demands increase (see News Report 18.1).

Soil erosion can be compensated for in the short run by using more fertilizer, increasing irrigation, and by planting higher-yielding strains. But the potential yield from prime agricultural land will drop by as much as 20% over the next 20 years if only moderate erosion continues. One study tabulated the market value of lost nutrients and other variables in the most comprehensive soil-erosion study to date. The sum of direct damage (to agricultural land) and indirect damage (to streams, society's infrastructure, and human health) was estimated at more than $25 billion a year in the United States and hundreds of billions of dollars worldwide. (Of course, this is a controversial assessment in the agricultural industry.) The cost

News Report 18.1

Soil Is Slipping Through Our Fingers

- The U.S. General Accounting Office estimates that from 3 to 5 million acres of prime farmland are lost each year in the United States through mismanagement or conversion to nonagricultural uses. About half of all cropland in the United States and Canada is experiencing excessive rates of soil erosion—these countries are two of the few that monitor loss of topsoil. Worldwide, about one-third of potentially farmable land has been lost to erosion, much of that in the past 40 years.

- The Canadian Environmental Advisory Council estimated that the organic content of cultivated prairie soils has declined by as much as 40% compared with noncultivated native soils. In Ontario and Québec, losses of organic content increased to as

much as 50%, and losses are even higher in the Atlantic Provinces, which were naturally low in organic content before cultivation.

- A 1995 study completed at Cornell University concluded that soil erosion is a major environmental threat to the sustainability and productive capacity of agriculture worldwide. Studies by David Montgomery and others published in 2007 further reinforce that there is a crisis (see D. Montgomery, *Dirt: The Erosion of Civilization*, Berkeley: University of California Press, 2007).

- Since 1950, the rate of soil loss from farmable land continues at 5 to 6 million hectares (about 12 to 15 million acres) per year—560 million hectares, 1380 million acres,

to date (World Resources Institute and UNEP, 1997).

- The causes for degraded soils, in order of severity, include: overgrazing, vegetation removal, agricultural activities, overexploitation, and industrial and bioindustrial use (UNEP, 1997).

- The world's human population is growing at the rate of 6.6 million people a month (net increase), increasing the demand for food and agricultural productivity. Especially significant is that proportion of the global population that is adopting a meat-centered diet as American tastes and food outlets spread worldwide—as we see in the next chapter, meat production is an inefficient use of the grain supply.

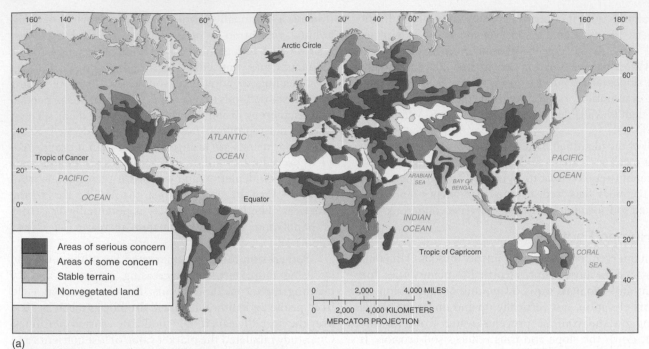

FIGURE 18.8 Soil degradation.
(a) Approximately 1.2 billion hectares (3.0 billion acres) of Earth's soils suffer degradation through erosion caused by human misuse and abuse. (b) Typical loss through soil sheet erosion on a northeastern Wisconsin farm. One millimeter of soil lost from an acre weighs about 5 tons. [(a) *A Global Assessment of Soil Degradation,* adapted from UNEP, International Soil Reference and Information Centre, "Map of Status of Human-Induced Soil Degradation," Nairobi, Kenya; (b) photo by D. P. Burnside/Photo Researchers, Inc.]

to bring erosion under control in the United States is estimated at approximately $8.5 billion, or about 30 cents on every dollar of damage and loss. Figure 18.8 maps regions of soil loss.

Our spatial impact on precious soil resources across each agricultural region should be recognized and protective measures taken. There is a need for cooperative international action. Critical to sustaining soil fertility is a long-term assessment of the *cost* of continued behavior—that is, business as usual—compared to the *benefits* of mitigating damage and preserving the market value of prime soils through conservation actions—a cost-benefit analysis. In sustaining natural systems, the benefits of doing things right far outweigh the continued costs of doing things wrong, over the long term. A recent article by scientist David Montgomery well summarizes our predicament:

Recent compilations of data from around the world show that soil erosion under conventional agriculture exceeds both rates of soil production and geological

erosion rates by several orders of magnitude. Consequently, modern agriculture—and therefore global society—faces a fundamental question over the upcoming centuries. Can an agricultural system capable of feeding a growing population safeguard both soil fertility and the soil itself?*

Soil Classification

Classification of soils is complicated by the continuing interaction of the physical properties and processes just discussed. This interaction creates thousands of distinct soils, with well over 15,000 soil series identified in the United States and Canada alone. Not surprisingly, different classification systems are in use worldwide. The United States, Canada, the United Kingdom, Germany, Australia, Russia, and the United Nations Food and Agricultural

---

*D. R. Montgomery, "Is agriculture eroding civilization's foundation?" *GSA Today,* October 2007, p. 4.

Organization each have their own soil classification system. Each system reflects the environment of its country. For example, the National Soil Survey Committee of Canada developed a system suited to its great expanses of boreal forest, tundra, and cool climatic regimes.

Soil Taxonomy

The U.S. soil classification system, *Soil Taxonomy—A Basic System of Soil Classification for Making and Interpreting Soil Surveys*, was published in 1975 and revised in a second edition in 1999. Soil scientists refer to it as **Soil Taxonomy**. Over the years, various revisions and clarifications in the system were published in *Keys to the Soil Taxonomy*, now in its 10th edition (2006, **http://soils.usda.gov/technical/classification/tax_keys/**), which includes all the revisions to the 1975 *Soil Taxonomy*. Much of the information in this chapter is derived from these two keystone publications. Two soil orders were added to the original 10: Andisols (volcanic soils) in 1990 and Gelisols (cold and frozen soils) in 1998. Soil properties and morphology (appearance, form, and structure) actually seen in the field are key to the Soil Taxonomy system. Thus, it is open to addition, change, and modification as the sampling database grows. The system recognizes the importance of interactions between humans and soils and the changes that humans have introduced, both purposely and inadvertently.

The classification system divides soils into six categories, creating a hierarchical sorting system (Table 18.2). The smallest, most detailed category is the soil series, which ideally includes only one polypedon but may include adjoining polypedons. In sequence from smallest category to the largest, the Soil Taxonomy recognizes *soil series, soil families, soil subgroups, soil great groups, soil suborders,* and *soil orders*.

Pedogenic Regimes   Prior to the Soil Taxonomy system, **pedogenic regimes** were used to describe soils. These regimes keyed specific soil-forming processes to climatic regions. Although each pedogenic process may be active in several soil orders and in different climates, we discuss them within the soil order where they commonly occur. Such climate-based regimes are convenient for relating climate and soil processes. However, *the Soil Taxonomy system recognizes the great uncertainty and inconsistency in basing soil classification on such climatic variables*. Aspects of several pedogenic processes are discussed with appropriate soil orders:

- **Laterization**: a leaching process in humid and warm climates, discussed with Oxisols
- **Salinization**: a process that concentrates salts in soils in climates with excessive potential evapotranspiration (POTET) rates, discussed with Aridisols
- **Calcification**: a process that produces an illuviated accumulation of calcium carbonates in continental climates, discussed with Mollisols and Aridisols
- **Podzolization**: a process of soil acidification associated with forest soils in cool climates, discussed with Spodosols
- **Gleization**: a process that includes an accumulation of humus and a thick, water-saturated gray layer of clay beneath, usually in cold, wet climates and poor drainage conditions

Diagnostic Soil Horizons

To identify a specific soil series within the Soil Taxonomy, the U.S. Natural Resources Conservation Service describes diagnostic horizons in a pedon. A *diagnostic horizon* reflects a distinctive physical property (color, texture, structure, consistence, porosity, moisture) or a dominant soil process (discussed with the soil types).

In the solum (A, E, and B horizons), two diagnostic horizons may be identified: the epipedon and the diagnostic subsurface. The presence or absence of either of these diagnostic horizons usually distinguishes a soil for classification.

- The **epipedon** (literally, "over the soil") is the diagnostic horizon at the surface where most of the rock structure has been destroyed. It may extend downward through the A horizon, even including all or part of an illuviated B horizon. It is visibly darkened by organic matter and sometimes is leached of minerals. Excluded from the epipedon are alluvial deposits, eolian deposits, and cultivated areas, because soil-forming processes have lacked the time to erase these relatively short-lived characteristics.
- The **diagnostic subsurface horizon** originates below the surface at varying depths. It may include part of the A or B horizon or both. Many diagnostic subsurface horizons have been identified.

The 12 Soil Orders of the Soil Taxonomy

At the heart of the Soil Taxonomy are 12 general soil orders, listed in Table 18.3. Their worldwide distribution is shown in Figure 18.9 and in individual maps with each description. Please consult this table and these maps as you read the following descriptions. Because the Soil Taxonomy evaluates each soil order on its own characteristics, there is no priority to the classification. However, you will find a progression in this discussion, for the 12 orders are arranged loosely by latitude, beginning with Oxisols along the equator as in Chapters 10 (climates) and 20 (terrestrial biomes).

Table 18.2   U.S. Soil Taxonomy

Soil Category	Number of Soils Included
Orders	12
Suborders	47
Great groups	230
Subgroups	1,200
Families	6,000
Series	15,000

FIGURE 18.9 Soil Taxonomy.
Worldwide distribution of the Soil
Taxonomy's 12 soil orders.
[Adapted from maps prepared
by World Soil Resources Staff,
Natural Resources Conservation
Service, USDA, 1999, 2006.]

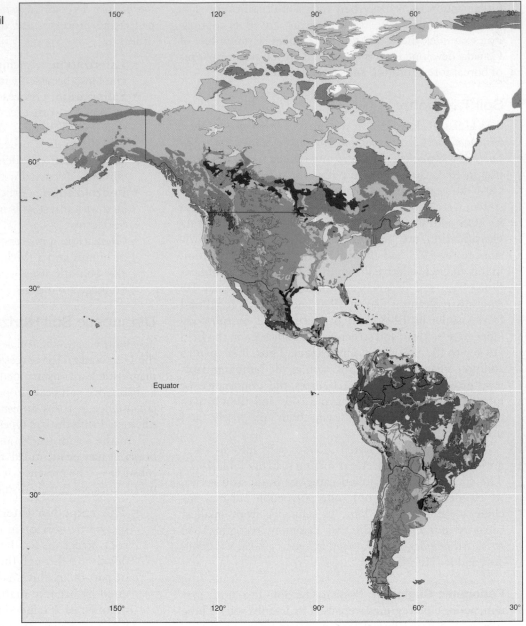

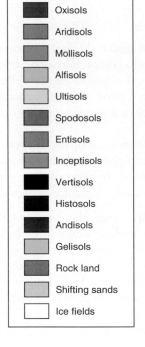

- Oxisols
- Aridisols
- Mollisols
- Alfisols
- Ultisols
- Spodosols
- Entisols
- Inceptisols
- Vertisols
- Histosols
- Andisols
- Gelisols
- Rock land
- Shifting sands
- Ice fields

Oxisols The intense moisture, temperature, and uniform daylength of equatorial latitudes profoundly affect soils. These generally old landscapes, exposed to tropical conditions for millennia or hundreds of millennia, are deeply developed. Soil minerals are highly altered (except in certain newer volcanic soils in Indonesia—the Andisols). Oxisols are among the most mature soils on Earth. Distinct horizons usually are lacking where these soils are well drained (Figure 18.10a). Related vegetation is the luxuriant and diverse tropical and equatorial rain forest. Oxisols include five suborders.

Oxisols (tropical soils) are so called because they have a distinctive horizon of iron and aluminum oxides. The concentration of oxides results from heavy precipitation, which leaches soluble minerals and soil constituents

from the A horizon. Typical Oxisols are reddish (from the iron oxide) or yellowish (from the aluminum oxides), with a weathered claylike texture, sometimes in a granular soil structure that is easily broken apart. The high degree of eluviation removes basic cations and colloidal material to lower illuviated horizons. Thus, Oxisols are low in CEC (cation-exchange capacity) and fertility, except in regions augmented by alluvial or volcanic materials.

To have the lush rain forests in the same regions as soils poor in inorganic nutrients seems an irony. However, this forest system relies on the recycling of nutrients from soil organic matter to sustain fertility, although this nutrient recycling ability is quickly lost when the ecosystem is disturbed. Consequently, Oxisols have a diagnostic

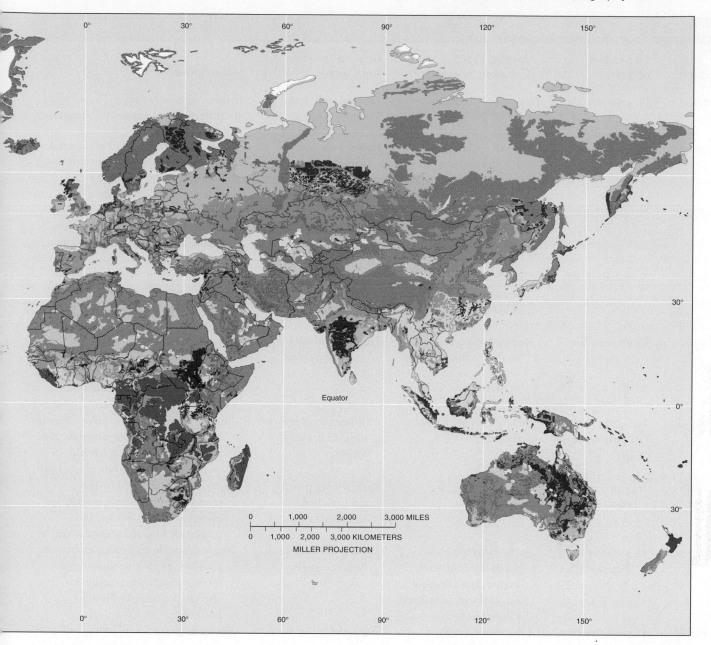

subsurface horizon that is highly weathered, contains iron and aluminum oxides, is at least 30 cm (12 in.) thick, and lies within 2 m (6.5 ft) of the surface (see Figure 18.10).

Figure 18.11 illustrates *laterization*, the leaching process that operates in well-drained soils in warm, humid, tropical and subtropical climates. If Oxisols are subjected to repeated wetting and drying, an *ironstone hardpan* develops (this is a hardened soil layer in the lower A or in the B horizon—iron-rich and humus-poor clay with quartz and other minerals). It is called a *plinthite* (from the Greek *plinthos*, meaning "brick"). This form of soil, also called a *laterite*, can be quarried in blocks and used as a building material (Figure 18.12).

Agricultural activities can be conducted with care in these soils. Early cultivation practices, called *slash-and-*

burn shifting cultivation, were adapted to these soil conditions and formed a unique style of crop rotation. The scenario went like this: People in the tropics cut down (slashed) and burned the rain forest in small tracts and then cultivated the land using stick-and-hoe tools, planting maize, beans, and squash. Mineral nutrients in the organic material and short-lived fertilizer input from the fire ash would quickly be exhausted. After several years, the soil lost fertility through leaching by intense rainfall, so the people shifted cultivation to another tract and repeated the process. After many years of moving from tract to tract, the group returned to the first patch to begin the cycle again. This practice protected the limited fertility of the soils somewhat, allowing periods of recovery.

(text continued on page 590)

Table 18.3 Soil Taxonomy Soil Orders

Order	Derivation of Term	Marbut, 1938* (Canadian System**)	General Location and Climate	Description
Oxisols	Fr. *oxide*, "oxide" Gr. *oxide*, "acid or sharp"	Latosols, lateritic soils	Tropical soils; hot, humid areas	Maximum weathering of Fe and Al and eluviation, continuous plinthite layer
Aridisols	L. *aridos*, "dry"	Reddish desert, gray desert, sierozems	Desert soils; hot, dry areas	Limited alteration of parent material, low climate activity, light color, low humus content, subsurface illuviation of carbonates
Mollisols	L. *mollis*, "soft"	Chestnut, chernozem (Chernozemic)	Grassland soils; subhumid, semiarid lands	Noticeably dark with organic material; humus rich; base saturation; high, friable surface with well-structured horizons
Alfisols	Invented syllable	Gray-brown podzolic, degraded chernozem (Luvisol)	Moderately weathered forest soils; humid temperate forests	B horizon high in clays, moderate to high degree of base saturation, illuviated clay accumulation, no pronounced color change with depth
Ultisols	L. *ultimus*, "last"	Red-yellow podzolic, reddish yellow lateritic	Highly weathered forest soils; subtropical forests	Similar to Alfisols, B horizon high in clays, generally low amount of base saturation, strong weathering in subsurface horizons, redder than Alfisols
Spodosols	Gr. *spodos* or L. *spodus*, "wood ash"	Podzols, brown podzolic (Podzol)	Northern conifer forest soils; cool, humid forests	Illuvial B horizon of Fe/Al clays, humus accumulation, without structure, partially cemented, highly leached, strongly acid, coarse texture of low bases
Entisols	Invented syllable from *recent*	Azonal soils, tundra	Recent soils, profile undeveloped, all climates	Limited development, inherited properties from parent material, pale color, low humus, few specific properties, hard and massive when dry
Inceptisols	L. *inceptum*, "beginning"	Ando, subarctic brown forest lithosols, some humic gleys (Brunisol, Cryosol with permafrost, Gleysol wet)	Weakly developed soils; humid regions	Intermediate development, embryonic soils but few diagnostic features, further weathering possible in altered or changed subsurface horizons
Vertisols	L. *verto*, "to turn"	Grumusols, (1949) tropical black clays	Expandable clay soils; subtropics, tropics; sufficient dry period	Forms large cracks on drying, self-mixing action, contains >30% in swelling clays, light color, low humus content
Histosols	Gr. *histos*, "tissue"	Peat, muck, bog (Organic)	Organic soils, wet places	Peat or bog, >20% organic matter, much with clay >40 cm thick, surface, organic layers, no diagnostic horizons
Andisols	L. *ando*, "volcanic ash"	—	Areas affected by frequent volcanic activity (formerly within Inceptisols and Entisols)	Volcanic parent materials, particularly ash and volcanic glass; weathering and mineral transformation important; high CEC and organic content, generally fertile
Gelisols	L. *gelatio*, "freezing"	Formerly Inceptisols and Entisols (Cryosols, some Brunisols)	High latitudes in Northern Hemisphere, southern limits near tree line	Permafrost within 100 cm of the soil surface; evidence of cryoturbation (frost churning) and/or an active layer; patterned-ground

*C.F. Marbut, USDA soil scientist, developed the first American system of soil classification in the 1930s, first published in the USDA *Yearbook of Agriculture* in 1938.

**For comparison, soil orders from the Canadian System of Soil Classification (CSSC), last revised in 1998.

(a)

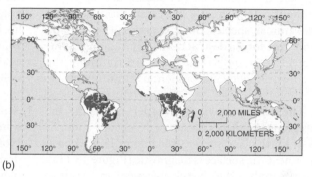

(b)

FIGURE 18.10 Oxisols.
(a) Deeply weathered Oxisol profile in central Puerto Rico. (b) General map showing worldwide distribution of these tropical soils. [Photo from the Marbut Collection, Soil Science Society of America.]

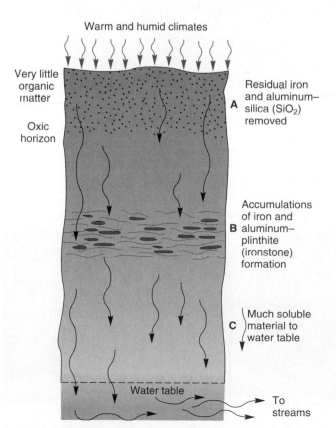

FIGURE 18.11 Laterization.
The laterization process is characteristic of moist tropical and subtropical climate regimes.

FIGURE 18.12 Oxisols are used for bricks.
Here plinthite is being quarried in India for building material. Inset photo shows a close-up of plinthite bricks. [Photos by Henry D. Foth; inset courtesy of F. H. Benroth, Guy Smith Memorial Slide Collection.]

The invasion of foreign plantation interests, development by local governments, vastly increased population pressures, and conversion of vast forest tracts to pasturage disrupted this orderly land rotation. Permanent tracts of cleared land, taken out of the former rotation mode, put tremendous pressure on the remaining forest and brought disastrous consequences. When Oxisols are disturbed, soil loss can exceed a thousand tons per square kilometer per year, not to mention the greatly increased extinction rate of plant and animal species that accompanies such soil depletion and rain forest destruction. The regions dominated by the Oxisols and rain forests are rightfully the focus of much worldwide environmental attention.

Aridisols The largest single soil order occurs in the world's dry regions. **Aridisols** (desert soils) occupy approximately 19% of Earth's land surface (see Figure 18.9). A pale, light soil color near the surface is diagnostic (Figure 18.13a).

Not surprisingly, the water balance in Aridisol regions has periods of soil-moisture deficit and generally inadequate soil moisture for plant growth. High potential evapotranspiration and low precipitation produce very shallow soil horizons. Usually there is no period greater than 3 months when the soils have adequate moisture. Lacking water and therefore lacking vegetation, Aridisols also lack organic matter of any consequence. Low precipitation means infrequent leaching, yet Aridisols are leached easily when exposed to excessive water, for they lack a significant colloidal structure.

Salinization is common in Aridisols, resulting from excessive potential evapotranspiration rates in deserts and semiarid regions. Salts dissolved in soil water migrate to surface horizons and are deposited there as the water evaporates. These deposits appear as subsurface salty horizons, which will damage or kill plants when the horizons occur near the root zone. The salt accumulation associated with a desert playa is an example of extreme salinization (see Chapter 15).

Obviously, salinization complicates farming in Aridisols. The introduction of irrigation water may either waterlog poorly drained soils or lead to salinization. Nonetheless, vegetation does grow where soils are well drained and low in salt content. If large capital investments are made in water, drainage, and fertilizers, Aridisols possess much agricultural potential (Figures 18.13c and 18.14). In the Nile and Indus river valleys, for example, Aridisols are intensively farmed with a careful balance of these environmental factors, although thousands of acres of once-productive land, not so carefully treated, now sit idle and salt-encrusted. In California, the Kesterson

(a)

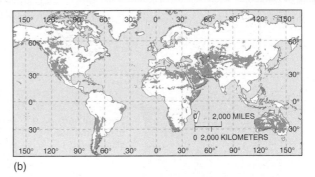

(b)

(c)

FIGURE 18.13 Aridisols.
(a) Soil profile from central Arizona. (b) General map showing worldwide distribution of these desert soils. (c) Irrigated Aridisols in the Imperial Valley, rows of Aloe plants. [Photos (a) the Marbut Collection, Soil Science Society of America; (c) Bobbé Christopherson.]

FIGURE 18.14 Asparagus in the desert.
The Coachella Valley of southeastern California is a desert in agricultural bloom; asparagus is growing in the foreground. Providing drainage for excessive water application in such areas often is necessary to prevent salinization of the rooting zone. [Photo by author.]

Wildlife Refuge was reduced to a toxic waste dump in the early 1980s by contaminated agricultural drainage. Focus Study 18.1 elaborates on the Kesterson tragedy.

Mollisols   Some of Earth's most significant agricultural soils are **Mollisols** (grassland soils). There are seven recognized suborders, which vary in fertility. The dominant diagnostic horizon is a dark, organic surface layer some 25 cm (10 in.) thick (Figure 18.15). As the Latin name implies (words using the same root, *mollis*, are *mollify* and *emollient*), Mollisols are soft, even when dry. They have

(a)

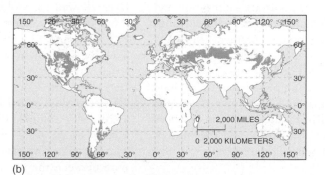

(b)

(c)

(d)

FIGURE 18.15 Mollisols.
(a) Soil profile in eastern Idaho, from calcareous loess related to the soils of the Palouse. (b) General map showing worldwide distribution of these grassland soils. (c) Mollisols in northwestern Iowa. (d) In eastern Washington State, wheat flourishes in the Palouse. These fertile Mollisols will be increasingly valuable in the future. [Photos (a) courtesy of Soil Science Division, University of Idaho; (c) and (d) by Bobbé Christopherson.]

Focus Study 18.1

Selenium Concentration in Western Soils

Irrigated agriculture has increased greatly since 1800, when only 8 million hectares (about 20 million acres) were irrigated worldwide. Today, approximately 255 million hectares (about 630 million acres) are irrigated, and this figure is on the increase (USDA, FAO, 1996). Representing about 16% of Earth's agricultural land, irrigated land accounts for nearly 36% of the harvest.

Two related problems common in irrigated lands are salinization and water logging, especially in arid lands that are poorly drained. In many areas, production has decreased and even ended because of salt buildup in the soils. Examples include areas along the Tigris and Euphrates rivers, the Indus River valley, sections of South America and Africa, and the western United States.

Irrigation in the West

About 95% of the irrigated acreage in the United States lies west of the 98th meridian. This region is increasingly troubled with salinization and waterlogging problems. In addition, at least nine sites in the West, particularly California's western San Joaquin Valley, are experiencing related contamination of a more serious nature—increasing selenium concentrations. Toxic effects of selenium were reported during the 1980s in some domestic animals grazing on grasses grown in selenium-rich soils in the Great Plains. In California, as parent materials weathered, selenium-rich alluvium washed into the semiarid valley, forming the soils that needed only irrigation water to become productive.

Drainage of agricultural wastewater poses a particular problem in semiarid and arid lands, where river discharge is inadequate to dilute and remove field runoff. One solution to prevent salt accumulations and water-clogged soils is to place field drains beneath the soil to collect gravitational water from fields that have been purposely overwatered to keep salts away from the effective rooting depth of the crops. Drainage tiles (perforated ceramic pipe) are laid in place, and the soil returned. The problem has been to find a place to dump the salt- and chemical-laden drainage water.

In the Imperial and Coachella valleys, the drainage water was sent to the Salton Sea (Figure 18.1.1)—a large inland sea formed by an irrigation accident in 1906 (a broken canal levee allowed the Colorado River to flow into the desert of southern California for 18 months). For the Wellton–Mohawk area, the contaminated drainage water was pumped back into the Colorado River just above the Mexican border. By the late 1950s, salty Colorado River water was killing every plant it touched in Mexico, triggering many years of dispute between the two countries. For the San Joaquin Valley of central California this problem of finding a "drain" triggered a 15-year controversy.

Death of Kesterson

Central California's potential drain outlets are to the ocean, San Francisco Bay, or the central valley. But all these suggested destinations failed to pass environmental-impact assessments under Environmental Policy Act requirements.

Nonetheless, by the late 1970s, about 80 miles of a drain were finished, even though no formal plan or adequate funding had been completed—a drain was built with no outlet! In the absence of any plan, large-scale irrigation continued, supplying the field drains with salty, selenium-laden runoff that made its way to the Kesterson National Wildlife Refuge in the northern portion of the San Joaquin Valley east of San Francisco. The unfinished drain abruptly stopped at the boundary to the refuge.

The selenium-tainted drainage took only 3 years to destroy the wildlife refuge, which was officially declared a contaminated toxic waste site (Figure 18.1.2). Aquatic life forms (e.g., marsh plants, plankton, and insects) had taken in the selenium, which then made its way up the food chain and into the diets of higher life forms in the refuge. According to U.S. Fish and Wildlife Service scientists, the toxicity moved through the food chain and genetically damaged and killed wildlife, including all varieties of birds that nested at Kesterson; approximately 90% of the exposed birds perished or were injured. Because this wildlife refuge was a major migration flyway and stopover point for birds from throughout the Western

FIGURE 18.1.1 Fields drain into canals.
Soil drainage canal collects contaminated water from field drains and directs it into the Salton Sea.
[Photo by author.]

FIGURE 18.1.2 Soil contamination in the wildlife refuge.
Salt-encrusted soil and plants in evidence at the contaminated Kesterson National Wildlife Refuge.
[Photo by Gary R. Zahm/DRK Photo.]

Hemisphere, the destruction of the refuge also violated several multinational wildlife protection treaties.

The field drains were sealed and removed in 1986, following a court order that forced the federal government to uphold existing laws. Irrigation water then immediately began backing up in the corporate farmlands, producing both waterlogging and selenium contamination.

Since 1985, more than 0.6 million hectares (1.5 million acres) of irrigated Aridisols and Alfisols have gone out of production in California, marking the end of several decades of irrigated farming in climatically marginal lands. Severe cutbacks in irrigated acreage no doubt will continue, underscoring the need to preserve prime farmlands in wetter regions and to understand essential soil processes.

Frustrated agricultural interests have asked the federal government to finish the drain, either to San Francisco Bay or to the ocean. However, neither option appears capable of passing an environmental-impact analysis. Another strategy is to allow irrigated lands to start pumping again into the former wildlife refuge because it is now a declared toxic dump site anyway! There are nine such threatened sites in the West; Kesterson was simply the first of these to fail from contamination. Such damage to a wildlife refuge presents a real warning to human populations—remember where we are, at the top of the food chain.

---

granular or crumbly peds, loosely arranged when dry. These humus-rich soils are high in basic cations (calcium, magnesium, and potassium) and have a high cation-exchange capacity and, therefore, high fertility. In soil moisture, these soils are intermediate between humid and arid.

Soils of the world's steppes and prairies belong to this soil group: the North American Great Plains, the Pampas of Argentina, and the region from Manchuria in China westward to Europe. Agriculture ranges from large-scale commercial grain farming to grazing along the drier portions of the soil order. With fertilization or soil-building practices, high crop yields are common.

The "fertile triangle" of Ukraine, Russia, and western portions of the Russian Federation is of this soil type. The region stretches from north of the Caspian Sea and Kazakstan westward in a widening triangle toward Central Europe (Figure 18.16). The soils in the fertile triangle are known as the *chernozem* in other classification systems. The remainder of the Russian landscape presents varied soils of lower fertility and productivity.

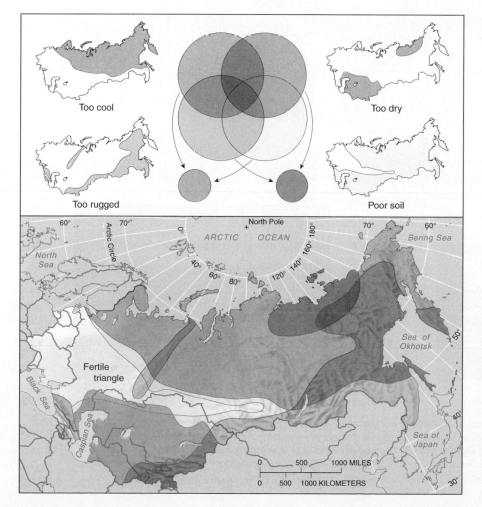

FIGURE 18.16 Soil fertility in Eurasia. Top: Four main factors influence soil fertility. Bottom: The fertile triangle of central Eurasia. The white area is highest in soil fertility (no negative factors). Areas shown with any color or color combination are limited for the reasons shown. The least-fertile area in Russia is northeastern Siberia. [After P. W. English and J. A. Miller, *World Regional Geography: A Question of Place*, p. 183, © 1989. Adapted by permission of John Wiley & Sons, Inc.]

In North America, the Great Plains straddle the 98th meridian, which is coincident with the 51-cm (20-in.) isohyet of annual precipitation—wetter to the east and drier to the west. The Mollisols here mark the historic division between the short- and tall-grass prairies (Figure 18.17).

Calcification is a soil process characteristic of some Mollisols and adjoining marginal areas of Aridisols. Calcification is the accumulation of calcium carbonate or magnesium carbonate in the B and C horizons. Calcification

by calcium carbonate ($CaCO_3$) forms a diagnostic subsurface horizon that is thickest along the boundary between dry and humid climates (Figure 18.18).

When cemented or hardened, these deposits are called *caliche*, or *kunkur*; they occur in widespread soil formations in central and western Australia, the Kalahari region of interior Southern Africa, and the High Plains of the west-central United States, among other places. (Soils dominated by the processes of calcification and salinization were formerly known as pedocals.)

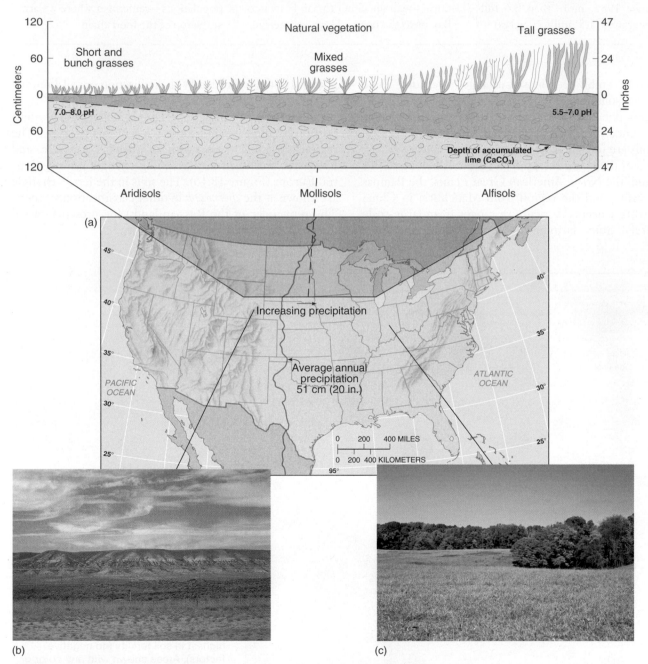

FIGURE 18.17 Soils of the Midwest.
(a) Aridisols (to the west), Mollisols (central), and Alfisols (to the east)—a soil continuum in the north-central United States and southern Canadian prairies. Graduated changes that occur in soil pH and the depth of accumulated lime are shown. (b) Bunch grasses and shallow soils of Wyoming. (c) Farmlands and Alfisols south of Bedford, Indiana. [(a) Illustration adapted from N. C. Brady, *The Nature and Properties of Soils,* 10th ed., © 1990 by Macmillan Publishing Company, adapted by permission; photos by (b) author and (c) Bobbé Christopherson.]

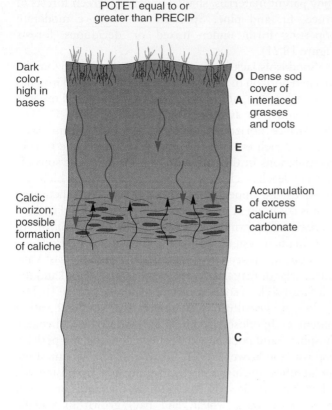

POTET equal to or
greater than PRECIP

Dark
color,
high in
bases

Calcic
horizon;
possible
formation
of caliche

O
A — Dense sod
cover of
interlaced
grasses
and roots

E

B — Accumulation
of excess
calcium
carbonate

C

FIGURE 18.18 Calcification in soil.
The calcification process in Aridisol/Mollisol soils occurs in
climatic regimes that have potential evapotranspiration equal
to or greater than precipitation.

Alfisols   Spatially the most widespread of the soil orders
are **Alfisols** (moderately weathered forest soils), extending
in five suborders from near the equator to high latitudes.
Representative Alfisol areas include Boromo and Burkina
Faso (interior western Africa); Fort Nelson, British Columbia; the states near the Great Lakes; and the valleys of central California. Most Alfisols are grayish brown to reddish
and are considered moist versions of the Mollisol soil
group. Moderate eluviation is present, as well as a subsurface horizon of illuviated clays and clay formation because
of a pattern of increased precipitation (Figure 18.19).

Alfisols have moderate to high reserves of basic
cations and are fertile. However, productivity depends on
moisture and temperature. Alfisols usually are supplemented by a moderate application of lime and fertilizer in
areas of active agriculture. Some of the best farmland in
the United States stretches from Illinois, Wisconsin, and
eastern Minnesota through Indiana, Michigan, and Ohio
to Pennsylvania and New York (see Figure 18.17c). This
land produces grains, hay, and dairy products. The soil
here is an Alfisol subgroup called Udalfs, which are characteristic of humid continental, hot summer climates.

The Xeralfs, another Alfisol subgroup, are associated
with the moist winter, dry summer pattern of the Mediterranean climate. These naturally productive soils are farmed
intensively for subtropical fruits, nuts, and special crops that
can grow in only a few locales worldwide (e.g., California
grapes, olives, citrus, artichokes, almonds, and figs).

(a)

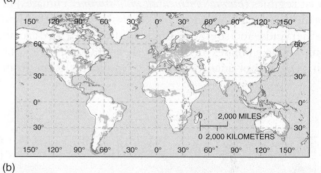

(b)

(c)

FIGURE 18.19 Alfisols.
**(a) Soil profile from northern Idaho, formed in loess.
(b) General map showing worldwide distribution of these
moderately weathered forest soils. (c) Cultivated Alfisols
farmland near Lompoc, California.** [Photos (a) courtesy of Soil
Science Division, University of Idaho; and (c) by Bobbé
Christopherson.]

Ultisols Farther south in the United States are the **Ultisols** (highly weathered forest soils) and their five suborders. An Alfisol might degenerate into an Ultisol, given time and exposure to increased weathering under moist conditions. These soils tend to be reddish because of residual iron and aluminum oxides in the A horizon (Figure 18.20).

The relatively high precipitation in Ultisol regions causes greater mineral alteration and more eluvial leaching than in other soils. Therefore, the level of basic cations is lower and the soil fertility is lower. Fertility is further reduced by certain agricultural practices and the effect of soil-damaging crops such as cotton and tobacco, which deplete nitrogen and expose soil to erosion. These soils respond well if subjected to good management—for example, crop rotation restores nitrogen, and certain cultivation practices prevent sheetwash and soil erosion. Peanut plantings assist in nitrogen restoration. Much needs to be done to achieve sustainable management of these soils.

Spodosols The **Spodosols** (northern coniferous forest soils) and their four suborders occur generally to the north and east of the Alfisols. They are in cold and forested moist regimes (*humid continental, mild summer* climates) in northern North America and Eurasia, Denmark, The Netherlands, and southern England. Because there are no comparable climates in the Southern Hemisphere, this soil type is not identified there. Spodosols form from sandy parent materials, shaded under evergreen forests of spruce, fir, and pine. Spodosols with more moderate properties form under mixed or deciduous forests (Figure 18.21).

Spodosols lack humus and clay in the A horizon. An eluviated horizon of sandy and leached of clays and iron lies in the A horizon, above a B horizon of illuviated organic matter and iron and aluminum oxides (Figure 18.21c). The surface horizon receives organic litter from base-poor, acid-rich evergreen trees, which contribute to acid accumulations in the soil. The solution in acidic soils effectively leaches clays, iron, and aluminum, which are passed to the upper diagnostic horizon. An ashen-gray color is common in these subarctic forest soils and is characteristic of a formation process called *podzolization*. In the Canadian system, Spodosols fall within the Podzolic Great Group, as seen in the temperate rain forests of Vancouver Island, British Columbia, in Figure 18.21d and the rich forest soils of south-central Norway in Figure 18.21e.

When agriculture is attempted, the low basic cation content of Spodosols requires the addition of nitrogen, phosphate, and potash (potassium carbonate), and perhaps crop rotation as well. A soil *amendment* such as limestone can significantly increase crop production by raising the pH of these acidic soils. For example, the yields of several crops (corn, oats, wheat, and hay) grown in specific Spodosols in New York State were increased up to a third with the application of 1.8 metric tonnes (2 tons) of limestone per 0.4 hectare (1.0 acre) during each 6-year rotation.

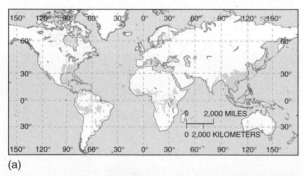

(a)

FIGURE 18.20 Ultisols.
(a) General map showing worldwide distribution of these highly weathered forest soils. (b) A type of Ultisol in central Georgia bearing its characteristic dark reddish color is planted with pecan trees. (c) Distinctive reddish soils and the invasive kudzu plant tell us we are in the American Southeast. [Photos by Bobbé Christopherson.]

(b)

(c)

(a)

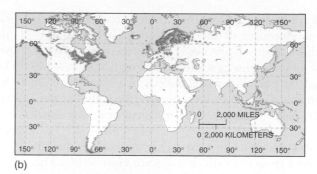

(b)

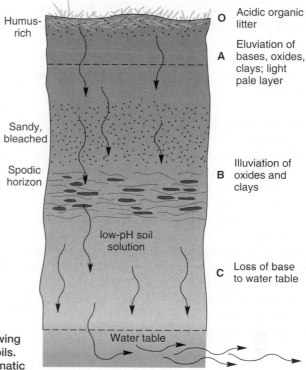

Cool and moist climate

Humus-rich

Sandy, bleached

Spodic horizon

low-pH soil solution

Water table

O Acidic organic litter

A Eluviation of bases, oxides, clays; light pale layer

B Illuviation of oxides and clays

C Loss of base to water table

(c)

FIGURE 18.21 Spodosols.
(a) Soil profile from northern New York. (b) General map showing worldwide distribution of these northern coniferous forest soils. (c) The podzolization process is typical in cool and moist climatic regimes. (d) Characteristic temperate forest and Spodosols in the cool, moist climate of central Vancouver Island. (e) Spodosols and farms growing feed grains in central Norway along a fjord.
[Photos (a) from the Marbut Collection, Soil Science Society of America; (d) and (e) by Bobbé Christopherson.]

(d)

(e)

Entisols The **Entisols** (recent, undeveloped soils) lack vertical development of their horizons. The five suborders of this soil group are based on differences in parent materials and climatic conditions, although the presence of Entisols is not climate-dependent, for they occur in many climates worldwide. Entisols are true soils, but they have not had sufficient time to generate the usual horizons.

Entisols generally are poor agricultural soils, although those formed from river silt deposits are quite fertile. The same conditions that inhibit complete development also prevent adequate fertility—too much or too little water, poor structure, and insufficient accumulation of weathered nutrients. Active slopes, alluvium-filled floodplains, poorly drained tundra, tidal mudflats, dune sands and erg (sandy) deserts, and plains of glacial outwash all are characteristic regions that have these soils. Figure 18.22 shows an Entisol in a desert climate where shales formed the parent material.

Inceptisols Since they have not reached a mature condition, **Inceptisols** (weakly developed soils) and their six suborders are inherently infertile. They are young soils, although they are more developed than the Entisols. Inceptisols include a wide variety of different soils, all having in common a lack of maturity with evidence of weathering just beginning. Inceptisols are associated with moist soil regimes and are regarded as eluvial because they demonstrate a loss of soil constituents throughout their profile but retain some weatherable minerals. This soil group has no distinct illuvial horizons. Inceptisols include most of the glacially derived till and outwash materials from New York down through the Appalachians and alluvium on the Mekong and Ganges floodplains.

Gelisols The **Gelisols** are cold and frozen soils and have three suborders, which represent inclusion of high-latitude (Canada, Alaska, Russia, Arctic Ocean islands, and the Antarctic Peninsula) and high-elevation (mountain) soil conditions (Figure 18.23a). Temperatures in these regions are at or below 0°C (32°F), making soil development a slow process and disturbances of the soil long-lasting. Gelisols can develop organic diagnostic horizons because cold temperatures slow decomposition of materials (Figure 18.23b and c). Characteristic vegetation is that of tundra, such as lichens, mosses, sedges, dwarf willows, and other plants adapted to the harsh cold and permafrost.

Gelisols are subject to *cryoturbation* (frost churning and mixing) in the freeze–thaw cycle in the active layer (see Chapter 17). This process disrupts soil horizons, pulling organic material to lower layers and rocky C-horizon material to the surface. Patterned-ground phenomena are possible under such conditions.

As we saw in Chapter 17, periglacial processes occupy about 20% of Earth's land surface, including permafrost (frozen ground) that underlies some 13% of these lands. Previously these soils were included in the Inceptisol, Entisol, and Histosol soil orders.

Andisols Areas of volcanic activity feature **Andisols** (volcanic parent materials), which have seven suborders. These soils formerly were classified as Inceptisols and Entisols, but in 1990 they were placed in this new order. Andisols are derived from volcanic ash and glass. Previous soil horizons frequently are found buried by ejecta from repeated eruptions. Volcanic soils are unique in their mineral content because they are recharged by eruptions. For an example, see Figure 19.28, which illustrates recovery and succession of pioneer species in the developing soil north of Mount St. Helens.

Weathering and mineral transformations are important in this soil order. Volcanic glass weathers readily into allophane (a noncrystalline aluminum silicate clay mineral that acts as a colloid) and oxides of aluminum and iron. Andisols feature a high CEC and high water-holding ability and develop moderate fertility, although phosphorus availability is an occasional problem. In Hawai'i, the fertile Andisol fields produce coffee, pineapple, macadamia nuts, and some sugar cane as important cash crops (Figure 18.24a). Andisol distribution is small in areal extent; however, such soils are locally important in the volcanic ring of fire surrounding the Pacific Rim. Feed crops and pastureland of fertile Andisols sustain a large sheep industry in Iceland. The annual fall roundup is a time when many ranchers combine forces over several weeks to bring in the herds (Figure 18.24b).

Vertisols Heavy clay soils are the **Vertisols** (expandable clay soils). They contain more than 30% swelling clays (clays that swell significantly when they absorb water), such as *montmorillonite*. They are located in regions experiencing highly variable soil-moisture balances through the seasons. These soils occur in areas of subhumid to semiarid moisture and moderate to high temperature. Vertisols frequently form under savanna and grassland vegetation in tropical and subtropical climates and are

FIGURE 18.22 Entisols.
A characteristic Entisol: young, undeveloped soil forming from a shale parent material in the Anza-Borrego desert, California. [Photo by Bobbé Christopherson.]

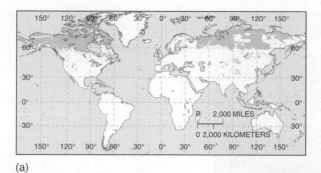

(a)

(b)

(c)

FIGURE 18.23 Gelisols.
(a) General map showing worldwide distribution of these cold and frozen soils. (b) Gelisols and tundra, green in the brief summer season on Spitsbergen Island. (c) Turning over the upper organic layer exposes the fibrous organic content and slow decomposition; permafrost is just below this layer that briefly thaws in summer, in the Arctic. [Photos by Bobbé Christopherson.]

(a)

(b)

FIGURE 18.24 Andisols in agricultural production.
Fertile Andisols planted in macadamia nut groves on the big island of Hawai'i. (b) Pasture and feed grains for raising sheep grow in Andisols in southwestern Iceland. [Photos (a) and (b) by Bobbé Christopherson.]

FIGURE 18.25 Vertisols.
(a) Soil profile in the Lajas Valley of Puerto Rico. (b) General map showing worldwide distribution of these expandable clay soils. (c) Vertisols in the Texas coastal plain, northeast of Palacios near the Tres Palacios River, planted with a commercial sorghum crop. Note the dark soil color indicative of Vertisols, wet and shiny from the rains of Tropical Storm Allison, June 2001. [Photos (a) from the Marbut Collection, Soil Science Society of America; and (c) by Bobbé Christopherson.]

sometimes associated with a distinct dry season following a wet season. Although widespread, individual Vertisol units are limited in extent.

Vertisol clays are black when wet, but not because of organics; rather, the blackness is due to specific mineral content. They range from brown to dark gray. These deep clays swell when moistened and shrink when dried. In the drying process, vertical cracks may form, as wide as 2–3 cm (0.8–1.2 in.) and up to 40 cm (16 in.) deep. Loose material falls into these cracks, only to disappear when the soil again expands and the cracks close. After many such cycles, soil contents tend to invert or mix vertically, bringing lower horizons to the surface (Figure 18.25).

Despite the fact that clay soils are plastic and heavy when wet, with little available soil moisture for plants, Vertisols are high in bases and nutrients and thus are some of the better farming soils where they occur. For example, they occur in a narrow zone along the coastal plain of Texas (Figure 18.25c) and in a section along the Deccan region of India. Vertisols often are planted with grain sorghums, corn, and cotton.

Histosols   Accumulations of thick organic matter can form **Histosols** (organic soils), including four suborders. In the midlatitudes, when conditions are right, beds of former lakes may turn into Histosols, with water gradually replaced by organic material to form a bog and layers of peat (Figure 18.26). (Lake succession and bog or marsh formation are discussed in Chapter 19.) Histosols also form in small, poorly drained depressions, with conditions ideal for significant deposits of sphagnum peat to form. This material can be cut, baled, and sold as a soil amendment. In some locales, dried peat has served for centuries as a low-grade fuel (see the chapter-opening photo of peat harvesting as well).

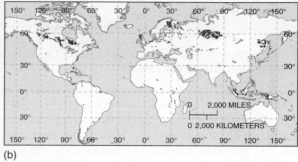

FIGURE 18.26 Histosols.
(a) A bog in coastal Maine, near Popham Beach State Park. (b) General map showing worldwide distribution of these organic soils. (c) Sphagnum peat soil profile and drying peat blocks on Mainland Island, north of Scotland. [Photos by Bobbé Christopherson.]

Summary and Review—The Geography of Soils

■ *Define* soil and soil science, and *describe* a pedon, polypedon, and typical soil profile.

Soil is the portion of the land surface in which plants can grow. It is a dynamic natural body composed of fine materials and contains both mineral and organic matter. **Soil science** is the interdisciplinary study of soils involving physics, chemistry, biology, mineralogy, hydrology, taxonomy, climatology, and cartography. *Pedology* deals with the origin, classification, distribution, and description of soil. *Edaphology* specifically focuses on the study of soil as a medium for sustaining the growth of plants.

The basic sampling unit used in soil surveys is the **pedon**. The **polypedon** is the soil unit used to prepare local soil maps and may contain many pedons. Each discernible layer in an exposed pedon is a **soil horizon**. The horizons are designated O (contains **humus**, a complex mixture of decomposed and synthesized organic materials), A (rich in humus and clay, darker), E (zone of **eluviation**, the removal of fine particles and minerals by water), B (zone of **illuviation**, the deposition of clays and minerals translocated from elsewhere), C (*regolith*, weathered bedrock), and R (bedrock). Soil horizons A, E, and B experience the most active soil processes and together are designated the **solum**.

soil (p. 575)
soil science (p. 576)
pedon (p. 576)
polypedon (p. 577)
soil horizon (p. 577)
humus (p. 577)
eluviation (p. 577)
illuviation (p. 577)
solum (p. 577)

1. Soils provide the foundation for animal and plant life and therefore are critical to Earth's ecosystems. Why is this true?
2. What are the differences among soil science, pedology, and edaphology?
3. Define polypedon and pedon, the basic units of soil.
4. Characterize the principal aspects of each soil horizon. Where does the main accumulation of organic material occur? Where does humus form? Explain the difference between the eluviated layer and the illuviated layer. Which horizons constitute the solum?

■ *Describe* soil properties including color, texture, structure, consistence, porosity, and soil moisture.

We use several physical properties to classify soils. Color suggests composition and chemical makeup. Soil texture refers to the size of individual mineral particles and the proportion of different sizes. For example, **loam** is a balanced mixture of sand, silt, and clay. Soil structure refers to the arrangement of soil peds, which are the smallest natural cluster of particles in a soil. The cohesion of soil particles to each other is called soil consistence. Soil porosity refers to the size, alignment, shape, and location of spaces in the soil. Soil moisture refers to water in the soil pores and its availability to plants.

loam (p. 578)

5. Soil color is identified and compared using what technique?
6. Define a soil separate. What are the various sizes of particles in soil? What is loam? Why is loam regarded so highly by agriculturists?

7. What is a quick, hands-on method for determining soil consistence?
8. Summarize the five soil-moisture regimes common in mature soils.

■ *Explain* basic soil chemistry, including cation-exchange capacity, and *relate* these concepts to soil fertility.

Particles of clay and organic material form negatively charged **soil colloids** that attract and retain positively charged mineral ions in the soil. The capacity to exchange ions between colloids and roots is called the **cation-exchange capacity (CEC)**. CEC is a measure of **soil fertility**, the ability of soil to sustain plants. Fertile soil contains organic substances and clay minerals that absorb water and retain certain elements needed by plants.

soil colloids (p. 581)
cation-exchange capacity (CEC) (p. 582)
soil fertility (p. 582)

9. What are soil colloids? How are they related to cations and anions in the soil? Explain cation-exchange capacity.
10. What is meant by the concept of soil fertility?

■ *Evaluate* the principal soil formation factors, including the human element.

Environmental factors that affect soil formation include parent materials, climate, vegetation, topography, and time. Human influence is having great impact on Earth's prime soils. Essential soils for agriculture and their fertility are threatened by mismanagement, destruction, and conversion to other uses. Much soil loss is preventable through the application of known technologies, improved agricultural practices, and government policies.

11. Briefly describe the contribution of the following factors and their effect on soil formation: parent material, climate, vegetation, landforms, time, and humans.
12. Explain some of the details that support the concern over loss of our most fertile soils. What cost estimates have been placed on soil erosion?

■ *Describe* the 12 soil orders of the Soil Taxonomy classification system, and *explain* their general occurrence.

The **Soil Taxonomy** classification system is used in the United States and is built around an analysis of various diagnostic horizons and 12 soil orders, as actually seen in the field. The system divides soils into six hierarchical categories: series, families, subgroups, great groups, suborders, and orders.

Specific soil-forming processes keyed to climatic regions (not a basis for classification) are called **pedogenic regimes**: **laterization** (leaching in warm and humid climates), **salinization** (collection of salt residues in surface horizons in hot, dry climates), **calcification** (accumulation of carbonates in the B and C horizons in drier continental climates), **podzolization** (soil acidification in forest soils in cool climates), and **gleization** (humus and clay accumulation in cold, wet climates with poor drainage).

The Soil Taxonomy system uses two diagnostic horizons to identify soil: the **epipedon**, or the surface soil, and the **diagnostic subsurface horizon**, or the soil below the surface at various depths. (For an overview and definition of the 12 soil orders, please refer to Table 18.3.) The 12 soil orders are **Oxisols** (tropical soils), **Aridisols** (desert soils), **Mollisols** (grassland soils), **Alfisols** (moderately weathered, temperate forest soils), **Ultisols** (highly weathered, subtropical forest soils), **Spodosols** (northern conifer forest soils), **Entisols** (recent, undeveloped soils), **Inceptisols** (weakly developed, humid-region soils), **Gelisols** (cold soils underlain by permafrost), **Andisols** (soils formed from volcanic materials), **Vertisols** (expandable clay soils), and **Histosols** (organic soils).

Soil Taxonomy (p. 585)
pedogenic regimes (p. 585)
laterization (p. 585)
salinization (p. 585)
calcification (p. 585)
podzolization (p. 585)
gleization (p. 585)
epipedon (p. 585)
diagnostic subsurface horizon (p. 585)
Oxisols (p. 586)
Aridisols (p. 590)
Mollisols (p. 591)
Alfisols (p. 595)
Ultisols (p. 596)
Spodosols (p. 596)
Entisols (p. 598)
Inceptisols (p. 598)
Gelisols (p. 598)
Andisols (p. 598)
Vertisols (p. 598)
Histosols (p. 600)

13. Summarize the basis of soil classification described in this chapter. What have soil scientists revised in the latest Soil Taxonomy classification system?

14. What is the basis of the Soil Taxonomy system? How many orders, suborders, great groups, subgroups, families, and soil series are there?

15. Define an epipedon and a diagnostic subsurface horizon. Give a simple example of each.

16. Locate each soil order on the world map and on the U.S. map as you give a general description of it.

17. How was slash-and-burn shifting cultivation, as practiced in the past, a form of crop and soil rotation and conservation of soil properties?

18. Describe the salinization process in arid and semiarid soils. What associated soil horizons develop?

19. Which of the soil orders are associated with Earth's most productive agricultural areas?

20. What is the significance to plants of the 51-cm (20-in.) isohyet in the Midwest relative to soils, pH, and lime content?

21. Describe the podzolization process associated with northern coniferous forest soils. What characteristics are associated with the surface horizons? What strategies might enhance these soils?

22. What former Inceptisols now form a new soil order? Describe these soils as to location, nature, and formation processes. Why do you think they were separated into their own order?

23. Why has a selenium contamination problem arisen in western U.S. soils? Explain the impact of agricultural practices, and tell why you think this is or is not a serious problem.

NetWork

The *Geosystems* Student Learning Center provides on-line resources for this chapter on the World Wide Web. To begin: Once at the Center, click on the cover of this textbook, scroll the Table of Contents menu, and select this chapter. You will find self-tests that are graded, review exercises, specific updates for items in the chapter, and in "Destinations" many links to interesting related pathways on the Internet. *Geosystems* Student Learning Center is found at **http://www.prenhall.com/christopherson/**.

Critical Thinking

A. Select a small soil sample from your campus or near where you live. Using the sections in this chapter on soil characteristics, properties, and formation, describe as completely as possible this sample, within these limited constraints. Using the general soil map and any other sources available (e.g., local agriculture extension agent, Internet, related department on campus), are you able to roughly place this sample in one of the soil orders?

B. Using the local phone directory or Internet, see if you can locate an agency or extension agent that provides information about local soils and advice for soil management. Is there a place where you can get soil tested?

C. Refer to "Critical Thinking" for Chapter 18 at the Geosystems Student Learning Center, item 1. Please complete the analysis of the three photographs for environments of deciduous forests of eastern North America, through evergreen forests, and into the arctic tundra. What soil order would you expect in these regions? Why?

An ancient forest ecosystem and community of mosses, ferns, heather, grasses, and lush undergrowth cover the forest floor and rock outcrop. The mixed birch forest and marine west coast climate is in central Scotland, near Loch Ness. [Photo by Bobbé Christopherson.]

Ecosystem Essentials

■ Key Learning Concepts

After reading the chapter, you should be able to:

- ■ **Define** ecology, biogeography, and the ecosystem concept.
- ■ **Describe** communities, habitats, and niches.
- ■ **Explain** photosynthesis and respiration, and **derive** net photosynthesis and the world pattern of net primary productivity.
- ■ **List** abiotic ecosystem components, and **relate** those components to ecosystem operations.
- ■ **Explain** trophic relationships in ecosystems.
- ■ **Relate** how biological evolution led to the biodiversity of life on Earth.
- ■ **Define** succession, and **outline** the stages of general ecological succession in both terrestrial and aquatic ecosystems.

Diversity is an impressive feature of the living Earth. The diversity of organisms is a response to the interaction of the atmosphere, hydrosphere, and lithosphere, all powered by solar energy. The evolution of life on Earth, in all its forms, is intertwined with the evolution of Earth's physical and chemical systems. This interaction produces a variety of conditions within which the biosphere exists. The diversity of life also results from the intricate interplay and coevolution of living organisms themselves. We, as part of this vast natural complex, seek to find our place and understand the feelings nature stimulates within us.

The physical beauty of nature is certainly among its most powerful appeals to the human animal. The complexity of the aesthetic response is suggested by its wide-ranging expression from the contours of a mountain landscape to the ambient colors of a setting Sun to the fleeting vitality of a breaching whale. Each exerts a powerful aesthetic impact on most people, often accompanied by feelings of awe at the extraordinary physical appeal and beauty of the natural world.*

The biosphere, the sphere of life and organic activity, extends from the ocean floor to about 8 km (5 mi) altitude into the atmosphere. The biosphere includes myriad ecosystems from simple to complex, each operating within general spatial boundaries. An **ecosystem** is a self-sustaining association of living plants and animals and their nonliving physical environment. Earth's biosphere itself is a collection of ecosystems within the natural boundary of the atmosphere and Earth's crust.

Natural ecosystems are open systems for both solar energy and matter, with almost all ecosystem boundaries functioning as transition zones rather than as sharp demarcations. Distinct ecosystems—for example, forests, seas, mountaintops, deserts, beaches, islands, lakes, and ponds—make up the larger whole.

Ecology is the study of the relationships between organisms and their environment and among the various ecosystems in the biosphere. The word *ecology*, developed by German naturalist Ernst Haeckel in 1869, is derived from the Greek *oikos* ("household," or "place to live") and *logos* ("study of"). **Biogeography** is the study of the distribution of plants and animals, the diverse spatial patterns they create, and the physical and biological processes, past and present, that produce Earth's species richness.

The degree to which modern society understands Earth's biogeography and conserves Earth's living legacy will determine the extent of our success as a species and the long-term survival of a habitable Earth:

The time is ripe to step up and expand current efforts to understand the great interlocking systems of air, water, and minerals nourishing the Earth. . . . Moreover, without vigorous action toward that goal, nations will be seriously handicapped in trying to cope with proven and suspected threats to ecosystems and to human health and welfare resulting from alterations in the cycles of carbon, nitrogen, phosphorus, sulfur, and related materials. . . . Society depends upon this life-support system of planet Earth.[†]

Here we are more than 28 years after this UNEP report and Gilbert White's statement, reading from the new UNEP assessment *Global Environmental Outlook 4*,

published October 25, 2007. In its *Summary for Policy Makers* (p. 12), the UNEP states:

Biodiversity decline and loss of ecosystem services continue to be a major threat to future development. The reduction in distribution and functioning of land, fresh water, and marine biodiversity is more rapid than at any time in human history. Ecosystems such as forests, wetlands, and drylands are being transformed and, in some cases, irreversibly degraded. Rates of species extinction are increasing. The great majority of well-studied species, including commercially important fish stocks, are declining in distribution or abundance or both. Genetic diversity of agricultural and other species is widely considered to be in decline.

What was learned in all these years? The essential challenge for us in geographic education is in the last topic sentence, "The degree to which modern society understands Earth's biodiversity. . . ." We must work to better understand and communicate these "Ecosystem Essentials."

Earth's most influential biotic (living) agents are humans. This is not arrogance; it is fact, because we powerfully influence every ecosystem on Earth. From the time humans first developed agriculture, the raising of animals, and the use of fire, we began a process that has led to our dominance over Earth's physical systems.

Since life arose on the planet, there have been six major extinctions. The fifth one was 65 million years ago, whereas the sixth is happening across the present decades. Of all these extinction episodes, this is the only one of biotic origin, caused for the most part by human activity. We are the greatest evolutionary force as we essentially dominate freshwater quantity and quality, croplands, ocean fishery production, nitrogen and other chemical budgets, the concentration of greenhouse gases, and the present extinction rate.

In this chapter: We explore ecosystems and the community, habitat, and niche concepts. Plants are the essential living component in the biosphere, translating solar energy into usable forms to energize life. The role of nonliving systems, including biogeochemical cycles, is examined. We cover the organization of living ecosystems along complex food chains and webs. The biodiversity of living organisms is seen as a product of biological evolution over the past 3.6+ billion years, approximately. Ecosystem stability and resilience, and how living landscapes change over space and time through the process of succession, are important. We examine climate-induced changes to planting guides. Included is coverage of the effects of global change on ecosystems and rates of succession. The chapter ends with a look at the Great Lakes environment and status.

Ecosystem Components and Cycles

An ecosystem is a complex of many variables, all functioning independently yet in concert, with complicated flows of energy and matter (Figure 19.1). The key is to

*S. R. Kellert, "The biological basis for human values of nature," in S. R. Kellert and E. O. Wilson, eds., *The Biophilia Hypothesis* (Washington, DC: Island Press, 1993), p. 49.
[†]G. F. White and M. K. Tolba, *Global Life Support Systems*, United Nations Environment Programme Information, No. 47 (Nairobi, Kenya: United Nations, 1979), p. 1.

(a)

(b)

FIGURE 19.1 The web of life.
(a) Study the spider web as you read the following quotation:

Life devours itself: everything that eats is itself eaten; everything that can be eaten is eaten; every chemical that is made by life can be broken down by life; all the sunlight that can be used is used.... The web of life has somany threads that a few can be broken without making it all unravel, and if this were not so, life could not have survived the normal accidents of weather and time, but still the snapping of each thread makes the whole web shudder, and weakens it.... You can never do just one thing: the effects of what you do in the world will always spread out like ripples in a pond.

(b) Abandoned Black-browed Albatross nests indicate a change in government fishing regulations in the surrounding ocean— "...snapping each thread makes the whole web shudder...." [(a) Photo by author; quotation from Friends of the Earth and Amory Lovins, The U.N. Stockholm Conference, *Only One Earth* (London: Earth Island Limited, 1972), p. 20. (b) Photos by Bobbé Christopherson.]

understand the linkages and interconnections within and between systems. As the quote in the caption says, "... The effects of what you do in the world will always spread out like ripples in a pond." The decrease in the Black-browed Albatross (*Diomedea melanophris*) colony evident from the empty nests in Figure 19.1b was the result of changes in regulations in the fishing industry around the Falkland Islands. An increase in quotas for the squid harvest, a primary food source for the albatross, led to the reduction in bird population.

An ecosystem includes both biotic (living) and abiotic (nonliving) components. Nearly all ecosystems depend on a direct input of solar energy; the few limited ones that exist in dark caves, in wells, or on the ocean floor depend on chemical reactions (chemosynthesis). Ecosystems are divided into subsystems, with the biotic portion composed of producers (plants), consumers (animals), and detritus feeders and decomposers (worms, mites, bacteria, fungi). The abiotic flows in an ecosystem include gaseous, hydrologic, and mineral cycles. Figure 19.2a illustrates these essential elements of an ecosystem. Figure 19.2b and c illustrates how biotic and abiotic ingredients operate together to form a delicate forest floor in the Northwest and the hearty lichens in the hostile environment of Antarctica.

Communities

A convenient biotic subdivision within an ecosystem is a community. A **community** is formed by interactions among populations of living animals and plants at a particular time. An ecosystem is the interaction of many communities with the abiotic physical components of its environment.

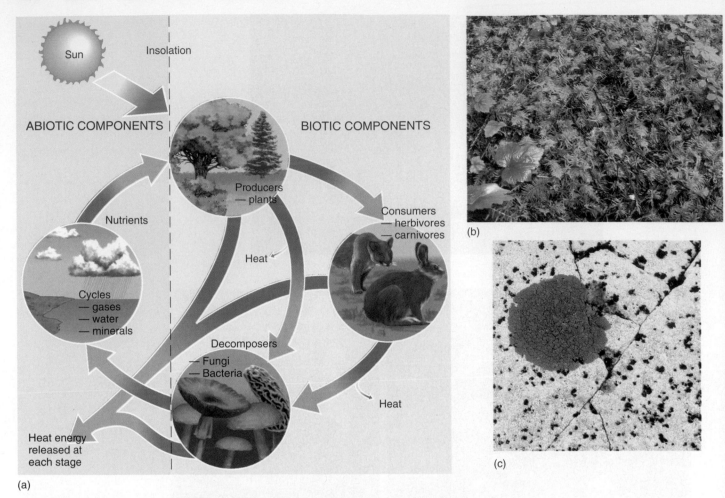

FIGURE 19.2 Biotic and abiotic components of ecosystems.
(a) Solar energy is the input that drives the biotic and abiotic components. Heat energy and biomass
are the outputs from the biosphere. (b) Biotic and abiotic ingredients operate together to form this
temperate forest-floor ecosystem in British Columbia, with mosses beginning the process of community
development. (c) Two species of lichen, perhaps more, function under extremely hostile climate
conditions in Antarctica. Each little indentation provides some advantage to the darker lichen. [Photos
by Bobbé Christopherson.]

Some examples help to clarify these concepts. In a forest ecosystem, a specific community may exist on the forest floor, whereas another community functions in the canopy of leaves high above (Figure 19.3). Not all forests appear familiar, as the dwarf willow forest in Greenland demonstrates (Figure 19.3c). Similarly, within a lake ecosystem, the plants and animals that flourish in the bottom sediments form one community, whereas those near the surface form another. A community is identified in several ways—by its physical appearance, the species present and the abundance of each, the complex patterns of their interdependence, and the trophic (feeding) structure of the community.

Within a community, two concepts are important: habitat and niche. **Habitat** is the type of environment in which an organism resides or is biologically adapted to live. In terms of physical and natural factors, most species have specific habitat requirements with definite limits and a specific regimen of sustaining nutrients.

Niche (French *nicher*, "to nest") refers to the function, or occupation, of a life form within a given community. It is the way an organism obtains and sustains the physical, chemical, and biological factors it needs to survive. A niche has several facets. Among these are a habitat niche, a trophic (food) niche, and a reproductive niche. For example, the Red-winged Blackbird (*Agelaius phoeniceus*) occurs throughout the United States and most of Canada in habitats of meadow, pastureland, and marsh. This species nests in blackberry tangles and thick vegetation in freshwater marshes, sloughs, and fields. Its trophic niche is weed seeds and cultivated seed crops throughout the year, and during the nesting season it adds insects to its diet—an aspect of its reproductive niche. These birds disperse seeds of many plants during their travels.

In a stable community, no niche is left unfilled. The *competitive exclusion principle* states that no two species can occupy the same niche (food or space) successfully in a stable community. Thus, closely related species are spatially separated. In other words, each species operates to reduce competition and to maximize its own reproduction rate—literally, species survival depends on successful reproduction. This strategy in turn leads to greater diversity as species shift and adapt to fill each niche. Figure 19.4

(a)

(b)

(c)

FIGURE 19.3 Different forest communities.
(a) Pioneer Mothers' Memorial Forest near Paoli, Indiana, is 36 hectares (88 acres) of old-growth forest, virtually undisturbed since around 1816, featuring black walnut, white oak, yellow poplar, white ash, and beech trees, among others. (b) Turtles bask on a log in a subtropical swamp (low, waterlogged ground) at Juniper Springs Recreation Area, Florida. (c) Fall colors highlight this dwarf willow forest in Greenland. Musk Oxen are foraging for food and drinking meltwater. The inset photo shows a 13-year-old willow, dated by the small annual growth segments below the leaf. Some dwarf willows can range to over 300 years in age. [Photos by Bobbé Christopherson.]

shows 11 different community environments and representative life forms.

Some species are *symbiotic*, an arrangement whereby two or more species exist together in an overlapping relationship. One type of symbiosis, *mutualism*, occurs when each organism benefits and is sustained over an extended period by the relation. For example, lichen (pronounced "liken") is made up of algae and fungi living together (Figure 19.2c). The alga is the producer and food source for the fungus, and the fungus provides structure and physical support. Their mutualism allows the two to occupy a niche in which neither could survive alone. Lichen developed from an earlier parasitic relationship in which the fungi broke into the algal cells. Today, the two organisms have evolved into a supportive harmony and symbiotic relationship. The partnership of corals and algae discussed in Chapter 16 is another example of mutualism in a symbiotic relationship.

By contrast, another form of symbiosis is a *parasitic* relationship, which may eventually kill the host, thus destroying the parasite's own niche and habitat. An example is parasitic mistletoe (*Phoradendron*), which lives on and may kill various kinds of trees. Some scientists are questioning whether our human society and the physical systems of Earth constitute a global-scale symbiotic relationship of mutualism, which is sustainable, or a parasitic one, which is unsustainable.

Plants: The Essential Biotic Component

Plants are the critical biotic link between solar energy and the biosphere. Ultimately, the fate of all members of the biosphere, including humans, rests on the success of plants and their ability to turn sunlight into food.

Land plants (and animals) became common about 430 million years ago, according to fossilized remains. **Vascular plants** developed conductive tissues and true roots for internal transport of fluid and nutrients. (*Vascular* is from a Latin word for "vessel-bearing," referring to the conducting cells.) Presently, about 270,000 species of plants are known to exist, and most are vascular. Many more species have yet to be identified. They represent a great untapped resource base. Only about 20 species of plants provide 90% of the world's food; just three—wheat, maize (corn), and rice—comprise half of the food supply. Plants are a major source of new medicines and chemical compounds that benefit humanity. Plants also are the core of healthy, functioning ecosystems that sustain all life.

(a) Elephant heads

(b) Western yellow-bellied racer

(c) Polar bear and cubs

(d) Svalbard reindeer

(e) Killdeer chicks

(f) Coral mushroom

(g) Desert scene, barrel cactus

(h) Leopard seal

(i) Alligator in swamp

(j) Tricolored Heron

(k) Lichens and flower

FIGURE 19.4 Plants and animals fit specific niches.
(a) Elephant heads (*Pedicularis groenlandica*), a wildflower, above 1800 m (6000 ft) in wet meadows. (b) A western yellow-bellied racer (*Coluber constrictor*) lives in prairies and feeds on insects, lizards, and mice. (c) A female polar bear (*Ursus maritimus*) with two 8-month-old cubs in the Arctic Ocean. (d) A Svalbard reindeer (*Rangifer tarandus platyrhynchus*) on Spitsbergen Island, Arctic Ocean, losing velvet from his antlers. (e) Killdeer chicks (*Charadrius vociferus*) begin life in an exposed rocky nest. (f) Coral mushroom (*Ramaria* sp.), a fungal decomposer, at work on organic matter. (g) A large barrel cactus flourishes in the rocky soils of the desert. (h) A leopard seal (*Hydrurga leptonyx*) resting on a small iceberg near the Antarctic Peninsula; the blood stains are from a recent meal. (i) An alligator (*Alligator mississippiensis*) hides in a mangrove swamp. (j) Tricolored Heron (*Egretta tricolor*) in wetlands. (k) Five kinds of lichen and sea thrift flower in Scotland. [Photos (a), (c), (d), (f), (h), (l), (j), (k) by Bobbé Christopherson; (b), (e), (g) by author.]

Leaves are solar-powered chemical factories, wherein photochemical reactions take place. Veins in the leaf bring in water and nutrient supplies and carry off the sugars (food) produced by photosynthesis. The veins in each leaf connect to the stems and branches of the plant and to the main circulation system.

Flows of carbon dioxide, water, light, and oxygen enter and exit the surface of each leaf (see Figure 1.4 in Chapter 1). Gases flow into and out of a leaf through small pores called **stomata** (singular: *stoma*), which usually are most numerous on the lower side of the leaf. Each stoma is surrounded by guard cells that open and close the pore, depending on the plant's changing needs. Water that moves through a plant exits the leaves through the stomata and evaporates from leaf surfaces, thereby assisting the plant's temperature regulation. As water evaporates from the leaves, a pressure deficit is created that allows atmospheric pressure to push water up through the plant all the way from the roots, in the same manner that a soda straw works. We can only imagine the complex operation of a 100-m (330-ft) tree!

Photosynthesis and Respiration

Powered by energy from certain wavelengths of visible light, **photosynthesis** unites carbon dioxide and hydrogen (hydrogen is derived from water in the plant). The term is descriptive: *photo-* refers to sunlight and *-synthesis* describes the "manufacturing" of starches and sugars through reactions within plant leaves. The process releases oxygen and produces energy-rich food for the plant.

The largest concentration of light-responsive, photosynthetic structures (known as *organelles*) in a leaf rests below the leaf's upper layers. These organelle units within cells are the *chloroplasts*, and within each resides a green, light-sensitive pigment called **chlorophyll**. Within this pigment, light stimulates photochemistry. Consequently, competition for light is a dominant factor in the formation of plant communities. This competition is expressed in the height, orientation, distribution, and structure of plants.

Only about one-quarter of the light energy arriving at the surface of a leaf is useful to the light-sensitive chlorophyll. Chlorophyll absorbs only the orange-red and violet-blue wavelengths for photochemical operations, and it reflects predominantly green hues (and some yellow). That is why trees and other vegetation look green.

Photosynthesis essentially follows this equation:

$$6CO_2 + 6H_2O + \text{Light} \rightarrow C_6H_{12}O_6 + 6O_2$$

(carbon dioxide) (water) (solar energy) (glucose, carbohydrate) (oxygen)

From the equation, you can see that photosynthesis removes carbon (in the form of CO_2) from Earth's atmosphere. The quantity is enormous: approximately 91 billion metric tonnes (100 billion tons) of carbon dioxide per year. Carbohydrates, the organic result of the photosynthetic process, are combinations of carbon, hydrogen, and oxygen. They can form simple sugars, such as *glucose* ($C_6H_{12}O_6$). Plants use glucose to build starches, which are more complex carbohydrates and the principal food stored in plants.

Plants store energy for later use. They consume this energy as needed through respiration by converting the carbohydrates to energy for their other operations. Thus, **respiration** is essentially a reverse of the photosynthetic process:

$$C_6H_{12}O_6 + 6O_2 \rightarrow 6CO_2 + 6H_2O + \text{energy}$$

(glucose carbohydrate) (oxygen) (carbon dioxide) (water) (heat energy)

In respiration, plants oxidize carbohydrates (stored energy), releasing carbon dioxide, water, and energy as heat. The overall growth of a plant depends on a surplus of carbohydrates beyond those lost through plant respiration. Figure 19.5 presents a simple schematic of this process that produces plant growth.

The *compensation point* is the break-even point between the production and consumption of organic material. Each leaf must operate on the production side of the compensation point, or else the plant eliminates it—something each of us has no doubt experienced with a houseplant that received inadequate water or light. The difference between photosynthetic production and respiration loss is *net photosynthesis*. The amount varies, depending on controlling environmental factors such as light, water, temperature, soil fertility, and the plant's site, elevation, and competition from other plants and animals.

The rapidly increasing carbon dioxide concentration in the atmosphere stimulates the photosynthetic processes. However, higher temperatures due to global warming occur more at night than during the day in many regions, resulting in higher respiration rates. One of the key scientific

FIGURE 19.5 How plants live and grow.
The balance between photosynthesis and respiration determines net photosynthesis and plant growth.
[Temperate rain forest photo by Bobbé Christopherson.]

findings is that this unnatural warming is emphasized during the night and during wintertime, whereas a natural warming cycle would be balanced day to night or summer to winter. Plants respire all the time, but at night respiration processes are the plant's main operation. Studies of tropical rain forests in Costa Rica determined that increased respiration rates had made the forest a net carbon gain to the atmosphere.

Plant productivity increases as light availability increases—up to a point. When the light level is too high, *light saturation* occurs and most plants actually reduce their output in response. Some plants are adapted to shade, whereas others flourish in full sunlight. Crops such as rice, wheat, and sugarcane do well with high light intensity. Figure 19.6 portrays the general energy budget for green plants, showing energy receipt, energy utilization, and disposition of net primary production—how the energy is spent.

Net Primary Productivity The net photosynthesis for an entire plant community is its **net primary productivity**. This is the amount of stored chemical energy that the community generates for the ecosystem. **Biomass** is the net dry weight of all this organic material and its stored chemical energy.

Net primary productivity is measured as fixed carbon per square meter per year. ("Fixed" means chemically bound into plant tissues.) Study the map and satellite images in Figure 19.7 and you can see that on land, net primary production tends to be highest between the Tropics of Cancer and Capricorn at sea level and decreases toward higher latitudes and elevation. Precipitation also affects productivity, as evidenced by the correlations of abundant precipitation with high productivity adjacent to the equator and reduced precipitation with low productivity in the subtropical deserts. Even though deserts receive high amounts of solar radiation, other controlling factors limit productivity, namely, water availability and soil conditions.

In the oceans, differing nutrient levels control and limit productivity. Regions with nutrient-rich upwelling currents off western coastlines generally are the most productive. The map in Figure 19.7 shows that the tropical oceans and areas of subtropical high pressure are quite low in productivity.

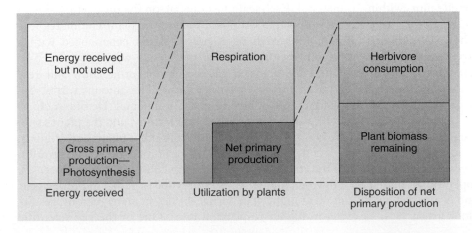

FIGURE 19.6 Energy budget of the biosphere.
Energy receipt, energy utilization, and energy disposition of net primary production by green plants.

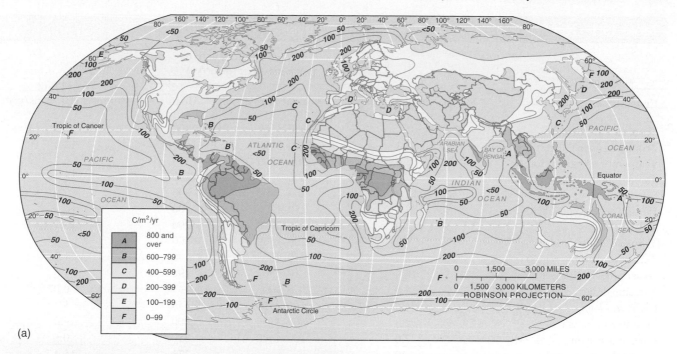

(a)

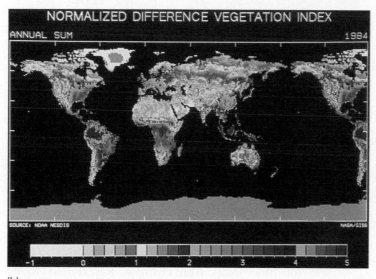

(b)

(c)

FIGURE 19.7 Net primary productivity.
(a) Worldwide net primary productivity in grams of carbon per square meter per year (approximate
values). (b) Normalized difference vegetation index. False coloration indicates bare ground in browns,
dense vegetation in blues. (c) *SeaWIFS* image of surface chlorophyll concentration for land and sea;
intensity of green color represents higher levels of chlorophyll. This is the first image of a continuous
record over a 3-year period for the oceans. [(a) After D. E. Reichle, *Analysis of Temperate Forest Ecosystems*
(Heidelberg: Springer, 1970). Adapted by permission. (b) NASA/GSFC. (c) *SeaWIFS* image by NASA/GSFC and
GeoEye, Dulles, VA. Used by permission.]

In temperate and high latitudes, the rate at which carbon is fixed by vegetation varies seasonally. It increases in
spring and summer as plants flourish with increasing solar
input and, in some areas, with more available (nonfrozen)
water, and it decreases in late fall and winter. Productivity
rates in the tropics are high throughout the year, and
turnover in the photosynthesis–respiration cycle is faster,
exceeding by many times the rates experienced in a desert
environment or in the far northern limits of the tundra. A
lush hectare (2.5 acres) of sugarcane in the tropics might
fix 45 metric tonnes (50 tons) of carbon in a year, whereas

desert plants in an equivalent area might achieve only 1%
of that amount.

Table 19.1 lists various ecosystems, their net primary
productivity, and an estimate of net total biomass
worldwide—170 billion metric tons of dry organic matter
per year. Compare the various ecosystems, especially
cultivated land (in *italics*), with most of the natural communities. Net productivity is generally regarded as the most
important aspect of any type of community, and the distribution of productivity over Earth's surface is an important
subject of biogeography.

Table 19.1 Net Primary Productivity and Plant Biomass on Earth

Ecosystem	Area $(10^6 \text{ km}^2)^a$	Net Primary Productivity per Unit Area $(\text{g/m}^2\text{/yr})^b$		World Net Biomass $(10^9\text{/tons/yr})^c$
		Normal Range	Mean	
Tropical rain forest	17.0	1000–3500	2200	37.4
Tropical seasonal forest	7.5	1000–2500	1600	12.0
Temperate evergreen forest	5.0	600–2500	1300	6.5
Temperate deciduous forest	7.0	600–2500	1200	8.4
Boreal forest	12.0	400–2000	800	9.6
Woodland and shrubland	8.5	250–1200	700	6.0
Savanna	15.0	200–2000	900	13.5
Temperate grassland	9.0	200–1500	600	5.4
Tundra and alpine region	8.0	10–400	140	1.1
Desert and semidesert scrub	18.0	10–250	90	1.6
Extreme desert, rock, sand, ice	24.0	0–10	3	0.07
Cultivated land	*14.0*	*100–3500*	*650*	*9.1*
Swamp and marsh	2.0	800–3500	2000	4.0
Lake and stream	2.0	100–1500	250	0.5
Total continental	**149**	—	773	**115.17**
Open ocean	332.0	2–400	125	41.5
Upwelling zones	0.4	400–1000	500	0.2
Continental shelf	26.6	200–600	360	9.6
Algal beds and reefs	0.6	500–4000	2500	1.6
Estuaries	1.4	200–3500	1500	2.1
Total marine	**361.0**	—	152	**55.0**
Grand total	**510.0**	—	333	**170.17**

Source: R. H. Whittaker, *Communities and Ecosystems* (Heidelberg: Springer, 1975), p. 224. Reprinted by permission.

[a]1 km^2 = 0.38 mi^2.

[b]1 g per m^2 = 8.92 lb per acre.

[c]1 metric ton (10^6 g) = 1.1023 tons.

Abiotic Ecosystem Components

Critical in each ecosystem are the flow of energy and the cycling of nutrients and water in life-supporting systems. These abiotic (nonliving) components set the stage for ecosystem operations.

Light, Temperature, Water, and Climate Solar energy powers ecosystems, so the pattern of solar energy receipt is crucial. Solar energy enters an ecosystem by way of photosynthesis, and heat energy is dissipated from the system at many points. Of the total energy intercepted at Earth's surface and available for work, only about 1.0% is actually fixed by photosynthesis as carbohydrates in plants.

The duration of Sun exposure is the *photoperiod*. Along the equator, days are essentially 12 hours long year-round; however, with increasing distance from the equator, seasonal effects become pronounced. Plants have adapted their flowering and seed germination to seasonal changes in insolation. Some seeds germinate only when daylength reaches a certain number of hours. A plant that responds in the opposite manner is the poinsettia (*Euphorbia pulcherrima*), which requires at least 2 months of 14-hour nights to start flowering.

Monarch butterflies (*Danaus plexippus*) migrate some 3600 km (2235 mi) from eastern North America to wintering forests in Mexico. The butterflies navigate using an internal time-compensated Sun compass to maintain their southwestward track while they compensate for time-of-day Sun changes. Researchers found receptors in their eyes sensitive to ultraviolet light that play the key role in accurate navigation. For Monarchs, ultraviolet light is an important abiotic component.

Other components are important to ecosystem processes. Air and soil temperatures determine the rates at which chemical reactions proceed. Significant temperature factors are seasonal variation and duration and the pattern of minimum and maximum temperatures.

Operations of the hydrologic cycle and water availability depend on precipitation and evaporation rates and their seasonal distribution. Water quality—its mineral content, salinity, and levels of pollution and toxicity—is important. Also, regional climates affect the pattern of vegetation and ultimately influence soil development. All of these factors work together to form the limits for ecosystems in a given location.

Figure 19.8 illustrates the general relationship among temperature, precipitation, and vegetation. Can you identify the characteristic vegetation type and related temperature and moisture regime that fit areas you know well,

FIGURE 19.8 How temperature and precipitation affect ecosystems.
(a) Climate controls (wetness, dryness, warmth, cold) and ecosystem types; (b) subtropical desert near Death Valley, California;
(c) Sonoran desert of Arizona; (d) cold desert of north-central Nevada; (e) dry tundra of East Greenland; (f) El Yunque tropical rain
forest of Puerto Rico; (g) deciduous trees in fall colors in Ohio; (h) needleleaf forest near Independence Pass, Colorado; and
(i) moist tundra of Spitsbergen, Arctic Ocean. [Photos (c, d) by author; (f) photo by Tom Bean; (b), (e), (g), (h), (i) by Bobbé Christopherson.]

such as a place you have lived or an area around the school you now attend?

Life Zones Alexander von Humboldt (1769–1859), an explorer, geographer, and scientist, observed that plants and animals recur in related groupings wherever similar conditions occur in the abiotic environment. He described the similarities and dissimilarities of vegetation and the associated uneven distribution of various organisms. After several years of study in the Andes Mountains of Peru, he described a distinct relation between elevation and plant communities as his *life zone concept*. As he climbed the mountains, he noticed that the experience was similar to that of traveling away from the equator toward higher latitudes (Figure 19.9a).

This zonation of plants with altitude is noticeable on any trip from lower valleys to higher elevations. Each **life zone** possesses its own temperature, precipitation, and insolation relations and therefore its own biotic communities.

The Grand Canyon in Arizona provides a good example. Plants and animals of the inner gorge at the bottom of the canyon (600 m, or 2000 ft, in elevation) are characteristic of the lower Sonoran Desert of northern Mexico. However, communities similar to those of southern Canadian forests dominate the north rim of the canyon (2100 m, or 7000 ft, in elevation). On the summits of the nearby San Francisco Mountains (3600 m, or 12,000 ft, in elevation), the vegetation is similar to that of the arctic tundra of northern Canada. Needleleaf forest and alpine tundra in the Colorado Rockies at high elevation are pictured in Figure 19.9b and c.

Other Factors Beyond these general conditions, each ecosystem further produces its own *microclimate*, which is specific to individual sites. For example, in a forest the insolation reaching the ground is reduced and shade from its own trees blankets the forest floor. A pine forest cuts light by 20%–40%, whereas a birch–beech forest reduces it by as much as 50%–75%. Forests also are about 5%

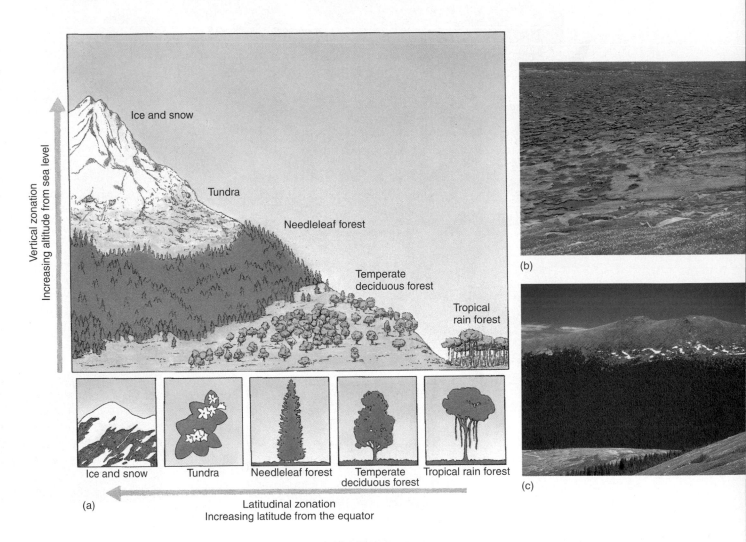

FIGURE 19.9 Vertical and latitudinal zonation of plant communities.
(a) Progression of plant community life zones with changing elevation or latitude. (b) Alpine tundra ecosystem in the Colorado Rockies. (c) The timberline for a needleleaf forest in the Rockies. The line of trees marks the limit of highest continuous forest, whereas the area above it marks the zone above which no trees grow. [Photos by Bobbé Christopherson.]

News Report 19.1

Earth's Magnetic Field—An Abiotic Factor

The fact that birds and bees can detect Earth's magnetic field and use it for finding direction is well established. Small amounts of magnetically sensitive particles in the skull of the bird and the abdomen of the bee provide compass directions for bird migration and for alerting the hive to the location of the latest nectar find. For migrating birds, the timing and accuracy of an exact flight path is important, for the birds know the nutritional places they need to visit to build their fat storage for long journeys and where each "pit stop" is located along the way.

A team of biologists from the University of North Carolina, Catherine and Kenneth Lohmann, has found that sea turtles can detect magnetic fields of different strengths and the inclination (angle) of these magnetic fields. This means that the turtles have a built-in navigation system that helps them know where they are on Earth (like our global positioning system, which requires multimillion-dollar satellites). Using magnetic field strengths and inclination, the turtles evidently are aware of their global position—similar to knowing their latitude and longitude. This magnetic map is evidently imprinted in their brain.

Loggerhead sea turtles hatch in Florida, crawl into the water, and spend the next 70 years traveling thousands of miles between North America and Africa around the subtropical high-pressure gyre in the Atlantic Ocean (see Figure 6.13). The females return to near where they were hatched to lay their eggs! The researchers think that the hatchlings are imprinted with magnetic data unique to the beach where they hatched and then develop a more global sense as they live a life swimming across the ocean.

---

more humid than nonforested landscapes, have moderated temperatures (warmer winters and cooler summers), and experience reduced winds.

Slope orientation and Sun exposure are important, too, for they translate into differences in temperature and moisture efficiency, especially in middle and higher latitudes. With all other factors equal, slopes facing away from the Sun's rays tend to be moister and more vegetated than slopes facing toward the Sun. In the Northern Hemisphere, these moister slopes face north. Such highly localized *microecosystems* are evident where changes in exposure and moisture occur related to geographic orientation.

Even Earth's magnetic field, another abiotic component, plays an interesting role in ecosystems. Read about turtle and animal navigation in News Report 19.1.

Elemental Cycles

The most abundant natural elements in living matter are hydrogen (H), oxygen (O), and carbon (C). Together, these elements make up more than 99% of Earth's biomass; in fact, all life (organic molecules) contains hydrogen and carbon. In addition, nitrogen (N), calcium (Ca), potassium (K), magnesium (Mg), sulfur (S), and phosphorus (P) are significant nutrients, elements necessary for the growth of a living organism.

Several key chemical cycles function in nature. Oxygen, carbon, and nitrogen each have *gaseous cycles*, part of which are in the atmosphere. Other elements have *sedimentary cycles*, which principally involve mineral and solid phases; major elements that have these cycles include phosphorus, calcium, potassium, and sulfur. Some elements combine gaseous and sedimentary cycles. The recycling of gases and sedimentary (nutrient) materials forms Earth's **biogeochemical cycles**, so called because they involve chemical reactions necessary for growth and development of living systems. The chemical elements themselves recycle over and over again in life processes.

Oxygen and Carbon Cycles   We consider these two cycles together because they are so closely intertwined through photosynthesis and respiration (Figure 19.10). The atmosphere is the principal reserve of available oxygen. Larger reserves of oxygen exist in Earth's crust, but they are unavailable, being chemically bound with other elements, especially the silicate (SiO_2) and carbonate (CO_3) mineral families. Unoxidized reserves of fossil fuels and sediments also contain oxygen.

The oceans are enormous pools of carbon—about 42,900 billion metric tons. (Metric tons times 1.10 equals short tons, or 47,190 billion tons.) However, all of this carbon is bound chemically in carbon dioxide, calcium carbonate, and other compounds. The ocean initially absorbs carbon dioxide by means of the photosynthesis carried on by phytoplankton; it becomes part of the living organisms and is fixed in certain carbonate minerals, such as limestone ($CaCO_3$). In Chapter 16 we saw that excessive absorption of CO_2 from the atmosphere by the ocean is lowering pH. This condition of acidification makes it harder for plankton, corals, and other organisms to maintain calcium carbonate skeletons.

The atmosphere, which is the integrating link in the cycle, contains only about 700 billion tons of carbon (as carbon dioxide) at any moment. This is far less carbon than is stored in fossil fuels and oil shales (13,200 billion

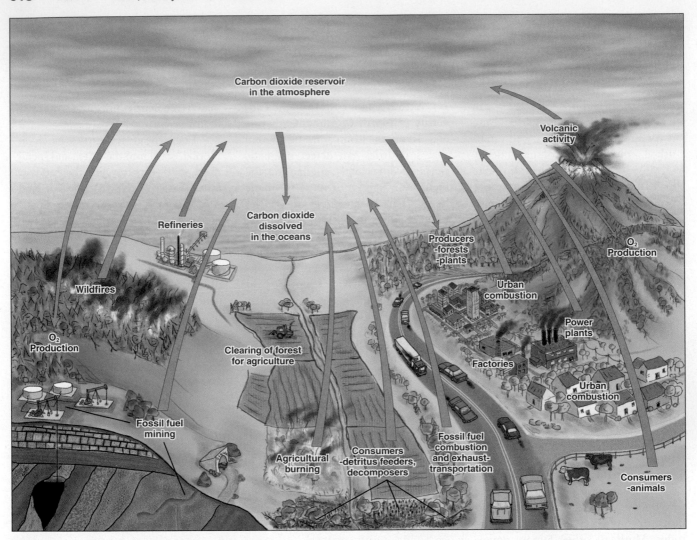

FIGURE 19.10 The carbon and oxygen cycles.
Carbon is fixed through photosynthesis and begins its passage through ecosystem operations. Respiration by living organisms, burning of forests and grasslands, and the combustion of fossil fuels releases carbon to the atmosphere. These cycles are greatly influenced by human activities.

ANIMATION

The Global
Carbon Cycle

metric tons, as hydrocarbon molecules) or in living and dead organic matter (2500 billion metric tonnes, as carbohydrate molecules). Carbon dioxide is released into the atmosphere by the respiration of plants and animals, volcanic activity, and fossil-fuel combustion by industry and transportation.

The carbon dumped into the atmosphere by human activity constitutes a vast geochemical experiment, using the real-time atmosphere as a laboratory. Annually, we are adding carbon to the atmospheric pool in an amount 400% greater than 1950 levels. Global emissions of carbon from fossil fuels continues to increase, with some 6.88 billion metric tons (7.56 billion tons) of carbon added to the atmosphere in 2005—compare this to the 1.47 billion metric tons (1.61 billion tons) in 1950, or 4.74 billion metric tons (5.21 billion tons) in 1980. About 50% of the carbon emitted since the beginning of the Industrial Revolution and not absorbed by oceans and organisms remains in the atmosphere, enhancing Earth's natural greenhouse effect, and thus global warming.

Nitrogen Cycle Nitrogen, which accounts for 78.084% of each breath we take, is the major constituent of the atmosphere. Nitrogen also is important in the makeup of organic molecules, especially proteins, and therefore is essential to living processes. A simplified view of the nitrogen cycle is portrayed in Figure 19.11.

This vast atmospheric reservoir of nitrogen is not directly accessible to most organisms. *Nitrogen-fixing bacteria* provide the key link to life. They live principally in the soil and are associated with the roots of certain plants—for example, the *legumes* such as clover, alfalfa, soybeans, peas, beans, and peanuts. Colonies of these bacteria reside in nodules on the legume roots and chemically combine the nitrogen from the air in the form of nitrates (NO_3) and ammonia (NH_3). Plants use these chemically bound forms of nitrogen to produce their own organic matter. Anyone or anything feeding on the plants thus ingests the nitrogen. Finally, the nitrogen in the organic wastes (sewage, manure) of these consuming organisms is freed by

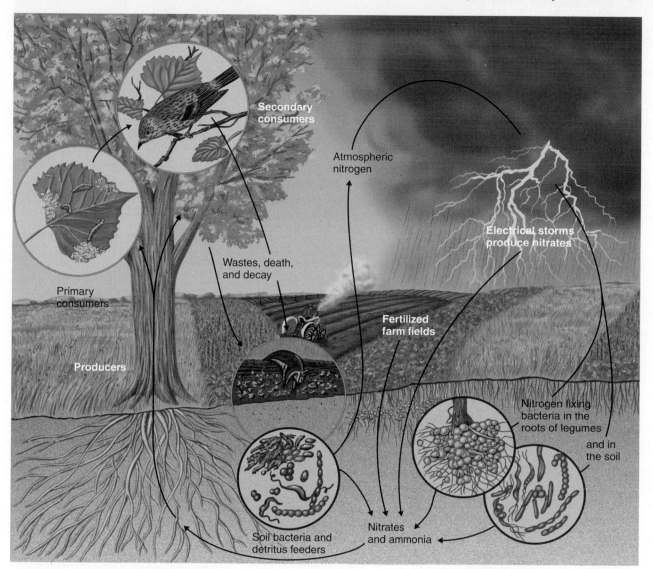

FIGURE 19.11 The nitrogen cycle.
The atmosphere is the essential reservoir of gaseous nitrogen. Atmospheric nitrogen gas is chemically fixed by bacteria to produce ammonia. Lightning and forest fires produce nitrates, and fossil-fuel combustion forms nitrogen compounds that are washed from the atmosphere by precipitation. Plants absorb nitrogen compounds and produce organic material. [From T. Audesirk and G. Audesirk, *Biology: Life on Earth*, 4th ed., Figure 45-10, p. 903. © 1996 Prentice Hall Inc. Used by permission.]

ANIMATION   The Nitrogen Cycle

denitrifying bacteria, which recycle nitrogen back to the atmosphere.

To improve agricultural yields, many farmers use synthetic inorganic fertilizers, as opposed to soil-building organic fertilizers (manure and compost). Inorganic fertilizers are chemically produced through artificial nitrogen fixation at factories using the Haber–Bosch process developed in 1913, a process that takes inert nitrogen gas and hydrogen to form ammonia (NH_3) or the ammonium ion NH_4^+, a reactive form of nitrogen that plants can use. The annual production of inorganic fertilizers far exceeds the ability of natural denitrification systems—and the present production of synthetic fertilizers now is doubling every 8 years; some 1.82 million metric tons (2 million tons) are produced per week, worldwide. Humans presently fix more nitrogen

per year than all terrestrial sources combined. The crossover point when anthropogenic sources of fixed nitrogen exceeded the normal range of naturally fixed nitrogen was 1970.

This surplus of usable nitrogen accumulates in Earth's ecosystems. Some is present as excess nutrients, washed from soil into waterways and eventually to the ocean. This excess nitrogen load begins a water pollution process that feeds an excessive growth of algae and phytoplankton, increases biochemical oxygen demand, diminishes dissolved oxygen reserves, and eventually disrupts the aquatic ecosystem. In addition, excess nitrogen compounds in air pollution are a component in acid deposition, further altering the nitrogen cycle in soils and waterways. For more on this problem, specific to North America, read News Report 19.2.

News Report 19.2

The Dead Zone

The flush of agricultural runoff and fertilizers, animal manure, sewage, and other wastes is carried by the Mississippi River to the Gulf, causing huge spring blooms of phytoplankton. By summer, the biological oxygen demand of bacteria feeding on the decay exceeds the dissolved oxygen, and hypoxia develops, killing any fish that ventures into the area. These low-oxygen conditions act as a limiting factor on marine life. Figure 19.2.1 shows the dead zone, a region of oxygen-depleted (hypoxic) water off the coast of Louisiana in the Gulf of Mexico.

The Mississippi drainage system handles the runoff for 41% of the continental United States. When the 1993 Midwest floods hit, the river's nutrient discharge was such that the size of the dead zone doubled to 17,500 km² (6755 mi²). From 2002 on, it has expanded to more than 22,000 km²

(8500 mi²). The agricultural, feedlot, and fertilizer industries dispute the connection between their nutrient input and the dead zone. Following Hurricane Katrina in 2005, the dead-zone region was overwhelmed by the discharge carrying all the sewage, chemical toxins, oil spills, household pesticides, animal wastes and carcasses, and nutrients. Full disclosure of the contents of the "muck" has yet to be made and total volume will never be known. The temporary plume from this event tracked around the Florida peninsula and northward up the Eastern seaboard (see suggested plot in Figure 19.2.1)

In other parts of the world, the connection is established. In Sweden and Denmark, a concerted effort to reduce nutrient flows into their rivers reversed hypoxic conditions in the Kattegat Strait (between the Baltic and North Seas). And, with the fall of

the U.S.S.R. and state agriculture in 1990, fertilizer use is down more than 50%, and for the first time in recent decades the Black Sea is getting several months' break from year-around hypoxia from river discharge inlets.

As with most environmental situations, the cost of mitigation is cheaper than the cost of continued damage to marine ecosystems. A cut in nitrogen inflow upstream of 20% to 30% is estimated to increase dissolved oxygen levels by more than 50% in the dead zone region of the Gulf. One government estimate placed the application of excess nitrogen fertilizer at 20% more than needed in Iowa, Illinois, and Indiana. The initial step to resolving this issue might be to mandate applying only the levels of fertilizer needed—a savings in overhead costs for agricultural interests—and, secondly, action to begin dealing with animal wastes in the basin.

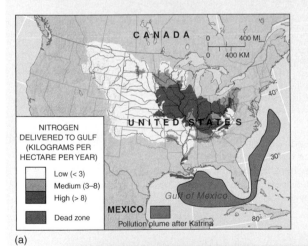

(a)

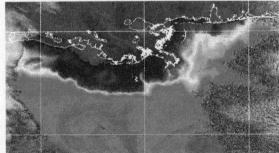

(b)

(c)

FIGURE 19.2.1 The Gulf Coast dead zone. (a) Mapped total nitrogen originating in the upstream watershed that feeds the Mississippi River system. (b) The flush of nutrients from the Mississippi River drainage basin enriches the offshore waters in the Gulf of Mexico, causing huge phytoplankton blooms in early spring (red areas). (c) By late summer, oxygen-depleted waters (hypoxic) dominate what is known as the *dead zone*. [(a) Adapted from R. B. Alexander et al., "Effect of stream channel size on the delivery of nitrogen to the Gulf of Mexico," *Nature* (February 17, 2000): 761; (b) *SeaWiFS* image from NASA/GSFC, courtesy of GeoEye, Dulles, VA.]

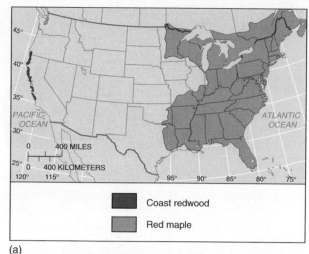

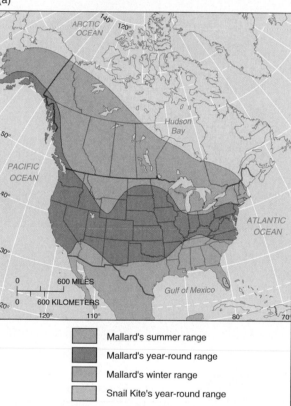

FIGURE 19.12 Limiting factors affect the distribution of every plant and animal species.
(a) Distributions of coast redwood and red maple demonstrate the effect of limiting factors, (b) as do the distributions of the Mallard Duck and the Snail Kite across North America. The Mallard is a generalist and feeds widely. In contrast, the range of its food, a single type of snail, limits the Snail Kite.

Limiting Factors

The term **limiting factor** identifies the one physical or chemical abiotic component that most inhibits biotic operations, through either its lack or its excess. Here are a few examples:

- Low temperatures limit plant growth at high elevations.

- Lack of water limits growth in a desert.
- Excess water limits growth in a bog.
- Changes in salinity levels affect aquatic ecosystems.
- Lack of iron in ocean surface environments limits photosynthetic production.
- Low phosphorus content of soils limits plant growth.
- General lack of active chlorophyll above 6100 m (20,000 ft) limits primary productivity.

In most ecosystems, precipitation is the limiting factor. However, temperature, light levels, and soil nutrients certainly affect vegetation patterns.

Each organism possesses a range of tolerance for each limiting factor in its environment. This fact is illustrated in Figure 19.12, which shows the geographic range for two tree species and two bird species. The coast redwood (*Sequoia sempervirens*) is limited to a narrow section of the Coast Ranges in California, covering barely 9500 km² (less than 4000 mi²), concentrated in areas that receive necessary summer advection fog. On the other hand, the red maple (*Acer rubrum*) thrives over a large area under varying conditions of moisture and temperature, thus demonstrating a broader tolerance to environmental variations.

The Mallard Duck (*Anas platyrhynchos*) and the Snail Kite (*Rostrhamus sociabilis*) also demonstrate a variation in tolerance and range (Figure 19.12b). The Mallard, a generalist, feeds from widely diverse sources, is easily domesticated, and is found throughout most of North America in at least one season of the year. By contrast, the Snail Kite is a specialist that feeds only on one specific type of snail. This single food source, then, is its limiting factor. Note its small habitat area near Lake Okeechobee in Florida.

Biotic Ecosystem Operations

The abiotic components of energy, atmosphere, water, weather, climate, and minerals make up the life support for the biotic components of each ecosystem. The flow of energy, cycling of nutrients, and trophic (feeding) relations determine the nature of an ecosystem. As energy cascades through this process-flow system, it is constantly replenished by the Sun. But nutrients and minerals cannot be replenished from an external source, so they constantly cycle within each ecosystem and through the biosphere. Let us examine these biotic operations.

Producers, Consumers, and Decomposers

Organisms that are capable of using carbon dioxide as their sole source of carbon are *autotrophs* (self-feeders), or **producers**. These are the plants. They chemically fix carbon through photosynthesis. Organisms that depend on producers as their carbon source are *heterotrophs* (feed on others), or **consumers**. Generally, these are animals (Figure 19.13).

Autotrophs are the essential producers in an ecosystem—capturing light energy and converting it to

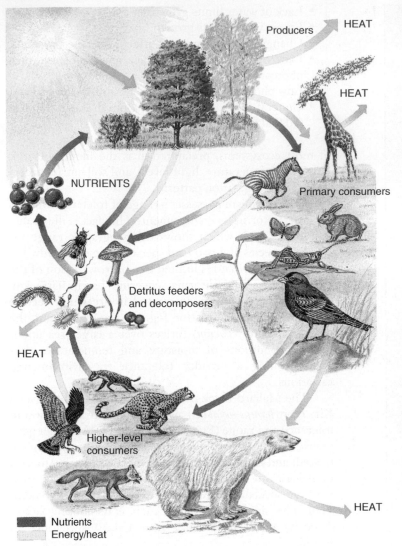

NUTRIENTS

Producers

HEAT

HEAT

Primary consumers

Detritus feeders and decomposers

HEAT

Higher-level consumers

HEAT

■ Nutrients
■ Energy/heat

(a)

(b)

(c)

 ANIMATION Building a Food Web

FIGURE 19.13 Energy, nutrient, and food pathways in the environment.
(a) The flow of energy, cycling of nutrients, and trophic (feeding) relationships portrayed for a generalized ecosystem. The operation is fueled by radiant energy supplied by sunlight and first captured by the plants. (b) A bearded seal on a bergy bit in the Arctic Ocean. (c) A solitary male polar bear drags its seal feast across pack ice. (d) A mother polar bear and cubs eat a meal on an iceberg after mom caught a seal; birds work on the leftovers. [From T. Audesirk and G. Audesirk, *Biology: Life on Earth*, 4th ed., Figure 45-1, p. 892. © 1996 Prentice Hall Inc. Used by permission. Photos (b), (c), and (d) by Bobbé Christopherson.]

chemical energy, incorporating carbon, forming new plant tissue and biomass, and freeing oxygen. Solar energy enters each food chain through the producer plant or producer phytoplankton, subsequently flowing through higher and higher levels of consumers.

From the producers, which manufacture their own food, energy flows through the system along a circuit called the **food chain**, reaching consumers and eventually *detritivores.* Organisms that share the same basic foods are said to be at the same *trophic level.* Ecosystems generally are structured in a **food web**, a complex network of interconnected food chains. In a food web, consumers participate in several different food chains, comprising both *strong interactions* and *weak interactions* between species in the food web.

In Figure 19.13b a bearded seal (*Erignathus barbatus*), with body fat exceeding 30%, rests on a bergy bit, or small iceberg, in Arctic waters where it feeds on fish and clams. Figure 19.13c shows a male polar bear (*Ursus maritimus*) hauling its bearded seal kill onto the ice. Polar bears are the dominant Arctic predator, a marine mammal. In 19.13d, on the pack ice many kilometers from land we see a mother and her 6-month-old cubs-of-the-year feeding on a seal catch. Even though the cubs still nurse they share the kill with mom. The polar bear consumes most of the seal except for the bones and intestines, which are quickly eaten by birds scavenging the leftovers. Note the Glaucous Gulls and Ivory Gulls in the photo.

Figure 19.14 illustrates simple terrestrial and oceanic food chains. Primary consumers feed on producers. Because producers are always plants, the primary consumer is an **herbivore**, or plant eater, like the krill. A **carnivore** is a secondary consumer and primarily eats meat, as with the leopard seal and polar bear. A tertiary consumer eats primary and secondary consumers and is referred to as the "top carnivore" in the food chain. A consumer that feeds on both producers (plants) and consumers (meat) is an **omnivore**—a category occupied by humans, among others.

In the Arctic region, the walrus (*Odobenus rosmarus*) has a special diet and adaptation to obtain it (Figure 19.14c). They feed on clams in the shallows near land, brushing aside sand and sediment with their whiskers, pulling the clam into their mouths, and, with a powerful sucking action, pulling out the meat. Walrus will occasionally eat squid and fish if needed to sustain their huge calorie requirement. In the photo a female walrus places a protective flipper on her pup on a precarious piece of pack ice.

Detritivores (detritus feeders and decomposers) are the final link in the endless chain. Detritivores renew the entire system by releasing simple inorganic compounds and nutrients with the breaking down of organic materials. Detritus refers to all the dead organic debris—remains, fallen leaves, and wastes—that living processes leave. Detritus feeders—worms, mites, termites, centipedes, snails, crabs, and even vultures, among others—work like an army to consume detritus and excrete nutrients that fuel an ecosystem. **Decomposers** are primarily bacteria and fungi that digest organic debris outside their bodies and absorb and release nutrients in the process. This metabolic work of microbial

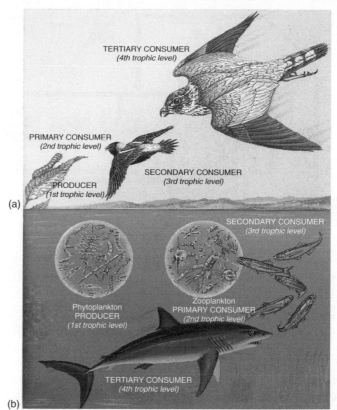

(a)

(b)

(c)

FIGURE 19.14 Food chains.
(a) A simplified terrestrial food chain, from leaf to caterpillar to Bobolink to a fast-moving Merlin Hawk. (b) A simplified aquatic food chain from phytoplankton to zooplankton to schooling fish and to a mako shark. (c) A female walrus and her pup rest on the pack ice. [(a) and (b) From T. Audesirk and G. Audesirk, *Biology: Life on Earth*, 4th ed., Figure 45-4, p. 895. © 1996 Prentice Hall Inc. Used by permission. Photo (c) by Bobbé Christopherson]

ANIMATION   Building a Food Web

decomposers produces the rotting that breaks down detritus. Detritus feeders and decomposers, although different in operation, have a similar function in an ecosystem.

Examples of Complex Food Webs

An example of a complex community is the oceanic food web that includes krill, a primary consumer (Figure 19.15a). Krill (*Euphausia sp.*, at least a dozen species) is a shrimplike

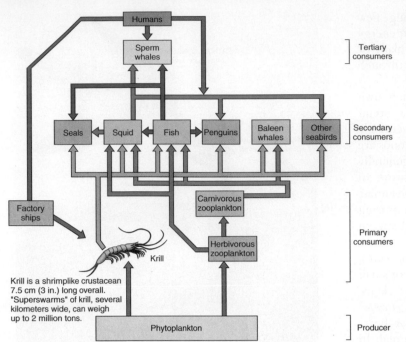

Building a Food Web

Midlatitude Productivity

FIGURE 19.15 A complex food web. The food web in Antarctic waters, from phytoplankton producers (bottom) through various consumers. Phytoplankton began this chain using photosynthesis to store solar energy. Krill and other herbivorous zooplankton eat the phytoplankton. Krill in turn are consumed by the next trophic level.

crustacean that is a major food for an interrelated group of organisms, including whales, fish, seabirds, seals, and squid in the Antarctic region. All of these organisms participate in numerous other food chains as well, some consuming and some being consumed.

Phytoplankton began this chain by harvesting solar energy in photosynthesis. *Herbivorous zooplankton* such as krill and other organisms eat phytoplankton. Consumers eat krill at the next trophic level. Because krill are a protein-rich, plentiful food, increasingly factory ships, such as those from Japan and Russia, seek them out. The annual krill harvest currently surpasses a million tons, principally as feed for chickens and livestock and as protein for human consumption. Many of the Antarctic-dwelling birds depend on krill and on fish that eat krill.

The impact on the food web of further increases in the krill harvest is uncertain, for little is known about the krill reproduction rate. Their long life span (greater than 8 years) and slow growth rate limit the ultimate extent of the resource. In addition, the possible effect on krill of increased ultraviolet radiation resulting from the seasonal "hole" in the ozone layer above Antarctica is under investigation.

Another example of a complex food web comes from eastern North America and a temperate deciduous forest. As with the krill food web, the diagram is simplified from the actual complexity of nature (Figure 19.16). From our discussion find the primary producers and then locate the primary, secondary, and tertiary consumers in this community. Do you find the detritivores (worms are shown) playing a role as well? Bacteria act as decomposers in the system; do you find any other decomposers in the illustration?

Efficiency in a Food Web

Any assessment of world food resources depends on the level of consumer being targeted. Let us use humans as an example. Many people can be fed if wheat is eaten

directly. However, if the grain is first fed to cattle (herbivores) and then we eat the beef, the yield of available food energy is cut by 90% (810 kg of grain is reduced to 82 kg of meat); far fewer people can be fed from the same land area (Figure 19.17).

In terms of energy, only about 10% of the kilocalories (food Calories, not heat calories) in plant matter survives from the primary to the secondary trophic level. When humans consume meat instead of grain, there is a further loss of biomass and added inefficiency. More energy is lost to the environment at each progressive step in the food chain. You can see that an omnivorous diet such as that of an average North American and European is quite expensive in terms of biomass and energy.

Food web concepts are becoming politicized as world food issues grow more critical. Today, approximately half of the cultivated acreage in the United States and Canada is planted for animal consumption—beef and dairy cattle, hogs, chickens, and turkeys. Livestock feed includes approximately 80% of the annual corn and nonexported soybean harvest. In addition, some lands cleared of rain forest in Central and South America were converted to pasture to produce beef for export to restaurants, stores, and fast-food outlets in developed countries. Thus, lifestyle decisions and dietary patterns in North America and Europe are perpetuating inefficient food webs, not to mention the destruction of valuable resources, both here and overseas.

Ecological Relations

A study of food webs is a study of who eats what and where they do the eating. Figure 19.18 charts the summertime distribution of populations in two ecosystems, grassland and temperate forest. The stepped population pyramid is characteristic of summer conditions in such ecosystems.

FIGURE 19.16 Temperate forest food webs.
A simplified illustration of food webs in a temperate forest. Examine the art and identify primary producers, and primary, secondary, and tertiary consumers. Featured are strong interactions, but many weak interactions not illustrated hold such natural communities together, affirming the hypothesis that complexity is stabilizing and not weakening the community.
[Jay Withgott and Scott Brennan, *Environment, The Science Behind the Stories*, San Francisco: Benjamin Cummings, © 2007, Fig. 6.11, p. 159. Used by permission.]

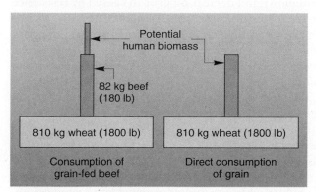

FIGURE 19.17 Efficiency and inefficiency in biomass consumption.
Biomass pyramids illustrate a great difference in efficiency between direct and indirect consumption of grain.

You can see the decreasing number of organisms supported at successively higher trophic levels. The base of the temperate forest pyramid is narrow, however, because most of the producers are large, highly productive trees and shrubs, which are outnumbered by the consumers they can support.

One approach to ecological study is to analyze a community's metabolism. *Metabolism* is the way in which a community uses energy and produces food for continued operation (the sum of all its chemical processes). In 1957, Howard T. Odum completed one of the best-known studies of this type for Silver Springs, Florida.

Figure 19.19a, an adaptation of his work, shows that the energy source for the community is insolation. The amount of light absorbed by the plants for photosynthesis

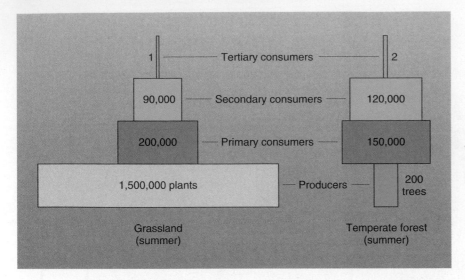

FIGURE 19.18 Population distribution in two ecosystems.
Ecological pyramids for 0.1 hectare (0.25 acre) of land show the difference in producers and consumers between grassland and forest in the summer. [From E. P. Odum, *Fundamentals of Ecology*, 3rd ed., with permission of Brooks/Cole, an imprint of the Wadsworth Group, Thomson Learning.]

is 24% of the total amount arriving at the study site. Of the absorbed amount, only 5% is transformed into the community's gross photosynthetic production—and this is only 1.2% of the total insolation input. The net plant production (gross production minus respiration) is 42.4% of the gross production, or 8833 kilocalories per square meter per year. That amount then moves through the food chain of herbivores and carnivores, with the top carnivores receiving only 0.24% of the net plant production of the system.

You can see that, at each trophic level, some of the biomass flows to the decomposers. The largest portion of the biomass is exported from the system in the form of community respiration. In addition, a relatively small portion leaves the system as organic particles by downstream export out of this system. The biomass pyramid shown in Figure 19.19b portrays the same community from the viewpoint of a *standing crop* (all the biomass in a particular environment at one time), which is another way of examining the existing biomass at each trophic level.

Concentration of Pollution in Food Chains

When chemical pesticides are applied to an ecosystem of producers and consumers, the food web concentrates some of these chemicals. Many chemicals are degraded or diluted in air and water and thus are rendered relatively harmless. Other chemicals, however, are long-lived, stable, and soluble in the fatty tissues of consumers. They become increasingly concentrated at each higher trophic level. Figure 19.20 illustrates this *biological amplification*, or *magnification*.

DDT is one such chemical. As an insecticide, it saved millions of human lives in the tropics by killing the mosquitoes that spread malaria. Crop-destroying insects also were brought under control. However, the persistence of the chemical in the environment and its destruction of established food webs brought many

unwanted consequences—death and illness to animals and humans alike. Although banned in some countries (the United States banned it in 1973), it still is widely used in less-developed countries. In addition, some other organic and synthetic chemicals, radioactive debris, and heavy metals such as lead and mercury become concentrated in food webs.

Thus, pollution in a food web can efficiently poison the organism at the top. Many species are threatened in this manner, and, of course, humans are at the top of many food chains and can ingest concentrated chemicals in what is consumed, amplified in this way.

Ecosystems, Evolution, and Succession

Far from being static, Earth's ecosystems have been dynamic (vigorous and energetic) and ever-changing from the beginning of life on the planet. Over time, communities of plants and animals have adapted, evolved, and in turn shaped their environments to produce great diversity. A constant interplay exists between increasing growth in a community and decreasing growth caused by resistance factors that create limits and disruptions (Figure 19.21). Each ecosystem is constantly adjusting to changing conditions and disturbances in the struggle to survive. The concept of change is key to understanding ecosystem stability.

The nature of change in natural ecosystems can range from gradual transitions from one equilibrium to another, balancing the factors in Figure 19.20, to extreme catastrophes such as an asteroid impact or severe volcanism, among other events. In the evolution of life, or the physical and chemical evolution of Earth's systems, such sudden changes created a punctuated equilibrium. Following each of the five major extinctions of life over the past 450 million years, there were great bursts

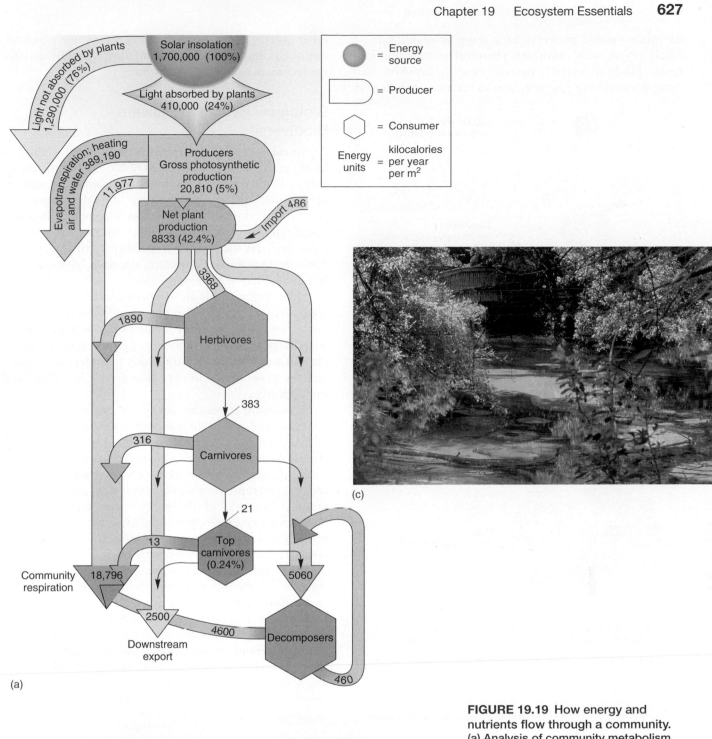

(a)

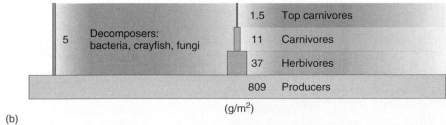

(b)

(c)

FIGURE 19.19 How energy and nutrients flow through a community. (a) Analysis of community metabolism. **(b)** Standing crop distribution in grams per square meter for Silver Springs, Florida. **(c)** A similar nearby swampy forest ecosystem. [Adapted from E. P. Odum, *Fundamentals of Ecology*, 3rd ed., with permission of Brooks/Cole, an imprint of the Wadsworth Group, Thomson Learning. (c) Photo by Bobbé Christopherson.]

of diversity and proliferation of new and different species, a process known as *speciation*. Although the extinctions (noted in Chapter 11, Figure 11.1) temporarily limited biodiversity, each created evolutionary

opportunities into which life diversified into differentiated populations.

For most of the last century, scientists thought that an undisturbed ecosystem—whether forest, grassland, aquatic, or other—would progress to a stage of equilibrium, a stable point, with maximum chemical storage and biomass. Modern research has determined, however, that ecosystems do not progress to some static conclusion. In other words, there is no "balance of nature"; instead, think of a constant interplay of physical, chemical, and living factors in a dynamic equilibrium.

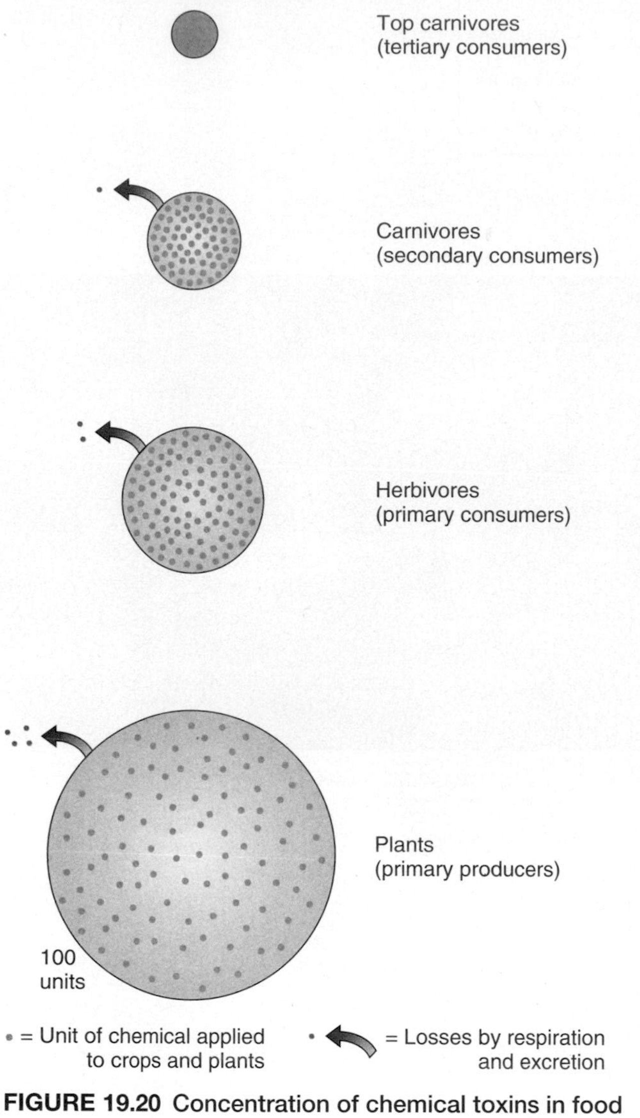

Biological Evolution Delivers Biodiversity

A critical aspect of ecosystem stability and vitality is **biodiversity**, or species richness of life (a combination of *bio*logical and *diversity*). Biodiversity includes the number of different species and quantity of each species, the genetic diversity in species (number of genetic characteristics), and ecosystem and habitat diversity. A working estimate places the number of species presently populating Earth at 13.6 million; however, scientists have classified less than 2 million of these.

The origin of this diversity is embodied in the *theory of evolution*. (The definition of a *scientific theory* is given in Focus Study 1.1.) A theory is based on several extensively tested hypotheses and represents a broad general principle and a powerful tool with which to understand nature and the general laws in operation across the planet and the Universe.

Evolution states that single-cell organisms adapted, modified, and passed along inherited changes to multicellular organisms. The genetic makeup of successive generations was shaped by environmental factors, physiological functions, and behaviors that created a greater rate of survival and reproduction, traits that were passed along—literally, "survival of the species that reproduces the most." A species refers to a population that can reproduce sexually and produce viable offspring. By definition, this means reproductive isolation from other species.

Each inherited trait is guided by a *gene* that is a segment of the primary genetic material called *DNA* (deoxyribonucleic acid) that resides in the nest of chromosomes in every cell nucleus. These traits are passed to successive generations, especially those traits that are most successful at maximizing niches different from those of other species or traits that help the species adapt to environmental changes. Such differential reproduction and adaptation passes along successful genes or groups of genes in a process of genetic favoritism

Top carnivores
(tertiary consumers)

Carnivores
(secondary consumers)

Herbivores
(primary consumers)

Plants
(primary producers)

100 units

• = Unit of chemical applied to crops and plants

↖ = Losses by respiration and excretion

FIGURE 19.20 Concentration of chemical toxins in food chains.
Chemical residues are passed along a simple food chain to the top carnivores, where the accumulation becomes concentrated. This phenomenon is *biological magnification.*

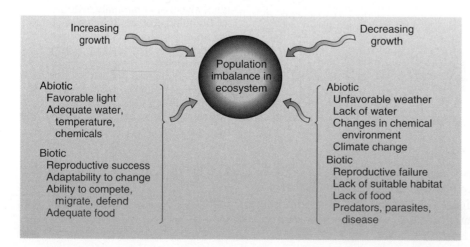

Increasing growth

Decreasing growth

Population imbalance in ecosystem

Abiotic
 Favorable light
 Adequate water, temperature, chemicals

Biotic
 Reproductive success
 Adaptability to change
 Ability to compete, migrate, defend
 Adequate food

Abiotic
 Unfavorable weather
 Lack of water
 Changes in chemical environment
 Climate change
Biotic
 Reproductive failure
 Lack of suitable habitat
 Lack of food
 Predators, parasites, disease

FIGURE 19.21 Factors controlling the population balance of an ecosystem.
Potential growth factors and resistance factors ultimately determine the population of an ecosystem.

called *natural selection*. The process of evolution follows generation after generation and tracks the passage of inherited characteristics that were successful—the failures passed into extinction. Thus, it is not an understatement to say that today's humans are the result of billions of years of positive change and affirmative natural selection.

New mixes in the *gene pool*, or all genes possessed by individuals in a given population, involve *mutations*. Mutation is a process that occurs when a random action, or perhaps an error as the DNA reproduces, forms new genetic material and inserts new traits into the inherited stream. Geography comes into play, as spatial effects are important. A species may disperse through migration to different environments or may be separated by natural *vicariance* (fragmentation of the environment) events that establish natural barriers. Such isolation allows natural selection to evolve new sets of genes. Imagine the migrations across ice bridges or land connections at times of low sea level, or the slowly evolving rise of a mountain range, or the formation of an isthmus such as Panama connecting continents and separating oceans. To restate, the physical and chemical evolution of Earth's systems is closely linked to the biological evolution of life.

The deciphering of complete species *genomes*, or the total genetic identity of an organism, is a scientific triumph and is disclosing a complex history and geography of each species—where species originated, how species are dispersed, what diseases and traumas were of consequence. Analysis of the human genome has indicated our origins and provided a road map of how we got to where we are. Thus, the intricate interplay that led to present biodiversity across myriad ecosystems, both terrestrial and aquatic, is here in Chapters 19 and 20.

All life falls into one of five major groups, called *kingdoms*. Focus Study 20.1 and Table 20.1 in the next chapter provide a specific count of the tremendous diversity of life within the five kingdoms. These kingdoms are classified based on source of nutrition, cellular structure (lacking a nucleus–the *prokaryotes*, or having a membrane-enclosed nucleus–the *eukaryotes*), and numbers of cells. The five kingdoms of life are:

- *Monera Kingdom:* Unicellular prokaryotes, such as bacteria; absorbing food from their environment, some with the capability to photosynthesize (e.g., cyanobacteria).
- *Protista Kingdom:* Usually unicellular eukaryotes, protozoans, such as amoebas; includes slime molds, some with photosynthesizing ability, and some mobile members with simple cilia for propulsion.
- *Fungi Kingdom:* Multicellular eukaryotes, fungi, mushrooms, bread molds, with specialized cells for reproduction and feeding.
- *Plant Kingdom:* Multicellular eukaryotes, dominated by flowering plants including trees, not mobile; photosynthesis for food production, interrelationships with animals for nutrient supplies and pollination; mosses,

liverworts, ferns, coniferous trees, and food plants, among others.
- *Animal Kingdom:* Multicellular eukaryotes, ingest food, many specialized organs and tissues, specialized cells, sensory ability and mobile; includes vertebrates, worms, mollusks, among other types; largest group of animals are invertebrates such as anemones.

Ecosystem Stability and Diversity

In an ecosystem the tendency of species populations to remain stable (*inertial stability*) does not necessarily foster the ability to recover from change (*resilience*). Several of the dramatic asteroid impact episodes that triggered partial extinctions over the past 440 million years tested communities of plants and animals and their resilience. Think of resilience as an ecosystem's ability to recover from disturbance, its ability to absorb disturbance up to some threshold point before it snaps to a different set of relations, thus altering some logical ecological progression.

Examples of communities with high inertial stability include a redwood forest, a pine forest at a high elevation, and a tropical rain forest near the equator. But, cleared forest tracts recover slowly and therefore have poor resilience. Figure 19.22 shows how clear-cut tracts of former forest have drastically altered microclimatic conditions, making regrowth of the same species difficult. Full exposure to sunlight and wind will make recovery difficult. In contrast, a midlatitude grassland is low in stability; yet, when burned, its resilience is high because the community recovers rapidly.

In the context of ecosystem stability and resilience, consider the elimination of a full section of forest ecosystem, shown in Figure 19.22, on private land in southern Oregon, south-southeast of Crater Lake. Note the high road density, the inadequate buffers along watercourses to prevent sedimentation of streams, the drag lines and skid trails to each pickup spot that pockmark the landscape. On nearby U.S. Forest Service land, the rules are clear: a full 95-m (312-ft) tree buffer is left on either side of a stream, road densities are minimized, and harvesting regimes include partial removals. Scientists have concluded that for sustainable forests, the sole objective should never be the complete removal of timber.

Greater biodiversity in an ecosystem results in greater stability and greater productivity. In other words, the strength of **biodiversity** is that it spreads risk over the entire community, because several food sources exist at each trophic level. Therefore, biodiversity matters. Consider more than just the diversity of species; rather, think of all the positive interactions of species embodied in the *facilitation concept*. Facilitation describes how a diverse community manages nutrient dynamics among the populations, and pest resistance, and resilience to events such as fire. News Report 19.3 presents exciting confirmation of biodiversity.

The present loss of biodiversity is irreversible, and yet, here we are in the midst of the biosphere's sixth major

(a)

(b)

FIGURE 19.22 Poor practices disrupt stable forest communities.
(a) An example of clear-cut timber harvesting that has devastated a stable forest community and produced drastic changes in microclimatic conditions. (b) A logged pine forest and disruptive logging road. About 10% of the Northwest's old-growth forests remain, as identified through satellite images and GIS analysis. [Photos (a) by Bobbé Christopherson; (b) by author.]

extinction episode. Extinction is final no matter how complex the organism or how long it lived since it evolved. E. O. Wilson, the famed biologist, stated:

> Biological diversity—the full sweep from ecosystems to species within ecosystems, thence to genes within species—is in trouble. Mass extinctions are commonplace, especially in tropical regions where most of the biodiversity occurs. Among the more recent are more than half the exclusively freshwater fishes of peninsular Asia, half of the fourteen birds of the Philippines island of Cebu, and more than ninety plant species growing on a single mountain ridge in Ecuador. In the United States an estimated 1 percent of all species have been extinguished; another 32 percent are imperiled.*

Agricultural Ecosystems   Humans simplify communities by eliminating biodiversity, and in this way we place more ecosystems at risk of harmful change and perhaps failure. An artificially produced monoculture community, such as a field of wheat, is singularly vulnerable to failure owing to weather or attack from insects or plant disease. In some regions, simply planting multiple crops brings more stability to the ecosystem. This is an important principle of sustainable agriculture, a movement to make farming more ecologically compatible.

A modern agricultural ecosystem is vulnerable to failure not only because of its lack of ecological diversity but also because it creates an enormous demand for energy, chemical pesticides and herbicides, artificial fertilizer, and

*E. O. Wilson, *Consilience, The Unity of Knowledge* (New York: Knopf, 1998), p. 292.

irrigation water. The practice of harvesting and removing biomass from the land interrupts the cycling of materials into the soil (Figure 19.23). This net loss of nutrients must be artificially replenished. The type of plantation agriculture in Figure 19.23b is depleting soil nutrients over time; this farming practice requires subsidies of chemical fertilizers and herbicides to eliminate competing vegetation.

Climate Change   Obviously, the distribution of plant species is affected by climate change. Many species have survived wide climate swings in the past. Consider the beginning of the Tertiary Period, 75 million years ago. Warm, humid conditions and tropical forests dominated the land as far north as southern Canada, pines grew in the Arctic, and deserts were few. Then, between 15 and 50 million years ago, deserts began developing in the southwestern United States. Mountain-building processes produced higher elevations, creating rain-shadow aridity that affected plant distribution.

Recall, too, that the movement of Earth's tectonic plates created climate changes important in the evolution and distribution of plants and animals (see Figure 11.16a–e). Europe and North America were joined in Pangaea and positioned near the equator, where vast swamps formed (the site of coal deposits today). The southern mass of Gondwana was extensively glaciated as it drifted at high latitudes in the Southern Hemisphere. Glaciation left matching glacial scars and specific plant and animal distributions across South America, Africa, India, and Australia. The diverse and majestic dinosaurs dispersed with the drifting continents.

The key question for our future is: As temperature patterns change, how fast can plants either adapt to new

News Report 19.3

Experimental Prairies Confirm the Importance of Biodiversity

Little of the original midlatitude prairies remains. Only a few widespread areas of unaltered prairie are protected. Iowa was once 80% prairie, and now 99.9% of the prairie is gone, converted to other uses. One such plot of 240 acres (97 ha) is in Howard County, Iowa, south of the Minnesota border. The Hayden Prairie State Preserve sits quietly, almost missed by passing cars as they kick up pale dust from the gravel county road (Figure 19.3.1). Grasses and shooting star flowers (*Dodecatheon sp.*), and a blocked former road and trails, mark this pristine place in Iowa.

Field experiments have confirmed an important scientific assumption: that greater biological diversity in an ecosystem leads to greater stability, productivity, and soil nutrient use within that ecosystem. For instance, during a drought, some species of plants will be damaged. In a diverse ecosystem, however, other species with deeper roots and better water-obtaining ability will thrive.

Exciting research on natural prairie ecosystems and biodiversity, the role of fire, and climate-change impacts is ongoing at Konza Prairie LTER (Long-Term Ecological Research) in the Flint Hills of northeastern Kansas. Kansas State University and the Nature Conservancy operate this native tall-grass prairie research station (see **http://climate.konza. ksu.edu/**).

Ecologist David Tilman at the University of Minnesota tracked the operation of 147 grassland plots, each 3 m (9.8 ft) square. Species diversity was carefully controlled on each plot (Figure 19.3.2). Plots were sown with different numbers of native North American prairie plant seeds and cared for by a team of 50 people. The results demonstrated that the plots with a more diverse plant community were able to retain and use nutrients more efficiently than the plots with less diversity. This efficiency reduced the loss of soil nitrogen through leaching, thus increasing soil fertility.

Biodiversity improved the nutrient dynamic in the prairie ecosystem.

Greater plant diversity led to both higher productivity and better resource utilization. Also, total plant cover was found to increase with species richness. Tilman and his colleagues affirm:

This extends the earlier results to the field, providing direct evidence that the current rapid loss of species on Earth, and management practices that decrease local biodiversity, threaten ecosystem productivity and the sustainability of nutrient cycling. Observational, laboratory, and now field experimental evidence supports the hypothesis that biodiversity influences ecosystem productivity, sustainability, and stability.*

*D. Tilman, D. Wedin, and J. Knops, "Productivity and sustainability influenced by biodiversity in grassland ecosystems," *Nature* 379 (February 22, 1996): 720. Also see the summary in *Science* 271 (March 15, 1996): 1497.

FIGURE 19.3.1 One of the few remaining tracts of prairie.
The Hayden Prairie State Preserve in Howard County, Iowa, is 240 acres of tallgrass prairie named after Dr. Ada Hayden, a former botany professor at Iowa State University. [Photo by Bobbé Christopherson.]

FIGURE 19.3.2 Biodiversity experiment.
Experimental plots are planted with native North American prairie grasses. A total of 147 plots were sown with random seed selections of 1, 2, 4, 6, 8, 12, or 24 species. Experimental results confirm many assumptions regarding the value of rich biodiversity. [Photo by David Tilman/University of Minnesota, Twin Cities Campus.]

(a)

(b)

(c)

(d)

FIGURE 19.23 Agricultural ecosystems.
(a) The Sacramento Valley of California is one of the most intensively farmed regions of the world. Field crops including irrigated rice, an irrigation canal, and the Sacramento River are in the scene. (b) A pine tree plantation (farm) in Georgia, principally for the production of wood pulp—the forest ecosystem is reduced to efficient rows for clear-cutting when mature. (c) Modern mechanized farming and a freshly mowed field of oats near Dokka, Norway. (d) Highly fragmented, fertile agricultural land along the north German plain. [All photos by Bobbé Christopherson.]

conditions or migrate through succession (location change) to remain within their shifting specific habitats? Satellite observations indicate rapid changes in vegetation patterns in northern latitudes since 1980. Global warming is reducing spring snow cover, allowing spring greening to occur earlier, up to 3 weeks sooner in some high-latitude locations. Likewise, the onset of fall is delayed to a later date than previously experienced. Nina Leopold Bradley is studying the Wisconsin setting her father, Aldo Leopold, made famous in his book *A Sand County Almanac* (Oxford University Press, 1949). He kept detailed records of all things dealing with nature and seasonal change on his tract of land. Her findings show that more than one-third of animal and plant species start spring activities 4 weeks earlier than they did 50 years ago.

Current global change is occurring rapidly, at the rate of decades instead of millions of years. Thus, we see die-out and succession of different species along disadvantageous habitat margins. The displaced species may colonize new regions made more hospitable by climate change. Also shifting will be agricultural lands that produced wheat, corn, soybeans, and other commodities. Society will have to adapt to different crops growing in different places.

A study completed by biologist Margaret Davis on North American forests suggests that trees will have to respond quickly as temperatures increase. Changes in the climate inhabited by certain species could shift 100–400 km (60–250 mi) during the next 100 years. Some species, such as the sugar maple, may migrate northward, disappearing from the United States except for Maine, and moving into eastern Ontario and Québec. Davis prepared a map showing the possible impact of increasing temperatures on the distribution of beech and hemlock trees (Figure 19.24).

Changes in the geographical distributions of plant and animal species in response to future greenhouse warming threaten to reduce biotic diversity.... The risk posed by CO_2-induced warming depends on the distances that regions of suitable climate are displaced northward [in the Northern Hemisphere] and on the rate of displacement.... If the change occurs too rapidly for colonization of newly available regions, population sizes may fall to critical levels, and extinction will occur.*

*M. B. Davis and C. Zabinski, "Changes in the geographical range resulting from greenhouse warming: Effects on biodiversity in forests," in R. L. Peters and T. E. Lovejoy, eds., *Global Warming and Biological Diversity* (New Haven, CT: Yale University Press, 1992), p. 297.

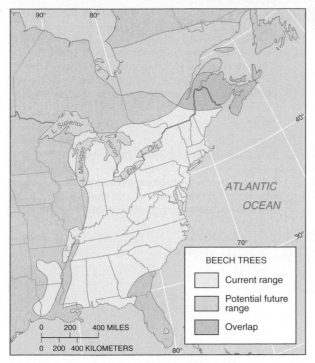

(a)

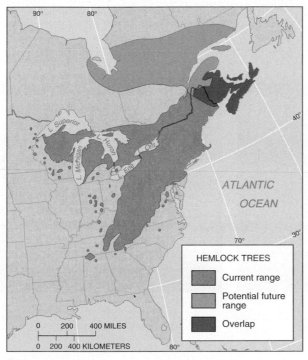

(b)

FIGURE 19.24 Present and predicted distribution of two tree species.
(a) Beech trees and (b) hemlock trees in North America. The potential future range shown for each reflects expected climate change resulting from a doubling of carbon dioxide. [After M. B. Davis and C. Zabinski, "Changes in the geographical range resulting from greenhouse warming: Effects on biodiversity in forests," in R. L. Peters and T. E. Lovejoy, eds., *Global Warming and Biological Diversity* (New Haven, CT: Yale University Press, 1992), p. 301.]

Another sign of the impact of climate change on ecosystems is the updating of planting and gardening guides so that they better fit actual changing climatic patterns. A comparison of a 2006 Hardiness Zone Map with the USDA 1990 map illustrates such change in a dramatic way. Each zone is based on the low temperature reached, with the zones set at 10° increments—a zone 3 has a low of –30°F to –40°F (–34°C to –40°C), a zone 4 increases by 10 F° (5.6 C°), and so forth. Based on NOAA and National Climatic Data center input from 5000 stations, the Arbor Day Foundation produced the new plant hardiness map. Figure 19.25 presents both the 1990 and 2006 maps and a map that portrays the differences, or zone changes, that occurred between the two. For instance, Ohio, Indiana, and Illinois shifted from Zone 5 to the warmer Zone 6.

Ecological Succession

Ecological succession occurs when newer communities (usually more complex) replace older communities of plants and animals (usually simpler)—a change of species composition. Each successive community of species modifies the physical environment in a manner suitable for a later community of species. Changes apparently move toward a more mature condition, although disturbances are common and constantly disrupt the sequence.

Traditionally, it was assumed that plants and animals formed a predictable *climax community*—a stable, self-sustaining, and functioning community with balanced birth, growth, and death—but scientists have abandoned this notion. Contemporary conservation biology, biogeography, and ecology assume nature to be in constant adaptation and nonequilibrium.

Rather than thinking of an ecosystem as a uniform set of communities, think of ecosystems as a patchwork mosaic of habitats—each striving to achieve an optimal range and low environmental stress. This is the study of *patch dynamics*, or disturbed portions of habitats. Within an ecosystem, individual patches may arise only to fail later. The smaller the patch, the faster species loss occurs. Succession is the interaction among patches. Most ecosystems are in actuality made up of patches of former landscapes. Biodiversity in some respects is the result of such patch dynamics.

Given the complexity of natural ecosystems, there may be several stages, or a *polyclimax* condition, with adjoining ecosystems or patches within ecosystems, each at different stages in the same environment. Mature communities are properly thought of as being in dynamic equilibrium. Or at times, these communities may be out of phase and in nonequilibrium with the immediate physical environment because of the usual lag time in their adjustment.

Succession often requires an initiating disturbance. Examples include windstorms, severe flooding, a volcanic eruption, a devastating wildfire, or an agricultural practice such as prolonged overgrazing (Figure 19.26). When existing organisms are disturbed or removed, new communities can emerge. At such times of nonequilibrium

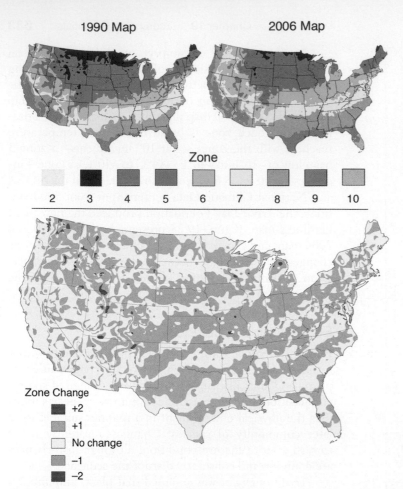

1990 Map   2006 Map

Zone

2	3	4	5	6	7	8	9	10

Zone Change
■ +2
■ +1
□ No change
■ −1
■ −2

FIGURE 19.25 Climate change impacts Hardiness Zones.
The Arbor Day Foundation revised the 1990 USDA Hardiness Zone map used to guide farmers and gardeners. Climate changes necessitated the 2006 revision. Compare the two maps, then check the differences map and "zone change." Locate your home and your college towns. Did these places get reclassified? [Maps courtesy of Arbor Day Foundation and USDA; see http://www.arborday.org/media/map_change.cfm.]

Find your home town/city and campus town/city on this "Zone Change" map and on the 1990 and 2006 maps. What zone changes do you note for each location?

     Home: _____
     Campus: _____

(a)

(b)

(c)

(d)

FIGURE 19.26 Disturbances in ecosystems alter succession patterns.
(a) River flooding along the American River, California. (b) Community development proceeds in a burned forest, progressing perhaps by chance, as species with an adaptive edge, such as fireweed (*Epilobium* sp.), succeed. (c) Unwise practices cause extensive soil erosion on a hog farm in Iowa, 1999; a disrupted prairie landscape. (d) Drought-stressed Coulter pines are attacked by bark beetles, altering relationships in the ecosystem, now ripe for wildfire in southern California. [Photos (a), (b), (d) by Bobbé Christopherson; (c) by USDA/NRCS.]

transition, the interrelationships among species produce elements of chance, and species having an adaptive edge will succeed in the competitive struggle for light, water, nutrients, space, time, reproduction, and survival.

Land and water experience different forms of succession. We first look at terrestrial succession, which is characterized by competition for sunlight, and then at aquatic succession, characterized by progressive changes in nutrient levels.

Terrestrial Succession

An area of bare rock or a disturbed site with no vestige of a former community can be a place for **primary succession**, the beginning of an ecosystem. Examples include new surfaces created by mass movement of land, areas exposed by a retreating glacier, cooled lava flows and volcanic eruption landscapes, surface mining and clear-cut logging scars, or an area of sand dunes. Plants taking hold on new lava flows from the Kīlauea volcano, Hawai'i, illustrate primary succession in Figure 19.27. In terrestrial ecosystems, succession begins with early species that form a **pioneer community**. You can see these plants taking hold in the photo. Primary succession often begins with lichens, mosses, and ferns growing on bare rock. These early inhabitants prepare the way for further succession to grasses, shrubs, and trees.

More common in nature is **secondary succession**, which begins if some aspects of a previously functioning community are present. An area where the natural community has been destroyed or disturbed, but where the underlying soil remains intact, may experience secondary succession.

Examples of secondary succession are most of the areas affected by the Mount St. Helens eruption and blast in 1980 (Figure 19.28). Some soils, young trees, and plants were protected under ash and snow, so community development began almost immediately after the event. About 38,450 hectares (95,000 acres) of trees were blown down and burned, however, and as the photos show, the effects of the eruption were devastating and lingering. Of course, the areas completely destroyed near the Mount St. Helens volcano or those buried beneath the massive landslide north of the mountain became candidates for primary succession.

As succession progresses, soil develops and a different set of plants and animals with different niche requirements may adapt. Further niche expansion follows as the community matures. Succession, which is really the developmental process forming new communities, is a dynamic set of interactions with sometimes unpredictable outcomes, steered by sometimes random, unpredictable events that trigger a threshold leap to a new set of relations—the basis of *dynamic ecology*.

Succession Through the Ice Age   Glaciation is a large-scale disturbance that well illustrates how succession operates. As Earth cycled through glacial and interglacial ages (Chapter 17), the ecological succession in all ecosystems was affected repeatedly. Imagine the milder midlatitude climate at the beginning of an interglacial warming, with lush herb and shrub vegetation slowly giving way to pioneer trees of birch, aspen, and pine. Winds and animals dispersed seeds, and changing communities readily spread in response to changing conditions. During the warm interglacial, trees increased shade and the organic content of soil increased. Deciduous trees such as oak, elm, and ash spread freely on well-drained soils; willow, cottonwood, and alder grew on poorly drained land.

As climates cooled with the approach of the next glacial, soils became more acidic, so podzolization (cool, moist, forested) processes dominated, helped by the acidity of spruce and fir trees. Ecosystems slowly returned to the vegetation community that existed at the beginning of the interglacial. The increased cold produced open regions of disturbed ecosystems. Freezing and the advance of ice disrupted highly acidic soils.

The Holocene (the past 10,000 years) is apparently atypical because of the development of human societies and agriculture. Civilization is creating an artificial and accelerated succession, similar in some ways to the closing centuries of an interglacial. The challenge for biogeographers is to understand such long-term developmental change principles, given the accelerated pace at which change is occurring in the industrial and postindustrial era.

Wildfire and Fire Ecology   Fire and its effects are one of Earth's significant ecosystem and economic processes. In the United States during 2000, about 7.4 million acres, more than double the annual average, burned in wildfires—an incredible 92,000 fires. In subsequent years, wildfires consumed the following: 2001, 3.6 million acres; 2002, 7.1 million acres; 2003, 3.8 million acres; 2005, about 8.4 million acres (4.4 million acres of this in Alaska); and 2006, 9.4 million acres; 2007 is at 9.1 million acres (see

FIGURE 19.27 Primary succession.
Ferns take hold as primary succession on new lava flows that came from the Kīlauea volcano, Hawai'i. [Photo by Bobbé Christopherson.]

(a) 1979
(b) 1980

(c) 1983
(d) 1983

(e) 1983

(f) 1983

1999

1999

1999

1999

Landsat-7 8/22/1999

FIGURE 19.28 The pace of change in the region of Mount St. Helens. (a) The area north of the volcano before the eruption (1979) and (b) after the eruption (1980) experienced profound change in all ecosystems. A series of four photo comparisons made in 1983 and 1999 of matching scenes illustrate the slow recovery of secondary succession: (c) and (d) at Meta Lake; (e) Spirit Lake panorama view (note the mat of floating logs after 19 years); (f) detail of the destroyed forest community. [(a) and all 1983 photos by author; (b) and all 1999 photos by Bobbé Christopherson; *Landsat-7* 8/22/1999 satellite image courtesy of NASA/GSFC.]

National Interagency Fire Center at **http://www.nifc. gov/fire_info/nfn.htm**).

Over the past 50 years, the role of fire in ecosystems has been the subject of much scientific research and experimentation. Today, fire is recognized as a natural component of most ecosystems and not the enemy of nature that it was once thought. In fact, in many forests, undergrowth and surface litter is purposely burned in controlled "cool fires" to remove fuel and prevent catastrophic and destructive "hot fires." Forestry experts have learned that when fire-prevention strategies are rigidly followed they can lead to the abundant undergrowth accumulation that fuels major fires.

Modern society's demand for fire prevention to protect property goes back to European forestry of the 1800s. Fire prevention became an article of faith for forest managers in North America. But, in studies of the longleaf pine forest that stretches in a wide band from the Atlantic coastal plain to Texas, fire was discovered to be an integral part of regrowth following lumbering. In fact, seed dispersal of some pine species, such as the knobcone pine, does not occur unless assisted by a forest fire. Heat from the fire opens the cones, releasing seeds so they can fall to the ground for germination. Also, these fire-disturbed areas quickly recover with protein-rich woody growth, young plants, and a stimulated seed production that provides abundant food for animals.

The science of **fire ecology** imitates nature by recognizing fire as a dynamic ingredient in community succession. Controlled ground fires, deliberately set to prevent accumulation of forest undergrowth, now are widely regarded as a wise forest management practice and are used across the country. After the highly visible Yellowstone National Park fires, an outcry was heard from forestry and recreational interests (Figure 19.29). Critics called fire ecology practices the government's "let it burn" policy.

FIGURE 19.29 Yellowstone burns and succession; wildfires strike and recovery in Montana and southern California.
Yellowstone National Park in northwestern Wyoming: (a) 1988 wildfire progresses up a slope; (b) trees and shrubs return to Yellowstone with secondary succession. (c) The drama of wildfire unfolds in the Bitterroot National Forest in Montana, August 2000, as elk stand in the Bitterroot River attempting to survive the conflagration. (d) Fire-adapted chaparral vegetation in southern California a few months after a wildfire, as recovery begins. Note the stump sprouts emerging. [Photos (a) by Joe Peaco/National Park Service; (b) and (d) by Bobbé Christopherson; (c) by John McColgan, BLM Alaska Fire Service.]

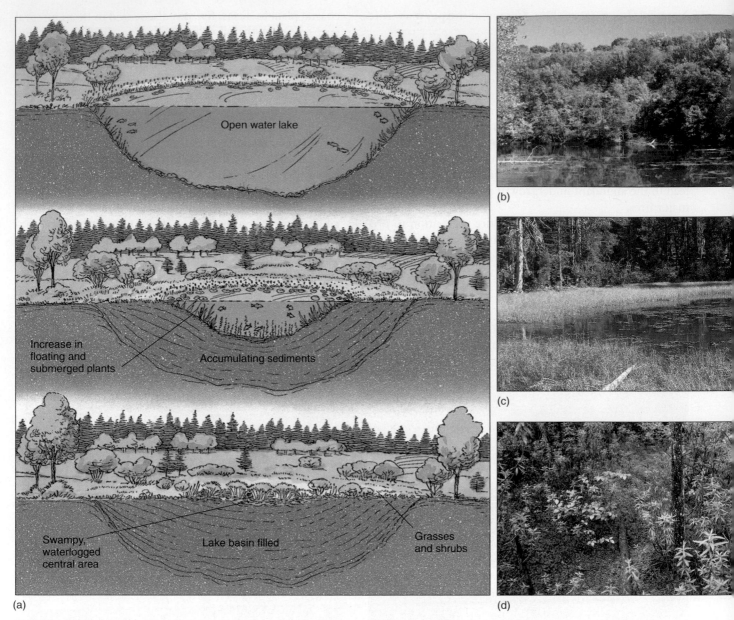

FIGURE 19.30 Idealized lake–bog–meadow succession in temperate conditions.
(a) What begins as a lake gradually fills with organic and inorganic sediments, which successively shrink the area of the pond. A bog forms, then a marshy area, and finally a meadow completes the successional stages. (b) Spring Mill Lake, Indiana. (c) Aquatic succession in a mountain lake. (d) The Richmond Nature Park and its Lulu Island Bog was ocean just 8000 years ago. Frasier River sediment created the mudflats and set the stage for the evolution of this sphagnum bog with acidic soils that "quakes" when you walk on it. Plants in the bog include moss, shore pine, hemlock, blueberry, salal shrubs, and Labrador tea, among others. [Photos by Bobbé Christopherson.]

In its final report on the 1988 Yellowstone fires, a government interagency task force concluded, "an attempt to exclude fire from these lands leads to major unnatural changes in vegetation . . . as well as creating fuel accumulation that can lead to uncontrollable, sometimes very damaging wildfire." Thus, federal land managers and scientists reaffirmed their stand that fire ecology is a fundamentally sound concept.

An increasingly serious problem is emerging as people seek to live in the wildlands near cities. Urban development has encroached on forests, and the added suburban landscaping has created new fire hazards.

Uncontrolled wildfires now can destroy homes and threaten public safety. People living in these areas demand fire protection that actually causes undergrowth to accumulate and worsens the risk. In 2003 and again in 2007, wildfires devastated communities in southern California, especially where development was in the "wilds" of chaparral country. In 2007, fire destroyed more than 2000 homes, and over 500,000 acres burned in two-dozen wildfires. Strong Santa Ana winds (see Chapter 6 for description) added the blowtorch effect, taking embers miles from their ignition source. Chapter 21 opens with a satellite view of southern California at the height of the inferno.

The connection between the record wildfires in the American West since 2000 and climate change was reported in *Science* in 2006:

> Higher spring and summer temperatures and earlier snowmelt are extending the wildfire season and increasing the intensity of wildfires in the western U.S....Here, we show that large wildfire activity increased suddenly and markedly in the mid-1980s, with higher large-wildfire frequency. Longer wildfire durations, and longer wildfire seasons...[are] strongly associated with increased spring and summer temperatures and an earlier spring snowmelt.[*]

Aquatic Succession

Ecosystems occur in water as well as on land. Lakes, estuaries, ponds, and shorelines are complex ecosystems. The concepts of terrestrial succession that you just read about—stability and resilience, biodiversity, community succession—apply to open water, shoreline, and watershed systems as well.

A lake or pond is really a temporary feature on the landscape, when viewed across geologic time. Lakes and ponds exhibit successional stages as they fill with nutrients and sediment and as aquatic plants take root and grow. This growth captures more sediment and adds organic debris to the system (Figure 19.30). This gradual enrichment in water bodies is known as eutrophication (from the Greek *eutrophos*, meaning "well nourished").

In moist climates, a floating mat of vegetation grows outward from the shore to form a bog. Cattails and other marsh plants become established, and partially decomposed organic material accumulates in the basin, with additional vegetation bordering the remaining lake surface. Vegetation and soil and a meadow may fill in as water is displaced;

willow trees follow, and perhaps cottonwood trees; and eventually the lake may evolve into a forest community.

The stages in lake succession are named for their nutrient levels: *oligotrophic* (low nutrients), *mesotrophic* (medium nutrients), and *eutrophic* (high nutrients). Greater primary productivity and resultant decreases in water transparency mark each stage, so that photosynthesis becomes concentrated near the surface. Energy flow shifts from production to respiration in the eutrophic stage, with oxygen demand exceeding oxygen availability.

Nutrient levels also vary spatially: Oligotrophic conditions occur in deep water, whereas eutrophic conditions occur along the shore, in shallow bays, or where sewage, fertilizer, or other nutrient inputs occur. Even large bodies of water may have eutrophic areas along the shore. As society dumps sewage, agricultural runoff, and pollution in waterways, the nutrient load is enhanced beyond the cleansing ability of natural biological processes. The result is *cultural eutrophication*, which hastens succession in aquatic systems.

Research has uncovered an apparent opposite pattern to the lake–bog succession described, in areas that have been recently deglaciated. The new study points to the reality that succession is poorly understood in aquatic ecosystems. Scientists examined lakes in the Glacier Bay, Alaska, area and found that they became more dilute, acidic, and unproductive during the past 10,000 years; in other words, less eutrophic. Successional changes in the surrounding vegetation and soils seem to be linked to the aquatic ecosystem. These limnology (the study of lakes and other inland water) studies signify that in cool, moist, temperate rain forest climates recovering from glacial retreat, different factors are at work in lakes created by continental glaciation.

Focus Study 19.1 examines the Great Lakes of North America and their geographically and biologically diverse aquatic ecosystems, interrelated with the surrounding terrestrial ecosystems. This is the largest lake system on Earth and is jointly managed by the United States and Canada.

[*]A. L. Westerling, H. G. Hidalgo, et al., "Warming and earlier spring increase Western U.S. forest wildfire activity," *Science* 313, no. 5789 (August 2006): 927–27, 940–43.

Focus Study 19.1

The Great Lakes

The basins of the Great Lakes are gifts of the last ice age to North America (see Figure 17.28 in Chapter 17). The glaciers advanced and retreated over this region, excavating the basins for five large lakes. This international waterway of lakes and connecting rivers has played an important role in the history of Canada and the United States. Today, about 10% of the U.S. population and 25% of Canada's population live in the Great Lakes drainage basin (Figure 19.1.1). Major agricultural regions surround the lakes, as do many centers of industrial activity. Tourism, sport fishing, and maritime commerce are also important.

Society asks much of the Great Lakes: to dilute wastes from cities and industry, to dissipate thermal pollution from power plants, to provide municipal drinking water and irrigation water, and to sustain unique and varied ecosystems—open lake, coastal shore, coastal marsh, lakeplain (former lake bed), inland wetlands, and inland

(continued)

Focus Study 19.1 (continued)

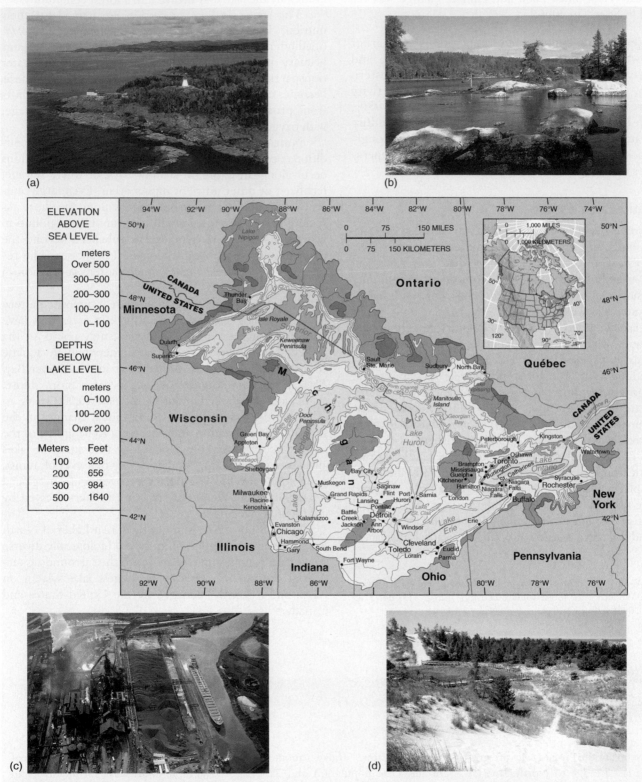

FIGURE 19.1.1 The Great Lakes Basin.
Elevations above sea level, depths below lake level, and principal urban areas are shown. The map outline is the limit of the drainage basin for the Great Lakes system. (a) The rocky shores of Lake Superior, Otter Island lighthouse. (b) Forests along Lake Ontario shores. (c) The Cuyahoga River enters Lake Erie at Cleveland. (d) Sand dunes along the southern shore of Lake Michigan. Note the wooden pedestrian walkway to prevent erosion of sand by human traffic. [(a) Photo by Carl R. Sams, II/Peter Arnold, Inc.; (b) photo by Wayne Lankinen/DRK Photo; (c) photo by Alex S. MacLean/Peter Arnold, Inc.; (d) photo by Bobbé Christopherson. Map courtesy of Environment Canada, U.S. EPA, and Brock University cartography.]

terrestrial (upland areas of forest, prairie, and barrens). Here is a brief profile of this important hydrological and ecological resource. (See Environment Canada's Great Lakes home page at **http://www.on.ec.gc.ca/greatlakes/** and NOAA's Great Lakes Research Lab site at **http://www.glerl.noaa.gov/**.)

Geography and Physical Characteristics

The Great Lakes—Superior, Michigan, Huron, Erie, and Ontario—contain 18% of the total volume of all freshwater lakes in the world, some 23,000 km³ (5500 mi³) of water. Their combined surface area covers 244,000 km² (94,000 mi²), a little less than the state of Wyoming. The entire drainage basin embraces 528,000 km² (204,000 mi²), or an area about the size of Manitoba.

As we tour the lakes, it is useful to follow along in Figure 19.1.1 and to look at their profile in Figure 19.1.2a. Lake Superior is the highest in elevation, highest in latitude, deepest, and largest in the system. It drains through St. Mary's River into Lake Huron. Lake Michigan, the only lake of the five that lies entirely within the United States, is at the same level as Lake Huron because it is joined by the wide connection through the Straits of Mackinac. (This is why, in Figure 19.1.2, we combine these two lakes on one hydrograph.)

The St. Clair River, Lake St. Clair, and the Detroit River carry water on to Lake Erie, the shallowest lake in the system. Compared to an average water retention time of 191 years in Lake Superior, Lake Erie has the shortest retention time of 2.6 years.

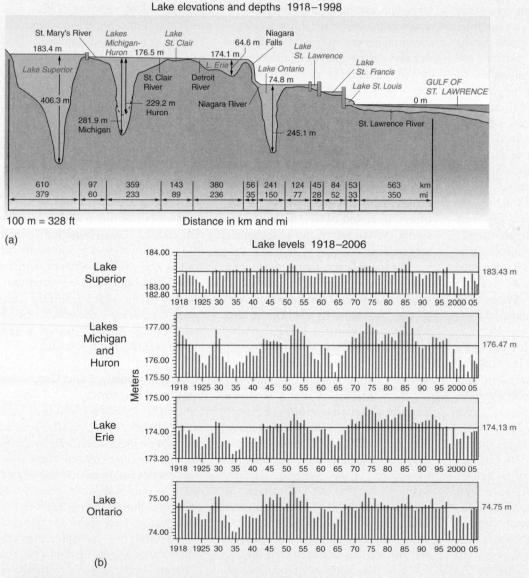

(a)

(b)

FIGURE 19.1.2 Great Lakes elevation profile and hydrographs.
(a) Average depth and lake surface levels, 1918–1998. (b) Annual hydrographs for each of the Great Lakes, 1918–2006. Both figures were prepared using the International Great Lakes Datum of 1985. [Data courtesy of the Canadian Hydrographic Service, Central Region, and the International Coordinating Committee on the Great Lakes Basin Hydraulic and Hydrographic Data.]

(continued)

Focus Study 19.1 (continued)

The Niagara River, plunging dramatically over Niagara Falls, transports water into Lake Ontario. Lake Ontario is drained by the St. Lawrence River, which carries the entire discharge of the Great Lakes system to the Gulf of St. Lawrence and eventually into the North Atlantic Ocean. The entire basin extends over 10° of latitude (41° N to 51° N) and 18° of longitude (75° W to 93° W). (For additional Great Lakes information, see http://www.seagrant.wisc.edu/outreach/.)

Because of the large size of the overall Great Lakes system, its associated climate, soils, and topography vary widely. Dominating the north are colder microthermal climates, exposed portions of the Canadian Shield bedrock, and vast stands of conifers and acidic soils. To the south, warmer mesothermal climates and fertile, glacially deposited soils provide a vast agricultural base. Farms and urbanization replaced previous stands of mixed forest. Virtually none of the original land cover is undisturbed.

Prevailing winds from the west and seasonal change mix air masses from different source regions, producing variable weather. The presence of the Great Lakes basin strongly influences passing air masses and weather systems, producing lake-effect snowfall to the east (see the dramatic satellite image of this in Figure 8.5). Precipitation over the drainage basin feeds the lake storage as part of the renewable hydrologic cycle. Figure 19.1.2 illustrates Great Lakes water levels from 1918 to 2006. Water levels were above average during 1996 and 1997. During 1998 through 2001, water levels lowered to below-average levels and a 45-year low, reflecting near-drought conditions across the region. However, through 2003 and 2006 lake-level trends rose toward the 1985 Datum, only to lower after 2005.

Short-term lake levels vary from winter to summer; they are higher in summer after snowmelt and the arrival of maximum summer precipitation. Over the long term, highest lake levels occur during years of heavier precipitation and during times of cooler temperatures, which reduce evaporation. Wind also affects lake levels. A wind setup (wind-driven water) occurs along the downwind shoreline, as water is pushed higher onshore.

The Great Lakes Ecosystem The Great Lakes ecosystem should be thought of as young and a fruitful laboratory for ecological studies. These lakes are not simply large bodies of water with a uniform mixture. A stratification occurs in them related to density and temperature differences. In the summer, the surface and shallow waters are warmed and become less dense so that the lakes develop a sharp stratification. The warm surface water and light penetration in the upper layers support most biotic production and adequate dissolved oxygen. This layering affects water quality because it can prevent mixing of pollution and other effluents with cooler bottom water during the summer months.

As the fall season matures, surface water cools and sinks, displacing deeper water and creating a turnover of the lake mass. By midwinter, the temperature from the surface to the bottom is uniformly around 4°C (39°F), the point at which water is densest); temperatures at the surface are near freezing. A trend in Lake Superior water temperature is a 2.5 C° (4.5 F°) increase since 1979, higher than the 1.5 C° (2.7 F°) increase in regional air temperatures. It is difficult to imagine Lake Superior surface water at 23.9°C (75°F), which it reached in July 2007. Scientists are analyzing the impact of these temperature increases on lake ecosystems and any relation to climate change.

The lakes support a food chain of producers and consumers, as does any aquatic ecosystem. Native fish populations have been greatly affected by human activities: overfishing, introduced nonnative species, pollution from excess nutrients, toxic contamination of fish, and disruption of spawning habitats.

Peak commercial fishing occurred in the 1880s. The fisheries declined to their lowest levels as water pollution peaked in the 1960s and early 1970s. The U.S. and Canadian governments implemented strong pollution control programs, working in concert with many citizens, industries, and private organizations. These efforts brought the lakes into the present period of recovery. Today, lake trout (in Lake Superior), sturgeon, herring, smelt, alewife, splake, yellow perch, walleye, and white bass are fished. However, health advisories are occasionally issued warning of danger in consuming fish of certain species, sizes, and locales.

Coastline for the five lakes totals 18,000 km (11,000 mi) and is diverse: major dunes and sandy beaches, bedrock shorelines, gravel beaches, and adjoining coastal marsh systems. Marshes can be found at the far western end of Lake Superior in the St. Louis River estuary and at the far eastern end of Lake Ontario and its coastal lagoon. Here, as elsewhere, marshes occupy that key interface between the land and the water, storing and cycling organic material and nutrients into aquatic ecosystems.

Human Activities, Land Use, Loss, and Recovery

Figure 19.1.3 portrays land use in the Great Lakes Basin. Each activity has its own impact on lake systems. Runoff from agricultural areas contains chemicals, nutrients, and eroded soil. The pulp and paper industry was a major polluter but has improved as more was learned about the dangers of pollution; for example, mercury contamination was halted in the 1970s and better disposal methods for wastes were implemented. The forests were not always treated in a sustainable way in terms of reforestation, and

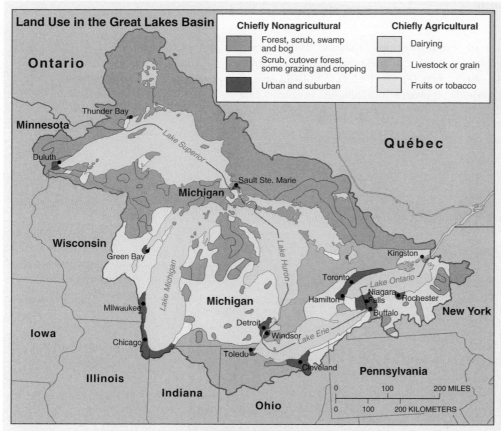

Land Use in the Great Lakes Basin

Chiefly Nonagricultural
- Forest, scrub, swamp and bog
- Scrub, cutover forest, some grazing and cropping
- Urban and suburban

Chiefly Agricultural
- Dairying
- Livestock or grain
- Fruits or tobacco

FIGURE 19.1.3 Land use in the Great Lakes Basin.
Generalized distribution map of land use, nonagricultural and agricultural, in the Great Lakes Basin.
[Map courtesy of Environment Canada, U.S. EPA, and Brock University cartography.]

today forests may be regarded as a diminished resource in the basin. Recovery efforts are slow to make up for bad past practices, but progress is underway.

Lake Erie was the first of the Great Lakes to show severe effects of *cultural eutrophication*, for it is the shallowest and warmest of the five lakes. About a third of the overall population in the Great Lakes Basin lives in the drainage of Lake Erie, making it the major recipient of sewage effluent from treatment plants. Over the years, dissolved oxygen levels dropped as biochemical oxygen demand increased, caused by sewage-treatment discharges and algal decay. Algae flourished in these eutrophic conditions and coated the beaches, turning the lake a greenish brown.

The most infamous incident in the 1960s was the Cuyahoga River (through Cleveland, Ohio) catching fire because it was so contaminated. In 1972, Canada and the United States signed the Great Lakes Water Quality Agreement and the cleanup was under way. The government, spurred by public opinion, was moved to action, knowing that one of these priceless lakes was actually being lost. One example of the success of these efforts is that the input of phosphorus was reduced by 90% (an element with high biological activity that dramatically stimulates unwanted plant growth in aquatic ecosystems).

Concern for the health of the entire Great Lakes ecosystem remains necessary because of the biological amplification of poisons in the food chain. For example, organic chemicals such as PCBs accumulate to dangerous levels. PCBs (polychlorinated biphenyls) are highly viscous, inert fluids used in electrical transformers and hydraulic systems, and they

are carcinogenic. Samples reported by Environment Canada show PCBs in phytoplankton at 0.0025 ppm, in a rainbow smelt at 1.04 ppm, in a lake trout at 4.83 ppm, and, higher in the food chain, a herring gull eggshell at 124 ppm! Later additions to the 1972 agreement reduced toxic levels, organic chemicals, and heavy metals entering the lakes. Since humans are at the end of the amplification process, it is in our own best interest to help the continuing recovery of these lakes.

An ecosystem approach to management of the Great Lakes Basin is the key to recovery. This approach considers the physical, chemical, and biological components that constitute the total system. Thus, a herring gull eggshell becomes an indicator of chemical pollution, blue-green algal blooms denote hastening eutrophication, and a loss of species richness and

(continued)

Focus Study 19.1 (continued)

diversity exposes the destruction of habitat. This is where geographic principles of spatial analysis, an integrative tool such as geographic information systems (GIS), and remote-sensing capabilities come into play (Figure 19.1.4). A holistic, spatial approach to ecological problems is a marked departure from the past, when pollution problems were considered locally and separately.

FIGURE 19.1.4 Orbital view of the Great Lakes.
The Great Lakes region in true color, made March 6, 2000, by sensors aboard *Terra*.
[*Terra* MODIS sensor image courtesy of NASA/MODIS Land Science Team.]

Summary and Review—Ecosystem Essentials

■ *Define* ecology, biogeography, and the ecosystem concept.

Earth's biosphere is unique in the Solar System; its ecosystems are the essence of life. **Ecosystems** are self-sustaining associations of living plants and animals and their nonliving physical environment. **Ecology** is the study of the relationships between organisms and their environment and among the various ecosystems in the biosphere.

   Biogeography is the study of the distribution of plants and animals and the diverse spatial patterns they create. Biogeographers trace communities across the ages, considering plate tectonics and ancient dispersal of plants and animals.

   ecosystem (p. 606)
   ecology (p. 606)
   biogeography (p. 606)

1. What is the relationship between the biosphere and an ecosystem? Define *ecosystem*, and give some examples.
2. What does biogeography include? Describe its relationship to ecology.
3. Briefly summarize what ecosystem operations imply about the complexity of life.

■ *Describe* communities, habitats, and niches.

A **community** is formed by the interactions among populations of living animals and plants at a particular time. Within a community, a **habitat** is the specific physical location of an organism—its address. A **niche** is the function or operation of a life form within a given community—its profession.

   community (p. 607)
   habitat (p. 608)
   niche (p. 608)

4. Define a community within an ecosystem.
5. What do the concepts of habitat and niche involve? Relate them to some specific plant and animal communities.
6. Describe symbiotic and parasitic relationships in nature. Draw an analogy between these relationships and human societies on our planet. Explain.

■ *Explain* photosynthesis and respiration, and *derive* net photosynthesis and the world pattern of net primary productivity.

As plants evolved, the **vascular plants** developed conductive tissues. **Stomata** on the underside of leaves are the portals through which the plant participates with the atmosphere and hydrosphere. Plants (primary producers) perform **photosynthesis** as sunlight stimulates a light-sensitive pigment called **chlorophyll**. This process produces food sugars and oxygen to drive biological processes. **Respiration** is essentially the reverse of photosynthesis and is the way the plant derives energy by oxidizing carbohydrates. **Net primary productivity** is the net photosynthesis (photosynthesis minus respiration) of an entire community. This stored chemical energy is the **biomass** that the community generates, or the net dry weight of organic material.

vascular plants (p. 609)
stomata (p. 611)
photosynthesis (p. 611)
chlorophyll (p. 611)
respiration (p. 611)
net primary productivity (p. 612)
biomass (p. 612)

7. Define a vascular plant. How many plant species are there on Earth?
8. How do plants function to link the Sun's energy to living organisms? What is formed within the light-responsive cells of plants?
9. Compare photosynthesis and respiration and the derivation of net photosynthesis. What is the importance of knowing the net primary productivity of an ecosystem and how much biomass an ecosystem has accumulated?
10. Briefly describe the global pattern of net primary productivity.

■ *List* abiotic ecosystem components, and *relate* those components to ecosystem operations.

Light, temperature, water, and the cycling of gases and nutrients constitute the life-supporting abiotic components of ecosystems. Elevation and latitudinal position on Earth create a variety of physical environments. The zonation of plants with altitude is the **life zone concept** and is visible as you travel between different elevations.

Life is sustained by **biogeochemical cycles**, through which circulate the gases and sedimentary materials (nutrients) necessary for growth and development of living organisms. The environment may inhibit biotic operations, either through lack or excess. These **limiting factors** may be physical or chemical in nature.

life zone (p. 616)
biogeochemical cycles (p. 617)
limiting factor (p. 621)

11. What are the principal abiotic components in terrestrial ecosystems?
12. Describe what Alexander von Humboldt found that led him to propose the life zone concept. What are life zones? Explain the interaction among elevation, latitude, and the types of communities that develop.
13. What are biogeochemical cycles? Describe several of the essential cycles.
14. What is a limiting factor? How does it function to control the spatial distribution of plant and animal species?

■ *Explain* trophic relationships in ecosystems.

Trophic refers to the feeding and nutrition relations in an ecosystem. It represents the flow of energy and cycling of nutrients. **Producers**, which fix the carbon they need from carbon dioxide,

are the plants, including phytoplankton in aquatic ecosystems. **Consumers**, generally animals including zooplankton in aquatic ecosystems, depend on the producers as their carbon source. Energy flows from the producers through the system along a **food chain** circuit. Within ecosystems, the feeding interrelationships are complex and arranged in a complex network of interconnected food chains in a **food web**.

The primary consumer is an **herbivore**, or plant eater. A **carnivore** (meat eater) is a secondary consumer. A consumer that eats both producers and consumers is an **omnivore**—a role occupied by humans. **Detritivores** are detritus feeders and decomposers that release simple inorganic compounds and nutrients as they break down organic materials. Detritus feeders, including worms, mites, termites, and centipedes, process debris. **Decomposers** are the bacteria and fungi that process organic debris outside their bodies and absorb nutrients in the process—the rotting that breaks down detritus. Biomass and population pyramids characterize the flow of energy and the numbers of producers, consumers, and decomposers operating in an ecosystem.

producers (p. 621)
consumers (p. 621)
food chain (p. 623)
food web (p. 623)
herbivore (p. 623)
carnivore (p. 623)
omnivore (p. 623)
detritivore (p. 623)
decomposer (p. 623)

15. What role is played in an ecosystem by producers and consumers?
16. Describe the relationship among producers, consumers, and detritivores in an ecosystem. What is the trophic nature of an ecosystem? What is the place of humans in a trophic system?
17. What are biomass and population pyramids? Describe how these models help explain the nature of food chains and communities of plants and animals.
18. Follow the flow of energy and biomass through the Silver Springs, Florida, ecosystem. Describe the pathways in Figures 19.13a and 19.19a.

■ *Relate* how biological evolution led to the biodiversity of life on Earth.

Evolution states that single-cell organisms adapted, modified, and passed along inherited changes to multicellular organisms. The genetic makeup of successive generations was shaped by environmental factors, physiological functions, and behaviors that created a greater rate of survival and reproduction that were passed along through natural selection.

Biodiversity (a combination of *bio*logical and *diversity*) includes the number of different species and quantity of each species, the genetic diversity within species (number of genetic characteristics), and ecosystem and habitat diversity. The greater the biodiversity within an ecosystem, the more stable and resilient it is and the more productive it will be. Modern agriculture often creates a nondiverse monoculture that is particularly vulnerable to failure.

evolution (p. 628)
biodiversity (p. 629)

19. Referring back to Focus Study 1.1, define the scientific method and a theory.

20. Construct a simple model to show the progressive stages that lead to the development of a theory, such as the theory of evolution.

■ *Define* succession, and *outline* the stages of general ecological succession in both terrestrial and aquatic ecosystems.

Ecological succession describes the process whereby older communities of plants and animals are replaced by newer communities that are usually more complex. Past glacial and interglacial climatic episodes have created a long-term succession. An area of bare rock and soil with no trace of a former community can be a site for **primary succession**. The initial community that occupies an area of early succession is a **pioneer community**. **Secondary succession** begins in an area that has a vestige of a previously functioning community in place. Rather than progressing smoothly to a definable stable point, ecosystems tend to operate in a dynamic condition, with succeeding communities overlapping in time and space.

Wildfire is one external factor that disrupts a successional community. Aquatic ecosystems also experience community succession, as exemplified by the eutrophication of a lake ecosystem.

The science of **fire ecology** has emerged in an effort to understand the natural role of fire in ecosystem maintenance and succession.

ecological succession (p. 633)
primary succession (p. 635)
pioneer community (p. 635)
secondary succession (p. 635)
fire ecology (p. 637)

21. What is meant by ecosystem stability?
22. How does ecological succession proceed? What are the relationships between existing communities and new, pioneer communities?
23. Assess the impact of climate change on natural communities and ecosystems. Consider the plant Hardiness Zone changes or western wildfire conditions, for example.
24. Discuss the concept of fire ecology in the context of the Yellowstone National Park fires of 1988. What were the findings of the government task force?
25. How is the spread of urbanization into wilderness areas related to increasing wildfire risk?
26. Summarize the process of succession in a body of water. What is meant by cultural eutrophication?

NetWork

The *Geosystems* Student Learning Center provides on-line resources for this chapter on the World Wide Web. To begin: Once at the Center, click on the cover of this textbook, scroll the Table of Contents menu, and select this chapter. You will find self-tests that are graded, review exercises, specific updates for items in the chapter, and in "Destinations" many links to interesting related pathways on the Internet. *Geosystems* Student Learning Center is found at **http://www.prenhall.com/christopherson/**.

Critical Thinking

A. This chapter states, "Some scientists are questioning whether our human society and the physical systems of Earth constitute a global-scale symbiotic relationship of mutualism (sustainable) or a parasitic one (nonsustainable)." Referring to the definition of these terms, what is your response to the statement? How do you equate our planetary economic system with the need to sustain life-supporting natural systems? Mutualism? Parasitism?

B. Over the next several days as you travel between home and campus, a job, or other areas, observe the landscape. What types of ecosystem disturbances do you see? Imagine that several acres in the same area you are viewing had escaped any disruptions for a century or more. Describe what some of these ecosystems and communities might be like.

Fields of rapeseed (*Brassica napus*) are in bloom on coastal farms, northeastern Scotland. Grown for vegetable oil (known as the modern canola oil), animal feed, and increasingly as a source for biofuel, rapeseed outproduces soy beans or corn in oil per volume of crop. The abundant nectar produced makes it popular with honey bees. [Photo by Bobbé Christopherson.]

Terrestrial Biomes

■ Key Learning Concepts

After reading the chapter, you should be able to:

- ■ **Define** the concept of biogeographic realms of plants and animals, and **define** ecotone, terrestrial ecosystem, and biome.

- ■ **Define** six formation classes and the life-form designations, and **explain** their relationship to plant communities.

- ■ **Describe** 10 major terrestrial biomes, and **locate** them on a world map.

- ■ **Relate** human impacts, real and potential, to several of the biomes.

n 1874, Earth was viewed as offering few limits to human enterprise. But visionary American diplomat and conservationist George Perkins Marsh perceived the need to manage and to conserve the environment:

We have now felled forest enough everywhere, in many districts far too much. Let us now restore this one element of material life to its normal proportions, and devise means of maintaining the permanence of its relations to the fields, the meadows, and the pastures, to the rain and the dews of heaven, to the springs and rivulets with which it waters the Earth.*

Unfortunately, his warning was ignored. The clearing of the forests is perhaps the most fundamental change wrought by humans in our transformation of Earth. The development and expansion of civilization was fueled and built on the consumption of forests and other natural resources. The natural "work" supplied by Earth's physical, chemical, and biological systems has an estimated annual worth of some $35 trillion to the global economy.

*G. P. Marsh, *Physical Geography as Modified by Human Action* (New York: Scribner's, 1874), pp. 385–386.

649

Today, we see that Earth's natural systems do pose limits to human activity. Scientists are measuring ecosystems carefully, and they are building elaborate computer models to simulate our evolving real-time human-environment experiment—in particular, shifting patterns of environmental factors (temperatures and changing frost periods; precipitation timing and amounts; air, water, and soil chemistry; and a redistribution of nutrients). We know from history that the effect of climate change on ecosystems will be significant. More than 15 years ago two Harvard researchers assessed the impact of these alterations on ecosystems:

Based on more than a decade of research, it is obvious that the CO_2-rich atmosphere of our future will have direct and dramatic effects on the composition and operation of ecosystems. According to the best scientific evidence, we see no reason to be sanguine [optimistic] about the response of these habitats to our changing environment.*

Compare this to UNEP's 2007 *Global Environmental Outlook–GEO4* quote from their *GEO4 Executive Summary* (p. 4):

Human life and all other species depend on healthy ecosystems. But current biodiversity changes, the fastest in human history, mean losses are restricting future development options. About 60% of the ecosystem services that have been previously assessed are degraded or used unsustainably. Species are becoming extinct at rates which are 100 times faster than the rate shown in the fossil record, because of land-use changes, habitat loss, overexploitation of resources, pollution, and the spread of invasive alien species.

*F. A. Bazzaz and E. D. Fajer, "Plant life in a CO_2-rich world," *Scientific American* 256, no. 1 (January 1992): 74.

What will the quotations about Earth's environment be like 135 years from now, about the same time interval since G. P. Marsh made his assessment? Humans now have become the most powerful biotic agent on Earth, influencing all ecosystems on a planetary scale. We simply can't view physical geography in some academic compartment. Rather, we are studying systems in dynamic equilibrium, changing through time. This is the challenge for physical geography and the other geospatial sciences, to provide information that might be applied in some ways to benefit the future of natural systems.

In this chapter: We explore Earth's major terrestrial ecosystems, conveniently grouped into 10 biomes. Each biome is an idealized assemblage because much alteration of the natural environment has occurred. The discussion covers biome appearance, structure, and location; the present conditions of related plants, animals, and environments; and the state of biodiversity. Table 20.1 summarizes most of the information about Earth's physical systems that is presented through the pages of this text. Through biomes, we synthesize the physical geography of place and region.

Biogeographic Realms

Figure 20.1 shows Earth's biogeographic realms. The map illustrates the botanical (plant) realms worldwide and generally overlaps with the traditional zoological (animal) realms. Each realm contains many distinct ecosystems that distinguish it from other realms.

A **biogeographic realm** is a geographic region where a group of plant and animal species evolved and traditionally has been used as a basis for establishing terrestrial ecosystem regions. As you can see, these realms correspond generally to continents. The main topographic barrier that separates these realms is the ocean. Every species attempts to migrate worldwide according to its niche requirements, reproductive success, and competition

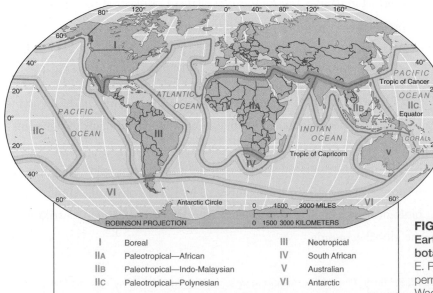

FIGURE 20.1 Biogeographic realms.
Earth's biogegraphic realms roughly based on botanical realms (after R. D. Good, 1947). [From E. P. Odum, *Fundamentals of Ecology*, 3rd ed., with permission of Brooks/Cole, an imprint of the Wadsworth Group, Thomson Learning.]

I	Boreal	III	Neotropical
IIA	Paleotropical—African	IV	South African
IIB	Paleotropical—Indo-Malaysian	V	Australian
IIC	Paleotropical—Polynesian	VI	Antarctic

FIGURE 20.2 The unique Australian realm.
The flora and fauna of Australia form a special assemblage of communities. [Photo by Mark Newman/Photo Researchers, Inc.]

with other species but is constrained by climatic and topographic barriers. Recognition that such distinct regions of flora (plants) and fauna (animals) exist was an early beginning for *biogeography* as a discipline.

The Australian realm is unique, giving rise to 450 species of *Eucalyptus* among its plants and 125 species of marsupials, including four of the six orders (animals such as kangaroos, that carry their young in pouches, where gestation is completed; Figure 20.2). Also, monotremes, egg-laying mammals such as the platypus, add further distinctiveness to this realm. Australia's unique flora and fauna is the result of its early isolation from the other continents. During critical evolutionary times, Australia drifted away from Pangaea (see Chapter 11) and never again was reconnected by a land bridge, even when sea level lowered during repeated glacial ages. New Zealand's isolation from Australia also undoubtedly explains why no native marsupials exist in New Zealand.

Transition Zones

Alfred Wallace (1823–1913), the first scholar of *zoogeography*, used the stark contrast in animal species among islands in present-day Indonesia to delimit a boundary between the Paleotropical and Australian realms (Figure 20.1). He thought that deep water in the straits had completely prevented species crossover. However, he later recognized that a boundary line actually is a wide transition zone, where one region grades into the other. Boundaries between natural systems are "zones of shared traits." Thus, despite distinctions among individual realms and our discussion of biomes, it is best to think of their borders as transition zones of mixed identity and composition, rather than as rigidly defined boundaries.

A boundary transition zone between adjoining ecosystem regions is an **ecotone**. Because ecotones are defined by different physical factors, they vary in width—think of a

transition zone instead of a line. Climatic ecotones usually are more gradual than physical ecotones, where differences in soil or topography sometimes form abrupt boundaries. An ecotone between prairies and northern forests may occupy many kilometers of land. The ecotone is an area of tension, as similar species of plants and animals compete for the resource base.

Scientists have identified these boundary and edge areas as generating biodiversity (a species richness) as adaptation and natural selection favor the population with reproductive success. Organisms may compete, yet they may act to facilitate other organisms in a positive interaction in whatever strategies affect rates of ecosystem productivity, such as pollinating insects. A process of *coevolution* is possible where interacting populations in the tension zone can lead to genetic changes in each population (*microevolution*). Edges and ecotones are dynamic areas in the evolution of ecosystems and communities.

Terrestrial Ecosystems

A **terrestrial ecosystem** is a self-sustaining association of land-based plants and animals and their abiotic environment, characterized by specific plant formation classes. Plants are the most visible part of the biotic landscape, and they are key members of Earth's terrestrial ecosystems. In their growth, form, and distribution, plants reflect Earth's physical systems: its energy patterns; atmospheric composition; temperature and winds; air masses; water quantity, quality, and seasonal timing; soils; regional climates; geomorphic processes; and ecosystem dynamics. As discussed in the previous chapter, aquatic ecosystems are likewise important and result from the interaction of physical systems (see News Report 20.1).

A brief review of the ways that living organisms relate to the environment and to each other is helpful at this point. Interacting populations of plants and animals in an area form a *community*. An *ecosystem* involves the interplay between a community of plants and animals and their abiotic physical environment. Each plant and animal occupies an area in which it is biologically suited to live—its *habitat*—and within that habitat it performs a basic operational function—its *niche* (see Chapter 19).

A **biome** is a large, stable, terrestrial ecosystem characterized by specific plant and animal communities. Each biome usually is named for its *dominant vegetation*, because that is the single most easily identified feature. We can generalize Earth's wide-ranging plant species into six broad biomes: *forest, savanna, grassland, shrubland, desert,* and *tundra*. Because plant distributions respond to environmental conditions and reflect variations in climate and soil, the world climate map in Figure 10.5 in Chapter 10 is a helpful reference for this chapter. The biome concept can also be applied to aquatic ecosystems: polar seas, temperate seas, tropical seas, seafloor, shoreline, and coral reef.

More specific characteristics are used for the structural classification of plants themselves. *Life-form designations* are based on the outward physical properties of

Table 20.1 Major Terrestrial Biomes and Their Characteristics

Biome and Ecosystems (map symbol)	Vegetation Characteristics	Soil Orders (Soil Taxonomy and CSSC)	Climate Type	Annual Precipitation Range	Temperature Patterns	Water Balance
Equatorial and Tropical Rain Forest (ETR) Evergreen broadleaf forest Selva	Leaf canopy thick and continuous; broadleaf evergreen trees, vines (lianas), epiphytes, tree ferns, palms	Oxisols Ultisols (on well-drained uplands)	Tropical	180–400 cm (>6cm/mo)	Always warm (21°–30°C; avg. 25°C)	Surpluses all year
Tropical Seasonal Forest and Scrub (TrSF) Tropical monsoon forest Tropical deciduous forest Scrub woodland and thorn forest	Transitional between rain forest and grasslands; broadleaf, some deciduous trees; open parkland to dense undergrowth; acacias and other thorn trees in open growth	Oxisols Ultisols Vertisols (in India) Some Alfisols	Tropical Monsoon savanna	130–200 cm (>40 rainy days during 4 driest months)	Variable, always warm (>18°C)	Seasonal surpluses and deficits
Tropical Savanna (TrS) Tropical grassland Thorn tree scrub Thorn woodland	Transitional between seasonal forests, rain forests, and semiarid tropical steppes and desert; trees with flattened crowns, clumped grasses, and bush thickets; fire association	Alfisols (dry: Ultalfs) Ultisols Oxisols	Tropical Savanna	9–150 cm, seasonal	No cold-weather limitations	Tends toward deficits, therefore fire- and drought-susceptible
Midlatitude Broadleaf and Mixed Forest (MBME) Temperate broadleaf Midlatitude deciduous Temperate needleleaf	Mixed broadleaf and needleleaf trees; deciduous broadleaf, losing leaves in winter; southern and eastern evergreen pines demonstrate fire association	Ultisols Some Alfisols (Podzols, red and yellow)	Humid Subtropical warm summer Humid Continental warm summer	75–150 cm	Temperate with cold season	Seasonal pattern with summer maximum PRECIP and POTET (PET); no irrigation needed
Needleleaf Forest and Montane Forest (NF/MF) Taiga Boreal forest Other montane forests and highlands	Needleleaf conifers, mostly evergreen pine, spruce, fir; Russian larch, a deciduous needleleaf	Spodosols Histosols Inceptisols Alfisols (Boralfs: cold) (Gleysols) (Podzols)	Subarctic Humid Continental cool summer	30–100 cm	Short summer, cold winter	Low POTET (PET), moderate PRECIP, moist soils, some waterlogged and frozen in winter; no deficits
Temperate Rain Forest (TeR) West coast forest Coast redwoods (U.S.)	Narrow margin of lush evergreen and deciduous trees on windward slopes; redwoods, tallest trees on Earth	Spodosols Inceptisols (mountainous environs) (Podzols)	Marine west coast	150–500 cm	Mild summer and mild winter for latitude	Large surpluses and runoff
Mediterranean Shrubland (MSh) Sclerophyllous shrubs Australian eucalyptus forest	Short shrubs, drought adapted, tending to grassy woodlands and chaparral	Alfisols (Xeralfs) Mollisols (Xerolls) (Luvisols)	Mediterranean dry summer	25–65 cm	Hot, dry summers, cool winters	Summer deficits, winter surpluses
Midlatitude Grasslands (MGr) Temperate grassland Sclerophyllous shrub	Tallgrass prairies and shortgrass steppes, highly modified by human activity; major areas of commercial grain farming; plains, pampas, and veld	Mollisols Aridisols (Chernozemic)	Humid subtropical Humid Continental hot summer	25–75 cm	Temperate continental regimes	Soil moisture utilization and recharge balanced; irrigation and dry farming in drier areas
Warm Desert and Semidesert (DBW) Subtropical desert and scrubland	Bare ground graduating into xerophytic plants including succulents, cacti, and dry shrubs	Aridisols Entisols (sand dunes)	Desert Arid		Average annual temperature, around 18°C, highest temperatures on Earth	Chronic deficits, irregular precipitation events, PRECIP POTET (PET)
Cold Desert and Semidesert (DBC) Midlatitude desert, scrubland, and steppe	Cold desert vegetation includes short grass and dry shrubs	Aridisols Entisols	Steppe Semiarid	2–25 cm	Average annual temperature around 18°C	PRECIP >½ POTET (PET)
Arctic and Alpine Tundra (AAT)	Treeless; dwarf shrubs, stunted sedges, mosses, lichens, and short grasses; alpine, grass meadows	Gelisols Histosols Entisols (permafrost) (Organic) (Cryosols)	Tundra Subarctic very cold	15–180 cm	Warmest months only 2 or 3 mo above freezing	Not applicable most of the year, poor drainage in summer
Ice			Ice sheet ice cap			

News Report 20.1

Aquatic Ecosystems and the LME Concept

Oceans, estuaries, and freshwater bodies (streams and lakes) are home to aquatic ecosystems. Poor understanding of aquatic ecosystems has led to the demise of some species, such as herring in the Georges Bank prime fishing area of the Atlantic, and an overall decline in fisheries worldwide. One analytical tool is the designation of certain areas as large marine ecosystems (LMEs, **http://woodsmoke.edc.uri.edu/Portal/**). Kenneth Sherman and Lewis Alexander, scientists with the U.S. Marine Fisheries Service Laboratory and the University of Rhode Island, respectively, conceived the LME concept to encourage resource planners, managers, and researchers to consider complete ecosystems, not just a targeted species.

The Large Marine Ecosystems Strategy is a worldwide scientific and political effort to understand these aquatic ecosystems and move closer to sustainability of these resources. Involved is the World Conservation Union (IUCN), the Intergovernmental Oceanographic Commission of UNESCO (IOC) and other United Nations agencies, and the National Oceanic and Atmospheric Administration (NOAA).

An LME is a distinctive oceanic region having unique organisms, ocean-floor topography, currents, areas of nutrient-rich upwelling circulation, or areas of significant predation, including human. Examples of identified LMEs include the Gulf of Alaska, California Current, Gulf of Mexico, Northeast Continental Shelf, and the Baltic and Mediterranean Seas. Some 62 LMEs, each encompassing more than 200,000 km² (77,200 mi²), are presently defined worldwide.

The largest government-protected area in North America is the Monterey Bay National Marine Sanctuary, established in 1992 within the California Current LME. (The Florida Keys Marine Sanctuary is second largest.) Stretching along 645 km (400 mi) of coastline and covering 13,500 km² (5300 mi²), the Monterey Bay sanctuary is home to 27 species of marine mammals, 94 species of shorebirds, 345 species of fish, and the largest sampling of invertebrates in any one place in the Pacific. This represents one of the most species-rich and diverse marine communities on Earth (see Monterey Bay Aquarium at **http://www.mbayaq.org/**). In the United States, NOAA and The National Marine Sanctuaries administer these protected areas. The web site is at **http://sanctuaries.noaa.gov/**.

---

individual plants or the general form and structure of a vegetation cover. These physical life forms include:

- *Trees* (large woody main trunk, perennial; usually exceeds 3 m, or 10 ft, in height)
- *Lianas* (woody climbers and vines)
- *Shrubs* (smaller woody plants; stems branching at the ground)
- *Herbs* (small plants without woody stems; includes grasses and other nonwoody vascular plants)
- *Bryophytes* (nonflowering, spore-producing plants; includes mosses and liverworts)
- *Epiphytes* (plants growing above the ground on other plants, using them for support)
- *Thallophytes* (lack true leaves, stems, or roots; include bacteria, fungi, algae, lichens)

These plant assemblages in their general biomes are divided into more-specific vegetation units called **formation classes**, which refer to the structure and appearance of dominant plants in a terrestrial ecosystem. Examples are equatorial rain forest, northern needleleaf forest, Mediterranean shrubland, and arctic tundra. Each formation class includes numerous plant communities, and each community includes innumerable plant habitats. Within those habitats, Earth's diversity is expressed in 270,000 plant species. Despite this intricate complexity, we can generalize Earth's numerous formation classes into 10 global terrestrial biome regions, as portrayed in Figure 20.4 and detailed in Table 20.1.

Now let us go on a tour of Earth's biomes, keeping in mind that they synthesize all we have learned about the atmosphere, hydrosphere, lithosphere, and biosphere in the pages of this text. Here we bring it all together.

Earth's Major Terrestrial Biomes

Few natural communities of plants and animals remain; most biomes have been greatly altered by human intervention. Thus, the "natural vegetation" identified on many biome maps reflects *ideal potential* mature vegetation, given the environmental characteristics in a region. Even in somewhat remote, pristine Norway, the former needleleaf and montane forest today is a mix of second- and third-growth forests, farmlands, and altered landscapes (Figure 20.3). Even though human practices have greatly altered these ideal forms, it is valuable to study the natural (undisturbed) biomes to better understand the environment and to assess the extent of human-caused alteration.

Knowing the ideal in a region guides us to a closer approximation of natural vegetation in the plants we

FIGURE 20.3 Needleleaf forest landscape modified by human activity.
Forests and farms sweep down to meet the long, narrow Vagavatnet Lake near the town of Vagamo
in south-central Norway, at 62° N latitude. [Photo by Bobbé Christopherson.]

introduce. In the United States and Canada, we humans are perpetuating a type of *transition community*, somewhere between grasslands and a forest—an artificial successional stage, produced by large-scale interruption and disturbance. We plant trees and lawns and then must invest energy, water, and capital to sustain such artificial modifications. Or we graze animals and plant crops that perpetuate this disturbed transitional state.

When these activities cease, natural succession recovers, and we see the land slowly return to its various vegetation-cover potentials. Some irreversible damage might occur when plants, animals, and organisms are knowingly or accidentally brought into an ecosystem or biome in which they are not native. These exotic species pose an increasing concern to scientists and society—a subject of News Report 20.2.

News Report 20.2

Alien Invaders of Exotic Species

The title above sounds like that of a cult science fiction movie; instead, it refers to a real problem in the integrity of many ecosystems, both aquatic and terrestrial. Animal and plant species brought or somehow getting into biomes outside their native home is an important subject in biogeography and invasion biology. Niche takeovers by nonnative species can prove damaging to otherwise healthy ecosystems. These intruders are known as *exotic species*, or *alien* or *nonnative species*.

The Institute for Biological Invasions at the University of Tennessee at **http://invasions.bio.utk.edu/bio_invasions/index.html** is dedicated to studying these invasive plants, animals, and other organisms such as zebra

mussels. Some states post warning signs at their borders (Figure 20.2.1a) and others offer web sites, such as Minnesota's "Prohibited and Noxious Plants by Scientific Name" (**http://www.dnr.state.mn.us/invasives/index.html**).

The African "killer bees" are in the headlines whenever an attack occurs, as are brown tree snakes in Guam, zebra mussels in the Great Lakes, and Eurasian cheatgrass in the Utah desert, to name but a few examples. Brought from Africa to the central coast of Brazil in 1957 to increase honey production, killer bees now have interbred with native bees and range to southern California, Arizona, New Mexico, Texas, and Puerto Rico. More than 1000 people have died

from their attacks. The dangers are in their mass response to disturbances and that some people are allergic to even one bee sting. This is a classic case of geographical diffusion of an exotic species.

Probably 90% of exotic species fail when they try to move into established niches in a community. The 10% that succeed, some 4500 species documented in a 1993 Office of Technology Assessment report, prove damaging to as much as one-fifth of ecosystems they invade. Figure 20.2.1b, 1c, 1d, and e shows four such nonnative species: purple loosestrife, kudzu, gorse, and zebra mussels.

Purple loosestrife (*Lythrum salicaria*) was introduced from Europe in the 1800s as a desired ornamental and

(a)

(b)

(c)

(d)

(e)

FIGURE 20.2.1 Exotic species.
(a) A sign in Wallowa County, Washington State, makes a plea for awareness in this agricultural country, listing noxious (nonnative, often problematic) weeds at the county line. (b) Invasive purple loosestrife near Lake Michigan and the Indiana Dunes National Lakeshore, Indiana (foreground and left center). (c) Kudzu overruns pasture and forest in western Georgia. (d) Gorse (yellow-flowering shrub), intended for barriers between pastures, is loose across the landscape in West Point, Falkland Islands. (e) Zebra mussels literally cover most hard surfaces in the Great Lakes. [Photos by (a), (b), (c), and (d) Bobbé Christopherson; (e) David M. Dennis, Maxximages.com.]

had some medicinal applications. The plant's seeds also arrived in ships that used soil for ballast. A hardy perennial, it got loose and invaded wetlands across the eastern portions of the United States and Canada, through the upper Midwest, and as far west as Vancouver Island, British Columbia, replacing plants on which native wildlife depend (Figure 20.2.1b). The plant is known to infect drier landscapes as well and poses a potential threat to agriculture.

The infamous kudzu (*Pueraria montana*) was brought from Japan in 1876 for cattle feed, erosion control, and as an ornamental. It has spread to east Texas, southern Pennsylvania, and across the South into Florida. The plant can grow 0.3 m (1 ft) on a hot, humid summer day and overtake

forests and structures. Serious effort is going into finding some practical use for the prodigious plant, from pulp (treeless paper), to recipes for cooking, to the telling of new kudzu jokes.

Gorse (*Ulex*) was introduced into the timber-poor Falkland Islands as a barrier plant, like a substitute fence. Its thorny stickers and dense growth were meant to keep grazing sheep and a few cattle under control. The plant spread beyond this intended use (Figure 20.2.1d).

Zebra mussel (*Dreissena polymorpha*) adults are up to 3.8 cm (1.5 in.) long and have shells with alternating brown and black colored stripes. The females can produce more than 1 million eggs a year. Zebra mussels are native to Europe and Western Russia and were discovered in the Great Lakes in

1988 in ballast water dumped from oceangoing ships. Microscopic larvae travel in the water. Also, they attach to boat hulls, fishing gear and nets, and boat equipment. Zebra mussels foul beaches, disrupt natural food webs, destroy native mussels, clog water systems, and lead to fish and wildlife die-offs (Figure 20.2.1e). They have spread throughout the Great Lakes and most of the Mississippi River and are now in other rivers and inland lakes.

In learning about biomes, we find out about native plant, animal, and organism locations and the unique place each species occupies. We must avoid knowingly or perhaps unconsciously adding to an escalating problem by bringing exotic species into nonnative sites where they do not naturally belong.

FIGURE 20.4 The 10 major global terrestrial biomes.

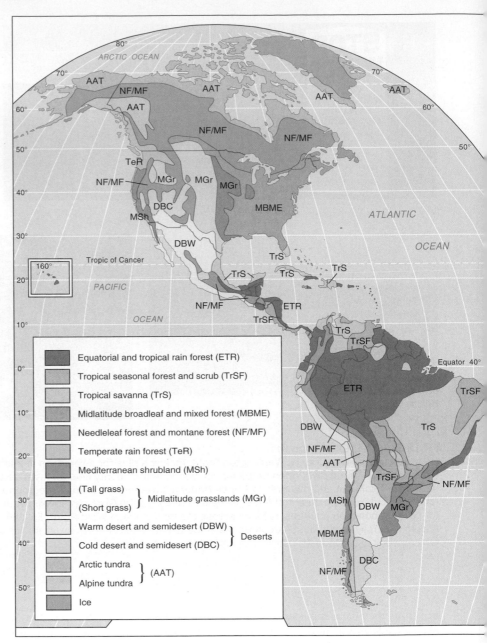

The legend reads:

Equatorial and tropical rain forest (ETR)
Tropical seasonal forest and scrub (TrSF)
Tropical savanna (TrS)
Midlatitude broadleaf and mixed forest (MBME)
Needleleaf forest and montane forest (NF/MF)
Temperate rain forest (TeR)
Mediterranean shrubland (MSh)
(Tall grass) } Midlatitude grasslands (MGr)
(Short grass)
Warm desert and semidesert (DBW) } Deserts
Cold desert and semidesert (DBC)
Arctic tundra } (AAT)
Alpine tundra
Ice

The global distribution of Earth's major terrestrial biomes, based on vegetation formation classes, is portrayed in Figure 20.4. Table 20.1 describes each biome on the map and summarizes other pertinent information gathered from throughout *Geosystems*.

Now is a good time to refer to "The Living Earth" composite image that appears inside the front cover of this text. A computer artist using hundreds of thousands of satellite images produced this cloudless view of Earth in natural color typical of a local summer day. Compare the map in Figure 20.4 with this remarkable composite image and see what correlations you can make. Then compare the biomes with population distribution as indicated by the nighttime lighting display also on the inside front cover.

Equatorial and Tropical Rain Forest

Earth is girdled with a lush biome—the **equatorial and tropical rain forest**. In a climate of consistent year-round daylength (12 hours), high insolation, average annual temperatures around 25°C (77°F), and plentiful moisture, plant and animal populations have responded with the most diverse expressions of life on the planet.

As Figure 20.4 shows, the Amazon region, also called the *selva*, is the largest tract of equatorial and tropical rain forest. In addition, rain forests cover equatorial regions of Africa, Indonesia, the margins of Madagascar and Southeast Asia, the Pacific coast of Ecuador and Colombia, and the east coast of Central America, with small discontinuous patches elsewhere. The cloud forests of western Venezuela are such tracts

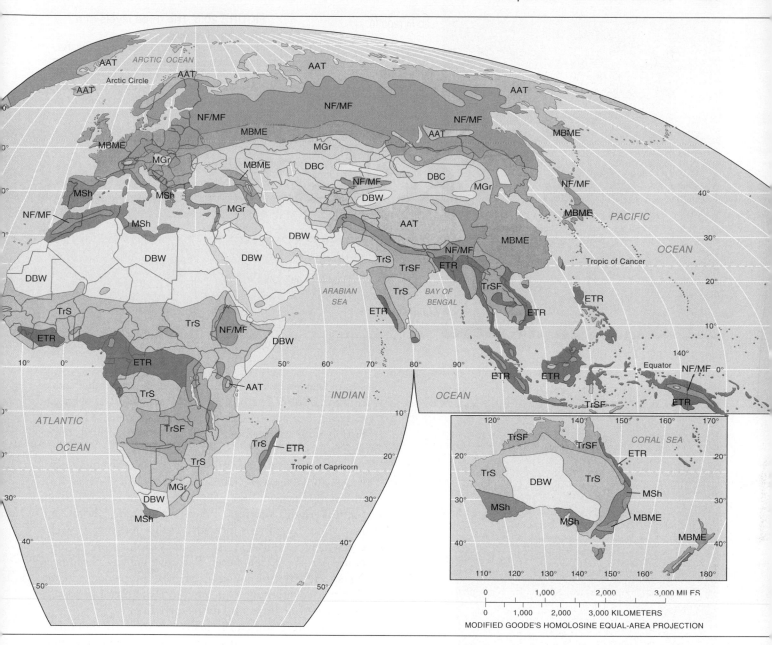

MODIFIED GOODE'S HOMOLOSINE EQUAL-AREA PROJECTION

of rain forest at high elevation, perpetuated by high humidity and cloud cover. Undisturbed tracts of rain forest are rare.

Rain forests represent approximately one-half of Earth's remaining forests, occupying about 7% of the total land area worldwide. This biome is stable in its natural state, resulting from the long-term residence of these continental plates near equatorial latitudes and their escape from glacial activity.

Rain forests feature ecological niches that are distributed vertically rather than horizontally because of the competition for light. Biomass in a rain forest is concentrated high up in the canopy, that dense mass of overhead leaves. The canopy is filled with a rich variety of plants and animals. *Lianas* (vines) stretch from tree to tree,

entwining them with cords that can reach 20 cm (8 in.) in diameter. *Epiphytes* flourish there too—plants such as orchids, bromeliads, and ferns live entirely aboveground, supported physically but not nutritionally by the structures of other plants. Calm conditions on the forest floor make wind pollination difficult, so pollination is by insects, other animals, and self-pollination.

The rain forest canopy forms three levels—see Figure 20.5. The upper level is not continuous, but features emergent tall trees whose high crowns rise above the middle canopy. This upper level appears as a broken *overstory* of tall trees breaking through a middle canopy that is nearly continuous (Figure 20.5b). The broad leaves block much of the light and create a darkened *understory* area and forest floor. The lower level is composed of

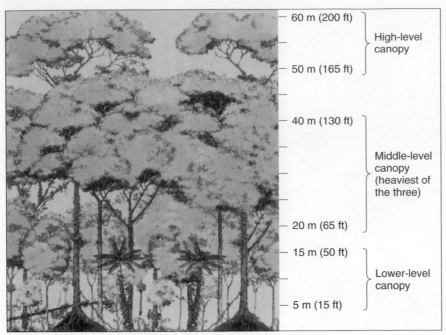

High-level canopy
— 60 m (200 ft)
— 50 m (165 ft)

Middle-level canopy (heaviest of the three)
— 40 m (130 ft)
— 20 m (65 ft)

Lower-level canopy
— 15 m (50 ft)
— 5 m (15 ft)

(a)

FIGURE 20.5 The three levels of a rainforest canopy.
(a) Lower, middle, and high levels of the rain forest. (b) Undisturbed rain forest on the Peninsula de Osa in Costa Rica. Note a few high-level canopy trees that extend above the dense, continuous cover of the middle canopy. [Photo by Barbara Cushman Rowell/Mountain Light Photography, Inc.]

(b)

seedlings, ferns, bamboo, and the like, leaving the litter-strewn ground level in deep shade and fairly open.

Aerial photographs of a rain forest, or views along riverbanks covered by dense vegetation, or the false Hollywood-movie imagery of the jungle make it difficult to imagine the shadowy environment of the actual rainforest floor, which receives only about 1% of the sunlight arriving at the canopy (Figure 20.6a). The constant moisture, rotting fruit and moldy odors, strings of thin roots and vines dropping down from above, windless air, and echoing sounds of life in the trees together create a unique environment.

The smooth, slender trunks of rainforest trees are covered with thin bark and buttressed by large wall-like flanks that grow out from the trees to brace the trunks (Figure 20.6b). These buttresses form angular, open enclosures, a ready habitat for various animals. There are usually no branches for at least the lower two-thirds of the tree trunks.

The wood of many rainforest trees is extremely hard, heavy, and dense—in fact, some species will not even float in water. Exceptions are balsa and a few others, which are light. Varieties of trees include mahogany, ebony, and rosewood. Logging is difficult because individual species are widely scattered; a species may occur only once or twice per square kilometer. Selective cutting is required for species-specific logging, whereas pulpwood production takes everything. Conversion of the forest to pasture usually is done by setting destructive fires.

Rainforest soils, principally Oxisols, are essentially infertile, yet they support rich vegetation. The trees have adapted to these soil conditions with root systems able to capture nutrients from litter decay at the soil surface. High precipitation and temperatures work to produce deeply weathered and leached soils, characteristic of the laterization process, with a claylike texture sometimes breaking up into a granular structure. Oxisols lack nutrients and colloidal material. With much investment in fertilizers, pesticides, and machinery, these soils can be productive.

The animal and insect life of the rain forest is diverse, ranging from small decomposers (bacteria) working the surface to many animals living exclusively in the upper stories of the trees. These tree dwellers are referred to as *arboreal*, from the Latin for "tree," and include sloths, monkeys, lemurs, parrots, and snakes. Beautiful birds of many colors, tree frogs, lizards, bats, and a rich insect community that includes more than 500 species of butterflies are found in rain forests. Surface animals include pigs (bushpigs and the giant forest hog in Africa, wild boar and

(a)

(b)

FIGURE 20.6 The rain forest.
(a) Equatorial rain forest is thick along the Amazon River where light breaks through to the surface, producing a rich gallery of vegetation. (b) The rainforest floor in Corcovado, Costa Rica, with typical buttressed trees and lianas. [Photos by (a) Wolfgang Kaehler Photography; (b) Frank S. Balthis.]

bearded pig in Asia, and peccary in South America), species of small antelopes (bovids), and mammalian predators (the tiger in Asia, jaguar in South America, and leopard in Africa and Asia).

The present human assault on Earth's rain forests has put this diverse fauna and the varied flora at risk. It also jeopardizes an important recycling system for atmospheric carbon dioxide and a potential source of valuable pharmaceuticals and many types of new foods—and so much is still unknown and undiscovered.

Deforestation of the Tropics

More than half of Earth's original rain forest is gone, cleared for pasture, timber, fuel wood, and farming. Worldwide, an area nearly the size of Wisconsin is lost each year (169,000 km²; 65,000 mi²) and about a third more is disrupted by selective cutting of canopy trees, a damage that occurs along the *edges* of deforested areas.

When orbiting astronauts look down on the rain forests at night, they see thousands of human-set fires. During the day, the lower atmosphere in these regions is choked with the smoke. These fires are used to clear land for agriculture, which is intended to feed the domestic population as well as to produce accelerating cash exports of beef, rubber, coffee, soybeans, and other commodities. Edible fruits are not abundant in an undisturbed rain forest, but cultivated clearings produce bananas (plantains), mangos, jack fruit, guava, and starch-rich roots such as manioc and yams.

Because of the poor soil fertility, the cleared lands are quickly exhausted under intensive farming and are then generally abandoned in favor of newly burned and cleared lands, unless fertility is maintained artificially. The dominant trees require from 100 to 250 years to reestablish themselves after major disturbances. Once cleared, the former forest becomes a mass of low bushes

intertwined with vines and ferns, slowing the return of the forests.

Figure 20.7a–c shows satellite false-color images of a portion of western Brazil called Rondônia, recorded in June 1975 (*Landsat 2*), August 1986 (*Landsat 5*), and June 1992 (*Landsat 4*) and in (d) true color in June 2001 (*Terra*). These images give a sense of the level of rainforest destruction in progress. You can clearly see encroachment along new roads branching from highway BR364. The *edges* of every road and cleared area represent a significant portion of the species habitat disturbance, population dynamic changes, and carbon losses to the atmosphere—perhaps as much as 1.5 times more impact occurs along these edges than in the tracts of clear-cut logging. The hotter, drier, and windier edge conditions penetrate the forest up to 100 m. Figure 20.7e shows a tract of former rain forest that has just been burned to begin the clearing and road-building process.

In 1998–1999, despite all the efforts to introduce sustainable forestry practices, deforestation was 33,926 km² (13,100 mi²) in 2 years. Since 1972, the losses totaled more than the area of the north-central states of Minnesota, Iowa, Missouri, North and South Dakota, Nebraska, and Kansas, noted on the map in Figure 20.7f. The loss in 2000 was 19,836 km² (7660 mi²), an area roughly equivalent to Connecticut, Rhode Island, and Delaware combined.

Estimates based on satellite imagery and ground surveys place the losses for the 3 years 2001–2003 of an area equal to slightly less than the states of Vermont, New Hampshire, and Massachusetts combined. The losses in 2004–2005 are about half the area of Maine. In 1970, only about 1% of the Brazilian Amazon had been deforested, yet here we see lost since 1972 forest area more than the area of France.

This is a highly charged issue, as can be seen by the growth of Brazil's cattle industry that uses these new pasture lands. Cattle herds will increase from 25 million head (1990) to a forecasted 200 million head by 2010. Brazil is

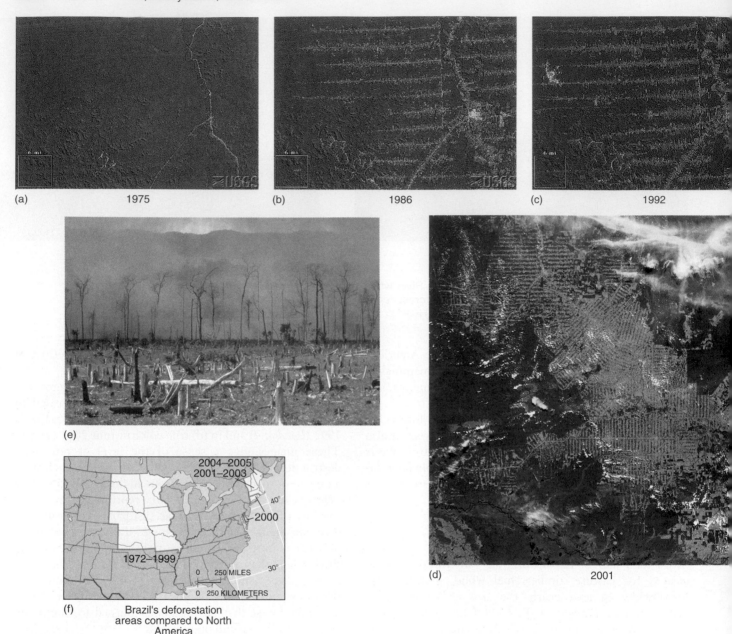

(a) 1975

(b) 1986

(c) 1992

(e)

(f) Brazil's deforestation areas compared to North America

2004–2005
2001–2003
2000
1972–1999

0 250 MILES
0 250 KILOMETERS

(d) 2001

FIGURE 20.7 Amazon rainforest losses over three decades.
Satellite images of a portion of western Brazil called Rondônia, recorded in false color in (a) June 1975 (*Landsat 2*), (b) August 1986 (*Landsat 5*), and (c) June 1992 (*Landsat 4*) and in true color (d) June 1, 2001 (the area of the other three images is in the upper left). Highway BR364, the main artery in the region, passes through the towns of Jaru and Ji-Paraná near the center of image (d) at about 62° W 11° S. The branching pattern of feeder roads encroaches on the rain forest. (e) Surface view of burning rain forest. (f) Deforested areas in Brazil, cleared between 1972–1999, 2000, 2001–2003, and 2004–2005, compared to North America. [*Landsat* images courtesy of NASA/USGS EROS Data Center; *Terra* image courtesy of NASA/MODIS Land Rapid Response Team; (e) photo by George Holton/Photo Researchers, Inc.]

a major beef exporter and consumer of its own rainforest resource base.

The United Nations Food and Agricultural Organization (FAO) estimates that if this destruction to rain forests continues unabated, they will be completely removed by 2050. By continent, rainforest losses are estimated at more than 50% in Africa, more than 40% in Asia, and 40% in Central and South America. Among many references, see Tropical Rainforest Coalition at **http://www.rainforest.org/**, World Resources Institute publications on forests at **http://www.globalforestwatch.org/english/index.htm**, the Rainforest Information Centre at **http://www.rainforestinfo.org.au/welcome.htm**, or the Rainforest Action Network at **http://www.ran.org/**.

Focus Study 20.1 looks more closely at efforts to curb increasing rates of species extinction and loss of Earth's biodiversity, much of which is directly attributable to the loss of rain forests.

Focus Study 20.1

Biodiversity and Biosphere Reserves

When the first European settlers landed on the Hawaiian Islands in the late 1700s, they counted 43 species of birds. Today, 15 of those species are extinct and 19 more are threatened or endangered, with only one-fifth of the original species relatively healthy. Humans have great impact on animals and plants and global biodiversity.

Relatively pristine habitats around the world are being lost at unprecedented rates as an expanding human population converts them to agriculture, forestry, and urban centers. As these habitats are altered, untold numbers of species are disappearing before they have been recognized, much less studied, and the functioning of entire ecosystems is threatened. This loss of biodiversity, at the very time when the value of biotic resources is becoming widely recognized, has made it strikingly clear that current strategies for conservation are failing dismally.*

As more is learned about Earth's ecosystems and their related communities, more is known of their value to civilization and our interdependence with them. Natural ecosystems are a major source of new foods, new chemicals, new medicines, and specialty woods, and of course they are indicators of a healthy, functioning biosphere.

International efforts are under way to study and preserve specific segments of the biosphere; among others: the UN Environment Programme, World Resources Institute, The World Conservation Union, Natural World Heritage Sites, Wetlands of International Importance, and the Nature Conservancy. The International Union for Conservation of Nature and Natural Resources (IUCN) Red List of global endangered species provide an important focus. (See IUCN Red List at http://www.iucnredlist.org/, the IUCN at http://www.iucn.org/, the endangered species home page of the Fish and Wildlife Service at http://www.fws.gov/endangered/, and the World Wildlife Fund at http://www.panda.org/.) The motivation to set aside natural sanctuaries is directly related to concern over the increase in the rate of species extinctions. We are facing a loss of genetic diversity that may be unparalleled in Earth's history, even compared with the major extinctions over the geologic record.

Species Threatened—Examples

The IUCN in 2006 listed the polar bear (*Ursus maritimus*) as *Vulnerable* to extinction, a "Threatened" category on their Endangered Species List—the *Red Book*. In spring 2007, during the 90-day comment period concerning the bears' status, the Fish and Wildlife Service received more than 600,000 letters concerning the inclusion of polar bears under the Endangered Species Act. Research released by the USGS in September 2007 predicted that with the loss of Alaskan, Canadian, and Russian sea-ice habitat, some two-thirds of the world's polar bears will die off by 2050. The remaining 7500 bears will be struggling (Figure 20.1.1). The Fish and Wildlife Service opened a second comment period over these findings. Any action society takes now will hopefully result in a slowing of such sea-ice loss and the buying of a little adaptation time.

Black rhinos (*Diceros bicornis*) and white rhinos (*Ceratotherium simum*) in Africa exemplify species in jeopardy. Rhinos once grazed over much of the savanna grasslands and woodlands. Today, they survive only in protected

FIGURE 20.1.1 The polar bear.
A male polar bear, snow on his face, seems to be looking directly at us with questions in his eyes, wondering about who we are. Polar bear vision is about the same quality as human sight; however, they have an extraordinary sense of smell. One study predicts that 67% of the remaining polar bear population of 23,000 will be gone by 2050 because of climate change and loss of sea ice—that will leave 7590 bears across the Arctic region. [Photo by Bobbé Christopherson.]

NOTEBOOK

Polar Bear Photo Gallery
Polar Bear Movie - Mom and Cubs on the Park Ice

*T. D. Sisk, A. E. Launer, K. R. Switky, and P. R. Ehrlich, "Identifying extinction threats," *Bioscience* 44 (October 1994): 592.

(continued)

Focus Study 20.1 (continued)

districts in heavily guarded sanctuaries and are threatened even there (Figure 20.1.2). In 1998, 2599 black rhinos remained, 96% fewer than the 70,000 in 1960. Populations remained statistically stable through the last half of the 1990s; this count is supported by South Africa's guarding of about 50% of the remaining population. A slow recovery is underway as numbers of black rhinos rose to 3,600 in 2003 (source: IUCN/SSC/African Rhino Specialist Group). This is not the case with the western black rhino, a subspecies, which is down to its last few individuals.

There were 11 white rhinos (northern subspecies) in existence in 1984, increasing to an estimated 29 by 1998 (Congo, 25; Côte d'Ivoire, 4), although the IUNC suggests that in 2005 there might actually be only 5 to 10 of these rhinos left. Political unrest in the Congo and the area of the Garamba National Park has led to the deaths of an unknown number of these few. The southern white rhino population reached 11,000 by 2002. Rhinoceros horn sells for $29,000 per kilogram for use as a supposed aphrodisiac. These large land mammals are nearing extinction and will

Table 20.1.1   Known and Estimated Species on Earth

Categories of Living Organisms	Number of Known Species	Estimated Number of Species High (,000)	Estimated Number of Species Low (,000)	Working Estimate (,000)	Accuracy
Viruses	4,000	1,000	50	400	Very poor
Bacteria	4,000	3,000	50	1000	Very poor
Fungi	72,000	27,000	200	1500	Moderate
Protozoa	40,000	200	60	200	Very poor
Algae	40,000	1,000	150	400	Very poor
Plants	270,000	500	300	320	Good
Nematodes	25,000	1,000	100	400	Poor
Arthropods					
Crustaceans	40,000	200	75	150	Moderate
Arachnids	75,000	1,000	300	750	Moderate
Insects	950,000	100,000	2,000	8,000	Moderate
Mollusks	70,000	200	100	200	Moderate
Chordates	45,000	55	50	50	Good
Others	115,000	800	200	250	Moderate
Total	**1,750,000**	**111,655**	**3,635**	**13,620**	**Very poor**

Source: United Nations Environment Programme, *Global Biodiversity Assessment* (Cambridge: Cambridge University Press, 1995), Table 3.1–2, p. 118.

survive only as a dwindling zoo population. The limited genetic pool that remains complicates further reproduction.

Table 20.1.1 summarizes the numbers of known and estimated species on Earth. Scientists have classified only 1.75 million species of plants and animals of an estimated 13.6 million overall; this figure represents an increase in what scientists once thought to be the diversity of life on Earth. And those yet-to-be-discovered species represent a potential future resource for society—the wide range of estimates places the expected species count between a low of 3.6 million and a high of 111.7 million.

Estimates of annual species loss range between 1000 and 30,000, although this range might be conservative. The possibility exists that more than half of Earth's present species could be extinct within the next 100 years. The effects of pollution, climate change, loss of wild habitat, excessive grazing, poaching, and collecting are at the root of such devastation. Approximately 60% of the extinctions are attributable to the clearing and loss of rain forests alone.

FIGURE 20.1.2 The rhinoceros in Africa.
Black rhino and young in Tanzania, escorted by Oxpecker birds. These rhinos are quite nearsighted (they can see clearly only up to 10 m); the birds act as the rhino's early-warning system for disturbances in the distance. [Photo by Stephen J. Krasemann/DRK Photo.]

The Question of Medicine and Food

Wheat, maize (corn), and rice—just three grains—fulfill about 50% of human planetary food demands. About 7000 plant species have been gathered for food throughout human history, but more than 30,000 plant species have edible parts. Undiscovered potential food resources are in nature waiting to be found and developed. Biodiversity, if preserved, provides us a potential cushion for all future food needs, but only if species are identified, inventoried, and protected. The same is true for pharmaceuticals.

Nature's biodiversity is like a full medicine cabinet. Since 1959, 25% of all prescription drugs were originally derived from higher plants. In preliminary surveys, 3000 plants have been identified as having anticancer properties. The rosy periwinkle (*Catharanthus roseus*) of Madagascar contains two alkaloids that combat two forms of cancer. (Alkaloids are compounds found in certain plants that help the plant defend against insects and are potentially significant as human medicines; examples are atropine, quinine, and morphine.) Yet, less than 3% of flowering plants have been examined for alkaloid content. It defies common sense to throw away the medicine cabinet before we even open the door to see what is inside.

Biosphere Reserves

Formal natural reserves are a possible strategy for slowing the loss of biodiversity and protecting this resource base. Setting up a *biosphere reserve* involves principles of *island biogeography*. Island communities are special places for study because of their spatial isolation and the relatively small number of species present. Islands resemble natural experiments, because the impact of individual factors, such as civilization, can be more easily assessed on islands than over larger continental areas.

Studies of islands also can assist in the study of mainland ecosystems, for in many ways a park or biosphere reserve, surrounded by modified areas and artificial boundaries, is like an island. Indeed, a biosphere reserve is conceived as an ecological island in the midst of change. The intent is to establish a core in which genetic material is protected from outside disturbances, surrounded by a buffer zone that is, in turn, surrounded by a transition zone and experimental research areas. An important variable to consider in setting aside a biosphere reserve is any change that might occur in temperature and precipitation patterns as a result of global change. A carefully considered reserve could end up outside its natural range in several decades as ecotones shift. Scientists predict that new, undisturbed reserves will not be possible in a decade, because pristine areas will be gone. The biosphere reserve program is coordinated by the Man and the Biosphere (MAB) Programme of UNESCO (**http://www.unesco.org/mab/**). More than 480 biosphere reserves are now operated voluntarily in 100 countries.

Not all protected areas are ideal bioregional entities. Some are simply imposed on existing park space and some remain in the planning stage, although they are officially designated. Some of the best biosphere reserve examples have been in operation since the late 1970s and range from the struggling Everglades National Park in Florida, to the Changbai Reserve in China, to the Tai Forest on the Côte d'Ivoire (Ivory Coast), to the Charlevoix Biosphere Reserve in Québec, Canada. Added to these efforts is the work of the Nature Conservancy, which acquires land for preservation as a valuable part of the reserve (**http://nature.org/**).

The ultimate goal, about half achieved, is to establish at least one reserve in each of the 194 distinctive biogeographic communities presently identified. UNESCO and the World Conservation Monitoring Centre compile the United Nations' list of national parks and protected areas. Presently in these biogeographic communities, 6930 areas covering 657 million hectares (1.62 billion acres) are designated in some protective form.

Tropical Seasonal Forest and Scrub

A varied biome on the margins of the rain forest is the **tropical seasonal forest and scrub**, which occupies regions of low and erratic rainfall. The shifting intertropical convergence zone (ITCZ) brings precipitation with the seasonally shifting high Sun of summer and then a season of dryness with the low Sun of winter, producing a seasonal pattern of moisture deficits, some leaf loss, and dry-season flowering. The term *semideciduous* applies to some of the broadleaf trees that lose their leaves during the dry season.

Areas of this biome have fewer than 40 rainy days during their four consecutive driest months, yet heavy monsoon downpours characterize their summers (see Chapter 6, especially Figure 6.20). The climates *tropical monsoon* and *tropical savanna* apply to these transitional communities between rain forests and tropical grasslands.

Portraying such a varied biome is difficult. In many areas humans disturb the natural biome so that the savanna grassland adjoins the rain forest directly. The biome does include a gradation from wetter to drier areas: monsoonal forests, to open woodlands and scrub woodland, to thorn forests, to drought-resistant scrub species (Figure 20.8). In South America, an area of transitional tropical deciduous forest surrounds an area of savanna in southeastern Brazil and portions of Paraguay.

The monsoonal forests average 15 m (50 ft) high with no continuous canopy of leaves, graduating into orchard-like parkland with grassy openings or into areas choked by dense undergrowth. In more open tracts, a common tree is the acacia, with its flat top and usually thorny stems.

(a)

(b)

FIGURE 20.8 Kenyan landscapes.
Two views of the open thorn forest and savanna near and in the Samburu Reserve, Kenya, representing a broad ecotone between the two biomes. [Photos by (a) Gael Summer; (b) Stephen J. Krasemann/DRK Photo.]

These trees have branches that look like an upside-down umbrella (spreading and open skyward), as do trees in the tropical savanna.

Local names are given to these communities: the *caatinga* of the Bahia State of northeastern Brazil, the *chaco* area of Paraguay and northern Argentina, the *brigalow* scrub of Australia, and the *dornveld* of southern Africa. Figure 20.4 shows areas of this biome in Africa, extending from eastern Angola through Zambia to Tanzania and Kenya; in southeast Asia and portions of India, from interior Myanmar through northeastern Thailand; and in parts of Indonesia.

The trees throughout most of this biome make poor lumber, but some, especially teak, may be valuable for fine cabinetry. In addition, some of the plants with dry-season adaptations produce usable waxes and gums, such as carnauba and palm-hard waxes. Animal life includes the koalas and cockatoos of Australia and the elephants, large cats, rodents, and ground-dwelling birds in other occurrences of this biome.

Tropical Savanna

The **tropical savanna** is large expanses of grassland, interrupted by trees and shrubs. This is a transitional biome between the tropical forests and semiarid tropical steppes and deserts. The savanna biome also includes treeless tracts of grasslands, and in dry savannas, grasses grow discontinuously in clumps, with bare ground between them. The trees of the savanna woodlands are characteristically flat-topped, in response to light and moisture regimes.

Savannas covered more than 40% of Earth's land surface before human intervention but were especially modified by human-caused fire. Fires occur annually throughout the biome. The timing of these fires is important. Early in the dry season they are beneficial and increase tree cover; late in the season they are very hot and kill trees and seeds. Savanna trees are adapted to resist the "cooler" fires.

Forests and elephant grasses averaging 5 m (16 ft) high once penetrated much farther into the dry regions, for they are known to survive there when protected. Savanna grasslands are much richer in humus than the wetter tropics and are better drained, thereby providing a better base for agriculture and grazing. Sorghums, wheat, and groundnuts (peanuts) are common commodities.

Tropical savannas receive their precipitation during less than 6 months of the year, when they are influenced by the shifting ITCZ. The rest of the year they are under the drier influence of shifting subtropical high-pressure cells. Savanna shrubs and trees are frequently *xerophytic*, or drought-resistant, with various adaptations to protect them from the dryness: small, thick leaves; rough bark; or leaf surfaces that are waxy or hairy.

Africa has the largest area of this biome, including the famous Serengeti Plains of Tanzania and Kenya and the Sahel region, south of the Sahara. Sections of Australia, India, and South America also are part of the savanna biome. Some of the local names for these lands include the *Llanos* in Venezuela, stretching along the coast and inland east of Lake Maricaibo and the Andes; the *Campo Cerrado* of Brazil and Guiana; and the *Pantanal* of southwestern Brazil.

Particularly in Africa, savannas are the home of large land mammals (zebra, giraffe, buffalo, gazelle, wildebeest, antelope, rhinoceros, elephant). They graze on savanna grasses (Figure 20.9) or feed upon the grazers themselves (lion, cheetah). Birds include the Ostrich, Martial Eagle (largest of all eagles), and Secretary Bird. Many species of venomous snakes as well as the crocodile are present in this biome.

However, in our lifetime we may see the reduction of these animal herds to zoo stock only, because of poaching and habitat losses, unless preservation efforts succeed. The loss of the rhino is discussed in Focus Study 20.1. Establishment of large tracts of savanna as biosphere reserves is critical for the preservation of this biome and its associated fauna.

(a)

(b)

FIGURE 20.9 Animals and plants of the Serengeti Plains.
(a) Savanna landscape of the Serengeti Plains, with wildebeest, zebras, and thorn forest. (b) Near Samburu, Kenya, Grevy's zebra and reticulated giraffe forage. [Photos by (a) Stephen F. Cunha; (b) Galen Rowell/Mountain Light Photography, Inc.]

Midlatitude Broadleaf and Mixed Forest

Moist continental climates support a mixed forest in areas of warm to hot summers and cool to cold winters. This **midlatitude broadleaf and mixed forest** biome includes several distinct communities in North America, Europe, and Asia. Along the Gulf of Mexico relatively lush evergreen broadleaf forests occur. Northward are mixed deciduous and evergreen needleleaf stands associated with sandy soils and burning. When areas are given fire protection, broadleaf trees quickly take over. Pines (longleaf, shortleaf, pitch, loblolly) predominate in the southeastern and Atlantic coastal plains. Into New England and westward in a narrow belt to the Great Lakes, white and red pines and eastern hemlock are the principal evergreens, mixed with deciduous oak, beech, hickory, maple, elm, chestnut, and many others (Figure 20.10a).

These mixed stands contain valuable timber, but their distribution has been greatly altered by human activity. Native stands of white pine in Michigan and Minnesota were removed before 1910; only later reforestation sustains their presence today. In northern China these forests have almost disappeared as a result of centuries of harvest. The forest species that once flourished in China are similar to species in eastern North America: oak, ash, walnut, elm, maple, and birch.

This biome is quite consistent in appearance from continent to continent and at one time represented the principal vegetation of the *humid subtropical hot summer* and *marine west coast* climatic regions of North America, Europe, and Asia.

A wide assortment of mammals, birds, reptiles, and amphibians is distributed throughout this biome. Representative animals (some migratory) include red fox, white-tail deer, southern flying squirrel, opossum, bear, and a great variety of birds, including Tanager and Cardinal (Figure 20.10b). To the west of this biome are the grasslands and to the north poorer soils and colder climates

(a)

(b)

FIGURE 20.10 The mixed broadleaf forest.
(a) A mixed broadleaf forest in the Midwest. (b) A young white-tailed deer grazes in the undergrowth of a mixed forest. [Photos by (a) Bobbé Christopherson; (b) John Shaw/Tom Stack & Associates.]

FIGURE 20.11 Climatic shifts impact the Midwest.
The relative future climatic conditions of Illinois in response to
changes in temperature and precipitation forecasted by 2090.
[After Canadian General Circulation Model (CGCM2) and *Climate
Change Impacts on the United States*, U.S. Global Change
Research Program, 2000.]

**FIGURE 20.12 Boreal forest of Canada (boreal means
"northern").** [Photo by author.]

that favor stands of coniferous trees and a gradual transition
to the northern needleleaf forests.

As with all terrestrial biomes, climate change is causing biome boundaries, the ecotone transition zones that
surround them, to shift. The Hardiness Zone planting
maps in Chapter 19, Figure 19.25, illustrated the changes
over just 16 years—change is underway. The climate
forecast models predict these transitions. For example,
Figure 20.11 illustrates the "migration" of Illinois this century based on changes in temperature and precipitation,
summer conditions in particular. By 2030, Illinois' summers will be similar to those of the Missouri–Arkansas border; by 2090, Illinois' climate will be similar to east Texas.
In turn, east Texas will be like north-central Mexico.

Simply stated, a physical geography text written 20 to
70 years from now will have to redo the biome maps to
reflect these climatic shifts. Imagine the change underway
for farm crops, pollinating insects, bird populations, cities
and energy-use patterns, and water resources.

Needleleaf Forest and Montane Forest

Stretching from the east coast of Canada and the Atlantic
provinces westward to Alaska and continuing from Siberia
across the entire extent of Russia to the European Plain is
the northern **needleleaf forest** biome, also called the
boreal forest (Figure 20.12). A more open form of boreal
forest, transitional to arctic and subarctic regions, is the
taiga. The Southern Hemisphere, lacking *humid microthermal* climates except in mountainous locales, has no such
biome. But **montane forests** of needleleaf trees exist
worldwide at high elevation.

Boreal forests of pine, spruce, fir, and larch occupy
most of the subarctic climates on Earth that are dominated
by trees. Although these forests are similar in formation,
individual species vary between North America and
Eurasia. The larch (*Larix*) is interesting because it is the
rare needleleaf tree that loses its needles in the winter
months, perhaps as a defense against the extreme cold of

its native Siberia (see the Verkhoyansk climograph and
photograph in Figure 10.21, Chapter 10). Larches also
occur in North America.

The Sierra Nevada, Rocky Mountains, Alps, and
Himalayas have similar forest communities occurring at
lower latitudes. Douglas fir and white fir grow in the western mountains in the United States and Canada. Economically, these forests are important for lumbering, with trees
for lumber occurring in the southern margins of the biome
and for pulpwood throughout the middle and northern
portions. Present logging practices and the sustainability
of these yields are issues of increasing controversy.

In the Sierra Nevada montane forests of California, the
giant sequoias naturally occur in 70 isolated groves. These
trees are Earth's largest living things (in terms of biomass),
although they began as a small seed (Figure 20.13a).
Some of these *Sequoia gigantea* exceed 8 m in diameter
(28 ft) and 83 m (270 ft) in height. The largest of these is
the General Sherman tree in Sequoia National Park
(Figure 20.13c); it is estimated to be 3500 years old. The
bark is fibrous, a half-meter thick, and lacks resins, so it
effectively resists fire. Imagine the lightning strikes
and fires that must have moved past the Sherman tree in
35 centuries! Standing among these giant trees is an over-
whelming experience and creates a sense of the majesty of
the biosphere.

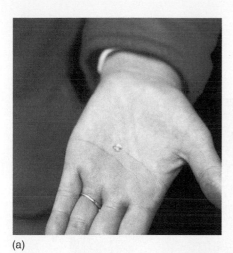
(a)

(b)

(c)

FIGURE 20.13 Sequoia: the seed, the seedling, the giant.
(a) A *Sequoia gigantea* seed. About 300 seeds are in each sequoia cone. (b) Seedling at approximately 50 years of age. (c) The General Sherman tree in Sequoia National Park, California. This tree is probably wider than a standard classroom. The first branch is 45 m (150 ft) off the ground and is 2 m (6.5 ft) in diameter! [All photos by author.]

Certain regions of the northern needleleaf biome experience permafrost conditions, discussed in Chapter 17. When coupled with rocky and poorly developed soils, these conditions generally limit the existence of trees to those with shallow rooting systems. The summer thaw of surface-active layers results in muskeg (moss-covered) bogs and Histosols (organic) soils of poor drainage and stability. Soils of the taiga are typically Spodosols (Podzolic, subject to podzolization), characteristically acidic and leached of humus and clays. Global warming is altering conditions in the high latitudes, causing increased melting and depth in the permafrost active layer. By the late 1990s, some affected forests were beginning to die in response to waterlogging in soils.

Representative fauna in this biome include wolf, elk, moose (the largest in the deer family), bear, lynx, beaver, wolverine, marten, small rodents, and migratory birds during the brief summer season (Figure 20.14). Birds include hawks and eagles, several species of grouse, Pine Grosbeak, Clark's Nutcracker, and several owls. About 50 species of insects particularly adapted to the presence of coniferous trees inhabit the biome.

Temperate Rain Forest

Lush forests at middle and high latitudes are in the **temperate rain forest** biome. In North America, it occurs only along narrow margins of the Pacific Northwest (Figure 20.15a). Some similar types of forest exist in southern China, small portions of southern Japan, New Zealand, and a few areas of southern Chile.

This biome contrasts with the diversity of the equatorial and tropical rain forest in that only a few species make up the bulk of the trees. The rain forest of the Olympic Peninsula in Washington State is a mixture of broadleaf and needleleaf trees, huge ferns, and thick undergrowth. Precipitation approaching 400 cm (160 in.) per year on

FIGURE 20.14 Animals of the northern needleleaf forest. A male elk (*Cervus canadensis*) in a snowstorm looking for any available vegetation to eat during winter among needleleaf trees. [Photo by Steven K. Huhtala.]

the western slopes, moderate air temperatures, summer fog, and an overall maritime influence produce this moist, lush vegetation community. Animals include bear, badger, deer, wild pig, wolf, bobcat, and fox. The trees are home to numerous bird species, including the famous Spotted Owl (Figure 20.15b).

The tallest trees in the world occur in this biome—the coastal redwoods (*Sequoia sempervirens*). Their distribution is shown on the map in Figure 19.12a in Chapter 19. These trees can exceed 1500 years in age and typically range in height from 60 to 90 m (200 to 300 ft), with some exceeding 100 m (330 ft). Virgin stands of other representative trees, such as Douglas fir, spruce, cedar, and hemlock, have been reduced by timber harvests to a few remaining valleys in Oregon and Washington, less than 10% of the original

(a)

(b)

FIGURE 20.15 Temperate rain forest.
(a) An ancient temperate rain forest in the Pacific Northwest in the Gifford Pinchot National Forest features huge old-growth Douglas fir, redwoods, cedars, and a mix of deciduous trees, ferns, and mosses. Only a small percentage of these old-growth forests remain. (b) The Northern Spotted Owl is an indicator species of the temperate rain forest. [Photos by (a) Bobbé Christopherson; (b) Greg Vaughn/Tony Stone Images.]

forest. Replanting and secondary-growth forests are predominant. In similar forests in Chile, large-scale timber harvests and new mills began operations in 2000. U.S. corporations are shifting logging operations to these forests in the Chilean Lake District and northern Patagonia.

Many studies by the U.S. Forest Service and others noted the failing ecology of these forest ecosystems and suggested that timber management plans should include ecosystem preservation as a priority. The ultimate solution must be one of economic and ecological synthesis rather than continuing conflict and forest losses—the forests can't all be cut down, nor can they all be preserved. The opposing sides in this conflict between logging and ecological interests should combine their efforts in a sustainable forestry model.

Mediterranean Shrubland

The **Mediterranean shrubland** biome, also referred to as a temperate shrubland, occupies those regions poleward of the shifting subtropical high-pressure cells. As those cells shift poleward with the high summer Sun, they cut off available storm systems and moisture. Their stable high-pressure presence produces the characteristic *Mediterranean dry summer* and establishes conditions conducive to fire (Figure 19.26d in Chapter 19).

Plant ecologists think that this biome is well adapted to frequent wildfires, for many of its characteristically deep-rooted plants have the ability to resprout from their roots after a fire. Earlier stands of evergreen woodlands (holm or evergreen oak) no longer dominate.

The dominant shrub formations that occupy these regions are stunted and able to withstand hot-summer drought. The vegetation is *sclerophyllous* (from *sclero*, for "hard," and *phyllos*, for "leaf"). It averages a meter or two in height and has deep, well-developed roots, leathery leaves, and uneven low branches.

Typically, the vegetation varies between woody shrubs covering more than 50% of the ground and grassy woodlands covering 25%–60% of the ground. In California, the Spanish word *chaparro* for "scrubby evergreen" gives us the word **chaparral** for this vegetation type (Figure 20.16). This scrubland includes species such as manzanita, toyon, red bud, ceanothus, mountain mahogany, blue and live oaks, and the dreaded poison oak.

A counterpart to the California chaparral in the Mediterranean region is the *maquis*, which includes live and cork oak trees (source of cork) as well as pine and olive trees. The overall similar appearance of the California and Spanish oak savannas is demonstrated in Figure 10.16c and d in Chapter 10. In Chile, such a region is the *mattoral*;

FIGURE 20.16 Mediterranean chaparral.
Chaparral vegetation associated with the Mediterranean dry-summer climate in southern California. [Photo by Bobbé Christopherson.]

FIGURE 20.17 Farming in the grasslands of North America.
An autumn sunset bathes these Alberta, Canada, prairie farms. Warmer average temperatures are expanding the growing season in Canada and melting permafrost soil in the far north. Conditions of drought are a concern. [Photo by John Eastcott/Yva Momatiuk, DRK Photo.]

in southwestern Australia, *mallee scrub*. In Australia, the bulk of the eucalyptus species is sclerophyllous in form and structure in whichever climate it occurs.

As described in Chapter 10, Mediterranean climates are important in commercial agriculture for subtropical fruits, vegetables, and nuts, with many food types produced only in this biome (e.g., artichokes, olives, almonds). Larger animals, such as several types of deer, are grazers and browsers, with coyote, wolf, and bobcat as predators. Many rodents, other small animals, and a variety of birds also proliferate.

Midlatitude Grasslands

Of all the natural biomes, the **midlatitude grasslands** are the most modified by human activity. Here are the world's "breadbaskets"—regions that produce bountiful grain (wheat and corn), soybeans, and livestock (hogs and cattle). Figure 20.17 shows a typical midlatitude grassland under cultivation. In these regions, the only naturally occurring trees were deciduous broadleaf trees along streams and other limited sites. These regions are called grasslands because of the predominance of grasslike plants before human intervention:

> In this study of vegetation, attention has been devoted to grass because grass is the dominant feature of the Plains and is at the same time an index to their history. Grass is the visible feature which distinguishes the Plains from the desert. Grass grows, has its natural habitat, in the transition area between timber and desert. . . . The history of the Plains is the history of the grasslands.*

*From *The Great Plains*, ©1931 by Walter Prescott Webb, published by Ginn and Company, Needham Heights, MA., p. 32.

In North America, tallgrass prairies once rose to heights of 2 m (6.5 ft) and extended westward to about the 98th meridian, with shortgrass prairies in the drier lands farther west. The 98th meridian is roughly the location of the 51-cm (20-in.) isohyet, with wetter conditions to the east and drier conditions to the west (see Figure 18.17 in Chapter 18).

The deep, tough sod of these grasslands posed problems for the first European settlers, as did the climate. The self-scouring steel plow, introduced in 1837 by John Deere, allowed the interlaced grass sod to be broken apart, freeing the soils for agriculture. Other inventions were critical to opening this region and solving its unique spatial problems: barbed wire (the fencing material for a treeless prairie); well-drilling techniques developed by Pennsylvania oil drillers but used for water wells; windmills for pumping; and railroads to conquer the distances.

Few patches of the original prairies (tall grasslands) or steppes (short grasslands) remain within this biome. For prairies alone, the reduction of natural vegetation went from 100 million hectares (250 million acres) down to a few areas of several hundred hectares each. A state-designated 10-hectare (25-acre) remnant of the "prairie" is located 8 km north of Ames, Iowa, and the Hayden Prairie Preserve, also in Iowa, is pictured in News Report 19.3. The map in Figure 20.4 shows the natural location of these former prairie and steppe grasslands.

Outside North America, the Pampas of Argentina and Uruguay and the grasslands of Ukraine are characteristic midlatitude grassland biomes. In each region of the world where these grasslands have occurred, human development of them was critical to territorial expansion.

This biome is the home of large grazing animals, including deer, pronghorn, and bison (the almost complete annihilation of the latter is part of American history, Figure 20.18). Grasshoppers and other insects feed on the grasses and crops as well, and gophers, prairie dogs, ground squirrels, Turkey Vultures, grouse, and Prairie

FIGURE 20.18 Bison graze on the grasslands of Montana's National Bison range. [Photo by Lowell Georgia/CORBIS.]

Chickens are on the land. Predators include the coyote, nearly extinct black-footed ferret, badger, and birds of prey—hawks, eagles, and owls.

Deserts

Earth's **desert biomes** cover more than one-third of its land area, as is obvious in Figure 20.4. In Chapter 15, we examined desert landscapes, and in Chapter 10, desert climates. On a planet with such a rich biosphere, the deserts stand out as unique regions of fascinating adaptations for survival. Much as a group of humans in the desert might behave with short supplies, plant communities also compete for water and site advantage. Some desert plants, called *ephemerals*, wait years for a rainfall event, at which time their seeds quickly germinate and the plants develop, flower, and produce new seeds, which then rest again until the next rainfall event. The seeds of some xerophytic species open only when fractured by the tumbling, churning action of flash floods cascading down a desert arroyo, and, of course, such an event produces the moisture that a germinating seed needs.

Perennial desert plants employ other adaptive features to cope with the desert. Unique xerophytic features include

- Long, deep taproots (example, the mesquite); succulence (thick, fleshy, water-holding tissue, such as that of cacti),
- Spreading root systems to maximize water availability, waxy coatings and fine hairs on leaves to retard water loss, leafless conditions during dry periods (example, palo verde and ocotillo),
- Reflective surfaces to reduce leaf temperatures,
- Tissue that tastes bad to discourage herbivores.

The creosote bush *(Larrea tridentata)* sends out a wide pattern of roots and contaminates the surrounding soil with toxins that prevent the germination of other creosote seeds, possible competitors for water. When a creosote bush dies, surrounding plants or germinating seeds will occupy the abandoned site, but they must wait for infrequent rains to remove the toxins, according to one hypothesis.

The faunas of both warm and cold deserts are limited by the extreme conditions and include few resident large animals. Exceptions are the camel, which can lose up to 30% of its body weight in water without suffering (for humans, a 10%–12% loss is dangerous), and the desert bighorn sheep, in scattered populations in inaccessible mountains and places such as the Grand Canyon along rocky outcrops and cliffs but not at the rim. In an effort to reestablish bighorn populations, several states are transplanting the animals to their former ranges. Some representative desert animals are the ring-tail cat, kangaroo rat, lizards, scorpions, and snakes. Most of these animals are quite secretive and become active only at night, when

temperatures are lower. In addition, various birds have adapted to desert conditions and available food sources—for example, Roadrunners, Thrashers, Ravens, Wrens, Hawks, Grouse, and Nighthawks.

We are witnessing an unwanted expansion of the desert biome, discussed in Chapter 15 and Figure 15.25. This process, known as *desertification*, is now a worldwide phenomenon along the margins of semiarid and arid lands. Desertification is due principally to poor agricultural practices (overgrazing and agricultural activities that abuse soil structure and fertility), improper soil-moisture management, erosion and salinization, deforestation, and the ongoing climatic change.

Earth's deserts are subdivided into desert and semi-desert associations, to distinguish those with expanses of bare ground from those covered by xerophytic plants of various types. The two broad associations are further separated into warm deserts, principally tropical and subtropical, and cold deserts, principally midlatitude.

Warm Desert and Semidesert   Earth's **warm desert and semidesert** biomes are caused by the presence of dry air and low precipitation from subtropical high-pressure cells. These areas are very dry, as evidenced by the Atacama Desert of northern Chile, where only a minute amount of rain has ever been recorded—a 30-year annual average of only 0.05 cm (0.02 in.)! Like the Atacama, the true deserts are under the influence of the descending, drying, and stable air of high-pressure systems from 8 to 12 months of the year.

Remember that these dry regions are defined by low amounts of precipitation that fail to satisfy high amounts of potential evapotranspiration. Deserts receive precipitation that is less than one-half of potential evapotranspiration. *Semiarid steppe* climates receive precipitation that is more than one-half of annual potential evapotranspiration. A few of the subtropical deserts—such as those in Chile, Western Sahara, and Namibia—are right on the seacoast and are influenced by cool offshore ocean currents. As a result, these true deserts experience summer fog that mists the plant and animal populations with needed moisture.

Vegetation ranges from almost none in the arid deserts to numerous xerophytic shrubs, succulents, and thorn tree forms. The lower Sonoran Desert of southern Arizona is a warm desert (Figure 20.19). This desert landscape features the unique saguaro cactus (*Carnegiea gigantea*), which, if undisturbed, grows to many meters in height and up to 200 years in age. First blooms do not appear until it is 50 to 75 years old!

The equatorward margin of the subtropical high-pressure cell is a region of transition to savanna, thorn tree, scrub woodland, and tropical seasonal forest. Poleward of the warm deserts, the subtropical cells shift to produce the Mediterranean dry-summer regime along west coasts and may grade into cool deserts elsewhere.

and scrub vegetation were actually dry shortgrass regions in the past, before extensive grazing altered their ecology. The deserts in the upper Great Basin today are the result of more than a century of such cultural practices.

Arctic and Alpine Tundra

High Latitude Animals Photo Gallery
Geographic Scenes: East Greenland
Photo Gallery

NOTEBOOK

The **arctic tundra** is found in the extreme northern area of North America, around the coast of Greenland, and in Russia, bordering on the Arctic Ocean and generally north of the 10°C (50°F) isotherm for the warmest month (Figure 20.20). The photos in Figure 10.20 and 10.22 in Chapter 10 show tundra lands. Daylength varies greatly

(a)

FIGURE 20.19 Sonoran Desert scene.
Characteristic vegetation in the Lower Sonoran Desert west of Tucson, Arizona (32° N; elevation 900 m, or 2950 ft). [Photo by author.]

Cold Desert and Semidesert The **cold desert and semidesert** biomes tend to occur in middle latitudes. Here, seasonal shifting of subtropical high pressure is of some influence less than 6 months of the year. Interior locations are dry because of their distance from moisture sources or their location in rain-shadow areas on the lee side of mountain ranges, such as the Sierra Nevada of the western United States, the Himalayas, and the Andes. The combination of interior location and rain-shadow positioning produces the cold deserts of the Great Basin of western North America.

Winter snows occur in the cold deserts, but generally they are light. Summers are hot, with highs from 30°C to 40°C (86°F to 104°F). Nighttime lows, even in the summer, can cool 10–20 C° (18–36 F°) from the daytime high. The dryness, generally clear skies, and sparse vegetation lead to high radiative heat loss and cool evenings. Many areas of these cold deserts that are covered by sagebrush

(b)

FIGURE 20.20 Arctic tundra.
(a) Tundra on the Kamchatka Peninsula, Russia. The Uzon Caldera is in the center-background mountain range. (b) Lush by tundra standards, flowers, grasses, mosses, and dwarf willow flourish in the cold climate of the Arctic. [Photos by (a) Wolfgang Kaehler Photography; (b) Bobbé Christopherson.]

throughout the year, seasonally changing from almost continuous day to continuous night. Winters in this biome, the *tundra* climate classification, are long and cold; cool summers are brief. The region, except for a few portions of Alaska and Siberia, was covered by ice during all of the Pleistocene glacial episodes.

Intensely cold continental polar air masses and stable high-pressure anticyclones govern tundra winters. A growing season of sorts lasts only 60–80 days, and even then frosts can occur at any time. Vegetation is fragile in this flat, treeless world; soils are poorly developed periglacial surfaces, which are underlain by permafrost. In the summer months, only the surface horizons thaw, thus producing a mucky surface of poor drainage. Roots can penetrate only to the depth of thawed ground, usually about a meter. The surface is shaped by freeze-thaw cycles that create the frozen-ground phenomena discussed in Chapter 17.

Tundra vegetation consists of low, ground-level herbaceous plant species, such as sedges, mosses, arctic meadow grass, snow lichen, and some woody dwarf willow, as shown in Figure 19.3c and here in Figure 20.20b. Owing to the short growing season, some perennials form flower buds one summer and open them for pollination the next. Animals of the tundra include musk ox, caribou, reindeer, rabbit, ptarmigans, lemmings, other small rodents (important food for the larger carnivores), wolves, foxes, weasels, Snowy Owls, polar bears (in nearby marine and pack ice environments), and, of course, mosquitoes. The tundra is an important breeding ground for geese, swans, and other waterfowl.

Alpine tundra is similar to arctic tundra, but it can occur at lower latitudes because it is associated with high elevation. This biome usually is described as above the timberline, that elevation above which trees cannot grow. Timberlines increase in elevation equatorward in both hemispheres. Alpine tundra communities occur in the Andes near the equator, the White Mountains and Sierra

FIGURE 20.21 Alpine tundra conditions.
An alpine tundra and grazing mountain goats near Mount Evans, Colorado, at 3660 m (12,000 ft) elevation. [Photo by Bobbé Christopherson.]

of California, the American and Canadian Rockies, the Alps, and Mount Kilimanjaro of equatorial Africa, as well as in mountains from the Middle East to Asia.

Alpine meadows feature grasses, herbaceous annuals (small plants), and stunted shrubs, such as willows and heaths. Because alpine locations are frequently windy sites, many plants appear to have been sculpted by the wind. Alpine tundra can experience permafrost conditions. Characteristic fauna include mountain goats, bighorn sheep, elk, and voles (Figure 20.21).

Because the tundra biome is of such low productivity, it is fragile. Disturbances such as tire tracks, hydroelectric projects, and mineral exploitation leave marks that persist for hundreds of years. As development continues, the region will face even greater challenges from oil spills, contamination, and disruption (News Report 20.3).

News Report 20.3

ANWR Faces Threats

Planning continues for oil exploration and development of the Arctic National Wildlife Refuge (ANWR) on Alaska's North Slope. A 1998 USGS assessment disclosed that the oil under the Arctic Refuge coastal plain is likely held in many small reservoirs. Recovering this oil would require alteration of the Arctic National Wildlife Refuge landscape, a point disputed in the 2001 *National Energy Policy*, which

recommended oil exploration and development in ANWR. However, the U.S. Senate defeated that plan in 2002 as a tentative step toward ANWR protection. Political and corporate pressure is fairly constant to exploit these expected reserves of fossil fuels.

This pristine wilderness is above the Arctic Circle, bordering on the Beaufort Sea and adjoining the Yukon Territory of Canada (Figure 20.3.1).

The refuge area sustains almost 200,000 caribou, polar and grizzly bears, musk oxen, and wolves. Some have referred to it as "America's Serengeti," given the annual migration of hundreds of thousands of large animals.

The Porcupine caribou herd, for example, is a population of tundra caribou that inhabit Alaska, Yukon, and Northwest Territories (Figure 20.3.2). Most of the herd migrates to the

Alaska coastal plain to calve in early June. The calving area is fairly small and 80%–85% of the herd use it year after year. Because the caribou come from areas both within ANWR and from areas in Yukon and Northwest Territories, this is a transboundary issue. Oil exploration is not permitted in these areas because petroleum development can be devastating to arctic wildlife. The 1002 Area is 1.5 million coastal acres with potential oil and gas resources, yet critical animal habitat for caribou and polar bears (**http://energy. usgs.gov/alaska/anwr.html**). The debate is ongoing; see **http:// justice.uaa. alaska.edu/rlinks/environment/ak_ anwr.html**.

An ironic situation is that the thawing permafrost and deepening active layer triggered by rising temperatures are reducing the number of days oil exploration heavy equipment can operate on lands just outside ANWR. Previously, 200 days of work were possible, as the ground remained frozen and supported the heavy trucks and rigs. Now with regional-scale warming, such travel is restricted to just 100 days. Clearly, economic ventures should be weighed against the ecosystem itself, its limitations and its uniqueness, and a total assessment of all costs to the consumer.

FIGURE 20.3.1 Arctic National Wildlife Refuge.
In Alaska's Arctic National Wildlife Refuge, Mount Chamberlin, the second-highest peak of the Brooks Range, overlooks tundra in the foreground. Is this region destined for petroleum exploration and development or for continued preservation as wilderness? [Photo by Scott T. Smith.]

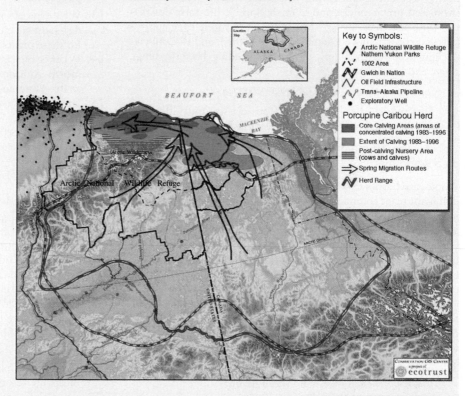

FIGURE 20.3.2 Caribou herd range, ANWR, migration routes, and 1002 area.
Areas of caribou migration routes and calving grounds. [Courtesy of the Alaska Center for the Environment.]

Summary and Review—Terrestrial Biomes

■ *Define* the concept of biogeographic realms of plants and animals, and *define* ecotone, terrestrial ecosystem, and biome.

The interplay among supporting physical factors within Earth's ecosystem determines the distribution of plant and animal communities and Earth's biodiversity. A **biogeographic realm** of plants and animals is a geographic region in which a group of species evolved.

This recognition was a start at understanding distinct regions of flora and fauna and the broader pattern of terrestrial ecosystems. A boundary transition zone adjoining ecosystems is an **ecotone**.

A **terrestrial ecosystem** is a self-sustaining association of plants and animals and their abiotic environment that is characterized by specific plant formation classes. A **biome** is a large, stable ecosystem characterized by specific plant and animal communities. Biomes carry the name of the dominant vegetation because it is the most easily identified feature: forest, savanna, grassland, shrubland, desert, tundra.

> biogeographic realm (p. 650)
> ecotone (p. 651)
> terrestrial ecosystem (p. 651)
> biome (p. 651)

1. Reread the three opening quotations and analysis in this chapter. What clues do they give you to the path ahead for Earth's forests and ecosystems? Is our future direction controllable? Sustainable? Explain.
2. What is a biogeographic realm? How is the world subdivided according to botanical (plant) types?
3. Describe a transition zone between two ecosystems. How wide is an ecotone? Explain.
4. Define *biome*. What is the basis of the designation?

■ *Define* six formation classes and the life-form designations, and *explain* their relationship to plant communities.

Biomes are divided into more specific vegetation units called **formation classes**. The structure and appearance of the vegetation is described: rain forest, needleleaf forest, Mediterranean shrubland, arctic tundra, and so forth. Specific life-form designations include trees, lianas, shrubs, herbs, bryophytes, epiphytes (plants growing aboveground on other plants), and thallophytes (lacking true leaves, stems, or roots, including bacteria, fungi, algae, and lichens).

> formation classes (p. 653)

5. Distinguish between formation classes and life-form designations as a basis for spatial classification.

■ *Describe* 10 major terrestrial biomes, and *locate* them on a world map.

Biomes are Earth's major terrestrial ecosystems, each named for its dominant plant community. The 10 major biomes are generalized from numerous formation classes that describe vegetation. Ideally, a biome represents a mature community of natural vegetation. In reality, few undisturbed biomes exist in the world, for most have been modified by human activity. Many of Earth's plant and animal communities are experiencing an accelerated rate of change that could produce dramatic alterations within our lifetime.

For an overview of Earth's 10 major terrestrial biomes and their vegetation characteristics, soil orders, climate-type designations, annual precipitation range, temperature patterns, and water balance characteristics, please review Table 20.1.

> equatorial and tropical rain forest (p. 656)
> tropical seasonal forest and scrub (p. 663)
> tropical savanna (p. 664)
> midlatitude broadleaf and mixed forest (p. 665)
> needleleaf forest (p. 666)
> boreal forest (p. 666)
> taiga (p. 666)
> montane forest (p. 666)
> temperate rain forest (p. 667)
> Mediterranean shrubland (p. 668)
> chaparral (p. 668)
> midlatitude grasslands (p. 669)
> desert biomes (p. 670)
> warm desert and semidesert (p. 670)
> cold desert and semidesert (p. 671)
> arctic tundra (p. 671)
> alpine tundra (p. 672)

6. Using the integrative chart in Table 20.1 and the world map in Figure 20.4, select any two biomes and study the correlation of vegetation characteristics, soil, moisture, and climate with their spatial distribution. Then, contrast the two using each characteristic.
7. Describe the equatorial and tropical rain forests. Why is the rainforest floor somewhat clear of plant growth? Why are logging activities for specific species so difficult there?
8. What issues surround the deforestation of the rain forest? What is the impact of these losses on the rest of the biosphere? What new threat to the rain forest has emerged?
9. What do *caatinga*, *chaco*, *brigalow*, and *dornveld* refer to? Explain.
10. Describe the role of fire or fire ecology in the tropical savanna biome and the midlatitude broadleaf and mixed forest biome.
11. Why does the northern needleleaf forest biome not exist in the Southern Hemisphere, except in mountainous regions? Where is this biome located in the Northern Hemisphere, and what is its relationship to climate type?
12. In which biome do we find Earth's tallest trees? Which biome is dominated by small, stunted plants, lichens, and mosses?
13. What type of vegetation predominates in the Mediterranean dry-summer climates? Describe the adaptations necessary for these plants to survive.
14. What is the significance of the 98th meridian in terms of North American grasslands? What types of inventions enabled humans to cope with the grasslands?
15. Describe some of the unique adaptations found in a desert biome.
16. What is desertification (review from Chapter 15 and this chapter)? Explain its impact.
17. What physical weathering processes are specifically related to the tundra biome? What types of plants and animals are found there?

■ *Relate* human impacts, real and potential, to several of the biomes.

The equatorial and tropical rainforest biome is undergoing rapid deforestation. Because the rain forest is Earth's most diverse biome and is important to the climate system, such losses are creating great concern among citizens, scientists, and nations. Efforts are underway worldwide to set aside and protect remaining representative sites within most of Earth's principal biomes.

18. What is the relationship between island biogeography and biosphere reserves? Describe a biosphere reserve. What are the goals?
19. Compare the map in Figure 20.4 with the composite satellite image inside the front cover of this text. What correlations can you make between the local summertime portrait of Earth's biosphere and the biomes identified on the map?
20. From the background in News Report 20.3, what is your assessment of ANWR? Using credible Internet sources, what is the present status relative to ongoing, but halted, efforts to begin oil and gas exploration in Area 1002?
21. As an example of shifting climate impacts, we tracked temperature and precipitation conditions for Illinois in Figure 20.11. What impacts do you think such climate change will have in the United States and in other countries? See Critical Thinking item "B."

NetWork

Critical Thinking

A. Given the information presented in this chapter about deforestation in the tropics, assess the present situation. What are the main issues? What natural assets are at stake? Natural resources versus sovereign state rights? How are global biodiversity and greenhouse warming related to these issues? What is the perspective of the less-developed countries that possess most of the rain forest? What is the perspective of the more-developed countries and their transnational corporations and millions of environmentally active citizens? What kind of action plan would you develop to accommodate all parties? How would you proceed?

B. Using Figure 20.4 (biomes), Figure 10.5 (climates), Figures 10.2 and 9.5 (precipitation), and Figure 8.2 (air masses), and the printed graphic scales on these maps, consider the following hypothetical. Assume a northward climatic shift in the United States and Canada of 500 km (310 mi) (in other words, imagine moving North America 500 km south, to simulate climatic categories shifting north). Describe your analysis of conditions through the Midwest from Texas to the prairies of Canada. Describe your analysis of conditions from New York through New England and into the Maritime provinces. What economic dislocations and relocations do you envision?

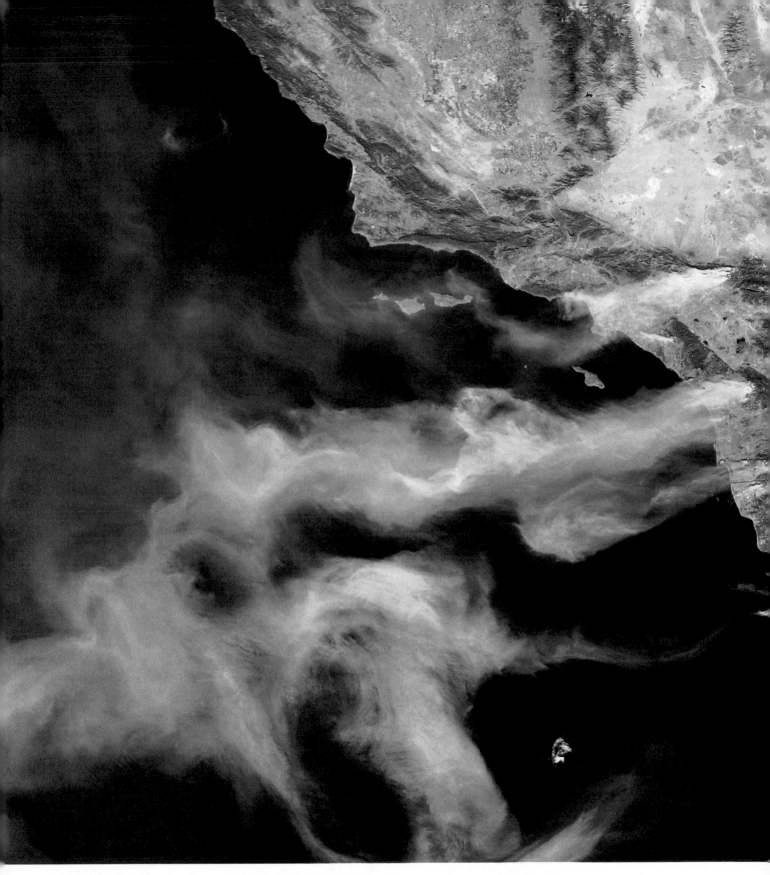

Twelve large fires were burning on October 23, 2007, as *Terra* passed overhead. Wildfire devastated Mediterranean chapparel and forest, homes, and businesses across southern California. The smoke, particulates, and carbon dioxide streamed out over the Pacific Ocean. Lost were more than 2000 homes and 800 other structures; more than 500,000 acres burned, with final fires not over until late November. Strong Santa Ana winds added the blowtorch effect, taking embers miles from their ignition source. The *Terra* satellite image captures the Salton Sea and irrigated desert farms; the Colorado River and Hoover Dam/Lake Mead are in the upper-right corner. This entire region is in a prolonged drought and experiencing higher-than-normal temperatures. [*Terra* true-color image MODIS Rapid Response Team, GSFC/NASA.]

21

Earth and the Human Denominator

■ Key Learning Concepts

After reading the chapter, you should be able to:

- ■ **Determine** an answer to Carl Sagan's question, "Who speaks for Earth?"

- ■ **Describe** the growth in human population, and **speculate** on possible future trends.

- ■ **Analyze** "An Oily Bird," and **relate** your analysis to energy consumption patterns in the United States and Canada.

- ■ **List** the subjects of environmental agreements, conventions, and protocols, and **relate** them to physical geography and Earth systems science.

- ■ **List** and **discuss** 12 paradigms for the twenty-first century.

- ■ **Appraise** your place in the biosphere, and **realize** your relation to Earth systems.

During my space flight, I came to appreciate my profound connection to the home planet and the process of life evolving in our special corner of the Universe, and I grasped that I was part of a vast and mysterious dance whose outcome will be determined largely by human values and actions.*

E arth can be observed from profound vantage points, as this astronaut experienced on the 1969 *Apollo IX* mission. Since November 2000, the International Space Station (ISS) has operated at an

*Rusty Sweickart, "Our backs against the bomb, our eyes on the stars," *Discovery*, July 1987, p. 62.

677

FIGURE 21.1 Our orbital perspective—Earth's Space Station.
The International Space Station (ISS) as it looked in August 2007. The ISS began scientific operations in November 2000. The station orbits at a speed of 28,000 kmph (17,400 mph), orbiting Earth in 91.6 min., 15.7 times a day. [Photo of ISS courtesy of NASA.]

orbital altitude of about 400 km (250 mi), in the upper reaches of Earth's thermosphere. Just imagine, a 16-nation coalition has built a scientific workstation, in orbit, that is 44.5 m long, 78 m wide, and 27.5 m tall (146 ft by 256 ft by 90 ft), and it is not yet finished (Figure 21.1). The 219-metric-ton (482,345-lb) complex has had more than 120 astronauts and scientists aboard in 392 m³ (14,000 ft³) of pressurized work space.

Several hundred scientific investigations and experiments are completed or underway to help us better understand Earth and life systems through research in the unique space environment. All life-support supplies must be brought from Earth. Water recycling and purification techniques provide the bulk of the water needed. Oxygen is electrolytically derived from water and supplemented by compressed oxygen. Enormous solar panel arrays generate electricity. Without plants, carbon dioxide must be scrubbed from the air the astronauts breathe.

If completed as planned, the ISS will weigh 455 metric tons (1 million lb) and have the cabin volume of a 747 jumbo jet. This seems like an important fulfillment of our human nature: to question, to explore, to discover, to test the limits of the known, and to push beyond to the unknown. (See **http://www.nasa.gov/mission_pages/station/main/index.html** for more information and updates.)

Our vantage point in this book is that of physical geography. We examine Earth's many systems: its energy, atmosphere, winds, ocean currents, water, weather, climate, endogenic and exogenic systems, soils, ecosystems, and biomes. This exploration has led us to an examination of the planet's most abundant large animal, *Homo sapiens*.

We stand in the first decade of the twenty-first century. This century will be an adventure for the global society, historically unparalleled in experimenting with Earth's life-supporting systems. You will spend the majority of your life in this century. What preparations and "future thinking" can you do to understand all that is to occur?

In his 1980 book and public television series *Cosmos*, the late astronomer Carl Sagan asked:

> What account would we give of our stewardship of the planet Earth? We have heard the rationales offered by

the nuclear superpowers. We know who speaks for the nations. But who speaks for the human species? Who speaks for Earth?*

Indeed, who does speak for Earth? We might answer: Perhaps we physical geographers, and other scientists who have studied Earth and know the operations of the global ecosystem, should speak for Earth. However, some might say that questions of technology, environmental politics, and future thinking belong outside of science and that our job is merely to learn how Earth's processes work and to leave the spokesperson's role to others. Biologist-ecologist Marston Bates addressed this line of thought in 1960:

> Then we came to humans and their place in this system of life. We could have left humans out, playing the ecological game of "let's pretend humans don't exist." But this seems as unfair as the corresponding game of the economists, "let's pretend nature doesn't exist." The economy of nature and the ecology of humans are inseparable and attempts to separate them are more than misleading, they are dangerous. Human destiny is tied to nature's destiny and the arrogance of the engineering mind does not change this. Humans may be a very peculiar animal, but they are still a part of the system of nature.†

A fact of life is that Earth's more-developed countries (MDCs), through their economic dominance, speak for the billions who live in less-developed countries (LDCs). The gross state product (GSP) of California's 36.5 million people was $1.4 trillion in 2004, which was about half the gross domestic product (GDP) of China and its 1.325 billion people—that's $38,000 per capita GSP in California, compared to $2100 per capita GDP in China. The scale of disparity on our home planet is difficult to comprehend.

The fate of traditional modes of life may rest in some distant financial capital. But, economics aside, the reality

---

*C. Sagan, *Cosmos* (New York: Random House, 1980), p. 329.
†M. Bates, *The Forest and the Sea* (New York: Random House, 1960), p. 247.

FIGURE 21.2 Lands distant from the global centers of power. These lands were depopulated under Stalin when they were part of the former USSR. Only in the last decade or so have people returned to this traditional rural landscape in the Pamirs of Tajikistan, beginning life again, so distant from the more-developed countries. [Photo by Stephen F. Cunha.]

is that the remote lands of Siberia are linked by Earth systems to the Pampas of Argentina, to the Great Plains in North America, and to those harvesting grain by hand in the remote Pamirs of Tajikistan (Figure 21.2).

To understand these linkages among Earth's myriad systems is our quest in *Geosystems* (Figure 21.3). We explored the atmosphere, hydrosphere, lithosphere, and biosphere, synthesizing operating systems into the web of

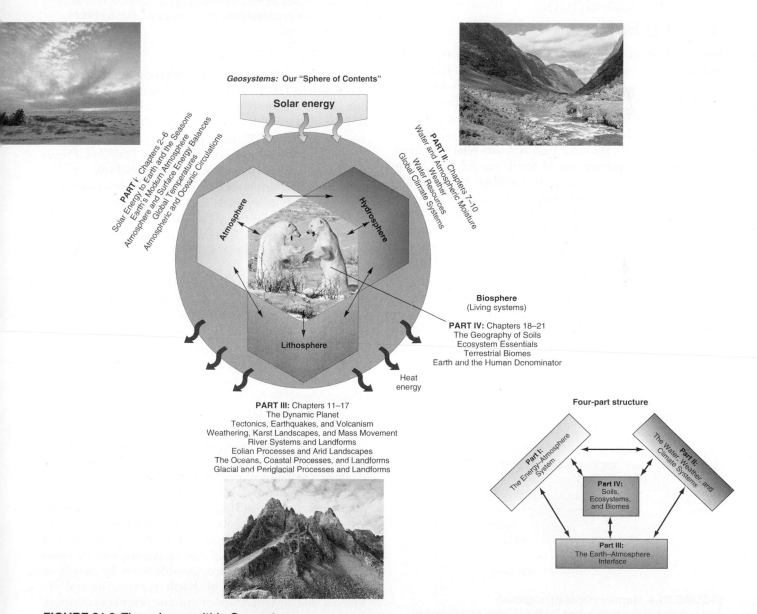

FIGURE 21.3 The spheres within *Geosystems.*
Through this text we covered the atmosphere, hydrosphere, lithosphere, and the synthesizing biosphere—the culmination of life-sustaining interactions. The systems approach shows us the flow of energy and matter and the sequence of events through time and across space. [All photos by Bobbé Christopherson.]

life. We now see global linkages among physical and living systems and how actions in one place can affect change elsewhere. For example, as described in Chapter 19, scientists are studying the links between climate change and the record wildfires this decade in the western United States (see chapter-opening satellite image).

In this sense, Earth can be compared to a spaceship. In the same way the members of a crew aboard the Space Station are inextricably linked to each other's lives and survival, we, too, are each connected through the operation of planetary systems. Growing international environmental awareness in the public sector is gradually prodding governments into action. People, when informed, generally favor environmental protection and public health progress over economic interests.

The Human Count and the Future

Because human influence is pervasive, we consider the totality of our impact the *human denominator*. Just as the denominator in a fraction tells how many parts a whole is divided into, so the growing human population and the increasing demand for resources and its rising planetary impact suggest how much the whole Earth system must adjust. Yet, Earth's resource base remains relatively fixed.

The human population of Earth passed 6 billion in August 1999, adding another 625 million by mid-2007; more people are alive today than at any previous point in the planet's long history, unevenly arranged in 192 countries and numerous colonies. In a year about 70 million more people are added—this is 192,000 a day, or a new U.S. population every 4.2 years. Over the span of human history these billion-mark milestones are occurring at closer intervals (Figure 21.4).

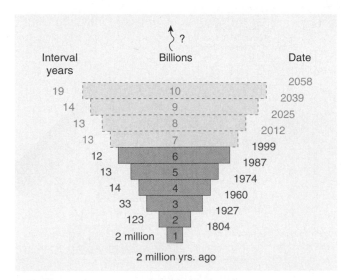

FIGURE 21.4 Human population growth.
The number of years required for the human population to add 1 billion to the count grows shorter and shorter. The twenty-first century should mark a slowing of this history of growth if policy actions are taken. Note the population forecasts for the next half-century.

FIGURE 21.5 Dhaka, Bangladesh.
The Bangladesh capital typifies the plight of the less-developed world as millions converge on cities to find work and more modern lifestyles. [Photo by Bruce Bander/Photo Researchers, Inc.]

Thirty-seven percent of Earth's population live in just two countries (20% in China and 17% in India—2.46 billion people combined). Moreover, we are a young planetary population, with some 28% of those now alive under the age of 15 (2007 data from Population Reference Bureau and U.S. Bureau of the Census *POPClock Projection:* http://www.census.gov/cgi-bin/popclock).

Virtually all new population growth is in the less-developed countries, which now possess 81.5%, or about 5.4 billion, of the total population; the MDCs have the other 18.5%, or 1.23 billion (Figure 21.5). Table 21.1 illustrates this natural increase principle and the differences between the MDCs and LDCs.

If you consider only the population count, the MDCs do not have a population growth problem. In fact, some European countries are actually declining in growth or are near replacement levels. However, people in the developed world produce a greater impact on the planet per person and therefore constitute a population impact crisis. An equation expresses this impact concept:

$$\text{Planetary impact } (I) = P \cdot A \cdot T$$

where P is *population*; A is *affluence*, or consumption per person; and T is *technology*, or the level of environmental impact per unit of production. The United States and Canada, with about 5% of the world's population, produce more than 22% ($14.2 trillion in 2004) of the world's gross domestic product and use more than double the energy per capita of Europeans, more than 7 times that of Latin Americans, 10 times that of Asians, and 20 times that of Africans. Therefore, consideration by people in the MDCs of the state of Earth systems, natural resources, and sustainability of current practices is critical. Earth systems science is here to provide a high level of planetary monitoring and analysis to assist this process.

The concept of an individual's "footprint" has emerged—ecological footprint, carbon footprint, lifestyle

Table 21.1   Planetary Population Increase Rates

	Birth Rate	Death Rate	Natural Increase (%)	Population Growth 2007–2050 (%)
World—2007	**21**	**9**	**1.2**	40%
More developed	**11**	**10**	**0.1**	3%
United States	14	8	0.6	39%
Japan	9	9	0.0	–26%
United Kingdom	12	10	0.3	13%
Less developed	**23**	**8**	**1.5**	49%
LDC exclu. China	**27**	**9**	**1.8**	61%
Mexico	21	5	1.7	24%
Nigeria	43	18	2.5	95%
India	24	8	1.6	55%

Source: World Population Data Sheet 2007. Population Reference Bureau, Washington, DC, http://www.prb.org.

footprint. These consider what your affluence and technology, the "A" and the "T" in the equation, cost planetary systems. Footprint assessment is grossly simplified, but it can give you an idea of your impact and even an estimate of how many planets it would take to sustain that lifestyle and economy if everyone lived like you. Your author has taken action to neutralize the carbon footprint for the production and printing of these textbooks, and the publisher used sustainable-forestry green paper for *Geosystems* to lessen ecological impacts. There are many Internet links to try for footprints; here is a sample:

http://www.earthday.net/Footprint/index.asp
http://www.footprintnetwork.org/
http://www.ecologicalfootprint.org/
http://www.carbonfootprint.com/calculator.html
https://www.econeutral.com/carboncalculator.html
http://www.ucsusa.org/publications/greentips/whats-your-carb.html

The MDCs not only should effect change in their own planetary impact but also provide direction for the LDCs, which have a right to move along the road of progress and improve their living conditions. Let us examine this idea of global impact.

An Oily Bird

At first glance, the chain of events that exposes wildlife to oil contamination seems to stem from a technological problem. An oil tanker splits open at sea and releases its petroleum cargo, which is moved by ocean currents toward shore, where it coats coastal waters, beaches, and animals. In response, concerned citizens mobilize and try to save as much of the spoiled environment as possible (Figure 21.6). But the real problem goes far beyond the physical facts of the spill.

In Prince William Sound off the southern coast of Alaska, in clear weather and calm seas, the *Exxon Valdez*, a

FIGURE 21.6   An oily bird.
A Western Grebe contaminated with oil from the *Exxon Valdez* tanker accident in Prince William Sound, Alaska. This oily bird is the result of a long chain of events and mistakes. [Photo by Geoffrey Orth/SIPA Press.]

single-hulled supertanker operated by Exxon Corporation, struck a reef in 1989. The tanker spilled 42 million liters (11 million gallons) of oil. It took only 12 hours for the *Exxon Valdez* to empty its contents, yet a complete cleanup is impossible, and costs and private claims have exceeded $15 billion. Scientists are still finding damage and residual oil spill. Eventually, more than 2400 km (about 1500 mi) of sensitive coastline was ruined for years to come, affecting three national parks and eight other protected areas (Figure 21.7).

The death toll of animals was massive: At least 5000 sea otters died, or about 30% of resident otters; about 300,000 birds and uncounted fish, shellfish, plants, and aquatic microorganisms also perished. Sublethal effects, namely mutations, now are appearing in fish. The Pacific herring is still in significant decline, as are the harbor seals; other species are in recovery, such as the Bald Eagle

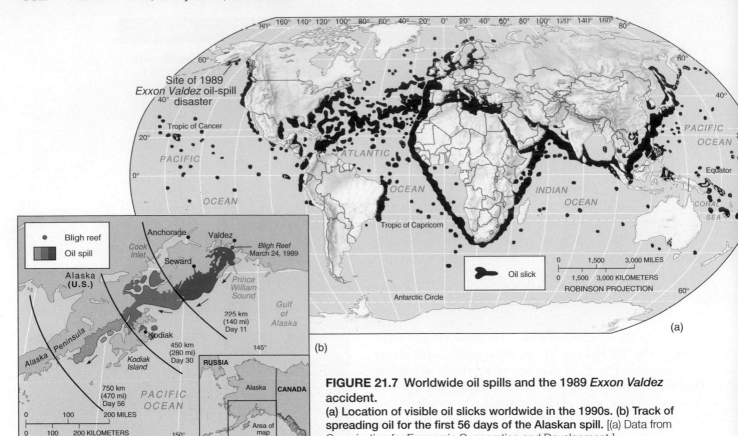

FIGURE 21.7 Worldwide oil spills and the 1989 *Exxon Valdez* accident.
(a) Location of visible oil slicks worldwide in the 1990s. (b) Track of spreading oil for the first 56 days of the Alaskan spill. [(a) Data from Organization for Economic Cooperation and Development.]

and Common Murre. Almost two decades later, oil remains in mudflats and marsh soils beneath rocks.

The immediate problem of cleaning oil off a bird symbolizes a national and international concern with far-reaching spatial significance. And while we search for answers, oil slicks continue their contamination. There is an average of 27 accidents a day, 10,000 a year worldwide, ranging from a few disastrous spills to numerous small ones. There have been 50 spills equal to the *Exxon Valdez* or larger since 1970. In addition to oceanic oil spills, people improperly dispose of crankcase oil from their automobiles in a volume that annually exceeds all of these tanker spills.

The immediate effect of global oil spills on wildlife is contamination and death. But the issues involved are much bigger than dead birds. Let's ask some fundamental questions about the *Exxon Valdez* accident:

- Why was the oil tanker there in the first place?
- Why are only a few of the 28 tankers that traverse the Prince William Sound of the safer double-hulled design?
- Why is petroleum imported into the continental United States in such enormous quantities? Is the demand for petroleum products based on real need and the operation of efficient systems?
- The U.S. demand for oil is higher per capita than the demand of any other country. Efficiencies and

mileage ratings in the U.S. transportation sector have not improved for cars, trucks, and SUVs (light trucks). Why? And what about proven hybrid gas–electric technology?

Yet, hypocrisy is apparent in our outrage over energy politics and oil-spill accidents. We continue to consume gasoline at record levels in inefficient vehicles, thus creating the demand for oil imports. The task of physical geography is to analyze all the spatial aspects of these events in the environment and the ironies that are symbolized by an oily bird.

With the shrinking Arctic Ocean permanent and seasonal sea ice, countries and corporations are lining up to exploit the open water and resources (Figure 21.8). Freighters and tankers are being designed wider than the Panama or Suez canals for shipping a decade or two from now through the newly ice-free Northwest and Northeast passages. The exhaust particulates, soot, and smoke and wastewater discharges will damage the fragile Arctic environment. These pollutants will darken the remaining ice shelves and Greenland's ice sheet, forcing lower albedos, thus increasing insolation absorption and accelerating the melting of remaining ice inventories. Think of the impact from physical geography's perspective of this situation using elements from the chapters you have studied. What should be on the planning table for discussion? Commerce or albedo forcing?

FIGURE 21.8 Record low Arctic Ocean sea-ice extent, 2007.
Global warming and wind patterns are combining to reduce the Arctic Ocean sea ice. The record low was September 14, 2007 (see Chapter 5). What the image doesn't show is the degraded thickness of the multiyear ice. Not only did the atmosphere record increased warmth but the Arctic Ocean warmed as well. [Image courtesy of Scientific Visualization Studio, GSFC/NASA.]

The Need for Cooperation

The idea for a global meeting on the environment was put forward at the 1972 U.N. Conference on the Human Environment held in Stockholm. The U.N. General Assembly in 1987 achieved a landmark in global planning by agreeing to hold the Earth Summit. *Our Common Future*, a book written at the time, set the tone for the 1992 Earth Summit:

> The Earth is one but the world is not. We all depend on one biosphere for sustaining our lives. Yet each community, each country, strives for survival and prosperity with little regard for its impact on others. Some consume the Earth's resources at a rate that would leave little for future generations. Others, many more in number, consume far too little and live with the prospect of hunger, squalor, disease, and early death.*

The setting for the 1992 Earth Summit in Rio de Janeiro was ironic, for many of the very problems discussed at the conference were evident on the city's streets, with their abundant air pollution, water pollution, toxins, noise, wealth and poverty, and daily struggle for health and education.

Five agreements were written at the Earth Summit: the Climate Change Framework; the Biological Diversity Treaty; the Management, Conservation, and Sustainable Development of All Types of Forests; the Earth Charter (a nonbinding statement of 27 environmental and economic principles); and Agenda 21 for Sustainable Development. A product of the Earth Summit was the

*World Commission on Environment and Development, *Our Common Future* (Oxford, UK: Oxford University Press, 1987), p. 27.

United Nations Framework Convention on Climate Change (FCCC), which led to a series of Conference of the Parties (COP) meetings—COP–13 was in Bali, Indonesia, December 2007, and COP–14 is in The Hague, Netherlands, in 2008.

The Kyoto Protocol, agreed to in 1997 to reduce global carbon emissions, was finalized in 2001, in Marrakech, Morocco, at COP-7. Following Russian ratification, the Kyoto Protocol and Rulebook became international law in March 2005, without United States or Australian participation.

Asking whether the Earth Summit "succeeded" or "failed" is the wrong question. The occurrence of this largest-ever official gathering is a remarkable accomplishment. From the Earth Summit emerged a new organization—the U.N. Commission on Sustainable Development—to oversee the promises made in the five agreements. This momentum lead to a second Earth Summit in 2002 (http://www.earthsummit2002.org/) in Johannesburg, South Africa, that completed an agenda including climate change, freshwater, gender issues, global public goods, HIV/AIDS, sustainable finance, and the five Rio Conventions.

A clear conclusion is that members of society must work to move the solutions for environmental and developmental problems off the bench and into play. We need international cooperation to consider our symbiotic relations with each other and with Earth's resilient, yet fragile, life-support systems (News Report 21.1).

A critical corollary to these international efforts is the linkage of academic disciplines. A positive step in that direction is the Earth systems science approach, embodied in physical geography, that synthesizes content from across the disciplines to create a holistic perspective. Exciting progress toward an integrated understanding of Earth's physical and biological systems is in progress. Such a multidisciplinary effort began in 2000 at a Ministerial Meeting of the Arctic Council, when the Arctic Climate Impact Assessment (ACIA) was launched. The ACIA final report, *Climate Change in the Arctic*, was released in November 2004 (Figure 21.9). The effort involved 300 scientists, eight countries, and six circumpolar indigenous peoples' organizations. For a look at this international triumph of cooperative science and summary of *Key Findings*, see http://www.acia.uaf.edu/.

Founded in 1904, the Association of American Geographers (AAG) is a scientific and educational society of 9500 geographers and other members from across the globe, dedicated to the advancement of geographic science. Many scientific/academic professional organizations, societies, and associations, in addition to many states and cities, have called for action on climate change. The AAG took a step toward the future at its annual meeting in Chicago in 2006 and passed a resolution requesting action on climate change—the *AAG Resolution Requesting Action on Climate Change* (http://www.aag.org/).

More than a decade ago, *Time* magazine gave Earth an interesting honor and this issue is today a time capsule

(a)

(b)

FIGURE 21.9 Arctic Climate Impact Assessment Symposium.
(a) Dr. Robert Corell, chair of the ACIA Integration Team, welcomes nearly 400 scientists, managers, government and nongovernmental representatives, student researchers, and indigenous people of the Arctic to a first-ever Arctic climate symposium, held in Reykjavík, Iceland, November 2004. The ACIA released their climate impact assessment to the Arctic Council and to a global audience. (b) The logo of the Arctic Climate Impact Assessment. [Photo (a) by Bobbé Christopherson.]

FIGURE 21.10 Earth made the cover of *Time.*
More than a decade ago, global concerns about environmental impacts prompted *Time* magazine to deviate from its 60-year tradition of naming a prominent citizen as its person of the year, instead naming Earth the "Planet of the Year." The magazine devoted 33 pages to Earth's physical and human geography. Importantly, *Time* also offered positive policy strategies for consideration. [January 2, 1989, issue, Copyright © 1989 The *Time* Inc. Magazine Company. Reprinted by permission.]

of Earth-systems analysis and activist suggestions (Figure 21.10). As if a victim of short-lived fame, we wonder where Earth awareness fits into the global scheme of things today. Is there a follow-up in pop culture to making the cover of *Time?*

Twelve Paradigms for the Twenty-First Century

As we conclude, a brief list of the dominant themes and patterns of concern for the twenty-first century seems appropriate. We hope these *paradigms* will provide a useful framework for brainstorming and discussion of the central issues that will affect us in the new century, listed here in no particular order. The paradigms:

Twelve Paradigms for the Twenty-First Century
1. Population increases in the less-developed countries
2. Planetary impact per person (on the biosphere and resources; $I = P \cdot A \cdot T$), ecological and carbon footprint
3. Feeding the world's population
4. Global and national disparities of wealth and resource allocation
5. Status of women and children (health, welfare, rights)
6. Global climate change (temperatures, sea level, weather and climate patterns, disease, and diversity)
7. Energy supplies and energy demands; renewables and demand management
8. Loss of biodiversity (habitats, genetic wealth, and species richness)
9. Pollution of air, surface water (quality and quantity), groundwater, oceans, and land
10. The persistence of wilderness (biosphere reserves and biodiversity hot spots)
11. Globalization versus cultural diversity
12. Conflict resolution

News Report 21.1

Gaia Hypothesis Triggers Debate

Some scientists and philosophers view Earth as one vast, self-regulating organism. The concept is one of global symbiosis, or mutualism. This controversial concept is the *Gaia hypothesis* (Gaia was the Earth-Mother goddess in ancient mythology). It was proposed in 1979 by James Lovelock, a British astronomer and inventor, and elaborated by American biologist Lynn Margulis.

Gaia is the ultimate synergistic relationship, in which the whole greatly exceeds the sum of the individual interacting components. The hypothesis contends that life processes control and shape Earth's inorganic physical and chemical processes, with the ecosphere so interactive that a very small mass can affect a very large mass. Thus, Lovelock and Margulis think that the material environment and the evolution of species are tightly joined; as species evolve through natural selection, they (including us) in turn affect their environment. The

present oxygen-rich composition of the atmosphere is given as proof of this coevolution of living and nonliving systems.

From the perspective of physical geography, the Gaia hypothesis permits a view of all Earth and the spatial interrelations among systems. In fact, such a perspective is necessary for analyzing specific environmental issues. Many variables interact synergistically, producing both wanted and unwanted results.

One disturbing aspect of this unity is that any biotic threat to the operation of an ecosystem tends to move toward extinction itself. This trend preserves the system overall. Earth systems operation and feedback naturally tend to eliminate offensive members. The degree to which humans represent a planetary threat, then, becomes a topic of great concern, for Earth (Gaia) will prevail, regardless of the outcome of the human experiment.

The maladies of Gaia do not last long in terms of her life span. Anything that makes the world uncomfortable to live in tends to induce the evolution of those species that can achieve a new and more comfortable environment. It follows that, if the world is made unfit by what we do, there is the probability of a change in regime to one that will be better for life but not necessarily better for us.*

The debate is vigorous regarding the true applicability of this hypothesis to nature, or whether it is true science at all. Regardless, it remains philosophically intriguing in its portrayal of the relationship between humans and Earth and stimulates discussion.

*J. Lovelock, *The Ages of Gaia—A Biography of Our Living Earth* (New York: Norton, 1988), p. 178.

Who Speaks for Earth?

Geographic awareness and education is an increasingly positive force on Earth. The National Geographic Society conducts an annual National Geography Bee to promote geography education and global awareness to millions of 6th- to 8th-graders, their parents, and teachers. The National Geography World Championship (formerly the International Geography Olympiad) is now an annual event (Figure 21.11). There are presently 60 geographic alliances in 47 states that coordinate geographic education among teachers and students at all levels: K–12, community college, college, and university. People are learning more about Earth–human relations.

Yet, ideological and ethical differences still remain within society. This dichotomy was addressed by biologist Edward O. Wilson:

The evidence of swift environmental change calls for an ethic uncoupled from other systems of belief. Those committed by religion to believe that life was put on Earth in one divine stroke will recognize that we are destroying the Creation; and those who perceive biodiversity to be the product of blind evolution will

FIGURE 21.11 National Geographic World Championship. Students from across the globe meet every two years for the National Geographic World Championship, sponsored by the National Geographic Society, with Alex Trebek of *Jeopardy!* as moderator. Here Alex congratulates finalist teams from three countries. The contest is an indicator of growing international geographic awareness. [Photo by O. Louis Mazzatenta, National Geographic Society.]

FIGURE 21.12 View of Earth from Saturn.
The *Cassini–Huygens* spacecraft was on the far side of Saturn so the Sun lit the rings and atmosphere in backlighting. The surprise catch was a glimpse of Earth as a little, pale blue dot, some 1.23 billion km (791 million mi) distant. It was September 19, 2006 on Earth. [Image courtesy of NASA/JPL/Space Science Institute.]

agree. . . . Defenders of both premises seem destined to gravitate toward the same position on conservation. . . . For what, in the final analysis, is morality but the command of conscience seasoned by a rational examination of consequences? . . . An enduring environmental ethic will aim to preserve not only the health and freedom of our species, but access to the world in which the human spirit was born.*

In an ideal sense, we hold all of Earth's systems in common. Now, what about Sagan's question to us? The late Carl Sagan asked, "Who speaks for Earth?" He answered with this perspective:

We have begun to contemplate our origins: starstuff pondering the stars; organized assemblages of ten billion billion billion atoms considering the evolution of atoms; tracing the long journey by which, here at least, consciousness arose. Our loyalties are to the species and the planet. We speak for Earth. Our obligation to survive is owed not just to ourselves but also to that Cosmos, ancient and vast, from which we spring.†

The *Cassini–Huygens* spacecraft on its Saturn mission provided us a dramatic perspective on our Home Planet. In September 2006, JPL scientists directed the craft so that Saturn was backlit by the Sun, a view of the rings that we never see from Earth. The small pale dot to the left of Saturn (see arrow), just inside the bright ring, is us! All that we studied in the text and that we are trying to understand is on that small, blue dot (Figure 21.12).

*E. O. Wilson, *The Diversity of Life* (Cambridge, MA: Harvard University Press, 1992), p. 351.

†C. Sagan, *Cosmos* (New York: Random House, 1980), p. 345.

May we all perceive our spatial importance within Earth's ecosystems and do our part to maintain a life-supporting and sustaining Earth for ourselves and countless generations in the future. "For the polar bears and all the grandchildren who will enjoy them, may there be sea ice."

The *Geosystems* Student Learning Center provides on-line resources for this chapter on the World Wide Web. To begin: Once at the Center, click on the cover of this textbook, scroll the Table of Contents menu, and select this chapter. You will find self-tests that are graded, review exercises, specific updates for items in the chapter, and in "Destinations" many links to interesting related pathways on the Internet. *Geosystems* Student Learning Center is found at **http://www.prenhall.com/christopherson/**.

Critical Thinking and Learning

A. What part do you think technology, politics, and thinking about the future should play in science courses?

B. Assess population growth issues: the count, the impact per person, and future projection. What strategies do you see as important?

C. According to the discussion in the chapter, what worldwide factors led to the *Exxon Valdez* accident? Describe the complexity of that event from a global perspective. In your analysis, examine both supply-side (corporations and utilities) and demand-side (consumers) issues, as well as environmental and strategic factors. And, what about the oily bird?

D. What is meant by the Gaia hypothesis? Describe several concepts from this text that might pertain to this hypothesis.

E. Relate the content of the various chapters in this text to the integrative Earth systems science concept. Which chapters help you to better understand Earth–human relations and human impacts?

F. After examining the list of 12 paradigm issues for the twenty-first century, suggest items that need to be added to the list, omitted from the list, or expanded in coverage. Rearrange and organize the list as needed to match your concerns and sense of priorities as to importance.

G. This chapter states that we already know many of the solutions to the problems we face. Why do you think these solutions are not being implemented at a faster pace?

H. Who speaks for Earth?

Appendix A

Maps in This Text and Topographic Maps

Maps Used in This Text

Geosystems uses several map projections: Goode's homolosine, Robinson, and Miller cylindrical, among others. Each was chosen to best present specific types of data. **Goode's homolosine projection** is an interrupted world map designed in 1923 by Dr. J. Paul Goode of the University of Chicago. Rand McNally *Goode's Atlas* first used it in 1925. Goode's homolosine equal-area projection (Figure A.1) is a combination of two oval projections (*homolo*graphic and *sin*usoidal projections).

Two equal-area projections are cut and pasted together to improve the rendering of landmass shapes. A *sinusoidal projection* is used between 40° N and 40° S latitudes. Its central meridian is a straight line; all other meridians are drawn as sinusoidal curves (based on sine-wave curves) and parallels are evenly spaced. A *Mollweide projection*, also called a *homolographic projection*, is used from 40° N to the North Pole and from 40° S to the South Pole. Its central meridian is a straight line; all other meridians are drawn as elliptical arcs and parallels are unequally spaced—farther apart at the equator, closer together poleward. This technique of combining two projections preserves areal size relationships, making the projection excellent for mapping spatial distributions when interruptions of oceans or continents do not pose a problem.

We use Goode's homolosine projection throughout this book. Examples include the world climate map and smaller climate type maps in Chapter 10, topographic regions and continental shields maps (Figures 12.3 and 12.4), world karst map (Figure 13.14), world sand regions and loess deposits (Figures 15.12 and 15.14), and the terrestrial biomes map in Chapter 20.

Another projection we use is the **Robinson projection**, designed by Arthur Robinson in 1963 (Figure A.2). This projection is neither equal area nor true shape, but is a compromise between the two. The North and South poles appear as lines slightly more than half the length of the equator; thus higher latitudes are exaggerated less than on other oval and cylindrical projections. Some of the Robinson maps employed include the latitudinal geographic zones map in Chapter 1 (Figure 1.12), the daily net radiation map (Figure 2.11), the world temperature range map in Chapter 5 (Figure 5.19), the maps of lithospheric plates of crust and volcanoes and earthquakes in Chapter 11 (Figures 11.17 and 11.20), and the global oil spills map in Chapter 21 (Figure 21.7).

Another compromise map, the **Miller cylindrical projection**, is used in this text (Figure A.3). Examples of this projection include the world time zone map (Figure 1.17), global temperature maps in Chapter 5 (Figures 5.14 and 5.17), two global pressure maps in Figure 6.10, and global soil maps and soil order maps in Chapter 18 (Figure 18.9). This projection is neither true shape nor true area, but is a compromise that avoids the severe scale distortion of the Mercator. The Miller projection frequently appears in world atlases. The American Geographical Society presented Osborn Miller's map projection in 1942.

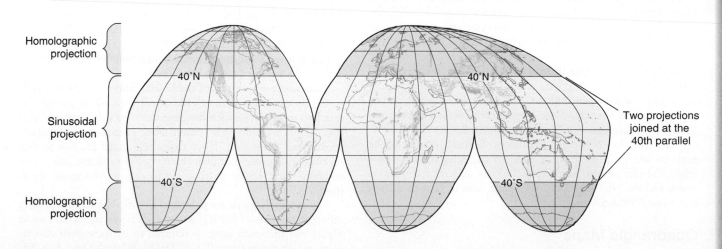

FIGURE A.1 Goode's homolosine projection.
An equal-area map. [Copyright by the University of Chicago. Used by permission of the University of Chicago Press.]

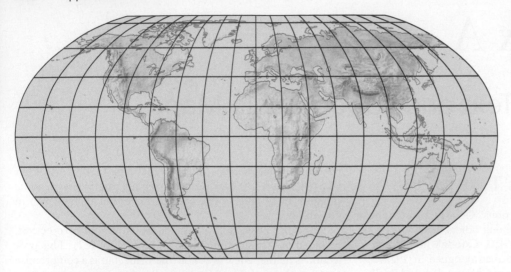

FIGURE A.2 Robinson projection.
A compromise between equal area and true shape. [Developed by Arthur H. Robinson, 1963.]

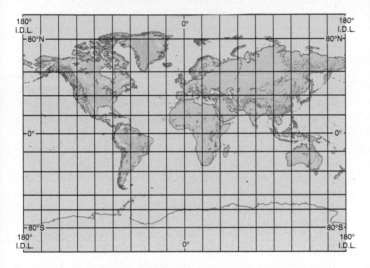

FIGURE A.3 Miller cylindrical projection.
A compromise map projection between equal area and true shape. [Developed by Osborn M. Miller, American Geographical Society, 1942.]

Mapping, Quadrangles, and Topographic Maps

The westward expansion across the vast North American continent demanded a land survey for the creation of accurate maps. Maps were needed to subdivide the land and to guide travel, exploration, settlement, and transportation. In 1785, the Public Lands Survey System began surveying and mapping government land in the United States. In 1836, the Clerk of Surveys in the Land Office of the Department of the Interior directed public-land surveys. The Bureau of Land Management replaced this Land Office in 1946. The actual preparation and recording of survey information fell to the U.S. Geological Survey (USGS), also a branch of the Department of the Interior (see **http://www.usgs.gov/pubprod/maps.html**).

In Canada, National Resources Canada conducts the national mapping program. Canadian mapping includes base maps, thematic maps, aeronautical charts, federal topographic maps, and the National Atlas of Canada, now in its fifth edition (see **http://atlas.nrcan.gc.ca/**).

Quadrangle Maps

The USGS depicts survey information on quadrangle maps, so called because they are rectangular maps with four corner angles. The angles are junctures of parallels of latitude and meridians of longitude rather than political boundaries. These quadrangle maps utilize the Albers equal-area projection, from the conic class of map projections.

The accuracy of conformality (shape) and scale of this base map is improved by the use of not one but two standard parallels. (Remember from Chapter 1 that standard lines are where the projection cone touches the globe's surface, producing greatest accuracy.) For the conterminous United States (the "lower 48"), these parallels are 29.5° N and 45.5° N latitude (noted on the Albers projection shown in Figure 1.21c). The standard parallels shift for conic projections of Alaska (55° N and 65° N) and for Hawai'i (8° N and 18° N).

Because a single map of the United States at 1:24,000 scale would be more than 200 m wide (more than 600 ft), some system had to be devised for dividing the map into a manageable size. Thus, a quadrangle system using latitude and longitude coordinates developed. Note that these maps are not perfect rectangles, because meridians converge toward the poles. The width of quadrangles narrows noticeably as you move north (poleward).

Quadrangle maps are published in different series, covering different amounts of Earth's surface at different scales. You see in Figure A.4 that each series is referred to by its angular dimensions, which range from 1° × 2° (1:250,000 scale) to 7.5′ × 7.5′ (1:24,000 scale). A map that is one-half a degree (30′) on each side is called a "30-minute quadrangle," and a map one-fourth of a degree (15′) on each side is a "15-minute quadrangle" (this was the

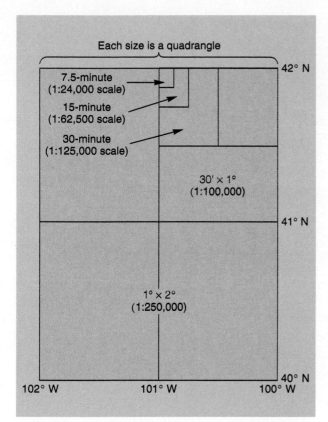

FIGURE A.4 Quadrangle system of maps used by the USGS.

USGS standard size from 1910 to 1950). A map that is one-eighth of a degree (7.5′) on each side is a 7.5-minute quadrangle, the most widely produced of all USGS topographic maps and the standard since 1950. The progression toward more-detailed maps and a larger-scale map standard through the years reflects the continuing refinement of geographic data and new mapping technologies.

The USGS National Mapping Program completed coverage of the entire country (except Alaska) on 7.5-minute maps (1 in. to 2000 ft, a large scale). It takes 53,838 separate 7.5-minute quadrangles to cover the lower 48 states, Hawai‘i, and the U.S. territories. A series of smaller-scale, more-general 15-minute topographic maps offer Alaskan coverage.

In the United States, most quadrangle maps remain in English units of feet and miles. The eventual changeover to the metric system requires revision of the units used on all maps, with the 1:24,000 scale eventually changing to a scale of 1:25,000. However, after completing only a few metric quads, the USGS halted the program in 1991. In Canada, the entire country is mapped at a scale of 1:250,000, using metric units (1.0 cm to 2.5 km). About half the country also is mapped at 1:50,000 (1.0 cm to 0.50 km).

Topographic Maps

The most popular and widely used quadrangle maps are **topographic maps** prepared by the USGS. An example of such a map is a portion of the Cumberland, Maryland, quad shown in Figure A.5. You will find topographic maps throughout *Geosystems* because they portray landscapes so effectively. As examples, see Figures 13.15 and 13.16, karst landscapes and sinkholes near Orleans, Indiana, and Winter Park, Florida;

FIGURE A.5 An example of a topographic map from the Appalachians.
Cumberland, MD, PA, WV 7.5-minute quadrangle topographic map prepared by the USGS. Note the water gap through Haystack Mountain. In the *Applied Physical Geography*, 7/e, lab manual this topographic map is accessible through Google Earth™ mapping services where you experience the map as a 3-D landscape and maneuver with your computer to see the topography at any angle or detail you choose.

(a)

(b)

FIGURE A.6 Topographic map of a hypothetical landscape.
(a) Perspective view of a hypothetical landscape. (b) Depiction of that landscape on a topographic
map. The contour interval on the map is 20 feet (6.1 m). [After the U.S. Geological Survey.]

Figure 14.8, river drainage patterns; Figure 14.22, river meander scars; Figure 15.18, an alluvial fan in Montana; Figure 17.3, glaciers in Alaska; and Figure 17.16, drumlins in New York.

A **planimetric map** shows the horizontal position (latitude/longitude) of boundaries, land-use aspects, bodies of water, and economic and cultural features. A highway map is a common example of a planimetric map.

A topographic map adds a vertical component to show topography (configuration of the land surface), including slope and relief (the vertical difference in local landscape elevation). These fine details are shown through the use of elevation contour lines (Figure A.6). A *contour line* connects all points at the same elevation. Elevations are shown above or below a vertical datum, or reference level, which usually is mean sea level. The contour interval is the vertical distance in elevation between two adjacent contour lines (20 ft, or 6.1m in Figure A. 6b).

The topographic map in Figure 6b shows a hypothetical landscape, demonstrating how contour lines and intervals depict slope and relief, which are the three-dimensional aspects of terrain. The pattern of lines and the spacing between them indicate slope. The steeper a slope or cliff, the closer together the contour lines appear—in the figure, note the narrowly spaced contours that represent the cliffs to the left of the

highway. A wider spacing of these contour lines portrays a more gradual slope, as you can see from the widely spaced lines on the beach and to the right of the river valley.

In Figure A.7 are the standard symbols commonly used on these topographic maps. These symbols and the colors used are standard on all USGS topographic maps: black for human constructions, blue for water features, brown for relief features and contours, pink for urbanized areas, and green for woodlands, orchards, brush, and the like.

The margins of a topographic map contain a wealth of information about its concept and content. In the margins of topographic maps, you find the quadrangle name, names of adjoining quads, quad series and type, position in the latitude-longitude and other coordinate systems, title, legend, magnetic declination (alignment of magnetic north) and compass information, datum plane, symbols used for roads and trails, the dates and history of the survey of that particular quad, and more.

Topographic maps may be purchased directly from the USGS or Centre for Topographic Information, NRC (http://maps.nrcan.gc.ca/). Many state geological survey offices, national and state park headquarters, outfitters, sports shops, and bookstores also sell topographic maps to assist people in planning their outdoor activities.

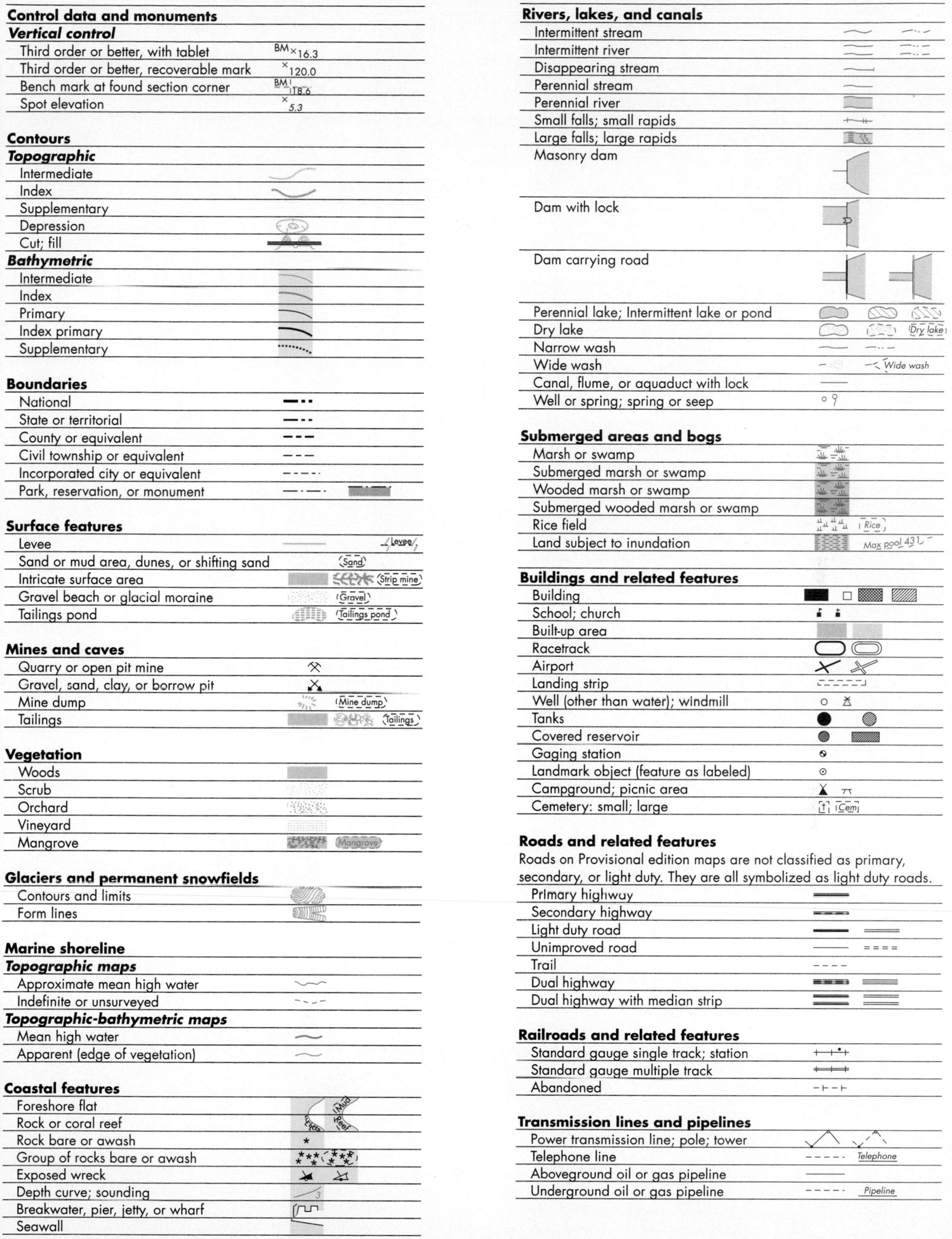

Control data and monuments

Vertical control

Third order or better, with tablet	BM ×16.3
Third order or better, recoverable mark	× 120.0
Bench mark at found section corner	BM 18.6
Spot elevation	× 5.3

Contours

Topographic

Intermediate	
Index	
Supplementary	
Depression	
Cut; fill	

Bathymetric

Intermediate	
Index	
Primary	
Index primary	
Supplementary	

Boundaries

National	
State or territorial	
County or equivalent	
Civil township or equivalent	
Incorporated city or equivalent	
Park, reservation, or monument	

Surface features

Levee	(Levee)
Sand or mud area, dunes, or shifting sand	(Sand)
Intricate surface area	(Strip mine)
Gravel beach or glacial moraine	(Gravel)
Tailings pond	(Tailings pond)

Mines and caves

Quarry or open pit mine	
Gravel, sand, clay, or borrow pit	
Mine dump	(Mine dump)
Tailings	(Tailings)

Vegetation

Woods	
Scrub	
Orchard	
Vineyard	
Mangrove	(Mangrove)

Glaciers and permanent snowfields

Contours and limits	
Form lines	

Marine shoreline

Topographic maps

Approximate mean high water	
Indefinite or unsurveyed	

Topographic-bathymetric maps

Mean high water	
Apparent (edge of vegetation)	

Coastal features

Foreshore flat	
Rock or coral reef	
Rock bare or awash	
Group of rocks bare or awash	
Exposed wreck	
Depth curve; sounding	3
Breakwater, pier, jetty, or wharf	
Seawall	

Rivers, lakes, and canals

Intermittent stream	
Intermittent river	
Disappearing stream	
Perennial stream	
Perennial river	
Small falls; small rapids	
Large falls; large rapids	
Masonry dam	
Dam with lock	
Dam carrying road	
Perennial lake; Intermittent lake or pond	
Dry lake	(Dry lake)
Narrow wash	
Wide wash	(Wide wash)
Canal, flume, or aquaduct with lock	
Well or spring; spring or seep	

Submerged areas and bogs

Marsh or swamp	
Submerged marsh or swamp	
Wooded marsh or swamp	
Submerged wooded marsh or swamp	
Rice field	(Rice)
Land subject to inundation	Max pool 431

Buildings and related features

Building	
School; church	
Built-up area	
Racetrack	
Airport	
Landing strip	
Well (other than water); windmill	
Tanks	
Covered reservoir	
Gaging station	
Landmark object (feature as labeled)	
Campground; picnic area	
Cemetery: small; large	(Cem)

Roads and related features

Roads on Provisional edition maps are not classified as primary, secondary, or light duty. They are all symbolized as light duty roads.

Primary highway	
Secondary highway	
Light duty road	
Unimproved road	
Trail	
Dual highway	
Dual highway with median strip	

Railroads and related features

Standard gauge single track; station	
Standard gauge multiple track	
Abandoned	

Transmission lines and pipelines

Power transmission line; pole; tower	
Telephone line	Telephone
Aboveground oil or gas pipeline	
Underground oil or gas pipeline	Pipeline

FIGURE A.7 Standardized topographic map symbols used on USGS maps.
English units still prevail, although a few USGS maps are in metric. [From USGS, Topographic Maps, 1969.]

A.5

Appendix B

The Köppen Climate Classification System

The Köppen climate classification system was designed by Wladimir Köppen (1846–1940), a German climatologist and botanist, and is widely used for its ease of comprehension. The basis of any empirical classification system is the choice of criteria used to draw lines on a map to designate different climates. Köppen–Geiger climate classification uses *average monthly temperatures, average monthly precipitation,* and *total annual precipitation* to devise its spatial categories and boundaries. But we must remember that boundaries really are transition zones of gradual change. The trends and overall patterns of boundary lines are more important than their precise placement, especially with the small scales generally used on world maps.

Take a few minutes and examine the Köppen system, the criteria, and the considerations for each of the principal climate categories. The modified Köppen–Geiger system has its drawbacks, however. It does not consider winds, temperature extremes, precipitation intensity, quantity of sunshine, cloud cover, or net radiation. Yet the system is important because its correlations with the actual world are reasonable and the input data are standardized and readily available.

Köppen's Climatic Designations

Figure 10.4 in Chapter 10 shows the distribution of each of Köppen's six climate classifications on the land. This generalized map shows the spatial pattern of climate. The Köppen system uses capital letters (A, B, C, D, E, H) to designate climatic categories from the equator to the poles. The guidelines for each of these categories are in the margin of Figure B.1.

Five of the climate classifications are based on thermal criteria:

A Tropical (equatorial regions)
C Mesothermal (Mediterranean, humid subtropical, marine west coast regions)
D Microthermal (humid continental, subarctic regions)
E Polar (polar regions)
H Highland (compared to lowlands at the same latitude, highlands have lower temperatures—recall the normal lapse rate—and more efficient precipitation due to lower moisture demand)

Only one climate classification is based on moisture as well:

B Dry (deserts and steppes)

Within each climate classification additional lowercase letters are used to signify temperature and moisture conditions. For example, in a tropical rain forest *Af* climate, the *A* tells us that the average coolest month is above 18°C (64.4°F, average for the month), and the *f* indicates that the weather is constantly wet, with the driest month receiving at least 6 cm (2.4 in.) of precipitation. (The designation *f* is from the German *feucht,* for "moist.") As you can see on the climate map, the tropical rain forest *Af* climate dominates along the equator and equatorial rain forest.

In a *Dfa* climate, the *D* means that the average warmest month is above 10°C (50°F), with at least one month falling below 0°C (32°F); the *f* says that at least 3 cm (1.2 in.) of precipitation falls during every month; and the *a* indicates a warmest summer month averaging above 22°C (71.6°F). Thus, a *Dfa* climate is a humid-continental, hot-summer climate in the microthermal category.

Köppen Guidelines

The Köppen guidelines and map portrayal are in Figure B.1. First, check the guidelines for a climate type, then the color legend for the subdivisions of the type, then check out the distribution of that climate on the map. You may want to compare this with the climate map in Figure 10.5 that presents causal elements that produce these climates.

Five of the climate classifications are based on thermal criteria:

A Tropical (equatorial regions)
C Mesothermal (Mediterranean, humid subtropical, marine west coast regions)
D Microthermal (humid continental, subarctic regions)
E Polar (polar regions)
H Highland (compared to lowlands at the same latitude, highlands have lower temperatures—recall the normal lapse rate—and more efficient precipitation due to lower moisture demand)

Only one climate classification is based on moisture as well:

B Dry (deserts and steppes)

FIGURE B.1 World climates and their guidelines according to the Köppen classification system.

Köppen Guidelines
Tropical Climates — A

Consistently warm with all months averaging above 18°C (64.4°F); annual water supply exceeds water demand.

Af — Tropical rain forest:
  f = All months receive precipitation in excess of 6 cm (2.4 in.).

Am — Tropical monsoon:
  m = A marked short dry season with 1 or more months receiving less than 6 cm (2.4 in.) precipitation; an otherwise excessively wet rainy season. ITCZ 6–12 months dominant.

Aw — Tropical savanna:
  w = Summer wet season, winter dry season; ITCZ dominant 6 months or less, winter water-balance deficits.

Mesothermal Climates — C

Warmest month above 10°C (50°F); coldest month above 0°C (32°F) but below 18°C (64.4°F); seasonal climates.

Cfa, Cwa — Humid subtropical:
  a = Hot summer; warmest month above 22°C (71.6°F).
  f = Year-round precipitation.
  w = Winter drought, summer wettest month 10 times more precipitation than driest winter month.

Cfb, Cfc — Marine west coast, mild-to-cool summer:
  f = Receives year-round precipitation.
  b = Warmest month below 22°C (71.6°F) with 4 months above 10°C.
  c = 1–3 months above 10°C.

Csa, Csb — Mediterranean summer dry:
  s = Pronounced summer drought with 70% of precipitation in winter.
  a = Hot summer with warmest month above 22°C (71.6°F).
  b = Mild summer; warmest month below 22°C.

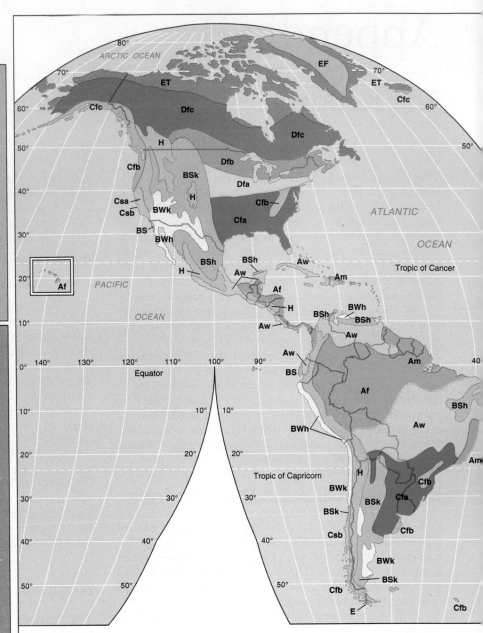

Microthermal Climates — D

Warmest month above 10°C (50°F); coldest month below 0°C (32°F); cool temperature-to-cold conditions; snow climates. In Southern Hemisphere, occurs only in highland climates.

Dfa, Dwa — Humid continental:
  a = Hot summer; warmest month above 22°C (71.6°F).
  f = Year-round precipitation.
  w = Winter drought.

Dfb, Dwb — Humid continental:
  b = Mild summer; warmest month below 22°C (71.6°F).
  f = Year-round precipitation.
  w = Winter drought.

Dfc, Dwc, Dwd — Subarctic:
Cool summers, cold winters.
  f = Year-round precipitation.
  w = Winter drought.
  c = 1–4 months above 10°C
  b = Coldest month below −38°C (−36.4°F), in Siberia only.

Dry Arid and Semiarid Climates — B

Potential evapotranspiration* (natural moisture demand) exceeds precipitation (natural moisture supply) in all B climates. Subdivisions based on precipitation timing and amount and mean annual temperature.

Earth's arid climates.

BWh — Hot low-latitude desert

BWk — Cold midlatitude desert
  BW = Precipitation less than 1/2 natural moisture demand.
  h = Mean annual temperature >18°C (64.4°F).
  k = Mean annual temperature <18°C.

Earth's semiarid climates.

BSh — Hot low-latitude steppe

BSk — Cold midlatitude steppe
  BS = Precipitation more than 1/2 natural moisture demand but not equal to it.
  h = Mean annual temperature >18°C.
  k = Mean annual temperature <18°C.

Polar Climates — E

Warmest month below 10°C (50°F); always cold; ice climates.

ET — Tundra:
Warmest month 0–10°C (32–50°F); precipitation exceeds small potential evapotranspiration demand*; snow cover 8–10 months.

EF — Ice cap:
Warmest month below 0°C (32°F); precipitation exceeds a very small potential evapotranspiration demand; the polar regions.

EM — Polar marine:
All months above −7°C (20°F), warmest month above 0°C; annual temperature ran <17 C° (30 F°).

*Potential evapotranspiration = the amount of water that would evaporate or transpire if it were available—the natural moisture deman in an environment; see Chapter 9.

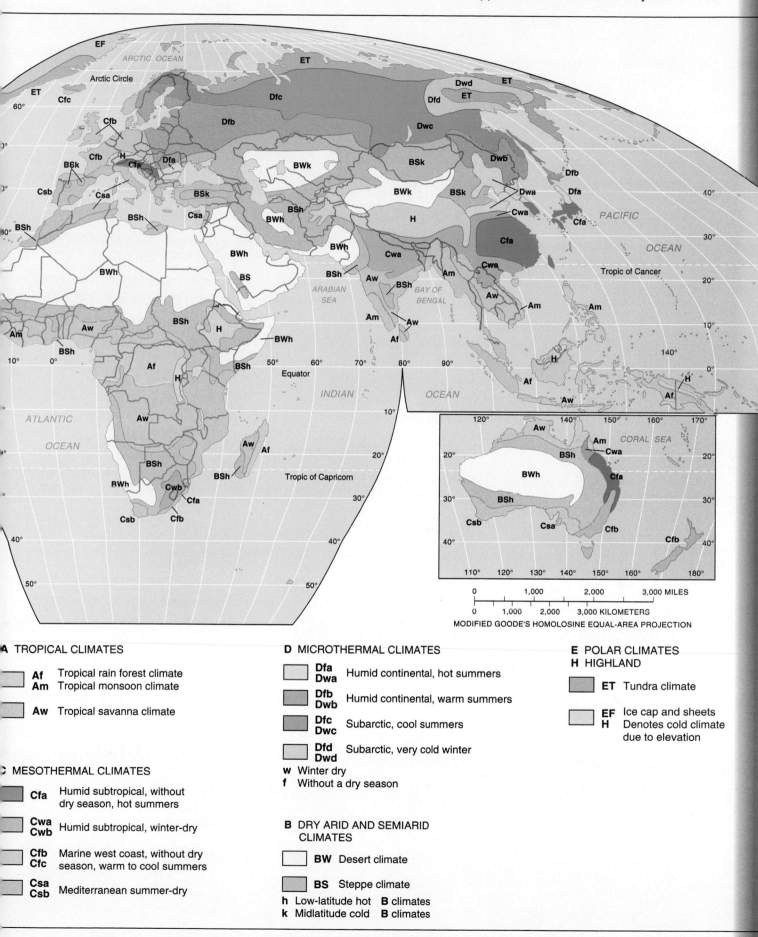

MODIFIED GOODE'S HOMOLOSINE EQUAL-AREA PROJECTION

A TROPICAL CLIMATES

Af	Tropical rain forest climate
Am	Tropical monsoon climate
Aw	Tropical savanna climate

C MESOTHERMAL CLIMATES

Cfa	Humid subtropical, without dry season, hot summers
Cwa **Cwb**	Humid subtropical, winter-dry
Cfb **Cfc**	Marine west coast, without dry season, warm to cool summers
Csa **Csb**	Mediterranean summer-dry

D MICROTHERMAL CLIMATES

Dfa **Dwa**	Humid continental, hot summers
Dfb **Dwb**	Humid continental, warm summers
Dfc **Dwc**	Subarctic, cool summers
Dfd **Dwd**	Subarctic, very cold winter

w Winter dry
f Without a dry season

B DRY ARID AND SEMIARID CLIMATES

BW	Desert climate
BS	Steppe climate

h Low-latitude hot **B** climates
k Midlatitude cold **B** climates

E POLAR CLIMATES
H HIGHLAND

ET	Tundra climate
EF **H**	Ice cap and sheets Denotes cold climate due to elevation

Appendix C

Common Conversions

Metric to English

Metric Measure	Multiply by	English Equivalent
Length		
Centimeters (cm)	0.3937	Inches (in.)
Meters (m)	3.2808	Feet (ft)
Meters (m)	1.0936	Yards (yd)
Kilometers (km)	0.6214	Miles (mi)
Nautical mile	1.15	Statute mile
Area		
Square centimeters (cm^2)	0.155	Square inches ($in.^2$)
Square meters (m^2)	10.7639	Square feet (ft^2)
Square meters (m^2)	1.1960	Square yards (yd^2)
Square kilometers (km^2)	0.3831	Square miles (mi^2)
Hectare (ha) (10,000 m^2)	2.4710	Acres (a)
Volume		
Cubic centimeters (cm^3)	0.06	Cubic inches ($in.^3$)
Cubic meters (m^3)	35.30	Cubic feet (ft^3)
Cubic meters (m^3)	1.3079	Cubic yards (yd^3)
Cubic kilometers (km^3)	0.24	Cubic miles (mi^3)
Liters (l)	1.0567	Quarts (qt), U.S.
Liters (l)	0.88	Quarts (qt), Imperial
Liters (l)	0.26	Gallons (gal), U.S.
Liters (l)	0.22	Gallons (gal), Imperial
Mass		
Grams (g)	0.03527	Ounces (oz)
Kilograms (kg)	2.2046	Pounds (lb)
Metric ton (tonne) (t)	1.10	Short ton (tn), U.S.
Velocity		
Meters/second (mps)	2.24	Miles/hour (mph)
Kilometers/hour (kmph)	0.62	Miles/hour (mph)
Knots (kn) (nautical mph)	1.15	Miles/hour (mph)
Temperature		
Degrees Celsius (°C)	1.80 (then add 32)	Degrees Fahrenheit (°F)
Celsius degree (C°)	1.80	Fahrenheit degree (F°)
Additional water measurements:		
Gallon (Imperial)	1.201	Gallon (U.S.)
Gallons (gal)	0.000003	Acre-feet

1 cubic foot per second per day = 86,400 cubic feet = 1.98 acre = feet

Additional Energy and Power Measurements

1 watt (W) = 1 joule/s

1 joule = 0.239 calorie

1 calorie = 4.186 joules

1 W/m^2 = 0.001433 cal/min

697.8 W/m^2 = 1 cal/cm^2min^{-1}

1 W/m^2 = 61.91 cal/cm^2month^{-1}

W/m^2 = 753.4 cal/cm^2year^{-1}

100W/m^2 = 75 kcal/cm^2year^{-1}

Solar constant:
1372 W/m^2
2 cal/cm^2min^{-1}

English to Metric

English Measure	Multiply by	Metric Equivalent
Length		
Inches (in.)	2.54	Centimeters (cm)
Feet (ft)	0.3048	Meters (m)
Yards (yd)	0.9144	Meters (m)
Miles (mi)	1.6094	Kilometers (km)
Statute mile	0.8684	Nautical mile
Area		
Square inches (in.^{2})	6.45	Square centimeters (cm^2)
Square feet (ft^2)	0.0929	Square meters (m^2)
Square yards (yd^2)	0.8361	Square meters (m^2)
Square miles (mi^2)	2.5900	Square kilometers (km^2)
Acres (a)	0.4047	Hectare (ha)
Volume		
Cubic inches (in.^{3})	16.39	Cubic centimeters (cm^3)
Cubic feet (ft^3)	0.028	Cubic meters (m^3)
Cubic yards (yd^3)	0.765	Cubic meters (m^3)
Cubic miles (mi^3)	4.17	Cubic kilometers (km^3)
Quarts (qt), U.S.	0.9463	Liters (l)
Quarts (qt), Imperial	1.14	Liters (l)
Gallons (gal), U.S.	3.8	Liters (l)
Gallons (gal), Imperial	4.55	Liters (l)
Mass		
Ounces (oz)	28.3495	Grams (g)
Pounds (lb)	0.4536	Kilograms (kg)
Short ton (tn), U.S.	0.91	Metric ton (tonne) (t)
Velocity		
Miles/hour (mph)	0.448	Meters/second (mps)
Miles/hour (mph)	1.6094	Kilometers/hour (kmph)
Miles/hour (mph)	0.8684	Knots (kn) (nautical mph)
Temperature		
Degrees Fahrenheit (°F)	0.556 (after subtracting 32)	Degrees Celsius (°C)
Fahrenheit degree (F°)	0.556	Celsius degree (C°)
Additional water measurements:		
Gallon (U.S.)	0.833	Gallons (Imperial)
Acre-feet	325,872	Gallons (gal)

Additional Notation

Multiples	Prefixes	
$1{,}000{,}000{,}000 = 10^9$	giga	G
$1{,}000{,}000 = 10^6$	mega	M
$1{,}000 = 10^3$	kilo	k
$100 = 10^2$	hecto	h
$10 = 10^1$	deka	da
$1 = 10^0$		
$0.1 = 10^{-1}$	deci	d
$0.01 = 10^{-2}$	centi	c
$0.001 = 10^{-3}$	milli	m
$0.000001 = 10^{-6}$	micro	μ

Glossary

The chapter in which each term appears **boldfaced** in parentheses is followed by a specific definition relevant to the term's usage in the chapter.

Aa (12) Rough, jagged, and clinkery basaltic lava with sharp edges. This texture is caused by the loss of trapped gases, a slow flow, and the development of a thick skin that cracks into the jagged surface.

Abiotic (1) Nonliving; Earth's nonliving systems of energy and materials.

Ablation (17) Loss of glacial ice through melting, sublimation, wind removal by deflation, or the calving of blocks of ice (see deflation).

Abrasion (14, 15, 17) Mechanical wearing and erosion of bedrock accomplished by the rolling and grinding of particles and rocks carried in a stream, removed by wind in a "sandblasting" action, or imbedded in glacial ice.

Absorption (4) Assimilation and conversion of radiation from one form to another in a medium. In the process, the temperature of the absorbing surface is raised, thereby affecting the rate and wavelength of radiation from that surface.

Active layer (17) A zone of seasonally frozen ground that exists between the subsurface permafrost layer and the ground surface. The active layer is subject to consistent daily and seasonal freeze-thaw cycles (see permafrost, periglacial).

Actual evapotranspiration (9) ACTET; the actual amount of evaporation and transpiration that occurs; derived in the water balance by subtracting the deficit (DEFIC) from potential evapotranspiration in the surface.

Adiabatic (7) Pertaining to the cooling of an ascending parcel of air through expansion or the warming of a descending parcel of air through compression, without any exchange of heat between the parcel and the surrounding environment.

Advection (4) Horizontal movement of air or water from one place to another (compare with convection).

Advection fog (7) Active condensation formed when warm, moist air moves laterally over cooler water or land surfaces, causing the lower layers of the air to be chilled to the dew-point temperature.

Aerosols (3) Small particles of dust, soot, and pollution suspended in the air.

Aggradation (14) The general building of land surface because of deposition of material; opposite of degradation. When the sediment load of a stream exceeds the stream's capacity to carry it, the stream channel becomes filled through this process.

Air (3) A simple mixture of gases (N, O, Ar, CO_2, and trace gases) that is naturally odorless, colorless, tasteless, and formless, blended so thoroughly that it behaves as if it were a single gas.

Air mass (8) A distinctive, homogeneous body of air that has taken on the moisture and temperature characteristics of its source region.

Air pressure (3, 6) Pressure produced by the motion, size, and number of gas molecules in the air and exerted on surfaces in contact with the air. Normal sea-level pressure, as measured by the height of a column of mercury (Hg), is expressed as 1013.2 millibars, 760 mm of Hg, or 29.92 inches of Hg. Air pressure can be measured with mercury or aneroid barometers (see listing for both).

Albedo (4) The reflective quality of a surface, expressed as the percentage of reflected insolation to incoming insolation; a function of surface color, angle of incidence, and surface texture.

Aleutian low (6) See subpolar low-pressure cell.

Alfisols (18) A soil order in the Soil Taxonomy. Moderately weathered forest soils that are moist versions of Mollisols, with productivity dependent on specific patterns of moisture and temperature; rich in organics. Most wide-ranging of the soil orders.

Alluvial fan (15) Fan-shaped fluvial landform at the mouth of a canyon; generally occurs in arid landscapes where streams are intermittent.

Alluvial terraces (14) Level areas that appear as topographic steps above a stream, created by the stream as it scours with renewed downcutting into its floodplain; composed of unconsolidated alluvium (see alluvium).

Alluvium (14) General descriptive term for clay, silt, sand, gravel, or other unconsolidated rock and mineral fragments transported by running water and deposited as sorted or semi-sorted sediment on a floodplain, delta, or stream bed.

Alpine glacier (17) A glacier confined in a mountain valley or walled basin, consisting of three subtypes: valley glacier (within a valley), piedmont glacier (coalesced at the base of a mountain, spreading freely over nearby lowlands), and outlet glacier (flowing outward from a continental glacier).

Alpine tundra (20) Tundra conditions at high elevation (see Arctic tundra).

Altitude (2) The angular distance between the horizon (a horizontal plane) and the Sun (or any point in the sky).

Altocumulus (7) Middle-level, puffy clouds that occur in several forms: patchy rows, wave patterns, a "mackerel sky," or lens-shaped "lenticular" clouds.

Andisols (18) A soil order in the Soil Taxonomy; derived from volcanic parent materials in areas of volcanic activity. A new order, created in 1990, of soils previously considered under Inceptisols and Entisols.

Anemometer (6) A device that measures wind velocity.

Aneroid barometer (6) A device that measures air pressure using a partially evacuated, sealed cell (see air pressure).

Angle of repose (13) The steepness of a slope that results when loose particles come to rest; an angle of balance between driving and resisting forces, ranging between 33° and 37° from a horizontal plane.

Antarctic Circle (2) This latitude (66.5° S) denotes the northernmost parallel (in the Southern Hemisphere) that experiences a 24-hour period of darkness in winter or daylight in summer.

Antarctic high (6) A consistent high-pressure region centered over Antarctica; source region for an intense polar air mass that is dry and associated with the lowest temperatures on Earth.

Anthropogenic atmosphere (3) Earth's future atmosphere, so named because humans appear to be the principal causative agent.

Anticline (12) Upfolded rock strata, in which layers slope downward from the axis of the fold, or central ridge (compare syncline).

Anticyclone (6) A dynamically or thermally caused area of high atmospheric pressure with descending and diverging airflows that rotate clockwise in the Northern Hemisphere and counterclockwise in the Southern Hemisphere (compare cyclone).

Aphelion (2) The point of Earth's greatest distance from the Sun in its elliptical orbit; reached on July 4 at a distance of 152,083,000 km (94.5 million mi); variable over a 100,000-year cycle (compare perihelion).

Aquiclude (9) A body of rock that does not conduct groundwater in usable amounts; an impermeable rock layer, related to an aquitard that slows but does not block water flow (compare aquifer).

Aquifer (9) A body of rock that conducts groundwater in usable amounts; a permeable layer of rock (compare aquiclude).

Aquifer recharge area (9) The surface area where water enters an aquifer to recharge the water-bearing strata in a groundwater system.

Arctic and alpine tundra (20) A biome in the northernmost portions of North America and northern Europe and Russia, featuring low ground-level herbaceous plants as well as some woody plants.

Arctic Circle (2) This latitude (66.5° N) denotes the southernmost parallel (in the Northern Hemisphere) that experiences a 24-hour period of darkness in winter or daylight in summer.

Arête (17) A sharp ridge that divides two cirque basins. Derived from "knife edge" in French, these form sawtooth and serrated ridges in glaciated mountains.

Aridisols (18) A soil order in the Soil Taxonomy; largest soil order. Typical of dry climates; low in organic matter and dominated by calcification and salinization.

Artesian water (9) Pressurized groundwater that rises in a well or a rock structure above the local water table; may flow out onto the ground without pumping (see potentiometric surface).

Asthenosphere (11) Region of the upper mantle just below the lithosphere; the least rigid portion of Earth's interior and known as the plastic layer, flowing very slowly under extreme heat and pressure.

Atmosphere (1) The thin veil of gases surrounding Earth, which forms a protective boundary between outer space and the biosphere; generally considered to extend out about 480 km (300 mi) from Earth's surface.

Aurora (2) A spectacular glowing light display in the ionosphere, stimulated by the interaction of the solar wind with oxygen and nitrogen gases and atoms at high latitudes, called aurora borealis in the Northern Hemisphere and aurora australis in the Southern Hemisphere.

Autumnal (September) equinox (2) The time around September 22–23 when the Sun's declination crosses the equatorial parallel (0° latitude) and all places on Earth experience days and nights of equal length. The Sun rises at the South Pole and sets at the North Pole (compare vernal equinox).

Available water (9) The portion of capillary water that is accessible to plant roots; usable water held in soil moisture storage (see capillary water).

Axial parallelism (2) Earth's axis remains aligned the same throughout the year (it "remains parallel to itself"); thus, the axis extended from the North Pole points into space always near Polaris, the North Star.

Axial tilt (2) Earth's axis tilts 23.5° from a perpendicular to the plane of the ecliptic (plane of Earth's orbit around the Sun).

Axis (2) An imaginary line, extending through Earth from the geographic North Pole to the geographic South Pole, around which Earth rotates.

Azores high (6) A subtropical high-pressure cell that forms in the Northern Hemisphere in the eastern Atlantic (see Bermuda high); associated with warm, clear water and large quantities of sargassum, or gulf weed, characteristic of the Sargasso Sea.

Backswamp (14) A low-lying, swampy area of a floodplain; adjacent to a river, with the river's natural levee on one side and higher topography on the other (see floodplain, yazoo tributary).

Bajada (15) A continuous apron of coalesced alluvial fans, formed along the base of mountains in arid climates; presents a gently rolling surface from fan to fan (see alluvial fan).

Barrier beach (16) Narrow, long, depositional feature, generally composed of sand, that forms offshore roughly parallel to the coast; may appear as barrier islands and long chains of barrier beaches (see barrier island).

Barrier island (16) Generally, a broadened barrier beach (see barrier beach).

Barrier spit (16) A depositional landform that develops when transported sand or gravel in a barrier beach or island is deposited in long ridges that are attached at one end to the mainland and partially cross the mouth of a bay.

Basalt (11) A common extrusive igneous rock, fine-grained, comprising the bulk of the ocean-floor crust, lava flows, and volcanic forms; gabbro is its intrusive form.

Base level (14) A hypothetical level below which a stream cannot erode its valley, and thus the lowest operative level for denudation processes; in an absolute sense, it is represented by sea level, extending under the landscape.

Basin and Range Province (15) A region of dry climates, few permanent streams, and interior drainage patterns in the western United States; a faulted landscape composed of a sequence of horsts and grabens.

Batholith (11) The largest plutonic form exposed at the surface; an irregular intrusive mass; it invades crustal rocks, cooling slowly so that large crystals develop (see pluton).

Bay barrier (16) An extensive barrier spit of sand or gravel that encloses a bay, cutting it off completely from the ocean and forming a lagoon; produced by littoral drift and wave action; sometimes referred to as a baymouth bar (see barrier spit, lagoon).

Beach (16) The portion of the coastline where an accumulation of sediment is in motion.

Beach drift (16) Material, such as sand, gravel, and shells, that is moved by the longshore current in the effective direction of the waves.

Beaufort wind scale (6) A descriptive scale for the visual estimation of wind speeds; originally conceived in 1806 by Admiral Beaufort of the British Navy.

Bed load (14) Coarse materials that are dragged along the bed of a stream by traction or by the rolling and bouncing motion of saltation; involves particles too large to remain in suspension (see traction, saltation).

Bedrock (13) The rock of Earth's crust that is below the soil and is basically unweathered; such solid crust sometimes is exposed as an outcrop.

Bergschrund (17) These form when a crevasse or wide crack opens along the headwall of a glacier; most visible in summer when covering snow is gone.

Bermuda high (6) A subtropical high-pressure cell that forms in the western North Atlantic (see Azores high).

Biodiversity (19) A principle of ecology and biogeography: the more diverse the species population in an ecosystem (in number of species, quantity of members in each species, and genetic content), the more risk is spread over the entire community, which results in greater overall stability, greater productivity, and increased use of nutrients, as compared to a monoculture of little or no diversity.

Biogeochemical cycle (19) One of several circuits of flowing elements and materials (carbon, oxygen, nitrogen, phosphorus, water) that combine Earth's biotic (living) and abiotic (nonliving) systems; the cycling of materials is continuous and renewed through the biosphere and the life processes.

Biogeographic realm (20) One of eight regions of the biosphere, each representative of evolutionary core areas of related flora (plants) and fauna (animals); a broad geographical classification scheme.

Biogeography (19) The study of the distribution of plants and animals and related ecosystems; the geographical relationships with their environments over time.

Biomass (19) The total mass of living organisms on Earth or per unit area of a landscape; also, the weight of the living organisms in an ecosystem.

Biome (20) A large terrestrial ecosystem characterized by specific plant communities and formations; usually named after the predominant vegetation in the region (see terrestrial ecosystem).

Biosphere (1) That area where the atmosphere, lithosphere, and hydrosphere function together to form the context within which life exists; an intricate web that connects all organisms with their physical environment.

Biotic (1) Living; referring to Earth's living system of organisms.

Blowout depression (15) Eolian (wind) erosion in which deflation forms a basin in areas of loose sediment. Diameter may range up to hundreds of meters (see deflation).

Bolson (15) The slope and basin area between the crests of two adjacent ridges in a dry region.

Boreal forest (20) See needleleaf forest.

Brackish (16) Descriptive of seawater with a salinity of less than 35‰; for example, the Baltic Sea (contrast to brine).

Braided stream (14) A stream that becomes a maze of interconnected channels laced with excess sediment. Braiding often occurs with a reduction of discharge that reduces a stream's transporting ability or with an increase in sediment load.

Breaker (16) The point where a wave's height exceeds its vertical stability and the wave breaks as it approaches the shore.

Brine (16) Seawater with a salinity of more than 35‰; for example, the Persian Gulf (contrast to brackish).

Calcification (18) The illuviated (deposited) accumulation of calcium carbonate or magnesium carbonate in the B and C soil horizons.

Caldera (12) An interior sunken portion of a composite volcano's crater; usually steep-sided and circular, sometimes containing a lake; also can be found in conjunction with shield volcanoes.

Capillary water (9) Soil moisture, most of which is accessible to plant roots; held in the soil by the water's surface tension and cohesive forces between water and soil (see also available water, field capacity, hygroscopic water, and wilting point).

Carbonation (13) A chemical weathering process in which weak carbonic acid (water and carbon dioxide) reacts with many minerals that contain calcium, magnesium, potassium, and sodium (especially limestone), transforming them into carbonates.

Carbon monoxide (3) An odorless, colorless, tasteless combination of carbon and oxygen produced by the incomplete combustion of fossil fuels or other carbon-containing substances; toxicity to humans is due to its affinity for hemoglobin, displacing oxygen in the bloodstream; CO.

Carnivore (19) A secondary consumer that principally eats meat for sustenance. The top carnivore in a food chain is considered a tertiary consumer (compare herbivore).

Cartography (1) The making of maps and charts; a specialized science and art that blends aspects of geography, engineering, mathematics, graphics, computer science, and artistic specialties.

Cation-exchange capacity (CEC) (18) The ability of soil colloids to exchange cations between their surfaces and the soil solution; a measured potential that indicates soil fertility (see soil colloid, soil fertility).

Chaparral (20) Dominant shrub formations of Mediterranean dry-summer climates; characterized by sclerophyllous scrub and short, stunted, tough forests; derived from the Spanish *chapparo*; specific to California (see Mediterranean shrubland).

Chemical weathering (13) Decomposition and decay of the constituent minerals in rock through chemical alteration of those minerals. Water is essential, with rates keyed to temperature and precipitation values. Chemical reactions are active at microsites even in dry climates. Processes include hydrolysis, oxidation, carbonation, and solution.

Chinook wind (8) North American term for a warm, dry, downslope airflow; characteristic of the rain-shadow region on the leeward side of mountains; known as föhn, or foehn, winds in Europe (see rain shadow).

Chlorofluorocarbons (CFC) (3) A manufactured molecule (polymer) made of chlorine, fluorine, and carbon; inert and possessing remarkable heat properties; also known as one of the halogens. After slow transport to the stratospheric ozone layer, CFCs react with ultraviolet radiation, freeing chlorine atoms that act as a catalyst to produce reactions that destroy ozone; manufacture banned by international treaties.

Chlorophyll (19) A light-sensitive pigment that resides within chloroplasts (organelles) in leaf cells of plants; the basis of photosynthesis.

Cinder cone (12) A volcanic landform of pyroclastics and scoria, usually small and cone-shaped and generally not more than 450 m (1500 ft) in height, with a truncated top.

Circle of illumination (2) The division between light and dark on Earth; a day–night great circle.

Circum-Pacific belt (12) A tectonically and volcanically active region encircling the Pacific Ocean; also known as the "ring of fire."

Cirque (17) A scooped-out, amphitheater-shaped basin at the head of an alpine glacier valley; an erosional landform.

Cirrus (7) Wispy, filamentous ice-crystal clouds that occur above 6000 m (20,000 ft); appear in a variety of forms, from feathery hairlike fibers to veils of fused sheets.

Classification (10) The process of ordering or grouping data or phenomena in related classes; results in a regular distribution of information; a taxonomy.

Climate (10) The consistent, long-term behavior of weather over time, including its variability; in contrast to weather, which is the condition of the atmosphere at any given place and time.

Climatic regions (10) An area of homogenous climate that features characteristic regional weather and air mass patterns.

Climatology (10) The scientific study of climate and climatic patterns and the consistent behavior of weather, including its variability and extremes, over time in one place or region; includes the effects of climate change on human society and culture.

Climograph (10) A graph that plots daily, monthly, or annual temperature and precipitation values for a selected station; may also include additional weather information.

Closed system (1) A system that is shut off from the surrounding environment so that it is entirely self-contained in terms of energy and materials; Earth is a closed material system (compare open system).

Cloud (7) An aggregate of tiny moisture droplets and ice crystals; classified by altitude of occurrence and shape.

Cloud-albedo forcing (4) An increase in albedo (the reflectivity of a surface) caused by clouds due to their reflection of incoming insolation.

Cloud-condensation nuclei (7) Microscopic particles necessary as matter on which water vapor condenses to form moisture droplets; can be sea salts, dust, soot, or ash.

Cloud-greenhouse forcing (4) An increase in greenhouse warming caused by clouds because they can act like insulation, trapping longwave (infrared) radiation.

Col (17) Formed by two headward-eroding cirques that reduce an arête (ridge crest) to form a high pass or saddlelike narrow depression.

Cold desert and semidesert (20) A type of desert biome found at higher latitudes than warm deserts. Interior location and rain-shadow locations produce these cold deserts in North America.

Cold front (8) The leading edge of an advancing cold air mass; identified on a weather map as a line marked with triangular spikes pointing in the direction of frontal movement (compare warm front).

Community (19, 20) A convenient biotic subdivision within an ecosystem; formed by interacting populations of animals and plants in an area.

Composite volcano (12) A volcano formed by a sequence of explosive volcanic eruptions; steep-sided, conical in shape; sometimes referred to as a stratovolcano, although composite is the preferred term (compare shield volcano).

Conduction (4) The slow molecule-to-molecule transfer of heat through a medium, from warmer to cooler portions.

Cone of depression (9) The depressed shape of the water table around a well after active pumping. The water table adjacent to the well is drawn down by the water removal.

Confined aquifer (9) An aquifer that is bounded above and below by impermeable layers of rock or sediment (see artesian water, unconfined aquifer).

Constant isobaric surface (6) An elevated surface in the atmosphere on which all points have the same pressure, usually 500 mb. Along this constant-pressure surface, isobars mark the paths of upper-air winds.

Consumer (19) Organism in an ecosystem that depends on producers (organisms that use carbon dioxide as their sole source of carbon) for its source of nutrients; also called a heterotroph (compare producer).

Consumptive use (9) A use that removes water from a water budget at one point and makes it unavailable further downstream (compare withdrawal).

Continental divide (14) A ridge or elevated area that separates drainage on a continental scale; specifically, that ridge in North America that separates drainage to the Pacific on the west side from drainage to the Atlantic and Gulf on the east side and to Hudson Bay and the Arctic Ocean in the north.

Continental drift (11) A proposal by Alfred Wegener in 1912 stating that Earth's landmasses have migrated over the past 225 million years from a supercontinent he called Pangaea to the present configuration; the widely accepted plate tectonics theory today (see plate tectonics).

Continental effect (5) A quality of regions that lack the temperature-moderating effects of the sea and that exhibit a greater range of minimum and maximum temperatures, both daily and annually (see marine, land–water heating differences).

Continental glacier (17) A continuous mass of unconfined ice, covering at least 50,000 km² (19,500 mi²); most extensive at present as ice sheets covering Greenland and Antarctica (compare alpine glacier).

Continental landmasses (12) The broadest category of landform, including those masses of crust that reside above or near sea level and the adjoining undersea continental shelves along the coastline; sometimes synonymous with continental platforms.

Continental shield (12) Generally, old, low-elevation heartland regions of continental crust; various cratons (granitic cores) and ancient mountains exposed at the surface.

Contour lines (Appendix A) Isolines on a topographic map that connect all points at the same elevation relative to a reference elevation called the vertical datum.

Convection (4) Transfer of heat from one place to another through the physical movement of air; involves a strong vertical motion (compare to advection).

Convectional lifting (8)   Air passing over warm surfaces gains buoyancy and lifts, initiating adiabatic processes.

Convergent lifting (8)   Airflows in conflict force lifting and displacement of air upward, initiating adiabatic processes.

Coordinated Universal Time (UTC) (1)   The official reference time in all countries, formerly known as Greenwich Mean Time; now measured by six primary standard atomic clocks, the time calculations are collected in Paris at the International Bureau of Weights and Measures (BIPM); the legal reference for time in all countries and broadcast worldwide.

Coral (16)   A simple, cylindrical marine animal with a saclike body that secretes calcium carbonate to form a hard external skeleton and, cumulatively, landforms called reefs; lives symbiotically with nutrient-producing algae; presently in a worldwide state of decline due to bleaching (loss of algae).

Core (11)   The deepest inner portion of Earth, representing one-third of its entire mass; differentiated into two zones—a solid-iron inner core surrounded by a dense, molten, fluid metallic-iron outer core.

Coriolis force (6)   The apparent deflection of moving objects (wind, ocean currents, missiles) from traveling in a straight path, in proportion to the speed of Earth's rotation at different latitudes. Deflection is to the right in the Northern Hemisphere and to the left in the Southern Hemisphere; maximum at the poles and zero along the equator.

Crater (12)   A circular surface depression formed by volcanism; built by accumulation, collapse, or explosion; usually located at a volcanic vent or pipe; can be at the summit or on the flank of a volcano.

Crevasse (17)   A vertical crack that develops in a glacier as a result of friction between valley walls, or tension forces of extension on convex slopes, or compression forces on concave slopes.

Cryosphere (1, 7, 10, 17, 20)   The frozen portion of Earth's waters, including ice sheets, ice caps and fields, glaciers, ice shelves, sea ice, and subsurface ground ice and frozen ground (permafrost).

Crust (11)   Earth's outer shell of crystalline surface rock, ranging from 5 to 60 km (3 to 38 mi) in thickness from oceanic crust to mountain ranges. Average density of continental crust is 2.7g/cm³, whereas oceanic crust is 3.0g/cm³.

Cumulonimbus (7)   A towering, precipitation-producing cumulus cloud that is vertically developed across altitudes associated with other clouds; frequently associated with lightning and thunder and thus sometimes called a thunderhead.

Cumulus (7)   Bright and puffy cumuliform clouds up to 2000 m in altitude (6500 ft).

Cyclogenesis (8)   An atmospheric process that describes the birth of a midlatitude wave cyclone, usually along the polar front. Also refers to strengthening and development of a midlatitude cyclone along the eastern slope of the Rockies, other north–south mountain barriers, and along the North American and Asian east coasts (see midlatitude cyclone, polar front).

Cyclone (6)   A dynamically or thermally caused area of low atmospheric pressure with converging and ascending airflows. Rotates counterclockwise in the Northern Hemisphere and clockwise in the Southern Hemisphere (compare anticyclone; see midlatitude cyclone, tropical cyclone).

Daylength (2)   Duration of exposure to insolation, varying during the year depending on latitude; an important aspect of seasonality.

Daylight saving time (1)   Time is set ahead 1 hour in the spring and set back 1 hour in the fall in the Northern Hemisphere. In the United States and Canada, time is set ahead on the second Sunday in March and set back on the first Sunday in November—except in Hawai‘i, Arizona, and Saskatchewan, which exempt themselves.

Debris avalanche (13)   A mass of falling and tumbling rock, debris, and soil; can be dangerous because of the tremendous velocities achieved by the onrushing materials.

Declination (2)   The latitude that receives direct overhead (perpendicular) insolation on a particular day; the subsolar point migrates annually through 47° of latitude between the Tropics of Cancer (23.5° N) and Capricorn (23.5° S).

Decomposer (19)   Bacteria and fungi that digest organic debris outside their bodies and absorb and release nutrients in an ecosystem (see detritivores).

Derechos (8)   Strong linear winds in excess of 26 m/sec (58 mph), associated with thunderstorms and bands of showers crossing a region.

Deficit (9)   DEFIC; in a water balance, the amount of unmet (unsatisfied) potential evapotranspiration; a natural water shortage (refer to POTET, or PE).

Deflation (15)   A process of wind erosion that removes and lifts individual particles, literally blowing away unconsolidated, dry, or non-cohesive sediments (see blowout depression).

Delta (14)   A depositional plain formed where a river enters a lake or an ocean; named after the triangular shape of the Greek letter delta, Δ.

Denudation (13)   A general term that refers to all processes that cause degradation of the landscape: weathering, mass movement, erosion, and transport.

Deposition (14)   The process whereby weathered, wasted, and transported sediments are laid down by air, water, and ice.

Desalination (9)   In a water resources context, the removal of organics, debris, and salinity from seawater through distillation or reverse osmosis to produce potable water.

Desert biome (20)   Arid landscapes of uniquely adapted dry-climate plants and animals.

Desertification (15)   The expansion of deserts worldwide, related principally to poor agricultural practices (overgrazing and inappropriate agricultural practices), improper soil-moisture management, erosion and salinization, deforestation, and the ongoing climatic change; an unwanted semipermanent invasion into neighboring biomes.

Desert pavement (15)   On arid landscapes, a surface formed when wind deflation and sheetflow remove smaller particles, leaving residual pebbles and gravels to concentrate at the surface; resembles a cobblestone street (see deflation, sheetflow).

Detritivores (19)   Detritus feeders and decomposers that consume, digest, and destroy organic wastes and debris. *Detritus feeders*—worms, mites, termites, centipedes, snails, crabs, and even vultures, among others—consume detritus and excrete nutrients and simple inorganic compounds that fuel an ecosystem. Compare with decomposers (see decomposers).

Dew-point temperature (7)   The temperature at which a given mass of air becomes saturated, holding all the water it can hold. Any further cooling or addition of water vapor results in active condensation.

Diagnostic subsurface horizon (18)   A soil horizon that originates below the epipedon at varying depths; may be part of the A and B horizons; important in soil description as part of the Soil Taxonomy.

Differential weathering (13)   The effect of different resistances in rock, coupled with variations in the intensity of physical and chemical weathering.

Diffuse radiation (4)   The downward component of scattered incoming insolation from clouds and the atmosphere.

Discharge (9)   The measured volume of flow in a river that passes by a given cross section of a stream in a given unit of time; expressed in cubic meters per second or cubic feet per second.

Dissolved load (14)   Materials carried in chemical solution in a stream, derived from minerals such as limestone and dolomite, or from soluble salts.

Downwelling current (6)   An area of the sea where a convergence or accumulation of water thrusts excess water downward; occurs, for example, at the western end of the equatorial current or along the margins of Antarctica (compare upwelling currents).

Drainage basin (14)   The basic spatial geomorphic unit of a river system; distinguished from a neighboring basin by ridges and highlands that form divides, marking the limits of the catchment area of the drainage basin.

Drainage density (14)   A measure of the overall operational efficiency of a drainage basin, determined by the ratio of combined channel lengths to the unit area.

Drainage pattern (14)   A distinctive geometric arrangement of streams in a region, determined by slope, differing rock resistance to weathering and erosion, climatic and hydrologic variability, and structural controls of the landscape.

Drawdown (9)   See cone of depression.

Drought (9)   Does not have a simple water-budget definition; rather, it can occur in at least four forms: *meteorological drought*, *agricultural drought*, *hydrologic drought*, and/or a *socioeconomic drought*.

Drumlin (17) A depositional landform related to glaciation that is composed of till (unstratified, unsorted) and is streamlined in the direction of continental ice movement—blunt end upstream and tapered end downstream with a rounded summit.

Dry adiabatic rate (DAR) (7) The rate at which an unsaturated parcel of air cools (if ascending) or heats (if descending); a rate of 10 C° per 1000 m (5.5 F° per 1000 ft) (see adiabatic; compare moist adiabatic rate).

Dune (15) A depositional feature of sand grains deposited in transient mounds, ridges, and hills; extensive areas of sand dunes are called sand seas.

Dust dome (4) A dome of airborne pollution associated with every major city; may be blown by winds into elongated plumes downwind from the city.

Dynamic equilibrium (1) Increasing or decreasing operations in a system demonstrate a trend over time, a change in average conditions.

Dynamic equilibrium model (13) The balancing act between tectonic uplift and erosion, between the resistance of crust materials and the work of denudation processes. Landscapes evidence ongoing adaptation to rock structure, climate, local relief, and elevation.

Earthquake (12) A sharp release of energy that sends waves traveling through Earth's crust at the moment of rupture along a fault or in association with volcanic activity. The moment magnitude scale (formerly the Richter scale) estimates earthquake magnitude; intensity is described by the Mercalli scale.

Earth systems science (1) An emerging science of Earth as a complete, systematic entity. An interacting set of physical, chemical, and biological systems that produce the processes of a whole Earth system. A study of planetary change resulting from system operations; includes a desire for a more quantitative understanding among components, rather than qualitative description.

Ebb tide (16) Falling or lowering tide during the daily tidal cycle (compare flood tide).

Ecological succession (19) The process whereby different and usually more complex assemblages of plants and animals replace older and usually simpler communities; communities are in a constant state of change as each species adapts to conditions. Ecosystems do not exhibit a stable point or successional climax condition as previously thought (refer to primary succession, secondary succession).

Ecology (19) The science that studies the relations between organisms and their environment and among various ecosystems.

Ecosphere (1) Another name for the biosphere.

Ecosystem (19, 20) A self-regulating association of living plants, animals, and their nonliving physical and chemical environments.

Ecotone (20) A boundary transition zone between adjoining ecosystems that may vary in width and represent areas of tension as similar species of plants and animals compete for the resources (see ecosystem).

Effusive eruption (12) A volcanic eruption characterized by low-viscosity basaltic magma and low-gas content, which readily escapes. Lava pours forth onto the surface with relatively small explosions and few pyroclastics; tends to form shield volcanoes (see shield volcano, lava, pyroclastics; compare explosive eruption).

Elastic-rebound theory (12) A concept describing the faulting process in Earth's crust, in which the two sides of a fault appear locked despite the motion of adjoining pieces of crust, but with accumulating strain they rupture suddenly, snapping to new positions relative to each other, generating an earthquake.

Electromagnetic spectrum (2) All the radiant energy produced by the Sun placed in an ordered range, divided according to wavelengths.

Eluviation (18) The removal of finer particles and minerals from the upper horizons of soil; an erosional process within a soil body (compare illuviation).

Empirical classification (10) A climate classification based on weather statistics or other data; used to determine general climate categories (compare genetic classification).

Endogenic system (11) The system internal to Earth, driven by radioactive heat derived from sources within the planet. In response, the surface fractures, mountain building occurs, and earthquakes and volcanoes are activated (compare exogenic system).

Entisols (18) A soil order in the Soil Taxonomy. Specifically lacks vertical development of horizons; usually young or undeveloped. Found in active slopes, alluvial-filled floodplains, poorly drained tundra.

Environmental lapse rate (3) The actual lapse rate in the lower atmosphere at any particular time under local weather conditions; may deviate above or below the normal lapse rate of 6.4 C° per 1000 m (3.5 F° per 1000 ft). (Compare normal lapse rate.)

Eolian (15) Caused by wind; refers to the erosion, transportation, and deposition of materials; spelled aeolian in some countries.

Epipedon (18) The diagnostic soil horizon that forms at the surface; not to be confused with the A horizon; may include all or part of the illuviated B horizon.

Equal area (1, Appendix A) A trait of a map projection; indicates the equivalence of all areas on the surface of the map, although shape is distorted (see map projections).

Equatorial and tropical rain forest (20) A lush biome of tall broadleaf evergreen trees and diverse plants and animals, roughly between 23.5° N and 23.5° S. The dense canopy of leaves is usually arranged in three levels.

Equatorial low-pressure trough (6) A thermally caused low-pressure area that almost girdles Earth, with air converging and ascending all along its extent; also called the intertropical convergence zone (ITCZ).

Erg desert (15) A sandy desert, or area where sand is so extensive that it constitutes a sand sea.

Erosion (14) Denudation by wind, water, or ice, which dislodges, dissolves, or removes surface material.

Esker (17) A sinuously curving, narrow deposit of coarse gravel that forms along a meltwater stream channel, developing in a tunnel beneath a glacier.

Estuary (14) The point at which the mouth of a river enters the sea, where freshwater and seawater are mixed; a place where tides ebb and flow.

Eustasy (7) Refers to worldwide changes in sea level that are not related to movements of land but rather to a rise and fall in the volume of water in the oceans.

Evaporation (9) The movement of free water molecules away from a wet surface into air that is less than saturated; the phase change of water to water vapor.

Evaporation fog (7) A fog formed when cold air flows over the warm surface of a lake, ocean, or other body of water; forms as the water molecules evaporate from the water surface into the cold, overlying air; also known as steam fog or sea smoke.

Evaporation pan (9) A weather instrument consisting of a standardized pan from which evaporation occurs, with water automatically replaced and measured; an evaporimeter.

Evapotranspiration (9) The merging of evaporation and transpiration water loss into one term (see potential and actual evapotranspiration).

Evolution (19) A theory that single-cell organisms adapted, modified, and passed along inherited changes to multicellular organisms. The genetic makeup of successive generations is shaped by environmental factors, physiological functions, and behaviors that created a greater rate of survival and reproduction and were passed along through natural selection.

Exfoliation dome (13) A dome-shaped feature of weathering, produced by the response of granite to the overburden removal process, which relieves pressure from the rock. Layers of rock slough ("sluff") off in slabs or shells in a sheeting process.

Exogenic system (11) Earth's external surface system, powered by insolation, which energizes air, water, and ice and sets them in motion, under the influence of gravity. Includes all processes of landmass denudation (compare endogenic system).

Exosphere (3) An extremely rarefied outer atmospheric halo beyond the thermopause at an altitude of 480 km (300 mi); probably composed of hydrogen and helium atoms, with some oxygen atoms and nitrogen molecules present near the thermopause.

Exotic stream (14) A river that rises in a humid region and flows through an arid region, with discharge decreasing toward the mouth; for example, the Nile River and the Colorado River.

Explosive eruption (12) A violent and unpredictable volcanic eruption, the result of magma that is thicker (more viscous), stickier, and higher in gas and silica content than that of an effusive eruption; tends to form blockages within a volcano; produces composite volcanic landforms (see composite volcano; compare effusive eruption).

Faulting (12) The process whereby displacement and fracturing occur between two portions of Earth's crust; usually associated with earthquake activity.

Feedback loop (1) Created when a portion of system output is returned as an information input, causing changes that guide further system operation (see negative feedback, positive feedback).

Field capacity (9) Water held in the soil by hydrogen bonding against the pull of gravity, remaining after water drains from the larger pore spaces; the available water for plants (see available water, capillary water).

Fire ecology (19) The study of fire as a natural agent and dynamic factor in community succession.

Firn (17) Snow of a granular texture that is transitional in the slow transformation from snow to glacial ice; snow that has persisted through a summer season in the zone of accumulation.

Firn line (17) The snow line that is visible on the surface of a glacier, where winter snows survive the summer ablation season; analogous to a snow line on land (see ablation).

Fjord (17) A drowned glaciated valley, or glacial trough, along a seacoast.

Flash flood (15) A sudden and short-lived torrent of water that exceeds the capacity of a stream channel; associated with desert and semiarid washes.

Flood (14) A high water level that overflows the natural riverbank along any portion of a stream.

Floodplain (14) A flat, low-lying area along a stream channel, created by and subject to recurrent flooding; alluvial deposits generally mask underlying rock.

Flood tide (16) Rising tide during the daily tidal cycle (compare ebb tide).

Fluvial (14) Stream-related processes; from the Latin *fluvius* for "river" or "running water."

Fog (7) A cloud, generally stratiform, in contact with the ground, with visibility usually reduced to less than 1 km (3300 ft).

Folding (12) The bending and deformation of beds of rock strata subjected to compressional forces.

Food chain (19) The circuit along which energy flows from producers (plants), which manufacture their own food, to consumers (animals); a one-directional flow of chemical energy, ending with decomposers.

Food web (19) A complex network of interconnected food chains (see food chain).

Formation class (20) That portion of a biome that concerns the plant communities only, categorized by size, shape, and structure of the dominant vegetation.

Friction force (6) The effect of drag by the wind as it moves across a surface; may be operative through 500 m (1600 ft) of altitude. Surface friction slows the wind and therefore reduces the effectiveness of the Coriolis force.

Frost action (13) A powerful mechanical force produced as water expands up to 9% of its volume as it freezes. Water freezing in a cavity in a rock can break the rock if it exceeds the rock's tensional strength.

Funnel cloud (8) The visible swirl extending from the bottom side of a cloud, which may or may not develop into a tornado. A tornado is a funnel cloud that has extended all the way to the ground (see tornado).

Fusion (2) The process of forcibly joining positively charged hydrogen and helium nuclei under extreme temperature and pressure; occurs naturally in thermonuclear reactions within stars, such as our Sun.

Gelifluction (17) Refers to soil flow in periglacial environments, a progressive, lateral movement; a type of solifluction formed under periglacial conditions of permafrost and frozen ground (see solifluction).

Gelisols (18) A new soil order in the Soil Taxonomy, added in 1998, describing cold and frozen soils at high latitudes or high elevations; characteristic tundra vegetation (see Canadian System of Soil Classification for similar types).

General circulation model (GCM) (10) Complex, computer-based climate model that produces generalizations of reality and forecasts of future weather and climate conditions. Complex GCMs (three-dimensional models) are in use in the United States and in other countries.

Genetic classification (10) A climate classification that uses causative factors to determine climatic regions; for example, an analysis of the effect of interacting air masses (compare empirical classification).

Geodesy (1) The science that determines Earth's shape and size through surveys, mathematical means, and remote sensing (see geoid).

Geographic information system (GIS) (1) A computer-based data processing tool or methodology used for gathering, manipulating, and analyzing geographic information to produce a holistic, interactive analysis.

Geography (1) The science that studies the interdependence and interaction among geographic areas, natural systems, processes, society, and cultural activities over space—a spatial science. The five themes of geographic education are location, place, movement, regions, and human-Earth relationships.

Geoid (1) A word that describes Earth's shape; literally, "the shape of Earth is Earth-shaped." A theoretical surface at sea level that extends through the continents; deviates from a perfect sphere.

Geologic cycle (11) A general term characterizing the vast cycling that proceeds in the lithosphere. It encompasses the hydrologic cycle, tectonic cycle, and rock cycle.

Geologic time scale (11) A depiction of eras, periods, and epochs that span Earth's history; shows both the sequence of rock strata and their absolute dates, as determined by methods such as radioactive isotopic dating.

Geomagnetic reversal (11) A polarity change in Earth's magnetic field. With uneven regularity, the magnetic field fades to zero, then returns to full strength but with the magnetic poles reversed. Reversals have been recorded 9 times during the past 4 million years.

Geomorphic threshold (13) The threshold up to which landforms change before lurching to a new set of relationships, with rapid realignments of landscape materials and slopes.

Geomorphology (13) The science that analyzes and describes the origin, evolution, form, classification, and spatial distribution of landforms.

Geostrophic wind (6) A wind moving between areas of different pressure along a path that is parallel to the isobars. It is a product of the pressure gradient force and the Coriolis force (see isobar, pressure gradient force, Coriolis force).

Geothermal energy (11, 12) The energy in steam and hot water heated by subsurface magma near groundwater. Geothermal energy literally refers to heat from Earth's interior, whereas *geothermal power* relates to specific applied strategies of geothermal electric or geothermal direct applications. This energy is used in Iceland, New Zealand, Italy, and northern California, among other locations.

Glacial drift (17) The general term for all glacial deposits, both unsorted (till) and sorted (stratified drift).

Glacial ice (17) A hardened form of ice, very dense in comparison to normal snow or firn.

Glacier (17) A large mass of perennial ice resting on land or floating shelflike in the sea adjacent to the land; formed from the accumulation and recrystallization of snow, which then flows slowly under the pressure of its own weight and the pull of gravity.

Glacier surge (17) The rapid, lurching, unexpected forward movement of a glacier.

Glacio-eustatic (7) Changes in sea level in response to changes in the amount of water stored on Earth as ice; the more water that is bound up in glaciers and ice sheets, the lower the sea level (compare eustasy).

Gleization (18) A process of humus and clay accumulation in cold, wet climates with poor drainage.

Global dimming (4) The decline in sunlight reaching Earth's surface owing to pollution, aerosols, and clouds.

Global positioning system (GPS) (1) Latitude, longitude, and elevation are accurately calibrated using this handheld instrument that calibrates radio signals from satellites.

Goode's homolosine projection (Appendix A) An equal-area projection formed by splicing together a sinusoidal and a homolographic projection.

Graben (12) Pairs or groups of faults that produce downward-faulted blocks; characteristic of the basins of the interior western United States (compare horst; see Basin and Range Province).

Graded stream (14) An idealized condition in which a stream's load and the landscape mutually adjust. This forms a dynamic equilibrium among erosion, transported load, deposition, and the stream's capacity.

Gradient (14) The drop in elevation from a stream's headwaters to its mouth, ideally forming a concave slope.

Granite (11) A coarse-grained (slow-cooling) intrusive igneous rock of 25% quartz and more than 50% potassium and sodium feldspars; characteristic of the continental crust.

Gravitational water (9) That portion of surplus water that percolates downward from the capillary zone, pulled by gravity to the groundwater zone.

Gravity (2) The mutual force exerted by the masses of objects that are attracted one to another and produced in an amount proportional to each object's mass.

Great circle (1) Any circle drawn on a globe with its center coinciding with the center of the globe. An infinite number of great circles can be drawn, but only one parallel is a great circle—the equator (compare small circle).

Greenhouse effect (4) The process whereby radiatively active gases (carbon dioxide, water vapor, methane, and CFCs) absorb insolation and emit the energy at longer wavelengths, which are retained longer, delaying the loss of infrared to space. Thus, the lower troposphere is warmed through the radiation and re-radiation of infrared wavelengths. The approximate similarity between this process and that of a greenhouse explains the name.

Greenwich Mean Time (GMT) (1) Former world standard time, now reported as Coordinated Universal Time (UTC) (see Coordinated Universal Time).

Ground ice (17) Subsurface water that is frozen in regions of permafrost. The moisture content of areas with ground ice may vary from nearly absent in regions of drier permafrost to almost 100% in saturated soils (compare permafrost).

Groundwater (9) Water beneath the surface that is beyond the soil-root zone; a major source of potable water.

Groundwater mining (9) Pumping an aquifer beyond its capacity to flow and recharge; an overuse of the groundwater resource.

Gulf Stream (5) A strong, northward-moving, warm current off the east coast of North America, which carries its water far into the North Atlantic.

Habitat (19, 20) A physical location to which an organism is biologically suited. Most species have specific habitat parameters and limits (compare to niche).

Hail (8) A type of precipitation formed when a raindrop is repeatedly circulated above and below the freezing level in a cloud, with each cycle freezing more moisture onto the hailstone until it becomes too heavy to stay aloft.

Hair hygrometer (7) An instrument for measuring relative humidity; based on the principle that human hair will change as much as 4% in length between 0% and 100% relative humidity.

Heat (3) The flow of kinetic energy from one body to another because of a temperature difference between them.

Herbivore (19) The primary consumer in a food web, which eats plant material formed by a producer (plant) that has photosynthesized organic molecules (compare carnivore).

Heterosphere (3) A zone of the atmosphere above the mesopause, 80 km (50 mi) in altitude; composed of rarefied layers of oxygen atoms and nitrogen molecules; includes the ionosphere.

Histosols (18) A soil order in the Soil Taxonomy. Formed from thick accumulations of organic matter, such as beds of former lakes, bogs, and layers of peat.

Homosphere (3) A zone of the atmosphere from Earth's surface up to 80 km (50 mi), composed of an even mixture of gases, including nitrogen, oxygen, argon, carbon dioxide, and trace gases.

Horn (17) A pyramidal, sharp-pointed peak that results when several cirque glaciers gouge an individual mountain summit from all sides.

Horst (12) Upward-faulted blocks produced by pairs or groups of faults; characteristic of the mountain ranges of the interior of the western United States (see graben and Basin and Range Province).

Hot spot (11) An individual point of upwelling material originating in the asthenosphere, or deeper in the mantle; tends to remain fixed relative to migrating plates; some 100 are identified worldwide, exemplified by Yellowstone National Park, Hawai'i, and Iceland.

Human–Earth relationships (1) One of the oldest themes of geography (the human-land tradition); includes the spatial analysis of settlement patterns, resource utilization and exploitation, hazard perception and planning, and the impact of environmental modification and artificial landscape creation.

Humidity (7) Water vapor content of the air. The capacity of the air to hold water vapor is mostly a function of the temperature of the air and the water vapor.

Humus (18) A mixture of organic debris in the soil worked by consumers and decomposers in the humification process; characteristically formed from plant and animal litter deposited at the surface.

Hurricane (8) A tropical cyclone that is fully organized and intensified in inward-spiraling rainbands; ranges from 160 to 960 km (100 to 600 mi) in diameter, with wind speeds in excess of 119 kmph (65 knots, or 74 mph); a name used specifically in the Atlantic and eastern Pacific (compare typhoon).

Hydration (13) A chemical weathering process involving water that is added to a mineral, which initiates swelling and stress within the rock, mechanically forcing grains apart as the constituents expand (contrast to hydrolysis).

Hydrology (14) The science of water, its global circulation, distribution, and properties, specifically water at and below Earth's surface.

Hydraulic action (14) The erosive work accomplished by the turbulence of water; causes a squeezing and releasing action in joints in bedrock; capable of prying and lifting rocks.

Hydrograph (14) A graph of stream discharge (in cms or cfs) over a period of time (minutes, hours, days, years) at a specific place on a stream. The relationship between stream discharge and precipitation input is illustrated on the graph.

Hydrologic cycle (9) A simplified model of the flow of water, ice, and water vapor from place to place. Water flows through the atmosphere, across the land, where it is also stored as ice, and within groundwater. Solar energy empowers the cycle.

Hydrolysis (13) A chemical weathering process in which minerals chemically combine with water; a decomposition process that causes silicate minerals in rocks to break down and become altered (contrast with hydration).

Hydrosphere (1) An abiotic open system that includes all of Earth's water.

Hygroscopic water (9) That portion of soil moisture that is so tightly bound to each soil particle that it is unavailable to plant roots; the water, along with some bound capillary water, that is left in the soil after the wilting point is reached (see wilting point).

Ice age (17) A cold episode, with accompanying alpine and continental ice accumulations, that has repeated roughly every 200 to 300 million years since the late Precambrian era (1.25 billion years ago); includes the most recent episode during the Pleistocene Ice Age, which began 1.65 million years ago.

Iceberg (17) Floating ice created by ice calving (a large piece breaking off) and floating adrift; a hazard to shipping because about nine-tenths of the ice is submerged and can be irregular in form.

Ice cap (17) A large, dome-shaped glacier, less extensive than an ice sheet (although it buries mountain peaks and the local landscape).

Ice field (17) The least extensive form of a glacier, with mountain ridges and peaks visible above the ice; less than an ice cap or ice sheet.

Icelandic low (6) See subpolar low-pressure cell.

Ice sheet (17) An enormous continuous continental glacier. The bulk of glacial ice on Earth covers Antarctica and Greenland in two ice sheets.

Ice wedge (17) Formed when water enters a thermal contraction crack in permafrost and freezes. Repeated seasonal freezing and melting of the water progressively expands the wedge.

Igneous rock (11) One of the basic rock types; it has solidified and crystallized from a hot molten state (either magma or lava). (Compare metamorphic rock, sedimentary rock.)

Illuviation (18) The downward movement and deposition of finer particles and minerals from the upper horizon of the soil; a depositional process. Deposition usually is in the B horizon, where accumulations of clays, aluminum, carbonates, iron, and some humus occur (compare eluviation; see calcification).

Inceptisols (18) A soil order in the Soil Taxonomy. Weakly developed soils that are inherently infertile; usually young soils that are weakly developed, although they are more developed than Entisols.

Industrial smog (3) Air pollution associated with coal-burning industries; it may contain sulfur oxides, particulates, carbon dioxide, and exotics.

Infiltration (9) Water access to subsurface regions of soil moisture storage through penetration of the soil surface.

Insolation (2) Solar radiation that is incoming to Earth systems.

Interception (9) Delays the fall of precipitation toward Earth's surface; caused by vegetation or other ground cover.

Internal drainage (14) In regions where rivers do not flow into the ocean, the outflow is through evaporation or subsurface gravitational flow. Portions of Africa, Asia, Australia, and the western United States have such drainage.

International Date Line (1) The 180° meridian, an important corollary to the prime meridian on the opposite side of the planet; established by the treaty of 1884 to mark the place where each day officially begins.

Intertropical convergence zone (ITCZ) (6) See equatorial low-pressure trough.

Ionosphere (3) A layer in the atmosphere above 80 km (50 mi) where gamma, X-ray, and some ultraviolet radiation is absorbed and converted into infrared and where the solar wind stimulates the auroras.

Isobar (6) An isoline connecting all points of equal atmospheric pressure.

Isostasy (11) A state of equilibrium in Earth's crust formed by the interplay between portions of the lithosphere and the asthenosphere and the principle of buoyancy. The crust depresses under weight and recovers with its removal, for example, the melting of glacial ice. The uplift is known as isostatic rebound.

Isotherm (5) An isoline connecting all points of equal temperature.

Jet contrails (4) *Condensation trails* produced by aircraft exhaust, particulates, and water vapor can form high cirrus clouds, sometimes called *false cirrus clouds*.

Jet stream (6) The most prominent movement in upper-level westerly wind flows; irregular, concentrated, sinuous bands of geostrophic wind, traveling at 300 kmph (190 mph). (See polar jet stream, subtropical jet stream.)

Joint (13) A fracture or separation in rock that occurs without displacement of the sides; increases the surface area of rock exposed to weathering processes.

Kame (17) A depositional feature of glaciation; a small hill of poorly sorted sand and gravel that accumulates in crevasses or in ice-caused indentations in the surface.

Karst topography (13) Distinctive topography formed in a region of chemically weathered limestone with poorly developed surface drainage and solution features that appear pitted and bumpy; originally named after the Krš Plateau in Slovenia.

Katabatic winds (6) Air drainage from elevated regions, flowing as gravity winds. Layers of air at the surface cool, become denser, and flow downslope; known worldwide by many local names.

Kettle (17) Forms when an isolated block of ice persists in a ground moraine, an outwash plain, or valley floor after a glacier retreats; as the block finally melts, it leaves behind a steep-sided hole that frequently fills with water.

Kinetic energy (3) The energy of motion in a body; derived from the vibration of the body's own movement and stated as temperature.

Köppen–Geiger climate classification (10, Appendix B) An empirical classification system that uses average monthly temperatures, average monthly precipitation, and total annual precipitation to establish regional climate designations.

Lagoon (16) An area of coastal seawater that is virtually cut off from the ocean by a bay barrier or barrier beach; also, the water surrounded and enclosed by an atoll.

Landfall (8) The location along a coast where a storm moves onshore.

Land–sea breeze (6) Wind along coastlines and adjoining interior areas created by different heating characteristics of land and water surfaces—onshore (landward) breeze in the afternoon and offshore (seaward) breeze at night.

Landslide (13) A sudden rapid downslope movement of a cohesive mass of regolith and/or bedrock in a variety of mass-movement forms under the influence of gravity; a form of mass movement.

Land–water heating difference (5) Differences in the degree and way that land and water heat, as a result of contrasts in transmission, evaporation, mixing, and specific heat capacities. Land surfaces heat and cool faster than water and have continentality, whereas water provides a marine influence.

Latent heat (7) Heat energy is stored in one of three states—ice, water, or water vapor. The energy is absorbed or released in each phase change from one state to another. Heat energy is absorbed as the latent heat of melting, vaporization, or evaporation. Heat energy is released as the latent heat of condensation and freezing (or fusion).

Latent heat of condensation (7) The heat energy released to the environment in a phase change from water vapor to liquid; under normal sea-level pressure, 540 calories are released from each gram of water vapor that changes phase to water at boiling, and 585 calories are released from each gram of water vapor that condenses at 20°C (68°F).

Latent heat of sublimation (7) The heat energy absorbed or released in the phase change from ice to water vapor or water vapor to ice—no liquid phase. The change from water vapor to ice is also called deposition.

Latent heat of vaporization (7) The heat energy absorbed from the environment in a phase change from liquid to water vapor at the boiling point; under normal sea-level pressure, 540 calories must be added to each gram of boiling water to achieve a phase change to water vapor.

Lateral moraine (17) Debris transported by a glacier that accumulates along the sides of the glacier and is deposited along these margins.

Laterization (18) A pedogenic process operating in well-drained soils that occur in warm and humid regions; typical of Oxisols. Plentiful precipitation leaches soluble minerals and soil constituents. Resulting soils usually are reddish or yellowish.

Latitude (1) The angular distance measured north or south of the equator from a point at the center of Earth. A line connecting all points of the same latitudinal angle is called a parallel (compare longitude).

Lava (11, 12) Magma that issues from volcanic activity onto the surface; the extrusive rock that results when magma solidifies (see magma).

Life zone (19) A zonation by altitude of plants and animals that form distinctive communities. Each life zone possesses its own temperature and precipitation relations.

Lightning (8) Flashes of light caused by tens of millions of volts of electrical charge heating the air to temperatures of 15,000° to 30,000°C.

Limestone (11) The most common chemical sedimentary rock (nonclastic); it is lithified calcium carbonate; very susceptible to chemical weathering by acids in the environment, including carbonic acid in rainfall.

Limiting factor (19) The physical or chemical factor that most inhibits biotic processes, either through lack or excess.

Lithification (11) The compaction, cementation, and hardening of sediments into sedimentary rock.

Lithosphere (1) Earth's crust and that portion of the uppermost mantle directly below the crust, extending down to about 70 km (45 mi). Some sources use this term to refer to the entire Earth.

Littoral drift (16) Transport of sand, gravel, sediment, and debris along the shore; a more comprehensive term, considers *beach drift* and *longshore drift* combined.

Littoral zone (16) A specific coastal environment; that region between the high-water line during a storm and a depth at which storm waves are unable to move sea-floor sediments.

Loam (18) A soil that is a mixture of sand, silt, and clay in almost equal proportions, with no one texture dominant; an ideal agricultural soil.

Location (1) A basic theme of geography dealing with the absolute and relative position of people, places, and things on Earth's surface.

Loess (15) Large quantities of fine-grained clays and silts left as glacial outwash deposits; subsequently blown by the wind great distances and redeposited as a generally unstratified, homogeneous blanket of material covering existing landscapes; in China, loess originated from desert lands.

Longitude (1) The angular distance measured east or west of a prime meridian from a point at the center of Earth. A line connecting all points of the same longitude is called a meridian.

Longshore current (16) A current that forms parallel to a beach as waves arrive at an angle to the shore; generated in the surf zone by wave action, transporting large amounts of sand and sediment (see littoral current, beach drift).

Lysimeter (9) A weather instrument for measuring potential and actual evapotranspiration; isolates a portion of a field so that the moisture moving through the plot is measured.

Magma (11) Molten rock from beneath Earth's surface; fluid, gaseous, under tremendous pressure, and either intruded into existing country rock or extruded onto the surface as lava (see lava).

Magnetosphere (2) Earth's magnetic force field, which is generated by dynamo-like motions within the planet's outer core; deflects the solar wind flow toward the upper atmosphere above each pole.

Mangrove swamp (16) A wetland ecosystem between 30° N or S and the equator; tends to form a distinctive community of mangrove plants (compare salt marsh).

Mantle (11) An area within the planet representing about 80% of Earth's total volume, with densities increasing with depth and averaging 4.5g/cm^3; occurs between the core and the crust; is rich in iron and magnesium oxides and silicates.

Map (1, Appendix A) A generalized view of an area, usually some portion of Earth's surface, as seen from above at a greatly reduced size (see scale and map projection).

Map projection (1, Appendix A) The reduction of a spherical globe onto a flat surface in some orderly and systematic realignment of the latitude and longitude grid.

Marine effect (5) A quality of regions that are dominated by the moderating effect of the ocean and that exhibit a smaller range of minimum and maximum temperature than continental stations (see continentality, land–water heating differences).

Mass movement (13) All unit movements of materials propelled by gravity; can range from dry to wet, slow to fast, small to large, and free-falling to gradual or intermittent.

Mass wasting (13) Gravitational movement of nonunified material downslope; a specific form of mass movement.

Meandering stream (14) The sinuous, curving pattern common to graded streams, with the energetic outer portion of each curve subjected to the greatest erosive action and the lower-energy inner portion receiving sediment deposits (see graded stream).

Mean sea level (MSL) (16) The average of tidal levels recorded hourly at a given site over a long period, which must be at least a full lunar tidal cycle.

Medial moraine (17) Debris transported by a glacier that accumulates down the middle of the glacier, resulting from two glaciers merging their lateral moraines; forms a depositional feature following glacial retreat.

Mediterranean shrubland (20) A major biome dominated by Mediterranean dry-summer climates and characterized by sclerophyllous scrub and short, stunted, tough forests (see chapparal).

Mercator projection (1) A true-shape projection, with meridians appearing as equally spaced straight lines and parallels appearing as straight lines that are spaced closer together near the equator. The poles are infinitely stretched, with the 84th north parallel and 84th south parallel fixed at the same length as that of the equator. It presents false notions of the size (area) of midlatitude and poleward landmasses but presents true compass direction (see rhumb line).

Mercury barometer (6) A device that measures air pressure using a column of mercury in a tube, one end of which is sealed, and the other end inserted in an open vessel of mercury (see air pressure).

Meridian (1) See longitude; a line designating an angle of longitude.

Mesocyclone (8) A large, rotating atmospheric circulation, initiated within a parent cumulonimbus cloud at midtroposphere elevation; generally produces heavy rain, large hail, blustery winds, and lightning; may lead to tornado activity.

Mesosphere (3) The upper region of the homosphere from 50 to 80 km (30 to 50 mi) above the ground; designated by temperature criteria; atmosphere extremely rarified.

Metamorphic rock (11) One of three basic rock types, it is existing igneous and sedimentary rock that has undergone profound physical and chemical changes under increased pressure and temperature. Constituent mineral structures may exhibit foliated or nonfoliated textures (compare igneous rock and sedimentary rock).

Meteorology (8) The scientific study of the atmosphere, including its physical characteristics and motions; related chemical, physical, and geological processes; the complex linkages of atmospheric systems; and weather forecasting.

Microclimatology (4) The study of local climates at or near Earth's surface, or that height above the Earth's surface where the effects of the surface are no longer of effect.

Midlatitude broadleaf and mixed forest (20) A biome in moist continental climates in areas of warm-to-hot summers and cool-to-cold winters; relatively lush stands of broadleaf forests trend northward into needleleaf evergreen stands.

Midlatitude cyclone (8) An organized area of low pressure, with converging and ascending airflow producing an interaction of air masses; migrates along storm tracks. Such lows or depressions form the dominant weather pattern in the middle and higher latitudes of both hemispheres.

Midlatitude grassland (20) The major biome most modified by human activity; so named because of the predominance of grasslike plants, although deciduous broadleafs appear along streams and other limited sites; location of the world's breadbaskets of grain and livestock production.

Mid-ocean ridge (11) A submarine mountain range that extends more than 65,000 km (40,000 mi) worldwide and averages more than 1000 km (600 mi) in width; centered along sea-floor spreading centers (see sea-floor spreading).

Milky Way Galaxy (2) A flattened, disk-shaped mass in space estimated to contain up to 400 billion stars; includes our Solar System.

Miller cylindrical projection (Appendix A) A compromise map projection that avoids the severe distortion of the Mercator projection (see map projection).

Mineral (11) An element or combination of elements that forms an inorganic natural compound; described by a specific formula and crystal structure.

Mirage (4) A refraction effect when an image appears near the horizon where light waves are refracted by layers of air at different temperatures (and consequently of different densities).

Model (1) A simplified version of a system, representing an idealized part of the real world.

Mohorovičić discontinuity, or Moho (11) The boundary between the crust and the rest of the lithospheric upper mantle; named for the Yugoslavian seismologist Mohorovičić; a zone of sharp material and density contrasts; also known as the Moho.

Moist adiabatic rate (MAR) (7) The rate at which a saturated parcel of air cools in ascent; a rate of 6 C° per 1000 m (3.3 F° per 1000 ft). This rate may vary, with moisture content and temperature, from 4 C° to 10 C° per 1000 m (2 F° to 6 F° per 1000 ft) (see adiabatic; compare dry adiabatic rate).

Moisture droplet (7) A tiny water particle that constitutes the initial composition of clouds. Each droplet measures approximately 0.002 cm (0.0008 in.) in diameter and is invisible to the unaided eye.

Mollisols (18) A soil order in the Soil Taxonomy. These have a mollic epipedon and a humus-rich organic content high in alkalinity. Some of the world's most significant agricultural soils are Mollisols.

Moment magnitude scale (12) An earthquake magnitude scale. Considers the amount of fault slippage, the size of the area that ruptured, and the nature of the materials that faulted in estimating the magnitude of an earthquake—an assessment of the seismic moment. Replaces the Richter scale (amplitude magnitude); especially valuable in assessing larger-magnitude events.

Monsoon (6) An annual cycle of dryness and wetness, with seasonally shifting winds produced by changing atmospheric pressure systems; affects India, Southeast Asia, Indonesia, northern Australia, and portions of Africa. From the Arabic word *mausim*, meaning "season."

Montane forest (20) Needleleaf forest associated with mountain elevations (see needleleaf forest).

Moraine (17) Marginal glacial deposits (lateral, medial, terminal, ground) of unsorted and unstratified material.

Mountain–valley breeze (6) A light wind produced as cooler mountain air flows downslope at night and as warmer valley air flows upslope during the day.

Movement (1) A major theme in geography involving migration, communication, and the interaction of people and processes across space.

Mudflow (13) Fluid downslope flows of material containing more water than earthflows.

Natural levee (14) A long, low ridge that forms on both sides of a stream in a developed floodplain; they are depositional products (coarse gravels and sand) of river flooding.

Neap tide (16) Unusually low tidal range produced during the first and third quarters of the Moon, with an offsetting pull from the Sun (compare spring tide).

Needleleaf forest (20) Forests of pine, spruce, fir, and larch, stretching from the east coast of Canada westward to Alaska and continuing from Siberia westward across the entire extent of Russia to the European Plain; called the *taiga* (a Russian word) or the *boreal forest*; principally in the microthermal climates. Includes montane forests that may be at lower latitudes at higher elevation.

Negative feedback (1) Feedback that tends to slow or dampen response in a system; promotes self-regulation in a system; far more common than positive feedback in living systems (see feedback loop, compare to positive feedback).

Net primary productivity (19) The net photosynthesis (photosynthesis minus respiration) for a given community; considers all growth and all reduction factors that affect the amount of useful chemical energy (biomass) fixed in an ecosystem.

Net radiation (NET R) (4) The net all-wave radiation available at Earth's surface; the final outcome of the radiation balance process between incoming shortwave insolation and outgoing longwave energy.

Niche (19, 20) The basic function, or occupation, of a life form within a given community; the way an organism obtains its food, air, and water.

Nickpoint (knickpoint) (14) The point at which the longitudinal profile of a stream is abruptly broken by a change in gradient; for example, a waterfall, rapids, or cascade.

Nimbostratus (7) Rain-producing, dark, grayish, stratiform clouds characterized by gentle drizzle.

Nitrogen dioxide (3) A noxious (harmful) reddish-brown gas produced in combustion engines; can be damaging to human respiratory tracts and to plants; participates in photochemical reactions and acid deposition.

Noctilucent cloud (3) A rare, shining band of ice crystals that may glow at high latitudes long after sunset; formed within the mesosphere, where cosmic and meteoric dust act as nuclei for the formation of ice crystals.

Normal fault (12) A type of geologic fault in rocks. Tension produces strain that breaks a rock, with one side moving vertically relative to the other side along an inclined fault plane (compare reverse fault).

Normal lapse rate (3) The average rate of temperature decrease with increasing altitude in the lower atmosphere; an average value of 6.4 C° per km, or 1000 m (3.5 F° per 1000 ft). (Compare environmental lapse rate.)

Occluded front (8) In a cyclonic circulation, the overrunning of a surface warm front by a cold front and the subsequent lifting of the warm air wedge off the ground; initial precipitation is moderate to heavy.

Ocean basin (12) The physical container (a depression in the lithosphere) holding an ocean.

Omnivore (19) A consumer that feeds on both producers (plants) and consumers (meat)—a role occupied by humans, among other animals (compare consumer, producer).

Open system (1) A system with inputs and outputs crossing back and forth between the system and the surrounding environment. Earth is an open system in terms of energy (compare closed system).

Orogenesis (12) The process of mountain building that occurs when large-scale compression leads to deformation and uplift of the crust; literally, the birth of mountains.

Orographic lifting (8) The uplift of a migrating air mass as it is forced to move upward over a mountain range—a topographic barrier. The lifted air cools adiabatically as it moves upslope; clouds may form and produce increased precipitation.

Outgassing (7) The release of trapped gases from rocks, forced out through cracks, fissures, and volcanoes from within Earth; the terrestrial source of Earth's water.

Outwash plain (17) Glacial stream deposits of stratified drift of meltwater-fed, braided, and overloaded streams; occurs beyond a glacier's morainal deposits.

Overland flow (9) Surplus water that flows across the land surface toward stream channels. Together with precipitation and subsurface flows, it constitutes the total runoff from an area.

Oxbow lake (14) A lake that was formerly part of the channel of a meandering stream; isolated when a stream eroded its outer bank forming a cutoff through the neck of the looping meander (see meandering stream). In Australia, known as a billabong (Aboriginal for "dead river").

Oxidation (13) A chemical weathering process in which oxygen dissolved in water oxidizes (combines with) certain metallic elements to form oxides; most familiar is the "rusting" of iron in a rock or soil (Ultisols, Oxisols), which produces a reddish-brown stain of iron oxide .

Oxisols (18) A soil order in the Soil Taxonomy. Tropical soils that are old, deeply developed, and lacking in horizons wherever well drained; heavily weathered, low in cation-exchange capacity, and low in fertility.

Ozone layer (3) See ozonosphere.

Ozonosphere (3) A layer of ozone occupying the full extent of the stratosphere (20 to 50 km, or 12 to 30 mi) above the surface; the region of the atmosphere where ultraviolet wavelengths of insolation are extensively absorbed and converted into heat.

Pacific high (6) A high-pressure cell that dominates the Pacific in July, retreating southward in the Northern Hemisphere in January; also known as the Hawaiian high.

Pahoehoe (12) Basaltic lava that is more fluid than aa. Pahoehoe forms a thin crust that forms folds and appears "ropy," like coiled, twisted rope.

Paleoclimatology (10, 17) The science that studies the climates, and the causes of variations in climate, of past ages, throughout historic and geologic time.

Paleolake (17) An ancient lake, such as Lake Bonneville or Lake Lahonton, associated with former wet periods when the lake basins were filled to higher levels than today.

Palsa (17) A rounded or elliptical mound of peat that contains thin perennial ice lenses rather than an ice core, as in a pingo. Palsas can be 2 to 30 m wide by 1 to 10 m high and usually are covered by soil or vegetation over a cracked surface.

PAN (3) See peroxyacetyl nitrate.

Pangaea (11) The supercontinent formed by the collision of all continental masses approximately 225 million years ago; named in the continental drift theory by Wegener in 1912 (see plate tectonics).

Parallel (1) See latitude; a line, parallel to the equator, that designates an angle of latitude.

Parent material (13) The unconsolidated material, from both organic and mineral sources, that is the basis of soil development.

Particulate matter (PM) (3) Dust, dirt, soot, salt, sulfate aerosols, fugitive natural particles, or other material particles suspended in air.

Paternoster lake (17) One of a series of small, circular, stair-stepped lakes formed in individual rock basins aligned down the course of a glaciated valley; named because they look like a string of rosary (religious) beads.

Patterned ground (17) Areas in the periglacial environment where freezing and thawing of the ground create polygonal forms of arranged rocks at the surface; can be circles, polygons, stripes, nets, and steps.

Pedogenic regime (18) A specific soil-forming process keyed to a specific climatic regime: laterization, calcification, salinization, and podzolization, among others; not the basis for soil classification in the Soil Taxonomy.

Pedon (18) A soil profile extending from the surface to the lowest extent of plant roots or to the depth where regolith or bedrock is encountered; imagined as a hexagonal column; the basic soil sampling unit.

Percolation (9) The process by which water permeates the soil or porous rock into the subsurface environment.

Periglacial (17) Cold-climate processes, landforms, and topographic features along the margins of glaciers, past and present; periglacial characteristics exist on more than 20% of Earth's land surface; includes permafrost, frost action, and ground ice.

Perihelion (2) That point of Earth's closest approach to the Sun in its elliptical orbit; occurs on January 3 at 147,255,000 km (91,500,000 mi); variable over a 100,000-year cycle (compare aphelion).

Permafrost (17) Forms when soil or rock temperatures remain below 0°C (32°F) for at least 2 years in areas considered periglacial; criterion is based on temperature and not on whether water is present (compare frozen ground; see periglacial).

Permeability (9) The ability of water to flow through soil or rock; a function of the texture and structure of the medium.

Peroxyacetyl nitrate (PAN) (3) A pollutant formed from photochemical reactions involving nitric oxide (NO) and volatile organic compounds (VOCs). PAN produces no known human health effect, but it is particularly damaging to plants.

Phase change (7) The change in phase, or state, among ice, water, and water vapor; involves the absorption or release of latent heat (see latent heat).

Photochemical smog (3) Air pollution produced by the interaction of ultraviolet light, nitrogen dioxide, and hydrocarbons; produces ozone and PAN through a series of complex photochemical reactions. Automobiles are the major source of the contributive gases.

Photogrammetry (1) The science of obtaining accurate measurements from aerial photos and remote sensing; used to create and to improve surface maps.

Photosynthesis (19) The process by which plants produce their own food from carbon dioxide and water, powered by solar energy. The joining of carbon dioxide and hydrogen in plants, under the influence of certain wavelengths of visible light; releases oxygen and produces energy-rich organic material, sugars, and starches (compare respiration).

Physical geography (1) The science concerned with the spatial aspects and interactions of the physical elements and processes that make up the environment: energy, air, water, weather, climate, landforms, soils, animals, plants, microorganisms, and Earth.

Physical weathering (13) The breaking and disintegrating of rock without any chemical alteration; sometimes referred to as mechanical or fragmentation weathering.

Pingo (17) Large areas of frozen ground (soil-covered ice) that develop a heaved-up, circular, ice-cored mound, rising above a periglacial landscape as water freezes into ice and expands; sometimes results from pressure developed by freezing artesian water that is injected into permafrost; occasionally exceeds 60 m in height (200 ft).

Pioneer community (19) The initial plant community in an area; usually found on new surfaces or those that have been stripped of life, as in beginning primary succession; including lichens, mosses, and ferns growing on bare rock.

Place (1) A major theme in geography, focused on the tangible and intangible characteristics that make each location unique; no two places on Earth are alike.

Plane of the ecliptic (2) A plane (flat surface) intersecting all the points of Earth's orbit.

Planetesimal hypothesis (2) Proposes a process by which early protoplanets formed from the condensing masses of a nebular cloud of dust, gas, and icy comets; a formation process now being observed in other parts of the galaxy.

Planimetric map (Appendix A) A basic map showing the horizontal position of boundaries, land-use activities, and political, economic, and social outlines.

Plateau basalt (12) An accumulation of horizontal flows formed when lava spreads out from elongated fissures onto the surface in extensive sheets; associated with effusive eruptions; also known as flood basalts (see basalt).

Plate tectonics (11) The conceptual model and theory that encompasses continental drift, sea-floor spreading, and related aspects of crustal movement; accepted as the foundation of crustal tectonic processes (see continental drift).

Playa (15) An area of salt crust left behind by evaporation on a desert floor usually in the middle of a desert or semiarid bolson or valley; intermittently wet and dry.

Pluton (11) A mass of intrusive igneous rock that has cooled slowly in the crust; forms in any size or shape. The largest partially exposed pluton is a batholith (see batholith).

Podzolization (18) A pedogenic process in cool, moist climates; forms a highly leached soil with strong surface acidity because of humus from acid-rich trees.

Point bar (14) In a stream the inner portion of a meander, where sediment fill is redeposited.

Polar easterlies (6) Variable weak, cold, and dry winds moving away from the polar region; an anticyclonic circulation.

Polar front (6) A significant zone of contrast between cold and warm air masses; roughly situated between 50° and 60° N and S latitude.

Polar high-pressure cells (6) A weak, anticyclonic, thermally produced pressure system positioned roughly over each pole; the region of the lowest temperatures on Earth (see Antarctic high).

Polypedon (18) The identifiable soil in an area, with distinctive characteristics differentiating it from surrounding polypedons that form the basic mapping unit; composed of many pedons (see pedon).

Porosity (9) The total volume of available pore space in soil; a result of the texture and structure of the soil.

Positive feedback (1) Feedback that amplifies or encourages responses in a system (compare to negative feedback, feedback loop).

Potential evapotranspiration (9) POTET, or PE; the amount of moisture that would evaporate and transpire if adequate moisture were available; it is the amount lost under optimum moisture conditions, the moisture demand.

Potentiometric surface (9) A pressure level in a confined aquifer, defined by the level to which water rises in wells; caused by the fact that the water in a confined aquifer is under the pressure of its own weight; also known as a piezometric surface. This surface can extend above the surface of the land, causing water to rise above the water table in wells in confined aquifers (see artesian water).

Precipitation (9) Rain, snow, sleet, and hail—the moisture supply; called PRECIP, or P, in the water balance.

Pressure gradient force (6) Causes air to move from an area of higher barometric pressure to an area of lower barometric pressure due to the pressure difference.

Primary succession (19) Succession that occurs among plant species in an area of new surfaces created by mass movement of land, areas exposed by a retreating glacier, cooled lava flows and volcanic eruption landscapes, surface mining and clear-cut logging scars, or an area of sand dunes, with no trace of a former community.

Prime meridian (1) An arbitrary meridian designated as 0° longitude, the point from which longitudes are measured east or west, at Greenwich, England, selected by international agreement in an 1884 treaty.

Process (1) A set of actions and changes that occur in some special order; analysis of processes is central to modern geographic synthesis.

Producer (19) Organism (plant) in an ecosystem that uses carbon dioxide as its sole source of carbon, which it chemically fixes through photosynthesis to provide its own nourishment; also called an autotroph (compare consumer).

Pyroclastic (12) An explosively ejected rock fragment launched by a volcanic eruption; sometimes described by the more general term tephra.

Radiation fog (7) Formed by radiative cooling of a land surface, especially on clear nights in areas of moist ground; occurs when the air layer directly above the surface is chilled to the dew-point temperature, thereby producing saturated conditions.

Rain gauge (9) A weather instrument; a standardized device that captures and measures rainfall.

Rain shadow (8) The area on the leeward slope of a mountain range, where precipitation receipt is greatly reduced compared to the windward slope on the other side (see orographic lifting).

Reflection (4) The portion of arriving insolation that is returned directly to space without being absorbed and converted into heat and without performing any work (see albedo).

Refraction (4) The bending effect on electromagnetic waves that occurs when insolation enters the atmosphere or another medium; the same process by which a crystal, or prism, disperses the component colors of the light passing through it.

Region (1) A geographic theme that focuses on areas that display unity and internal homogeneity of traits; includes the study of how a region forms, evolves, and interrelates with other regions.

Regolith (13) Partially weathered rock overlying bedrock, whether residual or transported.

Relative humidity (7) The ratio of water vapor actually in the air (content) compared to the maximum water vapor the air could hold (capacity) at that temperature; expressed as a percentage (compare vapor pressure, specific humidity).

Relief (12) Elevation differences in a local landscape; an expression of local height difference of landforms.

Remote sensing (1) Information acquired from a distance, without physical contact with the subject; for example, photography, orbital imagery, or radar.

Respiration (19) The process by which plants use food to derive energy for their operations; essentially, the reverse of the photosynthetic process; releases carbon dioxide, water, and heat energy into the environment (compare photosynthesis).

Reverse fault (12) Compressional forces produce strain that breaks a rock so that one side moves upward relative to the other side; also called a thrust fault (compare normal fault).

Revolution (2) The annual orbital movement of Earth about the Sun; determines the length of the year and the seasons.

Rhumb line (1) A line of constant compass direction, or constant bearing, which crosses all meridians at the same angle; a portion of a great circle.

Richter scale (12) An open-ended, logarithmic scale that estimates earthquake magnitude; designed by Charles Richter in 1935; now replaced by the moment magnitude scale (see moment magnitude scale).

Ring of fire (12) See circum-Pacific belt.

Robinson projection (Appendix A) A compromise (neither equal area nor true shape) oval projection developed in 1963 by Arthur Robinson.

Roche moutonnée (17) A glacial erosion feature; an asymmetrical hill of exposed bedrock; displays a gently sloping upstream side that has been smoothed and polished by a glacier and an abrupt, steep downstream side.

Rock (11) An assemblage of minerals bound together, or sometimes a mass of a single mineral.

Rock cycle (11) A model representing the interrelationships among the three rock-forming processes: igneous, sedimentary, and metamorphic; shows how each can be transformed into another rock type.

Rockfall (13) Free-falling movement of debris from a cliff or steep slope, generally falling straight or bounding downslope.

Rossby wave (6) An undulating horizontal motion in the upper-air westerly circulation at middle and high latitudes.

Rotation (2) The turning of Earth on its axis, averaging about 24 hours in duration; determines day–night relation; counterclockwise when viewed from above the North Pole and from above the equator, west to east or eastward.

Salinity (16) The concentration of natural elements and compounds dissolved in solution, as solutes; measured by weight in parts per thousand (‰) in seawater.

Salinization (18) A pedogenic process that results from high potential evapotranspiration rates in deserts and semiarid regions. Soil water is drawn to surface horizons, and dissolved salts are deposited as the water evaporates.

Saltation (14, 15) The transport of sand grains (usually larger than 0.2 mm, or 0.008 in.) by stream or wind, bouncing the grains along the ground in asymmetrical paths.

Salt marsh (16) A wetland ecosystem characteristic of latitudes poleward of the 30th parallel (compare to mangrove swamp).

Sand sea (15) An extensive area of sand and dunes; characteristic of Earth's erg deserts (contrast with *reg* deserts).

Saturation (7) Air that is holding all the water vapor that it can hold at a given temperature, known as the dew-point temperature.

Scale (1) The ratio of the distance on a map to that in the real world; expressed as a representative fraction, graphic scale, or written scale.

Scarification (13) Human-induced mass movements of Earth materials, such as large-scale open-pit mining and strip mining.

Scattering (4) Deflection and redirection of insolation by atmospheric gases, dust, ice, and water vapor; the shorter the wavelength, the greater the scattering; thus skies in the lower atmosphere are blue.

Scientific method (1) An approach that uses applied common sense in an organized and objective manner; based on observation, generalization, formulation, and testing of a hypothesis, and ultimately the development of a theory.

Sea-floor spreading (11) As proposed by Hess and Dietz, the mechanism driving the movement of the continents; associated with upwelling flows of magma along the worldwide system of mid-ocean ridges (see mid-ocean ridge).

Secondary succession (19) Succession that occurs among plant species in an area where vestiges of a previously functioning community

are present; an area where the natural community has been destroyed or disturbed, but where the underlying soil remains intact.

Sediment (13) Fine-grained mineral matter that is transported and deposited by air, water, or ice.

Sedimentary rock (11) One of three basic rock types; formed from the compaction, cementation, and hardening of sediments derived from other rocks (compare igneous rock, metamorphic rock).

Seismic wave (11) The shock wave sent through the planet by an earthquake or underground nuclear test. Transmission varies according to temperature and the density of various layers within the planet; provides indirect diagnostic evidence of Earth's internal structure.

Seismograph (12) A device that measures seismic waves of energy transmitted throughout Earth's interior or along the crust.

Sensible heat (3) Heat that can be measured with a thermometer; a measure of the concentration of kinetic energy from molecular motion.

Sheet flow (14) Surface water that moves downslope in a thin film as overland flow, not concentrated in channels larger than rills.

Sheeting (13) A form of weathering associated with fracturing or fragmentation of rock by pressure release; often related to exfoliation processes (see exfoliation dome).

Shield volcano (12) A symmetrical mountain landform built from effusive eruptions (low-viscosity magma); gently sloped, gradually rising from the surrounding landscape to a summit crater; typical of the Hawaiian Islands (compare to effusive eruption; composite volcano).

Sinkhole (13) Nearly circular depression created by the weathering of karst landscapes with subterranean drainage; also known as a doline in traditional studies; may collapse through the roof of an underground space (see karst topography).

Sleet (8) Freezing rain, ice glaze, or ice pellets.

Sling psychrometer (7) A weather instrument that measures relative humidity using two thermometers—a dry bulb and a wet bulb—mounted side-by-side.

Slipface (15) On a sand dune, formed as dune height increases above 30 cm (12 in.) on the leeward side at an angle at which loose material is stable—its angle of repose (30 to 34°).

Slope (13) A curved, inclined surface that bounds a landform.

Small circle (1) A circle on a globe's surface that does not share Earth's center; for example, all parallels of latitude other than the equator (compare to great circle).

Snow line (17) A temporary line marking the elevation where winter snowfall persists throughout the summer; seasonally, the lowest elevation covered by snow during the summer.

Soil (18) A dynamic natural body made up of fine materials covering Earth's surface in which plants grow, composed of both mineral and organic matter.

Soil colloid (18) A tiny clay and organic particle in soil; provides chemically active sites for mineral ion adsorption (see cation-exchange capacity).

Soil creep (13) A persistent mass movement of surface soil where individual soil particles are lifted and disturbed by the expansion of soil moisture as it freezes or by grazing livestock or digging animals.

Soil fertility (18) The ability of soil to support plant productivity when it contains organic substances and clay minerals that absorb water and certain elemental ions needed by plants through adsorption (see cation-exchange capacity, CEC).

Soil horizon (18) The various layers exposed in a pedon; roughly parallel to the surface and identified as O, A, E, B, C, and R (bedrock).

Soil-moisture recharge (9) Water entering available soil storage spaces.

Soil-moisture storage (9) STRGE, the retention of moisture within soil; it is a savings account that can accept deposits (soil-moisture recharge) or experiences withdrawals (soil-moisture utilization) as conditions change.

Soil-moisture utilization (9) The extraction of soil moisture by plants for their needs; efficiency of withdrawal decreases as the soil storage is reduced.

Soil science (18) Interdisciplinary science of soils. Pedology concerns the origin, classification, distribution, and description of soil. Edaphology focuses on soil as a medium for sustaining higher plants.

Soil Taxonomy (18) A soil classification system based on observable soil properties actually seen in the field; published in 1975 by the U.S.

Soil Conservation Service and revised in 1990 and 1998 by the Natural Resources Conservation Service to include 12 soil orders.

Soil-water budget (9) An accounting system for soil moisture using inputs of precipitation and outputs of evapotranspiration and gravitational water.

Solar constant (2) The amount of insolation intercepted by Earth on a surface perpendicular to the Sun's rays when Earth is at its average distance from the Sun; a value of $1370W/m^2$, or 1.968 calories/cm^2, per minute; averaged over the entire globe at the thermopause.

Solar wind (2) Clouds of ionized (charged) gases emitted by the Sun and traveling in all directions from the Sun's surface. Effects on Earth include auroras, disturbance of radio signals, and possible influences on weather.

Solifluction (17) Gentle downslope movement of a saturated surface material (soil and regolith) in various climatic regimes where temperatures are above freezing (compare gelifluction).

Solum (18) A true soil profile in the pedon; ideally, a combination of O, A, E, and B horizons (see pedon).

Spatial (1) The nature or character of physical space, as in an area; occupying or operating within a space. Geography is a spatial science; spatial analysis its essential approach.

Spatial analysis (1) The examination of spatial interactions, patterns, and variations over area and/or space; a key integrative approach of geography.

Specific heat (5) The increase of temperature in a material when energy is absorbed; water has a higher specific heat (can store more heat) than a comparable volume of soil or rock.

Specific humidity (7) The mass of water vapor (in grams) per unit mass of air (in kilograms) at any specified temperature. The maximum mass of water vapor that a kilogram of air can hold at any specified temperature is termed its maximum specific humidity (compare vapor pressure, relative humidity).

Speed of light (2) Specifically, 299,792 kilometers per second (186,282 miles per second), or more than 9.4 trillion kilometers per year (5.9 trillion miles per year)—a distance known as a light-year; at light speed Earth is 8 minutes and 20 seconds from the Sun.

Spheroidal weathering (13) A chemical weathering process in which the sharp edges and corners of boulders and rocks are weathered in thin plates that create a rounded, spheroidal form.

Spodosols (18) A soil order in the Soil Taxonomy classification that occurs in northern coniferous forests; best developed in cold, moist, forested climates; lacks humus and clay in the A horizons, with high acidity associated with podzolization processes.

Spring tide (16) The highest tidal range, which occurs when the Moon and the Sun are in conjunction (at new Moon) or in opposition (at full Moon) stages (compare neap tide).

Squall line (8) A zone slightly ahead of a fast-advancing cold front, where wind patterns are rapidly changing and blustery and precipitation is strong.

Stability (7) The condition of a parcel, whether it remains where it is or changes its initial position. The parcel is stable if it resists displacement upward, unstable if it continues to rise.

Stationary front (8) A frontal area of contact between contrasting air masses that shows little horizontal movement; winds in opposite direction on either side of the front flow parallel along the front.

Steady-state equilibrium (1) The condition that occurs in a system when rates of input and output are equal and the amounts of energy and stored matter are nearly constant around a stable average.

Stomata (19) A small opening on the undersides of leaves through which water and gasses pass.

Storm surge (8) A large quantity of seawater pushed inland by the strong winds associated with a tropical cyclone.

Storm tracks (8) Seasonally shifting paths followed by migrating low-pressure systems.

Stratified drift (17) Sediments deposited by glacial meltwater that appear sorted; a specific form of glacial drift (compare till).

Stratigraphy (11) A science that analyzes the sequence, spacing, geophysical and geochemical properties, and spatial distribution of rock strata.

Stratocumulus (7) A lumpy, grayish, low-level cloud, patchy with sky visible, sometimes present at the end of the day.

Stratosphere (3) That portion of the homosphere that ranges from 20 to 50 km (12.5 to 30 mi) above Earth's surface, with temperatures ranging from $-57°C$ and $-70°F$ at the tropopause to $0°C$ ($32°F$) at the stratopause. The functional ozonosphere is within the stratosphere.

Stratus (7) A stratiform (flat, horizontal) cloud generally below 2000 m (6500 ft).

Strike-slip fault (12) Horizontal movement along a faultline, that is, movement in the same direction as the fault; also known as a transcurrent fault. Such movement is described as right lateral or left lateral, depending on the relative motion observed across the fault (see transform fault).

Subduction zone (11) An area where two plates of crust collide and the denser oceanic crust dives beneath the less dense continental plate, forming deep oceanic trenches and seismically active regions.

Sublimation (7) A process in which ice evaporates directly to water vapor or water vapor freezes to ice (deposition).

Subpolar low-pressure cell (6) A region of low pressure centered approximately at 60° latitude in the North Atlantic near Iceland and in the North Pacific near the Aleutians, as well as in the Southern Hemisphere. Airflow is cyclonic; it weakens in summer and strengthens in winter (refer to cyclone).

Subsolar point (2) The only point receiving perpendicular insolation at a given moment—the Sun directly overhead (see declination).

Subtropical high-pressure cell (6) One of several dynamic high-pressure areas covering roughly the region from 20° to 35° N and S latitudes; responsible for the hot, dry areas of Earth's arid and semiarid deserts (refer to anticyclone).

Sulfate aerosols (3) Sulfur compounds in the atmosphere, principally sulfuric acid; principal sources relate to fossil fuel combustion; scatters and reflects insolation.

Sulfur dioxide (3) A colorless gas detected by its pungent odor; produced by the combustion of fossil fuels, especially coal, that contain sulfur as an impurity; can react in the atmosphere to form sulfuric acid, a component of acid deposition.

Summer (June) solstice (2) The time when the Sun's declination is at the Tropic of Cancer, at 23.5° N latitude, June 20–21 each year (compare to winter solstice).

Sunrise (2) That moment when the disk of the Sun first appears above the horizon.

Sunset (2) That moment when the disk of the Sun totally disappears.

Sunspot (2) Magnetic disturbances on the surface of the Sun, occurring in an average 11-year cycle; related flares, prominences, and outbreaks produce surges in solar wind.

Surface creep (15) A form of eolian transport that involves particles too large for saltation; a process whereby individual grains are impacted by moving grains and slide and roll.

Surplus (SURPL) (9) The amount of moisture that exceeds potential evapotranspiration; moisture oversupply when soil moisture storage is at field capacity; extra or surplus water.

Suspended load (14) Fine particles held in suspension in a stream. The finest particles are not deposited until the stream velocity nears zero.

Swell (16) Regular patterns of smooth, rounded waves in open water; can range from small ripples to very large waves.

Syncline (12) A trough in folded strata, with beds that slope toward the axis of the downfold (compare anticline).

System (1) Any ordered, interrelated set of materials or items existing separate from the environment or within a boundary; energy transformations and energy and matter storage and retrieval occur within a system.

Taiga (20) See needleleaf forest.

Talik (17) An unfrozen portion of the ground that may occur above, below, or within a body of discontinuous permafrost or beneath a body of water in the continuous region, such as a deep lake; may extend to bedrock and noncryotic soil under large deep lakes.

Talus slope (13) Formed by angular rock fragments that cascade down a slope along the base of a mountain; poorly sorted, cone-shaped deposits.

Tarn (17) A small mountain lake, especially one that collects in a cirque basin behind risers of rock material or in an ice-gouged depression.

Temperate rain forest (20)   A major biome of lush forests at middle and high latitudes; occurs along narrow margins of the Pacific Northwest in North America, among other locations; includes the tallest trees in the world.

Temperature (5)   A measure of sensible heat energy present in the atmosphere and other media; indicates the average kinetic energy of individual molecules within a substance.

Temperature inversion (3)   A reversal of the normal decrease of temperature with increasing altitude; can occur anywhere from ground level up to several thousand meters; functions to block atmospheric convection and thereby trap pollutants.

Terminal moraine (17)   Eroded debris that is dropped at a glacier's farthest extent.

Terrane (12)   A migrating piece of Earth's crust, dragged about by processes of mantle convection and plate tectonics. Displaced terranes are distinct in their history, composition, and structure from the continents that accept them.

Terrestrial ecosystem (20)   A self-regulating association characterized by specific plant formations; usually named for the predominant vegetation and known as a biome when large and stable (see biome).

Thermal equator (5)   The isoline on an isothermal map that connects all points of highest mean temperature.

Thermohaline circulation (6)   Deep ocean currents produced by differences in temperature and salinity with depth; Earth's deep currents.

Thermokarst (17)   Topography of hummocky, irregular relief marked by cave-ins, bogs, small depressions, and pits formed as ground ice melts; an erosion process caused by ground ice melting; not related to solution processes and chemical weathering associated with limestone (karst).

Thermopause (2, 3)   A zone approximately 480 km (300 mi) in altitude that serves conceptually as the top of the atmosphere; an altitude used for the determination of the solar constant.

Thermosphere (3)   A region of the heterosphere extending from 80 to 480 km (50 to 300 mi) in altitude; contains the functional ionosphere layer.

Threshold (1)   A moment in which a system can no longer maintain its character, so it lurches to a new operational level, which may not be compatible with previous conditions.

Thrust fault (12)   A reverse fault where the fault plane forms a low angle relative to the horizontal; an overlying block moves over an underlying block.

Thunder (8)   The violent expansion of suddenly heated air, created by lightning discharges, which send out shock waves as an audible sonic bang.

Tide (16)   A pattern of daily oscillations in sea level produced by astronomical relations among the Sun, the Moon, and Earth; experienced in varying degrees around the world (see neap and spring tides).

Till (17)   Direct ice deposits that appear unstratified and unsorted; a specific form of glacial drift (compare stratified drift).

Till plain (17)   A large, relatively flat plain composed of unsorted glacial deposits behind a terminal or end moraine. Low-rolling relief and unclear drainage patterns are characteristic.

Tombolo (16)   A landform created when coastal sand deposits connect the shoreline with an offshore island outcrop or sea stack.

Topographic map (Appendix A)   A map that portrays physical relief through the use of elevation contour lines that connect all points at the same elevation above or below a vertical datum, such as mean sea level.

Topography (12)   The undulations and configurations, including its relief, that give Earth's surface its texture, portrayed on topographic maps.

Tornado (8)   An intense, destructive cyclonic rotation, developed in response to extremely low pressure; generally associated with mesocyclone formation.

Total runoff (9)   Surplus water that flows across a surface toward stream channels; formed by sheet flow, combined with precipitation and subsurface flows into those channels.

Traction (14)   A type of sediment transport that drags coarser materials along the bed of a stream (see bed load).

Trade wind (6)   Wind from the northeast and southeast that converges in the equatorial low-pressure trough, forming the intertropical convergence zone.

Transform fault (11)   A type of geologic fault in rocks. An elongated zone along which faulting occurs between mid-ocean ridges; produces a relative horizontal motion with no new crust formed or consumed; strike-slip motion is either left or right lateral (see strike-slip fault).

Transmission (4)   The passage of shortwave and longwave energy through space, the atmosphere, or water.

Transparency (5)   The quality of a medium (air, water) that allows light to easily pass through it.

Transpiration (9)   The movement of water vapor out through the pores in leaves; the water is drawn by the plant roots from soil-moisture storage.

Transport (14)   The actual movement of weathered and eroded materials by air, water, and ice.

Tropical cyclone (8)   A cyclonic circulation originating in the tropics, with winds between 30 and 64 knots (39 and 73 mph); characterized by closed isobars, circular organization, and heavy rains (see hurricane and typhoon).

Tropical savanna (20)   A major biome containing large expanses of grassland interrupted by trees and shrubs; a transitional area between the humid rain forests and tropical seasonal forests and the drier, semiarid tropical steppes and deserts.

Tropical seasonal forest and scrub (20)   A variable biome on the margins of the rain forests, occupying regions of lesser and more erratic rainfall; the site of transitional communities between the rain forests and tropical grasslands.

Tropic of Cancer (2)   The parallel that marks the farthest north the subsolar point migrates during the year; 23.5° north latitude (see Tropic of Capricorn; summer or June solstice).

Tropic of Capricorn (2)   The parallel that marks the farthest south the subsolar point migrates during the year; 23.5° south latitude (see Tropic of Cancer; winter or December solstice).

Tropopause (3)   The top zone of the troposphere defined by temperature; wherever −57°C (−70°F) occurs.

Troposphere (3)   The home of the biosphere; the lowest layer of the homosphere, containing approximately 90% of the total mass of the atmosphere; extends up to the tropopause; occurring at an altitude of 18 km (11 mi) at the equator, 13 km (8 mi) in the middle latitudes, and at lower altitudes near the poles.

True shape (1)   A map property showing the correct configuration of coastlines, a useful trait of conformality for navigational and aeronautical maps, although areal relationships are distorted (see map projection; compare equal area).

Tsunami (16)   A seismic sea wave, traveling at high speeds across the ocean, formed by sudden motion in the sea floor, such as a sea-floor earthquake, submarine landslide, or eruption of an undersea volcano.

Typhoon (8)   A tropical cyclone in excess of 65 knots (74 mph) that occurs in the western Pacific; same as a hurricane except for location.

Ultisols (18)   A soil order in the Soil Taxonomy. Features highly weathered forest soils, principally in the humid subtropical climatic classification. Increased weathering and exposure can degenerate an Alfisol into the reddish color and texture of these more humid to tropical Ultisols. Fertility is quickly exhausted when Ultisols are cultivated.

Unconfined aquifer (9)   An aquifer that is not bounded by impermeable strata. It is simply the zone of saturation in water-bearing rock strata with no impermeable overburden, and recharge is generally accomplished by water percolating down from above.

Undercut bank (14)   In streams, a steep bank formed along the outer portion of a meandering stream; produced by lateral erosive action of a stream; sometimes called a cutbank (compare to point bar).

Uniformitarianism (11)   An assumption that physical processes active in the environment today are operating at the same pace and intensity that has characterized them throughout geologic time; proposed by Hutton and Lyell (compare to catastrophism).

Upslope fog (7)   Forms when moist air is forced to higher elevations along a hill or mountain and is thus cooled (compare valley fog).

Upwelling current (6)   An area of the sea where cool, deep waters, which are generally nutrient rich, rise to replace the vacating water, as occurs along the west coasts of North and South America (compare to downwelling current).

Urban heat island (4)   An urban microclimate that is warmer on average than areas in the surrounding countryside because of the interaction of solar radiation and various surface characteristics.

Valley fog (7) The settling of cooler, more dense air in low-lying areas; produces saturated conditions and fog.

Vapor pressure (7) That portion of total air pressure that results from water vapor molecules, expressed in millibars (mb). At a given dew-point temperature, the maximum capacity of the air is termed its saturation vapor pressure.

Vascular plant (19) A plant having internal fluid and material flows through its tissues; almost 250,000 species exist on Earth.

Ventifact (15) A piece of rock etched and smoothed by eolian erosion—abrasion by windblown particles.

Vernal (March) equinox (2) The time around March 20–21 each year when the Sun's declination crosses the equatorial parallel and all places on Earth experience days and nights of equal length. The Sun rises at the North Pole and sets at the South Pole (compare to autumnal equinox).

Vertisols (18) A soil order in the Soil Taxonomy. Features expandable clay soils; composed of more than 30% swelling clays. Occurs in regions that experience highly variable soil moisture balances through the seasons.

Volatile organic compounds (3) Compounds, including hydrocarbons, produced by the combustion of gasoline, from surface coatings and from electric utility combustion; participates in the production of PAN through reactions with nitric oxides.

Volcano (12) A mountainous landform at the end of a magma conduit, which rises from below the crust and vents to the surface. Magma rises and collects in a magma chamber deep below, erupting effusively or explosively and forming composite, shield, or cinder-cone volcanoes.

Warm desert and semidesert (20) A desert biome caused by the presence of subtropical high-pressure cells; dry air and low precipitation.

Warm front (8) The leading edge of an advancing warm air mass, which is unable to push cooler, passive air out of the way; tends to push the cooler, underlying air into a wedge shape; identified on a weather map as a line with semicircles pointing in the direction of frontal movement (compare to cold front).

Wash (15) An intermittently dry streambed that fills with torrents of water after rare precipitation events in arid lands.

Waterspout (8) An elongated, funnel-shaped circulation formed when a tornado exists over water.

Water table (9) The upper surface of groundwater; that contact point between the zone of saturation and aeration in an unconfined aquifer (see zone of aeration, zone of saturation).

Wave (16) An undulation of ocean water produced by the conversion of solar energy to wind energy and then to wave energy; energy produced in a generating region or a stormy area of the sea.

Wave-cut platform (16) A flat or gently sloping tablelike bedrock surface that develops in the tidal zone, where wave action cuts a bench that extends from the cliff base out into the sea.

Wave cyclone (8) See midlatitude cyclone.

Wavelength (2) A measurement of a wave; the distance between the crests of successive waves. The number of waves passing a fixed point in 1 second is called the frequency of the wavelength.

Wave refraction (16) A bending process that concentrates wave energy on headlands and disperses it in coves and bays; the long-term result is coastal straightening.

Weather (8) The short-term condition of the atmosphere, as compared to climate, which reflects long-term atmospheric conditions and extremes. Temperature, air pressure, relative humidity, wind speed and direction, daylength, and Sun angle are important measurable elements that contribute to the weather.

Weathering (13) The processes by which surface and subsurface rocks disintegrate, or dissolve, or are broken down. Rocks at or near Earth's surface are exposed to physical and chemical weathering processes.

West Antarctic ice sheet (10) A vast grounded ice mass held back by the Ross, Ronne, and Filcher ice shelves in Antarctica, drained by several active ice streams, such as the active Pine Island Glacier.

Westerlies (6) The predominant surface and aloft windflow pattern from the subtropics to high latitudes in both hemispheres.

Western intensification (6) The piling up of ocean water along the western margin of each ocean basin, to a height of about 15 cm (6 in.); produced by the trade winds that drive the oceans westward in a concentrated channel.

Wetland (16) A narrow, vegetated strip occupying many coastal areas and estuaries worldwide; highly productive ecosystems with an ability to trap organic matter, nutrients, and sediment.

Wilting point (9) That point in the soil-moisture balance when only hygroscopic water and some bound capillary water remains. Plants wilt and eventually die after prolonged stress from a lack of available water.

Wind (6) The horizontal movement of air relative to Earth's surface; produced essentially by air pressure differences from place to place; its direction is influenced by the Coriolis force and surface friction.

Wind vane (6) A weather instrument used to determine wind direction; winds are named for the direction from which they originate.

Winter (December) solstice (2) That time when the Sun's declination is at the Tropic of Capricorn, at 23.5° S latitude, December 21–22 each year. The day is 24 hours long south of the Antarctic Circle. The night is 24 hours long north of the Arctic Circle (compare to Summer [June] solstice).

Withdrawal (9) Sometimes called *offstream use*, the removal of water from the natural supply, after which it is used for various purposes and then is returned to the water supply.

Wrangellia terrane (12) One of many terranes that became cemented together to form present-day North America and the Wrangell Mountains, arriving from approximately 10,000 km (6200 mi) away; a former volcanic island arc and associated marine sediments.

Yardang (15) A streamlined rock structure formed by deflation and abrasion; appears elongated and aligned with the most effective wind direction.

Yazoo tributary (14) A small tributary stream draining alongside a floodplain; blocked from joining the main river by its natural levees and elevated stream channel (see backswamp).

Zone of aeration (9) A zone above the water table that has air in its pore spaces and may or may not have water.

Zone of saturation (9) A groundwater zone below the water table in which all pore spaces are filled with water.

Index

LICENSE AGREEMENT
Geosystems, 7e Student Animations CD
Robert W. Christopherson
ISBN: 0-13-605038-7
© 2009 Pearson Education, Inc.
Pearson Prentice Hall
Pearson Education, Inc.
Upper Saddle River, NJ 07458
Pearson Prentice Hall™ is a trademark of Pearson Education, Inc.

YOU SHOULD CAREFULLY READ THE TERMS AND CONDI-TIONS BEFORE USING THE CD-ROM PACKAGE. USING THIS CD-ROM PACKAGE INDICATES YOUR ACCEPTANCE OF THESE TERMS AND CONDITIONS.

Pearson Education, Inc. provides this program and licenses its use. You assume responsibility for the selection of the program to achieve your intended results, and for the installation, use, and results obtained from the program. This license extends only to use of the program in the United States or countries in which the program is marketed by author-ized distributors.

LICENSE GRANT
You hereby accept a nonexclusive, nontransferable, permanent license to install and use the program ON A SINGLE COMPUTER at any given time. You may copy the program solely for backup or archival purposes in support of your use of the program on a single computer. You may not modify, translate, disassemble, decompile, or reverse engineer the program, in whole or in part.

TERM
The License is effective until terminated. Pearson Education, Inc. reserves the right to terminate this License automatically if any provi-sion of the License is violated. You may terminate the License at any time. To terminate this License, you must return the program, including documentation, along with a written warranty stating that all copies in your possession have been returned or destroyed.

LIMITED WARRANTY
Pearson Education Inc.'s entire liability and your exclusive remedy shall be: 1. the replacement of any CD-ROM not meeting Pearson Education, Inc.'s "LIMITED WARRANTY" and that is returned to Pearson Education, or 2. if Pearson Education is unable to deliver a replacement CD-ROM that is free of defects in materials or workmanship, you may terminate this agreement by returning the program.

IN NO EVENT WILL PEARSON EDUCATION, INC. BE LIABLE TO YOU FOR ANY DAMAGES, INCLUDING ANY LOST PROF-ITS, LOST SAVINGS, OR OTHER INCIDENTAL OR CONSE-QUENTIAL DAMAGES ARISING OUT OF THE USE OR INABILITY TO USE SUCH PROGRAM EVEN IF PEARSON EDUCATION, INC. OR AN AUTHORIZED DISTRIBUTOR HAS BEEN ADVISED OF SUCH DAMAGES, OR FOR ANY CLAIM BY ANY OTHER PARTY.

SOME STATES DO NOT ALLOW FOR THE LIMITATION OR EXCLUSION OF LIABILITY FOR INCIDENTAL OR CONSE-QUENTIAL DAMAGES, SO THE ABOVE LIMITATION OR EXCLUSION MAY NOT APPLY TO YOU.

GENERAL
You may not sublicense, assign, or transfer the license of the program. Any attempt to sublicense, assign or transfer any of the rights, duties, or obligations hereunder is void. This Agreement will be governed by the laws of the State of New York. Should you have any questions concern-ing this Agreement, you may contact Pearson Education, Inc. by writing to:

ESM Media Development
Higher Education Division
Pearson Education, Inc.
1 Lake Street
Upper Saddle River, NJ 07458

Should you have any questions concerning technical support, you may write to:
New Media Production
Higher Education Division
Pearson Education, Inc.
1 Lake Street
Upper Saddle River, NJ 07458

YOU ACKNOWLEDGE THAT YOU HAVE READ THIS AGREE-MENT, UNDERSTAND IT, AND AGREE TO BE BOUND BY ITS TERMS AND CONDITIONS. YOU FURTHER AGREE THAT IT IS THE COMPLETE AND EXCLUSIVE STATEMENT OF THE AGREEMENT BETWEEN US THAT SUPERSEDES ANY PRO-POSAL OR PRIOR AGREEMENT, ORAL OR WRITTEN, AND ANY OTHER COMMUNICATIONS BETWEEN US RELATING TO THE SUBJECT MATTER OF THIS AGREEMENT.

PROGRAM INSTRUCTIONS
Windows
Inserting this CD-ROM will automatically start the application. If you need to manually start the application, follow the following steps:
• Insert the "Geosystems Student Animations CD" into your CD-ROM drive.
• Browse to the "Geosys7e" CD-ROM using your Windows Explorer window.
• Double-click the "Geosys.exe" application file.

Macintosh
Inserting this CD-ROM will automatically start the application. If you need to manually start the application, follow the following steps:
• Insert the "Geosystems Student Animations CD" into your CD-ROM drive.
• Double-click the "Geosys7e" CD icon on your desktop.
• Double-click the "Geosys" application file.

MINIMUM SYSTEM REQUIREMENTS
Windows
233 MHz Intel Pentium processor
Windows 98/ME/2000/NT4-sp3/XP
16-bit Sound Card
32 MB or more of available RAM
800x600 monitor resolution set to 16-bit color
Mouse or other pointing device
12x CD-ROM drive
Optional components
Active Internet connection
Netscape NS7 or Microsoft IE5.5, IE6 or Firefox 1.x

Macintosh
PowerPC G3 233 MHz processor or better
Operating System OSX
32 MB or more of available RAM
800x600 monitor resolution set to 16-bit color
Mouse or other pointing device
12x CD-ROM drive
Optional components
Active Internet connection
Netscape NS7 or Microsoft IE5.2 or Safari 1.x or Firefox 1.x

Support Information
If you are having problems with this software, visit our support site at http://247.prenhall.com. E-mail support is available 24 hours a day, 7 days a week. Chat support is available at hours specified on the site. Our technical staff will need to know certain things about your system in order to help us solve your problems more quickly and efficiently. You should have the following information ready:
• Textbook ISBN
• CD-ROM ISBN
• corresponding product and title
• computer make and model
• Operating System (Windows or Macintosh) and Version
• RAM available
• hard disk space available
• Sound card? Yes or No
• printer make and model
• network connection
• detailed description of the problem, including the exact wording of any error messages

NOTE: Pearson does not support and/or assist with the following:
• third-party software (i.e. Microsoft including Microsoft Office suite, Apple, Borland, etc.)
• homework assistance
• Textbooks and CD-ROMs purchased used are not supported and are non-replaceable. To purchase a new CD-ROM, contact Pearson Individual Order Copies at 1-800-282-0693

World – Physical

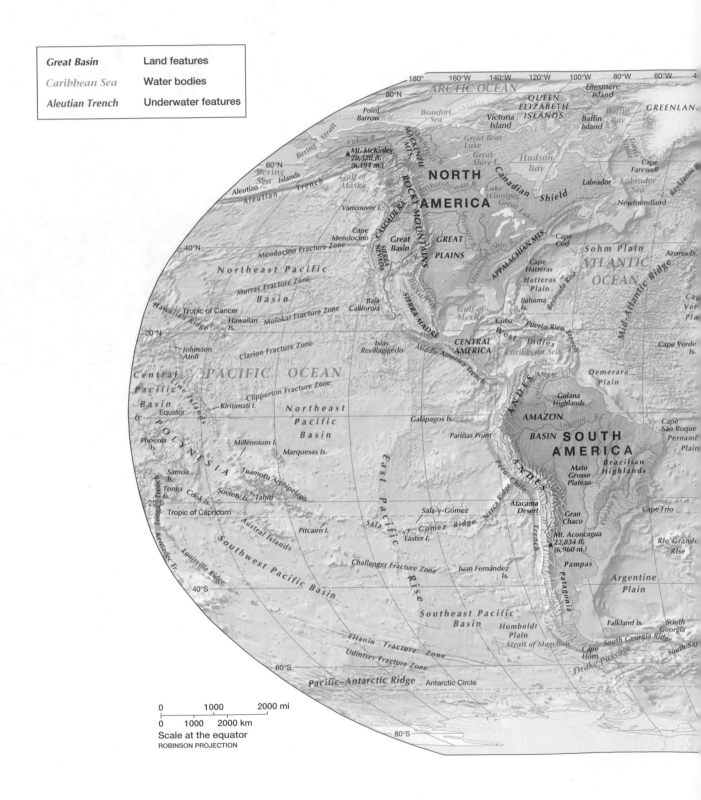

Great Basin	Land features
Caribbean Sea	Water bodies
Aleutian Trench	Underwater features

ARCTIC OCEAN

Point Barrow
Beaufort Sea
Victoria Island
QUEEN ELIZABETH ISLANDS
Baffin Island
Ellesmere Island
Baffin Bay
GREENLAN

80°N

Bering Strait
Yukon R.
Mt. McKinley 20,320 ft. (6,194 m.)
MACKENZIE MTS.
Great Bear Lake
Great Slave L.
Hudson Bay
Davis Strait
Cape Farewell
Labrador
Labrador Sea
Reykjanes Ridge

60°N
Bering Sea
Aleutian Islands
Aleutian Trench
Gulf of Alaska
ROCKY MOUNTAINS
NORTH AMERICA
Saskatchewan R.
Lake Winnipeg
Great Lakes
Canadian Shield
Newfoundland

Vancouver I.
CASCADE RA.
Missouri R.
Cape Cod
Sohm Plain

Cape Mendocino
Mendocino Fracture Zone
SIERRA NEVADA
Great Basin
GREAT PLAINS
Ohio R.
APPALACHIAN MTS.
Cape Hatteras
ATLANTIC OCEAN
Azores Is.

40°N
Northeast Pacific
Murray Fracture Zone
Basin
Colorado R.
SIERRA MADRE
Hatteras Plain
Bermuda Rise
Mid-Atlantic Ridge

Hawaiian Ridge
Tropic of Cancer
Baja California
Gulf of Mexico
Bahama Is.
Ca Ver Pla

Hawaiian Is.
Molokai Fracture Zone
Cuba
West Indies
Cape Verde Is.

20°N
Johnston Atoll
Clarion Fracture Zone
Islas Revillagigedo
CENTRAL AMERICA
Puerto Rico Trench
Caribbean Sea
Middle America Trench

Central Pacific Basin
Line Islands
PACIFIC OCEAN
Demerara Plain

Kiritimati I.
Clipperton Fracture Zone
Northeast Pacific Basin
Galápagos Is.
ANDES
Guiana Highlands
Orinoco R.
AMAZON

Equator 0°
POLYNESIA
Phoenix Is.
Pariñas Point
BASIN SOUTH AMERICA
Amazon R.
Cape São Roque
Pernaml Plain

Millennium I.
Marquesas Is.
Samoa Is.
Tonga Is.
Cook Is.
Society Is.
Tahiti
Tuamotu Archipelago
Brazilian Highlands
Mato Grosso Plateau

20°S
Tropic of Capricorn
Tonga Trench
Austral Islands
Pitcairn I.
Sala-y-Gómez
Sala-y-Gómez Ridge
Easter I.
East Pacific Rise
Peru-Chile Trench
Nazca Ridge
Atacama Desert
Gran Chaco
Cape Frio

Kermadec Tr.
Louisville Ridge
Southwest Pacific Basin
Challenger Fracture Zone
Juan Fernández Is.
Mt. Aconcagua 22,834 ft. (6,960 m.)
Pampas
Patagonia
Rio Grande Rise
Argentine Plain

40°S
Southeast Pacific Basin
Humboldt Plain
Strait of Magellan
Falkland Is.
South Georgia
South Georgia Ridge
South Sar

Eltanin Fracture Zone
Udintsev Fracture Zone
Cape Horn
Drake Passage

60°S
Pacific-Antarctic Ridge
Antarctic Circle

80°S

0   1000   2000 mi
0   1000   2000 km
Scale at the equator
ROBINSON PROJECTION

180° 160°W 140°W 120°W 100°W 80°W 60°W 4